MW01618016

Interactive SCIENCE

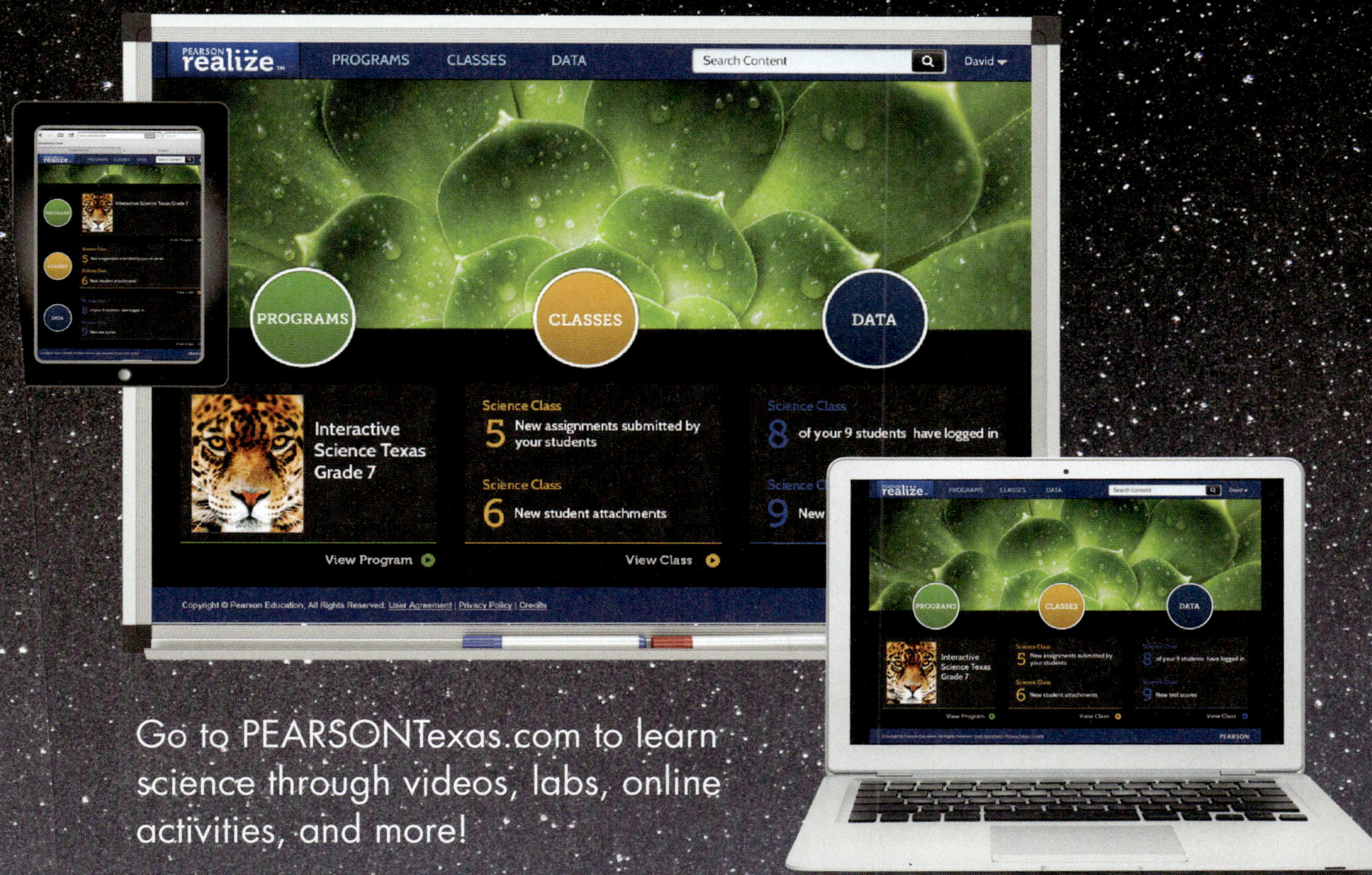

PEARSON Texas.com

Grade 7acher's Edition and Resource • Volume 1

Glenvie\linois • Boston, Massachusetts • Chandler, Arizona • Upper Saddle River, New Jersey

ALWAYS RNING

PEARSON

You're an author!

As you write in this science book, your answers and personal discoveries will be recorded for you to keep, making this book unique to you. That is why you are one of the primary authors of this book.

In the space below, print your name, school, town, and state. Then write a short autobiography that includes your interests and accomplishments.

Your Photo

YOUR NAME

SCHOOL

TOWN

AUTOBIOGRAPHY

Acknowledgments appear on pages C-1–C-4, which constitute an extension of this copyright page.

ON THE COVER

Jaguar

Panthera onca, commonly know the jaguar, stalks prey and prep to pounce. The genus *Panthera*, w also includes lions, tigers, and ards, is often referred to as the roari cats. The jaguar is the only mer of *Panthera* found in the Amer—in areas from Mexico to South Aıa— but its population numbers arall.

PEARSON

ISBN-13: 978-0-13-324434-2
ISBN-10: 0-13-324434-2
1 2 3 4 5 6 7 8 9 10 V011 17 16 15 14 13

Program Authors

DON BUCKLEY, M.Sc.
Director of Technology & Innovation,
The School at Columbia University, New York, New York
Don Buckley has transformed learning spaces, textbooks, and media resources so that they work for students and teachers. He has advanced degrees from leading European universities, is a former industrial chemist, published photographer, and former consultant to MoMA's Education Department. He also teaches a graduate course at Columbia Teacher's College in Educational Technology and directs the Technology and Innovation program at the school. He is passionate about travel, architecture, design, change, the future, and innovation.

ZIPPORAH MILLER, M.A.Ed.
Associate Executive Director for Professional Development Programs and Conferences, National Science Teachers Association, Arlington, Virginia
Mrs. Miller is currently the associate executive director for professional development programs and conferences at NSTA. She provides professional development and e-learning opportunities to science educators nationwide. She is a former K–12 science supervisor and STEM coordinator for the Prince George's County Public School District in Maryland. During her tenure there, she served as teacher, STEM coordinator, principal, and administrator. Mrs. Miller is passionate about providing quality educational opportunities to all students.

MICHAEL J. PADILLA, Ph.D.
Eugene P. Moore School of Education, Clemson University, Clemson, South Carolina
A former middle school teacher and a leader in middle school science education, Dr. Michael Padilla has served as president of the National Science Teachers Association and as a writer of the 1996 National Science Education Standards. He is a professor of science education at Clemson University. Having started with Pearson as lead author of the *Science Explorer* series, Dr. Padilla has inspired the team in developing *Interactive Science*, which promotes student inquiry and meets the needs of today's students.

KATHRYN THORNTON, Ph.D.
Professor, Mechanical & Aerospace Engineering,
University of Virginia,
Charlottesville, Virginia
Selected by NASA in May 1984, Dr. Kathryn Thornton is a veteran of four space flights. She has logged more than 975 hours in space, including more than 21 hours of extravehicular activity. As an author on the *Scott Foresman Science* series, Dr. Thornton's enthusiasm for science has inspired teachers around the globe.

MICHAEL E. WYSESSION, Ph.D.
Associate Professor of Earth and Planetary Science,
Washington University, St. Louis, Missouri
An author on more than 50 scientific publications, Dr. Wysession was awarded the prestigious Packard Foundation Fellowship and Presidential Faculty Fellowship for his research in geophysics. Dr. Wysession is an expert on Earth's inner structure and has mapped various regions of Earth using seismic tomography. He is known internationally for his work in geoscience education and research, and was an author of the Next Generation Science Standards.

Planet Diary Author

JACK HANKIN
Science/Mathematics Teacher,
The Hilldale School, Daly City, California
Founder, Planet Diary Web site
Mr. Hankin is the creator and writer of Planet Diary, a science current events Web site. Mr. Hankin is passionate about bringing science news and environmental awareness into classrooms.

ELL Consultant

JIM CUMMINS, Ph.D.
Professor and Canada Research Chair, Curriculum, Teaching and Learning Department at the University of Toronto
Dr. Cummins's research focuses on literacy development in multilingual schools and the role technology plays in learning across the curriculum. *Interactive Science* incorporates research-based principles for integrating language with the teaching of academic content based on Dr. Cummins's work.

Reading Consultant

HARVEY DANIELS, Ph.D.
Professor of Secondary Education,
University of New Mexico,
Albuquerque, New Mexico
Dr. Daniels is an international consultant to schools, districts, and educational agencies. He has authored or coauthored 13 books on language, literacy, and education. His most recent works are *Comprehension and Collaboration: Inquiry Circles in Action* and *Subjects Matter: Every Teacher's Guide to Content-Area Reading.*

Contributing Writers

Edward Aguado, Ph.D.
Professor, Department of Geography
San Diego State University
San Diego, California

Elizabeth Coolidge-Stolz, M.D.
Medical Writer
North Reading, Massachusetts

Donald L. Cronkite, Ph.D.
Professor of Biology
Hope College
Holland, Michigan

Jan Jenner, Ph.D.
Science Writer
Talladega, Alabama

Linda Cronin Jones, Ph.D.
Associate Professor of Science and Environmental Education
University of Florida
Gainesville, Florida

T. Griffith Jones, Ph.D.
Clinical Associate Professor of Science Education
College of Education
University of Florida
Gainesville, Florida

Andrew C. Kemp, Ph.D.
Teacher
Jefferson County Public Schools
Louisville, Kentucky

Matthew Stoneking, Ph.D.
Associate Professor of Physics
Lawrence University
Appleton, Wisconsin

R. Bruce Ward, Ed.D.
Senior Research Associate
Science Education Department
Harvard-Smithsonian Center for Astrophysics
Cambridge, Massachusetts

Content Reviewers

Paul D. Beale, Ph.D.
Department of Physics
University of Colorado at Boulder
Boulder, Colorado

Jeff R. Bodart, Ph.D.
Professor of Physical Sciences
Chipola College
Marianna, Florida

Joy Branlund, Ph.D.
Department of Earth Science
Southwestern Illinois College
Granite City, Illinois

Marguerite Brickman, Ph.D.
Division of Biological Sciences
University of Georgia
Athens, Georgia

Bonnie J. Brunkhorst, Ph.D.
Science Education and Geological Sciences
California State University
San Bernardino, California

Michael Castellani, Ph.D.
Department of Chemistry
Marshall University
Huntington, West Virginia

Charles C. Curtis, Ph.D.
Research Associate Professor of Physics
University of Arizona
Tucson, Arizona

Alice K. Hankla, Ph.D.
The Galloway School
Atlanta, Georgia

Mark Henriksen, Ph.D.
Physics Department
University of Maryland
Baltimore, Maryland

Chad Hershock, Ph.D.
Center for Research on Learning and Teaching
University of Michigan
Ann Arbor, Michigan

Jeremiah N. Jarrett, Ph.D.
Department of Biology
Central Connecticut State University
New Britain, Connecticut

Scott L. Kight, Ph.D.
Department of Biology
Montclair State University
Montclair, New Jersey

Jennifer O. Liang, Ph.D.
Department of Biology
University of Minnesota–Duluth
Duluth, Minnesota

Candace Lutzow-Felling, Ph.D.
State Arboretum of Virginia & Blandly Experimental Farm
Boyce, Virginia

Joseph F. McCullough, Ph.D.
Physics Program Chair
Cabrillo College
Aptos, California

Heather Mernitz, Ph.D.
Department of Physical Science
Alverno College
Milwaukee, Wisconsin

Sadredin C. Moosavi, Ph.D.
Department of Earth and Environmental Sciences
Tulane University
New Orleans, Louisiana

David L. Reid, Ph.D.
Department of Biology
Blackburn College
Carlinville, Illinois

Scott M. Rochette, Ph.D.
Department of the Earth Sciences
SUNY College at Brockport
Brockport, New York

Karyn L. Rogers, Ph.D.
Department of Geological Sciences
University of Missouri
Columbia, Missouri

Laurence Rosenhein, Ph.D.
Department of Chemistry
Indiana State University
Terre Haute, Indiana

Sara Seager, Ph.D.
Department of Planetary Sciences and Physics
Massachusetts Institute of Technology
Cambridge, Massachusetts

Tom Shoberg, Ph.D.
Missouri University of Science and Technology
Rolla, Missouri

Patricia Simmons, Ph.D.
North Carolina State University
Raleigh, North Carolina

William H. Steinecker, Ph.D.
Research Scholar
Miami University
Oxford, Ohio

Paul R. Stoddard, Ph.D.
Department of Geology and Environmental Geosciences
Northern Illinois University
DeKalb, Illinois

John R. Wagner, Ph.D.
Department of Geology
Clemson University
Clemson, South Carolina

Jerry Waldvogel, Ph.D.
Department of Biological Sciences
Clemson University
Clemson, South Carolina

Donna L. Witter, Ph.D.
Department of Geology
Kent State University
Kent, Ohio

Edward J. Zalisko, Ph.D.
Department of Biology
Blackburn College
Carlinville, Illinois

Museum of Science.

Special thanks to the Museum of Science, Boston, Massachusetts, and Ioannis Miaoulis, the Museum's president and director, for serving as content advisors for the technology and design strand in this program.

Teacher Reviewers

Herb Bergamini
The Northwest School
Seattle, Washington

David R. Blakely
Arlington High School
Arlington, Massachusetts

Jane E. Callery
Capital Region Education Council
Hartford, Connecticut

Jeffrey C. Callister
Former Earth Science Instructor
Newburgh Free Academy
Newburgh, New York

Colleen Campos
Cherry Creek Schools
Aurora, Colorado

Dan Gabel
Consulting Teacher, Science
Montgomery County Public Schools
Montgomery County, Maryland

Wayne Goates
Kansas Polymer Ambassador
Intersociety Polymer Education Council (IPEC)
Wichita, Kansas

Katherine Bobay Graser
Mint Hill Middle School
Charlotte, North Carolina

Darcy Hampton
Science Department Chair
Deal Middle School
Washington, D.C.

Sean S. Houseknecht
Elizabethtown Area Middle School
Elizabethtown, Pennsylvania

Tanisha L. Johnson
Prince George's County Public Schools
Lanham, Maryland

Karen E. Kelly
Pierce Middle School
Waterford, Michigan

Dave J. Kelso
Manchester Central High School
Manchester, New Hampshire

Beverly Crouch Lyons
Career Center High School
Winston-Salem, North Carolina

Angie L. Matamoros, Ed.D.
ALM Consulting
Weston, Florida

Corey Mayle
Durham Public Schools
Durham, North Carolina

Keith W. McCarthy
George Washington Middle School
Wayne, New Jersey

Timothy McCollum
Charleston Middle School
Charleston, Illinois

Bruce A. Mellin
Cambridge College
Cambridge, Massachusetts

John Thomas Miller
Thornapple Kellogg High School
Middleville, Michigan

Randy Mousley
Dean Ray Stucky Middle School
Wichita, Kansas

Yolanda O. Peña
John F. Kennedy Junior High School
West Valley, Utah

Kathleen M. Poe
Fletcher Middle School
Jacksonville Beach, Florida

Judy Pouncey
Thomasville Middle School
Thomasville, North Carolina

Vickki Lynne Reese
Mad River Middle School
Dayton, Ohio

Bronwyn W. Robinson
Director of Curriculum
Algiers Charter Schools Association
New Orleans, Louisiana

Shirley Rose
Lewis and Clark Middle School
Tulsa, Oklahoma

Linda Sandersen
Sally Ride Academy
Whitefish Bay, Wisconsin

Roxanne Scala
Schuyler-Colfax Middle School
Wayne, New Jersey

Patricia M. Shane, Ph.D.
Associate Director
Center for Mathematics & Science Education
University of North Carolina at Chapel Hill
Chapel Hill, North Carolina

Bradd A. Smithson
Science Curriculum Coordinator
John Glenn Middle School
Bedford, Massachusetts

Sharon Stroud
Consultant
Colorado Springs, Colorado

Master Teacher Board

Emily Compton
Park Forest Middle School
Baton Rouge, Louisiana

Georgi Delgadillo
East Valley School District
Spokane Valley, Washington

Treva Jeffries
Toledo Public Schools
Toledo, Ohio

James W. Kuhl
Central Square Middle School
Central Square, New York

Bonnie Mizell
Howard Middle School
Orlando, Florida

Joel Palmer, Ed.D.
Mesquite Independent School District
Mesquite, Texas

Leslie Pohley
Largo Middle School
Largo, Florida

Susan M. Pritchard, Ph.D.
Washington Middle School
La Habra, California

Anne Rice
Woodland Middle School
Gurnee, Illinois

Richard Towle
Noblesville Middle School
Noblesville, Indiana

Texas Content Reviewers

Serena R. Aldrich, Ph.D.
Department of Geography
Blinn College
Bryan, Texas

Scott Cordell
Amarillo Independent School District
Amarillo, Texas

Diane I. Doser, Ph.D.
Department of Geological Sciences
University of Texas
El Paso, Texas

Rick Duhrkopf, Ph.D.
Department of Biology
Baylor University
Waco, Texas

Joseph C. Hill, Ph.D.
Department of Geography and Geology
Sam Houston State University
Huntsville, Texas

Christopher B. Martin, Ph.D.
Department of Chemistry and Biochemistry
Lamar University
Beaumont, Texas

Diana Mason, Ph.D.
Department of Chemistry (retired)
University of North Texas
Denton, Texas

Suzanne Nesmith, Ph.D.
Department of Curriculum and Instruction
Baylor University
Waco, Texas

Hilary Clement Olson, Ph.D.
Institute for Geophysics
Dept. of Petroleum and Geosystems Engineering
The University of Texas at Austin
Austin, Texas

Elizabeth Polito, MS
Institute for Geophysics
The University of Texas at Austin
Austin, Texas

John R. Villarreal, Ph.D.
Department of Chemistry
The University of Texas–Pan American
Edinburg, Texas

Built for Texas

Texas Interactive Science covers 100% of the Texas Essential Knowledge and Skills for Science. Built on feedback from Texas educators, *Interactive Science* focuses on what is important to Texas teachers and students, and creates a personal, relevant, and engaging classroom experience.

Pearson would like to say a special *Thank You* to all of the teachers from all over Texas who have helped guide the development of this program since its inception.

CONTENTS Volume 1

Chapter 1
Scientific Thinking

Chapter 2
Introduction to Cells

Enter the Lab zone for hands-on inquiry. For a complete list of labs go to PearsonTexas.com.

Lab Investigations: Keeping Flowers Fresh • Design and Build a Microscope

Inquiry Warm-Ups: How Keen Are Your Senses? • Where Is the Safety Equipment in Your School? • What Can You See?

Quick Labs: Classifying Objects • Scientific Inquiry • Models in Nature • Light Sources • Classifying Animals • Comparing Cells • Observing Cells • Gelatin Cell Model • Tissues, Organs, Organ Systems • What Is a Compound? • What's That Taste? • Effect of Concentration on Diffusion

Chapter 3

Cell Processes and Energy

Chapter 4

Genetics: The Science of Heredity

Lab zone®

Lab Investigations: Exhaling Carbon Dioxide • Make the Right Call

Inquiry Warm-Ups: Measuring Calories • What's the Chance?

Quick Labs: Energy From the Sun • Observing Mitosis • Inferring the Parent Generation • Is It All in the Genes? • Chromosomes and Inheritance • Comparing Meiosis and Mitosis

CONTENTS Volume 1

Chapter 5
Plant Structure, Function, and Response

Enter the Lab zone for hands-on inquiry. For a complete list of labs go to PearsonTexas.com.

Lab Investigations: Investigating Stomata • Fruit Basket • A Look Beneath the Skin

Inquiry Warm-Ups: Can a Plant Respond to Touch? • How Is Your Body Organized? • How Do Muscles Work?

Quick Labs: Local Plant Diversity • Will Mosses Absorb Water? • Modeling Flowers • Watching Roots Grow • Seasonal Changes • Observing Cells and Tissues • Working Together Act I • Working Together Act II • Working to Maintain Balance • The Skeleton • Observing Joints • Observing Muscle Tissue • Sweaty Skin

Chapter 6
Introduction to the Human Body

Volume 2

Chapter 7

Managing Materials in the Body

Lab zone

Lab Investigations: Nutrient Identification • Modeling Negative Feedback • The Skin as a Barrier

Inquiry Warm-Up: Where Does Digestion Start?

Quick Labs: Direction of Blood Flow • Modeling Respiration • Perspiration • How Does Your Knee React? • Making Models: Hormones • Looking at Hormone Levels • Egg-cellent Protection

Chapter 8

Controlling Body Processes

CONTENTS Volume 2

Chapter 9

Ecosystems and Biomes

Chapter 10

Organisms and the Environment

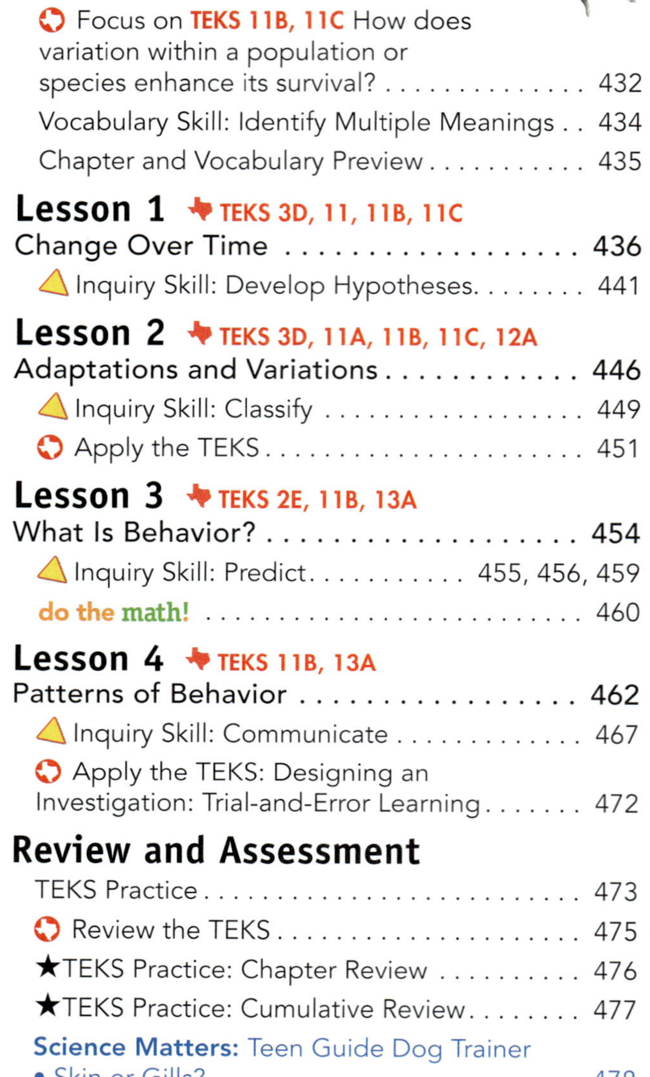

Enter the Lab zone for hands-on inquiry. For a complete list of labs go to PearsonTexas.com.

Lab Investigations: Last Remains • From Microhabitats to Biomes • Investigating Bird and Mammal Bones • Bird Beak Adaptations • Insect Behavior

Inquiry Warm-Ups: How Much Rain Is That? • How Do Living Things Vary? • Eggs-tra Protection • Communicating Without Words

Quick Labs: Observing Decomposition • Considering Fungi as Decomposers • Playing Nitrogen Cycle Roles • Dissolved Oxygen • Grocery Gene Pool and Potato Famine • Primary or Secondary? • Identifying Vertebrates Using Dichotomous Keys • Learning New Words

Chapter 11

Fresh Water

Lab Investigations: Water From Trees • Field Testing a Body of Water • Sand Hills

Inquiry Warm-Ups: Groundwater Pollution • How Can You Change Land?

Quick Labs: Water on Earth • Modeling a Watershed • Soil Percolation • Testing Water • Raindrops Falling • Shaping a Coastline • Desert Pavement • Local Landscapes • How Does Soil Type Affect Plant Growth? • What Happened Here? • Humans and Erosion

Chapter 12

Weathering, Erosion, and Deposition

CONTENTS Volume 2

Chapter 13

Space Exploration

Enter the Lab zone for hands-on inquiry. For a complete list of labs go to PearsonTexas.com.

Lab Investigation: Build a Space Exploration Vehicle

Inquiry Warm-Ups: What Is the Greenhouse Effect? • Using Space Science

Quick Labs: Water on Earth • Liquid Water and Water Vapor • What Is Work?

Untamed Science created this captivating video series for Interactive SCIENCE featuring a unique segment for every chapter of the program.

Chapter 1	**Mimicking Nature**
Chapter 2	**Touring Hooke's Crib**
Chapter 3	**Yum...Eating Solar Energy**
Chapter 4	**Where'd You Get Those Genes?**
Chapter 5	**Amazing Plant Defenses**
Chapter 6	**Feeling Just Spine, Thank You**
Chapter 7	**Blood Lines**
Chapter 8	**Think Fast!**
Chapter 9	**Give Me That Carbon!**
Chapter 10	**Why Would a Fish Have Red Lips?**
Chapter 11	**Water Cyclists**
Chapter 12	**Carving a Canyon**
Chapter 13	**Skeletons in Space**

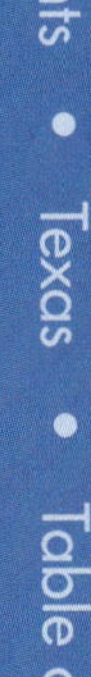

Dear Family Member,

As your child's science teacher, I am looking forward to helping your child learn about science. Because I know that you want your child to be successful, I offer these suggestions so that you can help your child gain proficiency in science.

- Your child's textbook is very different from most—it's meant for students to write in it. Therefore, it is a record of learning. Look through lessons your child has completed recently, and be sure to ask lots of questions. One of the best ways for students to check on their learning is to explain it to someone else.
- Ask your child about homework assignments and check that he or she has completed them.
- Help your child collect materials and information for school activities.
- Encourage computer literacy. Advise your child to use computers in school or at the library. If you have a home computer, help your child do research online.

In this year of study, your child will explore the skills and tools that scientists use to study the natural world. Your child will also learn about cells, genetics, human body systems, how organisms interact with their environments, Earth systems, and our solar system.

I encourage you to stay involved in your child's learning. By all means, visit the classroom during open house or make an appointment with me if you have questions.

Cordially,

____

Science Teacher

To access activities and interactive lessons online, go to **www.PearsonTexas.com.**

Estimados familiares:

Soy el maestro de ciencias de su hijo y me dará mucho gusto ayudarlo a aprender sobre las ciencias. Sé que ustedes desean que su hijo tenga un buen desempeño académico, por eso les ofrezco estas sugerencias para que lo ayuden a dominar las ciencias.

- El libro de texto de su hijo es muy diferente a los demás: tiene como objetivo que su hijo escriba en el libro. Por esa razón, es un registro de aprendizaje. Fíjense en las lecciones que ha terminado recientemente y asegúrense de hacerle muchas preguntas. Una buena manera de que el estudiante repase lo que ha aprendido es que se lo explique a otras personas.
- Pregúntenle a su hijo sobre la tarea que se le asigna y asegúrense de que la complete.
- Ayúdenlo a reunir materiales e información relacionados con las actividades escolares.
- Anímenlo a adquirir destrezas con la computadora, y a usar computadoras en la escuela o en la biblioteca. Si tienen una computadora en casa, ayúdenlo a hacer investigaciones en línea.

Este año, su hijo va a explorar las destrezas e instrumentos que usan los científicos para estudiar la naturaleza. También aprenderá sobre las células, la genética, los sistemas del cuerpo humano, la manera en que los organismos interactúan con su medio ambiente, los sistemas terrestres y nuestro sistema solar.

Los invito a que participen en el proceso de aprendizaje de su hijo. Pueden visitar el salón de clases durante las horas de visita o hacer una cita para hablar conmigo si tienen dudas.

Cordialmente,

____

Maestro de Ciencias

Para ver actividades y lecciones interactivas en línea, visiten **www.PearsonTexas.com.**

PEARSON Texas

Interactive SCIENCE

Built for Texas Powered for You

PEARSON Texas.com

Power of Connections

- It's Their Science
- Connect What They Do to What They Read and See
- Leverage ELPS Strategies Across All Content Areas

Power of Data

- Assign and Differentiate
- Focus, Apply, and Review the **TEKS**
- Track **TEKS** Mastery
- Use **TEKS** Data

Power of Flexibility

- Access **TEKS** Lessons
- Search by Keyword or **TEKS**
- Align to Your District Framework
- Edit Program Resources

Power of

Engage students with Untamed Science Videos to introduce each chapter.

It's Their Science.

Engage with the page! Connect and interact with science content with the Texas Interactive Science write-in Student Edition and Lab Manual.

Connections

Pearson Flipped Videos for Science deepen students' **TEKS** understanding.

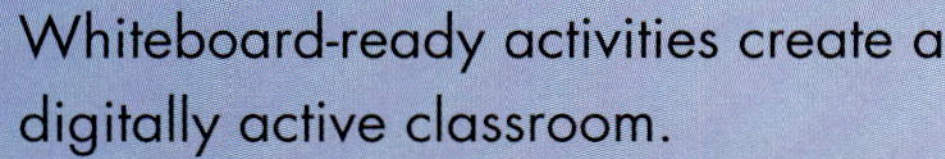

Whiteboard-ready activities create a digitally active classroom.

Access all labs online at PEARSONTexas.com.

Connect what they do to what they read and see.

Power of

Assign and Differentiate

Individual

Group

Whole Class

Assign Leveled Content Based on Performance

At PEARSONTexas.com, differentiated resources include:

- Student eText with Audio
- Directed and Open Inquiry Options
- Texas exam-formatted Benchmark Assessments
- Spanish resources available

Focus, Apply, and Review the TEKS and ELPS

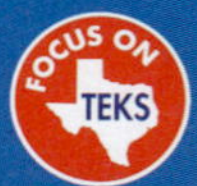

Focus on the TEKS with the Focus Question.

Apply content to new situations from Texas and beyond

Review the TEKS and practice with Texas exam-formatted questions.

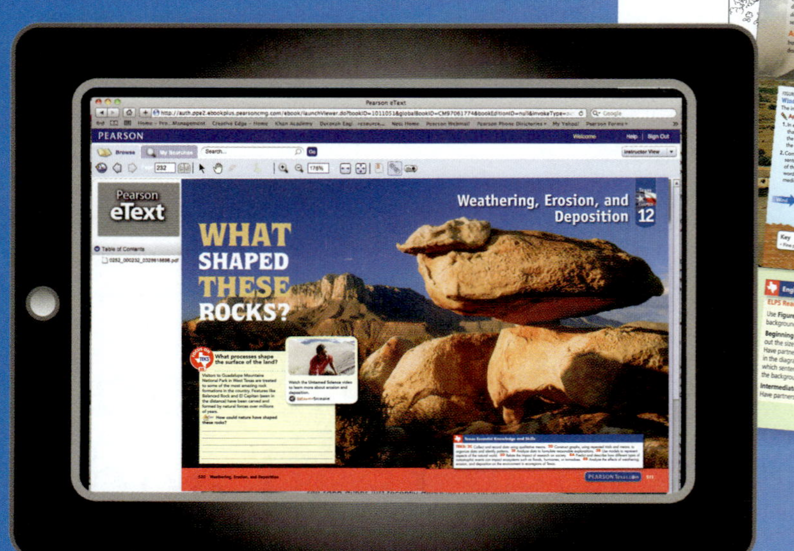

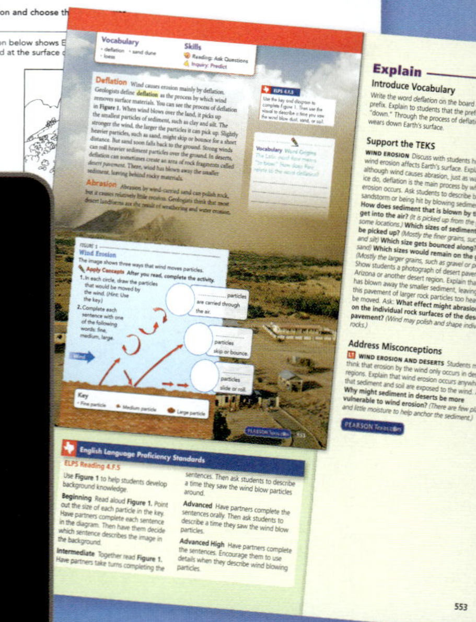

Data

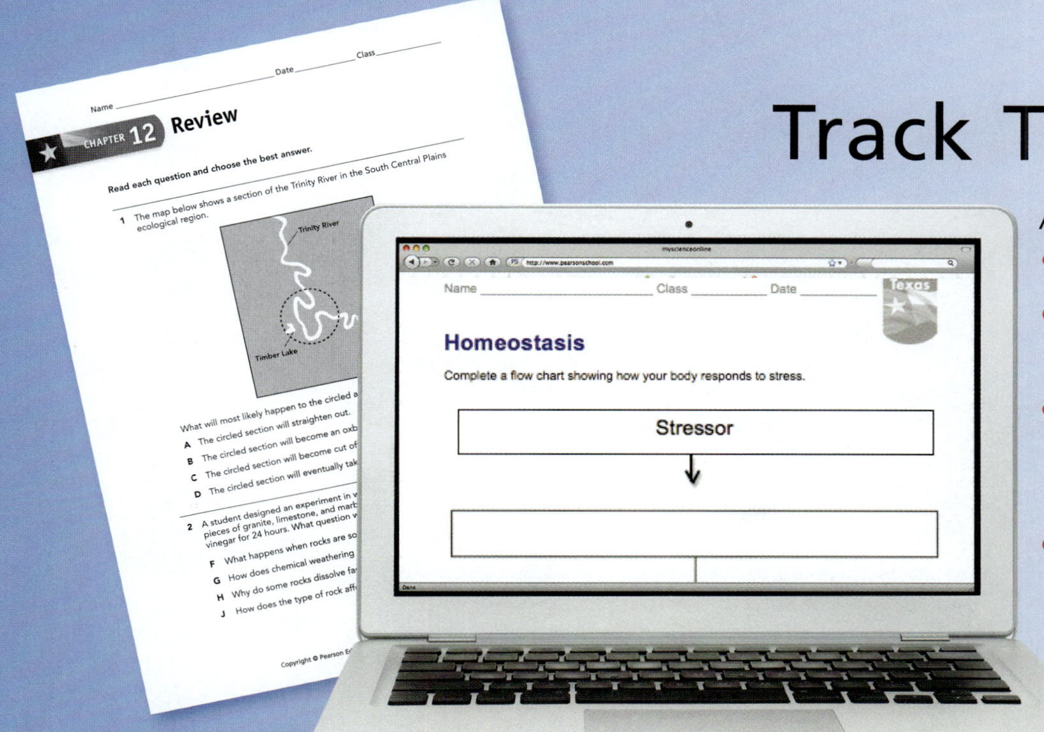

Track TEKS Mastery

At PEARSONTexas.com, you can:

- Identify what students already know.
- Check what students know at the end of each topic and chapter.
- Provide automated practice on skills and/or content, based on student performance.
- Predict your students' exam readiness with benchmark assessments throughout the year.

Use TEKS Data

PEARSONTexas.com brings you the power of data in an easy-to-view format. Features include:

- Up-to-the-minute standards mastery indicators
- Individual and class progress
- Usage data that shows how much time students are spending in the online course

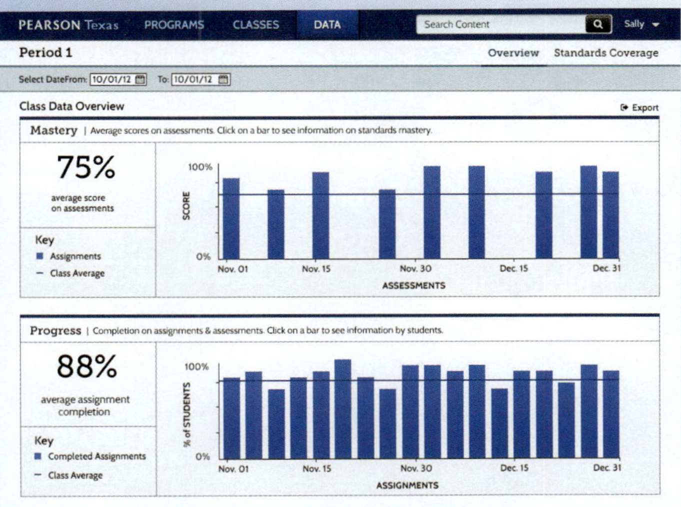

Power of

Access TEKS Lessons

Rich-media lessons that cover 100% of the **TEKS** and **ELPS**.

Search by Keyword or TEKS

Use powerful searching capabilities to easily find and add content to digital lessons. Search by Keyword, TEKS, or Resource Type.

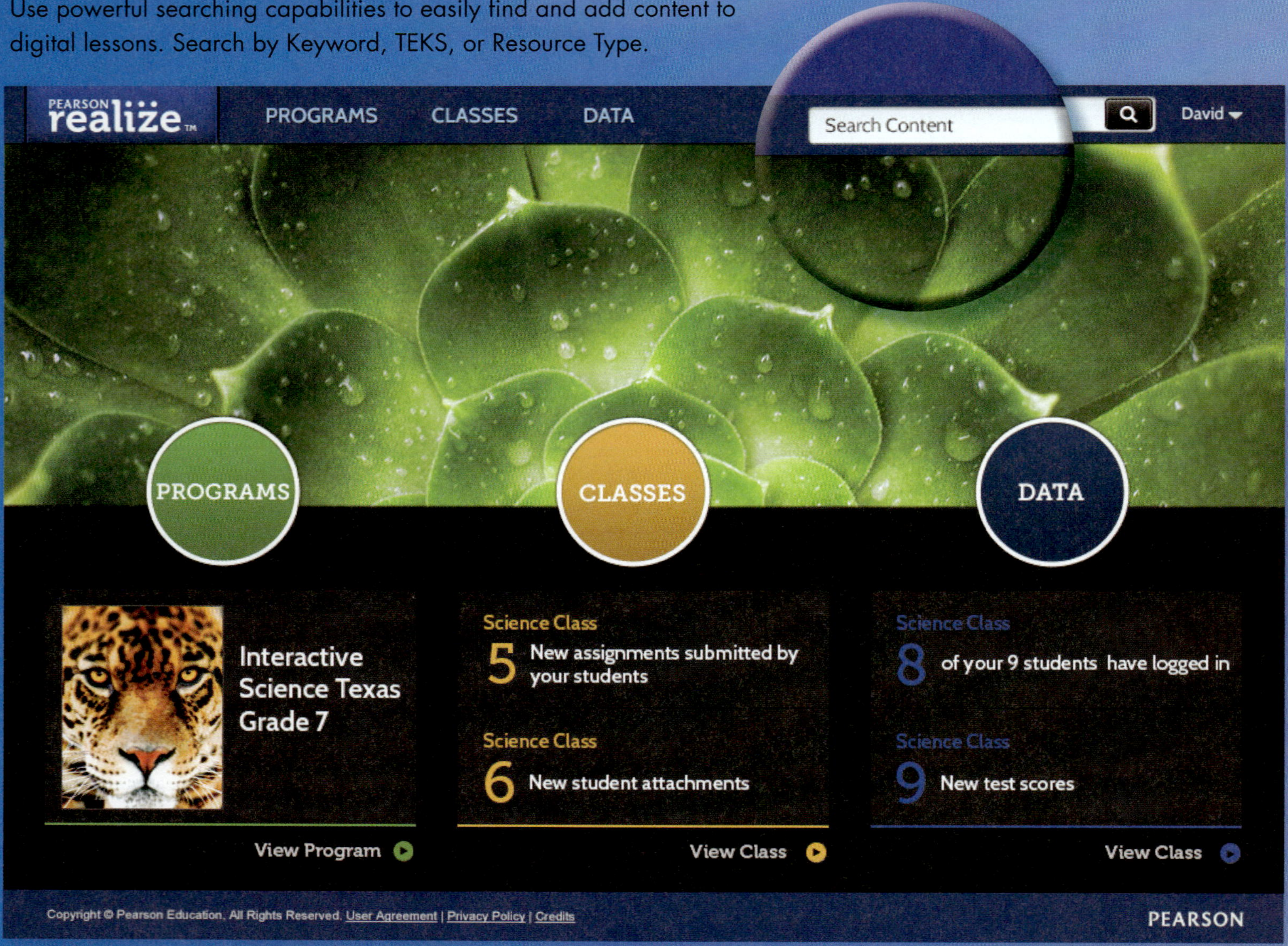

PEARSON realize™ PROGRAMS CLASSES DATA

Search Content

David

PROGRAMS

CLASSES

DATA

Interactive Science Texas Grade 7

View Program

Science Class
5 New assignments submitted by your students

Science Class
6 New student attachments

View Class

Science Class
8 of your 9 students have logged in

Science Class
9 New test scores

View Class

PEARSON

Flexibility

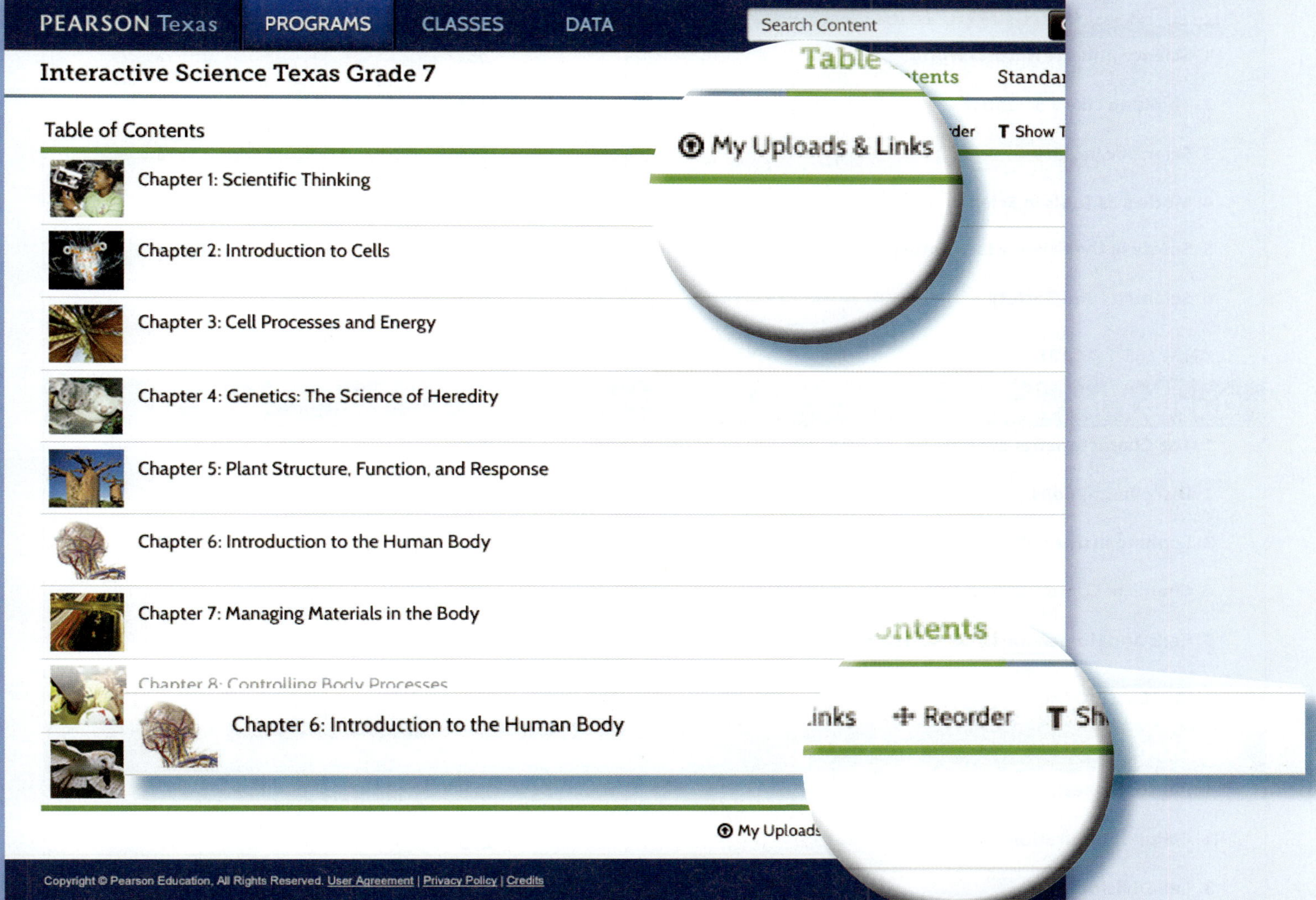

Upload Your Content

Upload district-created or your own content with PEARSONTexas.com.

Align to Your District Framework

Re-sequence chapter and/or lesson content to match district-level curriculum guides or your scope and sequence preference.

Edit Program Resources

Edit assessments and blackline masters, including labs, to meet the needs of each learner in your classroom.

Pacing Guide

The guide on these pages suggests the core time to spend on each lesson of each chapter in the **Texas Grade 7 *Interactive Science*** book. Time allotments include lab activities, but exclude assessment and activities or projects you may choose to add.

Chapter Title / Lesson Title	Periods (approx.)	Blocks (approx.)
Chapter 1 Scientific Thinking		
1 Science and the Natural World	2–3	1–1 ½
2 Thinking Like a Scientist	2–3	1–1 ½
3 Scientific Inquiry	3–4	1 ½–2
4 Models as Tools in Science	2–3	1–1 ½
5 Safety in the Science Laboratory	2–3	1–1 ½
6 Scientists and Society	1–2	½–1
CHAPTER 1 TOTAL	**12–18**	**6–9**
Chapter 2 Introduction to Cells		
1 The Characteristics of Life	1–2	½–1
2 Discovering Cells	2–3	1–1 ½
3 Looking Inside Cells	3–4	1 ½–2
4 Chemical Compounds in Cells	1–2	½–1
5 Cells and Homeostasis	1–2	½–1
CHAPTER 2 TOTAL	**8–13**	**4–6 ½**
Chapter 3 Cell Processes and Energy		
1 Photosynthesis	2–3	1–1 ½
2 Cellular Respiration	2–3	1–1 ½
3 Cell Division	2–3	1–1 ½
CHAPTER 3 TOTAL	**6–9**	**3–4 ½**

Chapter Title / Lesson Title	Periods (approx.)	Blocks (approx.)
Chapter 4 Genetics: The Science of Heredity		
1 What Is Heredity?	2–3	1–1 ½
2 Probability and Heredity	2–3	1–1 ½
3 Patterns of Inheritance	2–3	1–1 ½
4 Chromosomes and Inheritance	3–4	1 ½–2
CHAPTER 4 TOTAL	**9–13**	**4 ½–6 ½**
Chapter 5 Plant Structure, Function, and Response		
1 What Is a Plant?	2–3	1–1 ½
2 Plant Structures	3–4	1 ½–2
3 Plant Responses and Growth	1–2	½–1
CHAPTER 5 TOTAL	**6–9**	**3–4 ½**
Chapter 6 Introduction to the Human Body		
1 Body Organization	2–3	1–1 ½
2 System Interactions	2–3	1–1 ½
3 Homeostasis	2–3	1–1 ½
4 The Skeletal System	2–3	1–1 ½
5 The Muscular System	2–3	1–1 ½
6 The Integumentary System	1–2	½–1
CHAPTER 6 TOTAL	**11–17**	**5 ½–8 ½**
Chapter 7 Managing Materials in the Body		
1 The Digestive System	3–4	1 ½–2
2 The Circulatory System	3–4	1 ½–2
3 The Respiratory System	2–3	1–1 ½
4 The Excretory System	2–3	1–1 ½
CHAPTER 7 TOTAL	**10–14**	**5–7**

Chapter Title / Lesson Title	Periods (approx.)	Blocks (approx.)
Chapter 8 Controlling Body Processes		
1 The Nervous System	3–4	1 ½–2
2 The Endocrine System	2–3	1–1 ½
3 The Male and Female Reproductive Systems	2–3	1–1 ½
4 Pregnancy and Birth	2–3	1–1 ½
5 The Body's Defenses	3–4	1 ½–2
CHAPTER 8 TOTAL	**12–17**	**6–8 ½**
Chapter 9 Ecosystems and Biomes		
1 Energy Flow in Ecosystems	2–3	1–1 ½
2 Cycles of Matter	2–3	1–1 ½
3 Biomes	3–4	1 ½–2
4 Aquatic Ecosystems	1–2	½–1
5 Biodiversity	3–4	1 ½–2
6 Changing Ecosystems	1–2	½–1
CHAPTER 9 TOTAL	**12–18**	**6–9**
Chapter 10 Organisms and the Environment		
1 Change Over Time	3–4	1 ½–2
2 Adaptations and Variations	2–3	1–1 ½
3 What Is Behavior?	2–3	1–1 ½
4 Patterns of Behavior	3–4	1 ½–2
CHAPTER 10 TOTAL	**10–14**	**5–7**
Chapter 11 Fresh Water		
1 Water on Earth	2–3	1–1 ½
2 Surface Water	2–3	1–1 ½
3 Water Underground	2–3	1–1 ½
4 Water Pollution and Solutions	2–3	1–1 ½
CHAPTER 11 TOTAL	**8–12**	**4–6**

Chapter Title / Lesson Title	Periods (approx.)	Blocks (approx.)
Chapter 12 Weathering, Erosion, and Deposition		
1 Weathering and Mass Movement	3–4	1 ½–2
2 Water Erosion	3–4	1 ½–2
3 Wave and Glacial Erosion	2–3	1–1 ½
4 Wind Erosion	1–2	½–1
5 Weathering, Erosion, and Deposition in Texas	2–3	1–1 ½
6 Ecosystems and Catastrophic Events	1–2	½–1
CHAPTER 12 TOTAL	**12–18**	**6–9**
Chapter 13 Space Exploration		
1 Our Solar System	2–3	1–1 ½
2 Working and Living in Space	2–3	1–1 ½
3 Space Technology	1–2	½–1
CHAPTER 13 TOTAL	**5–8**	**2½–4**
GRADE 7 TOTAL	**121–180**	**60½–90**

TEKS at a Glance

The **Texas *Interactive Science*** program provides comprehensive coverage and assessment of the **Texas Essential Knowledge and Skills (TEKS).** Use this chart to track where the TEKS for Grade 7 are addressed throughout the essential components of the program.

KEY: **green** = in-depth coverage; **SE** = Student Edition; **LM** = Laboratory Manual; **TE** = Teacher's Edition

Texas Essential Knowledge and Skills	Where You Will Find It
7.1 Scientific investigation and reasoning. The student, for at least 40% of the instructional time, conducts laboratory and field investigations following safety procedures and environmentally appropriate and ethical practices. The student is expected to:	
7.1A Demonstrate safe practices during laboratory and field investigations as outlined in the Texas Safety Standards.	**SE:** 37–41, 50 **LM:** **xii–xiii,** 21, **50,** 72, 101, 103, **136,** 147, 157, 166, 173 **TE:** 39
7.1B Practice appropriate use and conservation of resources, including disposal, reuse, or recycling of materials.	**SE:** 431 **LM:** **xiii-xiv, 73,** 79, **94, 97, 154,** 178
7.2 Scientific investigation and reasoning. The student uses scientific inquiry methods during laboratory and field investigations. The student is expected to:	
7.2A Plan and implement comparative and descriptive investigations by making observations, asking well-defined questions, and using appropriate equipment and technology.	**SE:** 19, 203, 512 **LM:** **xx, xxi–xxiii,** 3, 17, 35, 43, **58, 59,** 86, 96, **109, 110, 111, 133,** 138, 157, 166, 195
7.2B Design and implement experimental investigations by making observations, asking well-defined questions, formulating testable hypotheses, and using appropriate equipment and technology.	**SE:** 19, 20, 21, 22, 26, 49, 472, 611 **LM:** **xx, xxi–xxiii,** 3, 5, 72, **103, 147, 149** **TE:** 193, 458
7.2C Collect and record data using the International System of Units (SI) and qualitative means such as labeled drawings, writing, and graphic organizers.	**SE:** **23,** 73, 146, 166, 173, 254, 255, 472, **612–619** **LM:** **53, 65,** 104, 126, **201** **TE:** 93, 195, 227, 380, 387, 401B, 405
7.2D Construct tables and graphs, using repeated trials and means, to organize data and identify patterns.	**SE:** 7, **23,** 126–127, 269, 343, 472, 530, **626, 627–633** **LM:** 6, 41, **43,** 119, **139, 170,** 178 **TE:** 25, 185
7.2E Analyze data to formulate reasonable explanations, communicate valid conclusions supported by the data, and predict trends.	**SE:** 7, **23, 24–25,** 88, 94, 125, 126–127, 166, 182, 254–255, 297, 304, 460, 486, 501, 512, 530, 570–571, 587 **LM:** 8, 26, **34,** 51, 55, **125,** 135, 150, **153,** 159, 168, 199 **TE:** 65B, 89B, 292, 295B, 361B, 389B, 451B, 511B, 543B, 569B
7.3 Scientific investigation and reasoning. The student uses critical thinking, scientific reasoning, and problem solving to make informed decisions and knows the contributions of relevant scientists. The student is expected to:	
7.3A In all fields of science, analyze, evaluate, and critique scientific explanations by using empirical evidence, logical reasoning, and experimental and observational testing, including examining all sides of scientific evidence of those scientific explanations, so as to encourage critical thinking by the student.	**SE:** 11–13, 14–17, 46–47, 48, 55, 304 **LM:** **xxiii–xxv,** 64, 65, 90, **114, 115, 124, 136, 150, 151,** 152, **160** **TE:** 445B
7.3B Use models to represent aspects of the natural world such as human body systems and plant and animal cells.	**SE:** 9, 17, **29,** 30–31, 32-35, 80, 166, 217, 132, 330, 361, 362, 493, 497, 604–605 **LM:** 9, 21, 22, 45, 47, 48, 49, 63, 79, 91, 94, 96, 100, 101, 103, 118, 130, **138,** 170, 171, 186, 187, 188, 189, 204, 210 **TE:** 9, 35, 35B, 81, 141, 149B, 159, 161, 163, 165, 165B, 181, 189, 193, 195, 198, 299, 397, 493, 503B, 510, 531, 551
7.3C Identify advantages and limitations of models such as size, scale, properties, and materials.	**SE:** 9, **29, 32–33,** 35, 49, 80, 362, 605 **LM:** 9, **63, 98** **TE:** 35B
7.3D Relate the impact of research on scientific thought and society, including the history of science and contributions of scientists as related to the content.	**SE:** **43–44,** 45, **50,** 54, 55, 66, **68–69,** 95, 100, 101, **143, 167,** 172, 263, 286–287, 310, 311, 336, **351,** 355, **364,** 368, 445, 450, 473, 518, 519, 600, 601, 603, **607,** 610 **TE:** 17B, 27B, 45B, 73B, 117B, 147, 219B, 277B, 459

Texas Essential Knowledge and Skills	Where You Will Find It
7.4 Science investigation and reasoning. The student knows how to use a variety of tools and safety equipment to conduct science inquiry. The student is expected to:	
7.4A Use appropriate tools to collect, record, and analyze information, including life science models, hand lens, stereoscopes, microscopes, beakers, Petri dishes, microscope slides, graduated cylinders, test tubes, meter sticks, metric rulers, metric tape measures, timing devices, hot plates, balances, thermometers, calculators, water test kits, computers, temperature and pH probes, collecting nets, insect traps, globes, digital cameras, journals/notebooks, and other equipment as needed to teach the curriculum.	**SE:** 8, **29,** 70–73, 78–79, 637, 638–639, 640 **LM: xv, xvi, xvii, xviii, xix,** 13, **14, 15,** 16, 18, **29, 31, 32, 33, 34, 51, 65, 66, 67, 68, 69, 83, 87, 88,** 89, **104,** 106, **109, 111, 115, 116, 117,** 119, **121, 129, 131,** 132, **135, 138,** 140, 141, **146, 148, 152, 155, 157,** 158, **159, 161, 162, 163, 164,** 167, 168, **173, 174, 175, 177, 179, 185, 193,** 199, **200, 201** **TE:** 195
7.4B Use preventative safety equipment, including chemical splash goggles, aprons, and gloves, and be prepared to use emergency safety equipment, including an eye/face wash, a fire blanket, and a fire extinguisher.	**SE:** 37–41 **LM: xi, xii, 10, 24, 27, 28, 29,** 72, 109, 132, **162,** 163, 171, 187, 188, 192, 199 **TE:** 39
7.5 Matter and energy. The student knows that interactions occur between matter and energy. The student is expected to:	
7.5A Recognize that radiant energy from the Sun is transformed into chemical energy through the process of photosynthesis.	**SE:** 91, 107–108, **109–111,** 115, 117, 128 **LM:** 27
7.5B Demonstrate and explain the cycling of matter within living systems such as in the decay of biomass in a compost bin.	**SE: 382–389,** 424, 431, 488–489 **LM:** 107, **117, 118**
7.5C Diagram the flow of energy through living systems, including food chains, food webs, and energy pyramids.	**SE:** 32, 33, 108, 129, **378–379, 380–381** **LM:** 112 **TE:** 381B
7.6 Matter and energy. The student knows that matter has physical and chemical properties and can undergo physical and chemical changes. The student is expected to:	
7.6A Identify that organic compounds contain carbon and other elements such as hydrogen, oxygen, phosphorus, nitrogen, or sulfur.	**SE: 85,** 86, 87, **89,** 94, **96,** 270 **LM:** 23
7.6B Distinguish between physical and chemical changes in matter in the digestive system.	**SE: 272–275, 277,** 305 **LM:** 84 **TE:** 225
7.6C Recognize how large molecules are broken down into smaller molecules such as carbohydrates can be broken down into sugars.	**SE:** 86, 114, 128, 270, 271, **272–275,** 304, **305** **LM:** 25
7.7 Force, motion, and energy. The student knows that there is a relationship among force, motion, and energy. The student is expected to:	
7.7A Contrast situations where work is done with different amounts of force to situations where no work is done such as moving a box with a ramp and without a ramp, or standing still.	**SE:** 248, 254–255, **592–593,** 606 **LM: 202**
7.7B Illustrate the transformation of energy within an organism such as the transfer from chemical energy to heat and thermal energy in digestion.	**SE:** 109–111, 113–114, 116–117, **115,** 126–127, **128,** 132 **LM:** 28, 38
7.7C Demonstrate and illustrate forces that affect motion in everyday life such as emergence of seedlings, turgor pressure, and geotropism.	**SE:** 193, **191,** 197, 201, 202–203, 204, 205 **LM: 65, 66**
7.8 Earth and space. The student knows that natural events and human activity can impact Earth systems. The student is expected to:	
7.8A Predict and describe how different types of catastrophic events impact ecosystems such as floods, hurricanes, or tornadoes.	**SE:** 564, **565–569,** 570–**571**, 574, 578 **LM:** 191 **TE:** 566
7.8B Analyze the effects of weathering, erosion, and deposition on the environment in ecoregions of Texas.	**SE:** 524–525, 526–529, 537–543, 546–547, 556–559, **560–561, 562,** 563, 570–571, **574,** 579 **LM:** 185, 186, 187, 188, 189, 190, 192, 210 **TE:** 563B, 561
7.8C Model the effects of human activity on groundwater and surface water in a watershed.	**SE: 501–503, 504–507,** 510–511, 514, 515, 518 **LM:** 155, **162, 172** **TE:** 501, 507

Texas Essential Knowledge and Skills	Where You Will Find It
7.9 Earth and space. The student knows components of our solar system. The student is expected to:	
7.9A Analyze the characteristics of objects in our solar system that allow life to exist such as the proximity of the Sun, presence of water, and composition of the atmosphere.	**SE:** **584–591,** 606, 607 **LM:** 193, 199 **TE:** 591
7.9B Identify the accommodations, considering the characteristics of our solar system, that enabled manned space exploration.	**SE:** 594–595, **596–599,** 604–605, 606, 607, 610, 611 **LM:** 195, 203 **TE:** 597, 599B
7.10 Organisms and environments. The student knows that there is a relationship between organisms and the environment. The student is expected to:	
7.10A Observe and describe how different environments, including microhabitats in schoolyards and biomes, support different varieties of organisms.	**SE:** 54, **391, 392–401,** 402–405, 422–423, **425,** 438, 439 **LM:** 119, **121, 122, 123,** 125, 207 **TE:** 377
7.10B Describe how biodiversity contributes to the sustainability of an ecosystem.	**SE:** 54, 60, 406, **407,** 408, 409–411, 412–415, **426** **LM:** 126
7.10C Observe, record, and describe the role of ecological succession such as in a microhabitat of a garden with weeds.	**SE:** **416–419,** 420–421, 423, 426, 430 **LM:** **128,** 191 **TE:** 421B
7.11 Organisms and environments. The student knows that populations and species demonstrate variation and inherit many of their unique traits through gradual processes over many generations. The student is expected to:	
7.11A Examine organisms or their structures such as insects or leaves and use dichotomous keys for identification.	**SE:** 64–65, 95, 179–180, 184–185, 190–191, 195, **192–193,** 208, 437, 438, 439, **452,** 453, 473 **LM:** 12, **59,** 63, **144,** 148 **TE:** 185, 453
7.11B Explain variation within a population or species by comparing external features, behaviors, or physiology of organisms that enhance their survival such as migration, hibernation, or storage of food in a bulb.	**SE:** 438, 443–444, 445, **446–449,** 450, 451, 458–460, 463, 464–467, 468, 471, **473,** 474, 475, 479 **LM:** 50, 129, 130, 141, 207 **TE:** 185B, 443, 448, 453B
7.11C Identify some changes in genetic traits that have occurred over several generations through natural selection and selective breeding such as the Galapagos Medium Ground Finch *(Geospiza fortis)* or domestic animals.	**SE:** 439, **440, 441, 442–445,** 450, **451** **LM:** 140 **TE:** 450
7.12 Organisms and environments. The student knows that living systems at all levels of organization demonstrate the complementary nature of structure and function. The student is expected to:	
7.12A Investigate and explain how internal structures of organisms have adaptations that allow specific functions such as gills in fish, hollow bones in birds, or xylem in plants.	**SE:** **181–183,** 184–185, 186–191, **204,** 448, 449, 479 **LM:** 51, 53, 58, **131–136** **TE:** 183
7.12B Identify the main functions of the systems of the human organism, including the circulatory, respiratory, skeletal, muscular, digestive, excretory, reproductive, integumentary, nervous, and endocrine systems.	**SE:** 218, **219,** 221–222, 223, 224, 225, 226, 227, 234–235, **237–238,** 239–240, **245–246,** 247, 248, **251–252,** 253, 256, **257, 258,** 262, **272, 277, 278–279,** 284–285, **288–291,** 292–295, **297,** 298–299, 300–301, **302, 303,** 305, **306,** 307, **317–318,** 324, 325–327, **329,** 330–333, 334, 335, **337–341,** 356, 357, 362, **363, 364,** 369 **LM:** 72, 76, 77, 78, 79, 80, 81, 82, 83, 84, 90, 91, 92, 93, 94, 96, 103, 204 **TE:** 227B
7.12C Recognize levels of organization in plants and animals, including cells, tissues, organs, organ systems, and organisms.	**SE:** 83, 96, **180,** 194–195, **204, 215–218, 219,** 238, 239, 241, 246, 253, 256, 283, 284-–285, 290, 301, 319–320, 321, 331, 345, 363 **LM:** 22, 68, 70
7.12D Differentiate between structure and function in plant and animal cell organelles, including cell membrane, cell wall, nucleus, cytoplasm, mitochondrion, chloroplast, and vacuole.	**SE:** **75, 76, 77,** 78–79, **80, 81,** 92, 96, **97,** 215, 256 **LM:** 14, 21 **TE:** 64, 81, 124, 125B

Texas Essential Knowledge and Skills	Where You Will Find It
7.12E Compare the functions of a cell to the functions of organisms such as waste removal.	**SE:** **67,** 90–93, **96,** 128, 289, 291, 306 **LM:** 26 **TE:** 115
7.12F Recognize that according to cell theory all organisms are composed of cells and cells carry on similar functions such as extracting energy from food to sustain life.	**SE:** 62–63, 67, **68–69,** 82, **95,** 115, 118–119, 120–124, 125, 129 **LM:** 14, 15
7.13 Organisms and environments. The student knows that a living organism must be able to maintain balance in stable internal conditions in response to external and internal stimuli. The student is expected to:	
7.13A Investigate how organisms respond to external stimuli found in the environment such as phototropism and fight or flight.	**SE:** 61, 63, **197–198,** 199–200, 209, 229–233, 300–301, 318, 324, 455 **LM:** **64**, 65 **TE:** 63, 232, 327B, 457
7.13B Describe and relate responses in organisms that may result from internal stimuli such as wilting in plants and fever or vomiting in animals that allow them to maintain balance.	**SE:** 179, **191, 204, 205, 229–231, 233,** 234–235, 257, 258, 274, 303, 306, 318, 333, 334–335, 358-361, **363,** 455 **LM:** **99** **TE:** 235B, 335B, 361B
7.14 Organisms and environments. The student knows that reproduction is a characteristic of living organisms and that the instructions for traits are governed in the genetic material. The student is expected to:	
7.14A Define heredity as the passage of genetic instructions from one generation to the next generation.	**SE:** **138, 140,** 154–155, 167, 444 **LM:** 40, 42, 47 **TE:** 155B
7.14B Compare the results of uniform or diverse offspring from sexual reproduction or asexual reproduction.	**SE:** 154, 155, **164–165, 168,** 169 **LM:** 49 **TE:** 165
7.14C Recognize that inherited traits of individuals are governed in the genetic material found in the genes within chromosomes in the nucleus.	**SE:** 141, 154, 157–159, **160–161, 168** **LM:** 48

English Language Proficiency Standards

The **Texas *Interactive Science*** program provides comprehensive coverage and assessment of the **Texas English Language Proficiency Standards (ELPS).** Use this chart to track where the ELPS are addressed throughout the Student Edition and Teacher's Edition.

KEY: **green** = in-depth coverage; **SE** = Student Edition; **TE** = Teacher's Edition

Texas English Language Proficiency Standards	Where You Will Find It
Learning Strategies (1) Cross-curricular second language acquisition/learning strategies. The ELL uses language learning strategies to develop an awareness of his or her own learning processes in all content areas. In order for the ELL to meet grade-level learning expectations across the foundation and enrichment curriculum, all instruction delivered in English must be linguistically accommodated (communicated, sequenced, and scaffolded) commensurate with the student's level of English language proficiency. The student is expected to:	
1.A Use prior knowledge and experiences to understand meanings in English;	**SE:** **11, 229** **TE:** **11, 229,** 601
1.B Monitor oral and written language production and employ self-corrective techniques or other resources;	**TE:** 61, 297, 391, **470**
1.C Use strategic learning techniques such as concept mapping, drawing, memorizing, comparing, contrasting, and reviewing to acquire basic and grade-level vocabulary;	**SE:** **66** **TE:** **67**
1.D Speak using learning strategies such as requesting assistance, employing non-verbal cues, and using synonyms and circumlocution (conveying ideas by defining or describing when exact English words are not known);	**SE:** **37** **TE:** **37**
1.F Use accessible language and learn new and essential language in the process.	**TE:** 58, **104,** 136, 176, 212, 266, 372, 434, 482, 522
Listening (2) Cross-curricular second language acquisition/listening. The ELL listens to a variety of speakers including teachers, peers, and electronic media to gain an increasing level of comprehension of newly acquired language in all content areas. ELLs may be at the beginning, intermediate, advanced, or advanced high stage of English language acquisition in listening. In order for the ELL to meet grade-level learning expectations across the foundation and enrichment curriculum, all instruction delivered in English must be linguistically accommodated (communicated, sequenced, and scaffolded) commensurate with the student's level of English language proficiency. The student is expected to:	
2.C Learn new language structures, expressions, and basic and academic vocabulary heard during classroom instruction and interactions;	**SE:** **19, 138** **TE:** **19, 94,** 126, **139,** 245, 263, 311, **557,** 593, 611
2.D Monitor understanding of spoken language during classroom instruction and interactions and seek clarification as needed;	**SE:** **351** **TE:** 2, 179, 329, **351,** 570, **585**
2.E Use visual, contextual, and linguistic support to enhance and confirm understanding of increasingly complex and elaborated spoken language;	**SE:** **499** **TE:** **499**
2.I Demonstrate listening comprehension of increasingly complex spoken English by following directions, retelling or summarizing spoken messages, responding to questions and requests, collaborating with peers, and taking notes commensurate with content and grade-level needs.	**SE:** **382, 485, 491** **TE:** 5, 145, 184, **221,** 362, 375, **383,** 447, **485,** 488, **491,** 505

Texas English Language Proficiency Standards	Where You Will Find It
Speaking (3) Cross-curricular second language acquisition/speaking. The ELL speaks in a variety of modes for a variety of purposes with an awareness of different language registers (formal/informal) using vocabulary with increasing fluency and accuracy in language arts and all content areas. ELLs may be at the beginning, intermediate, advanced, or advanced high stage of English language acquisition in speaking. In order for the ELL to meet grade-level learning expectations across the foundation and enrichment curriculum, all instruction delivered in English must be linguistically accommodated (communicated, sequenced, and scaffolded) commensurate with the student's level of English language proficiency. The student is expected to:	
3.B Expand and internalize initial English vocabulary by learning and using high-frequency English words necessary for identifying and describing people, places, and objects, by retelling simple stories and basic information represented or supported by pictures, and by learning and using routine language needed for classroom communication;	**SE:** **107, 417** **TE:** **107, 417**
3.C Speak using a variety of grammatical structures, sentence lengths, sentence types, and connecting words with increasing accuracy and ease as more English is acquired;	**SE:** **288, 337** **TE:** **289, 337**
3.D Speak using grade-level content area vocabulary in context to internalize new English words and build academic language proficiency;	**SE:** **118, 197** **TE:** **119, 197**
3.E Share information in cooperative learning interactions;	**SE:** **403** **TE:** **403**
3.F Ask and give information ranging from using a very limited bank of high-frequency, high-need, concrete vocabulary, including key words and expressions needed for basic communication in academic and social contexts, to using abstract and content-based vocabulary during extended speaking assignments;	**SE:** **215,** 251, **463** **TE:** **215,** 251, **463**
3.G Express opinions, ideas, and feelings ranging from communicating single words and short phrases to participating in extended discussions on a variety of social and grade-appropriate academic topics;	**TE:** 43, 133, 209, 269, **304,** 314, **407,** 479, 519, 582
3.H Narrate, describe, and explain with increasing specificity and detail as more English is acquired.	**SE:** **524** **TE:** **525**
Reading (4) Cross-curricular second language acquisition/reading. The ELL reads a variety of texts for a variety of purposes with an increasing level of comprehension in all content areas. ELLs may be at the beginning, intermediate, advanced, or advanced high stage of English language acquisition in reading. In order for the ELL to meet grade-level learning expectations across the foundation and enrichment curriculum, all instruction delivered in English must be linguistically accommodated (communicated, sequenced, and scaffolded) commensurate with the student's level of English language proficiency. The student is expected to:	
4.C Develop basic sight vocabulary, derive meaning of environmental print, and comprehend English vocabulary and language structures used routinely in written classroom materials;	**SE:** **29, 90, 279, 344** **TE:** **29, 91, 279, 345**
4.D Use prereading supports such as graphic organizers, illustrations, and pretaught topic-related vocabulary and other prereading activities to enhance comprehension of written text;	**SE:** **113** **TE:** **113**
4.E Read linguistically accommodated content area material with a decreasing need for linguistic accommodations as more English is learned;	**TE:** 85, 334, **565**
4.F Use visual and contextual support and support from peers and teachers to read grade-appropriate content area text, enhance and confirm understanding, and develop vocabulary, grasp of language structures, and background knowledge needed to comprehend increasingly challenging language;	**SE:** **75, 157, 186, 237, 251, 437, 454, 545, 553** **TE:** **75, 157, 187, 237, 251,** 375, 391, **437, 455, 545, 553**
4.G Demonstrate comprehension of increasingly complex English by participating in shared reading, retelling or summarizing material, responding to questions, and taking notes commensurate with content area and grade level needs.	**SE:** **150, 317, 535** **TE:** **151, 317, 535**

Master Guide to Grade 7 Lab Activities

Texas *Interactive Science* provides you with an extensive bank of labs. To easily locate a lab activity, look for the LabZone icon on the write-in Student Edition page to find the name and location of the lab. You can also use the guide below to determine whether a particular lab can be found in the online Lab Manual or online resources. All labs can be found in editable format at **PearsonTexas.com.**

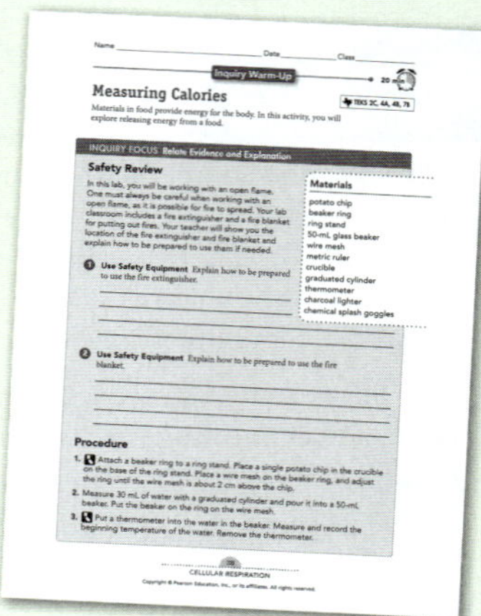

1 Every time you see a LabZone icon in the **Student Edition** it means there's an Inquiry activity allowing students to further explore lesson content. There are between 2 and 4 labs for every lesson.

Lab zone — Do the Inquiry Warm-Up *Measuring Calories.* Student Lab Manual, p. 28

3 Looking for even more opportunities to provide hands-on inquiry? All of the labs in the Lab Manual, plus an extensive additional group of lab activities, are available in an editable format. These additional labs delve further into the TEKS, reinforcing principles and refining student knowledge.

2 Activities in the Student Lab Manual have been specifically selected to provide your students with a robust, engaging exploration of the TEKS. These activities allow your students the opportunity to explore, investigate, demonstrate, and understand the TEKS through hands-on inquiry.

LESSON LAB ACTIVITIES	ONLINE LAB MANUAL	ONLINE RESOURCES
Chapter 1 Scientific Thinking		
Lesson 1 Science and the Natural World		
Is It Really True?		X
Classifying Objects	X	X
Lesson 2 Thinking Like a Scientist		
How Keen Are Your Senses?	X	X
Thinking Like a Scientist		X
Using Scientific Thinking		X
Lesson 3 Scientific Inquiry		
What's Happening?		X
Scientific Inquiry	X	X
Keeping Flowers Fresh	X	X
Theories and Laws		X

LESSON LAB ACTIVITIES	ONLINE LAB MANUAL	ONLINE RESOURCES
Lesson 4 Models as Tools in Science		
Scale Models		X
Making Models: Day and Night		X
Systems		X
Models in Nature	X	X
Lesson 5 Safety in the Science Laboratory		
Where Is the Safety Equipment in Your School?	X	X
Be Prepared		X
Just In Case		X
Lesson 6 Scientists and Society		
What Do Scientists Do?		X
Light Sources	X	X

LESSON LAB ACTIVITIES	ONLINE LAB MANUAL	ONLINE RESOURCES
Chapter 2 Introduction to Cells		
Lesson 1 The Characteristics of Life		
Is Yeast Alive or Not?		X
Modeling Trace Fossils		X
Classifying Animals	X	X
Lesson 2 Discovering Cells		
What Can You See?	X	X
Comparing Cells	X	X
Observing Cells	X	X
Design and Build a Microscope	X	X
Lesson 3 Looking Inside Cells		
How Large Are Cells?		X
Gelatin Cell Model	X	X
Tissues, Organs, Organ Systems	X	X
Lesson 4 Chemical Compounds in Cells		
Detecting Starch		X
What Is a Compound?	X	X
What's That Taste?	X	X
Lesson 5 Cells and Homeostasis		
Homeostasis		X
Effect of Concentration on Diffusion	X	X
Chapter 3 Cell Processes and Energy		
Lesson 1 Photosynthesis		
Where Does the Energy Come From?		X
Energy From the Sun	X	X
Looking at Pigments		X
Lesson 2 Cellular Respiration		
Measuring Calories	X	X
Exhaling Carbon Dioxide	X	X
Observing Fermentation		X
Lesson 3 Cell Division		
What Are the Yeast Cells Doing?		X
Observing Mitosis	X	X
Modeling Mitosis		X

LESSON LAB ACTIVITIES	ONLINE LAB MANUAL	ONLINE RESOURCES
Chapter 4 Genetics: The Science of Heredity		
Lesson 1 What Is Heredity?		
What Does the Father Look Like?		X
Observing Pistils and Stamens		X
Inferring the Parent Generation	X	X
Lesson 2 Probability and Heredity		
What's the Chance?	X	X
Coin Crosses		X
Make the Right Call	X	X
Lesson 3 Patterns of Inheritance		
Observing Traits		X
Patterns of Inheritance		X
Is It All in the Genes?	X	X
Lesson 4 Chromosomes and Inheritance		
Which Chromosome Is Which?		X
Chromosomes and Inheritance	X	X
Modeling the Genetic Code		X
Modeling Meiosis		X
Comparing Mitosis and Meiosis	X	X
Chapter 5 Plant Structure, Function, and Response		
Lesson 1 What Is a Plant?		
What Do Leaves Reveal About Plants?		X
Algae and Other Plants		X
Local Plant Diversity	X	X
Will Mosses Absorb Water?	X	X
Lesson 2 Plant Structures		
Which Plant Part Is It?		X
Investigating Stomata	X	X
Fruit Basket	X	X
Modeling Flowers	X	X
Lesson 3 Plant Responses and Growth		
Can a Plant Respond to Touch?	X	X
Watching Roots Grow	X	X
Seasonal Changes	X	X

LESSON LAB ACTIVITIES	ONLINE LAB MANUAL	ONLINE RESOURCES
Chapter 6 Introduction to the Human Body		
Lesson 1 Body Organization		
How Is Your Body Organized?	X	X
Observing Cells and Tissues	X	X
Lesson 2 System Interactions		
How Does Your Body Respond?		X
A Look Beneath the Skin	X	X
Working Together, Act I	X	X
Working Together, Act II	X	X
Lesson 3 Homeostasis		
Out of Balance		X
Working to Maintain Balance	X	X
Lesson 4 The Skeletal System		
Hard as a Rock?		X
The Skeleton	X	X
Observing Joints	X	X
Soft Bones?		X
Lesson 5 The Muscular System		
How Do Muscles Work?	X	X
Observing Muscle Tissue	X	X
Modeling How Skeletal Muscles Work		X
Lesson 6 The Integumentary System		
What Can You Observe About Skin?		X
Sweaty Skin	X	X
Chapter 7 Managing Materials in the Body		
Lesson 1 The Digestive System		
Where Does Digestion Start?	X	X
Nutrient Identification	X	X
As the Stomach Churns		X
Lesson 2 The Circulatory System		
Observing a Heart		X
Direction of Blood Flow	X	X
Do You Know Your A-B-Os?		X

LESSON LAB ACTIVITIES	ONLINE LAB MANUAL	ONLINE RESOURCES
Lesson 3 The Respiratory System		
How Big Can You Blow Up a Balloon?		X
Modeling Respiration	X	X
A Breath of Fresh Air		X
Lesson 4 The Excretory System		
How Does Filtering a Liquid Change the Liquid?		X
Kidney Function		X
Perspiration	X	X
Chapter 8 Controlling Body Processes		
Lesson 1 The Nervous System		
How Simple Is a Simple Task?		X
Ready or Not!		X
How Does Your Knee React?	X	X
Working Together		X
Lesson 2 The Endocrine System		
What's the Signal?		X
Making Models: Hormones	X	X
Modeling Negative Feedback	X	X
Lesson 3 The Male and Female Reproductive Systems		
What's the Big Difference?		X
Reproductive Systems		X
Looking at Hormone Levels	X	X
Lesson 4 Pregnancy and Birth		
Prenatal Growth		X
Way to Grow!		X
Egg-cellent Protection	X	X
Labor and Delivery		X
Lesson 5 The Body's Defenses		
Which Pieces Fit Together?		X
How Does a Disease Spread?		X
The Skin as a Barrier	X	X
Stuck Together		X

LESSON LAB ACTIVITIES	ONLINE LAB MANUAL	ONLINE RESOURCES
Chapter 9 Ecosystems and Biomes		
Lesson 1 Energy Flow in Ecosystems		
Where Did Your Dinner Come From?		X
Observing Decomposition	X	X
Last Remains	X	X
Lesson 2 Cycles of Matter		
Are You Part of a Cycle?		X
Following Water		X
Considering Fungi as Decomposers	X	X
Playing Nitrogen Cycle Roles	X	X
Lesson 3 Biomes		
How Much Rain Is That?	X	X
From Microhabitats to Biomes	X	X
Inferring Forest Climates		X
Lesson 4 Aquatic Ecosystems		
Where Does It Live?		X
Dissolved Oxygen	X	X
Lesson 5 Biodiversity		
How Much Variety Is There?		X
Modeling Keystone Species		X
Grocery Gene Pool and Potato Famine	X	X
Humans and Biodiversity		X
Lesson 6 Changing Ecosystems		
How Communities Change		X
Primary or Secondary	X	X
Change in a Microhabitat		X
Chapter 10 Organisms and the Environment		
Lesson 1 Change Over Time		
How Do Living Things Vary?	X	X
Selective Breeding		X
Adaptations of Birds		X
Lesson 2 Adaptations and Variations		
Eggs-tra Protection	X	X

LESSON LAB ACTIVITIES	ONLINE LAB MANUAL	ONLINE RESOURCES
Investigating Bird and Mammal Bones	X	X
Bird Beak Adaptations	X	X
Identifying Vertebrates Using Dichotomous Keys	X	X
Lesson 3 What Is Behavior?		
What Behaviors Can You Observe?		X
Insect Behavior	X	X
Learning New Words	X	X
Lesson 4 Patterns of Behavior		
Communicating Without Words	X	X
Modeling Animal Communication		X
One For All		X
Behavior Cycles		X
Chapter 11 Fresh Water		
Lesson 1 Water on Earth		
Where Does the Water Come From?		X
Water, Water Everywhere		X
Water on Earth	X	X
Water From Trees	X	X
Lesson 2 Surface Water		
Mapping Surface Waters		X
Modeling a Watershed	X	X
Modeling How a Lake Forms		X
Field Testing a Body of Water	X	X
Lesson 3 Water Underground		
Where Does the Water Go?		X
Soil Percolation	X	X
An Artesian Well		X
Lesson 4 Water Pollution and Solutions		
Groundwater Pollution	X	X
Testing Water	X	X
Getting Clean		X

LESSON LAB ACTIVITIES	ONLINE LAB MANUAL	ONLINE RESOURCES
Chapter 12 Weathering, Erosion, and Deposition		
Lesson 1 Weathering and Mass Movement		
How Does Gravity Affect Materials on a Slope?		X
Freezing and Thawing		X
Rusting Away		X
It's All on the Surface		X
Sand Hills	X	X
Lesson 2 Water Erosion		
How Does Moving Water Wear Away Rocks?		X
Raindrops Falling	X	X
Erosion Cube		X
Lesson 3 Wave and Glacial Erosion		
What Is Sand Made Of?		X
Shaping a Coastline	X	X
Surging Glaciers		X
Modeling Valleys		X
Lesson 4 Wind Erosion		
How Does Moving Air Affect Sediment?		X
Desert Pavement	X	X

LESSON LAB ACTIVITIES	ONLINE LAB MANUAL	ONLINE RESOURCES
Lesson 5 Weathering, Erosion, and Deposition in Texas		
How Can You Change Land?	X	X
Local Landscapes	X	X
How Does Soil Type Affect Plant Growth?	X	X
Lesson 6 Ecosystems and Catastrophic Events		
What Can Be Done About the Risk of Wildfire?		X
What Happened Here?	X	X
Humans and Erosion	X	X
Chapter 13 Space Exploration		
Lesson 1 Our Solar System		
What Is the Greenhouse Effect?	X	X
Build a Space Exploration Vehicle	X	X
Liquid Water and Water Vapor	X	X
Lesson 2 Working and Living in Space		
Space Wash		X
What Is Work?	X	X
What Do You Need to Survive in Space?		X
Lesson 3 Space Technology		
Using Space Science	X	X
Useful Satellites		X

Lab Notes

Scientific Thinking

Chapter TEKS Overview

This chapter focuses on TEKS 1A, 2A, 2B, 3A, and 4B. Students will explore the inquiry methods that scientists use to study the natural world. They will also learn about the importance of critical thinking and scientific reasoning, as well as the impact that scientific research has on scientific thought and society. Finally, students will learn about the importance of following safety guidelines in the lab and in the field.

Introduce the TEKS

Have students look at the image and read the Focus Question and description. Ask the students to infer what other things NASA astronauts might study in space. Have volunteers share their ideas with the class. Point out that the astronaut pictured is using a camera. Ask: **What might an astronaut record in space using a camera?** *(Sample answer: the actions or appearance of other astronauts, or growing plants or crystals)* **What other instruments might they use?** *(Sample answer: thermometer, computer, ruler, balance, scale)* **What questions would you like answered about space?** *(Sample answer: Do astronauts eat less in space? Do they use less energy?)*

Untamed Science Video

MIMICKING NATURE Before viewing, invite students to suggest ways in which technology has affected their lives. Then play the video. Lead a class discussion and make a list of questions that this video raises. You may wish to have students view the video again after they have completed the chapter to see if their questions have been answered.

Chapter at a Glance

CHAPTER PACING: 12–18 periods or 6–9 blocks

INTRODUCE THE CHAPTER: Use the Focus Question and the opening image to get students thinking about how scientists investigate the natural world. Activate prior knowledge and preteach vocabulary using the Getting Started pages.

Lesson 1: Science and the Natural World

Lesson 2: Thinking Like a Scientist

Lesson 3: Scientific Inquiry

Lesson 4: Models as Tools in Science

Lesson 5: Safety in the Science Laboratory

Lesson 6: Scientists and Society

ASSESSMENT OPTIONS:
Teacher's Edition: Lesson Quizzes, Texas End-of-Year Test Prep A and B
Online assessments

HOW CAN AN ASTRONAUT STUDY GRAVITY WHILE FLOATING?

FOCUS ON TEKS 2A, 2B, 3A

How do scientists investigate the natural world?

NASA studies how microgravity, or very little gravity, affects humans, plants, crystals, and liquids. For example, NASA has found that the muscles and bones of astronauts weaken during space missions. Plants grow in different directions, crystals grow larger, and water does not pour as it would on Earth, but falls out in spheres.

Infer What other ideas might NASA study in space?

Sample: NASA could study how fast the hair and teeth of astronauts grow during space missions.

Watch the **Untamed Science** video to learn more about science.

UntamedScience

Professional Development Note — From the Author

A critical part of adult thinking is the ability to design and carry out a controlled experiment. However, this seemingly simple activity, a part of our everyday lives, is not so simple developmentally for your students. When I was in graduate school, I studied the work of Jean Piaget in detail. I was surprised to find out that Piaget measured students' abilities to think abstractly by asking them to design science experiments. Research shows that adolescents cannot yet reason abstractly, at least independently. The experiments students perform in our classrooms provide useful practice for doing so. Your students will need many opportunities to practice their abilities with experimenting in order to get good at it. Be patient!

Michael Padilla

Texas CHAPTER 1

Scientific Thinking

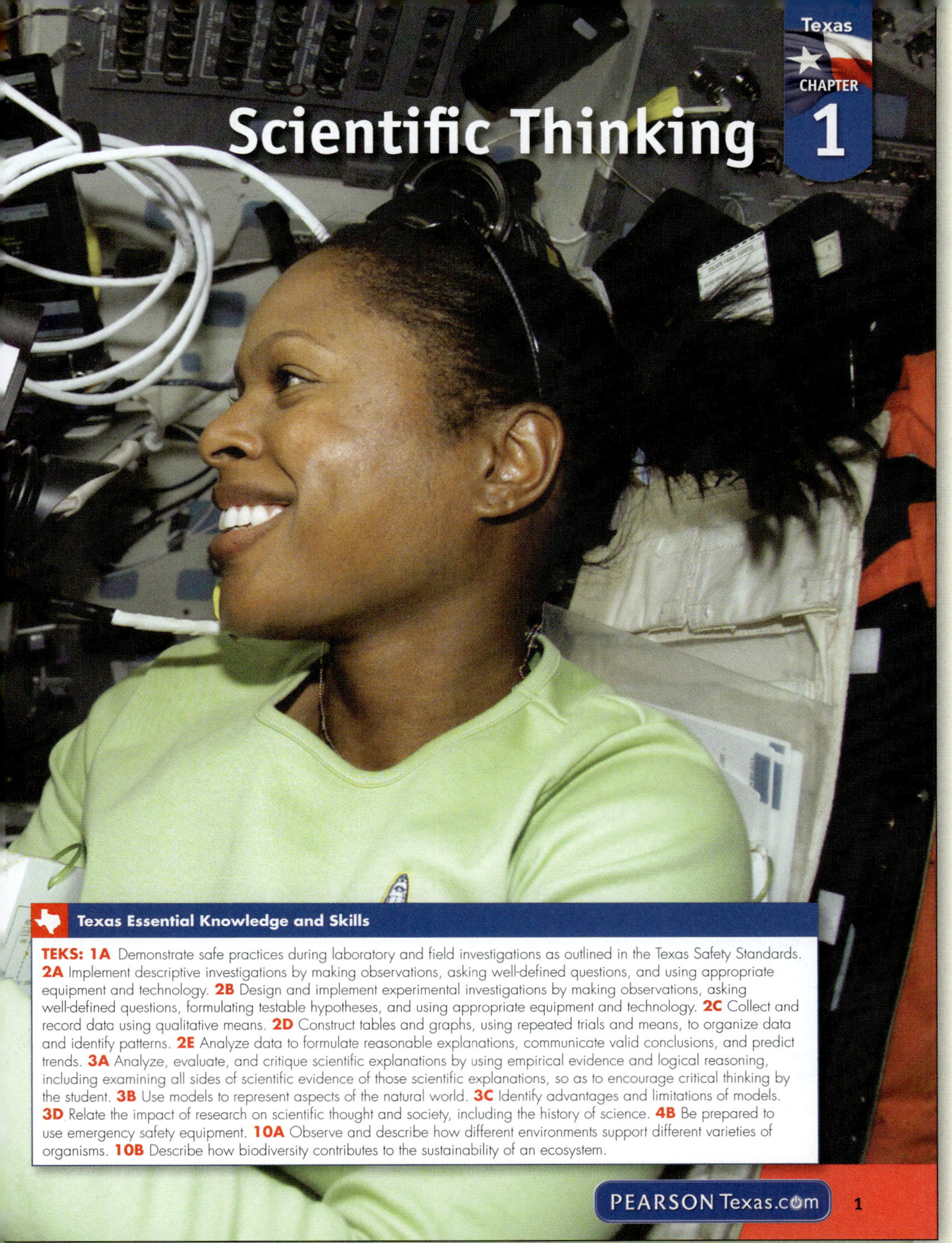

Texas Essential Knowledge and Skills

TEKS: 1A Demonstrate safe practices during laboratory and field investigations as outlined in the Texas Safety Standards. **2A** Implement descriptive investigations by making observations, asking well-defined questions, and using appropriate equipment and technology. **2B** Design and implement experimental investigations by making observations, asking well-defined questions, formulating testable hypotheses, and using appropriate equipment and technology. **2C** Collect and record data using qualitative means. **2D** Construct tables and graphs, using repeated trials and means, to organize data and identify patterns. **2E** Analyze data to formulate reasonable explanations, communicate valid conclusions, and predict trends. **3A** Analyze, evaluate, and critique scientific explanations by using empirical evidence and logical reasoning, including examining all sides of scientific evidence of those scientific explanations, so as to encourage critical thinking by the student. **3B** Use models to represent aspects of the natural world. **3C** Identify advantages and limitations of models. **3D** Relate the impact of research on scientific thought and society, including the history of science. **4B** Be prepared to use emergency safety equipment. **10A** Observe and describe how different environments support different varieties of organisms. **10B** Describe how biodiversity contributes to the sustainability of an ecosystem.

PEARSON Texas.com

Spotlight On Technology

Interactivity Make images come to life to help illustrate the work of some super scientists with an interactivity that asks students to engage with the content in an active, visual way.

Flipped Video for Science Show students a Pearson Flipped Video for Science for this chapter so that they can better understand science and the natural world, how to think like a scientist, and how to use models in science.

PEARSON Texas.com

Texas Essential Knowledge and Skills

1A Demonstrate safe practices during laboratory and field investigations as outlined in the Texas Safety Standards.

2A Plan and implement comparative and descriptive investigations by making observations, asking well-defined questions, and using appropriate equipment and technology.

2B Design and implement experimental investigations by making observations, asking well-defined questions, formulating testable hypotheses, and using appropriate equipment and technology.

2C Collect and record data using the International System of Units (SI) and qualitative means such as labeled drawings, writing, and graphic organizers.

2D Construct tables and graphs, using repeated trials and means, to organize data and identify patterns.

2E Analyze data to formulate reasonable explanations, communicate valid conclusions supported by the data, and predict trends.

3A In all fields of science, analyze, evaluate, and critique scientific explanations by using empirical evidence, logical reasoning, and experimental and observational testing, including examining all sides of scientific evidence of those scientific explanations, so as to encourage critical thinking by the student.

3B Use models to represent aspects of the natural world such as human body systems and plant and animal cells.

3C Identify advantages and limitations of models such as size, scale, properties, and materials.

3D Relate the impact of research on scientific thought and society, including the history of science and contributions of scientists as related to the content.

4B Use preventative safety equipment, including chemical splash goggles, aprons, and gloves, and be prepared to use emergency safety equipment, including an eye/face wash, a fire blanket, and a fire extinguisher.

10A Observe and describe how different environments, including microhabitats in schoolyards and biomes, support different varieties of organisms.

10B Describe how biodiversity contributes to the sustainability of an ecosystem.

The following **College and Career Readiness Standards** are covered in this chapter: **I.A.1, I.A.3, I.A.4, I.B.1, III.D.2**

Getting Started

Check Your Understanding

This activity assesses students' understanding of the process of posing questions and looking for evidence. After students have shared their answers, point out that science, or learning about the natural world, is a process that includes posing questions and gathering evidence.

Preteach Vocabulary Skills

Explain to students that related words share the same root. Often, these words differ in the affixes used with them. Learning to identify related word forms can make it easier to learn new vocabulary words. Also point out to students the forms of the verbs *manipulate, classify,* and *evaluate* on the next page. Point out that *manipulation, classification,* and *evaluation* are nouns related to these verbs.

Getting Started

Check Your Understanding

1. **Background** Read the paragraph below and then answer the question.

> Miki is in the **process** of preparing a stew for dinner at her campsite. After it is cooked, she sets the pot aside to cool. When she returns, the pot is empty. Immediately, she **poses** questions: Who ate the stew? What animals are active in the evening? She soon finds **evidence:** the pot cover, greasy spills, and a stinky smell. The thief is a skunk.

A **process** is a series of actions or events.

To **pose** is to put forward a question or a problem.

Facts, figures, or signs that help prove a statement are all pieces of **evidence.**

- How does the process of posing questions and looking for evidence help Miki solve the mystery of the missing stew?

  Her questions provide likely explanations to choose from.

Vocabulary Skill

Identify Related Word Forms Learn related forms of words to increase your vocabulary. The table below lists forms of words related to vocabulary terms.

Verb	Noun	Adjective
observe, *v.* to gather information using the senses	observation, *n.* facts learned by gathering information using the senses	observable, *adj.* able to be heard, seen, touched, tasted, or smelled
predict, *v.* to state or claim what will happen in the future	prediction, *n.* a statement or claim of what will happen in the future	predictable, *adj.* able to be predicted; behaving in a way that is expected

2. **Quick Check** **Complete the sentence with the correct form of the word.**

- It is difficult to predict ______ how much rain will fall.

2 Scientific Thinking

English Language Proficiency Standards

ELPS Listening 2.D.1

Have a discussion with students about the terms on page 3 to preview chapter vocabulary.

Beginning Have students practice asking for clarification of vocabulary terms. As you read the definitions and examples on page 3, model how to ask about unfamiliar words: *What does ______ mean?* Have students ask about unfamiliar words.

Intermediate Before you preview the terms on page 3, remind students that they can ask questions when they don't understand something. As you read the definitions and examples, have students ask questions about unfamiliar words or meanings.

Advanced Encourage students to ask questions about the meanings of the vocabulary terms and to use follow-up questions to confirm their understanding.

Advanced High Have students try to answer each other's questions about the vocabulary terms. Provide feedback.

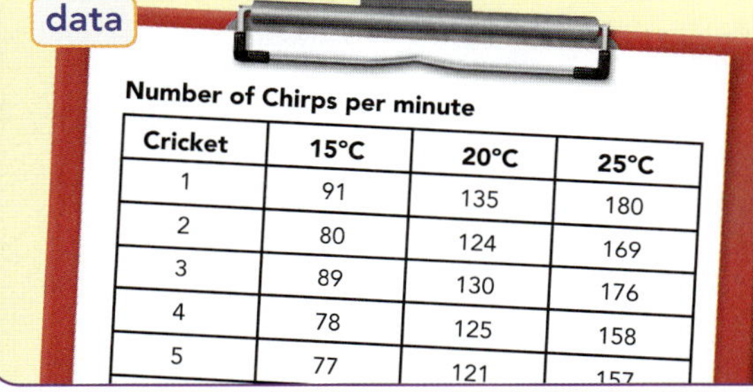
data

Number of Chirps per minute

Cricket	15°C	20°C	25°C
1	91	135	180
2	80	124	169
3	89	130	176
4	78	125	158
5	77	121	157

Chapter Preview

LESSON 1
- science • observing
- quantitative observation
- qualitative observation
- inferring • predicting
- classifying • evaluating • models
- Ask Questions
- Predict

LESSON 2
- skepticism • ethics
- personal bias • cultural bias
- experimental bias • objective
- subjective • deductive reasoning
- inductive reasoning
- Relate Cause and Effect
- Classify

LESSON 3
- scientific inquiry • hypothesis
- variable • manipulated variable
- responding variable
- controlled experiment
- data • SI • scientific theory
- scientific law
- Sequence
- Control Variables

LESSON 4
- system • input • process
- output • feedback
- Identify the Main Idea
- Make Models

LESSON 5
- safety symbol • field
- Summarize
- Observe

LESSON 6
- controversy
- Sequence
- Predict

Preview Vocabulary Terms

Have students work together to create a word wall to display the vocabulary terms for the chapter. Be sure to discuss and analyze each term before posting it on the wall. As the class progresses through the chapter, the words can be sorted and categorized in different ways.

L1 Have students look at the images on this page as you pronounce the vocabulary word. Have students repeat the word after you. Then read the definition below. Use the sample sentence in italics to clarify the meaning of the term.

observing *(ahb SURV ing)* The process of using one or more of our senses to gather information. *When Paul is noticing the color of a rock and measuring its mass, he is observing.*

subjective *(sub JEK tiv)* A situation in which personal feelings have entered into a decision or conclusion. *Since it was based on a dislike of water, Ms. Green's decision to travel by plane instead of boat was subjective.*

data *(DAY tuh)* Facts, figures, and other evidence gathered through observations. *Mike collected data on the thickness and texture of the tree bark.*

safety symbols (SAYF tee SIM buhls) Icons that alert you to possible sources of accidents in a lab investigation. *Safety symbols can warn you if you need to use special equipment to protect yourself during a lab.*

Academic Vocabulary

Each lesson includes key Academic Vocabulary. See also the Support All Readers box at the start of the lesson.

Lesson 1: predict, question

Lesson 2: cause, classify, effect, relate

Lesson 3: sequence, variables

Lesson 4: identify, model, main idea

Lesson 5: observe, summarize

Lesson 6: predict, sequence

Science and the Natural World

How do scientists investigate the natural world?

LESSON PACING:
2–3 periods or 1–$1\frac{1}{2}$ blocks

Lesson Vocabulary

- science
- observing
- quantitative observation
- qualitative observation
- inferring
- predicting
- classifying
- evaluating
- models

Lesson Objectives	TEKS	ELPS
Identify skills scientists use to learn about the natural world.	2D, 2E, 3B	2.I.4

Content Refresher

Skills and Attitudes Jane Goodall exemplifies the scientific skills and attitudes needed to investigate the natural world. Born and raised in England, Goodall first visited Kenya at the age of 22. She returned the following year and met with famed anthropologist Dr. Louis Leakey, who hired Goodall as his assistant. Recognizing Goodall's resourcefulness and patience—qualities needed for field research—Leakey hired Goodall to study wild chimpanzees in Tanzania. Goodall began her observations on the shore of Lake Tanganyika in 1960. Among the many important discoveries that Goodall made about chimpanzees was that these primates strip leaves off twigs to make a tool for finding termites in nests. Until this discovery, scientists thought that only humans made tools. Goodall also observed that chimps eat meat, disproving the widely accepted notion that they were herbivores. In 1977, she established the Jane Goodall Institute for Wildlife Research, Education, and Conservation. Goodall's field of biology is called ethology, which is the comparative study of animal behavior in the natural environment.

Texas Essential Knowledge and Skills

2D Construct tables and graphs, using repeated trials and means, to organize data and identify patterns.
2E Analyze data to formulate reasonable explanations, communicate valid conclusions supported by the data, and predict trends.
3B Use models to represent aspects of the natural world such as human body systems and plant and animal cells.

English Language Proficiency Standards

ELPS Listening 2.I.4 Demonstrate listening comprehension of increasingly complex spoken English by collaborating with peers commensurate with content and grade-level needs.

DIFFERENTIATED INSTRUCTION KEY
L1 Struggling Students or Special Needs
L2 On-Level Students L3 Advanced Students

LESSON PLANNER 1.1

Investigations and Activities	TEKS Review
My Planet Diary, **Student Edition,** p. 4 Inquiry: Inquiry Warm-Up, Is It Really True?, **PearsonTexas.com** Introduce Vocabulary, **Teacher's Edition,** p. 5 Teach Key Concepts, **Teacher's Edition,** p. 5 21st Century Learning, Interpersonal Skills, **Teacher's Edition,** p. 5 Support the TEKS, Interpret Observations, **Teacher's Edition,** p. 6 Address Misconceptions, Making Incorrect Inferences, **Teacher's Edition,** p. 6 21st Century Learning, Critical Thinking, **Teacher's Edition,** p. 6 Lead a Discussion, Make a Prediction, **Teacher's Edition,** p. 7 Do the Math!, **Student Edition,** p. 7 Differentiated Instruction, **Teacher's Edition,** p. 7 Lead a Discussion, Classifying, **Teacher's Edition,** p. 8 Inquiry: Teacher Demo, Classifying Plants, **Teacher's Edition,** p. 8 Lead a Discussion, Using Models, **Teacher's Edition,** p. 9 Teach With Visuals, **Teacher's Edition,** p. 9 Differentiated Instruction, **Teacher's Edition,** p. 9 Inquiry: Quick Lab, Classifying Objects, **Lab Manual,** p. 1	Apply the TEKS, Evaluating Scientific Explanations, **Student Edition,** p. 46 TEKS Practice, **Student Edition,** p. 48 TEKS Practice: Chapter and Cumulative Review, **Student Edition,** p. 52 Lesson 1.1, **TEKS Preparation and Study Guide Workbook,** p. 2 **SHORT ON TIME?** To do this lesson in approximately half the time, do the Activate Prior Knowledge activity. A discussion of the Key Concepts will familiarize students with the lesson content. Have students do the Quick Lab. The rest of the lesson can be completed by students independently.

These editable worksheets are available on **PearsonTexas.com.**
Print versions can be found in the **TEKS Preparation and Study Guide Workbook.**

Name ________ Date ________ Class ________

1.1 Science and the Natural World

Key Concept Summary

What Skills Do Scientists Use?

Science is a way of learning about the natural world. Science also includes all the knowledge gained by exploring the natural world. **Scientists use skills such as observing, inferring, predicting, classifying, evaluating, and making models to study the world.**

Observing means using one or more of your senses to gather information. It also means using tools to help your senses. A **quantitative observation** deals with numbers, or amounts. A **qualitative observation** deals with descriptions that cannot be expressed in numbers.

When you explain or interpret the things you observe, you are **inferring,** or making an inference. Inferring is not guessing. Inferences are based on reasoning from what you already know. **Predicting** means making a statement or claim about what will happen in the future based on past experience or evidence. Predictions and inferences are closely related. While inferences are attempts to explain what is happening or *has* happened, predictions are statements of claims about what *will* happen.

Classifying is the grouping together of items that are alike in some way. **Evaluating** involves comparing observations and data to reach a conclusion about them.

Models are representatives of processes or objects that are complex or too small or large to be observed. Some models can be touched, such as a map. Others are in the form of mathematical equations or computer programs.

2

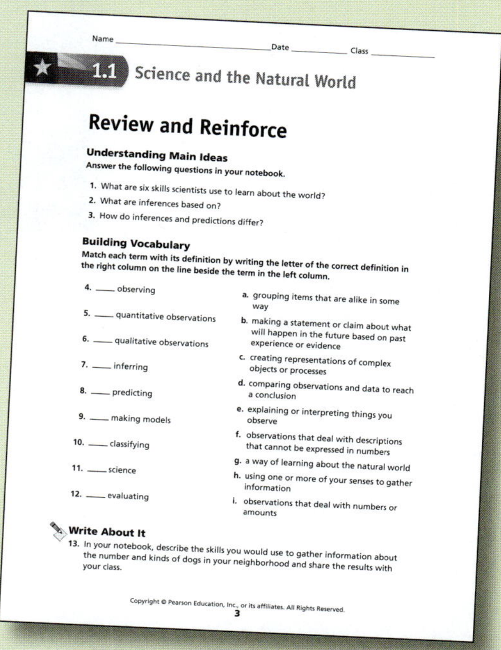

Name ________ Date ________ Class ________

1.1 Science and the Natural World

Review and Reinforce

Understanding Main Ideas

Answer the following questions in your notebook.

1. What are six skills scientists use to learn about the world?
2. What are inferences based on?
3. How do inferences and predictions differ?

Building Vocabulary

Match each term with its definition by writing the letter of the correct definition in the right column on the line beside the term in the left column.

4. ____ observing
5. ____ quantitative observations
6. ____ qualitative observations
7. ____ inferring
8. ____ predicting
9. ____ making models
10. ____ classifying
11. ____ science
12. ____ evaluating

a. grouping items that are alike in some way
b. making a statement or claim about what will happen in the future based on past experience or evidence
c. creating representations of complex objects or processes
d. comparing observations and data to reach a conclusion
e. explaining or interpreting things you observe
f. observations that deal with descriptions that cannot be expressed in numbers
g. a way of learning about the natural world
h. using one or more of your senses to gather information
i. observations that deal with numbers or amounts

Write About It

13. In your notebook, describe the skills you would use to gather information about the number and kinds of dogs in your neighborhood and share the results with your class.

3

Lexile Measure = 860L

LESSON 1.1

Science and the Natural World

Establish Learning Objective

After this lesson, students will be able to:

Identify skills scientists use to learn about the natural world.

Engage

Activate Prior Knowledge

MY PLANET DIARY Read *The Wild Chimpanzees of Gombe* with the class. Encourage students to describe experiences in science from previous years that involved observing and interpreting what they observed in nature. Ask: **What skills must a scientist like Jane Goodall possess to learn about the natural world?** *(Students may suggest skills like inferring and predicting. Accept all reasonable responses at this time.)*

Explore

Lab Resource: Inquiry Warm-Up

L3 **IS IT REALLY TRUE?** Students will design and conduct a scientific experiment to test whether a common belief is true or false. This Inquiry Warm-Up can be found online.

Science and the Natural World

What Skills Do Scientists Use?

TEKS 2D, 2E, 3B

MY PLANET DIARY

BIOGRAPHY

The Wild Chimpanzees of Gombe

The following words are from the writings of Jane Goodall, a scientist who studied wild chimpanzees in Africa for many years.

"Once, as I walked through thick forest in a downpour, I suddenly saw a chimp hunched in front of me. Quickly I stopped. Then I heard a sound from above. I looked up and there was a big chimp there, too. When he saw me he gave a loud, clear wailing *wraaaaah*—a spine-chilling call that is used to threaten a dangerous animal. To my right I saw a large black hand shaking a branch and bright eyes glaring threateningly through the foliage. Then came another savage *wraaaah* from behind...I was surrounded." Because Jane stood still, the chimps no longer felt threatened, so they went away.

Answer the question.

What is one advantage and one disadvantage of studying wild animals in their natural environment?

Sample: One advantage is that you can observe how they interact with their environment. One disadvantage is that you cannot control what the animals do.

Lab zone Do the Inquiry Wa[rm-Up] *Is It Really True?* Find the lab online

4 Scientific Thinking

SUPPORT ALL READERS

Lexile Measure = 860L **Lexile Word Count = 1211**

Prior Exposure to Content: Many students may have misconceptions on this topic

Academic Vocabulary: *predict, question*

Science Vocabulary: *science, observing, inferring, predicting*

Concept Level: Generally appropriate for most students in this grade

Preteach With: My Planet Diary "The Wild Chimpanzee of Gombe" and Figure 5 activity

Vocabulary
- science • observing • quantitative observation
- qualitative observation • inferring • predicting
- classifying • evaluating • models

Skills
Reading: Ask Questions
Inquiry: Predict

What Skills Do Scientists Use?

TEKS 2D, 2E, 3B In this section, you'll explore the different skills and tools that scientists use to investigate the natural world.

Jane Goodall trained herself to become a scientist, or a person who does science. **Science** is a way of learning about the natural world. Science also includes all the knowledge gained by exploring the natural world.  **Scientists use skills such as observing, inferring, predicting, classifying, evaluating, and making models to study the world.**

Observing

Observing means using one or more of your senses to gather information. It also means using tools, such as a microscope, to help your senses. By observing chimps like the one in **Figure 1,** Jane Goodall learned what they eat. She also learned what sounds chimps make and even what games they play.

Observations can be either quantitative or qualitative. A **quantitative observation** deals with numbers, or amounts. For example, seeing that you have 11 new text messages is a quantitative observation. A **qualitative observation** deals with descriptions that cannot be expressed in numbers. Noticing that a bike is blue or that a lemon tastes sour is a qualitative observation.

Ask Questions In the graphic organizer ask a *what, how,* or *why* question based on the text under Observing. As you read, write an answer to your question.

Thinking Like a Scientist

Question

Sample: What types of observations do scientists make?

Answer

Sample: Scientists make qualitative and quantitative observations.

FIGURE 1

Observing

A chimpanzee uses a rock as a tool to crack open a nut.

 **Observe** Write one quantitative observation and one qualitative observation about this chimp.

Sample: The chimp holds one rock in one hand (quantitative). The chimp is looking at the nut (qualitative).

Explain

Introduce Vocabulary

On the board, list all the vocabulary words including *observing* that end with *-ing,* and ask students to identify what these terms have in common. Remind students that *-ing* is a suffix used to express an action in the present without a subject doing the action (a present participle). Identify for students the root verbs of the listed terms.

Teach Key Concepts

Explain to students that in order to learn about the natural world, scientists rely on certain skills. These skills include observing, inferring, predicting, classifying, evaluating, and making models. Ask: **What do you do when you are observing?** *(I use my senses to get information.)* **What is the result of observing** *(An observation is made)* **What is a quantitative observation?** *(An observation that deals with a number, or amount)* **What is a qualitative observation?** *(An observation that deals with descriptions that cannot be expressed in numbers)*

Ask Questions Remind students that scientists often begin learning about the world around them by asking a question. Encourage students to ask questions about what they would like to find out about the skills scientists use.

21st Century Learning

INTERPERSONAL SKILLS Have student pairs observe the classroom and everything in it. Then have them work together to make a list of five qualitative and five quantitative observations. Have pairs exchange lists and check each other's list. Ask: **Did you correctly classify each of your observations by type? If not, how do you think you went wrong?** *(Accept all reasonable responses.)*

PEARSON Texas.com

English Language Proficiency Standards

ELPS Listening 2.1.4
Have students complete the Ask Questions activity on this page.

Beginning Have beginners listen as you read aloud the text and the Ask Questions activity. Ask what, how, or why questions based on the text. Encourage students to respond however they can—single words, phrases, gestures, or drawings.

Intermediate Read aloud with students the text and the Ask Questions activity. Have small groups complete the activity orally.

Advanced Have partners read the text and complete the Ask Questions activity together.

Advanced High Have students read the text and complete the Ask Questions activity. Then have students pair up to ask and answer the questions.

Texas Essential Knowledge and Skills

2D Construct tables and graphs, using repeated trials and means, to organize data and identify patterns.

2E Analyze data to formulate reasonable explanations, communicate valid conclusions supported by the data, and predict trends.

3B Use models to represent aspects of the natural world such as human body systems and plant and animal cells.

LESSON 1.1

Explain

Support the TEKS

INTERPRET OBSERVATIONS Tell students that making observations, or observing, is the first step in the process of learning about the world. The next step may be to explain or interpret these observations. Ask: **What observations did Jane Goodall make concerning chimpanzees and tree hollows?** *(She observed a chimp chew leaves, push them in a tree hollow, and then put them back into its mouth. She also observed the water on the leaves when pulled out of the tree hollow.)* **What inference did she make from her observations?** *(She inferred that there was water in the tree hollow.)* **What inference can you make about how the chimp used the leaves?** *(Sample: The chimp was using the leaves like a sponge to soak up and get water.)*

Address Misconceptions

L1 **MAKING INCORRECT INFERENCES** Explain that an inference made from an observation may be incorrect, even when the inference seems obvious. Point out that many people take vitamin C whenever they get a cold. Because they've quickly recovered from a cold when taking vitamin C in the past, they infer that the vitamin "cures" colds. Ask: **What is the observation?** *(The person quickly recovered from a cold after taking vitamin C.)* **What is the inference?** *(Vitamin C cured the cold.)* **How could you prove that taking vitamin C cures colds?** *(Carry out experiments testing whether people who took vitamin C recovered from colds faster than those who didn't.)* **If you skipped breakfast on a day that you aced a math test, would a reasonable inference be that skipping a meal caused your success?** *(Some students may say yes. Others will argue that skipping the meal had nothing to do with acing the test.)*

Elaborate

21st Century Learning

CRITICAL THINKING Give students this fictional scenario: A student is missing from his usual bus to school. When he finally arrives, he looks tired. Ask: **What might you infer from these observations?** *(Sample: Given these observations as well as past experience, I infer that the student overslept. Or the student may have walked to school.)*

Inferring One day, Jane watched as a chimp peered into a tree hollow. The chimp picked up a handful of leaves and chewed on them. Then, it took the leaves out of its mouth and pushed them into the hollow. When the chimp pulled the leaves out, Jane saw the gleam of water. The chimp then put the wet leaves back into its mouth. Jane reasoned that there was water in the tree. Jane made three observations. She saw the chimp pick up dry leaves, put them in the hollow, and then pull them out wet. But, Jane was not observing when she reasoned that there was water inside the tree. She was inferring. When you explain or interpret the things you observe, you are **inferring,** or making an inference. Inferring is not guessing. Inferences are based on reasoning from what you already know. They could also be based on assumptions you make about your observations. See what inferences you can make about the chimps in **Figure 2**.

FIGURE 2

Inferring

What can you infer about the chimps and the termite mound?

Complete the activities below.

1. **Observe** In the chart below, write two observations about the chimp on the left.
2. **Infer** Use the observations you wrote to make two related inferences.

Observation	Inference
The chimp is holding a stem.	The chimp is using the stem as a tool.
The chimp is sticking the stem in the hole.	There is something in the hole that the chimp wants.

Predicting Jane's understanding of chimp behavior grew over time. Sometimes, she could predict what a chimp would do next. **Predicting** means making a statement or a claim about what will happen in the future based on past experience or evidence.

By observing, Jane learned that when a chimp was frightened or angry its hairs stood on end. This response was sometimes followed by threatening gestures such as charging, throwing rocks, and shaking trees. Therefore, when Jane saw a chimp with its hair on end, she was able to predict that there was danger.

Predictions and inferences are closely related. While inferences are attempts to explain what is happening or *has* happened, predictions are statements of claims about what *will* happen. If you see a broken egg on the floor by a table, you might infer that the egg had rolled off the table. If, however, you see an egg rolling toward the edge of a table, you can predict that it's about to create a mess.

FIGURE 3

Predicting

Predictions are forecasts of what will happen next.

Predict Write a prediction about what this angry chimp might do next.

The chimp might charge, throw rocks, or shake trees.

do the math!

Chimp Diet in May	
Fruits	52%
Seeds	30%
Leaves	12%
Other foods	6%

Like all animals, chimps prefer to eat certain foods when they are available.

1. **Construct Graphs** Use the information in the table to create a bar graph.
2. Label the *x*-axis and the *y*-axis. Then write a title for the graph.
3. **Analyze Data** Did chimps feed more on seeds or leaves during May?

   They fed more on seeds than leaves.

4. **Draw Conclusions** What might chimps eat more of if fruits are not available in June?

   Sample: They might eat more foods that aren't fruits, unless another food source they prefer becomes available in June.

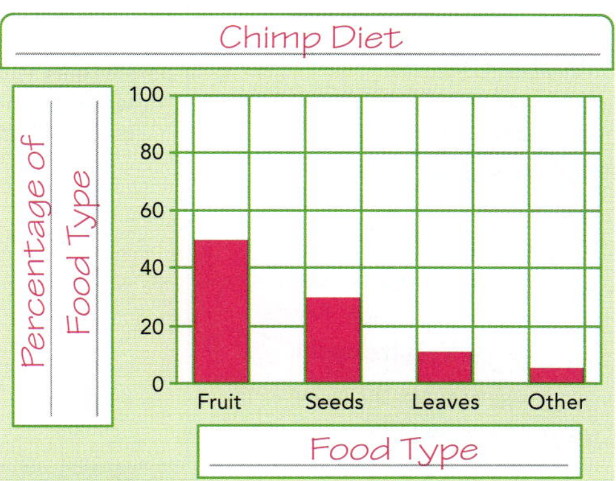

Differentiated Instruction

L1 Classify and Communicate Have students form small groups and ask them to create a poster that differentiates among observing, inferring, and predicting. Suggest that students give examples of each skill and use labeled drawings or pictures from magazines to illustrate them. Encourage groups to present their posters to the class and display them.

L3 Science Skills The scientist discussed in this section, Jane Goodall, has become world famous for her years of observing chimpanzees in Africa. Have students research this scientist's life and prepare an oral report or multimedia presentation to the class. They should include skills and methods she used in learning about the chimpanzees, Africa, and the natural world.

Explain

Lead a Discussion

MAKE A PREDICTION Tell students that scientists make predictions about what will happen in the future based on observations (or evidence) and prior experience. Ask: **What is predicting?** *(Making a statement or claim about what will happen in the future based on past experience or evidence)* **If on a sunny afternoon you observe a massive line of large, dark clouds quickly advancing, what prediction might you make?** *(Most students will predict that rain and even thunderstorms may soon arrive.)* **What would you base your prediction on?** *(On the observation of advancing clouds and on past experience with weather in the area)*

Predict Explain to students that when they predict, they use what they already know from past experience to state what is likely to happen in the future.

Elaborate

Do the Math!

L1 Remind students that on a bar or line graph, the *x*-axis is the horizontal axis and the *y*-axis is the vertical axis. Suggest students give each column in the table a descriptive head, and then use the heads to label the axes.

LESSON 1.1

Explain

Lead a Discussion

CLASSIFYING Tell students that one way scientists organize their observations is by classifying. Ask: **Jane grouped together information about Jomeo's feeding habits. What kind of information would she have classified in that grouping?** *(What kinds of foods Jomeo ate and the proportions of each kind of food, when Jomeo fed, whether Jomeo ate with other chimps or ate alone, and so on)* **How would this sort of classifying aid her in understanding chimpanzees?** *(By classifying behaviors, Jane could better compare and evaluate her observations of various chimps. She could also compare her observations of chimps with those of other scientists. She might also more easily see patterns and trends as she is evaluating.)*

Elaborate

Teacher Demo 

L1 CLASSIFYING PLANTS

Materials none

Time 20 minutes

Remind students that classifying involves sorting things into groups based on shared characteristics. Explain the basic difference between two common plant groups in your area, such as evergreen trees and deciduous trees. Then use the school grounds, surrounding area, or photographs to demonstrate the process of classifying each plant into one category or another.

Ask: **Why might classifying plants in a certain area like the school grounds be useful?** *(Sample: To determine what conditions they need for survival)*

Classifying What did chimps do all day? To find out, Jane's research team followed the chimps through the forest. They took detailed field notes about the chimps' behaviors. **Figure 4** shows some notes about Jomeo, an adult male chimp.

Suppose Jane had wanted to know how much time Jomeo spent feeding or resting that morning. She could have found out by classifying Jomeo's actions. **Classifying** is the grouping together of items that are alike in some way. Jane could have grouped together all the information about Jomeo's feeding habits or his resting behavior.

Evaluating Suppose Jane had found that Jomeo spent most of his time resting. What would this observation have told her about chimp behavior? Before Jane could have reached a conclusion, she would have needed to evaluate her observations. **Evaluating** involves comparing observations and data to reach a conclusion about them. For example, Jane would have needed to compare all of Jomeo's behaviors with those of other chimps to reach a conclusion. She would also need to have evaluated the resulting behavior data of Jomeo and the other chimps.

FIGURE 4

Classifying

By classifying the information related to a chimp's resting, climbing, or feeding, a scientist can better understand chimp behavior.

Classify Use the chart to classify the details from the field notes.

- 6:45 A.M. Jomeo rests in his nest. He lies on his back.
- 6:50 Jomeo leaves his nest, climbs a tree, and feeds on *viazi pori* fruits and leaves.
- 7:16 He wanders along about 175 m from his nest feeding on *budyankende* fruits.
- 8:08 Jomeo stops feeding, rests in a large tree, feeds on *viazi pori* fruits again.
- 8:35 He travels 50 m further, rests by a small lake.

Feeding	Resting	Changing Location
Jomeo eats *viazi pori* fruits, *budyankende* fruits, and leaves.	Sample: Jomeo sometimes lies on his back and likes to rest in big trees.	Sample: Jomeo moves often, traveling 225 m in the morning.

Using Models How far do chimps travel? Where do they go? Sometimes, Jane's research team followed a particular chimp for many days at a time. To show the chimp's movements, they might have made a model. **Models** are representations of processes or objects that are complex or too small or large to be observed. Using models, Jane and her team are able to study things that can't be observed directly and share information that would otherwise be difficult to explain. The model in **Figure 5** shows Jomeo's movements and behaviors during one day. Some models can be touched, such as a map. Others are in the form of mathematical equations or computer programs.

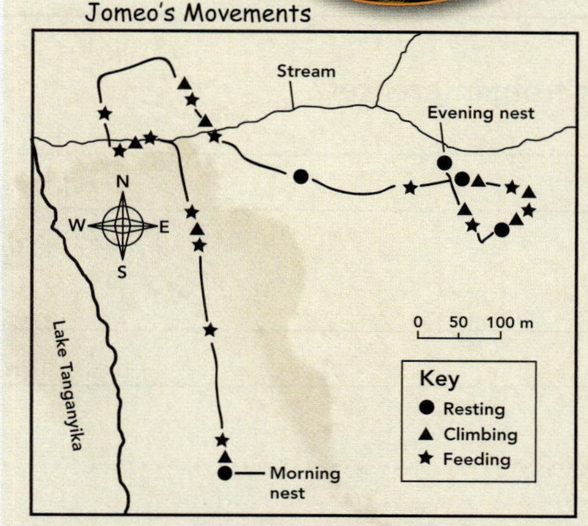

FIGURE 5

Using Models

This model shows Jomeo's movements and behaviors during one day.

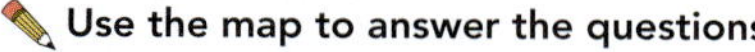

Use the map to answer the questions.

1. **Interpret Maps** How far did Jomeo travel during this day?
   Sample: Jomeo traveled about 1,400 m.
2. How many times did Jomeo stop to feed?
   He stopped 13 times.
3. How many times did Jomeo rest?
   He rested 5 times.

Do the Quick Lab *Classifying Objects.* Student Lab Manual, p. 1

Assess Your Understanding

TEKS 3B

1a. **Compare and Contrast** How do observations differ from inferences?
Observations are data gathered from the senses. Inferences are reasoned explanations of observations.

b. **Identify** Why do scientists use models?
Models allow you study things that are too complex, too large, or too small to observe directly.

got it?

- I get it! I know that scientists use skills such as observing, inferring, predicting, classifying, evaluating, and using models.
- I need extra help with See TE note.

Differentiated Instruction

L3 Other Scientific Skills Tell students that in addition to skills like inferring and predicting, scientists must possess skills related to general science topics and their particular discipline. For example, scientists must be computer literate. A microbiologist must be able to use a variety of microscopes. Have students form small groups and brainstorm or research other skills scientists must have.

L1 Design Models Provide students with newspapers, tape, glue, scissors, thin wire, and wire cutters. Challenge them to use just the materials provided to design a model of something in nature. Tell students they can use just one material or all the materials. Display their models in the classroom.

Explain

Lead a Discussion

USING MODELS Tell students that scientists often make models to study natural objects, events, and processes. Ask: **What is a model?** *(A representation of a complex object or process)* **What models can you point out in this classroom?** *(Sample answer: A globe, a map, a model of the human body, a labeled diagram of the solar system)* **How does using a model help scientists communicate information?** *(Using a model can communicate information among scientists that would otherwise be difficult to explain.)*

Teach With Visuals

Tell students to look at **Figure 5.** Ask: **From the map, what can you tell about the area in which the chimpanzees live?** *(The map shows that the area includes a stream and Lake Tanganyika.)* **What does the map show about Jomeo's day?** *(The chimp left its nest by Lake Tanganyika and traveled along the lake and then the stream before nesting along the stream in the evening.)* **Why is this map a model?** *(It is a representation of a complex process.)*

Elaborate

Lab Resource: Quick Lab 

L2 CLASSIFYING OBJECTS Students will observe and classify objects of different sizes, shapes, and colors. This Quick Lab can be found in the Student Lab Manual, p. 1, and online.

Evaluate

Assess Your Understanding

After students answer the questions, have them evaluate their understanding by completing the appropriate sentence.

RTI Response to Intervention

1a. If students need help contrasting observations and inferences, **then** help them review the definitions of the highlighted terms and discuss how they are different.

b. If students have trouble identifying an advantage of using a model, **then** have them review the highlighted term *models* on this page.

Name ______________________ Date __________ Class __________

Assess Your Understanding

Science and the Natural World

What Skills Do Scientists Use?

1a. COMPARE AND CONTRAST How do observations differ from inferences?

____

____

b. IDENTIFY Why do scientists use models?

____

____

got it?

○ **I get it!** Now I know that scientists use skills such as ______________________

____

____

○ **I need extra help with** ______________________

____

Place the outside corner, the corner away from the dotted line, in the corner of your copy machine to copy onto letter-size paper.

Name ______________________ Date ____________ Class ____________

Enrich

Science and the Natural World

Read the passage below. Then answer the questions that follow on a separate sheet of paper.

Scientists Needed!

Scientists need many skills, which you also use when you do scientific experiments and laboratory work. As you are working online one day, you come across a help wanted ad. It describes the skills needed for a particular scientific job.

> Wanted: A person who is observant and knows numbers and descriptions. Makes appropriate inferences from prior knowledge and predictions based on past experience. Can classify to organize information and make models of real objects and processes to better understand how they function. Good communication skills. Works well in a team atmosphere.

The application for the job includes the questions below.

1. **Think of an experiment you have done. What did you observe? Use both numbers and descriptions in your answer.**
2. **What can you infer from prior knowledge?**
3. **Predict what will happen based on past experience and what you have done in the experiment.**
4. **Is there anything you can classify?**
5. **Is there anything you can make a model of? Remember, a model can be a drawing or diagram.**
6. **Write a brief description of your experiment to communicate to others what you did.**
7. **Explain how you worked as part of a team.**

Name ______________________ Date ____________ Class ____________

Lesson Quiz

Science and the Natural World

Fill in the blank to complete each statement.

1. Grouping together all the things that are alike is called ______________.

2. A person who does ______________ learns about and explores the natural world.

3. When you state what you think will happen in the future, you are ______________.

4. A(n) ______________ observation deals with descriptions that cannot be expressed in numbers.

5. The skill of ______________ involves comparing observations and data to reach a conclusion about them.

6. You make a(n) ______________ when you interpret something you observe.

Write the letter of the correct answer on the line at the left.

7. ___ Which skill involves creating representations of complex objects or processes?
   - A classifying
   - B predicting
   - C making models
   - D evaluating

8. ___ Which of the following do scientists use when observing?
   - A only their senses
   - B only tools
   - C their senses and tools
   - D their tools and observations

9. ___ What kind of observation deals with numbers?
   - A qualitative
   - B quantitative
   - C sensory
   - D descriptive

10. ___ Which of the following is an example of a model?
    - A mathematical equation
    - B tool
    - C scientist
    - D observation

Place the outside corner, the corner away from the dotted line, in the corner of your copy machine to copy onto letter-size paper.

Science and the Natural World

Answer Key

Review and Reinforce

Find the worksheet in the Student Workbook.

1. observing, inferring, predicting, classifying, evaluating, making models
2. reasoning from what you already know or assumptions you make about your observations
3. Whereas inferences are attempts to explain what is happening or has happened, predictions are statements or claims about what will happen.

4. h **5.** i
6. f **7.** e
8. b **9.** c
10. a **11.** g
12. d

13. Sample: I would use the skills of observing to gather information, classifying to group dogs by type, evaluating to summarize my findings, and perhaps making a model to help me share my results.

Enrich

Answers will vary. Samples are given.

1. Six plants that received no sunlight drooped and died.
2. Since plants need sunlight for photosynthesis, without light they could not make food.
3. Plants that received the correct amounts of sunlight will grow.
4. Students could classify the plants as those that received sunlight (or water) and those that did not.
5. In this experiment, students might draw diagrams of plants that receive or do not receive sunlight.
6. I decided to see if plants that did not receive sunlight would grow and survive. Since I knew photosynthesis is necessary for plant growth, I hypothesized that the plants would not grow. My experiment proved my hypothesis.
7. I moved the plants away from sunlight while my partner measured the water and a third member of the team recorded how the plants looked every day.

Lesson Quiz

1. classifying **2.** science
3. predicting **4.** qualitative
5. evaluating **6.** inference
7. C **8.** C
9. B **10.** A

Thinking Like a Scientist

How do scientists investigate the natural world?

LESSON PACING:
2–3 periods or 1–1½ blocks

Lesson Vocabulary

- skepticism
- ethics
- personal bias
- cultural bias
- experimental bias
- objective
- subjective
- deductive reasoning
- inductive reasoning

Content Refresher

Science Attitudes Harold Urey's discovery of deuterium, a stable isotope of the element hydrogen, exemplifies creativity, an important science attitude. In 1931, Urey devised a method of using the fractional distillation of liquid hydrogen to concentrate heavy hydrogen isotopes. He also helped develop a method for separating hydrogen isotopes, and he conducted thorough investigations of the properties of those isotopes. His work earned Urey the Nobel Prize in Chemistry in 1934.

Urey's work in space science affords a notable example of other science attitudes—namely, curiosity, open-mindedness, and honesty. Urey had theorized that the moon's origin was independent from that of Earth and that the moon was older than Earth. He was one of the first scientists to examine the rocks that Apollo 11 brought back from the moon in 1969. His observations of these rocks indicated that his theory about the moon was wrong; on the basis of this new evidence, he readily revised his ideas, demonstrating his willingness to accept new information and change his own thinking.

Lesson Objectives	TEKS	ELPS
Describe the attitudes, or habits of mind, that are necessary for thinking scientifically, ethically, and without bias.	3A	1.A.1
Explain what scientific reasoning is and how scientists use it.	3A, 3D	

Texas Essential Knowledge and Skills

3A In all fields of science, analyze, evaluate, and critique scientific explanations by using empirical evidence, logical reasoning, and experimental and observational testing, including examining all sides of scientific evidence of those scientific explanations, so as to encourage critical thinking by the student.
3D Relate the impact of research on scientific thought and society, including the history of science and contributions of scientists as related to the content.

English Language Proficiency Standards

ELPS Learning Strategies 1.A.1 Use prior knowledge to understand meanings in English.

DIFFERENTIATED INSTRUCTION KEY
L1 Struggling Students or Special Needs
L2 On-Level Students L3 Advanced Students

LESSON PLANNER 1.2

Investigations and Activities	TEKS Review
My Planet Diary, **Student Edition,** p. 10 **Inquiry:** Inquiry Warm-Up, How Keen Are Your Senses?, **Lab Manual,** p. 2 Teach Key Concepts, **Teacher's Edition,** p. 11 Teach With Visuals, **Teacher's Edition,** p. 12 Address Misconceptions, Skepticism, **Teacher's Edition,** p. 12 Support the TEKS, Bias, **Teacher's Edition,** p. 12 Teacher to Teacher, **Teacher's Edition,** p. 12 Make Analogies, Bias Wall, **Teacher's Edition,** p. 13 Apply It!, **Student Edition,** p. 13 Differentiated Instruction, **Teacher's Edition,** p. 13 **Inquiry:** Quick Lab, Thinking Like a Scientist, **PearsonTexas.com**	Apply the TEKS, Evaluating Scientific Explanations, **Student Edition,** p. 46 TEKS Practice, **Student Edition,** p. 48 TEKS Practice: Chapter and Cumulative Review, **Student Edition,** p. 52 Lesson 1.2, **TEKS Preparation and Study Guide Workbook,** p. 4
Teach Key Concepts, **Teacher's Edition,** p. 14 Apply It!, **Student Edition,** p. 14 **Inquiry:** Build Inquiry, Objective and Subjective Reasoning, **Teacher's Edition,** p. 15 21st Century Learning, Creativity, **Teacher's Edition,** p. 15 Lead a Discussion, Deductive and Inductive Reasoning, **Teacher's Edition,** p. 16 21st Century Learning, Critical Thinking, **Teacher's Edition,** p. 16 Apply It!, **Student Edition,** p. 17 Differentiated Instruction, **Teacher's Edition,** p. 17 **Inquiry:** Quick Lab, Using Scientific Thinking, **PearsonTexas.com**	**SHORT ON TIME?** To do this lesson in approximately half the time, do the Activate Prior Knowledge activity. A discussion of the Key Concepts will familiarize students with the lesson content. Have students do the Quick Labs. The rest of the lesson can be completed by students independently.

These editable worksheets are available on **PearsonTexas.com.**
Print versions can be found in the **TEKS Preparation and Study Guide Workbook.**

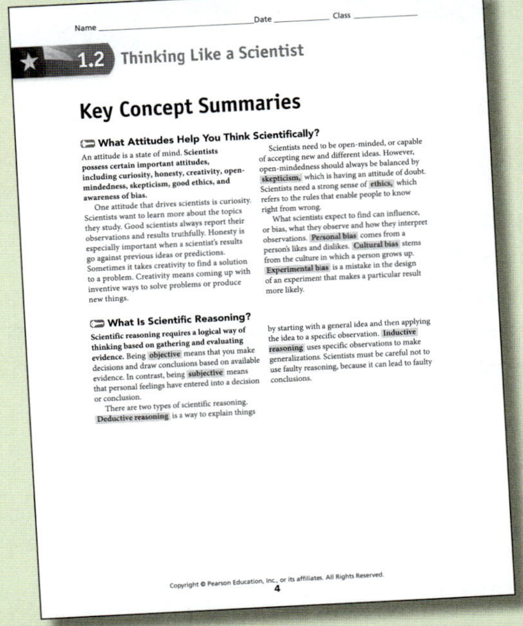

Name ______ Date ______ Class ______

1.2 Thinking Like a Scientist

Key Concept Summaries

What Attitudes Help You Think Scientifically?

An attitude is a state of mind. **Scientists possess certain important attitudes, including curiosity, honesty, creativity, open-mindedness, skepticism, good ethics, and awareness of bias.**

One attitude that drives scientists is curiosity. Scientists want to learn more about the topics they study. Good scientists always report their observations and results truthfully. Honesty is especially important when a scientist's results go against previous ideas or predictions. Sometimes it takes creativity to find a solution to a problem. Creativity means coming up with inventive ways to solve problems or produce new things.

Scientists need to be open-minded, or capable of accepting new and different ideas. However, open-mindedness should always be balanced by **skepticism,** which is having an attitude of doubt. Scientists need a strong sense of **ethics,** which refers to the rules that enable people to know right from wrong.

What scientists expect to find can influence, or bias, what they observe and how they interpret observations. **Personal bias** comes from a person's likes and dislikes. **Cultural bias** stems from the culture in which a person grows up. **Experimental bias** is a mistake in the design of an experiment that makes a particular result more likely.

What Is Scientific Reasoning?

Scientific reasoning requires a logical way of thinking based on gathering and evaluating evidence. Being **objective** means that you make decisions and draw conclusions based on available evidence. In contrast, being **subjective** means that personal feelings have entered into a decision or conclusion.

There are two types of scientific reasoning. **Deductive reasoning** is a way to explain things by starting with a general idea and then applying the idea to a specific observation. **Inductive reasoning** uses specific observations to make generalizations. Scientists must be careful not to use faulty reasoning, because it can lead to faulty conclusions.

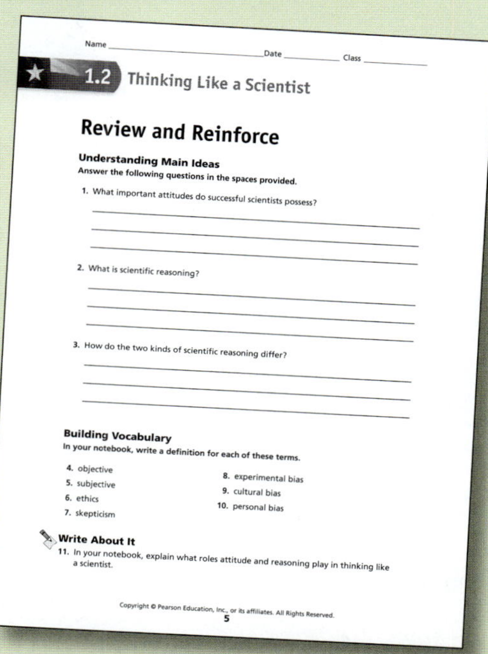

Name ______ Date ______ Class ______

1.2 Thinking Like a Scientist

Review and Reinforce

Understanding Main Ideas
Answer the following questions in the spaces provided.

1. What important attitudes do successful scientists possess?
2. What is scientific reasoning?
3. How do the two kinds of scientific reasoning differ?

Building Vocabulary
In your notebook, write a definition for each of these terms.

4. objective
5. subjective
6. ethics
7. skepticism
8. experimental bias
9. cultural bias
10. personal bias

Write About It

11. In your notebook, explain what roles attitude and reasoning play in thinking like a scientist.

Lexile Measure = 870L

LESSON 1.2

Thinking Like a Scientist

Establish Learning Objectives

After this lesson, students will be able to:

- Describe the attitudes, or habits of mind, that are necessary for thinking scientifically, ethically, and without bias.
- Explain what scientific reasoning is and how scientists use it.

Engage

Activate Prior Knowledge

MY PLANET DIARY Read *Incredible Inventions* with the class. Ask students to think about and describe a situation when they were curious about something, and their curiosity lead them to something they didn't expect. Ask: **Do you think is it helpful or harmful for a scientist to be curious? Why?** *(Some students may recognize that being curious leads scientists to ask numerous questions and discover new things.)*

Explore

Lab Resource: Inquiry Warm-Up 

L1 HOW KEEN ARE YOUR SENSES? Students will use their senses to observe an unexpected event. This Inquiry Warm-Up can be found in the Student Lab Manual, p. 2, and online.

Thinking Like a Scientist

- What Attitudes Help You Think Scientifically? TEKS 3A
- What Is Scientific Reasoning? TEKS 3A, 3D

my planet Diary — DISCOVERY

Incredible Inventions

Most scientific inventions are purposely created and result from curiosity, persistence, and years of hard work. However, some inventions have been accidentally discovered when their inventors were in the process of creating something else. While developing wallpaper cleaner, a type of clay was invented. A coil-shaped toy was originally designed as a spring to be used on ships. Instead of developing a substitute for synthetic rubber, toy putty was created. Self-stick notes, potato chips, and the hook and loop fasteners used on items such as clothing, shoes, and toys are also inventions that were discovered by accident. Like the inventors of these items, your curiosity may help you invent the next "big thing"!

Communicate Discuss the following questions with a partner. Write your answers below.

1. Why do you think it is important for scientists to be curious?
   Sample: Curiosity drives scientists to make discoveries.

2. What might you want to invent? Why?
   Sample: I want to invent a device that would help blind people know who was ringing the doorbell.

Do the Inquiry Warm-Up *How Keen Are Your Senses?* Student Lab Manual, p. 2

SUPPORT ALL READERS

Lexile Measure = 870L Lexile Word Count = 1554

Prior Exposure to Content: Many students may have misconceptions on this topic

Academic Vocabulary: *cause, classify, effect, relate*

Science Vocabulary: *skepticism, ethics, deductive reasoning*

Concept Level: Generally appropriate for most students in this grade

Preteach With: My Planet Diary "Incredible Inventions" and Figure 1 activity

Vocabulary
- skepticism
- ethics
- personal bias
- cultural bias
- experimental bias
- objective
- subjective
- deductive reasoning
- inductive reasoning

Skills
- Reading: Relate Cause and Effect
- Inquiry: Classify

What Attitudes Help You Think Scientifically?

TEKS 3A In this section, you'll learn about important attitudes that scientists need to possess.

Perhaps someone has told you that you have a good attitude. What does that mean? An attitude is a state of mind. Your actions say a lot about your attitude.  **Scientists possess certain important attitudes, including curiosity, honesty, creativity, open-mindedness, skepticism, good ethics, and awareness of bias.**

ELPS 1.A.1
Read aloud or listen to My Planet Diary on page 10. Discuss each question with a partner. Use what you know about inventing and what inventors do.

Curiosity One attitude that drives scientists is curiosity. Scientists want to learn more about the topics they study. **Figure 1** shows some things that may spark the interest of scientists.

Honesty Good scientists always report their observations and results truthfully. Honesty is especially important when a scientist's results go against previous ideas or predictions.

Creativity Whatever they study, scientists may experience problems. Sometimes, it takes creativity to find a solution. Creativity means coming up with inventive ways to solve problems or produce new things.

FIGURE 1
Curiosity
Curiosity helps scientists learn about the world around them.

Ask Questions For each image, write a question you are curious about in the boxes.

PEARSON Texas.com

Explain

Introduce Vocabulary

Explain that *personal bias* is one of several vocabulary words made up of more than one word. Discuss how the separate meaning of each word compares to its combined meaning.

Teach Key Concepts 

Explain to students that scientists must not only possess certain skills, but they must also possess certain attitudes. These attitudes include curiosity, honesty, creativity, open-mindedness, skepticism, good ethics, and awareness of bias. List these attitudes on the board. Ask: **What is an attitude?** *(A state of mind)* **What are some attitudes you possess?** *(Samples: optimism, pessimism, positivism, caution)* **What does it mean to be curious?** *(To wonder or want to learn more about something)* **What is involved in being honest?** *(Telling the truth)* **Why does it matter that a scientist be honest?** *(Without honesty, a scientist's observations and findings cannot be trusted by others or used to gain knowledge about the natural world.)* **What is creativity?** *(Being unique in the way you solve problems or produce things)* **How might creativity help a scientist?** *(It could lead the scientist to new information, ways of thinking about things, and solutions.)*

PEARSON Texas.com

LESSON 1.2

English Language Proficiency Standards

ELPS Learning Strategies 1.A.1

Have students read My Planet Diary on page 10 and answer the questions.

Beginning Define *invention* before reading aloud My Planet Diary. Have partners role-play scientists creating an invention. Help each pair name the invention and tell why they made it.

Intermediate Have students read My Planet Diary aloud with you. Ask students to name or draw examples of inventions. Then read aloud and discuss the questions. Provide sentence frames as needed: *A curious scientist* _____. *I want to invent a* _____ *because* _____.

Advanced Have pairs read aloud the text and discuss the questions. Encourage open-ended discussion, and prompt students to elaborate.

Advanced High Have partners read the text and discuss the questions. Then have students switch partners and present their answers.

Texas Essential Knowledge and Skills

3A In all fields of science, analyze, evaluate, and critique scientific explanations by using empirical evidence, logical reasoning, and experimental and observational testing, including examining all sides of scientific evidence of those scientific explanations, so as to encourage critical thinking by the student.

LESSON 1.2

Explain

Teach With Visuals

Tell students to look at **Figure 2.** Direct students' attention to the graphic organizer. Explain that a graphic organizer like this one can help them organize ideas and summarize concepts. Go through each attitude in the graphic organizer one by one, and ask students to describe a situation when they had that attitude. Ask: **Can you think of a time when you were open-minded?** *(Sample: A friend wanted to try a new kind of food and I agreed to try it with her.)* **When might a scientist need to be open-minded?** *(Sample: When another scientist interprets the same data differently.)* Repeat these same questions for each attitude in the graphic organizer.

Address Misconceptions

L1 SKEPTICISM Student may think that skepticism and open-mindedness are opposing attitudes. Explain that scientists should be both skeptical and open-minded. Without doubt, you can't distinguish what is really true from what is not. Without accepting new ideas, you can't gain knowledge about the world. Ask: **What things in science do you think people should be skeptical about?** *(Samples: Ideas about living things we can't observe directly; ideas not backed up by data)* **What things should they be open-minded about?** *(Samples: New ideas about organisms that are backed up with data)*

Support the TEKS

BIAS Explain to students that we all have past experiences, opinions, likes, and dislikes that affect the way we think. These influences can contribute to personal bias. Ask: **How can personal bias affect scientific results?** *(It can cause a scientist to misinterpret an observation or the results of an experiment.)* **Why is this something a scientist should avoid?** *(Because it can result in information that is not true or scientific)* **How can a scientist avoid personal bias in his or her work?** *(By being aware that it can happen and by analyzing his or her own interpretation)*

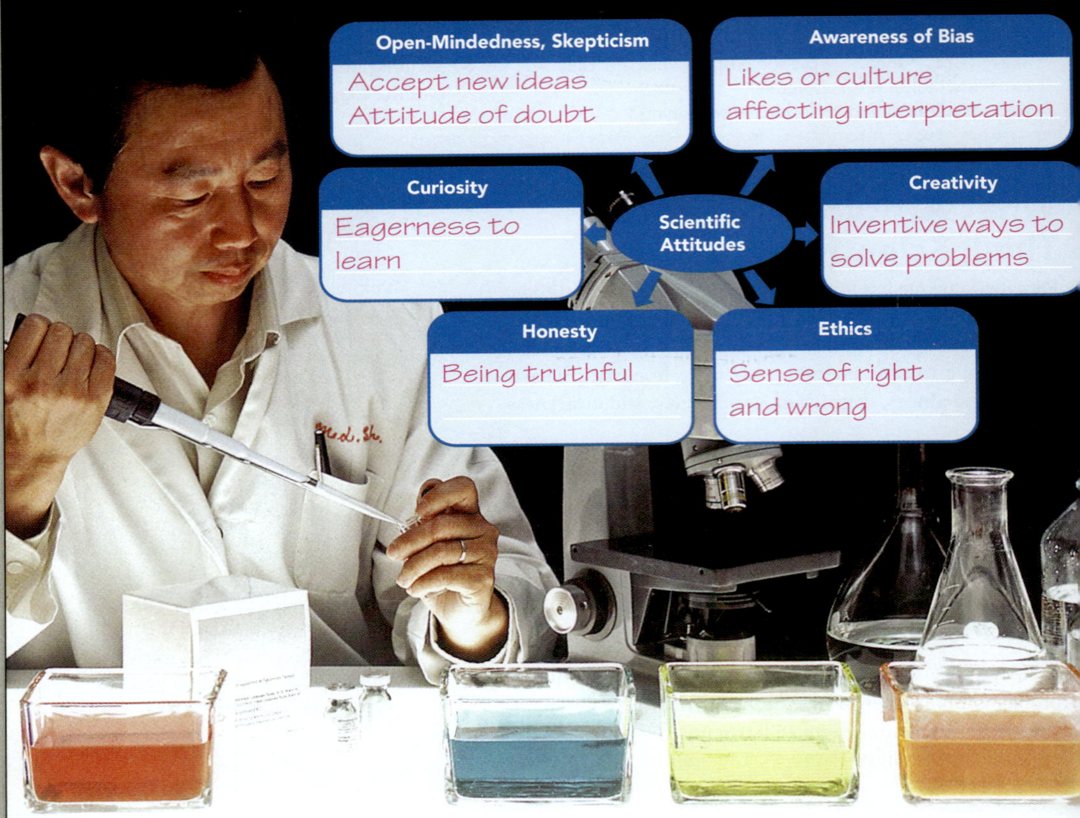

FIGURE 2

Attitudes of Scientists

This scientist is carefully conducting an experiment.

Summarize **After you have read the section *What Attitudes Help You Think Scientifically?*, write a summary of each scientific attitude in the graphic organizer.**

Open-Mindedness and Skepticism

Scientists need to be open-minded, or capable of accepting new and different ideas. However, open-mindedness should always be balanced by **skepticism,** which is having an attitude of doubt. Skepticism keeps a scientist from accepting ideas that may be untrue.

Ethics

Because scientists work with the natural world, they must be careful not to damage it. Scientists need a strong sense of **ethics,** which refers to the rules that enable people to know right from wrong. Scientists must consider all the effects their research may have on people and the environment. They make decisions only after considering the risks and benefits to living things or the environment. For example, scientists test medicine they have developed before the medicine is sold to the public. Scientists inform volunteers of the new medicine's risks before allowing them to take part in the tests. Look at **Figure 2** to review scientific attitudes.

Professional Development Note — Teacher to Teacher

Scientific Thinking If scientific thinking were intuitive, then everyone would do it. In reality, humans intuitively seek patterns and often jump to unsupported conclusions. Scientific thinking is an attempt to find reality unbiased by human fallacies. It is a difficult but rewarding process. Scientific thinking process involves observing, hypothesizing, predicting, as well as collecting and analyzing data. This recursive process is how all scientists incrementally learn new things and can be confident of what they know and what they still do not know.

Joel Palmer, Ed.D.

Mesquite ISD

Mesquite, Texas

Awareness of Bias What scientists expect to find can influence, or bias, what they observe and how they interpret observations. For example, a scientist might misinterpret the behavior of an animal because of what she already knows about animals.

There are different kinds of bias. **Personal bias** comes from a person's likes and dislikes. For instance, if you like the taste of a cereal, you might think everyone else should, too. **Cultural bias** stems from the culture in which a person grows up. For example, a culture that regards snakes as bad might overlook how well snakes control pests. **Experimental bias** is a mistake in the design of an experiment that makes a particular result more likely. For example, suppose you wanted to determine the boiling point of pure water. If your experiment uses water that has some salt in it, your results would be biased.

Relate Cause and Effect In the first paragraph, underline an example of bias. Then circle its effect.

apply it!

Matt likes cheese crackers best and thinks that most other students do too. So he observed what students bought at the vending machine during one lunch. Seven bought crackers, three bought nuts, and none bought raisins.

1. **Evaluate** Circle the evidence of personal bias.
2. **CHALLENGE** Describe the experimental bias.
   The cheese crackers are the cheapest option.

Do the Quick Lab *Thinking Like a Scientist.* Find the lab online.

Assess Your Understanding

TEKS 3A

1a. **Evaluate** What can bias a scientist's observations?
Expectations can bias a scientist's observations.

b. **Apply Concepts** Debbie discovered a new way to make pizza. What scientific attitude is this an example of?
It is an example of creativity.

got it?

O **I get it!** Now I know that attitudes that help you think scientifically are curiosity, honesty, creativity, open-mindedness, skepticism, ethics, and awareness of bias.

O I need extra help with See TE note.

13

Differentiated Instruction

L1 Attitude Cards Provide student pairs with index cards. Have them write each important attitude of scientists on a card and write the attitude's definition on the back. Have partners take turns drawing a card and testing each other on each attitude's definition. Students should also use each attitude in a sentence.

L3 Issues of Ethics Have students interview a local town official or scientist to identify an issue of science ethics in their community. Students should find out as much as they can about both sides of the issue, particularly why one or more actions are considered to be unethical. Have them take positions and hold a debate as to whether or not that action should be taken.

Elaborate

LESSON 1.2

Make Analogies

L1 BIAS WALL Tell students that personal bias or cultural bias is like a wall that blends in with its surroundings. You may not know it is there, yet it blocks part of your view and keeps you from seeing the entire picture of what is going on. Becoming aware of bias is the first step in removing it. Ask: **What is experimental bias like? Explain your answer.** *(Students might say that experimental bias is like a flaw in the make up of an appliance that keeps the appliance from working correctly.)*

Relate Cause and Effect Explain to students that a cause is an event that makes an effect happen. An effect happens as a result of a cause. The cause comes first and is followed by the effect.

Apply It!

L1 Before beginning the activity, review the three different kinds of bias described in the lesson. Remind students to look closely at the picture of the foods from the vending machine. Encourage them to speculate about how cultural bias might also affect Matt's results. *(Sample: Dried fruit and salted foods may not be common or popular in the students' culture.)*

Lab Resource: Quick Lab 

L2 THINKING LIKE A SCIENTIST Students will examine the biases they have about Earth and the sun. This Quick Lab can be found online.

Evaluate

Assess Your Understanding

After students answer the questions, have them evaluate their understanding by completing the appropriate sentence.

RTI Response to Intervention

1a. If students have trouble identifying what can bias a scientist's observations, **then** have them reread the information on awareness of bias.

b. If students need help classifying the scientific attitude described, **then** ask them what term best describes the state of mind of an artist who creates a new sculpture or an inventor who invents a new car that runs on air.

LESSON 1.2

Explain

Teach Key Concepts

Explain to students that scientists use a certain way of thinking called scientific reasoning. It is a logical form of thinking and is not based on emotions or opinions. Instead, it is based on researching through direct observation and then carefully evaluating the evidence that is gathered during research. Ask: **Why do scientists rely on research and evidence rather than their emotions or opinions?** *(Emotions or opinions vary from person to person, whereas direct observations do not. Direct observations can be repeated and result in facts being recorded.)* **What happens when you evaluate evidence?** *(You carefully look it over and think about it.)* **Which do you think is more likely to result in valid information—using emotions and stating opinions or using scientific reasoning?** *(Using scientific reasoning)*

Elaborate

Apply It!

L1 Before beginning the activity, have students restate the definitions of *objective reasoning* and *subjective reasoning* in their own words. Make sure students understand the difference between the two types of reasoning.

Classify Help students recall that when they classify, they group things together that are alike in some way. Remind students that they are classifying each example by placing a checkmark in the correct column. If they have difficulty with the classification, have a volunteer read each example aloud. Then ask: **Does this example involve feelings or an opinion?** If the answer is *yes,* tell students they should place a checkmark in the second column.

Texas Essential Knowledge and Skills

3A In all fields of science, analyze, evaluate, and critique scientific explanations by using empirical evidence, logical reasoning, and experimental and observational testing, including examining all sides of scientific evidence of those scientific explanations, so as to encourage critical thinking by the student.

3D Relate the impact of research on scientific thought and society, including the history of science and contributions of scientists as related to the content.

TEKS 3A, 3D In this section, you'll learn that scientific investigation and reasoning involve gathering, analyzing, evaluating, and critically examining evidence.

What Is Scientific Reasoning?

You use reasoning, or a logical way of thinking, when you solve word problems. Scientists use reasoning in their work, too. **Scientific reasoning requires a logical way of thinking based on gathering and evaluating evidence.** There are two types of scientific reasoning. Scientific reasoning can be deductive or inductive.

Scientific reasoning relies on gathering evidence during research and evaluating it, so it is a form of objective reasoning. Being **objective** means that you make decisions and draw conclusions based on available evidence. For example, scientists used to think chimps ate only plants. However, Jane Goodall's research showed that chimps eat meat. Based on this evidence, she concluded that chimps eat meat and plants.

In contrast, being **subjective** means that personal feelings have entered into a decision or conclusion. Personal opinions, values, and tastes are subjective because they are based on your feelings about something. For example, if you see a clear stream in the woods, you might take a drink because you think clear water is clean. However, you have not objectively tested the water's quality. The water might contain microorganisms you cannot see and be unsafe to drink.

apply it!

Classify Read the sentences below. Then decide if each example uses objective reasoning or subjective reasoning to reach a conclusion. Place a check mark in the corresponding column.

	Objective	Subjective
Jane Goodall saw a chimp chewing on wet leaves. She reasoned that chimps sometimes used leaves to drink water.	✔	
I like to run. I must be the fastest person in the class.		✔
Emily is 1.2 m tall. No one else in class is taller than 1 m. So, Emily is the tallest person in class.	✔	
I dislike dogs. Dogs must be the least friendly animals.		✔

Deductive Reasoning Scientists who study Earth think that the uppermost part of Earth's surface is made up of many sections they call plates. The theory of plate tectonics states that earthquakes should happen mostly where plates meet. There are many earthquakes in California. Therefore, California must be near a place where plates meet. This is an example of deductive reasoning. **Deductive reasoning** is a way to explain things by starting with a general idea and then applying the idea to a specific observation.

You can think about deductive reasoning as being a process. First, you state the general idea. Then you relate the general idea to the specific case you are investigating. Then you reach a conclusion. You can use this process in **Figure 3.** The process for the plate tectonics example is shown here.

- Earthquakes should happen mostly where plates meet.
- California has many earthquakes.
- California must be near a place where plates meet.

did you know?

Did you know that deductive reasoning is used by detectives? Sherlock Holmes, a fictional detective in the novels and short stories of Sir Arthur Conan Doyle, solved many mysteries using deductive reasoning.

FIGURE 3

Deductive Reasoning

Deductive reasoning occurs when a general idea is applied to a specific example and a conclusion is reached.

Apply Concepts **Apply each general idea to a specific example and then draw a conclusion.**

Dinner is always at 6 P.M.
Sample: It is 6 P.M.
So dinner is ready.

Classes end when the bell rings.
Sample: The bell just rang.
So my class is over.

Triangles have three sides.
Sample: This shape has three sides. This shape is a triangle.

15

Elaborate

Build Inquiry 

L1 OBJECTIVE AND SUBJECTIVE REASONING

Materials three flashlights of different colors

Time 10 minutes

Provide each group with three flashlights: one with the bulb removed, one with the batteries removed, and one with both working bulb and batteries. Tell students not to touch or turn on the flashlights until you instruct them to. First, have students decide which flashlight will work the best based solely on which one is the best color. Then tell students they can touch and examine the flashlights but not turn them on. This time, have groups decide which flashlight they think will work the best based solely on their observations and prior knowledge. Finally, have them turn on the flashlight to see if their reasoning was correct.

Ask: **How did you decide which flashlight would work the best the first time?** *(Based on the color we liked best)* **How did you decide which flashlight would work the best the second time?** *(Students will likely say they observed the flashlights and chose the only one that they knew from prior experience had all the parts needed to work.)* **When did you use objective reasoning? subjective reasoning?** *(The second time; the first time)*

21st Century Learning

CREATIVITY Have students compare and contrast the terms *objective* and *subjective*. Then have them write a diary entry about a time when they were objective and another entry about a time when they were subjective.

LESSON 1.2

Differentiated Instruction

L1 Logical Reasoning Students may have difficulty grasping the concept of what is logical and what is not. In the Key Concept statement, point out the word *logical*. Have a volunteer look up the work in a dictionary. Give examples of logical and illogical reasoning until students understand the difference.

L3 Deductive Reasoning Encourage students to read a novel or short story by Sir Arthur Conan Doyle about Sherlock Holmes. Have students identify examples of how Holmes uses deductive reasoning, summarizing examples for the class.

LESSON 1.2

Explain

Lead a Discussion

DEDUCTIVE AND INDUCTIVE REASONING Explain that both deductive reasoning and inductive reasoning are forms of logical, scientific reasoning. Each type follows a certain sequence of steps. Ask: **What three steps should you take in deductive reasoning?** *(1. State the general idea. 2. Relate the general idea to the specific case you are investigating. 3. Reach a conclusion.)* **What steps should you take in inductive reasoning?** *(1. Make specific observations of the case you are investigating. 2. Use the specific observations to reach a conclusion, or state the general idea.)* **What pattern do you notice when you compare the steps involved in deductive and inductive reasoning?** *(The steps in one occur in the reverse order to the other.)*

21st Century Learning

CRITICAL THINKING Remind students that there are different kinds of scientific reasoning. Ask: **How are deductive reasoning and inductive reasoning alike?** *(Both are logical ways of thinking or explaining things.)* **How are they different?** *(With deductive reasoning, you start with the general idea and apply it to a specific situation. With inductive reasoning, you start with specific observations and then make generalizations.)*

Inductive Reasoning Scientists also use inductive reasoning, which can be considered the opposite of deductive reasoning. **Inductive reasoning** uses specific observations to make generalizations. For example, suppose research shows that leaf-cutter ants appear to follow other ants along specific paths, as shown in **Figure 4**. The ants follow the paths to sources of food, water, and nest material. Then they return to their nests. These observations about the leaf-cutter ants are specific. From these specific observations you conclude that these ants must communicate to be able to always follow the same path. This conclusion is a generalization about the behavior of leaf-cutter ants based on your observations. Scientists frequently use inductive reasoning. They collect data and then reach a conclusion based on that data.

FIGURE 4

Scientific Reasoning

Leaf-cutter ants follow a chemical trail to find and harvest leaves.

Identify Look at the statements below. Write *D* next to the statements that use deductive reasoning. Write *I* next to the statements that use inductive reasoning.

1. Turtles have shells. They must use shells for protection. I
2. A puddle has frozen. It must be below 0°C outside. D
3. Because of gravity, everything that goes up must come down. D
4. Many birds fly toward the equator in fall. Birds prefer warm weather. I

16 Scientific Thinking

Faulty Reasoning Scientists must be careful not to use faulty reasoning, because it can lead to faulty conclusions. If you draw a conclusion based on too little data, your reasoning might lead you to the wrong general idea. For example, to conclude accurately that all ants communicate with each other, you would have to observe leaf-cutter ants and many other kinds of ants many times. In addition, based on observations of how leaf-cutter ants follow paths, you cannot conclude how they communicated. For example, you cannot say they follow the tiny footprints of the ants ahead of them. Such a conclusion would be a guess not based on observation.

Joy drew lines of symmetry on a square. She saw that a rectangle has four straight sides and four right angles, so she drew the same lines of symmetry on a rectangle.

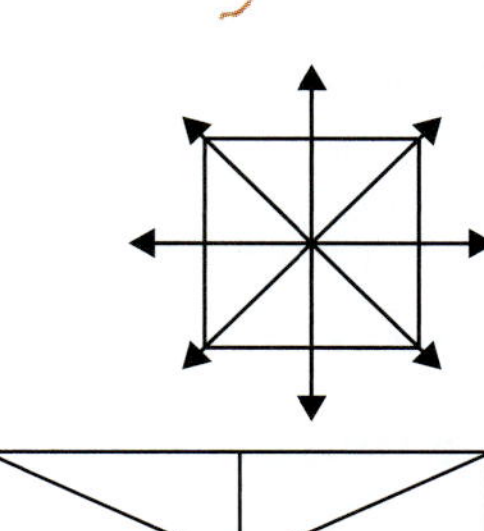

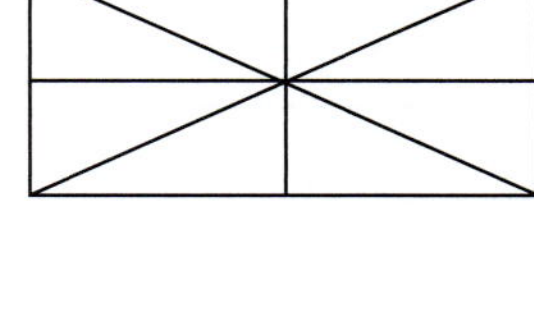

1 Use Models Fold a piece of rectangular notebook paper according to the lines of symmetry Joy drew on the rectangle. Are her lines of symmetry correct? Explain how you know.

No; a rectangular piece of notebook paper can only be symmetrically folded two ways.

2 Evaluate Underline Joy's reasoning for drawing the lines of symmetry on the rectangle. What other characteristic should Joy have considered?

She should have considered that a square has equal sides and a rectangle does not.

Lab zone® Do the Quick Lab *Using Scientific Thinking.* Find the lab online.

Assess Your Understanding

TEKS 3A

2a. Define Deductive reasoning uses a general idea to make a specific observation.

b. Analyze What is a cause of faulty reasoning?

Sample: drawing a conclusion based on too little data

got it?

O **I get it!** Now I know that scientific reasoning includes deductive and inductive reasoning.

O I need extra help with See TE note.

17

Differentiated Instruction

L1 Define and Model Reasoning On the board, draw a three-column chart with the heads *Deductive Reasoning, Inductive Reasoning,* and *Faulty Reasoning.* Elicit from students a definition of each kind of reasoning, and have volunteers fill in the table with their definitions. Then choose a situation in your school, such as the difference in temperature inside and outside your classroom. Give students an application of each of the three kinds of reasoning to that situation.

L3 Newspaper Article Have students write a newspaper article describing fictional or actual news involving scientific reasoning. Instruct them to identify the kind of reasoning exemplified in the article. Students can pool their articles into a newspaper.

Elaborate

Apply It!

L1 Before beginning the activity, explain to students that they will be looking for faulty reasoning. Tell them to look for subtle errors in Joy's reasoning as they read. Provide students with both square and rectangular scrap paper and have them draw lines on the pieces of paper as shown in the diagrams. Then have students fold the papers according to the lines and compare them.

Lab Resource: Quick Lab 

L2 USING SCIENTIFIC THINKING Students will conduct an experiment and then use inductive and deductive reasoning to analyze the results. This Quick Lab can be found online.

Evaluate

Assess Your Understanding

After students answer the questions, have them evaluate their understanding by completing the appropriate sentence.

RTI Response to Intervention

2a. If students cannot identify the type of reasoning, **then** have them reread the sentences that contain the boldface terms.

b. If students have difficulty identifying the cause of faulty reasoning, **then** hold a class discussion to help them review the information on faulty reasoning.

Name ______________________ Date __________ Class __________

Assess Your Understanding

Thinking Like a Scientist

What Attitudes Help You Think Scientifically?

1a. EVALUATE What can bias a scientist's observations?

____

b. APPLY CONCEPTS Debbie discovered a new way to make pizza. What scientific attitude is this an example of?

____

got it?

○ **I get it!** Now I know that attitudes that help you think scientifically are ____

○ **I need extra help with** ____

What Is Scientific Reasoning?

2a. DEFINE ______________ reasoning uses a general idea to make a specific observation.

b. ANALYZE What is a cause of faulty reasoning?

____

got it?

○ **I get it!** Now I know that scientific reasoning includes ____

○ **I need extra help with** ____

Name ______________________ Date ____________ Class ____________

Enrich

Thinking Like a Scientist

Scientists from different fields often work together toward a common goal. One of the greatest scientific discoveries of the twentieth century occurred as a result of this sort of cooperation. Read the passage and study the diagram below. Then answer the questions that follow on a separate sheet of paper.

Scientists Working Together

In 1953, two scientists were working together in a laboratory in England. One was the British scientist Francis Crick, who had been educated as a physicist. During World War II, he had worked on various projects including the use of radar. After the war, he became interested in biochemistry, which is a field of life science concerned with the chemistry of living things. The scientist working with Crick was an American, James Watson. He was a young biochemist from Chicago. The project that drew the attention of Watson and Crick concerned the structures of a chemical called DNA, which is found in living things. Many scientists thought that DNA had something to do with the inheritance of characteristics passed down from one generation to the next, but no one knew how this was accomplished.

DNA

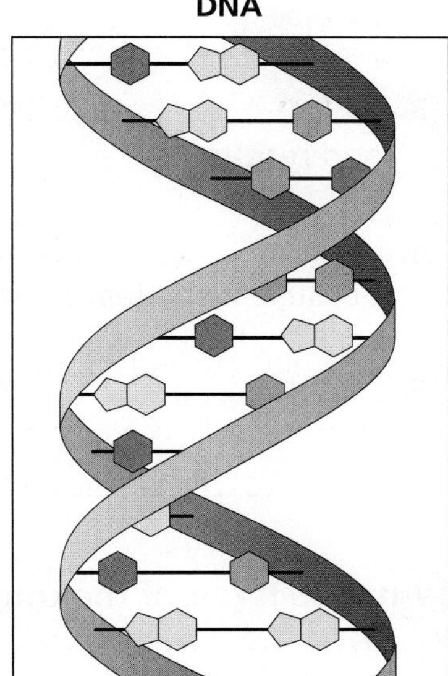

At another laboratory in England, two other scientists were also working with DNA. These scientists were the New Zealand physicist Maurice Wilkins and the British chemist Rosalind Franklin. Wilkins and Franklin were using X-rays to produce images of various chemicals, including DNA. One afternoon, Watson visited the college where Wilkins and Franklin worked. There, he saw an image of DNA that Franklin had produced. The image helped Watson and Crick determine the structure of DNA. From that image and other images of DNA, Watson and Crick built a model of the chemical, similar to the diagram on this page. Discovering the structure of DNA caused a revolution in life science. Now scientists could understand how characteristics were passed from one generation to the next.

1. What is DNA?
2. Look at the diagram of DNA. Describe its structure.
3. What kinds of scientists worked on the structures of DNA?
4. In what way did the work of Watson and Crick depend on the work of other scientists?
5. What was the significance of the work of these scientists?

Place the outside corner, the corner away from the dotted line, in the corner of your copy machine to copy onto letter-size paper.

Name ______________________ Date __________ Class __________

Lesson Quiz

Thinking Like a Scientist

If the statement is true, write *true.* If the statement is false, change the underlined word or words to make the statement true.

1. Inductive Deductive reasoning uses specific observations to make generalizations.

2. Skeptism A scientist's open-mindedness should always be balanced by bias.

3. true Good scientists use honesty when reporting their observations and results.

4. Faulty Scientists must be careful not to use inductive reasoning, because it can lead to faulty conclusions.

5. bias Personal ethics comes from a person's likes and dislikes.

6. true Scientific reasoning requires a logical way of thinking.

Write the letter of the correct answer on the line at the left.

7. ___ What kind of bias is a mistake in the design of an experiment that makes a particular result more likely?
   A deductive
   B cultural
   C personal
   D experimental

8. ___ Which attitude makes a scientist capable of accepting new and different ideas?
   A open-mindedness
   B skepticism
   C curiosity
   D creativity

9. ___ What are you being when you let personal feelings enter into a decision or conclusion?
   A inductive
   B deductive
   C subjective
   D objective

10. ___ Which attitude helps scientists to come up with inventive ways to solve problems or produce new things?
   A curiosity
   B creativity
   C good ethics
   D open-mindedness

Place the outside corner, the corner away from the dotted line, in the corner of your copy machine to copy onto letter-size paper.

Thinking Like a Scientist

Answer Key

Review and Reinforce

Find the worksheet in the Student Workbook.

1. Important attitudes that scientists possess include curiosity, honesty, creativity, open-mindedness, skepticism, good ethics, and awareness of bias.
2. Scientific reasoning is a logical way of thinking based on gathering and evaluating evidence.
3. Deductive reasoning is a way to explain things by starting with a general idea and then applying the idea to a specific observation. Inductive reasoning uses specific observations to make generalizations.
4. make decisions and draw conclusions based on available evidence
5. letting personal feelings enter into a decision or conclusion
6. the rules that enable people to know right from wrong
7. having an attitude of doubt
8. a mistake in the design of an experiment that makes a particular result more likely
9. bias that stems from the culture in which a person grows up
10. bias that comes from a person's likes and dislikes
11. To think scientifically, a scientist must possess certain important attitudes, including curiosity, honesty, creativity, open-mindedness, skepticism, good ethics, and awareness of bias. They must also use a logical way of thinking based on gathering and evaluating evidence, called scientific reasoning.

Enrich

1. DNA is a chemical found in living things that is involved in the inheritance of characteristics.
2. Sample: The structure of DNA is like a twisted ladder.
3. physicists, a chemist and a biochemist
4. Watson and Crick depended on the work of Wilkins, Franklin, and other scientists who produced images of the chemical DNA.
5. Discovering the structures of DNA caused a revolution in life science. With that structure, scientists could understand how characteristics are passed from one generation to the next.

Lesson Quiz

1. Inductive
2. skepticism
3. true
4. faulty
5. bias
6. true
7. D
8. A
9. C
10. B

Texas LESSON 3 Scientific Inquiry

How do scientists investigate the natural world?

LESSON PACING:
3–4 periods or $1\frac{1}{2}$–2 blocks

Lesson Vocabulary

- scientific inquiry • hypothesis • variable • manipulated variable
- responding variable • controlled experiment • data • SI
- scientific theory • scientific law

Content Refresher

A Process of Inquiry The process of scientific inquiry follows the traditional scientific method according to the relevant requirements of the investigation. Central to this process is the use of deductive reasoning to test hypotheses. A hypothesis is often a general statement from which a scientist predicts that specific events or data will result under certain conditions. The scientist then sets up an experiment to test whether the predicted results will occur. In practice, a hypothesis is rarely confirmed with just one experiment. The experiment must be repeated with the same results. The results of those experiments may then suggest other paths to follow. When a wide range of hypotheses are confirmed, an overarching scientific theory may be developed that explains a phenomenon. Scientists then consolidate observations into scientific laws, or predictions of what happens every time under a certain set of conditions.

Lesson Objectives	TEKS	ELPS
Recognize that scientific inquiry involves posing questions and developing hypotheses.	2A, 2B	2.C.3
Explain how to design and implement an experiment so that it uses sound scientific principles.	2B, 2C, 2D, 2E	
Differentiate between a scientific theory and a scientific law.	3A	

Texas Essential Knowledge and Skills

2A Plan and implement comparative and descriptive investigations by making observations, asking well-defined questions, and using appropriate equipment and technology.
2B Design and implement experimental investigations by making observations, asking well-defined questions, formulating testable hypotheses, and using appropriate equipment and technology.
2C Collect and record data using the International System of Units (SI) and qualitative means such as labeled drawings, writing, and graphic organizers.
2D Construct tables and graphs, using repeated trials and means, to organize data and identify patterns.
2E Analyze data to formulate reasonable explanations, communicate valid conclusions supported by the data, and predict trends.
3A In all fields of science, analyze, evaluate, and critique scientific explanations by using empirical evidence, logical reasoning, and experimental and observational testing, including examining all sides of scientific evidence of those scientific explanations, so as to encourage critical thinking by the student.

English Language Proficiency Standards

ELPS Listening 2.C.3 Learn basic vocabulary heard during classroom instruction and interactions.

DIFFERENTIATED INSTRUCTION KEY
L1 Struggling Students or Special Needs
L2 On-Level Students L3 Advanced Students

LESSON PLANNER 1.3

Investigations and Activities

My Planet Diary, **Student Edition,** p. 18
Inquiry: Inquiry Warm-Up, What's Happening?, **PearsonTexas.com**
Introduce Vocabulary, **Teacher's Edition,** p. 19
Teach Key Concepts, **Teacher's Edition,** p. 19
21st Century Learning, Critical Thinking, **Teacher's Edition,** p. 19
Inquiry: Quick Lab, Scientific Inquiry, **Lab Manual,** p. 3

Teach Key Concepts, **Teacher's Edition,** p. 21
Apply It!, **Student Edition,** p. 21
Differentiated Instruction, **Teacher's Edition,** p. 21
Inquiry: Teacher Demo, Observe Rate of Fall, **Teacher's Edition,** p. 22
Do the Math! **Student Edition,** p. 23
Differentiated Instruction, **Teacher's Edition,** p. 23
21st Century Learning, Accountability, **Teacher's Edition,** p. 25
Inquiry: Lab Investigation, Keeping Flowers Fresh, **Lab Manual,** p. 4

Teach Key Concepts, **Teacher's Edition,** p. 27
Teach With Visuals, **Teacher's Edition,** p. 27
Differentiated Instruction, **Teacher's Edition,** p. 27
Inquiry: Quick Lab, Theories and Laws, **PearsonTexas.com**

TEKS Review

Apply the TEKS, In a Scientist's Shoes, **Student Edition,** p. 26

Apply the TEKS, Evaluating Scientific Explanations, **Student Edition,** p. 46

TEKS Practice, **Student Edition,** p. 48

TEKS Practice: Chapter and Cumulative Review, **Student Edition,** p. 52

Lesson 1.3, **TEKS Preparation and Study Guide Workbook,** p. 6

SHORT ON TIME? To do this lesson in approximately half the time, do the Activate Prior Knowledge activity. A discussion of the Key Concepts will familiarize students with the lesson content. Use the Apply the TEKS activity to help students understand scientific inquiry. Have students do the Quick Labs. The rest of the lesson can be completed by students independently.

These editable worksheets are available on **PearsonTexas.com.**
Print versions can be found in the **TEKS Preparation and Study Guide Workbook.**

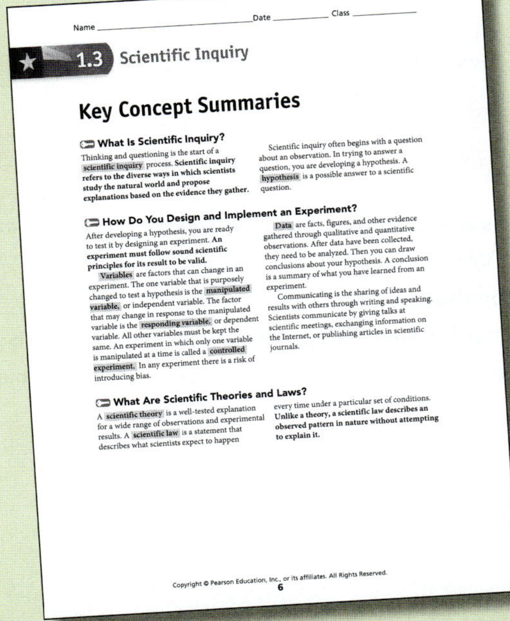
Name ____________ Date ______ Class ______

1.3 Scientific Inquiry

Key Concept Summaries

What Is Scientific Inquiry?

Thinking and questioning is the start of a scientific inquiry process. **Scientific inquiry refers to the diverse ways in which scientists study the natural world and propose explanations based on the evidence they gather.**

Scientific inquiry often begins with a question about an observation. In trying to answer a question, you are developing a hypothesis. A hypothesis is a possible answer to a scientific question.

How Do You Design and Implement an Experiment?

After developing a hypothesis, you are ready to test it by designing an experiment. **An experiment must follow sound scientific principles for its result to be valid.**

Variables are factors that can change in an experiment. The one variable that is purposely changed to test a hypothesis is the manipulated variable, or independent variable. The factor that may change in response to the manipulated variable is the responding variable, or dependent variable. All other variables must be kept the same. An experiment in which only one variable is manipulated at a time is called a controlled experiment. In any experiment there is a risk of introducing bias.

Data are facts, figures, and other evidence gathered through qualitative and quantitative observations. After data have been collected, they need to be analyzed. Then you can draw conclusions about your hypothesis. A conclusion is a summary of what you have learned from an experiment.

Communicating is the sharing of ideas and results with others through writing and speaking. Scientists communicate by giving talks at scientific meetings, exchanging information on the Internet, or publishing articles in scientific journals.

What Are Scientific Theories and Laws?

A scientific theory is a well-tested explanation for a wide range of observations and experimental results. A scientific law is a statement that describes what scientists expect to happen every time under a particular set of conditions. **Unlike a theory, a scientific law describes an observed pattern in nature without attempting to explain it.**

6

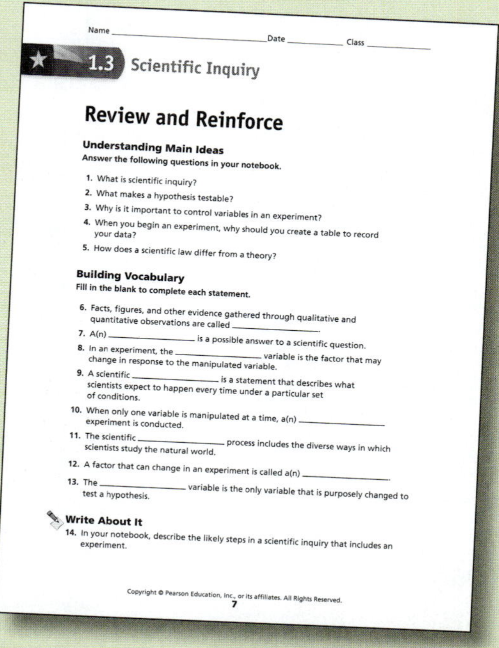
Name ____________ Date ______ Class ______

1.3 Scientific Inquiry

Review and Reinforce

Understanding Main Ideas
Answer the following questions in your notebook.

1. What is scientific inquiry?
2. What makes a hypothesis testable?
3. Why is it important to control variables in an experiment?
4. When you begin an experiment, why should you create a table to record your data?
5. How does a scientific law differ from a theory?

Building Vocabulary
Fill in the blank to complete each statement.

6. Facts, figures, and other evidence gathered through qualitative and quantitative observations are called ____________.
7. A(n) ____________ is a possible answer to a scientific question.
8. In an experiment, the ____________ variable is the factor that may change in response to the manipulated variable.
9. A scientific ____________ is a statement that describes what scientists expect to happen every time under a particular set of conditions.
10. When only one variable is manipulated at a time, a(n) ____________ experiment is conducted.
11. The scientific ____________ process includes the diverse ways in which scientists study the natural world.
12. A factor that can change in an experiment is called a(n) ____________.
13. The ____________ variable is the only variable that is purposely changed to test a hypothesis.

Write About It

14. In your notebook, describe the likely steps in a scientific inquiry that includes an experiment.

7

LESSON 1.3

Lexile Measure = 930L

Scientific Inquiry

Establish Learning Objectives

After this lesson, students will be able to:

- Recognize that scientific inquiry involves posing questions and developing hypotheses.
- Explain how to design and implement an experiment so that it uses sound scientific principles.
- Differentiate between a scientific theory and a scientific law.

Engage

Activate Prior Knowledge

MY PLANET DIARY Read *The Law of Falling Objects* with the class. Ask students to summarize and contrast Aristotle's idea and Galileo's idea. Ask: **Why do you think Aristotle's idea was accepted for so long?** *(Sampler: No one applied objective reasoning to the idea or conducted scientific experiments to test the idea.)*

Explore

Lab Resource: Inquiry Warm-Up

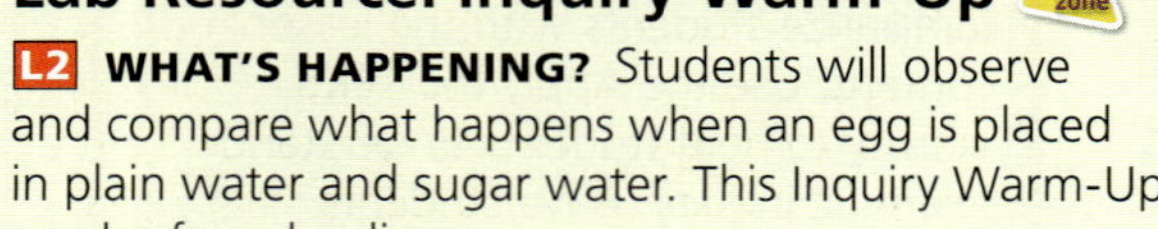

L2 **WHAT'S HAPPENING?** Students will observe and compare what happens when an egg is placed in plain water and sugar water. This Inquiry Warm-Up can be found online.

Scientific Inquiry

- What Is Scientific Inquiry? TEKS 2A, 2B
- How Do You Design and Implement an Experiment? TEKS 2B, 2C, 2D, 2E
- What Are Scientific Theories and Laws? TEKS 3A

my planet Diary

The Law of Falling Objects

Misconception: Heavier objects fall faster than lighter ones. This assumption is not true. They actually fall at the same rate, or with the same acceleration. The misconception was introduced by a philosopher named Aristotle and accepted for more than 2,000 years. But in the late 1500s, Galileo Galilei discovered something different—all free-falling objects fall with the same acceleration. To prove this, Galileo performed a number of experiments. Galileo's experiments involved rolling balls with different masses down a ramp called an inclined plane and making careful measurements.

Galileo and one of his acceleration experiments

MISCONCEPTION

Communicate Discuss the following questions with a partner. Write your answers below.

1. Why did Galileo perform experiments to see if all objects fall with the same acceleration?
   Galileo needed actual evidence to support his thinking.

2. Do you think a feather and a book that are dropped from the same height at the same time will hit the ground at the same time? Explain your answer in terms of Galileo's discovery.
   Sample: No, they won't. Galileo's discovery said they should, but his experiment did not show if the shape of a falling object matters.

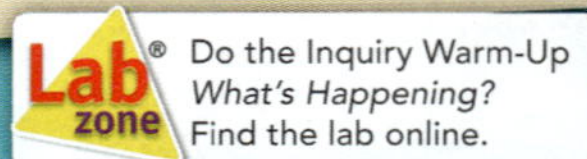

Do the Inquiry Warm-Up *What's Happening?* Find the lab online.

SUPPORT ALL READERS

Lexile Measure = 930L **Lexile Word Count = 1919**

Prior Exposure to Content: Many students may have misconceptions on this topic

Academic Vocabulary: *sequence, variables*

Science Vocabulary: *scientific inquiry, hypothesis, data, scientific theory*

Concept Level: Generally appropriate for most students in this grade

Preteach With: My Planet Diary "The Law of Falling Objects" and Figure 3 activity

Vocabulary

- scientific inquiry • hypothesis • variable
- manipulated variable • responding variable
- controlled experiment • data • SI
- scientific theory • scientific law

Skills

- Reading: Sequence
- Inquiry: Control Variables

What Is Scientific Inquiry?

Chirp, chirp, chirp. It is a hot summer night, and your bedroom windows are open. On most nights, the crickets chirp quietly, but not tonight. The noise from the crickets is almost deafening. Why do all the crickets in your neighborhood seem determined to keep you awake tonight? Your thinking and questioning is the start of the **scientific inquiry** process. **Scientific inquiry refers to the diverse ways in which scientists study the natural world and propose explanations based on the evidence they gather.** Some scientists investigate the natural world by making observations and drawing conclusions. Others conduct experiments in labs or in the field to test their ideas.

TEKS 2A, 2B In this section, you'll examine how scientists propose well-defined questions, develop hypotheses, and formulate explanations based on the evidence they gather.

ELPS 2.C.3

Read aloud or listen to the Posing Questions activity on this page. Discuss your observation and question with a small group. Tell the group whether your question is easy or difficult to investigate, and why.

Posing Questions Scientific inquiry often begins with a specific question about an observation. Your observation about the frequent chirping may lead you to ask a question: Why are the crickets chirping so much tonight? Questions come from your experiences, observations, and inferences. Curiosity plays a role, too. Because others may have asked similar questions, you should do research to find what information is already known about the topic before you go on with your investigation. Look at **Figure 1** to pose a well-defined question about an observation.

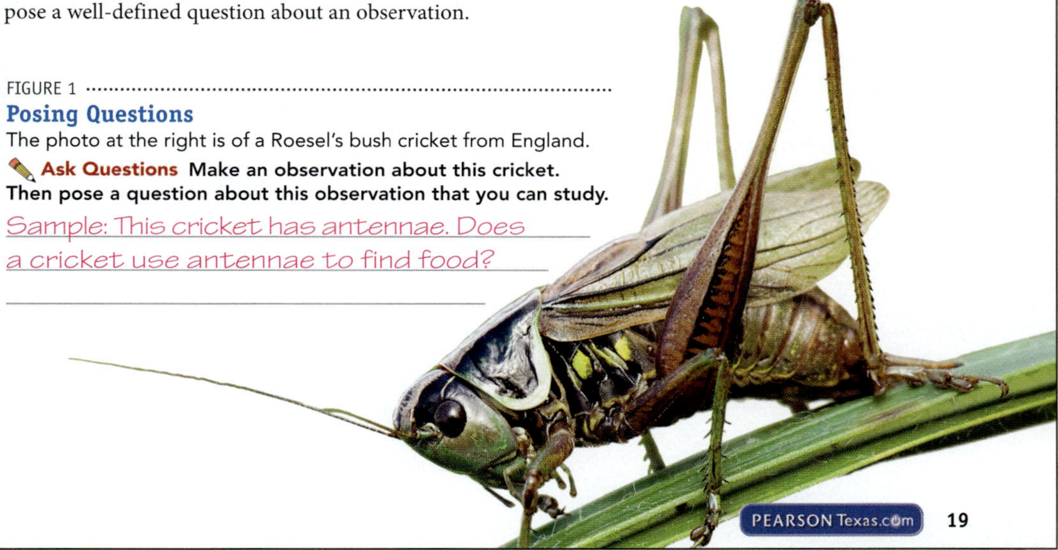

FIGURE 1

Posing Questions

The photo at the right is of a Roesel's bush cricket from England.

 **Ask Questions** **Make an observation about this cricket. Then pose a question about this observation that you can study.**

Sample: This cricket has antennae. Does a cricket use antennae to find food?

PEARSON Texas.com 19

Explain

Introduce Vocabulary

On the board, write the related word forms *inquire* and *inquiry*. Tell students that *inquire* is a verb that means "to seek information by questioning." *Inquiry* is a noun that means "a way of seeking information by questioning." Tell students they can use related word forms like these to help them infer and remember the meanings of vocabulary words.

Teach Key Concepts

Explain to students that scientists study the natural world through a process called scientific inquiry. This methodology involves making observations and collecting data. Scientists then use this evidence to propose explanations. Ask: **How do scientists often begin their process of inquiry?** *(By posing a question)* **Once they pose a question, why should doing research be their next step in a scientific inquiry?** *(Someone else may have already asked and answered the same question.)*

Elaborate

21st Century Learning

CRITICAL THINKING Tell students that a local biologist observes that a certain species of bird once common to your area has not been sighted for the past few years. Ask: **Given this observation, what scientific questions do you think the biologist would ask to begin a scientific inquiry?** *(Sample questions: Why has the bird disappeared from the area? Has a disease wiped out the population? Has the environment changed in some way?)*

PEARSON Texas.com

LESSON 1.3

English Language Proficiency Standards

ELPS Learning Strategies 2.C.3

Have students read and then complete **Figure 1.**

Beginning Have beginners listen as you read aloud the information in the figure. Point to and name each cricket feature. Then model how to complete the activity: look at the cricket; tell what you see; ask a question. Have partners complete the activity. Display question words as needed.

Intermediate Have students read the text aloud with you and name the cricket's features. Then have partners take turns asking questions. Provide question stems as needed.

Advanced As students work in pairs to complete the activity, encourage discussion of how they could find the answer to their questions.

Advanced High Have partners read the text and complete the activity by asking a question about each feature of the cricket.

Texas Essential Knowledge and Skills

2A Plan and implement comparative and descriptive investigations by making observations, asking well-defined questions, and using appropriate equipment and technology.

2B Design and implement experimental investigations by making observations, asking well-defined questions, formulating testable hypotheses, and using appropriate equipment and technology.

LESSON 1.3

Explain

Lead a Discussion

HYPOTHESIS SUMMARY Elicit from students the characteristics of a hypothesis, and then list them on the board. (*1. It is a possible answer to a scientific question. 2. It is not a fact. 3. It must be testable.*) Ask: **How does a hypothesis differ from a fact?** *(A fact has been confirmed as true through repeated observations, whereas a hypothesis may or may not be true.)* **One explanation for why crickets chirp more on hot summer nights is that they just feel like chirping more on those nights. Why is this explanation not a scientific hypothesis?** *(It is not testable.)* **What does it mean that a hypothesis is testable?** *(Researchers must be able to carry out investigations and gather evidence that will either support or disprove the hypothesis.)*

Elaborate

Lab Resource: Quick Lab 

L2 **SCIENTIFIC INQUIRY** Students will assess whether or not questions can be answered by scientific inquiry. This Quick Lab can be found in the Student Lab Manual, p. 3, and online.

Evaluate

Assess Your Understanding

After students answer the questions, have them evaluate their understanding by completing the appropriate sentence.

RTI Response to Intervention

1a. If students need help explaining why it is important to ask a well-defined question, **then** have them reread the section *Posing Questions*.

b. If students have difficulty developing another hypothesis, **then** lead a discussion about why students think crickets chirp.

Formulating a Hypothesis How could you answer your question about cricket chirping? In trying to answer the question, you are developing a hypothesis that you can test. A **hypothesis** (plural: *hypotheses*) is a possible answer to a scientific question. You may suspect that the hot temperatures affected the chirping. Your hypothesis would be that cricket chirping increases as a result of warmer air temperatures. Use **Figure 2** to practice developing a hypothesis.

A hypothesis is *not* a fact. In science, a fact is an observation that has been confirmed repeatedly. For example, that a cricket rubs its forelegs together to make the chirping noise is a fact. A hypothesis, on the other hand, is one possible answer to a question. For example, perhaps the crickets only seemed to be chirping more that night because there were fewer other sounds than usual.

In science, a hypothesis must be testable. Researchers must be able to carry out investigations and gather evidence that will either support or disprove the hypothesis. Disproven hypotheses are still useful, because they can lead to further investigations.

FIGURE 2

Formulating a Hypothesis

 **Develop Hypotheses** Write two hypotheses that might answer this student's question.

Hypothesis A	Hypothesis B
Sample: My digital music player needs to be recharged.	Sample: My digital music player got wet.

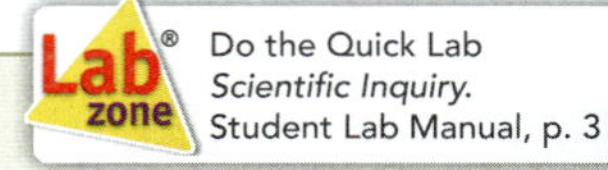

Assess Your Understanding

TEKS 2A, 2B

1a. Explain Why is it important to ask a well-defined question?

Sample: The question should be well-defined so that you can adequately answer it.

b. Develop Hypotheses What other hypothesis might explain why crickets chirp more frequently on some nights?

Sample: Crickets chirp more on nights when it doesn't rain.

got it?

○ **I get it!** Now I know that scientific inquiry is studying the natural world and posing explanations based on evidence.

○ I need extra help with See TE note.

Professional Development Note — Teacher to Teacher

Scientific Inquiry My students have studied the Scientific Method for at least two years by the time they are in my class, so I emphasize that there is a logical order to the process, not steps to memorize again. I ask them what they do if their locker doesn't open. I ask if they think, "Step 1, identify the problem; locker won't open. Step 2, make a hypothesis; I think it won't open because I am using my gym combination." Of course not! And yet, to open the locker, they must use the Scientific Method, but as a continual process, not step by step. Then I have each student write down a problem and solve it using the Scientific Method.

Mrs. Anne Rice
Woodland Middle School
Gurnee, Illinois

How Do You Design and Implement an Experiment?

TEKS 2B, 2C, 2D, 2E In this section, you'll learn that planning and implementing an experiment must follow certain scientific principles in order for the results to be accepted as valid.

After developing your hypothesis, you are ready to test it by designing an experiment. **An experiment must follow sound scientific principles for its results to be valid.** You know your experiment will involve counting cricket chirps at warm temperatures. But, how will you know how often a cricket would chirp at a low temperature? You cannot know unless you count other cricket chirps at low temperatures for comparison.

Controlling Variables To test your hypothesis, you will observe crickets at different air temperatures. All other **variables,** or factors that can change in an experiment, must be the same. This includes variables such as food and hours of daylight. By keeping these variables the same, you will know that any difference in cricket chirping is due to temperature alone.

The one variable that is purposely changed to test a hypothesis is the **manipulated variable,** or independent variable. The manipulated variable here is air temperature. The factor that may change in response to the manipulated variable is the **responding variable,** or dependent variable. The responding variable here is the number of cricket chirps.

apply it!

A student performs an experiment to determine whether 1 g of sugar or 1 g of salt dissolves more quickly in water.

1 **Control Variables** Identify the manipulated variable and the responding variable.
The manipulated variable is sugar or salt, and the responding variable is time.

2 **Identify** What are two other variables in this experiment?
Sample: Two other variables are the amount of water and the temperature of the water.

3 **Draw Conclusions** Write a hypothesis for this experiment.
Sample: Sugar dissolves faster in water than salt does.

Explain

Teach Key Concepts

Explain to students that one way scientists test a hypothesis is by designing and conducting experiments according to certain principles. By following these principles, they and other scientists can trust that the results of the experiment prove or disprove a hypothesis. After students have read the description of the cricket experiment, call on volunteers to define *variable, manipulated variable,* and *responding variable.* Ask: **What is the manipulated variable and the responding variable in this experiment?** *(Air temperature is the manipulated variable and number of cricket chirps is the responding variable.)*

Elaborate

Apply It!

L1 Review the definitions of *manipulated variable* and *responding variable* before beginning the activity.

Control Variables Explain that a variable is a factor that can change in a scientific investigation or experiment. When scientists control variables, they keep all but one of the factors the same. Have a volunteer read aloud the description of the experiment. Then ask students to identify any named or implied variables in the description. *(Sugar, salt, water, time)* If necessary, point out that the phrase "dissolves more quickly" points to the fact that time is a variable.

Differentiated Instruction

L3 **The Inquiry Process** Before students read this section, ask them to describe a process that they think would work best to find the answer to a question about the natural world. After they have read the section, ask them to compare the process they proposed with the process described in the text.

L1 **Questions and Hypotheses** Challenge students to write three scientific questions about today's weather and then develop a hypothesis for each question. Remind students that their hypotheses should be testable. Ask them to explain how they might test each of their hypotheses.

Texas Essential Knowledge and Skills

2B Design experimental investigations by making observations, asking well-defined questions, formulating testable hypotheses, and using appropriate equipment and technology.

2C Collect and record data using the International System of Units (SI) and qualitative means such as labeled drawings, writing, and graphic organizers.

2D Construct tables and graphs, using repeated trials and means, to organize data and identify patterns.

2E Analyze data to formulate reasonable explanations communicate valid conclusions supported by the data, and predict trends.

LESSON 1.3

Explain

Lead a Discussion

CONTROLLED EXPERIMENT Ask students what makes the cricket experiment described in the text a controlled experiment. Remind them that the hypothesis being tested is that cricket chirping increases as air temperature increases. Ask: **Why is this considered a controlled experiment?** *(Only one variable is being tested—air temperature. All other variables are controlled, or kept the same.)* **Why is only one variable manipulated in a controlled experiment?** *(So researchers can be sure which variable actually influenced the results.)*

Elaborate

Teacher Demo 

L2 OBSERVE RATE OF FALL

Materials 2 similar marbles of different mass, balance, towel, ruler

Time 15 minutes

Explain that one of the best-known experiments in history was conducted by Galileo. He showed that objects of different mass fall at the same rate and hit the ground at the same time. Use the balance to find the mass of each marble, and show students that one marble has more mass than the other. Have students speculate about which marble will hit the floor first if both are dropped from the same height at the same time. Then place both marbles on the edge of a desk and position a folded towel on the floor below. Have a volunteer in position to see which marble hits the floor first. Use the ruler to push both marbles off the desk at the same time. Repeat the procedure several times, allowing all students to observe.

Ask: **Which marble hit the floor first?** *(They both hit at the same time.)* Tell students that this experiment supports the hypothesis that objects fall at the same rate, no matter what their mass. Then ask: **What is the manipulated variable in this experiment?** *(Marble mass)* **What is the responding variable?** *(Rate of fall)* **What variables are controlled?** *(Height of fall, moment fall begins, landing area, shape and material of falling objects, air temperature)*

Setting Up a Controlled Experiment An experiment in which only one variable is manipulated at a time is called a **controlled experiment.** You decide to test the crickets at three different temperatures: 15°C, 20°C, and 25°C, as shown in **Figure 3.** All other variables are kept the same. Otherwise, your experiment would have more than one manipulated variable. Then there would be no way to tell which variable influenced your results.

Experimental Bias In any experiment there is a risk of introducing bias. For example, if you expect crickets to chirp more at 25°C, you may run experiments at just that temperature. Or, without meaning to, you might bias your results by selecting only the crickets that chirp the most often to test. Having a good sample size, or the number of crickets tested, is also important. Having too few crickets may bias your results because individual differences exist from cricket to cricket.

FIGURE 3

A Controlled Experiment

The manipulated variable in the experiment below is temperature.

Design Experiments **In the boxes, write the number of crickets you would test for this controlled experiment. On the lines below, write three other variables that must be kept the same.**

The type of container, the containers' contents, and the time that observations are made must be kept the same.

Collecting, Recording, and Analyzing Data

You decide to test five crickets, one at a time, at each temperature. You also decide to run multiple trials for each cricket. This is because a cricket may behave differently from one trial to the next. Before you begin your experiment, decide what observations you will make 1
and what data you will collect. **Data** are the facts, figures, and other evidence gathered through qualitative and quantitative means.

Qualitative means refers to collecting information about observations such as a color change in a chemical or the behavior of a cricket. To collect quantitative data, scientists use a version of the metric system called the International System of Units, or **SI,** as a standard system of measurement. To organize your data, you may want to make a data table. A data table provides you with an organized way to collect and record your observations. Decide what 2
3 your table will look like. Then you can start to collect your data.

After you have collected data, you need to analyze the data. 4
One tool that can help you analyze data is a graph. Graphs can be used to reveal patterns or predict trends in data. Sometimes, there is more than one interpretation for a set of data. For example, scientists agree that global temperatures have risen over the past 100 years. But they do not agree on how much they are likely to rise over the next 100 years.

Go to page 612 to learn more about SI Units. Go to pages 626–633 to learn more about tables and graphs.

Sequence Underline and number the steps involved in collecting and analyzing data.

do the math!

A data table helps you organize the information you collect in an experiment. Graphing the data may reveal any patterns in your data and help you predict trends.

1. **Read Graphs** Identify the manipulated variable and the responding variable.

   manipulated variable = temperature; responding variable = number of chirps per minute

2. **Analyze Data** As the temperature increases from 15°C to 25°C, what happens to the number of chirps per minute?

   It increases.

3. **Predict** How many chirps per minute would you expect when the temperature is 10°C?

   Sample: I would expect about 40 chirps per minute.

Number of Chirps per Minute

Cricket	15°C	20°C	25°C
1	91	135	180
2	80	124	169
3	89	130	176
4	78	125	158
5	77	121	157
Average	83	127	168

Average Chirps vs. Temperature

Chirps per Minute (0, 50, 100, 150, 200)

Temperature (°C) (0, 5, 10, 15, 20, 25, 30)

Differentiated Instruction

L1 Experimental Trials Students may not be familiar with the use of the word *trial* relative to an experiment. Explain that scientists usually repeat an experiment many times, always using the same procedure and variables. A trial is a repetition of an experiment. Trials help scientists check the accuracy of their results. Have students use *trial* in a sentence about an experiment.

L3 Cricket Experiment Challenge small groups of students to design an experiment to test this hypothesis: Crickets are more active at midnight than at noon. Emphasize that the experiment should have a manipulated variable and a responding variable, and all other variables should be controlled. Have students make a list of materials needed for their experiment and then research where such material can be obtained.

Explain

Lead a Discussion

ORGANIZING AND INTERPRETING DATA Tell students that a good controlled experiment yields quantitative observations that can be recorded in a table or displayed in a graph. Ask a volunteer to read the definition of *data*, and then direct students' attention to the table and graph shown. Ask: **What data were gathered in the cricket experiment?** *(The temperature and number of cricket chirps per minute)* **For what purpose were these data gathered?** *(To test the hypothesis that cricket chirping increases at higher temperatures)* Emphasize that recording data in a data table is a good way to organize data in preparation for drawing a conclusion about the hypothesis.

Sequence Explain to students that when they sequence, they put events in the order in which they occur. Sometimes objects can also be listed in order of their ascending or descending characteristics.

Address Misconceptions

L1 KINDS OF DATA Students may think that data only include numerical information. Explain that data are any kind of information gathered through observations, both quantitative and qualitative. Ask: **What kind of information is gained through quantitative observations?** *(Numbers or amounts)* **What kind of information is gained through qualitative observations?** *(Descriptions without numbers)* Provide students with examples of both qualitative data and quantitative data. Show them various data tables and graphs from experiments that have been published in scientific journals. Call on students to describe the kinds of data collected.

The Science Skills Practice and Support continues the lesson from this page by explaining more about the topics discussed. Teach the material to your students and assign the appropriate section for your students to read.

Elaborate

Do the Math!

L1 Point out that the table and graph show the same set of data from the cricket experiment described in the text, but in different ways. The kind of graph shown is a line graph, which is used to show the relationship between two variables. This graph shows how the responding variable (number of chirps per minute) changes in response to the manipulated variable (temperature). Tell students they can use a line graph to estimate values for conditions that were not tested in the experiment, such as the conditions described in Question 3.

LESSON 1.3

Explain

Lead a Discussion

A SUMMARY OF WHAT'S BEEN LEARNED Tell students that the purpose of gathering data from an experiment is to draw a conclusion about whether the data support or disprove the original hypothesis. Ask: **What was the hypothesis the cricket experiment was designed to test?** *(Cricket chirping increases at higher temperatures.)* **Was this hypothesis supported?** *(No, not entirely)* **What do you base your conclusion on?** *(An interpretation of the data shown in Figure 4)*

Teach With Visuals

Tell students to look at **Figure 4.** Make sure students understand that this table does not show the same data as the table and graph on the previous page. Both sets of data were collected using the same cricket experiment, but during different trials. Ask: **In this trial, at what temperature was the average number of chirps the greatest?** *(25°C)* **the least?** *(20°C)* **If the original hypothesis were correct, how would you expect the results of this trial to be different?** *(The least number of chirps would occur at 15°C.)* **What might account for the difference in results from the two trials?** *(During one of the trials, experimental bias may have occurred or the researchers may have made a mistake.)*

Drawing Conclusions Now you can draw conclusions about your hypothesis. A conclusion is a summary or a reasonable explanation of what you have learned from an experiment. To draw your conclusion, you must analyze your data objectively to see if they support or do not support your hypothesis. You also must consider whether you collected enough data.

You may decide that the data support your hypothesis. You conclude that the frequency of cricket chirping increases with temperature. Now, repeat your experiment to see if you get the same results. A conclusion or explanation is unreliable if it comes from an experiment with results that cannot be repeated. Many trials are needed before a hypothesis can be accepted as true.

Your data won't always support your hypothesis. When this happens, check your experiment for things that went wrong, or for improvements you can make. Then fix the problem and do the experiment again. If the experiment was done correctly the first time, your hypothesis was probably not correct. Propose a new hypothesis that you can test. Scientific inquiry usually doesn't end once an experiment is done. Often, one experiment leads to another.

Number of Chirps per Minute			
Cricket	15°C	20°C	25°C
1	98	100	120
2	92	95	105
3	101	93	99
4	102	85	97
5	91	89	98
Average	96	92	103

FIGURE 4

Drawing Conclusions

Sometimes the same experiment can produce very different data.

Answer the questions below.

1. **Interpret Tables** Look at the data in the table. Do the data support the hypothesis that crickets chirp more in warmer temperatures? Explain.
   No, because crickets chirp less on average at 20°C than at 15°C.
2. **Analyze Sources of Error** If the data in this table were yours, what might you do next? Explain.
   Sample: I would check if I made an error during the experiment. Then I would do the experiment again.
3. CHALLENGE Can you draw a conclusion from these data? Why or why not?
   No; the data do not show a clear relationship between the number of chirps and temperature.

Communicating Communicating is the sharing of ideas and results with others through writing and speaking. Scientists communicate by giving talks at scientific meetings, exchanging information on the Internet, or publishing articles in scientific journals.

When scientists share the results of their research, they describe their procedures so that others can repeat their experiments. It is important for scientists to wait until an experiment has been repeated many times before accepting a result. Therefore, scientists must keep accurate records of their methods and results. This way, scientists know that the result is accurate. Before the results are published, other scientists review the experiment for sources of error, such as bias, data interpretation, and faulty conclusions.

Sometimes, a scientific inquiry can be part of a huge project in which many scientists are working together around the world. For example, the Human Genome Project involved scientists from 18 different countries. The scientists' goal was to create a map of the information in your cells that makes you who you are. On such a large project, scientists must share their ideas and results regularly. Come up with ideas for communicating the results of your cricket experiment in **Figure 5.**

Vocabulary Identify Related Word Forms *Communication* is the noun form of the verb *communicate.* Write a sentence using the noun *communication.*

Sample: TV and the Internet are tools used for communication.

The Human Genome Project logo

BIOLOGY PHYSICS ETHICS INFORMATICS ENGINEERING CHEMISTRY

FIGURE 5

Communicating Results

Since the Human Genome Project touched upon many areas of science, communication was important.

Communicate Get together as a group and write three ways to share the results of your cricket experiment with other students.

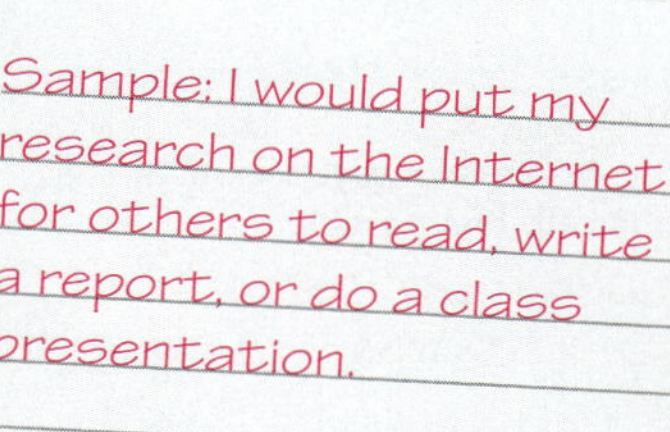

Sample: I would put my research on the Internet for others to read, write a report, or do a class presentation.

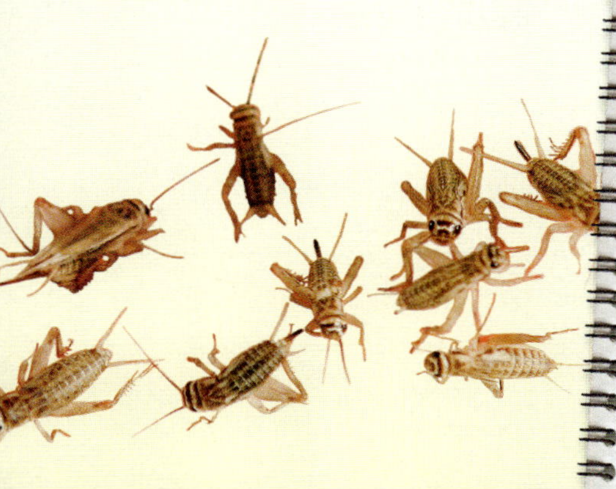

Differentiated Instruction

L1 Graph Weather Data Daily newspapers or online weather sites often have a table that shows the progression of air temperatures on the previous day. Provide students with a copy of one such table and have them make a line graph from those data. Then have them write a description of any pattern or trend they see in the temperatures during the day. Have them share their writing with a partner.

L3 Additional Inquiry Have students assume that the results of the cricket experiment supported the hypothesis. Have small groups or partners brainstorm additional questions based on those results. *(Samples: Do these results apply to all kinds of crickets? Does the rate of chirping decrease further at lower temperatures?)* Challenge groups to write a new hypothesis for one of the questions and describe an experiment that would test that hypothesis.

Elaborate

21st Century Learning

ACCOUNTABILITY Tell students that scientists depend on other scientists to communicate their findings fully and accurately. Scientists also have a responsibility to the general public to share findings that affect them. Ask: **What would happen if a space scientist discovered something new about the solar system but failed to report the discovery to other scientists?** *(Sample: Other scientists would continue their work without the new information, and the work might be wasted as a result.)* **When might it be important for a scientist to communicate the details of their experiments to the public?** *(Sample: When the public could be helped by doing so)*

LESSON 1.3

Explain

Apply the TEKS

Review the steps involved in conducting a controlled experiment. Ask: **What factors will you need to keep the same during the experiment?** *(Sample: the kind of paper, the number of sheets used for each shape, how far and where the sheets fall)* **What additional materials might you need to conduct this experiment?** *(A ruler and a stopwatch)* **At what point in the experiment would you interpret your data and draw a conclusion?** *(After collecting the data)*

Elaborate

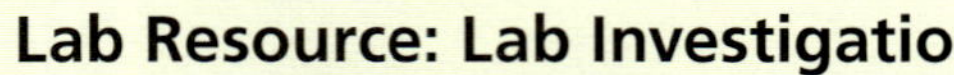

Lab Resource: Lab Investigation 

L2 KEEPING FLOWERS FRESH Students will explore how cut flowers can stay fresher for a longer period of time. This Lab Investigation can be found in the Student Lab Manual, p. 4, and online.

Evaluate

Assess Your Understanding

After students answer the questions, have them evaluate their understanding by completing the appropriate sentence.

RTI Response to Intervention

2a. If students have trouble identifying ways scientists communicate their results, **then** have them reread the section on communicating.

b. If students cannot explain how scientists investigate the natural world, **then** have them review the red headings in the lesson.

Texas Essential Knowledge and Skills

2A Plan comparative and descriptive investigations by making observations, asking well-defined questions, and using appropriate equipment and technology.

2B Design and implement experimental investigations by making observations, asking well-defined questions, formulating testable hypotheses, and using appropriate equipment and technology.

3A In all fields of science, analyze, evaluate, and critique scientific explanations by using empirical evidence, logical reasoning, and experimental and observational testing, including examining all sides of scientific evidence of those scientific explanations, so as to encourage critical thinking by the student.

APPLY THE TEKS 2A, 2B, 3A

In a Scientist's Shoes

How do scientists investigate the natural world?

FIGURE 6

When you think like a scientist, you develop hypotheses and design experiments to test them.

Design Experiments **Think like a scientist to find out which falls fastest: an unfolded sheet of paper, a sheet of paper folded in fourths, or a crumpled sheet of paper.**

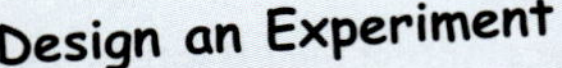

Design an Experiment

QUESTION *Which paper falls fastest?*

SCIENTIFIC ATTITUDES INVOLVED *Sample: Curiosity and creativity*

HYPOTHESIS *Sample: The crumpled paper will fall fastest.*

VARIABLES

Manipulated Variables *The shape of paper*

Responding Variables *How fast the paper falls*

Factors to Consider *Sample: The weight of the paper; the distance of the fall*

COLLECT DATA

Number of Trials *5 trials*

Units of Measure *Meters per second*

SCIENTIFIC SKILLS USED *Sample: Quantitative observations, evaluating*

NEXT STEPS *Sample: Redesign, communicate*

Do the Lab Investigation *Keeping Flowers Fresh.* Student Lab Manual, p. 4

Assess Your Understanding

TEKS 2A, 2B

2a. Identify Name two ways scientists communicate their results.

Sample: They talk at scientific meetings or exchange information on the Internet.

b. Apply Concepts How do scientists investigate the natural world?

Scientists use scientific skills, scientific reasoning, and the scientific process to answer questions about the world.

got it?

O **I get it!** Now I know that an experimental design must *follow sound scientific principles.*

O I need extra help with *See TE note.*

What Are Scientific Theories and Laws?

TEKS 3A In this section, you'll explore the difference between a scientific theory and a scientific law. You'll also learn how scientists use experimental and observational testing to develop, revise, or discard theories.

Sometimes, a large set of related observations can be connected by a single explanation. This explanation can lead to the development of a scientific theory. In everyday life, a theory can be an unsupported guess. Everyday theories are not scientific theories. A **scientific theory** is a well-tested explanation for a wide range of observations and experimental results. For example, according to the atomic theory, all substances are composed of particles called atoms. Atomic theory helps to explain many observations, such as why iron nails rust. Scientists accept a theory only when it can explain the important observations. If the theory cannot explain new observations, then the theory is changed or thrown out. In this way, theories are constantly being developed, revised, or discarded as more information is collected.

A **scientific law** is a statement that describes what scientists expect to happen every time under a particular set of conditions. **Unlike a theory, a scientific law describes an observed pattern in nature without attempting to explain it.** For example, the law of gravity states that all objects in the universe attract each other. Look at **Figure 7.**

FIGURE 7

A Scientific Law

According to the law of gravity, this parachutist will eventually land on Earth.

Apply Concepts Give another example of a scientific law.

Sample: Liquid water freezes into solid ice when its temperature falls to 0°C.

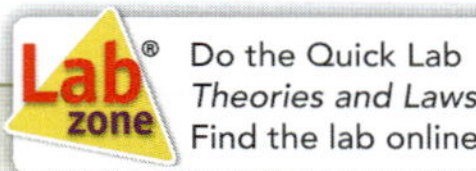

Do the Quick Lab *Theories and Laws.* Find the lab online.

Assess Your Understanding

got it?

○ **I get it!** Now I know that the difference between a scientific theory and a law is that a law describes an observed pattern, while a theory explains observations.

○ I need extra help with See TE note.

27

Differentiated Instruction

L1 Theory or Law Lead students in a discussion about the difference between a hypothesis, a scientific theory, and a scientific law. Call on students to define each term in their own words and give examples of each. Help students understand how these terms are related yet different.

L3 Research Theories and Laws Have students choose a subject area that interests them, such as geology or biology. They should then do research to identify a scientific theory and scientific law related to that subject area. Have students summarize the theory and law in their own words and read their summaries to the class.

Explain

Teach Key Concepts

Explain to students that the words *theory* and *law* have special meanings in science. Have a volunteer read aloud the definition of *scientific theory.* Discuss how that definition differs from the common meaning of the word. Then, have a volunteer read the definition of *scientific law,* and discuss how its definition differs from the common meaning. Explain that a scientific law summarizes a pattern observed in nature. It does not explain that pattern, as a scientific theory does. Ask: **Would a scientific law that is true in New York also be true in Alabama? Explain your answer.** *(Yes. Scientific laws hold true for every state, and throughout the universe. A scientific law is a rule of nature, not a law passed by a human legislature.)*

Teach With Visuals

Tell students to look at **Figure 7.** Ask: **According to the law of gravity, how do the parachutist and Earth affect each other?** *(They attract, or pull on, each other.)* **How does this relate to the motion of the parachutist?** *(Earth's pull on the parachutist is greater than the parachutist's pull on Earth, so the parachutist moves towards Earth.)*

Elaborate

Lab Resource: Quick Lab 

L1 THEORIES AND LAWS Students will take the first steps toward developing a scientific theory. This Quick Lab can be found online.

Evaluate

Assess Your Understanding

Have students evaluate their understanding by completing the appropriate sentence.

RTI Response to Intervention

If students need help distinguishing a scientific theory from a scientific law, **then** lead a class discussion about this topic, giving examples of both scientific theories and laws.

Texas Essential Knowledge and Skills

3A In all fields of science, analyze, evaluate, and critique scientific explanations by using empirical evidence, logical reasoning, and experimental and observational testing, including examining all sides of scientific evidence of those scientific explanations, so as to encourage critical thinking by the student.

Name ____________________ Date __________ Class __________

Assess Your Understanding

Scientific Inquiry

What Is Scientific Inquiry?

1a. **EXPLAIN** Why is it important to ask a well-defined question?

____

b. **DEVELOP HYPOTHESES** What other hypothesis might explain why crickets chirp more frequently on some nights?

____

How Do You Design and Implement an Experiment?

2a. **IDENTIFY** Name two ways scientists communicate their results.

____

____

b. **APPLY CONCEPTS** How do scientists investigate the natural world?

____

____

What Are Scientific Theories and Laws?

got it?

○ **I get it!** Now I know that the difference between a scientific theory and a law is that ______________________________

____

____

○ **I need extra help with** ______________________________

____

Name ______________________ Date ____________ Class ____________

Enrich

Scientific Inquiry

Henri Becquerel was a pioneer of scientific exploration. Read the passage. Then answer the questions that follow on a separate sheet of paper.

An Enlightening Discovery

In 1896, a French scientist named Henri Becquerel was experimenting with a sample of uranium salt to see if fluorescent substances give off X-rays. A fluorescent substance is one that glows when exposed to sunlight. Becquerel's experiment consisted of wrapping some photographic film in lightproof paper, placing a sample of uranium salt on top of the paper, and then leaving the setup out in the sun. His hypothesis was that uranium salt gives off X-rays when exposed to sunlight. He reasoned that the X-rays would pass through the paper and produce an image on the film.

When Becquerel developed the photographic film, he saw the image he was looking for—evidence, he thought, that X-rays had been produced by the uranium salt in sunlight. One trial was not enough, though, to support his hypothesis. He decided to repeat the experiment the next day. Much to his frustration, though, the next day was cloudy. Becquerel put the setup in a drawer and went on to do other things. A day later, he developed the film anyway, thinking there wouldn't be much of an image. Much to his amazement, he saw an image on the film just as clear as when the uranium salt had been left out in the sun.

Quite by accident, Becquerel had made a discovery of great importance. He realized that the uranium salt had given off an invisible "something" that could not be explained by sunlight hitting a fluorescent substance. Becquerel tested many more uranium compounds, and he drew the conclusion that uranium produced this mysterious "something." In time, scientists realized this "something" was a type of radiation.

1. **What was Becquerel's original hypothesis?**
2. **How was Becquerel convinced at first that his hypothesis was supported?**
3. **What observation caused Becquerel to pose another question?**
4. **What discovery did Becquerel make by accident?**
5. **Describe the pathway of scientific inquiry by which Henri Becquerel discovered that uranium gives off a type of radiation.**

Name ______________________ Date ____________ Class ____________

Lesson Quiz

Scientific Inquiry

If the statement is true, write *true.* If the statement is false, change the underlined word or words to make the statement true.

1. ___are___ Many trials are not needed before a hypothesis can be accepted as true.

2. ___true___ A conclusion is a summary of what is learned from an experiment.

3. ___True___ A factor that can change in an experiment is called a variable.

4. ___True___ A hypothesis is not the same as a fact.

5. ___data___ Facts and figures are examples of variables.

6. ___Theory___ A well-tested explanation for a wide range of observations and experimental results is known as a scientific inquiry.

Write the letter of the correct answer on the line at the left.

7. ___ The statement, "All objects in the universe attract each other," is an example of which of the following?
   - A scientific inquiry
   - B scientific theory
   - C scientific law
   - D controlled experiment

8. ___ Which of these is purposely changed during an experiment?
   - A hypothesis
   - B dependant variable
   - C responding variable
   - D manipulated variable

9. ___ Which of these is NOT an example of a way scientists communicate their results?
   - A taking out advertisements in the newspapers
   - B publishing articles in scientific journals
   - C giving talks at scientific meetings
   - D exchanging information on the Internet

10. ___ Which of these is a tool that can help you interpret data?
    - A theory
    - B variable
    - C hypothesis
    - D graph

Scientific Inquiry

Answer Key

Review and Reinforce

Find the worksheet in the Student Workbook.

1. Scientific inquiry refers to the diverse ways in which scientists study the natural world and propose explanations based on the evidence they gather.
2. Researchers must be able to carry out investigations and gather evidence that will either support or disprove a hypothesis.
3. If you don't control variables, there would be no way to know which variable explained your results.
4. A data table provides an organized way to collect and record data.
5. Unlike a theory, which is a well-tested explanation for a wide range of observations and experimental results, a scientific law describes an observed pattern in nature without attempting to explain it.

6. data	**7.** hypothesis
8. responding	**9.** law
10. controlled	**11.** inquiry
12. variable	**13.** manipulated

14. A scientific inquiry often begins with a question about an observation. To answer the question, a hypothesis is developed. A controlled experiment is planned and conducted to test the hypothesis. Data are collected and interpreted. A conclusion is drawn about the hypothesis. The results are communicated to others.

Enrich

1. Uranium salt gives off X-rays when exposed to sunlight.
2. When Becquerel developed film wrapped in lightproof paper that had been under uranium exposed to the sun, he saw an image.
3. The same image was produced when the setup was left in a drawer and not exposed to sunlight.
4. He discovered that uranium produces a type of radiation.
5. Becquerel developed a hypothesis and tested that hypothesis with an experiment. The results of that experiment led him to develop another hypothesis and another experiment, from which he drew a conclusion.

Lesson Quiz

1. are	**2.** true
3. true	**4.** true
5. data	**6.** theory
7. C	**8.** D
9. A	**10.** D

Models as Tools in Science

How do scientists investigate the natural world?

LESSON PACING:
2–3 periods or 1–1½ blocks

Lesson Vocabulary

- system
- input
- process
- output
- feedback

Lesson Objectives	TEKS	ELPS
Explain why models are used in science.	3B, 3C	4.C.2
Describe different types of systems and identify characteristics that all systems share.	3B	
Examine models of natural systems and identify their advantages and limitations.	3B, 3C	

Content Refresher

Systems Theory General systems theory is a framework by which any group of objects that work together to produce some result can be described. The theory was originally proposed in the 1920s by biologist and philosopher Ludwig von Bertalanffy, and it arose from the need to explain how organisms in ecosystems are interrelated. Born in a small village near Vienna in 1901, Von Bertalanffy proposed that a system is characterized by the interactions of its parts and the nonlinear nature of those interactions. His general systems theory was an interdisciplinary school of thought that attempted to provide alternatives to conventional models of organization. Systems theory has since been widely applied not just to biology but to many scientific and nonscientific fields including cybernetics, computer and information technology, psychology, economics, physics, geology, geography, sociology, and political science.

Texas Essential Knowledge and Skills

3B Use models to represent aspects of the natural world such as human body systems and plant and animal cells.
3C Identify advantages and limitations of models such as size, scale, properties, and materials.

English Language Proficiency Standards

ELPS Reading 4.C.2 Derive meaning of environmental print.

DIFFERENTIATED INSTRUCTION KEY
L1 Struggling Students or Special Needs
L2 On-Level Students L3 Advanced Students

LESSON PLANNER 1.4

Investigations and Activities	TEKS Review
My Planet Diary, **Student Edition,** p. 28 Inquiry: Inquiry Warm-Up, Scale Models, **PearsonTexas.com** Introduce Vocabulary, **Teacher's Edition**, p. 29 Teach Key Concepts, **Teacher's Edition,** p. 29 Teach With Visuals, **Teacher's Edition,** p. 29 Inquiry: Quick Lab, Making Models, **PearsonTexas.com**	Apply the TEKS, Evaluating Scientific Explanations, **Student Edition,** p. 46 TEKS Practice, **Student Edition,** p. 48 TEKS Practice: Chapter and Cumulative Review, **Student Edition,** p. 52 Lesson 1.4, **TEKS Preparation and Study Guide Workbook,** p. 8
Teach Key Concepts, **Teacher's Edition,** p. 30 Teach With Visuals, **Teacher's Edition,** p. 30 21st Century Learning, Critical Thinking, **Teacher's Edition,** p. 31 Apply It!, **Student Edition,** p. 31 Differentiated Instruction, **Teacher's Edition,** p. 31 Inquiry: Quick Lab, Systems, **PearsonTexas.com**	
Teach Key Concepts, **Teacher's Edition,** p. 32 Teach With Visuals, **Teacher's Edition,** p. 32 Support the TEKS, *Types of Models,* **Teacher's Edition,** p. 33 21st Century Learning, Communication, **Teacher's Edition,** p. 34 Inquiry: Build Inquiry, Earth Systems Model, **Teacher's Edition,** p. 35 Inquiry: Quick Lab, Models in Nature, **Lab Manual,** p. 9	**SHORT ON TIME?** To do this lesson in approximately half the time, do the Activate Prior Knowledge activity. A discussion of the Key Concepts will familiarize students with the lesson content. Have students do the Quick Labs. The rest of the lesson can be completed by students independently.

These editable worksheets are available on **PearsonTexas.com.**
Print versions can be found in the **TEKS Preparation and Study Guide Workbook.**

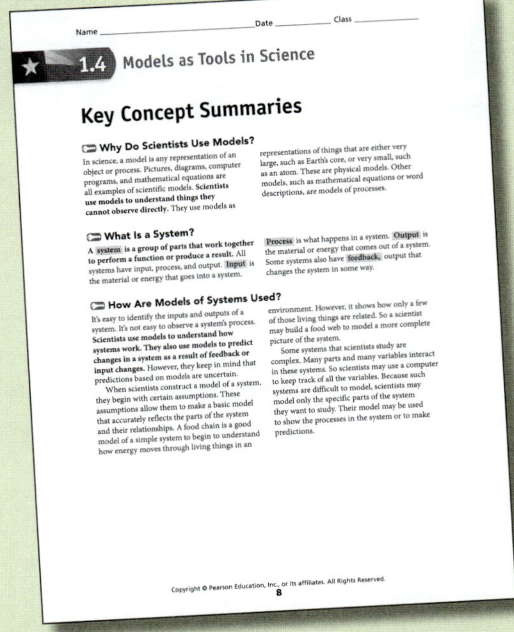

Name ________________ Date ________ Class ________

1.4 Models as Tools in Science

Key Concept Summaries

Why Do Scientists Use Models?

In science, a model is any representation of an object or process. Pictures, diagrams, computer programs, and mathematical equations are all examples of scientific models. **Scientists use models to understand things they cannot observe directly.** They use models as representations of things that are either very large, such as Earth's core, or very small, such as an atom. These are physical models. Other models, such as mathematical equations or word descriptions, are models of processes.

What Is a System?

A system is a group of parts that work together to perform a function or produce a result. All systems have input, process, and output. **Input** is the material or energy that goes into a system. **Process** is what happens in a system. **Output** is the material or energy that comes out of a system. Some systems also have **feedback,** output that changes the system in some way.

How Are Models of Systems Used?

It's easy to identify the inputs and outputs of a system. It's not easy to observe a system's process. **Scientists use models to understand how systems work. They also use models to predict changes in a system as a result of feedback or input changes.** However, they keep in mind that predictions based on models are uncertain.

When scientists construct a model of a system, they begin with certain assumptions. These assumptions allow them to make a basic model that accurately reflects the parts of the system and their relationships. A food chain is a good model of a simple system to begin to understand how energy moves through living things in an environment. However, it shows how only a few of those living things are related. So a scientist may build a food web to model a more complete picture of the system.

Some systems that scientists study are complex. Many parts and many variables interact in these systems. So scientists may use a computer to keep track of all the variables. Because such systems are difficult to model, scientists may model only the specific parts of the system they want to study. Their model may be used to show the processes in the system or to make predictions.

8

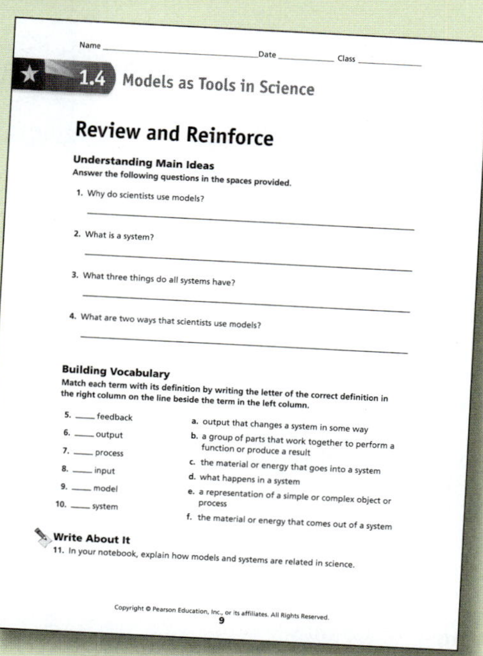

Name ________________ Date ________ Class ________

1.4 Models as Tools in Science

Review and Reinforce

Understanding Main Ideas

Answer the following questions in the spaces provided.

1. Why do scientists use models?
2. What is a system?
3. What three things do all systems have?
4. What are two ways that scientists use models?

Building Vocabulary

Match each term with its definition by writing the letter of the correct definition in the right column on the line beside the term in the left column.

5. ____ feedback	a. output that changes a system in some way
6. ____ output	b. a group of parts that work together to perform a function or produce a result
7. ____ process	c. the material or energy that goes into a system
8. ____ input	d. what happens in a system
9. ____ model	e. a representation of a simple or complex object or process
10. ____ system	f. the material or energy that comes out of a system

Write About It

11. In your notebook, explain how models and systems are related in science.

9

LESSON 1.4

Lexile Measure = 870L

Models as Tools in Science

Establish Learning Objectives

After this lesson, students will be able to:

- Explain why models are used in science.
- Describe different types of systems and identify characteristics that all systems share.
- Examine models of natural systems and identify their advantages and limitations.

Engage

Activate Prior Knowledge

MY PLANET DIARY Read *Flying Through Space* with the class. Ask students who have used computer simulations, including computer games that simulate an activity or event, to describe their experiences. Ask: **How was the simulation similar to and different from the real thing?** *(Sample: The sounds and sights were the same, but the physical sensations of touch and smell were not.)*

Explore

Lab Resource: Inquiry Warm-Up 

L3 **SCALE MODELS** Students will draw a scale model of their classroom and the objects in it. This Inquiry Warm-Up can be found online.

Models as Tools in Science

- Why Do Scientists Use Models? TEKS 3B, 3C
- What Is a System? TEKS 3B
- How Are Models of Systems Used? TEKS 3B, 3C

MY PLANET DIARY — FUN FACTS

Flying Through Space

You don't have to be an astronaut to experience what it's like to fly in space. Thanks to technological advances, space flight simulation software programs have been created. These programs range from simple and straightforward to detailed and complicated. Depending on which one you use, you can experience what it might feel like to fly to the moon, command a mission to Mars, and even explore other solar systems. If you've ever wondered what it's like to be an astronaut, now you have the chance to find out!

Read the following questions. Write your answers below.

1. Why would a flight simulation software program created today be more realistic than one that was created ten years ago?
   Technological advances have allowed designers to make the programs more realistic than they were in the past.

2. Would you be able to really fly in space if you knew how to use a space flight simulation software program? Explain.
   No, because astronauts go through years of intense training to learn how to fly in space.

Lab zone® Do the Inquiry Warm-Up *Scale Models.* Find the lab online.

Inside a flight simulator

SUPPORT ALL READERS

Lexile Measure = 870L **Lexile Word Count = 1215**

Prior Exposure to Content: Many students may have misconceptions on this topic

Academic Vocabulary: *identify, main idea, model*

Science Vocabulary: *system, feedback*

Concept Level: Generally appropriate for most students in this grade

Preteach With: My Planet Diary "Flying Through Space" and Figure 2 activity

Vocabulary
- system
- input
- process
- output
- feedback

Skills
 Reading: Identify the Main Idea
Inquiry: Make Models

Why Do Scientists Use Models?

"Who is that model on the cover?" "I still have that model car I built." The word *model* has many meanings. But, as with many words, *model* has a specific meaning in science. In science, a model is any representation of an object or process. Pictures, diagrams, computer programs, and mathematical equations are all examples of scientific models. Life-science models are often used to collect and analyze information about the structure and function of organisms.

Scientists use models to understand things they cannot observe directly. One advantage of using models is that scientists can study reasonable representations of things that are either very large, such as Earth's core, or very small, such as an atom. These kinds of models are physical models—drawings or three-dimensional objects. Other models, such as mathematical equations or word descriptions, are models of processes. Look at the models in **Figure 1**.

TEKS 3B, 3C In this section, you'll find out how scientists use models to represent aspects of the natural world they cannot observe directly.

ELPS 4.C.2
Give a tour of your classroom to a visitor or another student. Read and explain labels on any models, such as maps, globes, skeletons, or diagrams.

FIGURE 1

Two Science Models
Models may be three-dimensional objects or equations.

Explain Tell whether each of these models represents an object or a process and why each is useful.

Photosynthesis
Sample: It represents a process. You know the parts of the process and how they are related.

Carbon dioxide + Water $\xrightarrow{\text{sunlight}}$ Food + Oxygen

Sample: It represents an object. The model helps a person understand the heart's structure and function without having to examine a real heart.

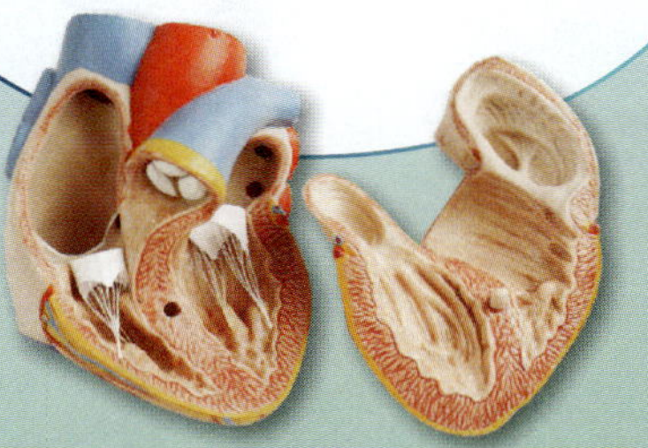

 Do the Quick Lab *Making Models.* Find the lab online.

Assess Your Understanding

got it?

- I get it! Now I know that scientists use models to understand things they cannot observe directly.
- I need extra help with See TE note.

English Language Proficiency Standards

ELPS Reading 4.C.2

Have students complete the ELPS activity on this page.

Beginning Have beginners listen as you read aloud the activity. Ask students to identify a classroom model. Display the model and ask questions about the labels as students point to or name answers.

Intermediate Read the activity together. Assign small groups to classroom models. Students take turns explaining the model and labels. Provide sentence frames: *This is ____. It shows ____. This part is ____.*

Advanced As partners complete the activity, have the "tourist" ask the guide questions about the model.

Advanced High Have small groups plan an entertaining tour of a classroom model. Encourage guides to think of ways to make the tour fun for visitors.

Explain

Introduce Vocabulary

Point out the compound vocabulary words *input* and *output*. Tell students that anything put *into* a system is *input,* and what comes *out* of a system is *output.*

Teach Key Concepts

Explain to students that scientists use models to understand things they cannot observe directly. For example, Earth scientists frequently use models to simulate changes in parts of Earth, such as in the atmosphere. You cannot fit the whole atmosphere inside a laboratory. Ask: **What are some models you have used to study science?** *(Students might have used human anatomical models, models of molecules, solar system models, or maps.)*

Teach With Visuals

Have students look at **Figure 1.** Make sure students understand that a model is a representation of an object, such as the human heart, or a process, such as photosynthesis. Ask: **How is an equation for photosynthesis an example of a model?** *(Sample: We cannot directly observe photosynthesis, so the equation represents what happens in the process.)*

Elaborate

Lab Resource: Quick Lab 

L2 **MAKING MODELS** Students will use a model to represent day and night on Earth. This Quick Lab can be found online.

Evaluate

Assess Your Understanding

Have students evaluate their understanding by completing the appropriate sentence.

RTI Response to Intervention

If students have trouble describing how scientists use models, **then** have them read the Key Concept statement.

PEARSON Texas.com

Texas Essential Knowledge and Skills

3B Use models to represent aspects of the natural world such as human body systems and plant and animal cells.

3C Identify advantages and limitations of models such as size, scale, properties, and materials.

LESSON 1.4

Explain

Teach Key Concepts

Explain to students that many things they are familiar with have individual parts that work together. Taken all together, the parts form a whole called a system. As a system, the individual parts either have a common function or they work together toward a common result. Ask: **If your body is a system, what are some parts that comprise it?** *(Samples: cells, organs, tissues, bones, respiratory system, nervous system)* **What function do these parts work together to perform?** *(To keep me alive and well)* **What is process?** *(What happens in a system)* **What are some processes that occur in your body system?** *(Samples: food gets digested, blood gets pumped, oxygen gets used, sights and sounds are detected and interpreted, waste is given off)* **What is input?** *(The material or energy that goes into a system)* **What input goes into your body system?** *(Samples: food, sensory stimuli, gases I breathe in, water)* **What is output?** *(The material or energy that comes out of a system)* **What output comes out of your body system?** *(Samples: waste, perspiration, gases I breathe out, sounds I make)*

Identify the Main Idea Tell students that every paragraph contains many ideas, but one idea is the most important. That idea is the main idea. As they read, encourage students to distinguish the main idea from other supporting information in a paragraph.

Teach With Visuals

Tell students to look at **Figure 2.** Explain that although a flashlight may seem to be a simple device, it is a system. Like other systems, a flashlight includes input, processes, output, and feedback. Have a volunteer take the flashlight apart. Call on volunteers to name and describe each of the parts of the flashlight. Show students a flashlight similar to the one pictured. Ask: **What function or result do these parts work together to perform or produce?** *(To give off light)* **What other devices can you think of that are systems with the same function or result?** *(Samples: electric lamps, oil lamps, lanterns, spotlights, light houses, street lights, car headlights)*

Texas Essential Knowledge and Skills

3B Use models to represent aspects of the natural world such as human body systems and plant and animal cells.

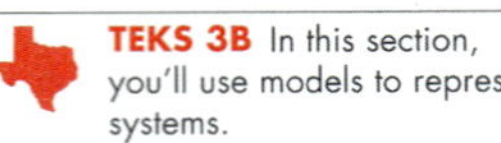

TEKS 3B In this section, you'll use models to represent systems.

What Is a System?

Many things you see and use are systems. For example, a toaster oven, your town's water pipes, and your bicycle are all systems. **A system is a group of parts that work together to perform a function or produce a result.**

Systems have common properties. All systems have input, process, and output. **Input** is the material or energy that goes into a system. **Process** is what happens in a system. **Output** is the material or energy that comes out of a system. In addition, some systems have feedback. **Feedback** is output that changes the system in some way. For example, the heating and cooling system in most homes has feedback. A sensor in the thermostat recognizes when the desired temperature has been reached. The sensor provides feedback that turns the system off temporarily. Look at **Figure 2** to see another example of a system.

Identify the Main Idea Circle the main idea in the second paragraph. Underline the details.

FIGURE 2

An Everyday System

In a flashlight, many parts work together as a system.

Apply Concepts Look at the flashlight and use what you know to fill in the chart.

	Flashlight
Parts of System	The parts are the batteries, the bulb, the bulb holder, the container, and the switch.
Input	The input is turning on the switch and starting chemical reactions in the batteries.
Process	The chemical energy in the batteries is converted into light energy.
Output	The output is light.

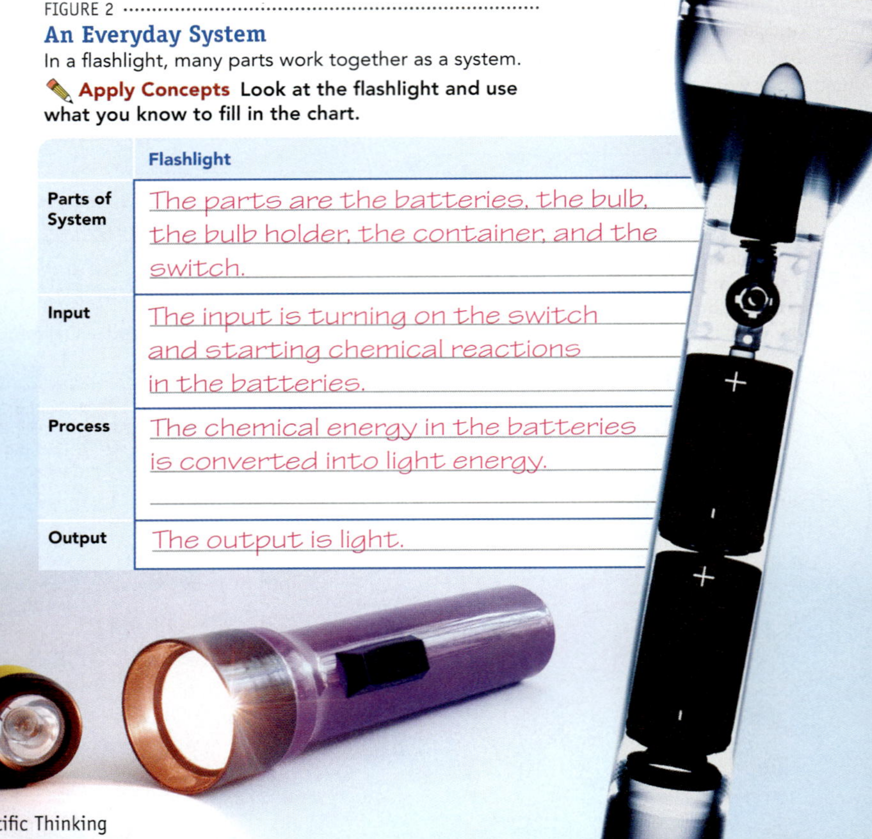

Sun, air, land, and water are the parts of a system that produce a sea breeze. During the day, the sun's energy heats both the land and the water. The land and water, in turn, heat the air above them. Air over the land becomes much warmer than the air over water. As the warmer air rises, the cooler air from over the water rushes in to replace it. A sea breeze is the result.

1 Use Models Identify the input, output, and process of the sea breeze system.

The input is the sun's energy. The output is the breeze. The process is the sun heating the land, which heats the air above it. The warm air then rises. The cooler air rushes into the empty space, resulting in the breeze.

2 CHALLENGE Which parts of this system will change after the sun sets? How will it change?

The input will change, because the sun's energy is not available to the area. The change in input will affect all other parts of the system.

Lab zone Do the Quick Lab *Systems.* Find the lab online.

Assess Your Understanding

TEKS 3B

1a. List What are the properties of a system?

The properties of a system are input, process, output, and sometimes feedback.

b. Apply Concepts A student uses a calculator to solve a math problem. Is this an example of a system? Explain your answer.

Yes; the input is the problem. The process is the calculation. The output is the answer.

got it?

O I get it! Now I know that a system is *a group of parts that work together to produce a specific function or result.*

O I need extra help with *See TE note.*

31

Differentiated Instruction

L1 Explain Models Have students give an example of a model that they are familiar with, such as a model of a molecule, a map or blueprint, or a model airplane. Then have students explain how the model is similar to and different from the object that it represents.

L3 Describe Systems Have students identify three examples of systems they are familiar with in their classroom, school, home, or neighborhood. Have them also identify the input, output, and process of each system. If possible, have students describe feedback in at least one of the systems.

21st Century Learning

CRITICAL THINKING Remind students that all systems have common properties, two of which are input and output. Ask: **How are input and output alike?** *(Both are properties that all systems have in common. Both are materials or energy.)* **How are input and output different?** *(Input is the material or energy that goes into a system, whereas output is the material or energy that comes out of a system.)*

Elaborate

Apply It!

L1 Before students begin the activity, have a volunteer read aloud the introductory text. Discuss the diagram and have another volunteer read aloud the labels. Make sure students understand the sequence of events and changes that produce a sea breeze. Then write the following terms on the board: *system, input, output, process.* Call on students to define each term in their own words.

Lab Resource: Quick Lab 

L2 SYSTEMS Students will make a lemon battery and describe how it works. This Quick Lab can be found online.

Evaluate

Assess Your Understanding

After students answer the questions, have them evaluate their understanding by completing the appropriate sentence.

RTI Response to Intervention

1a. If students need help listing the properties of a system, **then** suggest they reread the sentences that contain the boldface terms.

b. If students have trouble explaining whether or not a calculator is a system, **then** have them review **Figure 2.**

Explain

Teach Key Concepts

Explain to students that models are useful for understanding systems and how they work. One advantage os using models is that it allows scientists to predict how feedback and input will change a system. Ask: **What happens when you predict?** *(You use what you already know to make an inference about what is likely to happen.)* **Why is it possible to use a model to predict what is likely to happen in a system?** *(Since a model is a representation of the system and its parts, by studying the model you can learn about the system. Then you can use what you learn to make inferences about how the system is likely to change.)* **Are the predictions made using a model likely to always be correct? Why or why not?** *(No, since the model is based on certain assumptions that might not be true.)*

Teach With Visuals

Tell students to look at **Figures 3** and **4.** Review with students or explain to them the concepts shown in the illustrations. Ask: **What is a food chain?** *(A series of events in which one organism eats another and obtains energy)* **What is a food web?** *(The many overlapping food chains in an ecosystem)* **Why are food chains and food webs systems?** *(They are both groups of organisms that together produce a result—energy moving through an environment.)* Emphasize that a system is defined by the observer. A complex system, such as a food web, can be made up of several smaller, less complex systems.

Make Models Remind students that a model is something that represents an object, such as a planet, or a process, such as photosynthesis. A food chain is a model that represents the process by which energy is transferred through an ecosystem.

Texas Essential Knowledge and Skills

3B Use models to represent aspects of the natural world such as human body systems and plant and animal cells.

3C Identify advantages and limitations of models such as size, scale, properties, and materials.

TEKS 3B, 3C In this section, you'll learn how scientists use models of systems to represent and predict aspects of the natural world.

How Are Models of Systems Used?

It's easy to identify the materials and energy that make up the inputs and outputs of a system. It's not easy to observe a system's process. **Scientists use models to understand how systems work. They also use models to predict changes in a system as a result of feedback or input changes.** However, they keep in mind that predictions based on models are uncertain.

When scientists construct a model of a system, they begin with certain assumptions. These assumptions allow them to make a basic model that accurately reflects the parts of the system and their relationships. A scientist who wants to study how energy moves through living things in an environment might use a model called a food chain. A food chain is a series that shows who eats whom to obtain energy in an environment. The food chain shown in **Figure 3** assumes that hardhead catfish only eat Gulf killifish. Hardhead catfish actually eat many kinds of animals. However, the model still accurately reflects the relationship between the parts of a system.

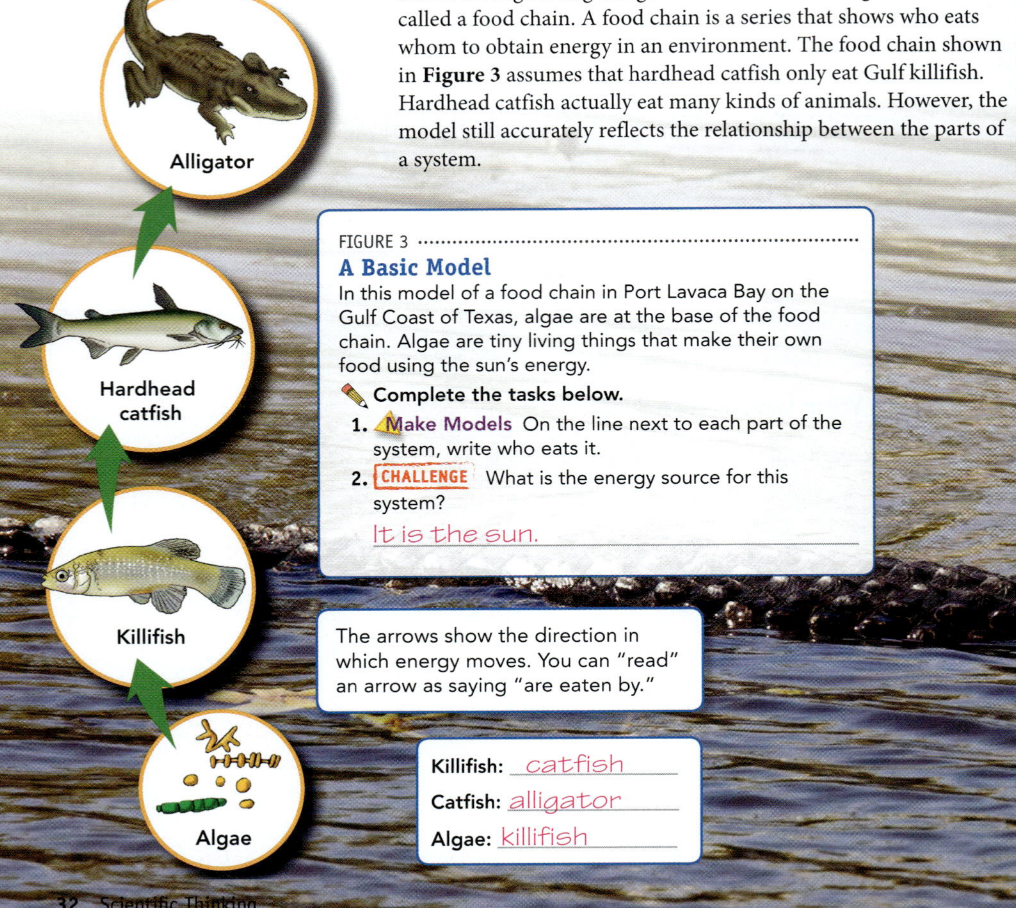

FIGURE 3

A Basic Model

In this model of a food chain in Port Lavaca Bay on the Gulf Coast of Texas, algae are at the base of the food chain. Algae are tiny living things that make their own food using the sun's energy.

Complete the tasks below.

1. **Make Models** On the line next to each part of the system, write who eats it.
2. **CHALLENGE** What is the energy source for this system?
   It is the sun.

The arrows show the direction in which energy moves. You can "read" an arrow as saying "are eaten by."

Killifish: catfish
Catfish: alligator
Algae: killifish

Modeling a Simple System A food chain is a good model to begin to understand how energy moves through living things in an environment. However, it has limitations because it shows how only a few of those living things are related. So a scientist may build a food web to model a more complete picture of the system. In **Figure 4** you can see a food web with many overlapping food chains. The food web is more detailed than one food chain. But it does not provide information about other factors, such as weather, that affect energy flow in the system.

FIGURE 4

A Model of a Simple System

This model of a Texas Gulf Coast food web contains overlapping food chains.

Interpret Diagrams Study the food web model. On the notebook page write two things you learned from this complex model.

33

Support the TEKS

TYPES OF MODELS Some students may think that a model has to be a physical, three-dimensional object, such as a model airplane. Be sure students understand that a model, especially one of a system, can be a two-dimensional diagram, such as the model of the food web. Ask: **What features of the food web make it an effective model?** *(Samples: It uses very little space to show the many organisms involved and it uses arrows to effectively show energy flow.)*

Lead a Discussion

L3 **MATHEMATICAL MODELS** Tell students that a model can also be a mathematical equation or computer model. A mathematical model can show not just where the energy flows in a food web, but how much is passed from one organism to the next. Using a computer to create such a model has many advantages. Ask: **How do you think a computer would be useful for managing a mathematical model of a energy transfer in a food web?** *(You can "plug in" different numbers to see how the energy transfer changes as conditions in the environment change.)*

Elaborate

21st Century Learning

L3 **CREATIVITY** Challenge small groups of students to make a model of a system. First, have them identify a natural system that interests them. Then have them decide what kind of model they will create and plan how to make it. Remind them that there are many different kinds of models, not just three-dimensional objects. Once you approve their plan, have students make the model and share it with the class.

Differentiated Instruction

L1 **Model Scavenger Hunt** Review the meanings of a model and a system with the class. Challenge student pairs to search the school for as many different kinds of models as that they can find. Have them make a list of the models they find. Then, have students share their lists with the class and tell why they think each model does or does not represent a system. Have them explain their answers, but allow them to modify their answers and the discussion continues.

LESSON 1.4

Explain

Lead a Discussion

MODEL STORMS Explain that the atmosphere is a complex system that affects society every day. Use the example of violent storms, which are part of the atmosphere, and discuss the impact they have on people. Ask: **How do storms like tornadoes and hurricanes affect people?** *(They cause damage to property and endanger the lives of people and organisms they depend on.)* **Why is it important to develop computer models of storm systems, such as tornadoes and hurricanes?** *(In order to predict the storm's strength and the path it will take; to understand something that it might be dangerous to observe directly)* **What might be one limitation of a model of a complex system like this?** *(Sample: Some information may be missing from the model, so the model may not fully explain the process it represents.)*

21st Century Learning

COMMUNICATION Display an example of a simple model and a more complex model for the same system. As a class, compare and contrast the models. Have students identify and discuss the advantages and limitations of each model.

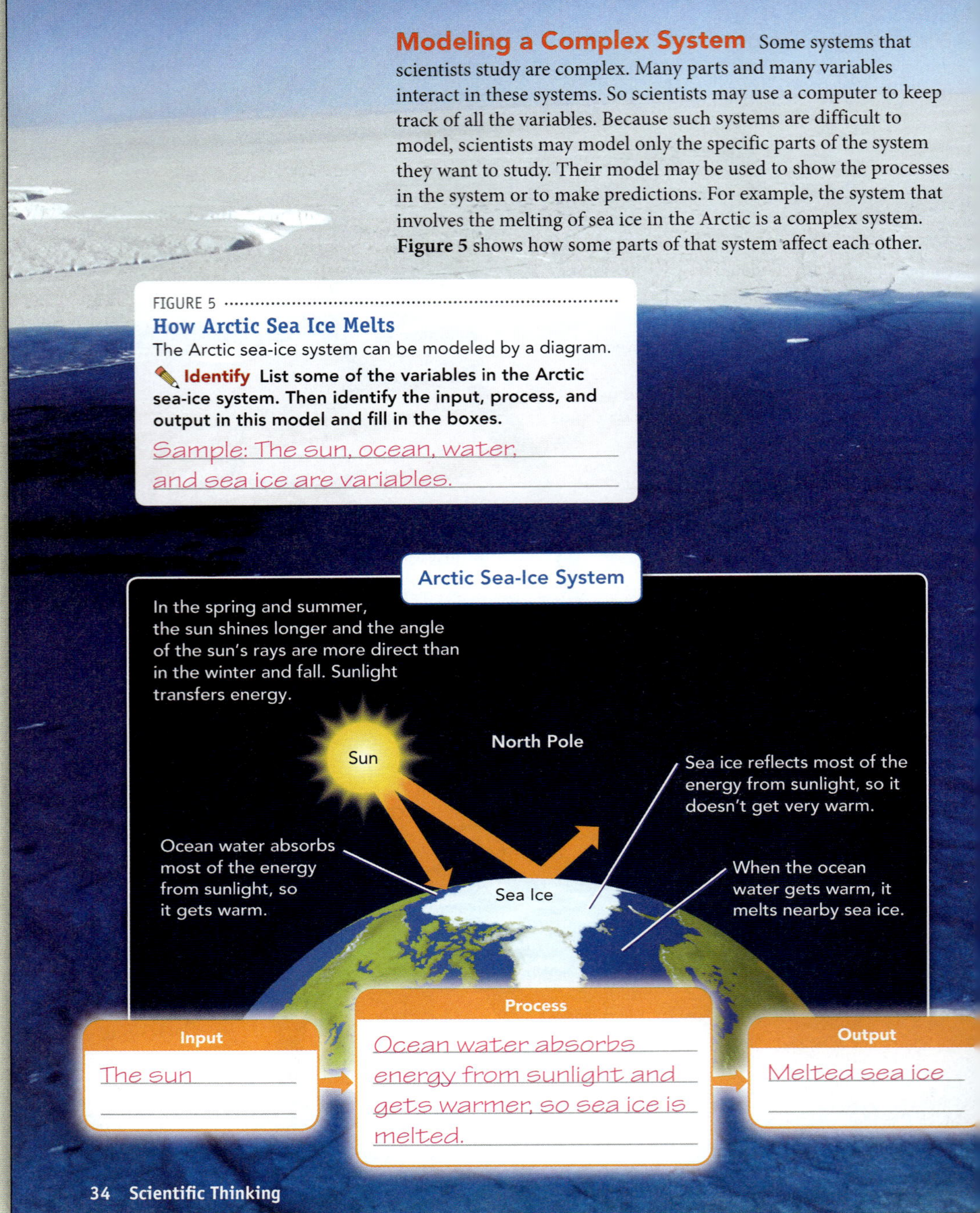

Modeling a Complex System Some systems that scientists study are complex. Many parts and many variables interact in these systems. So scientists may use a computer to keep track of all the variables. Because such systems are difficult to model, scientists may model only the specific parts of the system they want to study. Their model may be used to show the processes in the system or to make predictions. For example, the system that involves the melting of sea ice in the Arctic is a complex system. **Figure 5** shows how some parts of that system affect each other.

FIGURE 5

How Arctic Sea Ice Melts

The Arctic sea-ice system can be modeled by a diagram.

Identify List some of the variables in the Arctic sea-ice system. Then identify the input, process, and output in this model and fill in the boxes.

Sample: The sun, ocean, water, and sea ice are variables.

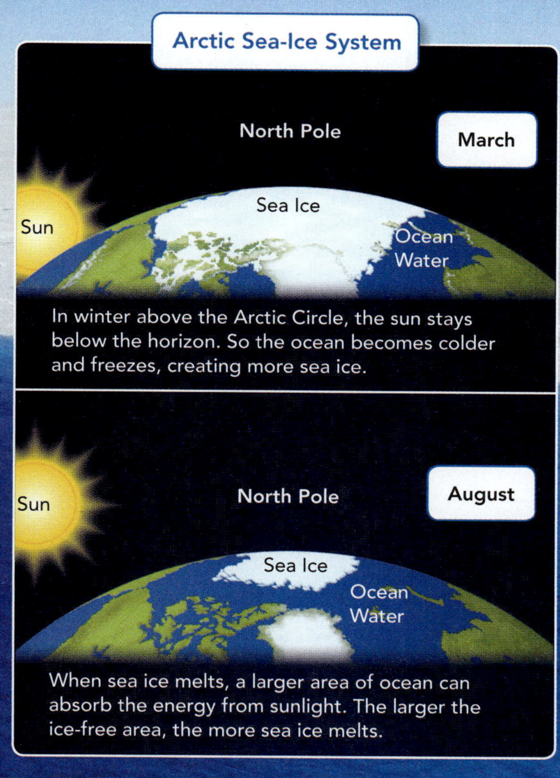

FIGURE 6

Arctic Sea-Ice System

These two diagrams show the amount of Arctic sea ice in March and August.

Answer the questions.

1. **Describe** In August, what is the feedback in this system?
   The ice that has melted allows more ice to melt.
2. **Explain** What causes changes in the system of melting Arctic sea ice to break the feedback in the system?
   The ocean absorbs less energy in winter.

Do the Quick Lab *Models in Nature.* Student Lab Manual, p. 9

Assess Your Understanding

TEKS 3B, 3C

2a. Explain Why do scientists use models?
They use them to understand how systems work and make predictions about how they might change.

b. Identify Why are models of complex systems limited in their accuracy?
Scientists may leave out parts of the system that they are not studying.

got it?

- I get it! Now I know that scientists use models of systems to understand how systems work and to predict changes in the output.
- I need extra help with See TE note.

35

Differentiated Instruction

L1 Compare and Contrast Models Show students a model of a simple machine, such as a crow bar, and one of the solar system. Discuss how one model shows a simple system and the other shows a complex system. Have students compare and contrast the two models and systems.

L3 Telephone Model Have students build and use a model telephone using two paper cups and string. Challenge them to do research to find out how the telephone system works and then explain to the class how the model represents a complex system of communication.

Elaborate

LESSON 1.4

Build Inquiry 

L2 EARTH SYSTEMS MODEL

Materials magazines, scissors, glue or tape, poster board, markers

Time 30 minutes

Explain to students that the Earth system is a system that has four main parts, called spheres: the atmosphere, the hydrosphere, the geosphere, and the biosphere. The atmosphere is Earth's outermost layer and is made up of gases. The geosphere includes the interior of Earth and the rocks and soil that make up Earth's solid surface. The hydrosphere is made up of all of Earth's water. The biosphere is Earth's living organisms. Have students form small groups to create a model of the Earth system in the form of a poster-size diagram. Suggest that they start by brainstorming how to represent each of the four main parts of the system, then use the materials provided to create their models.

Ask: **Is the Earth system a simple or complex system? Explain your answer.** *(It is a very complex system, since it consists of so very many parts, interactions, processes, inputs, and outputs.)* **Why might it be difficult to make predictions about the Earth system using only the model you created?** *(The system is so complex that too few of its parts are represented to make accurate predictions.)*

Lab Resource: Quick Lab 

L1 MODELS IN NATURE Students will build and evaluate a model of the solar system. This Quick Lab can be found in the Student Lab Manual, p. 9, and online.

Evaluate

Assess Your Understanding

After students answer the questions, have them evaluate their understanding by completing the appropriate sentence.

RTI Response to Intervention

2a. If students need help explaining why scientists use models, **then** have students reread the Key Concept statement for this section.

b. If students have difficulty explaining why models of complex system aren't completely accurate, **then** ask them if every part of the systems pictured in this section are represented and, if not, what parts might be missing.

Name ______________________ Date __________ Class __________

Assess Your Understanding

Models as Tools in Science

Why Do Scientists Use Models?

got it?

○ **I get it!** Now I know that scientists use models to ______________________

____

○ **I need extra help with** ______________________

____

What Is a System?

1a. LIST What are the properties of a system? ______________________

____

____

b. APPLY CONCEPTS A student uses a calculator to solve a math problem. Is this an example of a system? Explain your answer. ______________________

____

____

How Are Models of Systems Used?

2a. EXPLAIN Why do scientists use models? ______________________

____

____

b. IDENTIFY Why are models of complex systems limited in their accuracy? ______________________

____

____

Name ______________________ Date ____________ Class ____________

Enrich

Models as Tools in Science

A scientific model is a representation of a complex object or process. Read the passage and study the models at bottom and at right. Then use a separate sheet of paper to answer the questions that follow the passage.

A Scientific Model

A diagram is one example of a scientific model. A model can be very useful in providing information when people can't actually observe an object or process directly. Yet, often there is information missing from a model because a model is often made for the purpose of representing some specific characteristic of an object or process. Therefore other characteristics are not so important for the specific purpose of the model.

Model A and Model B are both models of the solar system that show the sun and the eight planets. Despite these similarities, the two models are quite different because they were made with different purposes in mind. Model A shows the relative sizes of the different planets. Model B shows the relative distances of the planets from the sun. Model A lacks the information of Model B, and Model B lacks the information of Model A. Both models, though, provide useful information about the solar system.

1. What is a scientific model?
2. Why is information often missing from a model?
3. What does Model A show about the solar system?
4. What information is missing from Model A?
5. What does Model B show about the solar system?
6. What information is missing from Model B?
7. Is either model better than the other? Explain your reasoning.

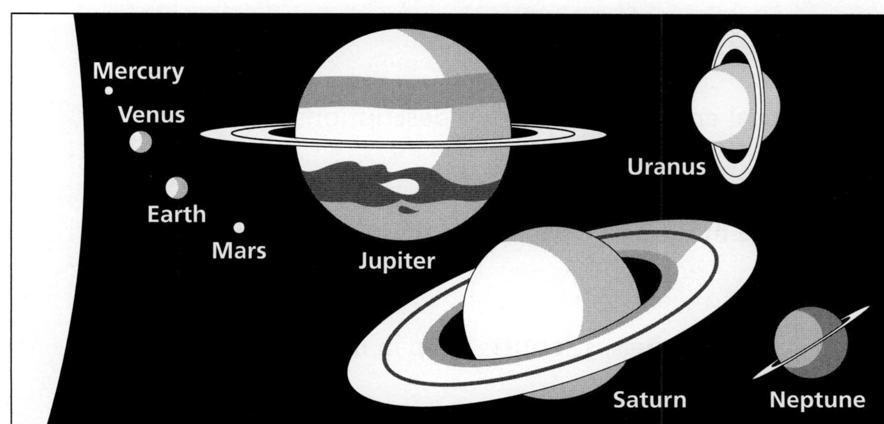

Model A

Model B

Name ______________________ Date ____________ Class ____________

Lesson Quiz

Models as Tools in Science

Write the letter of the correct answer on the line at the left.

1. ___ Which of the following do all systems have?
- A a model
- B variables
- C output
- D moving parts

2. ___ Which of the following is NOT a model?
- A an equation describing the force of gravity
- B a book on sharks
- C a drawing of the human digestive system
- D a large replica of an ant

3. ___ The material or energy that goes into a system is called
- A output
- B input
- C process
- D a variable

4. ___ Output that changes a system in some way is referred to as
- A feedback
- B variables
- C results
- D processes

If the statement is true, write *true*. If the statement is false, change the underlined word or words to make the statement true.

5. ________________ A process is a group of parts that work together to perform a function or produce a result.

6. ________________ Some systems have feedback.

7. ________________ Mathematical equations and word descriptions are examples of models of processes.

8. ________________ Scientists use models to understand things they can observe directly.

9. ________________ When they use models, scientists keep in mind that predictions based on models are certain.

10. ________________ What happens in a system is called input.

Place the outside corner, the corner away from the dotted line, in the corner of your copy machine to copy onto letter-size paper.

Models as Tools in Science

Answer Key

Review and Reinforce

Find the worksheet in the Student Workbook.

1. Scientists use models to understand things they cannot observe directly.
2. A system is a group of parts that work together to perform a function or produce a result.
3. All systems have input, process, and output.
4. Scientists use models to understand how systems work. They also use models to predict changes in a system as a result of feedback or input changes.
5. a
6. f
7. d
8. c
9. e
10. b
11. Scientists use models to understand how systems work and to predict changes in a system as a result of feedback or input changes.

Enrich

1. a representation of a complex object or process
2. Often a model is made for the purpose of representing some specific characteristic of an object or process.
3. It shows the relative sizes of the different planets.
4. Sample: Missing from the model are the relative distances of the planets from the sun.
5. It shows the relative distances of the planet from the sun.
6. Sample: Missing from the model are the relative sizes of the planets.
7. Sample: Neither model is better than the other. They each represent a different characteristic about the solar system.

Lesson Quiz

1. C
2. B
3. B
4. A
5. system
6. true
7. true
8. cannot
9. uncertain
10. process

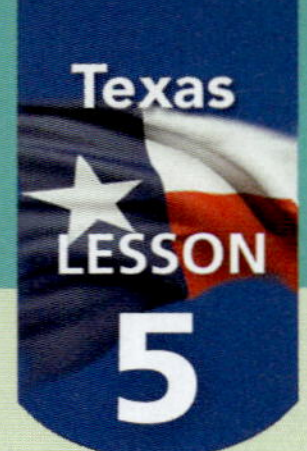

Safety in the Science Laboratory

How do scientists investigate the natural world?

LESSON PACING:
2–3 periods or 1–1½ blocks

Lesson Vocabulary

- safety symbol
- field

Content Refresher

Safety First Following safety precautions in the science laboratory cannot be overemphasized. In addition to having students follow safety cautions in their texts, you can take some actions that might further ensure safety. For instance, it is a good idea to lock the laboratory or the lab storeroom when you are not present. This will prevent the enterprising student from continuing an investigation when you are not there to supervise. Also consider having drills on what students should do if a substance catches fire or if there is an accidental chemical exposure or spill. You may want to post the phone numbers of the local poison control center in your laboratory.

Schedule frequent, periodic reviews of the safety and emergency equipment in your classroom to make sure it is in good working order. Many science laboratories have water, gas, and electricity valves in the room. As the instructor, it is necessary for you to know how to cut off those valves in the event of an emergency. You should also inspect the first aid supplies on a regular basis to make sure they are well stocked.

Lesson Objectives	TEKS	ELPS
Explain why it is important to prepare before carrying out investigations in the lab and in the field.	1A, 4B	1.D
Describe what you should do if an accident occurs.	1A, 4B	

Texas Essential Knowledge and Skills

1A Demonstrate safe practices during laboratory and field investigations as outlined in the Texas Safety Standards.
4B Use preventative safety equipment, including chemical splash goggles, aprons, and gloves, and be prepared to use emergency safety equipment, including an eye/face wash, a fire blanket, and a fire extinguisher.

English Language Proficiency Standards

ELPS Learning Strategies 1.D Speak using learning strategies such as requesting assistance, employing non-verbal cues, and using synonyms and circumlocution (conveying ideas by defining or describing when exact English words are not known).

DIFFERENTIATED INSTRUCTION KEY
L1 Struggling Students or Special Needs
L2 On-Level Students L3 Advanced Students

Investigations and Activities	TEKS Review
My Planet Diary for Texas, **Student Edition,** p. 36 Inquiry: Inquiry Warm-Up, Where Is the Safety Equipment in Your School?, **Lab Manual,** p. 10 Introduce Vocabulary, **Teacher's Edition,** p. 37 Teach Key Concepts, **Teacher's Edition,** p. 37 Teach With Visuals, **Teacher's Edition,** p. 37 Support the TEKS, *Math in Science,* **Teacher's Edition,** p. 38 Lead a Discussion, Staying Safe in the Laboratory, **Teacher's Edition,** p. 38 Address Misconceptions, Safety in All Labs, **Teacher's Edition,** p. 38 21st Century Learning, Critical Thinking, **Teacher's Edition,** p. 39 Inquiry: Teacher Demo, Demonstrate Safety Practices, **Teacher's Edition,** p. 39 Differentiated Instruction, **Teacher's Edition,** p. 39 Lead a Discussion, Staying Safe in the Field, **Teacher's Edition,** p. 40 Apply It!, **Student Edition,** p. 40 Inquiry: Quick Lab, Be Prepared, **PearsonTexas.com**	Apply the TEKS, Evaluating Scientific Explanations, **Student Edition,** p. 46 TEKS Practice, **Student Edition,** p. 48 TEKS Practice: Chapter and Cumulative Review, **Student Edition,** p. 52 Lesson 1.5, **TEKS Preparation and Study Guide Workbook,** p. 10
Teach Key Concepts, **Teacher's Edition,** p. 41 Differentiated Instruction, **Teacher's Edition,** p. 41 Inquiry: Quick Lab, Just in Case, **PearsonTexas.com**	**SHORT ON TIME?** To do this lesson in approximately half the time, do the Activate Prior Knowledge activity. A discussion of the Key Concepts will familiarize students with the lesson content. Have students do the Quick Labs. The rest of the lesson can be completed by students independently.

These editable worksheets are available on **PearsonTexas.com.**
Print versions can be found in the **TEKS Preparation and Study Guide Workbook.**

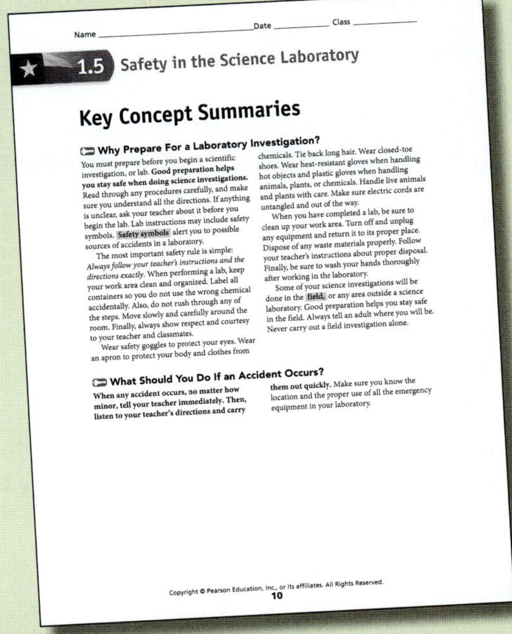

Name ______________________ Date __________ Class __________

1.5 Safety in the Science Laboratory

Key Concept Summaries

Why Prepare For a Laboratory Investigation?

You must prepare before you begin a scientific investigation, or lab. **Good preparation helps you stay safe when doing science investigations.** Read through any procedures carefully, and make sure you understand all the directions. If anything is unclear, ask your teacher about it before you begin the lab. Lab instructions may include safety symbols. **Safety symbols** alert you to possible sources of accidents in a laboratory.

The most important safety rule is simple: *Always follow your teacher's instructions and the directions exactly.* When performing a lab, keep your work area clean and organized. Label all containers so you do not use the wrong chemical accidentally. Also, do not rush through any of the steps. Move slowly and carefully around the room. Finally, always show respect and courtesy to your teacher and classmates.

Wear safety goggles to protect your eyes. Wear an apron to protect your body and clothes from chemicals. Tie back long hair. Wear closed-toe shoes. Wear heat-resistant gloves when handling hot objects and plastic gloves when handling animals, plants, or chemicals. Handle live animals and plants with care. Make sure electric cords are untangled and out of the way.

When you have completed a lab, be sure to clean up your work area. Turn off and unplug any equipment and return it to its proper place. Dispose of any waste materials properly. Follow your teacher's instructions about proper disposal. Finally, be sure to wash your hands thoroughly after working in the laboratory.

Some of your science investigations will be done in the **field,** or any area outside a science laboratory. Good preparation helps you stay safe in the field. Always tell an adult where you will be. Never carry out a field investigation alone.

What Should You Do If an Accident Occurs?

When any accident occurs, no matter how minor, tell your teacher immediately. Then, listen to your teacher's directions and carry them out quickly. Make sure you know the location and the proper use of all the emergency equipment in your laboratory.

10

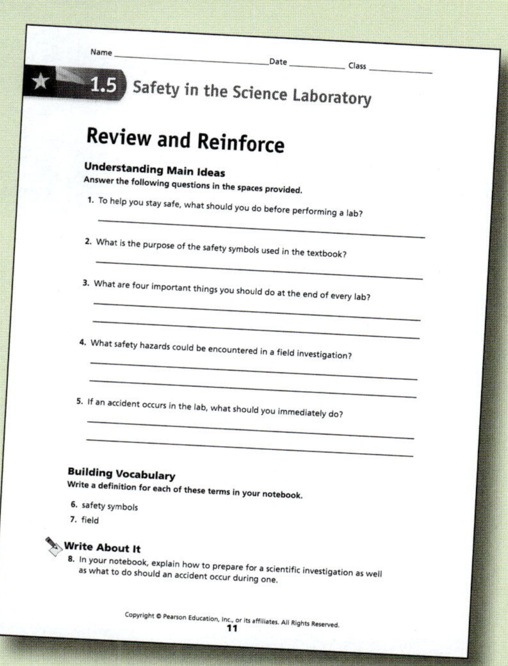

Name ______________________ Date __________ Class __________

1.5 Safety in the Science Laboratory

Review and Reinforce

Understanding Main Ideas
Answer the following questions in the spaces provided.

1. To help you stay safe, what should you do before performing a lab?
2. What is the purpose of the safety symbols used in the textbook?
3. What are four important things you should do at the end of every lab?
4. What safety hazards could be encountered in a field investigation?
5. If an accident occurs in the lab, what should you immediately do?

Building Vocabulary
Write a definition for each of these terms in your notebook.

6. safety symbols
7. field

Write About It
8. In your notebook, explain how to prepare for a scientific investigation as well as what to do should an accident occur during one.

11

Lexile Measure = 860L

LESSON 1.5

Safety in the Science Laboratory

Establish Learning Objectives

After this lesson, students will be able to:

Explain why it is important to prepare before carrying out investigations in the lab and in the field.

Describe what you should do if an accident occurs.

Engage

Activate Prior Knowledge

MY PLANET DIARY FOR TEXAS Read *Oil Refinery Explosion* with the class. Explain what an oil refinery is and discuss what the oil produced there is used for. Ask if students think the resulting products are worth risking lives to make. Ask: **In what other industries are safety precautions needed to prevent injuries and save lives?** *(Sample: In any industry where people work with dangerous chemicals, extreme heat or cold, heavy equipment, or vehicles)*

Explore

Lab Resource: Inquiry Warm-Up

L1 **WHERE IS THE SAFETY EQUIPMENT IN YOUR SCHOOL?** Students will familiarize themselves with school safety equipment. This Inquiry Warm-Up can be found in the Student Lab Manual, p. 10, and online.

Safety in the Science Laboratory

Why Prepare for a Laboratory Investigation?
TEKS 1A, 4B

What Should You Do if an Accident Occurs?
TEKS 1A, 4B

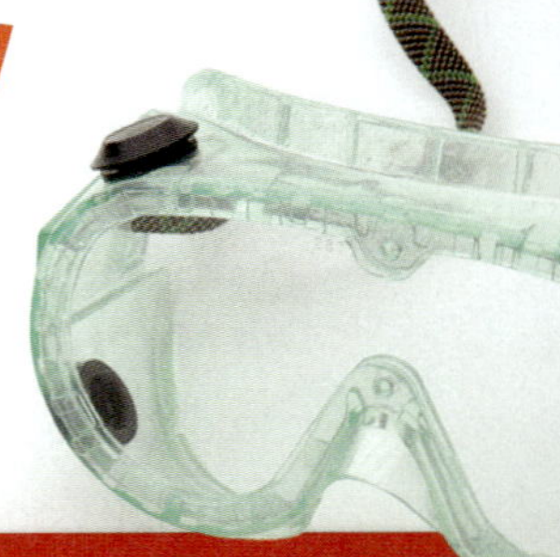

my planet Diary for TEXAS

DISASTER

Oil Refinery Explosion

On March 23, 2005, an explosion at an oil refinery in Texas took the lives of 15 people and wounded at least 170 others. Sadly, experts agree that this accident could have been prevented had safety codes not been ignored. Investigators found old, worn equipment and noticed that several repairs had not been made. One positive thing that has come out of this accident is a safety video made by the U.S. Chemical Safety Board. The video is based on the refinery accident and includes safety information about how to prevent such an accident from ever occurring again.

Read the following questions. Write your answers below.

1. How do investigators believe this accident might have been prevented?
It might have been prevented if the company had followed the safety codes.

2. What kind of information would you include in a safety video based on the accident?
Sample: I would list all the safety codes that were ignored by the refinery and suggest frequent inspections of a refinery.

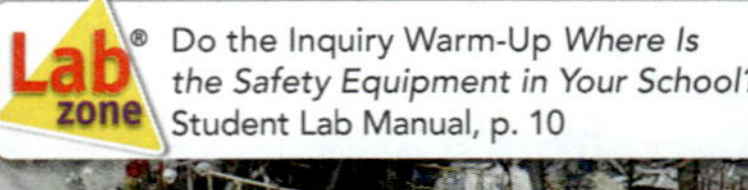

36 Scientific Thinking

Lexile Measure = 860L Lexile Word Count = 988

Prior Exposure to Content: Many students may have misconceptions on this topic

Academic Vocabulary: *observe, summarize*

Science Vocabulary: *safety symbol, field*

Concept Level: Generally appropriate for most students in this grade

Preteach With: My Planet Diary for Texas "Oil Refinery Explosion" and Figure 1 activity

Vocabulary
- safety symbol
- field

Skills
- Reading: Summarize
- Inquiry: Observe

Why Prepare for a Laboratory Investigation?

TEKS 1A, 4B In this section, you'll learn about safe lab practices and why they are important.

After hiking for many hours, you reach the campsite. You rush to set up your tent. Later that night, heavy rain falls. Water pours into the tent and soaks you. You look for a flashlight. Then you realize that you forgot to pack one.

Preparing for a Lab Just like for camping, you must prepare before you begin a laboratory investigation, or lab.  **Good preparation helps you stay safe when doing laboratory investigations.** You should prepare for a lab before you do it. Read through any procedures carefully, and make sure you understand all the directions. If anything is unclear, ask your teacher about it before you begin the lab. Investigations may include safety symbols like the ones you see in **Figure 1.** Safety symbols alert you to possible sources of accidents in a laboratory. Know where emergency equipment, such as fire extinguishers and fire blankets, are kept so that you know what to do when an accident occurs.

ELPS 1.D
With a partner, study and learn the Safety Symbols. For each symbol, tell the danger and the preventive action. Ask about those you don't understand.

FIGURE 1
Safety Symbols
Safety symbols identify how to work carefully and what safety equipment to use.

Apply Concepts In the notebook, list symbols that would appear for a lab investigation in which you measure the temperature of water as it heats to boiling.

Safety Symbols

Chemical Splash Goggles

Lab Apron

Breakage

Heat-Resistant Gloves

Heating

Poison

Physical Safety

Flames

No Flames

Measuring the Temperature of Water

Lab Safety

Sample: chemical splash goggles
lab apron
breakage
heat-resistant gloves
heating
flames

Explain

Introduce Vocabulary

Use *field* as an example of a word with multiple meanings. Students will be most familiar with the use of *field* to describe a stretch of open or cleared land, such as a farm field. Explain that many scientists do their work in the field, meaning any location outside the science lab.

Teach Key Concepts

Explain to students that the laboratory can be a dangerous place, and scientists take steps to ensure safety. Being prepared is one way scientists stay safe when doing a science investigation. Ask: **Why does good preparation help you stay safe in the lab?** *(If you prepare well, you know what equipment you will be using and the procedures you will be following. With that knowledge, you can take the proper precautions.)* **What should you do if you don't understand something that you're supposed to do in a lab activity?** *(Ask the teacher before beginning the activity.)*

Teach With Visuals

Tell students to look at **Figure 1.** Explain that in carrying out science investigations, they will see safety symbols like these that caution them about each investigation and its potential hazards. Then show them examples of laboratory investigations, and point out the symbols that are used. Remind students to make sure they are familiar with each symbol shown before beginning the investigation. Ask: **What does the "skull and crossbones" symbol mean when you see it in a laboratory procedure?** *(It warns you that poisonous chemicals are being used.)* **What does the "falling person" symbol stand for?** *(Physical safety)* **Why do you think pictures instead of words are used as safety symbols?** *(Pictures are more likely to be understood and remembered by everyone.)*

PEARSON Texas.com

English Language Proficiency Standards

ELPS Learning Strategies 1.D

Have students study and discuss the safety symbols in **Figure 1.**

Beginning Have beginners listen to and repeat the name of each symbol. Discuss meanings. Have partners match each symbol to the photos on pages 38–39. Students should tell or pantomime the danger for each symbol.

Intermediate Together read the ELPS activity and the names of the symbols. Have partners tell about each symbol: *The danger is ____. To protect yourself, ____.*

Advanced Have partners give instructions to go with each safety symbol. Have them begin each sentence with *Do* or *Don't*.

Advanced High After partners complete the ELPS activity, have them complete the activity in **Figure 1.**

Texas Essential Knowledge and Skills

1A Demonstrate safe practices during laboratory and field investigations as outlined in the Texas Safety Standards.

4B Use preventative safety equipment, including chemical splash goggles, aprons, and gloves, and be prepared to use emergency safety equipment, including an eye/face wash, a fire blanket, and a fire extinguisher.

LESSON 1.5

Explain

Support the TEKS

MATH IN SCIENCE Laboratory scientists require a logical approach to investigations in addition to strong math skills to produce accurate results safely. They need to number the steps of their experiments in an order that ensures safe and accurate results. Laboratory substances must be measured precisely to maintain controlled and safe conditions. Scientists carefully observe digital and analog equipment that registers experimental conditions, such as temperature, pressure, and electricity.

Mathematics and science are also integrally related in the natural world. Scientists studying nature may use math and science to track environmental conditions during natural disasters. From predicting storms to monitoring earthquakes and tsunamis, accurate mathematical calculations can help reduce casualties. Scientists have even developed Global Positioning Systems, based on geometry, which can pinpoint an emergency location or direct us home.

Lead a Discussion

STAYING SAFE IN THE LABORATORY Remind students that what they do in the lab can have serious consequences. As a result, their main concern should be to keep themselves and their classmates safe. Ask: **What is the most important safety rule in performing a lab?** *(Always follow your teacher's instructions and the directions exactly.)* **What is the last thing you should do at the end of a lab?** *(Wash your hands thoroughly.)* **Why is it important to dispose of waste materials properly at the end of a lab?** *(Chemicals must be disposed of properly because they can seep into the environment causing harm to people or other living things.)*

Address Misconceptions

L1 SAFETY IN ALL LABS Students may mistakenly think that they only need to follow safety procedures when they do chemistry experiments involving the use of chemicals, glassware, flames, or heat. Remind them that any science investigation can have potential hazards. For example, if they are measuring length around the classroom, they should move slowly to avoid tripping or bumping into others. Ask: **What should you do to stay safe while measuring the height of potted plants?** *(Wear an apron and wear plastic gloves to protect skin when handling plants.)* **What should you wear on your feet no matter what kind of lab you are doing?** *(Closed-toe shoes)*

FIGURE 2 

Safety in the Lab

Recognizing and preventing safety hazards are important skills to practice in the lab.

Complete the tasks.

1. **Make Models** In the empty boxes on each page, draw a safety symbol for wearing closed-toe shoes and one for tying back long hair.
2. **CHALLENGE** How might the student on this page protect himself from breathing in fumes from the flask or beakers?

   Sample: He should be working in a well-ventilated area.

Sample: A drawing of a sandal that has been crossed out.

Performing a Lab Whenever you do a science lab, your chief concern must be your safety and that of your classmates and your teacher. The most important safety rule is simple: *Always follow your teacher's instructions and the directions exactly.* Never try anything on your own without asking your teacher first.

Figure 2 shows a number of things that you can do to make your lab experience safe and successful. When performing a lab, keep your work area clean and organized. Label all containers so you do not use the wrong chemical accidentally. Also, do not rush through any of the steps. When you need to move around the room, move slowly and carefully so you do not trip or bump into another group's equipment. Finally, always show respect and courtesy to your teacher and classmates. Your teacher may refer you to the *Texas Safety Standards* for more information on safety in the laboratory.

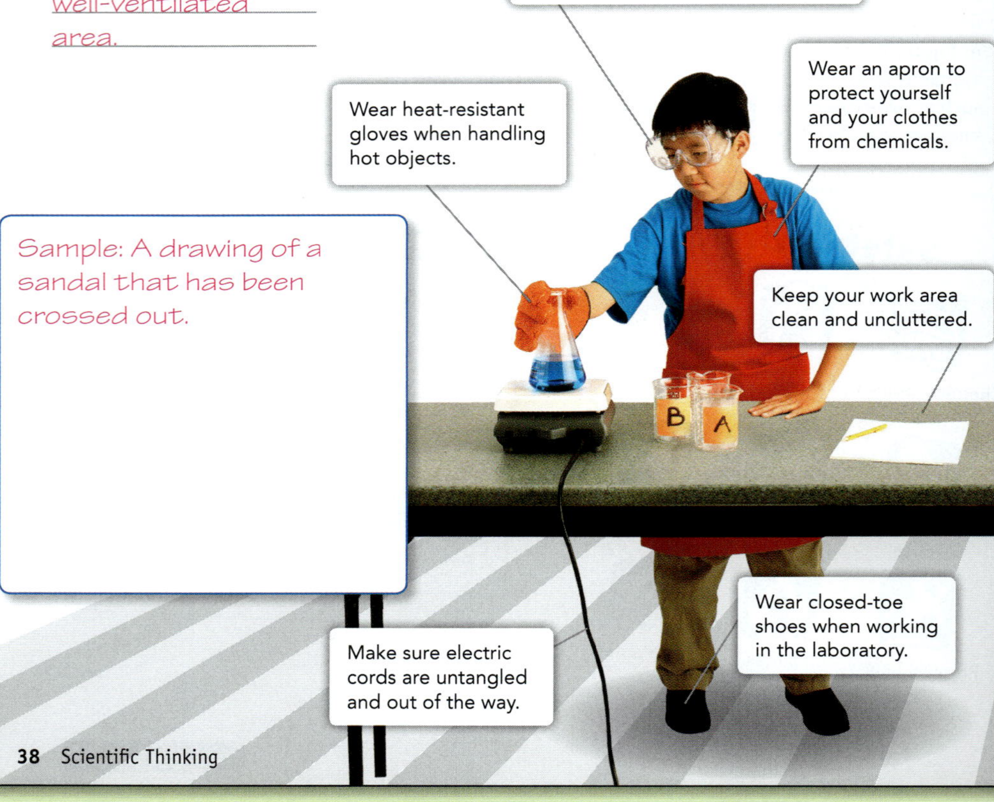

End-of-Lab Procedures There are important things you need to do at the end of every lab. When you have completed a lab, be sure to clean up your work area. Turn off and unplug any equipment and return it to its proper place. It is very important that you dispose of any waste materials properly. Some wastes should not be thrown in the trash or poured down the drain. Follow your teacher's instructions about proper disposal. Finally, be sure to wash your hands thoroughly after working in the laboratory.

Sample: A drawing of hair in a ponytail.

Summarize In the boxes provided, summarize the procedures you perform before, during, and after a lab-investigation.

Before
Read the directions before the lab. Ask questions if I don't understand them.

During
I follow the directions exactly. I label containers, move carefully, and am clean and neat.

After
I clean up my work area and turn off electric equipment. I dispose of waste properly and wash my hands.

Elaborate

21st Century Learning

CRITICAL THINKING To emphasize the importance of learning safe practices in the science laboratory, give students this fictional scenario: A student gets an idea for an investigation, so he stays late and works alone in the science laboratory. One material he is using in the experiment ignites. Ask: **Did the student violate any safety rules? If so, what?** *(Yes. You should never try anything on your own without asking your teacher first.)* **If the student had not been working alone, what is the first thing he should have done?** *(Tell the teacher.)*

Summarize Tell students that when they summarize, they briefly restate in their own words the main ideas of what they have read or heard.

Teacher Demo 

L1 DEMONSTRATE SAFETY PRACTICES

Materials chemical splash goggles, apron, heat-resistant gloves, plastic gloves, assorted lab equipment such as a hot plate, beaker, vegetable oil, balance, rock, and animal such as an earthworm

Time 15 minutes

Demonstrate safe practices in the science lab for students. Set up several mock investigations, involving heating vegetable oil, measuring the mass of a rock, and observing an animal. Show students the safe way to perform each task. Point out that you are wearing closed-toe shoes and note any situations in which long hair should be tied back. When appropriate, wear chemical splash goggles, an apron, heat-resistant gloves, or plastic gloves.

Ask: **What purpose do the chemical splash goggles serve?** *(To protect your eyes)* **What should you do with the oil when finished with it?** *(Ask your teacher how to dispose of it properly.)* **Once you clean up and dispose of waste, what else should you do?** *(Wash your hands.)*

Differentiated Instruction

L1 Communicate Have students use sketches, photographs, and short captions to create a visual display about safety in the laboratory and in the field. Tell students to draw ideas from the photograph in **Figure 2** as well as from the information they've learned about the safety symbols.

L3 Research Safe Practices Encourage interested students to ask an adult to accompany them to a science laboratory at a local college or university. They should ask for a copy of the lab's safety guidelines and inquire about safe practices. Ask these students to prepare a presentation to the class.

LESSON 1.5

Explain

Lead a Discussion

STAYING SAFE IN THE FIELD Explain to students that many scientific investigations are done in the field, where there may be many hazards. Ask: **What does it mean to do an investigation in the "field"?** *(The field can be any area outside a science laboratory, such as in the schoolyard, a forest, a park, or a beach.)* **Why should you never carry out a field investigation alone?** *(Hazards in the field are hard to plan for, and something might happen that could cause injury. If a person is hurt and alone, there is no one to assist or go for help.)*

Elaborate

Apply It!

L1 Before beginning the activity, discuss what is going on in the photograph. Ask volunteers to take turns describing one thing they observe in the photo. Point out that the students pictured correctly took the precaution of not going out in the field alone.

Observe Remind students that when they make observations of a photograph or diagram, they should take their time and thoroughly look over the entire image. They should be sure to read any labels or captions before reporting on what they observed.

Lab Resource: Quick Lab

L1 **BE PREPARED** Students will design a presentation for a fourth grade class about how to prepare for a lab. This Quick Lab can be found online.

Evaluate

Assess Your Understanding

After students answer the questions, have them evaluate their understanding by completing the appropriate sentence.

RTI Response to Intervention

1a. If students cannot explain why it is important to use safety equipment, **then** have them describe some potential dangers in a science laboratory.

b. If students have difficulty explaining why a field investigation takes more preparation, **then** ask students to contrast working in the lab with working outdoors in a forest or place they've never been to before.

Vocabulary Identify Multiple Meanings The noun *field* has several meanings. You have learned one meaning. Give two other meanings for *field*.

Sample: A field is an open area of land. A field is an area of study.

Safety in the Field Some of your science investigations will be done in the **field,** or any area outside a science laboratory. Just as in the laboratory, good preparation helps you stay safe. For example, there can be many safety hazards outdoors. You could encounter severe weather, traffic, wild animals, or poisonous plants. Whenever you set out to work in the field, you should always tell an adult where you will be. Never carry out a field investigation alone. Use common sense to avoid any potentially dangerous situations.

apply it!

These two students have not taken proper precautions to work in the field.

1 **Observe** Identify the clothing that is not appropriate for working in this field environment.

Sample: The students are wearing shorts and sandals.

2 **Draw Conclusions** Explain how one piece of clothing a student is wearing might expose the student to hazards in the field.

Sample: Wearing sandals exposes the feet to poisonous plants.

Lab zone® Do the Quick Lab *Be Prepared.* Find the lab online.

Assess Your Understanding

TEKS 1A, 4B

1a. Explain Why is it important to use safety equipment like chemical splash goggles?

Sample: The goggles protect your eyes from lab hazards such as chemicals and broken glass.

b. Make Generalizations Why would a field investigation take more preparation than a lab investigation?

There are more hazards to consider when preparing for a field investigation.

got it?

○ **I get it!** Now I know that the key to working safely in the lab and in the field is planning ahead and preparing.

○ I need extra help with See TE note.

What Should You Do if an Accident Occurs?

TEKS 1A, 4B In this section, you'll explore the first steps you should take if an accident occurs during a lab or field investigation.

Although you may have prepared carefully, at some point, an accident may occur. Would you know what to do? You should always start by telling an adult.

When any accident occurs, no matter how minor, tell your teacher immediately. Then listen to your teacher's directions and carry them out quickly. Make sure you know the location and the proper use of all the emergency equipment in your laboratory, including the fire extinguisher and fire blankets. Knowing safety and first-aid procedures beforehand will prepare you to handle accidents properly. **Figure 3** lists some emergency procedures.

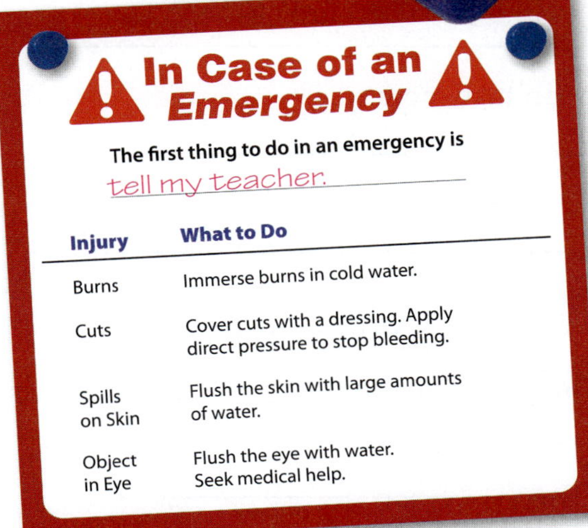

FIGURE 3

In Case of Emergency

These first-aid tips can help you in emergency situations in the lab.

Read and answer the questions.

1. **Review** Complete the sentence in the chart to identify the first step in responding to a lab emergency.
2. **Make Judgments** Suppose your teacher is involved in a lab accident. What should you do?
   Sample: Call the nurse's office. Ask the teacher in the next room to help.

In Case of an Emergency

The first thing to do in an emergency is tell my teacher.

Injury	What to Do
Burns	Immerse burns in cold water.
Cuts	Cover cuts with a dressing. Apply direct pressure to stop bleeding.
Spills on Skin	Flush the skin with large amounts of water.
Object in Eye	Flush the eye with water. Seek medical help.

Do the Quick Lab *Just in Case.* Find the lab online.

Assess Your Understanding

got it?

- **I get it!** Now I know that the first thing I should do in case of an accident is tell my teacher immediately.
- I need extra help with See TE note.

Differentiated Instruction

L1 Oral Presentation Call on students at random to explain what a student should do to stay safe in the field or when an accident occurs. After each student gives an answer, discuss responses as a group. Then allow each student to modify his or her thoughts.

L3 Write About Safety Ask students to write a tale of two students, one who follows safety guidelines and first-aid procedures after a lab accident and one who does not. Encourage students to read their stories to the class or post them on the class blog.

Explain

Teach Key Concepts 

Explain to students that accidents can occur in the laboratory, and they should know what to do if they are involved in a lab accident. Whether a major or a minor accident occurs, they should tell their teacher right away. Doing so can help prevent or lessen harm, because a teacher has been trained to take the correct actions in an emergency. Ask: **Once you tell your teacher, what is the next thing you should do?** *(Listen to your teacher's directions and carry them out quickly.)* **What should you do if, in preparing a liquid mixture in a beaker, you splashed some liquid on your skin?** *(Tell your teacher immediately.)* **What action should you and your teacher then take to try to prevent injury?** *(Flush the skin with large amounts of water)*

Elaborate

Lab Resource: Quick Lab 

L1 JUST IN CASE Students will learn more about how to respond in certain emergency situations. This Quick Lab can be found online.

Evaluate

Assess Your Understanding

Have students evaluate their understanding by completing the appropriate sentence.

RTI Response to Intervention

If students cannot identify the first thing to do after an accident, **then** review the Key Concept statement with them.

Texas Essential Knowledge and Skills

1A Demonstrate safe practices during laboratory and field investigations as outlined in the Texas Safety Standards.

4B Use preventative safety equipment, including chemical splash goggles, aprons, and gloves, and be prepared to use emergency safety equipment, including an eye/face wash, a fire blanket, and a fire extinguisher.

Name ______________________ Date ____________ Class ____________

Assess Your Understanding

Safety in the Science Laboratory

Why Prepare for a Laboratory Investigation?

1a. EXPLAIN Why is it important to use safety equipment like chemical splash goggles? ______

______

b. MAKE GENERALIZATIONS Why would a field investigation take more preparation than a lab investigation? ______

______

______

got*it*?

○ **I get it!** Now I know that the key to working safely in the lab and in the field is ______

______

○ **I need extra help with** ______

______

______

What Should You Do if an Accident Occurs?

got*it*?

○ **I get it!** Now I know that the first thing I should do in case of an accident is ______

______

○ **I need extra help with** ______

______

Place the outside corner, the corner away from the dotted line, in the corner of your copy machine to copy onto letter-size paper.

Name ______________________ Date ____________ Class ____________

Enrich

Safety in the Science Laboratory

Safety should always be the most important concern in the science laboratory. Read the passage. Then use a separate sheet of paper to answer the questions that follow.

A Day in the Labs

Mark, Juan, and Tina were eager to begin testing the properties of different chemicals. Juan placed a sample of chemical A in a test tube and began to heat it. Without goggles on, he peered into the test tube to see what was happening. Meanwhile, Mark was looking at a sample of liquid chemical B. It seemed to have an unusual odor, so he took a deep whiff of it. Then he put a drop of it on his finger to taste.

By this time, Juan was tired of heating chemical A, so he decided to heat a sample of chemical C—even though the instructions in his textbook stated that only chemical A was to be heated. As Juan tried his experiment, Tina began to observe the properties of chemical D by adding an acid to it. As she placed acid on the sample, she splashed some on her hand and then splashed more on the sleeve of her shirt. Then, leaving the top off the bottle of acid, she went on to test something else.

When the group was finished working, they left the chemical samples in the test tubes and put them away in a cupboard. One of the test tubes had a crack in it, but since it was not very large, they put that test tube away with the others. As Tina was leaving the lab, she noticed that she had a small cut on her finger. She decided that it was not important and did not mention it to anyone.

1. What did Juan do wrong as he was heating chemical A?

2. What mistake did Mark make with chemical B?

3. What was unsafe about Juan's moving on to chemical C?

4. What should Tina have done when she splashed acid on her hand and sleeve?

5. What did the group do wrong when leaving the lab?

6. What should Tina have done when she noticed the cut on her finger?

Name ______________________ Date ____________ Class ____________

Lesson Quiz

Safety in the Science Laboratory

Fill in the blank to complete each statement.

1. You should keep your lab work area clean and ____________.

2. Doing good ____________ ahead of time helps you stay safe during science investigations.

3. When an accident occurs, the first person you should tell is your ____________.

4. Be sure to wash your ____________ thoroughly after working in the laboratory.

5. Wear safety ____________ to protect your eyes from chemical splashes.

6. ______________________ are pictures that alert you to possible sources of accidents in an investigation.

Write the letter of the correct answer on the line at the left.

7. ___ Which of the following are you *least* likely to see represented by a safety symbol in the procedures of an investigation?

A poison
B time
C flames
D lab apron

8. ___ When doing a science investigation, what should be your chief concern?

A your own safety
B your classmates' safety
C both A and B
D neither A nor B

9. ___ Where do you conduct a science investigation in the field?

A in the science laboratory
B outside the science laboratory
C both A and B
D neither A nor B

10. ___ One way you can prepare for an accident in the lab is to know which of the following ahead of time?

A first-aid procedures
B investigation procedures
C location of aprons
D when to wear plastic gloves

Place the outside corner, the corner away from the dotted line, in the corner of your copy machine to copy onto letter-size paper.

Place the outside corner, the corner away from the dotted line, in the corner of your copy machine to copy onto letter-size paper.

Safety in the Science Laboratory

Answer Key

Review and Reinforce

Find the worksheet in the Student Workbook.

1. Read through any procedures carefully, and make sure you understand all the directions. If anything is unclear, ask you teacher about it before you begin the lab.
2. Safety symbols alert you to possible sources of accidents in an investigation. They identify how to work carefully and what safety equipment to use.
3. Clean up your work area. Turn off and unplug any equipment and return it to its proper place. Dispose of any waste materials properly according to your teacher's instructions. Be sure to wash your hands thoroughly.
4. severe weather, traffic, wild animals, or poisonous plants
5. Tell your teacher immediately. Then, listen to your teacher's directions and carry them out quickly.
6. symbols that alert you to possible sources of accidents in an investigation
7. any area outside a science laboratory
8. To prepare for an investigation, read through any procedures carefully, and make sure you understand all the directions. If anything is unclear, ask your teacher about it before you begin the lab. Also, pay attention to any safety symbols included in the lab directions. If an accident occurs during a lab investigation, tell your teacher immediately, and listen to and carry out your teacher's directions quickly.

Enrich

1. He should have been wearing goggles.
2. He should have avoided inhaling fumes directly. He should not have tasted or touched the chemical.
3. He should have followed the instructions in his book exactly.
4. She should have told the teacher, and she should have flushed the skin and sleeve with large amounts of water.
5. They should have disposed of wastes properly, and they should not have put a cracked test tube in a cupboard. They should also have washed their hands.
6. She should have told the teacher. She should also have covered the cut with a clean dressing and applied direct pressure to the wound to stop the bleeding.

Lesson Quiz

1. organized/uncluttered
2. preparation
3. teacher
4. hands
5. goggles
6. Safety symbols
7. B
8. C
9. B
10. A

Scientists and Society

How do scientists investigate the natural world?

LESSON PACING:
1–2 periods or $\frac{1}{2}$–1 block

Lesson Vocabulary

- controversy

Lesson Objectives	TEKS	ELPS
Explain how science and society impact one another.	3D	3.G.2

Content Refresher

Einstein and Society Albert Einstein's equation $E = mc^2$, which stands for "energy equals mass times the speed of light squared," might seem to be a work of science with little impact on society. Yet it provides the key to understanding some of the basic natural processes of the universe, including radioactivity and the formation of our universe.

In 1905, Einstein introduced his equation in a paper. In it, he proposed that his theory about the equation could be put to the test using radium, an element that emits 4,000 calories of heat per hour. Einstein thought that some of radium's mass was constantly converted to energy just as his equation indicated.

Today we know that Einstein was correct, and many technologies used in modern society take advantage of $E = mc^2$. For example, several elements, including radium, give off energetic particles as their atomic nuclei break down. This discovery made possible the development of Positron Emission Tomography (PET) scans by doctors and radiocarbon dating by archaeologists.

Texas Essential Knowledge and Skills

3D Relate the impact of research on scientific thought and society, including the history of science and contributions of scientists as related to the content.

English Language Proficiency Standards

ELPS Speaking 3.G.2 Express ideas ranging from communicating single words and short phrases to participating in extended discussions on a variety of social and grade-appropriate academic topics.

DIFFERENTIATED INSTRUCTION KEY

L1 Struggling Students or Special Needs

L2 On-Level Students **L3** Advanced Students

LESSON PLANNER 1.6

Investigations and Activities

My Planet Diary, **Student Edition,** p. 42

Inquiry: Inquiry Warm-Up, What Do Scientists Do?, **PearsonTexas.com**

Introduce Vocabulary, **Teacher's Edition,** p. 43

Teach Key Concepts, **Teacher's Edition,** p. 43

Make Analogies, Tug-of-War, **Teacher's Edition,** p. 43

Support the TEKS, Semmelweis and His Society, **Teacher's Edition,** p. 44

Inquiry: Build Inquiry, Sequence Events, **Teacher's Edition,** p. 44

Apply It!, **Student Edition,** p. 45

Differentiated Instruction, **Teacher's Edition,** p. 45

Inquiry: Quick Lab, Light Sources, **Lab Manual,** p. 11

TEKS Review

Apply the TEKS, Evaluating Scientific Explanations, **Student Edition,** p. 46

TEKS Practice, **Student Edition,** p. 48

TEKS Practice: Chapter and Cumulative Review, **Student Edition,** p. 52

Lesson 1.6, **TEKS Preparation and Study Guide Workbook,** p. 12

SHORT ON TIME? To do this lesson in approximately half the time, do the Activate Prior Knowledge activity. A discussion of the Key Concepts will familiarize students with the lesson content. Have students do the Quick Lab. The rest of the lesson can be completed by students independently.

These editable worksheets are available on **PearsonTexas.com.**
Print versions can be found in the **TEKS Preparation and Study Guide Workbook.**

Name ______ Date ______ Class ______

1.6 Scientists and Society

Key Concept Summary

How Does Society Affect the Work of Scientists?

Science discoveries have helped to bring cars, computers, phones, and other products to society. And today's scientists are working on tomorrow's discoveries. **The work that scientists do changes society. In turn, society influences the work of scientists.**

Sometimes scientific work conflicts with the beliefs of some people in society and society's leaders. As a result, a scientist's work causes a **controversy,** or a public disagreement between groups. Scientific controversies are not uncommon. Because scientists have defended their works, people today better understand the world.

In 1610, Galileo Galilei published his discoveries supporting the heliocentric model, a model of the universe in which Earth moves around the sun. This heliocentric model conflicted with the beliefs of his society's leaders. In 1616, all books that supported this model were banned. Galileo published another book that supported this model. As a result, he was tried, found guilty, and lived the rest of his life under house arrest. Galileo's stand eventually led to an acceptance of science as a way of explaining the natural world.

Ignaz Semmelweis, a doctor in the 1840s, observed that many women died of infections after giving birth in the hospital. After studying the problem, he suggested that doctors wash their hands before delivering babies. His solution to the problem was in conflict with common medical practices at the time. Therefore, Semmelweis lost his job at the hospital. Eventually, new discoveries about disease supported Semmelweis's ideas.

Rachel Carson was a biologist who wrote about science and nature. Her last book, *Silent Spring*, was published in 1962. It was about the effects of pesticides, such as DDT, on the environment. Pesticides are chemicals that farmers can use to kill insects that harm their crops. Carson used data to show how pesticides harmed animals, but many people disagreed with her ideas. Yet, in 1972, the government banned DDT, and *Silent Spring* played an important role in the government's decision.

12

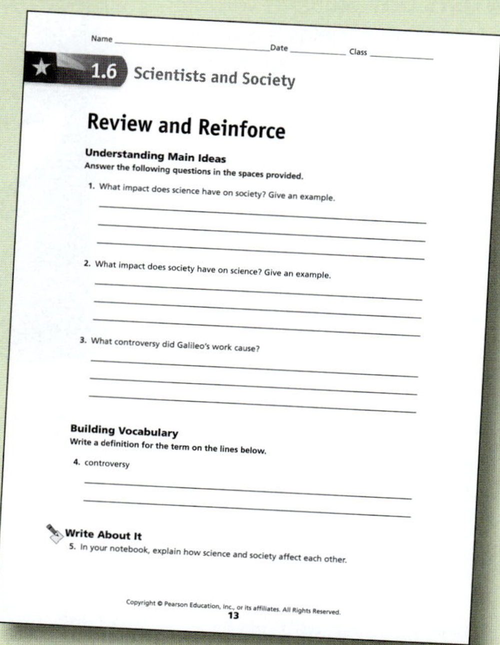

Name ______ Date ______ Class ______

1.6 Scientists and Society

Review and Reinforce

Understanding Main Ideas

Answer the following questions in the spaces provided.

1. What impact does science have on society? Give an example.

2. What impact does society have on science? Give an example.

3. What controversy did Galileo's work cause?

Building Vocabulary

Write a definition for the term on the lines below.

4. controversy

Write About It

5. In your notebook, explain how science and society affect each other.

13

Lexile Measure = 880L

LESSON 1.6

Scientists and Society

Establish Learning Objective

After this lesson, students will be able to:

Explain how science and society impact one another.

Engage

Activate Prior Knowledge

MY PLANET DIARY Read *Albert Einstein* with the class. Ask students to share what they already know about Einstein. Ask students if they have ever seen a T shirt, bumper sticker, or tote bag with $E = mc^2$ printed on it. Ask: **How do you think Einstein influenced society? What influence might society have had on Einstein?** *(Students may recognize that Einstein has become an icon of popular culture. They may suggest that the events occurring during his life influenced his work.)*

Explore

Lab Resource: Inquiry Warm-Up 

L1 **WHAT DO SCIENTISTS DO?** Students will infer how scientific work in a variety of fields has affected society. This Inquiry Warm-Up can be found online.

Texas LESSON 6

Scientists and Society

How Does Society Affect the Work of Scientists?

TEKS 3D

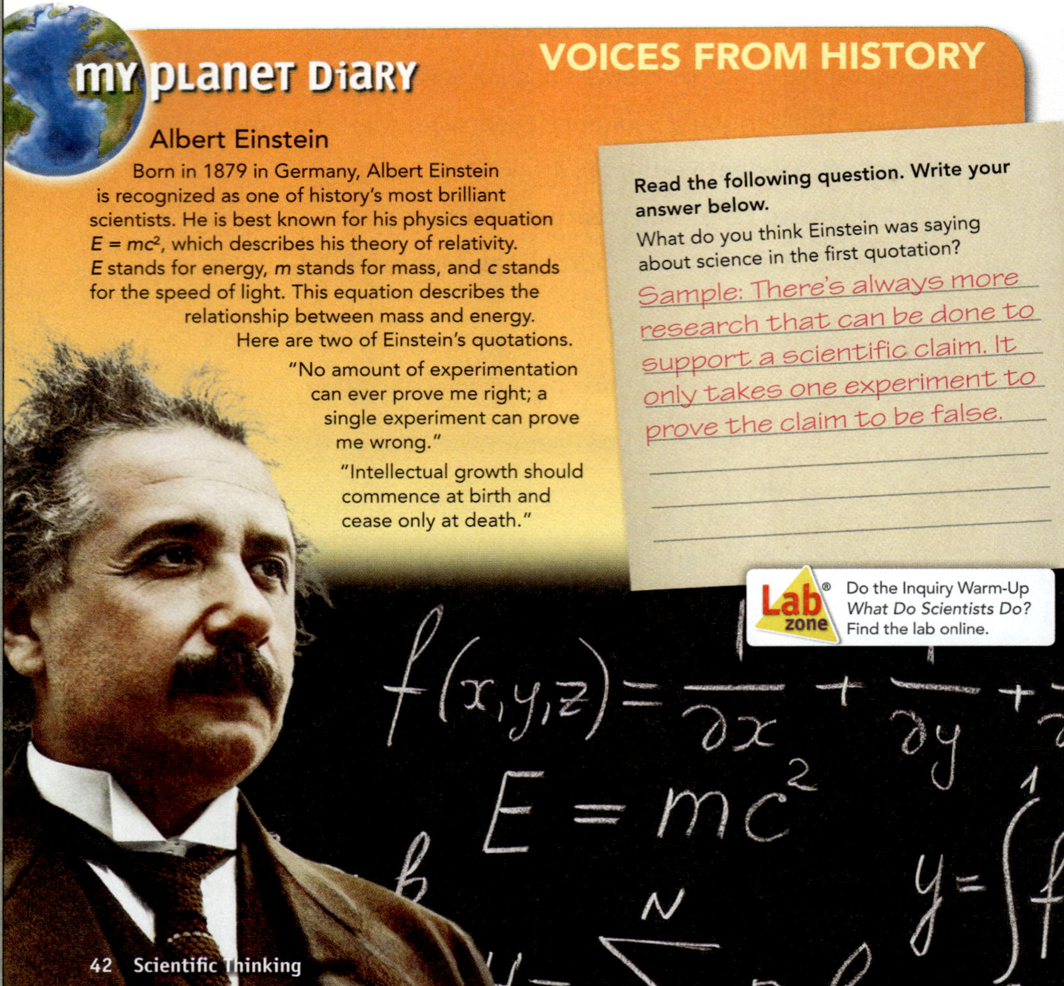

my planet Diary

VOICES FROM HISTORY

Albert Einstein

Born in 1879 in Germany, Albert Einstein is recognized as one of history's most brilliant scientists. He is best known for his physics equation $E = mc^2$, which describes his theory of relativity. E stands for energy, m stands for mass, and c stands for the speed of light. This equation describes the relationship between mass and energy.

Here are two of Einstein's quotations.

"No amount of experimentation can ever prove me right; a single experiment can prove me wrong."

"Intellectual growth should commence at birth and cease only at death."

Read the following question. Write your answer below.

What do you think Einstein was saying about science in the first quotation?

Sample: There's always more research that can be done to support a scientific claim. It only takes one experiment to prove the claim to be false.

Lab zone Do the Inquiry Warm-Up *What Do Scientists Do?* Find the lab online.

42 Scientific Thinking

SUPPORT ALL READERS

Lexile Measure = 880L **Lexile Word Count = 664**

Prior Exposure to Content: May be the first time students have encountered this topic

Academic Vocabulary: *predict, sequence*

Science Vocabulary: *controversy*

Concept Level: Generally appropriate for most students in this grade

Preteach With: My Planet Diary "Albert Einstein" and Figure 1 activity

Vocabulary
- controversy

Skills
- Reading: Sequence
- Inquiry: Predict

How Does Society Affect the Work of Scientists?

TEKS 3D In this section, you'll examine how the work that scientists do impacts society and how society can affect the work of scientists.

Science discoveries have helped to bring cars, computers, phones, and other products to society. And today's scientists are working on tomorrow's discoveries. **The work that scientists do changes society. In turn, society influences the work of scientists.**

Sometimes scientific research conflicts with the beliefs of a society or its leaders. When this happens, a scientist's work can cause a **controversy,** or a public disagreement between groups. Scientific controversies are not uncommon. By defending their work, scientists have helped people better understand the world.

Galileo Galilei For many years, Galileo Galilei observed the night sky with a telescope. In 1610, he published his research supporting the heliocentric model, a model of the universe in which Earth moves around the sun. This heliocentric model, shown at the right, conflicted with the beliefs of his society's leaders. In 1616, all books that supported this model were banned. In 1632, Galileo published another book that supported this model. As a result, he was tried, found guilty, and lived the rest of his life under house arrest. In time though, Galileo's work led to an acceptance of science as a way of explaining the natural world.

Galileo's diagram of the heliocentric model

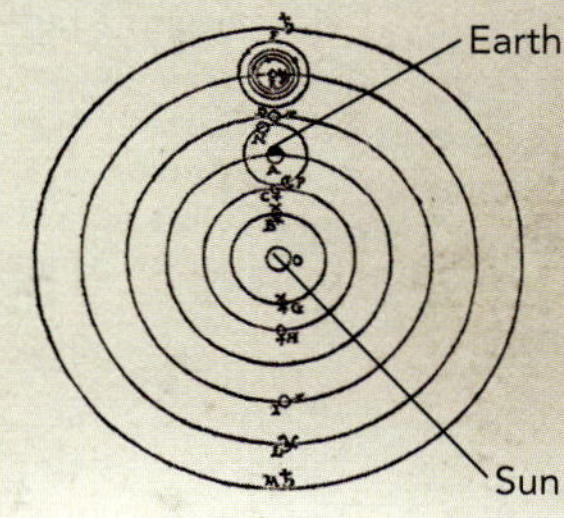

Sequence Write in your own words the order of events that led to Galileo's house arrest.

Event 1	Event 2	Event 3
Galileo published discoveries about the night sky.	His discoveries supported ideas that conflicted with those of society.	Books about the heliocentric model were banned.

Event 4	Event 5	Event 6
Galileo published another book about the heliocentric model.	Galileo was tried and found guilty.	Galileo was placed under house arrest.

PEARSON Texas.com 43

Explain

LESSON 1.6

Introduce Vocabulary

Explain that *controversy* comes from the Latin word *controversus,* meaning "turned against." Relate this meaning to the definition of *controversy* as a public disagreement.

Teach Key Concepts

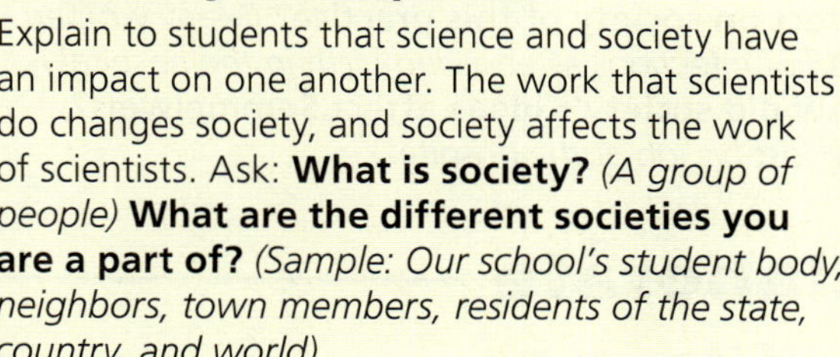

Explain to students that science and society have an impact on one another. The work that scientists do changes society, and society affects the work of scientists. Ask: **What is society?** *(A group of people)* **What are the different societies you are a part of?** *(Sample: Our school's student body, neighbors, town members, residents of the state, country, and world)*

Make Analogies

L1 TUG-OF-WAR Tell students that the interaction between science and society can be like a game of tug-of-war. When one side pulls on the rope, the other side is affected. Then when that side pulls on the rope or falls forward, the other side is affected. Ask: **What else is the interaction between science and society like?** *(Accept any reasonable responses that indicate a cause-and-effect or action-and-reaction relationship.)*

Sequence Explain that sequencing involves putting events in the order in which they occur or putting objects in order according to their characteristics.

PEARSON Texas.com

English Language Proficiency Standards

ELPS Speaking 3.G.2

Have students read My Planet Diary on page 42 and answer the question.

Beginning Have students listen as you read aloud My Planet Diary. Help students express ideas by scaffolding questions, for example: *Does a theory explain observations or tell what always happens? Can a scientist change a theory?*

Intermediate Together, read aloud My Planet Diary and discuss the question. If needed, review the term *scientific theory* on page 27.

Advanced Have pairs read aloud the text and discuss the questions. Encourage students to elaborate on their ideas.

Advanced High After students discuss the question, have them tell what they think Einstein was saying in the quotation about intellectual growth.

Texas Essential Knowledge and Skills

3D Relate the impact of research on scientific thought and society, including the history of science and contributions of scientists as related to the content.

LESSON 1.6

Explain

Support the TEKS

SEMMELWEIS AND HIS SOCIETY Tell students that scientific research can both help and harm members of society. Ask: **In Semmelweis's time, what scientific idea was harming people?** *(The common medical practice of doctors not washing their hands before delivering babies)* **What was the effect on society of this practice?** *(Many women died of infections after giving birth in the hospital.)* **How did society's ideas affect Semmelweis?** *(He lost his job at the hospital.)*

Elaborate

Build Inquiry 

L2 SEQUENCE EVENTS

Materials index cards, red and blue markers

Time 15 minutes

Review with students the impact that scientists Semmelweis and Carson had on society and that society had on them. Provide pairs of students with index cards. On the index cards, have them summarize the events related to these scientists as described in the text, writing one event per card. They should also draw individual red and blue arrows on several cards. Have students rearrange the cards in the order in which the events occurred. They should position a red arrow after a scientific event to point to the effect it had on society, and a blue arrow before a scientific event to point to the effect society had on science.

Ask: **What controversy was Semmelweis involved in?** *(His idea about doctors washing their hands before delivering babies was in conflict with common medical practices at the time.)* **What controversy was Carson involved in?** *(Many people disagreed with Carson's ideas about pesticides harming animals.)*

Ignaz Semmelweis Ignaz Semmelweis was a young doctor working in a hospital in Austria in the 1840s. He observed that many women died of infections after giving birth in the hospital. After researching the problem, he suggested that doctors wash their hands before delivering babies. As soon as doctors started doing this, death rates dropped. However, Semmelweis's solution to the problem was in conflict with common medical practices at the time. Therefore, Semmelweis lost his job at the hospital. Eventually, new discoveries about disease supported Semmelweis's ideas. Hand-washing before deliveries became routine, and many more mothers survived.

Rachel Carson Rachel Carson was a biologist who wrote about science and nature. Her last book, *Silent Spring*, was published in 1962. This book is about the effects of pesticides, such as DDT, on the environment. Pesticides are chemicals that farmers can use to kill insects that harm their crops. After World War II, farmers used pesticides and their harvests increased. However, large numbers of birds and other animals died as a result of pesticide use. Carson used data from her research to show how pesticides harmed animals, but many people disagreed with her ideas. Yet, in 1972, the government banned DDT, and *Silent Spring* played an important role in the government's decision. Eventually, most wildlife populations recovered from the harm DDT caused.

FIGURE 1

Effects of DDT

High concentrations of DDT in the environment caused birds to lay eggs with thin, fragile shells that would break easily.

Relate **Discuss with a partner why you think people were resistant to the research of Semmelweis and Carson despite being presented with data that supported their points. Then write your answer below.**

Sample: People were resistant to the ideas of Semmelweis and Carson because the data they had went against the thinking of society.

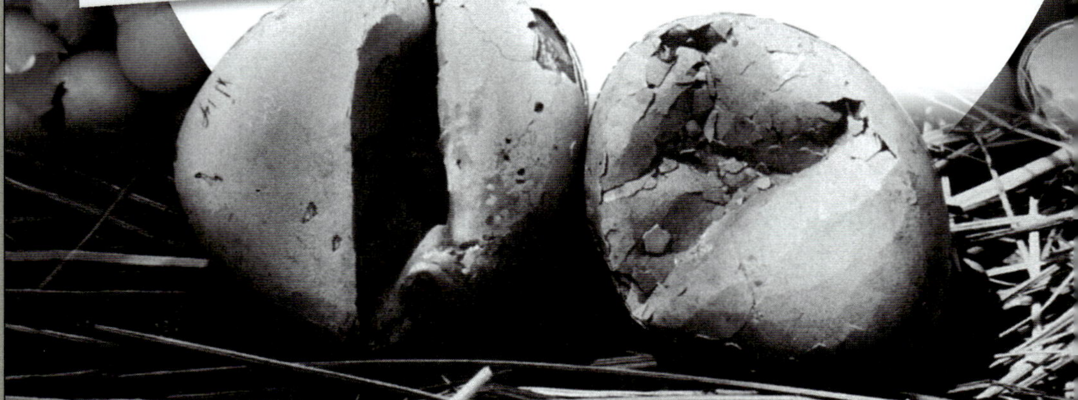

Suppose that fish in your town's river are dying. Scientists conclude that polluted water from a nearby factory is killing the fish. If it costs too much to fix the problem, the factory will have to close. Many people in town work in the factory. Many others work in fishing camps that attract tourists who come for the great fishing.

1 Predict Do you think each group below will reject or agree with the conclusion about why the fish are dying? Explain each answer.

Factory workers

Sample: They'll reject it because they will lose their jobs if the factory closes.

Fishing camp owners

Sample: They'll agree because if too many fish die, the tourists might stop visiting.

2 CHALLENGE What would be a cost and a benefit if the town offered to fix the problem for the factory?

Sample: A cost would be the money the town would have to pay for the cleanup. A benefit would be the clean water.

Lab zone Do the Quick Lab *Light Sources.* Student Lab Manual, p. 11

TEKS 3D

1a. **Define** What is a scientific controversy?

A disagreement between a scientist's work and a society's beliefs

b. **Relate** Why was there controversy over Semmelweis's research?

Washing your hands was not a common medical practice at the time.

got it?

O I get it! Now I know that science and society affect each other because science changes society and society influences science.

O I need extra help with See TE note.

45

Apply It!

L1 Before beginning the activity, review with students what they already know about water pollution, factories or other businesses closings, and recreational tourism in your area. Discuss how changes in water quality, businesses, and tourism have affected your local society.

Predict Tell students that when they predict, they use what they already know about a subject to infer what is most likely to happen in the future. Ask students to imagine they are factory workers or fishing camp employees to help them visual how members of each group might react.

Lab Resource: Quick Lab 

L1 **LIGHT SOURCES** Students will explore how the development of light technology has been shaped by science and society. This Quick Lab can be found in the Student Lab Manual, p. 11, and online.

Evaluate

Assess Your Understanding

After students answer the questions, have them evaluate their understanding by completing the appropriate sentence.

RTI Response to Intervention

1a. If students cannot define *scientific controversy,* **then** provide them with examples of other kinds of controversy with which they may be familiar.

b. If students have trouble explaining the controversy over Semmelweis's research, **then** reread the red section subtitled *Ignas Semmelweis* aloud as a class.

Differentiated Instruction

L1 **Public Service Announcement** Have students write and perform a public service announcement about the research of Galileo, Semmelweis, or Carson. The announcement should educate the audience about the scientific concepts and persuade the audience of their benefits to society.

L3 **Scientific Controversy** Have students do research to identify another scientist whose research caused a controversy. Have them write a short report about the related events and the scientist's effect on society. Students can add their reports to a class library.

Name ______________________ Date __________ Class __________

Assess Your Understanding

Scientists and Society

How Does Society Affect the Work of Scientists?

1a. DEFINE What is a scientific controversy?

____

____

b. RELATE Why was there controversy over Semmelweis's ideas?

____

____

got it?

○ **I get it!** Now I know that science and society affect each other because ______________

____

○ **I need extra help with** ______________________

____

Place the outside corner, the corner away from the dotted line, in the corner of your copy machine to copy onto letter-size paper.

Name ______________________ Date ____________ Class ____________

Enrich

Scientists and Society

Read the passage below. Then answer the questions that follow on a separate sheet of paper.

The Nobel Prizes in Science

Every year since 1901, the Nobel Pirize has been awarded for achievements in physics, chemistry, physiology or medicine, literature, and peace. These awards were established by Alfred Nobel, who was a scientist, inventor, entrepreneur, author, and pacifist. He was also the inventor of dynamite. Among the awards for science in 2009 were:

Nobel Prize for Physics
Charles K. Kao, Willard S. Boyle, George E. Smith
Fathers of fiber optics and digital imaging

Nobel Prize for Chemistry
Venkatrama Ramakrishnan, Thomas A Steitz, Ada E Yonath
Studies of the structure and function of the ribosome

Nobel Prize for Medicine
Elizabeth H. Blackburn, Carol W. Greider, Jack W. Szostak
Discovery of how chromosomes are protected by telomeres and the enzyme telomerase

The Nobel Prizes are to be awarded to those who "during the preceding year, shall have conferred the greatest benefit on mankind." No consideration is to be given to the nationality of the candidates. The process of nominating and selecting the winners is long and rigorous. Science prizes often are awarded to collaborators of two or three. The awards for science are for achievements "tested by time." So there is lag time between the discovery and the award.

1. **Choose one of the 2009 science awards and find out more about the achievement. Write a short summary.**

2. **Research two other science laureates (winners of the Nobel Prize) and describe their achievements briefly.**

3. **Write a short biography of Alfred Nobel. Explain why you think he gave much of his fortune to establish the Nobel Prizes.**

Name ______________________ Date __________ Class __________

Lesson Quiz

Scientists and Society

Fill in the blank to complete each statement.

1. The work that scientists do ________________ society.

2. A public disagreement between groups is called a(n) ________________.

3. The book *Silent Spring* was about the effects of ________________, such as DDT, on the environment.

4. ________________ influences the work of scientists.

5. Sometimes scientific work conflicts with the ________________ of some people in society and society's leaders.

6. ________________ was the Austrian doctor who suggested that doctors wash their hands before delivering babies.

Write the letter of the correct answer on the line at the left.

7. ___ Which scientist lived under house arrest as a result of publishing a book about his work on of the universe?
   - A Semmelweis
   - B Carson
   - C Galileo
   - D Einstein

8. ___ Which is NOT a product that science discoveries helped to bring to society?
   - A computers
   - B meteors
   - C phones
   - D cars

9. ___ Which event happened as a result of Rachel Carson's scientific work?
   - A More women died of infections after giving birth.
   - B Fewer women died of infections after giving birth.
   - C The government banned DDT.
   - D The government developed DDT.

10. ___ Which of the following describes the heliocentric model?
    - A The spread of infection is prevented by washing.
    - B Earth moves around the sun.
    - C The sun moves around Earth.
    - D Pesticides harm animals.

Place the outside corner, the corner away from the dotted line, in the corner of your copy machine to copy onto letter-size paper.

Place the outside corner, the corner away from the dotted line, in the corner of your copy machine to copy onto letter-size paper.

Scientists and Society

Answer Key

Review and Reinforce

Find the worksheet in the Student Workbook.

1. Sample: The work that scientists do changes society. Scientific discoveries have helped to bring products such as cars to society.
2. Sample: Society influences the work of scientists. Because books supporting Galileo's model of the universe were banned, he lived the rest of his life under house arrest.
3. Galileo's heliocentric model of the universe conflicted with the beliefs of his society's leaders, and books that supported this model were banned, including Galileo's book.
4. a public disagreement between groups
5. The work that scientists do changes society. Also, society influences the work of scientists. (Students should add details.)

Enrich

1. Answers will vary with choices of prize winners.
2. Answers will vary with choice of Nobel laureates.
3. Alfred Nobel was born in Stockholm, Sweden in 1833. In 1842, his family moved to St. Petersburg. In 1850, Nobel went to Paris to work in a laboratory. He then went on to Italy, Germany, and the United States. He started the Nobel Company which went bankrupt at the end of the Crimean War. In 1862, Nobel began experiments with nitroglycerin, eventually getting a patent for dynamite in 1867. Nobel became a wealthy man. His companies are in France, Germany, the United States, and Sweden. He wrote his will, establishing the Nobel Prizes in 1895.

Lesson Quiz

1. changes
2. controversy
3. pesticides
4. Society
5. beliefs
6. Semmelweis
7. C
8. B
9. C
10. B

CHAPTER 1

Scientific Investigation and Reasoning

This Apply the TEKS feature will enable students to apply the scientific investigation and reasoning skills in TEKS 3A. Students will analyze, evaluate, and critique scientific claims about the existence of the ivory-billed woodpecker, which was considered to be extinct or near extinct until 2004.

Evaluating Scientific Explanations

Have students read about the search for the ivory-billed woodpecker. You may wish to display a map showing the original range of the woodpecker. Explain to students that the ivory-billed woodpecker once thrived in the wetlands region of the southeastern U.S., including Texas. Human activity, including logging and hunting, resulted in extensive habitat loss and destruction, and the woodpecker vanished from sight after the early 1940s.

After students answer Question 1 and write a definition for *extinct,* have them develop criteria for meeting their definitions. Ask: **How much time do you think needs to pass before a species can be declared extinct?** *(Sample: At least ten years must pass without a sighting for a species to be declared extinct.)* **Who do you think should be responsible for declaring a species extinct?** *(Sample: A group of scientists or a government agency such as the U.S. Fish and Wildlife Service)*

Texas Essential Knowledge and Skills

3A In all fields of science, analyze, evaluate, and critique scientific explanations by using empirical evidence, logical reasoning, and experimental and observational testing, including examining all sides of scientific evidence of those scientific explanations, so as to encourage critical thinking by the student.

APPLY THE TEKS 3A

Evaluating Scientific Explanations

Everyone agrees that woolly mammoths are extinct because none have been spotted during recorded history. But how do scientists know when to declare a modern-day species extinct?

The ivory-billed woodpecker (*Campephilus principalis*) once lived in the southeast United States. The largest of the woodpecker species, it had a distinctive white pattern on its back and a unique call. The last confirmed sighting of the ivory-billed woodpecker was in 1944.

In 2005, a group of researchers announced the discovery of the woodpecker in eastern Arkansas. The announcement was based on a handful of reported sightings and a video clip recorded in 2004. Many ornithologists, scientists who study birds, were skeptical and thought the sightings were of another woodpecker species.

Researchers launched an intensive search for the woodpeckers, from southeast Texas to the Florida panhandle. But it seems to have turned out to be more like a wild goose chase. No conclusive signs of the ivory-billed woodpecker have been found.

▲ These still images, taken from the 2004 video, reportedly show the ivory-billed woodpecker in flight. The numbers indicate the frames in the video.

Answer the questions below.

1. **Form Operational Definitions** Write a clear and specific definition of the term *extinct*.
   Sample: Extinct means that every single member of a species has died.
2. **Use Empirical Evidence** Research more about the sightings of ivory-billed woodpeckers in the Southeast. Use reliable Internet or print sources to find more images of the woodpecker. If possible, view the 2004 video of the alleged woodpecker sighting.
3. **Evaluate** Examine all sides of the scientific evidence that you have found. What evidence is there that the ivory-billed woodpecker survives? What evidence supports the idea that it is extinct?
   Sample: Evidence that it survives: There have been occasional reports of sightings, as well as the video taken in 2004. Evidence that it is extinct: Many scientific surveys have failed to find a single bird, and most scientists think the reported sightings, including the video, are of another species.
4. **Critique Scientific Explanations** Think back to the definition you wrote for the term *extinct*. Does the evidence support the hypothesis that the ivory-billed woodpecker is not extinct?
   Sample: No; the evidence that some ivory-billed woodpeckers still live is not conclusive enough to conclude that the bird species survives.
5. **Make Judgments** People still continue to search for the woodpecker. What future evidence might cause you to revise your conclusion?
   Sample: If the bird were clearly caught on video or in a photo, then I might revise my conclusion.

Scientific Investigation and Reasoning, Cont.

You may want to provide students with additional information to help in their research for evidence in Question 2. Suggest that they search for the Cornell Lab of Ornithology, the academic research group that announced the existence of the ivory-billed woodpecker in 2005. Students may want to search for the video using the key words *Luneau video* (the video was recorded by David Luneau). In addition, the Texas Parks and Wildlife Department's official YouTube channel includes the short documentary *Chasing a Ghost,* a video about the search for the ivory-billed woodpecker in Texas.

Explain to students that the U.S. Fish and Wildlife Service has determined that there is not enough evidence to conclude that the bird in the Luneau video is a different species of woodpecker. As a result, the service has developed a recovery plan in case more definitive evidence comes to light in the future. Ask: **Why might some scientists and groups be hesitant to declare the woodpecker extinct rather than endangered?** *(Sample: If even a small population survives somewhere, it would not be protected.)* **How do you think the story of the ivory-billed woodpecker reflects the process of scientific inquiry?** *(Sample: Scientists gather and evaluate evidence, and then they draw conclusions based on their review of the evidence. In this case, different people and groups have drawn different conclusions based on the available evidence.)*

CHAPTER 1

TEKS Practice

Assess Understanding

Have students complete the answers to the TEKS Practice questions. Have a class discussion about what students find confusing. Write Key Concepts on the board to reinforce knowledge.

RTI Response to Intervention

2. If students have trouble defining the skill of classifying, **then** have them reread the sentence that contains the boldfaced term.

8. If students have trouble describing the three types of bias, **then** have them skim the lesson for the section on bias and reread it aloud with a partner.

Alternate Assessment

L1 DESIGN A GAME Have students work in small groups to design a game about scientific thinking and the inquiry process. Students can design a board game that requires players to answer questions in order to advance. Remind students to create rules, spinners, game pieces, and questions for their games. The questions for the game should include vocabulary terms and Key Concepts from the chapter. Students can exchange games with or play their own game against other groups.

PEARSON Texas.com

TEKS Practice

TEKS 1A, 2A, 2B, 2E, 3A, 3B, 3C, 3D, 4B

LESSON 1 Science and the Natural World

1. When you explain or interpret an observation, you are
   a. using models.
   b. classifying.
   c. inferring. (circled)
   d. predicting.

2. When scientists group observations that are alike in some way, they are *classifying.*

3. **Predict** How do scientists use observations to make predictions?
   They make predictions based on many observations with the same outcome.

4. **Infer** Suppose you come home to the scene below. What can you infer happened?

   Sample: The cat tried to catch a fish, fell into the water, and couldn't get out.

5. **Observe** What is a quantitative observation you might make in your school cafeteria?
   Sample: I might observe how many students buy lunch.

LESSON 2 Thinking Like a Scientist

6. The scientific attitude of having doubt is called
   a. open-mindedness.
   b. curiosity.
   c. honesty.
   d. skepticism. (circled)

7. When a person allows personal opinions, values, or tastes to influence a conclusion, that person is using *subjective* reasoning.

8. **Evaluate** Describe the three types of bias that can influence a science experiment.
   Sample: Personal bias comes from your own likes and dislikes. Cultural bias comes from opinions in your culture. Experimental bias comes from mistakes in the experiment's design.

9. **Draw Conclusions** Why is it important to report experimental results honestly, even when the results might be the opposite of the results you expect to see?
   Sample: The experimental results may show a scientific discovery.

10. **Write About It** A team of scientists is developing a new medicine to improve memory. Write about how the attitudes of curiosity, honesty, creativity, and open-mindedness help the scientists in their work. When might they need to think about ethics? How could bias influence their results?
   See TE rubric.

48 PEARSON Texas.com

Write About It 10 Assess student's writing using this rubric.

SCORING RUBRIC	SCORE 4	SCORE 3	SCORE 2	SCORE 1
Describe the four attitudes that help scientists.	Student correctly describes all four attitudes.	Student correctly describes some of the attitudes.	Student incorrectly describes all four attitudes.	Student does not describe any of the four attitudes.
Describes the influence of ethics and bias.	Student correctly describes the influence of both ethics and bias.	Student correctly describes the influence of either ethics or bias.	Student incorrectly describes the influence of both ethics and bias.	Student does not describe the influence of either ethics or bias.

LESSON 3 Scientific Inquiry

11. The facts, figures, and other evidence gathered through observations are called
  a. conclusions.
  (b.) data.
  c. predictions.
  d. hypotheses.

12. A scientific theory is a well-tested explanation for a wide range of observations.

13. **Explain** Why must a hypothesis be testable?
Sample: A hypothesis must be testable because you must be able to carry out an investigation to determine whether the hypothesis can be supported or disproved.

14. **Communicate** What are some ways that scientists communicate with each other?
Scientists can share their results by speaking at scientific meetings, exchanging information on the Internet, or publishing articles in scientific journals.

15. **Write About It** Suppose you want to find out which dog food your dog likes best. Write about the experiment you would design. What variables would you need to control? What kinds of data would you collect? How could you avoid experimental bias?
See TE rubric.

LESSON 4 Models as Tools in Science

16. Material or energy that goes into a system is
  a. output.
  (b.) input.
  c. feedback.
  d. process.

17. A complex system has many parts and variables.

18. **Use Models** You are an astronomer studying an exoplanet orbiting its host star. What are the advantages of using a computer model to study this planet and its star?
Sample: It is difficult to directly observe the planet and star because they are so far away. A model makes it easier to study them.

19. **Identify** Why are there more limitations to a simple model than a more complex one?
Sample: A complex model can represent more parts in the system. A simpler model may not show all the parts, so it might not be useful for making accurate predictions.

20. **Write About It** The output of the system below is text displayed on the screen. Describe the input and process that produces this output.

See TE rubric.

TEKS Practice, Cont.

CHAPTER 1

RTI Response to Intervention

14. If students need help with identifying ways scientists communicate with each other, **then** ask them what opportunities exist for researchers to share ideas in person, over long distances, and in writing.

18. If students cannot identify the advantages of using a model, **then** have them review the first Key Concept statement.

L3 WRITING IN SCIENCE Ask students to write a television interview with a scientist that explains to viewers how scientists investigate the natural world.

Write About It **15** Assess student's writing using this rubric.

SCORING RUBRIC	SCORE 4	SCORE 3	SCORE 2	SCORE 1
Describe experimental plan including variables and data	Student describes a valid experiment, with variables and data.	Student describes an experiment, with variables or data.	Student describes an experiment but no variables or data.	Student does not describe an experiment.
Explain how to avoid bias	Student correctly explain how to avoid bias.	Student partially explains how to avoid bias.	Student incorrectly explains how to avoid bias.	Student does not explain how to avoid bias.

Write About It **20** Assess student's writing using this rubric.

SCORING RUBRIC	SCORE 4	SCORE 3	SCORE 2	SCORE 1
Describe the input	Student correctly describes the input as typing.	Student fairly describes the input as typing.	Student incorrectly describes the input.	Student does not describe the input.
Describe the process	Student correctly describes the process.	Student fairly describes the process.	Student incorrectly describes the process.	Student does not describe the process.

CHAPTER 1

TEKS Practice, Cont.

RTI Response to Intervention

23. If students have trouble explaining why they should never bring food into a laboratory, **then** ask them what substances the food might accidentally come in contact with in a laboratory.

27. If students cannot define *controversy,* **then** have them review the highlighted term.

TEKS Practice

LESSON 5 Safety in the Science Laboratory

21. The outdoor area in which some of your scientific investigations will be done is called the

a. yard.
b. lawn.
c. park.
d. field. (circled)

22. Good preparation will help you stay safe when performing scientific investigations.

23. Make Judgments Why do you think that you should never bring food that is not part of an experiment into a laboratory?

Sample: It could become contaminated with materials in the lab, and the food might contaminate materials to be used in experiments.

24. Identify What are some examples of safety equipment that you use to help prevent injuries from happening in a lab?

Sample: Chemical splash goggles, aprons, and gloves are examples of safety equipment that help prevent injuries.

25. Apply Concepts Why is it important to read through the lab procedures before you do a lab?

Sample: It lets you make sure you understand the directions and know what equipment you might need to use.

LESSON 6 Scientists and Society

26. The scientist who is known for saying that Earth moves around the sun is

a. Rachel Carson.
b. Galileo Galilei. (circled)
c. Albert Einstein.
d. Ignaz Semmelweis.

27. A public disagreement between groups with different views is called a controversy.

28. Relate Identify a scientific controversy that you have seen reported on television, in a newspaper or magazine, or on the Internet.

Accept all reasonable answers, such as climate change.

29. Evaluate the Impact on Society How do you think Ignaz Semmelweis's research in the 1840s impacts modern medicine?

Accept all reasonable answers.

30. Write About It A company must clear the land of existing trees in order to erect a new building. Some people support the company because the expansion will provide new jobs. Other people are against it because they don't want the wildlife to be destroyed. On a separate sheet of paper, write which side of the controversy you would support and why.

See TE rubric.

Write About It 30 Assess student's writing using this rubric.

SCORING RUBRIC	SCORE 4	SCORE 3	SCORE 2	SCORE 1
Chooses a side in controversy and supports choice	Student provides strong, detailed support for chosen side in controversy.	Student provides adequate support for chosen side in controversy.	Student provides limited support for chosen side in controversy.	Student provides no support for chosen side in controversy.

How do scientists investigate the natural world?

31. Central Middle School is having problems with attendance during the winter. Many students are missing school due to illnesses. The principal wants to fix the problem, but she is not sure what to do. One idea is to install hand sanitizer dispensers in the classrooms.

Think about this problem scientifically. What is a possible hypothesis in this situation? What experiment could you design to test it? Mention at least three attitudes or skills that will be important in finding the answer.

Sample: The hypothesis could be, "Classroom hand sanitizer dispensers reduce winter absences." To test it, you could put dispensers in the sixth grade classrooms during the winter and not in the seventh grade classrooms. Then you could compare the sixth grade absences to the seventh grade absences. You would have to make careful quantitative observations and be curious and skeptical. See TE rubric.

2A, 2B, 3A

Interactive Science Chapter 1

Lesson 1
In Lesson 1.1 you learned that scientists use skills such as observing, inferring, predicting, classifying, evaluating, and using models to study the natural world.
TEKS: 2E, 3B

Lesson 2
In Lesson 1.2, you learned that scientists possess certain important attitudes, including curiosity, honesty, creativity, open-mindedness, skepticism, good ethics, and awareness of bias.
TEKS: 3A

Lesson 3
In Lesson 1.3, you learned that scientific inquiry refers to the ways in which scientists study the natural world and propose explanations based on the evidence they gather. An experiment must follow sound scientific principles for its results to be valid. You also learned the difference between a scientific theory and a scientific law.
TEKS: 2A, 2B, 2C, 3A

Lesson 4
In Lesson 1.4, you learned how models help scientists understand things they cannot observe directly. You also learned that a system is a group of parts that work together to produce a specific function or result.
TEKS: 3B, 3C

Lesson 5
In Lesson 1.5, you learned that good preparation helps you stay safe when doing science investigations. You also learned what to do when an accident occurs in the lab.
TEKS: 1A, 4B

Lesson 6
In Lesson 1.6, you learned that the work that scientists do impacts society. In turn, society influences the work of scientists.
TEKS: 3D

How do scientists investigate the natural world?
Assess student's response using this rubric.

SCORING RUBRIC	SCORE 4	SCORE 3	SCORE 2	SCORE 1
State a hypothesis and describe an experiment and at least three attitudes or skills.	Student states a hypothesis, describes an experiment, and mentions three or more attitudes or skills.	Student states a hypothesis, describes an experiment, but only mentions two attitudes or skills.	Student doesn't state a hypothesis or describes experiment, and mentions only one attitude or skill.	Student doesn't state a hypothesis, or describes an experiment, and doesn't mention any attitudes or skills.

TEKS Practice, Cont.

How do scientists investigate the natural world?

Students should be able to demonstrate understanding of how scientists think and investigate the natural world by answering this question. See the scoring rubric below.

Review the TEKS Chapter 1

REVIEW THE TEKS 2A, 2B, 3A

PARTNER REVIEW Have partners review the definitions of vocabulary terms and quiz each other. Students can read the Key Concept statements and leave out words for their partners to fill in. They also can change a statement so that it is false and then ask their partners to correct it.

CLASS ACTIVITY: FLOW CHART Help students develop one way to show how the information in this chapter is related. The way in which scientists learn about the natural world is a process involving different skills and procedures. As a class, make a poster-size or wall-size flow chart with interchangeable index cards to show the different paths of scientific inquiry. Have a class brainstorming session to identify the Key Concepts, key terms, and details. Then have volunteers write each one on index cards. Have students draw arrows on other index cards. Use the questions below to help students add to and organize the information on the cards. Then encourage students to arrange the cards and arrows several times, adding and removing cards as needed to represent variations in the process of scientific inquiry.

- What skills and attitudes do scientists use?
- What is scientific reasoning?
- How do you design and conduct an experiment?
- What are scientific theories and laws?
- How do scientists use models?
- Why is it important to follow safety procedures during an investigation?
- How do scientists and society impact one another?

CHAPTER 1

TEKS Practice: Chapter Review

Test-Taking Skills

INTERPRETING EXPERIMENTS Explain to students that some questions will ask them to interpret experiments. Remind students it is important to read the information on the procedure and the results carefully to make sure they understand what was tested in the experiment.

Question 1 TEKS 2B

The correct answer is C. The students purposely set up the samples so that they were the same for all factors except for temperature, so this is the manipulated variable.

If students chose A, explain that this factor is the same for both samples, so it is a control in the experiment.

If students chose B, explain that this factor is also the same for both samples, so it is a control in the experiment.

If students chose D, explain that this is the responding variable, because the student is watching to see whether it changes as a result of the manipulated variable.

Question 2 TEKS 2E

The correct answer is H. This is a way to observe color using numbers.

If students chose F, explain that this is a way to observe color using pictures rather than numbers.

If students chose G, explain that amount of sunlight is a quantitative observation, but it is not a direct observation of the grass color.

If students chose J, explain that this is a way to observe color using descriptions that do not involve numbers.

★ TEKS Practice: Chapter Review

Read each question and choose the best answer.

1 A student is investigating how earthworms respond to temperature. She places four earthworms in each of two clear plastic containers filled with the same amount of soil. She sets one container on a table. She sets the other container next to it in a bowl of ice water, as shown below. She observes the movement of both sets of worms for one hour.

Sample A: At room temperature

Sample B: In ice water

What is the independent, or manipulated, variable in this experiment?

A Amount of soil

B Number of earthworms

(C) Temperature of the soil

D Amount of earthworm movement

2 A student notices that the grass growing in different parts of a park is different colors. In some places the grass is dark green, while in other places it is light green. How might she make quantitative observations of the grass color?

F Use colored pencils to draw detailed pictures of both colors of grass in a science journal.

G Record whether the areas of dark green grass receive more or less sunlight.

(H) Collect samples of each color of grass and use a digital scale to measure the mass of each sample.

J Write down different words that describe shades of green and assign one to each sample of grass.

3 A student counted the number of students who rode his bus in the afternoons and recorded the data in the table below.

Number of Students on Bus Each Afternoon in September						
Sunday	**Monday**	**Tuesday**	**Wednesday**	**Thursday**	**Friday**	**Saturday**
					1	2
3	4 **Holiday**	5 **20**	6 **21**	7 **14**	8 **23**	9
10	11 **23**	12 **22**	13 **23**	14 **15**	15 **23**	16
17	18 **21**	19 **23**	20 **22**	21 **14**	22 **22**	23
24	25 **23**	26 **23**	27 **22**	28	29	30

Based on these data, which is the best prediction for the number of students who will ride the bus on September 28th?

A 5

(B) 15

C 21

D 44

4 A lab procedure instructs students to use a hot plate to heat water in a glass beaker. Then, it tells students to pick up the beaker and pour the heated water into another container. Which group of the following safety symbols likely appears in the instructions for this activity?

1  2  3  4 

F 1, 2, and 3

(G) 1, 2, and 4

H 1, 3, and 4

J All the above

If You Have Trouble With . . .				
Question	1	2	3	4
See Lesson	1.3	1.1	1.3	1.5
TEKS	7.2B	7.2E	7.2A	7.1A

TEKS Practice: Chapter Review, Cont.

Additional Assessment Resources

Teacher's Edition: Lesson Quizzes, Texas End-of-Year Test Prep A and B
Online assessments

Question 3 TEKS 2A

The correct answer is B. The data show a trend: On Thursdays, there are usually only 14 or 15 students riding the bus.

If students chose A, explain that this number is significantly lower than any previously observed number and is not a good prediction.

If students chose C, explain that the 28th is a Thursday, and this number does not match the trend for that day of the week in previously observed data.

If students chose D, explain that this number is significantly higher than any previously observed number and is not a good prediction.

Question 4 TEKS 1A

The correct answer is G. These safety icons warn of broken glass, electrical shocks, and hot objects.

If students chose F, explain that there is no danger from sharp objects from the described procedure, but the hot plate represents a danger from hot objects.

If students chose H, explain that there is no danger from sharp objects from the described procedure, but the there is potential danger from electric shock. The electric hot plate in this activity could shock someone if not used correctly.

If students chose J, explain that there is no danger from sharp objects from the described procedure, so it should not be included.

Science Matters

Focus on Texas

Have students read *Invasion!* Explain that an invasive species is a plant or animal—or even a microbe—that is not native to a particular ecosystem. Usually the organism is introduced into an area through human activity. Invasive species can have serious ecological and economic effects.

Point out that *Arundo* is native to western Asia and the Mediterranean region, and scientists believe it was brought to California in the early 1800s. The invasive plant is now found in over 25 states, including Texas.

The giant reed is a huge problem for the Lone Star state, especially along the Rio Grande. The plant develops quickly and forms thick mats where it grows. It crowds out native plants, causes flooding problems, increases fire risks, and harms or destroys habitats for native wildlife. Reinforce the idea that an invasive species can have ecological and economic benefits, but the threat posed by the invader often outweighs any of these benefits. Ask: **What are some of the beneficial uses of the giant reed?** *(The plant is used to make reeds for musical instruments and used as a fuel source.)* **What impact might this have on efforts to reduce the spread of the plant?** *(Sample: Some people and businesses might oppose efforts to reduce the spread of the plant or eradicate it altogether.)*

TEXAS SCIENCE MATTERS

Focus on Texas

Invasion!

TEKS 3D, 10A, 10B

Arundo is a plant that is commonly known as the "giant reed." It crowds out other plants and creates wildfire hazards where it grows. And it's a problem that is spreading through Texas.

Believed to be native to Asia and the Mediterranean, *Arundo* was brought to the United States in the early 1800s to help prevent erosion. The plant is also used for fuel and to make reeds for musical instruments. Unfortunately, the giant reed has invaded many places where it is not welcome.

Researchers at Texas A&M University are searching for ways to slow the plant's growth. Some biologists hope to control *Arundo* by releasing insects that feed on the plant in Europe and India. They are taking steps, however, to make sure that the insects won't create new problems. *Arundo* can be a very useful plant when controlled, and Texas scientists are working to keep it that way.

Research It With your classmates, investigate invasive plants that live nearby. The Texas Forest Service provides lists of non-native plants and maps of their locations. In a small group, choose an invasive plant to study in more detail. Prepare a multimedia report for your class that explains where the plant came from, how it got to the U.S., any ways the plant is useful to people or wildlife, any problems caused by the plant, and current methods to control the plant's spread.

▲ *Arundo* can grow to 9 meters tall.

54 Scientific Thinking

Quick Fact

The Texas Invasive Plant and Pest Council (TIPPC) was formally established in the state in 2008. The council is composed of state and federal agencies, non-profit organizations, academic institutions, and private businesses. The council's objectives include promoting awareness of invasive species in Texas; providing a forum for the exchange of information; and supporting research and conservation activities to reduce the impact of invasive species in the state.

Texas Essential Knowledge and Skills

3D Relate the impact of research on scientific thought and society, including the history of science and contributions of scientists as related to the content.

10A Observe and describe how different environments, including microhabitats in schoolyards and biomes, support different varieties of organisms.

10B Describe how biodiversity contributes to the sustainability of an ecosystem.

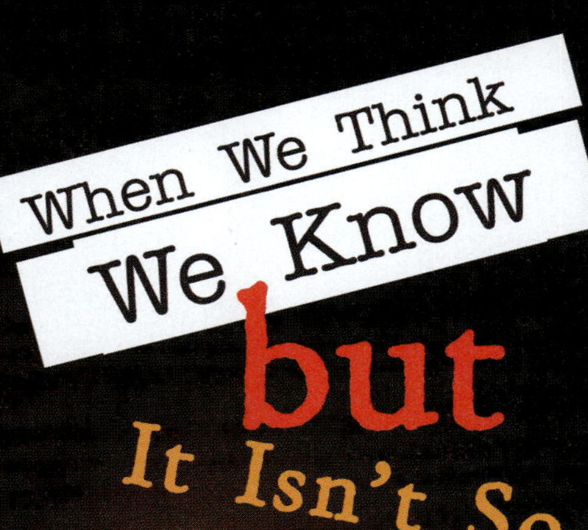

TEKS 3A, 3D

Science and History

Science is a way of thinking and learning about the world. It is not rigid or unchanging. In fact, scientists are constantly learning new things. And sometimes, they make mistakes! That's just what René Blondlot and his co-workers at Nancy University in France did in 1903.

X-rays had just been discovered. Scientists everywhere, including Blondlot, were experimenting with them. In a series of photographs taken in an experiment, Blondlot observed strange lights. He was convinced he had discovered another new form of radiation. He named his discovery the N-ray, in honor of Nancy University where he worked. Dozens of other scientists repeated his experiments expecting to see the lights. Some were convinced that they actually had seen them. But there was one very big problem—N-rays do not exist! Scientists who were skeptical did not see these lights and could not repeat the results of the experiment. It was soon discovered that when they looked for N-rays, Blondlot and his colleagues were just seeing what they expected and hoped to see.

This is a clear case of expectations influencing observations. Because these scientists did not avoid their bias, they made a big mistake.

Research It Research Robert W. Wood, the scientist who disproved the existence of N-rays. Write a paragraph summarizing how he came to this conclusion. What questions did he ask?

René Blondlot was a famous scientist who allowed his ambition to overrule his powers of observation.

Science and History

Have students read *When We Think We Know but It Isn't So.* Point out that Blondlot may have been mistaken about N-rays, but he conducted a great deal of other important scientific research. Tell students about Blondlot's work with electricity. He worked with mirrors to study the speed of electricity in a conductor. By comparing photographs of sparks made by different conductors, he learned that the speed of electricity in a conductor is close to that of light. He was also the first to measure the speed of radio waves. He discovered that the speed of radio waves is within 1% of the present value for the speed of light. This became important in confirming that light and radio waves are both forms of electromagnetic radiation.

Explain to students that Blondlot's observations are a good example of experimental bias. Tell students that experimental bias occurs when the experimenter strongly believes a certain result will occur. This can affect a scientist's observation of an experiment and cause him or her to see things that may not be true. N-rays can also be given as an example of pathological science, in which people trick themselves into believing false results based on their expectations. The term *pathological science* was first used by a chemist named Irving Langmuir, long after the N-ray incident.

As students research Wood's experiments with N-rays, have them note how Wood proved Blondlot's experimental bias.

Ask: **Why did Blondlot think he saw N-rays?** *(He wanted to discover a new kind of radiation.)* **How could you prevent experimental bias?** *(Sample: Design an experiment and have someone else perform it without telling them what you expect.)*

Texas Essential Knowledge and Skills

3A In all fields of science, analyze, evaluate, and critique scientific explanations by using empirical evidence, logical reasoning, and experimental and observational testing, including examining all sides of scientific evidence of those scientific explanations, so as to encourage critical thinking by the student.

3D Relate the impact of research on scientific thought and society, including the history of science and contributions of scientists as related to the content.

CHAPTER 2

Introduction to Cells

Chapter TEKS Overview

This chapter focuses on TEKS 12D and 12F. Students will differentiate the structures and functions in plant and animal cell organelles. They will also learn about the development of cell theory.

Introduce the TEKS

To pique student interest and introduce TEKS 12D and 12F, have students look at the image and read the Focus Question and description. Ask students to infer how they are similar to the deep-sea animal. Have volunteers explain their inferences. Point out that the cells in crustaceans and humans look and function in similar ways, even though their bodies look very different. Ask: **Have you seen other living things that look different from you?** *(Students will likely suggest other animals like dogs or birds, and may need to be reminded that plants and fungi are also alive.)* **What do all these living things have in common?** *(They all need food, they produce wastes, and they reproduce.)*

Untamed Science Video

TOURING HOOKE'S CRIB! Before viewing, invite students to discuss what they know about cells. Then play the video. Lead a class discussion and make a list of questions that this video raises. You may wish to have students view the video again after they have completed the chapter to see if their questions have been answered.

Chapter at a Glance

CHAPTER PACING: 8–13 periods or 4–6$\frac{1}{2}$ blocks

INTRODUCE THE CHAPTER: Use the Focus Question and the opening image to get students thinking about cells. Activate prior knowledge and preteach vocabulary using the Getting Started pages.

Lesson 1: The Characteristics of Life

Lesson 2: Discovering Cells

Lesson 3: Looking Inside Cells

Lesson 4: Chemical Compounds in Cells

Lesson 5: Cells and Homeostasis

ASSESSMENT OPTIONS:
Teacher's Edition: Lesson Quizzes, Texas End-of-Year Test Prep A and B
Online assessments

HOW ARE YOU LIKE THIS CREATURE?

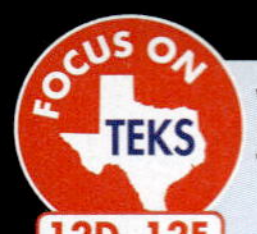

What structures are found in cells?

You sure don't see this sight when you look in the mirror! This deep-sea animal does not have skin, a mouth, or hair like yours. It's a young animal that lives in the Atlantic Ocean and may grow up to become a crab or shrimp. Yet you and this creature have more in common than you think.

Infer **What might you have in common with this young sea animal?**

Sample: We both have eyes and need to eat.

Watch the **Untamed Science** video to learn more about cells.

UntamedScience

Professional Development Note

From the Author

Visualization is one of the most difficult aspects of understanding cells, but the Internet can be very helpful to students in this visualization process. There are sites that provide cell models, both plant and animal, with all the parts highlighted with hot buttons. Other sites animate processes like mitosis and meiosis, with the capability to show the entire process in a continuous manner or to take the process in small, digestible steps. Certainly, there is a lot of "junk" on the Web. But there are also some great resources that, combined with a knowledgeable teacher, can produce a higher level of learning.

Michael Padilla

Texas CHAPTER 2

Introduction to Cells

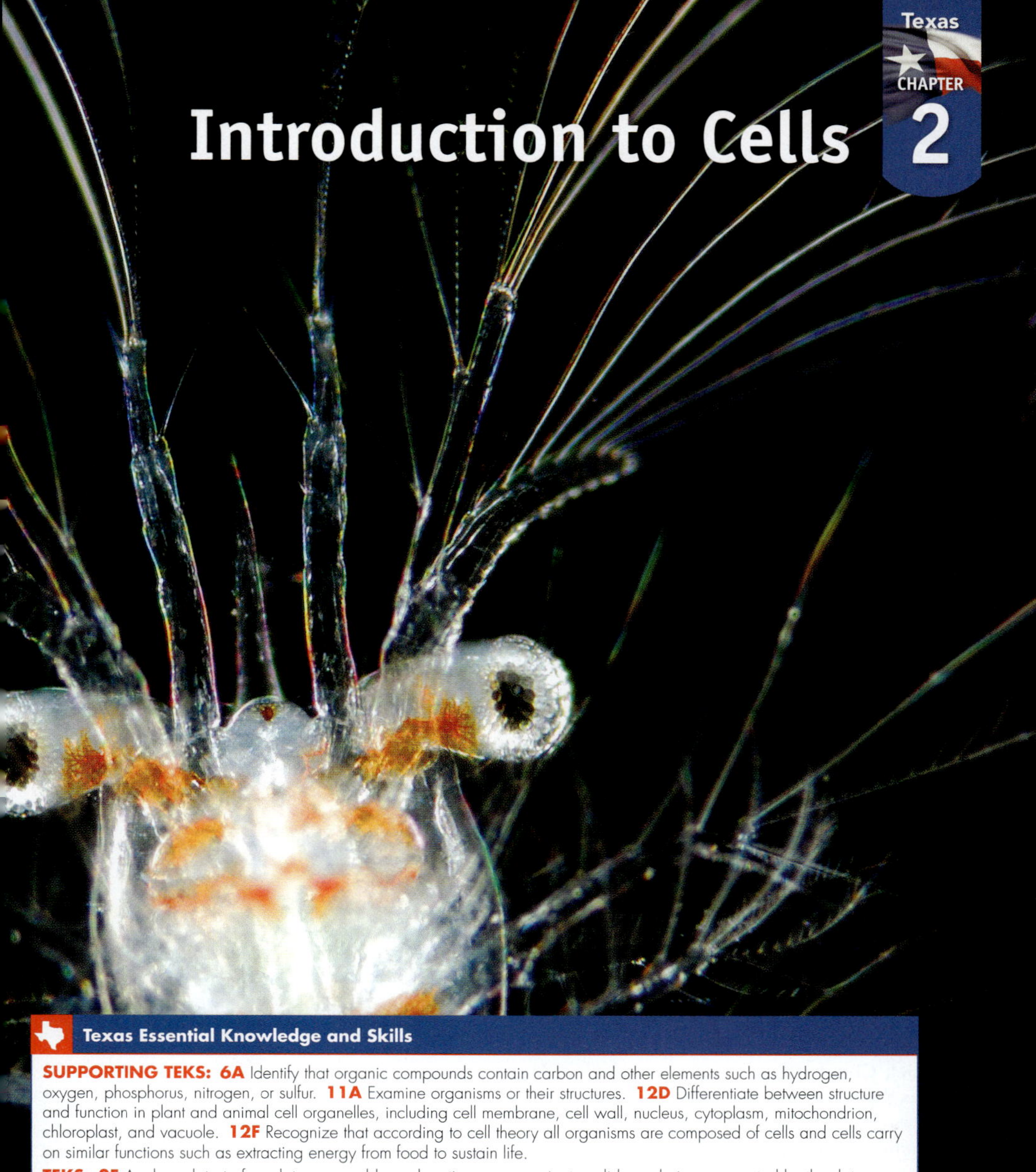

Texas Essential Knowledge and Skills

SUPPORTING TEKS: 6A Identify that organic compounds contain carbon and other elements such as hydrogen, oxygen, phosphorus, nitrogen, or sulfur. **11A** Examine organisms or their structures. **12D** Differentiate between structure and function in plant and animal cell organelles, including cell membrane, cell wall, nucleus, cytoplasm, mitochondrion, chloroplast, and vacuole. **12F** Recognize that according to cell theory all organisms are composed of cells and cells carry on similar functions such as extracting energy from food to sustain life.

TEKS: 2E Analyze data to formulate reasonable explanations, communicate valid conclusions supported by the data, and predict trends. **3B** Use models to represent aspects of the natural world such as plant and animal cells. **3D** Relate the impact of research on scientific thought and society, including the contributions of scientists as related to the content. **4A** Use appropriate tools to collect, record, and analyze information, including microscopes and metric rulers. **12C** Recognize levels of organization in plants and animals. **12E** Compare the functions of a cell to the functions of organisms such as waste removal. **13A** Investigate how organisms respond to external stimuli found in the environment.

PEARSON Texas.com

Spotlight On Technology

Virtual Lab Activity Set the tone for inquiry by having students engage in a quick online, inquiry-based experience to help them explore key questions about cells.

Flipped Video for Science Show students a Pearson Flipped Video for Science for this chapter so that they can better understand the characteristics of life, the cell theory, and the structure and function of cells.

PEARSON Texas.com

Texas Essential Knowledge and Skills

2E Analyze data to formulate reasonable explanations, communicate valid conclusions supported by the data, and predict trends.

3B Use models to represent aspects of the natural world such as human body systems and plant and animal cells.

3D Relate the impact of research on scientific thought and society, including the history of science and contributions of scientists as related to the content.

4A Use appropriate tools to collect, record, and analyze information, including life science models, hand lens, stereoscopes, microscopes, beakers, Petri dishes, microscope slides, graduated cylinders, test tubes, meter sticks, metric rulers, metric tape measures, timing devices, hot plates, balances, thermometers, calculators, water test kits, computers, temperature and pH probes, collecting nets, insect traps, globes, digital cameras, journals/notebooks, and other equipment as needed to teach the curriculum.

6A Identify that organic compounds contain carbon and other elements such as hydrogen, oxygen, phosphorus, nitrogen, or sulfur.

11A Examine organisms or their structures such as insects or leaves and use dichotomous keys for identification.

12C Recognize levels of organization in plants and animals, including cells, tissues, organs, organ systems, and organisms.

12D Differentiate between structure and function in plant and animal cell organelles, including cell membrane, cell wall, nucleus, cytoplasm, mitochondrion, chloroplast, and vacuole.

12E Compare the functions of a cell to the functions of organisms such as waste removal.

12F Recognize that according to cell theory all organisms are composed of cells and cells carry on similar functions such as extracting energy from food to sustain life.

13A Investigate how organisms respond to external stimuli found in the environment such as phototropism and fight or flight.

The following **College and Career Readiness Standards** are covered in this chapter: **I.B.1, I.D.3, I.E.1, VII.H.2**

Getting Started

Check Your Understanding

This activity assesses students' understanding of the tools used to study microscopic life. A hand lens magnifies an image beyond what the naked eye can see. The greater the magnification, the better you can see smaller things. In this example, the student could use a microscope, which would provide greater magnification than the hand lens.

Preteach Vocabulary Skills

Have students identify words on page 59 that have the same root but different prefixes. Discuss how the prefix changes the meaning of these words. *(Multicellular organisms have many cells but unicellular organisms are made of only one; large particles enter cells by endocytosis and leave them by exocytosis; lysosomes contain materials that break down molecules, while ribosomes contain ribose sugars.)*

Getting Started

Check Your Understanding

1. Background Read the paragraph below and then answer the question.

You heard that a pinch of soil can contain millions of **organisms,** and you decide to check it out. Many organisms are too small to see with just your eyes, so you bring a hand **lens.** You see a few organisms, but you think you would see more with greater **magnification.**

An **organism** is a living thing.

A **lens** is a curved piece of glass or other transparent material that is used to bend light.

Magnification is the condition of things appearing larger than they are.

- How does a hand lens help you see more objects in the soil than you can see with just your eyes?

  Sample: A hand lens magnifies an image.

Vocabulary Skill

Prefixes Some words can be divided into parts. A root is the part of the word that carries the basic meaning. A prefix is a word part that is placed in front of the root to change the word's meaning. The prefixes below will help you understand some of the vocabulary in this chapter.

Prefix	Meaning	Example
chroma-	color	chromatin, *n.* the genetic material in the nucleus of a cell, that can be colored with dyes
multi-	many	multicellular, *adj.* having many cells

2. Quick Check Circle the prefix in the boldface word below. What does the word tell you about the organisms?

- Fishes, insects, grasses, and trees are examples of **multicellular** organisms.

  They are made of many cells.

English Language Proficiency Standards

ELPS Learning Strategies 1.F

Support students as they create the personalized science glossary in Preview Vocabulary Terms on page 59.

Beginning Display and read aloud the terms before each lesson. Together, look for clues to meaning in images, labels, and captions. Give definitions and examples. Have students illustrate the terms in their glossaries.

Intermediate Before each lesson, display and read aloud the vocabulary terms. Discuss definitions and examples. Have small groups preview the text, look for clues to meaning, and add the terms to their glossaries.

Advanced Have partners preview the lesson, locate the terms, and discuss definitions. Then have them add the definitions to their glossaries.

Advanced High Have partners preview the text, locate the terms, and write definitions and sample sentences.

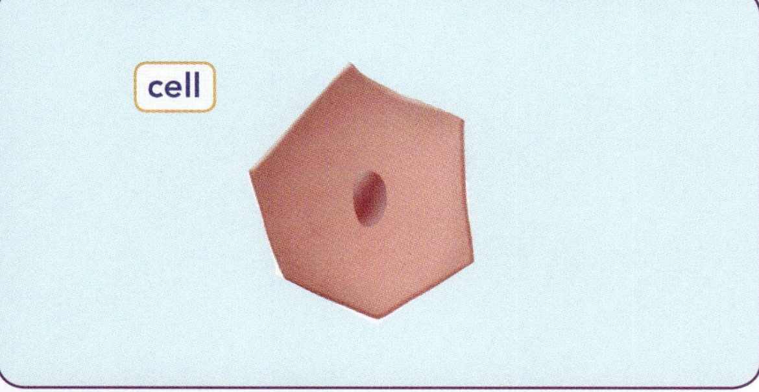

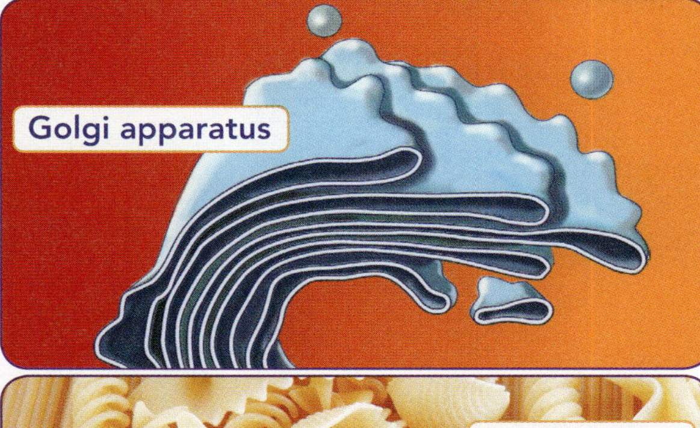

carbohydrate

Chapter Preview

LESSON 1

- organism • cell • unicellular
- multicellular • metabolism
- stimulus • response
- development
- asexual reproduction
- sexual reproduction • classification

Identify the Main Idea
Classify

LESSON 2

- microscope
- cell theory

Sequence
Measure

LESSON 3

- cell wall • nucleus
- organelle • ribosome
- cytoplasm • mitochondria
- endoplasmic reticulum
- Golgi apparatus • vacuole
- chloroplast • lysosome • tissue
- organ • organ system

Identify the Main Idea
Make Models

LESSON 4

- element • compound
- carbohydrate • lipid • protein
- enzyme • nucleic acid • DNA
- double helix

Compare and Contrast
Draw Conclusions

LESSON 5

- homeostasis
- diffusion
- active transport
- cell division

Compare and Contrast

Predict

Preview Vocabulary Terms

Have students create a personalized science glossary for the vocabulary terms in this chapter. In their glossaries, students should define each term and reference the pages in the chapter that define and explain the term. Encourage students to include drawings and diagrams that help explain the meaning of the terms and concepts.

L1 Have students look at the images on this page as you pronounce the vocabulary word. Have students repeat the word after you. Then read the definition below. Use the sample sentence in italics to clarify the meaning of the term.

organism *(AWR guhn izum)* A living thing. *A butterfly is an organism you might find near a pond.*

cell *(sel)* The basic unit of structure and function in living things. *A plant stem is built of millions of microscopic cells.*

Golgi apparatus *(GAWL jee ap uh RAT us)* A structure that packages and redistributes materials inside the cell. *The Golgi apparatus looks like a stack of flattened sacs.*

carbohydrate *(kahr boh HY drayt)* An energy-rich organic compound often used by organisms for food. *Pasta contains complex carbohydrates that your body breaks down for energy.*

Academic Vocabulary

Each lesson includes key Academic Vocabulary. See also the Support All Readers box at the start of the lesson.

Lesson 1: classify, identify, main idea

Lesson 2: measure, sequence

Lesson 3: identify, main idea, model

Lesson 4: compare, conclusion, contrast

Lesson 5: compare, contrast, predict

The Characteristics of Life

 **What structures are found in cells?**

LESSON PACING:
1–2 periods or $\frac{1}{2}$–1 block

Lesson Vocabulary

- organism • cell • unicellular • multicellular • metabolism
- stimulus • response • development • asexual reproduction
- sexual reproduction • classification

Content Refresher

Comparing Growth and Development Growth is an increase in the size of an organism or its parts as a result of ongoing mitotic division, which increases the number of cells. Development, which also results from cell division, involves changes in the structures and functions of cells. In plants and animals, development is a complex process that starts with a fertilized egg and ends with the emergence of an adult organism. The steps in development include fertilization, rapid cell division, and cell differentiation. During differentiation, groups of similar cells begin to differ in appearance from other such groups. The result is the formation of tissue and organ precursors—the first signs of an emerging plant or animal form.

Classification The system of classifying living things based on their structural similarities was developed in the eighteenth century by Swedish biologist Carl von Linné, known as Carolus Linnaeus. The system that Linnaeus devised included five hierarchical categories: kingdom, class, order, genus, and species; scientists later added the categories of phylum and family. Linnaeus recognized only plants and animals as living things. The five-kingdom scheme—which includes monera (prokaryotes), protists (single-cell organisms and some multicellular eukaryotes), fungi, plants, and animals—was developed by modern taxonomists. Scientists no longer rely on structural similarities for classification. Instead, they analyze the DNA codes of organisms, looking for similarities and differences.

Lesson Objectives	TEKS	ELPS
List the characteristics all living things share.	12E, 13A	1.B.1
Explain how living things are classified.	11A	

Texas Essential Knowledge and Skills

11A Examine organisms or their structures such as insects or leaves and use dichotomous keys for identification.
12E Compare the functions of a cell to the functions of organisms such as waste removal.
13A Investigate how organisms respond to external stimuli found in the environment such as phototropism and fight or flight.

English Language Proficiency Standards

ELPS Learning Strategies 1.B.1 Monitor oral language production and employ self-corrective techniques or other resources.

DIFFERENTIATED INSTRUCTION KEY
L1 Struggling Students or Special Needs
L2 On-Level Students L3 Advanced Students

Investigations and Activities	TEKS Review
My Planet Diary for Texas, **Student Edition,** p. 60 Inquiry: Inquiry Warm-Up, Is Yeast Alive or Not?, **PearsonTexas.com** Introduce Vocabulary, **Teacher's Edition,** p. 61 Teach Key Concepts, **Teacher's Edition,** p. 61 Teach With Visuals, **Teacher's Edition,** p. 61 Address Misconceptions, Plants Are Living Things, Too, **Teacher's Edition,** p. 61 Lead a Discussion, The Basic Unit of Living Things, **Teacher's Edition,** p. 62 Teach With Visuals, **Teacher's Edition,** p. 62 21st Century Learning, Critical Thinking, **Teacher's Edition,** p. 62 Inquiry: Teacher Demo, A Crystal Garden, **Teacher's Edition, p. 63** Differentiated Instruction, **Teacher's Edition,** p. 63 Inquiry: Quick Lab, Modeling Trace Fossils, **PearsonTexas.com**	Apply the TEKS, Comparing Elements, **Student Edition,** p. 94 TEKS Practice, **Student Edition,** p. 95 TEKS Practice: Chapter and Cumulative Review, **Student Edition,** p. 98 Lesson 2.1, **TEKS Preparation and Study Guide Workbook,** p. 16
Teach Key Concepts, **Teacher's Edition,** p. 64 Lead a Discussion, General and Specific Levels, **Teacher's Edition,** p. 64 Make Analogies, Organizing Time, **Teacher's Edition,** p. 64 Teacher to Teacher, **Teacher's Edition,** p. 64 Support the TEKS, Levels of Classification, **Teacher's Edition,** p. 65 Inquiry: Build Inquiry, Classification of Shoes, **Teacher's Edition, p. 65** Differentiated Instruction, **Teacher's Edition,** p. 65 Inquiry: Quick Lab, Classifying Animals, **Lab Manual,** p. 12	**SHORT ON TIME?** To do this lesson in approximately half the time, do the Activate Prior Knowledge activity. A discussion of the Key Concepts will familiarize students with the lesson content. Have students do the Quick Labs. The rest of the lesson can be completed by students independently.

These editable worksheets are available on **PearsonTexas.com.**
Print versions can be found in the **TEKS Preparation and Study Guide Workbook.**

Name ______ Date ______ Class ______

2.1 The Characteristics of Life

Key Concept Summaries

What Are the Characteristics of All Living Things?

All **organisms,** or living things, share six important characteristics. **All living things have a cellular organization, contain similar chemicals, use energy, respond to their surroundings, grow and develop, and reproduce.** All organisms are made up of cells. A **cell** is the basic unit of structure and function in an organism. Single-celled organisms, like bacteria, are **unicellular** organisms. Organisms composed of many cells are **multicellular.** The chemicals in cells include water, carbohydrates, proteins, and lipids. Nucleic acids are the genetic material of cells. The combination of reactions that break down and build up materials to provide a cell with energy is **metabolism.** A change in an organism's surroundings that causes the organism to react is called a **stimulus.** An organism reacts to a stimulus with an action or change in behavior called a **response. Development** is the process of change that occurs during an organism's life to produce a more complex organism. **Asexual reproduction** involves only one parent and produces offspring that are identical to the parent. **Sexual reproduction** involves two parents and combines their genetic material to produce a new organism that differs from both parents.

How Are Living Things Classified?

Taxonomists use organisms' structures and behaviors to classify them. **A domain is the broadest level of organization. Within a domain, there are kingdoms. Within kingdoms, there are phyla (FY luh; singular *phylum*). Within phyla are classes. Within classes are orders. Within orders are families. Each family contains one or more genera. Finally, each genus contains one or more species.** The more classification levels two organisms share, the more structures and characteristics they have in common and the more closely related they are.

16

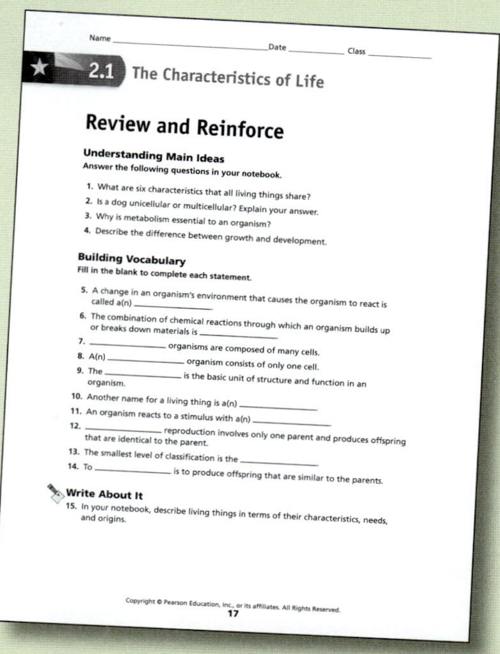

Name ______ Date ______ Class ______

2.1 The Characteristics of Life

Review and Reinforce

Understanding Main Ideas
Answer the following questions in your notebook.

1. What are six characteristics that all living things share?
2. Is a dog unicellular or multicellular? Explain your answer.
3. Why is metabolism essential to an organism?
4. Describe the difference between growth and development.

Building Vocabulary
Fill in the blank to complete each statement.

5. A change in an organism's environment that causes the organism to react is called a(n) ______.
6. The combination of chemical reactions through which an organism builds up or breaks down materials is ______.
7. ______ organisms are composed of many cells.
8. A(n) ______ organism consists of only one cell.
9. The ______ is the basic unit of structure and function in an organism.
10. Another name for a living thing is a(n) ______.
11. An organism reacts to a stimulus with a(n) ______.
12. ______ reproduction involves only one parent and produces offspring that are identical to the parent.
13. The smallest level of classification is the ______.
14. To ______ is to produce offspring that are similar to the parents.

Write About It

15. In your notebook, describe living things in terms of their characteristics, needs, and origins.

17

Lexile Measure = 950L

LESSON 2.1

The Characteristics of Life

Establish Learning Objectives

After this lesson, students will be able to:

- List the characteristics all living things share.
- Explain how living things are classified.

Engage

Activate Prior Knowledge

MY PLANET DIARY FOR TEXAS Read *Protecting Rare Species* with the class. Tell students that less than 5% of the jaguarundi's original habitat in Texas remains because much of it has been converted to grow crops and urban development. Part of the Wildlife Diversity program involves protecting existing shrub lands and restoring some of the habitat in the Rio Grande Valley by planting native trees and shrubs. Ask: **How will protecting existing shrub lands and planting native trees and shrubs in previously cleared areas affect the jaguarundi?** *(It will increase the area in which the jaguarundi can live.)*

Explore

Lab Resource: Inquiry Warm-Up

L1 IS YEAST ALIVE OR NOT? Students will form an operational definition of living thing by examining the characteristics of yeast. This Inquiry Warm-Up can be found online.

Classify Help students understand that when they classify, they group together items that are alike.

The Characteristics of Life

- What Are the Characteristics of All Living Things? TEKS 12E, 13A
- How Are Living Things Classified? TEKS 11A

MY PLANET DIARY for TEXAS

Protecting Rare Species

Tens of thousands of native plant and animal species call Texas home, some of which are found only in the state. The Wildlife Diversity program, part of the Texas Parks and Wildlife Department, is responsible for protecting the state's biological diversity by researching, restoring, and managing important habitats. For example, the jaguarundi is a small wildcat found in the southern tip of Texas. Due to its shrinking habitat, its population size has dropped and sightings of the cat have become extremely rare. One goal of the program is to identify rare species, such as the jaguarundi, and to determine ways to prevent them from disappearing from the state entirely.

SCIENCE STATS

Communicate Discuss the question with a group of classmates. Write your answer below.

Why do you think it is important to protect the biological diversity of Texas?

Sample: Many plants and animals are dependent on each other. When some of the animals or plants become rare, it can affect other plants and animals.

Do the Inquiry Warm-Up *Is Yeast Alive or Not?* Find the lab online.

The jaguarundi is a little bigger than a typical house cat.

SUPPORT ALL READERS

Lexile Measure = 950L **Lexile Word Count = 2106**

Prior Exposure to Content: Many students may have misconceptions on this topic

Academic Vocabulary: *classify, identify, main idea*

Science Vocabulary: *organism, cell, stimulus, reproduction*

Concept Level: Generally appropriate for most students in this grade

Preteach With: My Planet Diary for Texas "Protecting Rare Species" and Figure 2 activity

Vocabulary

• organism • cell • unicellular • multicellular • metabolism • stimulus • response • development • asexual reproduction • sexual reproduction • classification

Skills

Reading: Identify the Main Idea

Inquiry: Classify

What Are the Characteristics of All Living Things?

TEKS 12E, 13A In this section, you'll explore the characteristics that all living things have in common.

Quick! Think of some examples of living things, or **organisms.** You probably thought of yourself, a dog or cat, or a plant. But did you consider the furry mold on decaying fruit or the moss growing on the side of a tree? All are living things because they share important characteristics. **All living things have a cellular organization, contain similar chemicals, use energy, respond to their surroundings, grow and develop, and reproduce.**

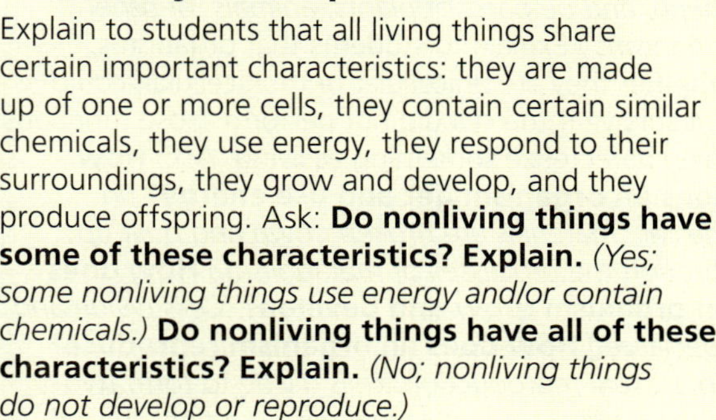

A

B

C

D

FIGURE 1

What Is Life?

Examine the photos. Then answer the questions.

1. **Classify** List the letter of the photo(s) that you think show living thing(s).
A, B, C, D

2. **Describe** What characteristics did you use to determine whether each thing shown was living or nonliving?
Accept all reasonable answers.

PEARSON Texas.com

English Language Proficiency Standards

ELPS Learning Strategies 1.B.1

Have students summarize the characteristics of living things on pages 62–63.

Beginning Have students listen as you read aloud each paragraph. Have partners complete the following sentence frame orally: *All living things* ______. Provide answer choices as needed.

Intermediate Together read aloud each paragraph. Have partners take turns summarizing the paragraph orally. Remind them to correct themselves and restate if they need to. Have them share a summary before continuing.

Advanced Have partners read the lesson and summarize each paragraph orally. Ask them to correct their answers if needed.

Advanced High Have partners read the lesson and summarize each paragraph orally. Ask them to correct their answers before writing them down.

Explain

Introduce Vocabulary

To help students understand the terms *unicellular* and *multicellular,* point to the root word *-cellular* and explain that it refers to cells. The prefix *uni-* means "one," so a unicellular organism is made up of just one cell. The prefix *multi-* means "many," so a multicellular organism is made up of many cells.

Teach Key Concepts

Explain to students that all living things share certain important characteristics: they are made up of one or more cells, they contain certain similar chemicals, they use energy, they respond to their surroundings, they grow and develop, and they produce offspring. Ask: **Do nonliving things have some of these characteristics? Explain.** *(Yes; some nonliving things use energy and/or contain chemicals.)* **Do nonliving things have all of these characteristics? Explain.** *(No; nonliving things do not develop or reproduce.)*

Teach With Visuals

Tell students to look at **Figure 1.** Ask: **Which of these images can you identify?** *(It is likely that the only image familiar to students is the butterfly.)* Tell students about the other images: (A) soap nuts from a soapberry tree, *Sapindus laurifolius;* (B) a Monarch butterfly, *Danaus plexippus;* (C) slime mold, *Diachea leucopodia;* and (D) rock lichens, orange: *Caloplaca sp.,* green: *Lecanova sp.,* and yellow: *Candelariella sp.*

Address Misconceptions

L1 PLANTS ARE LIVING THINGS, TOO Some students may not realize that plants are organisms, just as animals are. This misconception may arise from the fact that students do not understand how plants respond to their environment, grow and develop, and reproduce. Also, plant movement is not like animal movement and thus not always obvious. Ask: **What characteristics make plants living things?** *(Plants are made of cells, contain chemicals, use energy, respond to their surroundings, grow and develop, and reproduce.)*

PEARSON Texas.com

Texas Essential Knowledge and Skills

12E Compare the functions of a cell to the functions of organisms such as waste removal.

13A Investigate how organisms respond to external stimuli found in the environment such as phototropism and fight or flight.

LESSON 2.1

Explain

Lead a Discussion

THE BASIC UNIT OF LIVING THINGS Students are probably familiar with the concept of the cell. Have a discussion about what students know about cells. Ask: **How would you define a cell?** *(Samples: A cell is the basic unit of structure and function of all living things. A cell is the building block of life.)* **Are all cells alike? Explain.** *(Sample: No; cells can be large or small, have a well-defined shape or no shape, and be parts of plants, animals, or other organisms.)* Explain to students that organisms, whether they are unicellular or multicellular, could not survive if the cell did not perform those functions that characterize something as living. Ask: **How does an organism get and use energy?** *(In the cell, materials are broken down and built up through the processes of metabolism.)* **How does an organism grow and develop?** *(Cells divide and specialize.)* **How does an organism reproduce?** *(In asexual reproduction, cells divide to form two new identical cells. In sexual reproduction, special cells from two parents combine to produce a new organism different from both parents.)*

Teach With Visuals

Tell students to look at **Figure 2** and read the information about reproduction. Tell students that some plants reproduce asexually, such as tubers and bulbs. Ask: **How do most plants reproduce?** *(By sexual reproduction)* **What is necessary for sexual reproduction?** *(Two parents)* **What ways have you seen plants reproduce asexually, or with only one parent?** *(Students may be familiar with various forms of vegetative propagation, including rhizomes, bulbs, corms, tubers, and root propagation.)*

21st Century Learning

CRITICAL THINKING Have students reread the descriptions of the six characteristics of living things in **Figure 2.** Tell them that one characteristic is not necessary for the survival of an organism. Have them conclude which characteristic that is. *(Reproduction)*

Cellular Organization

All organisms are made of small building blocks called cells. A **cell,** like the one shown here, is the basic unit of structure and function in an organism. Organisms may be composed of only one cell or of many cells. For example, bacteria (bak TIHR ee uh) contain one cell and are called single-celled organisms or **unicellular** organisms. The single cell is responsible for carrying out all of the functions necessary to stay alive. Organisms that are composed of many cells are **multicellular.** For example, you are made of trillions of cells. In many multicellular organisms, the cells are specialized to do certain tasks. Specialized cells in your body, such as muscle and nerve cells, work together to keep you alive. Nerve cells carry messages to your muscle cells, making your body move.

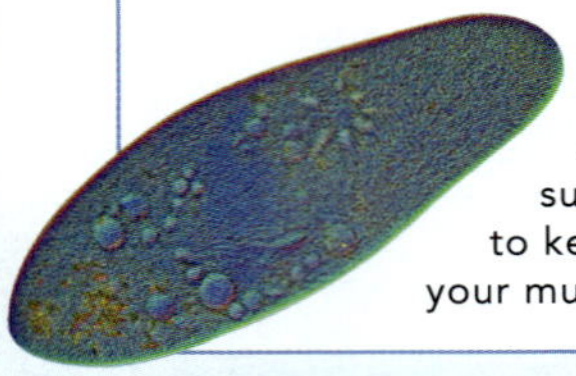

Identify the Main Idea Underline the sentence in the text that states the main idea of "Cellular Organization."

Characteristics of Living Things

The Chemicals of Life

The cells of living things are made of chemicals. The most abundant chemical in cells is water. Other chemicals, called carbohydrates (kahr boh HY drayts) are a cell's main energy source. Two other chemicals, proteins and lipids, are the building materials of cells, much as wood and bricks are the building materials of houses. Finally, nucleic (noo KLEE ik) acids are the genetic material of cells—the chemical instructions that cells need to carry out the functions of life.

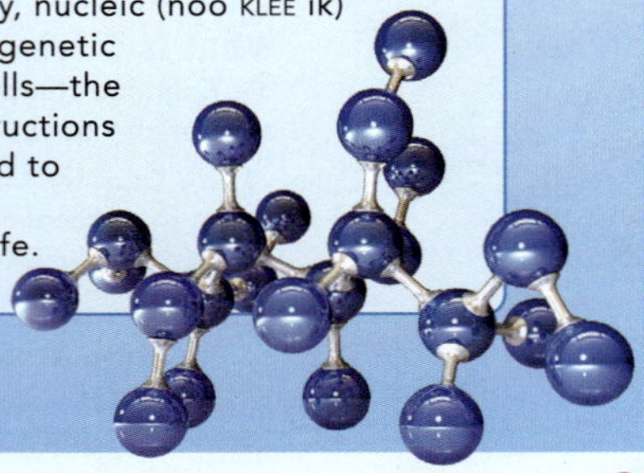

Energy Use

Organisms get energy from taking in and breaking down materials. The combination of chemical reactions through which an organism builds up or breaks down materials is called **metabolism.** The cells of organisms use energy to do what living things must do, such as grow and repair injured parts. An organism's cells are always hard at work. For example, as you read these words, not only are your eye and brain cells busy, but most of your other cells are working, too. Painted buntings, like the one shown above, need lots of energy to fly. An organism's metabolism can vary through its life span. Diet, exercise, hormones, and aging all affect metabolism.

FIGURE 2

Living Things

All living things share the same characteristics.

Make Judgments Which characteristic on these two pages do you think best identifies an object as a living thing? Explain your choice.

Sample: Reproduction; some nonliving things are also made of chemicals, use energy, or respond to their surroundings, but they cannot reproduce.

Response to Surroundings

If you've ever seen a plant in a sunny window, you may have observed that the plant's stems have bent so that the leaves face the sun. Like a plant bending toward the light, all organisms react to changes in their environment. A change in an organism's surroundings that causes the organism to react is called a **stimulus** (plural *stimuli*). Stimuli include changes in light, sound, and other factors.

An organism reacts to a stimulus with a **response**—an action or a change in behavior. For example, has someone ever knocked over a glass of water by accident during dinner, causing you to jump? The sudden spilling of water was the stimulus that caused your startled response.

Growth and Development

All living things grow and develop. Growth is the process of becoming larger. **Development** is the process of change that occurs during an organism's life. It produces a more complex organism, such as a tadpole developing into a frog. As they develop and grow, organisms use energy and make new cells.

Reproduction

Another characteristic of organisms is the ability to reproduce, or produce offspring that are similar to the parents. Organisms reproduce in different ways. **Asexual reproduction** involves only one parent and produces offspring that are identical to the parent. **Sexual reproduction** involves two parents and combines their genetic material to produce a new organism that differs from both parents. Mammals, birds, and most plants sexually reproduce. Dogs give birth to puppies that resemble their parents.

Lab zone® Do the Quick Lab *Modeling Trace Fossils*. Find the lab online.

Assess Your Understanding

TEKS 13A

1a. Review A change in an organism's surroundings is a (stimulus/response).

b. Infer A squirrel sitting in the grass runs away as you walk by. Which of the life characteristics explains the squirrel's behavior?

Response to surroundings

c. CHALLENGE A red oak doesn't move like a bat does, but both are living things. Why?

Sample: They have all of the characteristics of life. Although many living things move, movement is not a characteristic of life.

got it?

O I get it! Now I know that all living things *have a cellular organization, contain similar chemicals, use energy, respond to their surroundings, grow and develop, and reproduce.*

O I need extra help with *See TE note.*

63

Differentiated Instruction

L1 Demonstrating a Tropism Have students reread the material in **Figure 2** about the response of a plant to sunlight. Tell them that this response is called a *phototropism*. Have students work in groups to design an experiment to demonstrate phototropism. Encourage them to include a hypothesis and follow the scientific method.

L3 All About Tropisms The responses of plants to stimuli in their surroundings are called *tropisms*. Have students find out more about tropisms, including the different types and the chemical processes involved. Students might want to summarize their findings in a poster.

Elaborate

Teacher Demo 

L2 A CRYSTAL GARDEN

Materials crystal garden kit

Time 3 days

Review with students the difference between living and nonliving things. Construct a class crystal garden and have students observe crystal growth each day. (If you do cannot find a kit, instructions for growing a crystal garden are available on the Internet) Have them record their observations and list the six characteristics of living things. At the end of the three days, have students determine whether the crystals have each characteristic. Have students write a short paragraph to support their conclusions.

Ask: **Are the crystals alive?** *(No)* **What characteristics of life are the crystals missing?** *(Cellular organization, reproduction)* **What characteristics of life do the crystals appear to have?** *(Chemical composition, energy use, response to surroundings, growth and development)* **Can you list other objects that have some but not all of the characteristics of living things?** *(Samples: Fire, snowflake)*

Lab Resource: Quick Lab 

L1 MODELING TRACE FOSSILS Students will examine trace fossils and make inferences about the organisms that made them. This Quick Lab can be found online.

Evaluate

Assess Your Understanding

After students answer the questions, have them evaluate their understanding by completing the appropriate sentence.

RTI Response to Intervention

1a., b. If students cannot distinguish between a stimulus and a response, **then** have them reread the definitions of these terms and the example provided.

c. If students have trouble applying the characteristics of living things to a tree, **then** have them review **Figure 2.**

LESSON 2.1

Explain

Teach Key Concepts

Explain to students that the modern system of classification is hierarchical, meaning it consists of levels. There are eight levels, starting with domain, which is the largest and broadest, and ending with species, which is the smallest and most specific. Two organisms that share the most classification levels have more characteristics in common and are more closely related. Ask: **What are domains made up of?** *(Kingdoms)* **What are kingdoms made up of?** *(Phyla)* **What are the levels after phylum?** *(Class, order, family, genus, species)* Tell students that it will help them to remember the correct classification sequence from kingdom to species (largest to smallest) if they come up with a mnemonic device, or a sentence in which the first letter of each word represents a level, or taxon. Provide the following example: *Danish kings play cards on fat gold stools.* Have students create their own mnemonic devices.

Lead a Discussion

GENERAL AND SPECIFIC LEVELS Draw a series of eight concentric circles on the board. Label the outermost circle "Domain." Label the innermost circle "Species." Ask: **Working from the outside in, how should the other circles be labeled?** *(Kingdom, phylum, class, order, family, genus)* **Which is the most general level?** *(Domain)* **Which is the most specific level?** *(Species)*

Make Analogies

L1 ORGANIZING TIME Help students understand the levels of classification by inviting them to think about the organization of time using calendars and clocks. Ask: **How is the organization of time using calendars and clocks similar to the levels of classification of living things?** *(Larger units of time can be grouped into smaller and smaller units: years, months, weeks, days, hours, minutes, seconds.)*

21st Century Learning

CRITICAL THINKING Ask students to research information about the two extant subfamilies of *Felidae (Felinae* and *Pantherinae)* including the genera that make up these subfamilies. They can use this information to construct a family tree for *Felidae.* Encourage students to include examples of each genus in their diagrams.

Texas Essential Knowledge and Skills

11A Examine organisms or their structures such as insects or leaves and use dichotomous keys for identification.

TEKS 11A In this section, you'll study the major levels of classification that scientists use to organize living things and to show how closely they are related.

How Are Living Things Classified?

Taxonomists consider a great deal of information when classifying organisms, such as their structures and behaviors. The classification system that scientists use today is based on a series of many levels. **Classification** is the process of grouping things based on their similarities. To better understand the levels of classification, think about all the people in the United States. Then think about the people in Texas, and then those who live in your town. Finally, think about the people who live in your home. Each level down, the more you have in common with the other people.

The Major Levels of Classification Most biologists use the levels shown in **Figure 3** to classify organisms. An organism is placed into the broadest level possible, which is then divided into more and more specific levels.

A domain is the broadest level of organization. Within a domain, there are kingdoms. Within kingdoms, there are phyla (FY luh; singular *phylum*). Within phyla are classes. Within classes are orders. Within orders are families. Each family contains one or more genera. Finally, each genus contains one or more species. The more classification levels two organisms share, the more structures and characteristics they have in common and the more closely related they are.

FIGURE 3

Levels of Classification

The figure on the facing page shows how the levels of organization apply to *Sylvilagus aquaticus*, commonly known as a swamp rabbit.

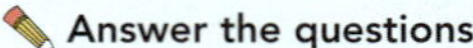

Answer the questions.

1. **Observe** List the characteristics that the organisms share at the kingdom level.
   Multicellullar, reproduce
2. **Observe** List the characteristics that the organisms share at the class level.
   Multicellular, fur, four legs
3. **Observe** List the characteristics that the organisms share at the genus level.
   Body shape, color of fur, long ears, large hind legs, cottony tails
4. **Draw Conclusions** How does the number of shared characteristics on your list change at each level? The closer to the species level, the longer the list of shared characteristics
5. **Interpret Diagrams** Wolves have more in common with (bees/rabbits). [rabbits circled]

Professional Development Note

Teacher to Teacher

Classification One of the most useful tools scientists have developed is the tool of classification. Classification, or categorization, is a way of grouping things by their similarities and differences. The separation of items into categories is based on specific characteristics. In 1735 Carl Linnaeus proposed a system for classifying living organisms. A modified form of the system is used today. This system divides organisms according to basic characteristics. For example, organisms that have a nucleus in their cells (eukaryotic) are in one category and organisms that do not have a nucleus is in their cells (prokaryotic) are in a different category. Each category tells us a number of things about all of the organisms that are placed in that category.

Joel Palmer, Ed.D.
Mesquite ISD
Mesquite, Texas

Levels of Classification

Domain Eukarya

Kingdom Animalia

Phylum Chordata

Class Mammalia

Order Lagomorpha

Family Leporidae

Genus *Sylvilagus*

Species *Sylvilagus aquaticus*

As you move down these levels of classification, the number of organisms decreases. The organisms that remain share more characteristics with one another and are more related.

Do the Quick Lab *Classifying Animals.* Student Lab Manual, p. 12

Assess Your Understanding

got it?

○ **I get it!** Now I know that the levels of classification are domain, kingdom, phylum, class, order, family, genus, species.

○ I need extra help with See TE note.

65

Differentiated Instruction

L1 The Largest and the Smallest Ask students to explain which classification level will always have the greatest number of different kinds of organisms and which level will always have the fewest number of different kinds of organisms. A correct explanation will indicate if students understand the concept of classification.

L3 Kitchen Classification Make sure students understand the concept of levels of classification by having them do the following activity. With a family member, students should go on a "classification hunt" in the kitchen. Tell students to look in the refrigerator, cabinets, and drawers to discover what classification systems their family uses to organize items. Remind students to identify the criteria used to classify the objects.

Support the TEKS

LEVELS OF CLASSIFICATION Direct students' attention to **Figure 3.** Have them look at the organisms shown in the top row and notice ways they are similar and ways they are different. Tell students that the five organisms farthest to the right are a tree, an amoeba, algae, a leech, and a bee. Ask: **Which organisms disappear at the second level? Why?** *(The tree, amoeba, and algae disappear because the second level is the animal kingdom and the tree, amoeba, and algae are not animals.)* **Which organisms disappear at the phylum level? Why?** *(The leech and the bee disappear because they are not chordates.)*

Elaborate

Build Inquiry 

L2 CLASSIFICATION OF SHOES

Materials paper, colored pencils or markers

Time 15 minutes

Have students work in pairs or groups of four. Tell students to carefully study **Figure 3.** Then tell them to develop a similar classification system for shoes. Encourage them to include as many levels as will be helpful, with a minimum of five. Remind students that their systems must be useful, easily understood by others, and easy to use.

Ask: **What is the purpose of your system?** *(Sample: To identify a specific type of shoe by its characteristics)* **Which level will have the most shoes?** *(Domain)* **Which will have the fewest shoes?** *(Species)*

Lab Resource: Quick Lab 

L1 CLASSIFYING ANIMALS Students will explore a leveled classification system. This Quick Lab can be found in the Student Lab Manual, p. 12, and online.

Evaluate

Assess Your Understanding

Have students evaluate their understanding by completing the appropriate sentence.

RTI Response to Intervention

If students have difficulty listing the levels of classification, **then** have them review **Figure 3** and identify the eight levels.

LESSON 2.1

Name ______________________ Date ____________ Class ____________

Assess Your Understanding

The Characteristics of Life

What Are the Characteristics of All Living Things?

1a. REVIEW A change in an organism's surroundings is a (stimulus/response).

b. INFER A squirrel sitting in the grass runs away as you walk by. Which of the life characteristics explains the squirrel's behavior? ______________________

____

c. CHALLENGE A red oak doesn't move like a bat does, but both are living things. Why?

____

____

____

____

got it?

○ **I get it!** Now I know that all living things ______________________

____

____

○ **I need extra help with** ______________________

____

How Are Living Things Classified?

got it?

○ **I get it!** Now I know that the levels of classification are ______________________

____

○ **I need extra help with** ______________________

____

Place the outside corner, the corner away from the dotted line, in the corner of your copy machine to copy onto letter-size paper.

Name ______________________ Date ____________ Class ____________

Enrich

The Characteristics of Life

Read the passage and study the diagrams below. Then use a separate sheet of paper to write a brief description of the conditions of each dish and explain why the bacteria appeared as they did.

Bacteria Counts

Many scientists grow bacteria in a laboratory for research using a specific process. A certain amount of bacteria are placed into a solution of water and nutrients the bacteria need as an energy source. After a period of a few days, a small amount of this solution is poured into a petri dish, a shallow round container with a cover, made from transparent plastic. Inside the petri dish is a layer of *agar*, usually a gelatinlike material that also contains nutrients on which bacteria can grow. The petri dishes are then put in a warm moist place for a week. At the end of a week, a scientist can place each dish under a large magnifying glass that has a grid drawn on it. She then counts the number of spots of bacteria growing on the agar in each of the squares. Each spot is actually a colony of bacteria. The scientist knows approximately how many bacteria are in a colony, so once she knows the number of colonies growing on a petri dish, she can calculate the number of bacteria present.

The drawings below show petri dishes growing bacteria and are labeled with the conditions the bacteria were grown in.

1.
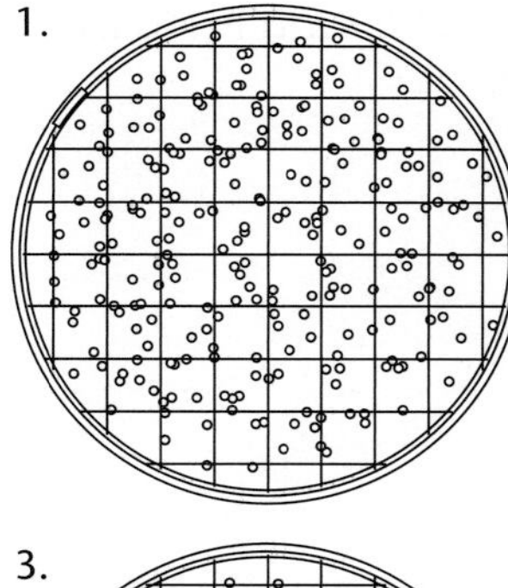
Growing conditions:
Fresh agar
37°C
moist environment
checked after 1 week

2.
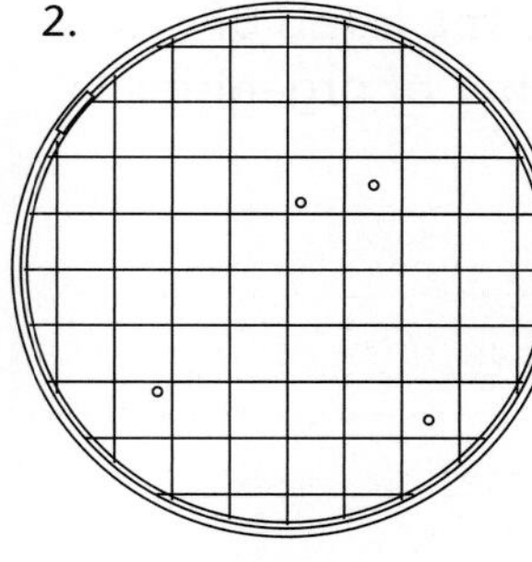
Growing conditions:
Fresh agar
0°C
moist environment
checked after 1 week

3.
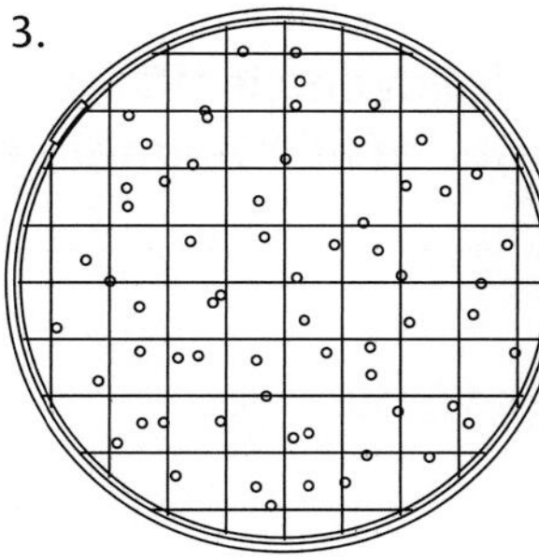
Growing conditions:
Old agar, past expiration date
37°C
moist environment
checked after 1 week

4.
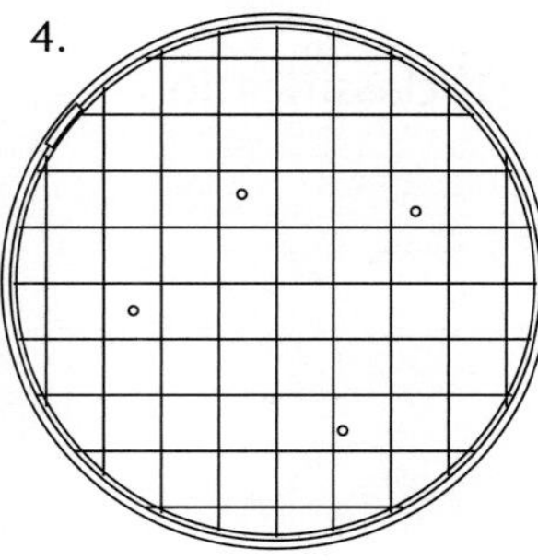
Growing conditions:
Fresh agar
37°C
moist environment
checked after 1 month

Name ______________________ Date ____________ Class ____________

Lesson Quiz

The Characteristics of Life

If the statement is true, write *true*. If the statement is false, change the underlined word or words to make the statement true.

1. ______________ Sexual reproduction involves only one parent.

2. ______________ The process of change during an organism's life that results in a more complex organism is called growth.

3. ______________ Bacteria, the most numerous organisms on Earth, are multicellular organisms.

4. ______________ Another name for living things is organisms.

5. ______________ Organisms break down and build up materials in a process called homeostasis.

6. ______________ In sexual reproduction, offspring are identical to the parent.

Write the letter of the correct answer on the line at the left.

7. ___ As you move down the levels of classification, the number of organisms in the level
   A increases
   B decreases
   C stays the same
   D increases, then decreases

8. ___ Which of the following is NOT a chemical of life?
   A mineral
   B water
   C nucleic acid
   D carbohydrate

9. ___ The broadest level of classification is the
   A kingdom
   B phylum
   C class
   D domain

10. ___ Which of the following is NOT a shared characteristic of all living things?
    A have wings
    B use energy
    C respond to their surroundings
    D reproduce

The Characteristics of Life

Answer Key

Review and Reinforce

Find the worksheet in the Student Workbook.

1. All living things have a cellular organization, contain similar chemicals, use energy, respond to their surroundings, grow and develop, and reproduce.
2. A dog is multicellular because it contains many cells that are specialized to do different tasks.
3. Metabolism creates usable energy for an organism to do things such as grow and repair injured parts.
4. Growth is the process of becoming larger. Development is the process of becoming more complex.
5. stimulus
6. metabolism
7. Multicellular
8. unicellular
9. cell
10. organism
11. response
12. Asexual
13. species
14. reproduce
15. All living things have a cellular organization, contain similar chemicals, use energy, respond to their surroundings, grow and develop, and reproduce. Living things arise from other living things through reproduction. Organisms metabolize the substances they take in and create substances such as water, carbohydrates, proteins, lipids and nucleic acids.

Enrich

1. The bacteria showed the best growth. They were provided with all of the nutrients necessary for growth and reproduction.
2. Bacterial growth was possibly poor due to the freezing temperature.
3. Only some colonies grew. Because the agar was old, it probably did not have as many nutrients needed for the growing bacteria.
4. Since one month passed, the scientists cannot be sure what happened. Perhaps the dish became too crowded and there were not enough nutrients for the colony of bacteria.

Lesson Quiz

1. Asexual	**2.** development
3. unicellular	**4.** true
5. metabolism	**6.** different from
7. B	**8.** A
9. D	**10.** A

Discovering Cells

 **What structures are found in cells?**

LESSON PACING:
2–3 periods or 1–1½ blocks

Lesson Vocabulary

- microscope
- cell theory

Content Refresher

Microscopes Most cells are far too small to be seen with the naked eye. So, even though every living thing is made of cells, the existence of cells was unknown until the invention of instruments that allowed people to see them. Early microscopes had relatively poor resolution because impurities in the glass used to make the lenses tended to split light into its constituent colors, thereby distorting the image. Modern microscopes use lenses made from clear glass, so resolving power is limited instead by the properties of the source of illumination.

To see the space between two separate, adjacent objects, light waves must pass between the objects without being scattered. If scattering occurs, the objects are visible but appear to be a single fuzzy object. Because of this physical limitation, the resolving power of light microscopes is no greater than half the wavelength of visible light—a useful magnification of about 1000×. Images can be made larger than this by photographic enlargement, but they will not reveal any more detail.

This limitation means that many biological structures cannot be resolved through a light microscope. Electron microscopes use a beam of electrons to illuminate samples; the electron beam has a much smaller wavelength than visible light, so its resolution power is much greater. Electron microscopes have a useful magnification of around 250,000×.

Lesson Objectives	TEKS	ELPS
Recognize and tell what cells are.	12E	1.C
Describe how scientists first observed cells and developed the cell theory.	3D, 12E, 12F	
Describe how microscopes produce magnified images.	4A	

Texas Essential Knowledge and Skills

3D Relate the impact of research on scientific thought and society, including the history of science and contributions of scientists as related to the content.
4A Use appropriate tools to collect, record, and analyze information, including life science models, hand lenses, stereoscopes, microscopes, beakers, Petri dishes, microscope slides, graduated cylinders, test tubes, meter sticks, metric rulers, metric tape measures, timing devices, hot plates, balances, thermometers, calculators, water test kits, computers, temperature and pH probes, collecting nets, insect traps, globes, digital cameras, journals/notebooks, and other equipment as needed to teach the curriculum.
12E Compare the functions of a cell to the functions of organisms such as waste removal.
12F Recognize that according to cell theory all organisms are composed of cells and cells carry on similar functions such as extracting energy from food to sustain life.

English Language Proficiency Standards

ELPS Learning Strategies 1.C Use strategic learning techniques such as concept mapping, drawing, memorizing, comparing, contrasting, and reviewing to acquire basic and grade-level vocabulary.

DIFFERENTIATED INSTRUCTION KEY
L1 Struggling Students or Special Needs
L2 On-Level Students L3 Advanced Students

LESSON PLANNER 2.2

Investigations and Activities	TEKS Review
My Planet Diary, **Student Edition,** p. 66 Inquiry: Inquiry Warm-Up, What Can You See?, **Lab Manual,** p. 13 Introduce Vocabulary, **Teacher's Edition,** p. 67 Teach Key Concepts, **Teacher's Edition,** p. 67 Inquiry: Quick Lab, Comparing Cells, **Lab Manual,** p. 14	Apply the TEKS, Comparing Elements, **Student Edition,** p. 94 TEKS Practice, **Student Edition,** p. 95
Teach Key Concepts, **Teacher's Edition,** p. 68 Support the TEKS, Hooke's Observations, **Teacher's Edition,** p. 68 Lead a Discussion, History of the Cell Theory, **Teacher's Edition,** p. 68 21st Century Learning, Creativity, **Teacher's Edition,** p. 69 Inquiry: Quick Lab, Observing Cells, **Lab Manual,** p. 15	TEKS Practice: Chapter and Cumulative Review, **Student Edition,** p. 98 Lesson 2.2, **TEKS Preparation and Study Guide Workbook,** p. 18
Teach Key Concepts, **Teacher's Edition,** p. 70 Lead a Discussion, Magnification, **Teacher's Edition,** p. 70 Teach With Visuals, **Teacher's Edition,** p. 71 Inquiry: Build Inquiry, Lenses and Magnification, **Teacher's Edition,** p. 71 Apply It!, **Student Edition,** p. 72 Inquiry: Teacher Demo, Resolution of Electron Microscopes, **Teacher's Edition,** p. 73 Differentiated Instruction, **Teacher's Edition,** p. 73 Inquiry: *Lab Investigation, Design and Build a Microscope, **Lab Manual,** p. 16	**SHORT ON TIME?** To do this lesson in approximately half the time, do the Activate Prior Knowledge activity. A discussion of the Key Concepts will familiarize students with the lesson content. Have students do the Quick Labs. The rest of the lesson can be completed by students independently.

These editable worksheets are available on **PearsonTexas.com.**
Print versions can be found in the **TEKS Preparation and Study Guide Workbook.**

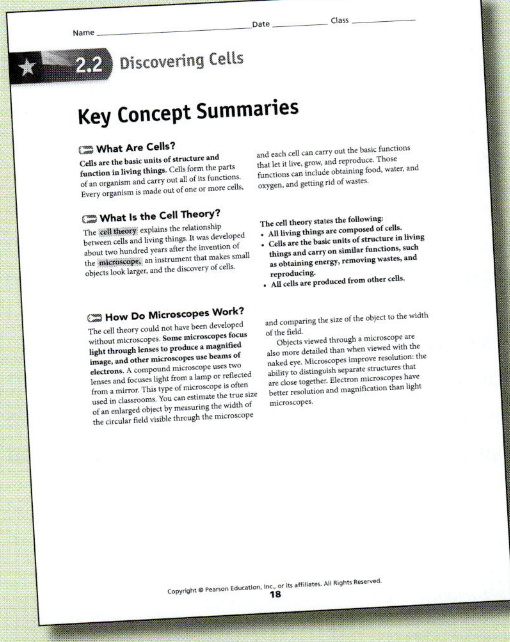
Name ____________ Date ______ Class ______

2.2 Discovering Cells

Key Concept Summaries

What Are Cells?

Cells are the basic units of structure and function in living things. Cells form the parts of an organism and carry out all of its functions. Every organism is made out of one or more cells, and each cell can carry out the basic functions that let it live, grow, and reproduce. Those functions can include obtaining food, water, and oxygen, and getting rid of wastes.

What Is the Cell Theory?

The **cell theory** explains the relationship between cells and living things. It was developed about two hundred years after the invention of the **microscope,** an instrument that makes small objects look larger, and the discovery of cells.

The cell theory states the following:
- **All living things are composed of cells.**
- **Cells are the basic units of structure in living things and carry on similar functions, such as obtaining energy, removing wastes, and reproducing.**
- **All cells are produced from other cells.**

How Do Microscopes Work?

The cell theory could not have been developed without microscopes. **Some microscopes focus light through lenses to produce a magnified image, and other microscopes use beams of electrons.** A compound microscope uses two lenses and focuses light from a lamp or reflected from a mirror. This type of microscope is often used in classrooms. You can estimate the true size of an enlarged object by measuring the width of the circular field visible through the microscope and comparing the size of the object to the width of the field.

Objects viewed through a microscope are also more detailed than when viewed with the naked eye. Microscopes improve resolution: the ability to distinguish separate structures that are close together. Electron microscopes have better resolution and magnification than light microscopes.

18

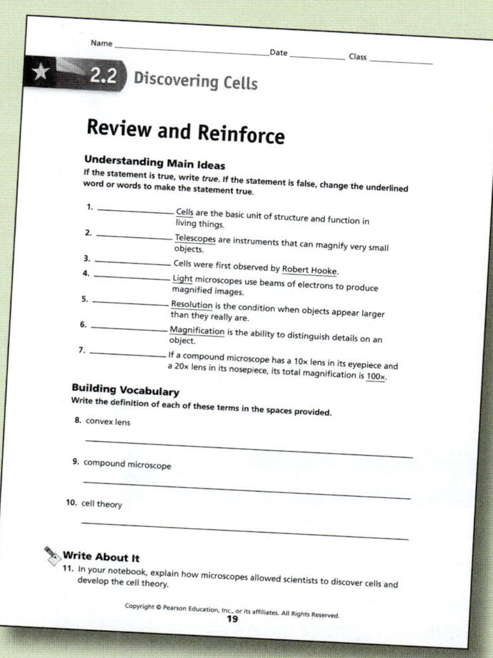
Name ____________ Date ______ Class ______

2.2 Discovering Cells

Review and Reinforce

Understanding Main Ideas

If the statement is true, write *true*. If the statement is false, change the underlined word or words to make the statement true.

1. ________ Cells are the basic unit of structure and function in living things.
2. ________ Telescopes are instruments that can magnify very small objects.
3. ________ Cells were first observed by Robert Hooke.
4. ________ Light microscopes use beams of electrons to produce magnified images.
5. ________ Resolution is the condition when objects appear larger than they really are.
6. ________ Magnification is the ability to distinguish details on an object.
7. ________ If a compound microscope has a 10× lens in its eyepiece and a 20× lens in its nosepiece, its total magnification is 100×.

Building Vocabulary

Write the definition of each of these terms in the spaces provided.

8. convex lens
9. compound microscope
10. cell theory

Write About It

11. In your notebook, explain how microscopes allowed scientists to discover cells and develop the cell theory.

19

LESSON 2.2

Lexile Measure = 940L

Discovering Cells

Establish Learning Objectives

After this lesson, students will be able to:

- Recognize and tell what cells are.
- Describe how scientists first observed cells and developed the cell theory.
- Describe how microscopes produce magnified images.

Engage

Activate Prior Knowledge

MY PLANET DIARY Read *Life at First Sight* with the class. Ask students if they have ever looked at objects through a magnifying glass or a microscope. Discuss how these instruments make objects appear larger. Ask: **Why is a microscope needed in order to study cells?** *(Cells are very small and cannot be seen with the unaided eye.)*

Explore

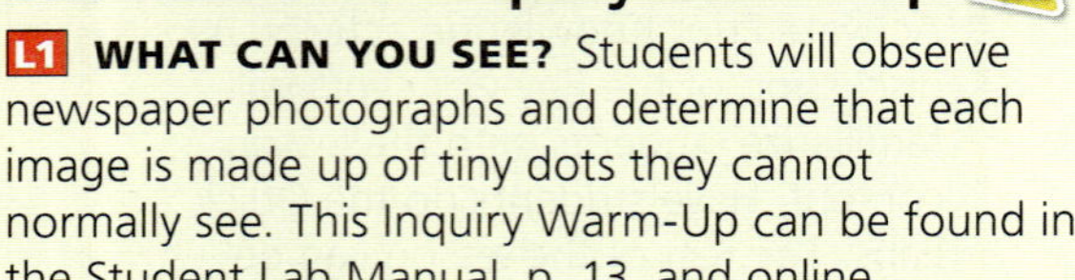

Lab Resource: Inquiry Warm-Up

L1 WHAT CAN YOU SEE? Students will observe newspaper photographs and determine that each image is made up of tiny dots they cannot normally see. This Inquiry Warm-Up can be found in the Student Lab Manual, p. 13, and online.

Texas Essential Knowledge and Skills

12E Compare the functions of a cell to the functions of organisms such as waste removal.

Discovering Cells

- What Are Cells?
  TEKS 12E
- What Is the Cell Theory?
  TEKS 3D, 12E, 12F
- How Do Microscopes Work?
  TEKS 4A

MY PLANET DIARY — VOICES FROM HISTORY

Life at First Sight

Anton van Leeuwenhoek was the first researcher to see bacteria under a microscope. In his journal, he described how he felt after discovering this new and unfamiliar form of life.

"For me . . . no more pleasant sight has met my eye than this of so many thousand of living creatures in one small drop of water."

Read the quote, and answer the question below.

Why do you think Leeuwenhoek was so excited about what he saw?

He saw living things smaller than anyone knew existed.

A modern view of bacteria similar to those seen by Leeuwenhoek

Do the Inquiry Warm-Up *What Can You See?* Student Lab Manual, p. 13

TEKS 12E In this section, you'll examine how the structures of cells relate to their functions.

ELPS 1.C

With a partner, fill in the "Needs of Cells" diagram on page 67. As you decide what to write on the arrows, use a sentence frame: *"The ____ needs to (take in/let out) ____."*

What Are Cells?

As you learned, all living things are made of cells. Cells form the parts of an organism and carry out all of its functions. **Cells are the basic units of structure and function in living things.**

Cells and Structure When you describe the structure of an object, you describe what it is made of and how its parts are put together. For example, the structure of a building depends on the way bricks, steel beams, or other materials are arranged. The structure of a living thing is determined by the amazing variety of ways its cells are put together.

SUPPORT ALL READERS

Lexile Measure = 940L **Lexile Word Count = 1581**

Prior Exposure to Content: Most students have encountered this topic in earlier grades

Academic Vocabulary: *measure, sequence*

Science Vocabulary: *microscope, cell theory*

Concept Level: Generally appropriate for most students in this grade

Preteach With: My Planet Diary "Life at First Sight" and Figure 1 activity

Vocabulary
- microscope
- cell theory

Skills
- Reading: Sequence
- Inquiry: Measure

FIGURE 1

Needs of Cells

A single cell has the same basic needs as an entire organism.

Classify On each blank arrow, write the name of a material that moves as shown.

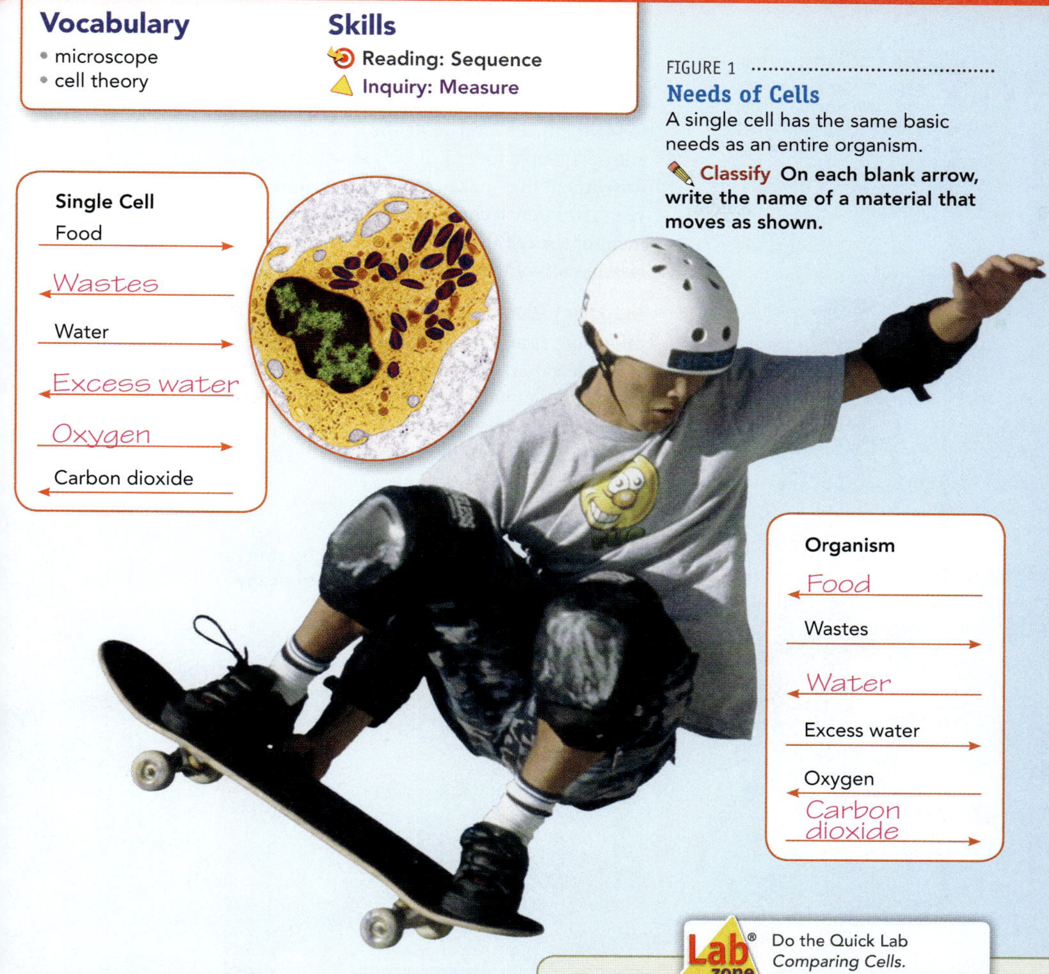

Cells and Function An organism's functions are the processes that enable it to live, grow, and reproduce. Those functions include obtaining oxygen, food, and water and getting rid of wastes. Cells are involved in all these functions. For example, cells in your digestive system absorb food. The food provides your body with energy and materials needed for growth. Cells in your lungs help you get oxygen. Your body's cells work together, keeping you alive. And for each cell to stay alive, it must carry out many of the same functions as the entire organism.

Lab zone: Do the Quick Lab *Comparing Cells.* Student Lab Manual, p. 14

Assess Your Understanding

got it? ·····

- I get it! Now I know that a cell is the basic unit of structure and function.
- I need extra help with See TE note.

Explain

Introduce Vocabulary

Write the term *microscope* on the board. Tell students that *micro-* means "small" and *scope* refers to an instrument used for looking at something. So a microscope is an instrument for looking at things that are small.

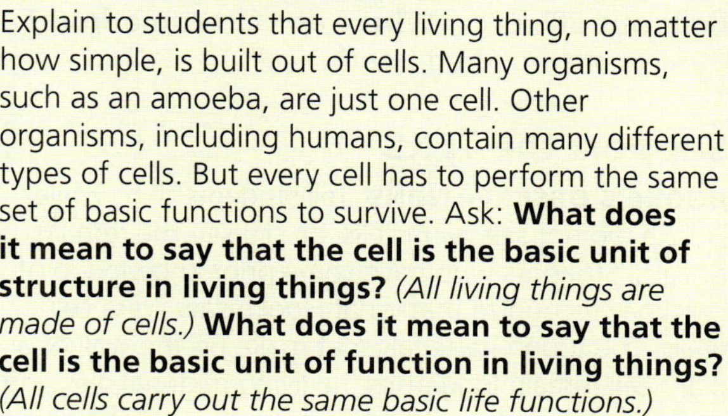

Teach Key Concepts

Explain to students that every living thing, no matter how simple, is built out of cells. Many organisms, such as an amoeba, are just one cell. Other organisms, including humans, contain many different types of cells. But every cell has to perform the same set of basic functions to survive. Ask: **What does it mean to say that the cell is the basic unit of structure in living things?** *(All living things are made of cells.)* **What does it mean to say that the cell is the basic unit of function in living things?** *(All cells carry out the same basic life functions.)*

Elaborate

Lab Resource: Quick Lab 

L2 **COMPARING CELLS** Students will observe the characteristics of plant and animal cells. This Quick Lab can be found in the Student Lab Manual, p. 14, and online.

Evaluate

Assess Your Understanding

Have students evaluate their understanding by completing the appropriate sentence.

RTI Response to Intervention

If students cannot explain what a cell is, **then** have them reread the Key Concept statement.

PEARSON Texas.com

English Language Proficiency Standards

ELPS Listening 1.C

Have students fill in the diagram for **Figure 1.**

Beginning Read aloud **Figure 1.** Name the cell and the organism. Model the first few examples. Have students repeat and point to the corresponding arrow in the diagram. Continue modeling until partners can continue on their own.

Intermediate Have students read **Figure 1.** Point out that the cell and the organism take in and let out the same things. Model how to complete the first two examples by comparing charts. Have partners orally complete the activity.

Advanced Have partners complete the activity in **Figure 1.**

Advanced High Have partners complete the activity in **Figure 1** and elaborate using words such as *obtain, absorb, get rid of,* and *remove.*

LESSON 2.2

LESSON 2.2

Explain

Teach Key Concepts

Explain to students that cells are the basic unit of structure and function in living things, and that every cell is descended from another cell. These ideas are the basis of the cell theory. Ask: **What is a theory?** *(A well-tested and widely accepted explanation of a natural phenomenon based on many observations and experimental results.)* **What are living things built out of?** *(Cells.)* **Where do cells come from?** *(All cells are produced by other cells.)*

Support the TEKS

HOOKE'S OBSERVATIONS Tell students that, while Hooke did not see living cells, he did see the remains of cells. Students who have some prior knowledge of cells may recognize that what Hooke saw were cell walls. The tough materials that make up the wall of a plant cell remain long after the living material dies and dries up. Ask: **Where did the cells Hooke observed come from?** *(The bark of a tree)* **What other product of a tree do you think is made up of the walls of cells that are no longer living?** *(Wood)*

Lead a Discussion

HISTORY OF THE CELL THEORY Remind students that no one knew cells existed until the microscope was invented. Point out that it took nearly 200 years after their discovery for scientists to determine they were the basic unit of structure and function in living things. Ask: **When were cells first discovered?** *(1663, by Robert Hooke)* **When was the cell theory developed?** *(Between 1838 and 1855)* Students may know that every person starts his or her development as a single cell called a zygote, or fertilized egg. All of the cells in the body were derived from that one cell. Ask: **What was Virchow's contribution to the cell theory?** *(The idea that all cells come from other cells)* **How is a zygote related to this idea?** *(The zygote is the cell that all the other cells in an individual descend from.)*

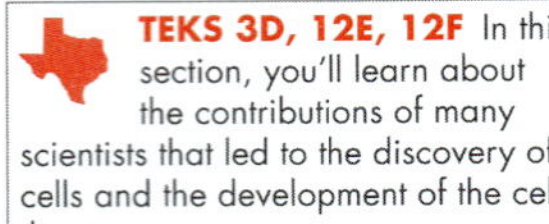
TEKS 3D, 12E, 12F In this section, you'll learn about the contributions of many scientists that led to the discovery of cells and the development of the cell theory.

What Is the Cell Theory?

Until the 1600s, no one knew cells existed because there was no way to see them. Around 1590, the invention of the first microscope allowed people to look at very small objects. A **microscope** is an instrument that makes small objects look larger. Over the next 200 years, this new technology revealed cells and led to the development of the cell theory. The **cell theory** is a widely accepted explanation of the relationship between cells and living things.

Seeing Cells English scientist Robert Hooke built his own microscopes and made drawings of what he saw when he looked at the dead bark of certain oak trees. Hooke never knew the importance of what he saw. A few years later, Dutch businessman Anton van Leeuwenhoek (LAY von hook) was the first to see living cells through his microscopes.

FIGURE 2

Growth of the Cell Theory

The cell theory describes how cells relate to the structure and function of living things. **Review** **Answer the questions in the spaces provided.**

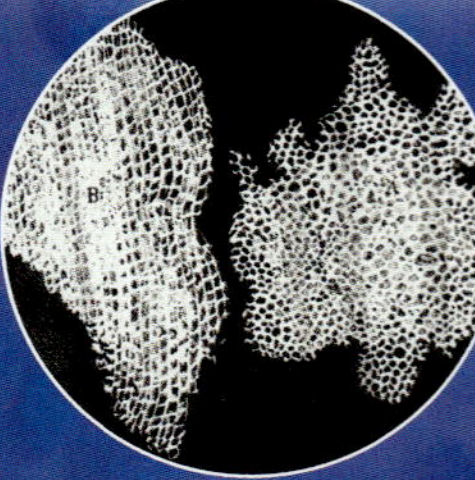
Hooke's drawing of cork

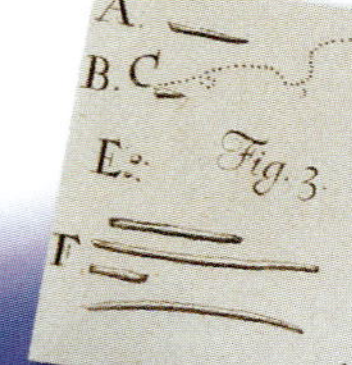

Drawing by Leeuwenhoek

Hooke's Microscope

In 1663, Robert Hooke used his microscope to observe a thin slice of cork. Cork, the bark of the cork oak tree, is made up of cells that are no longer alive. To Hooke, the empty spaces in the cork looked like tiny rectangular rooms. Therefore, Hooke called the empty spaces cells, which means "small rooms."

What was important about Hooke's work?

He was first to see the remains of cells. He gave "cells" their name.

Leeuwenhoek's Microscope

Leeuwenhoek built microscopes in his spare time. Around 1674, he looked at drops of lake water, scrapings from teeth and gums, and water from rain gutters. Leeuwenhoek was surprised to find a variety of single-celled organisms. He noted that many of them whirled, hopped, or shot through water like fast fish. He called these moving organisms animalcules, meaning "little animals."

What did Leeuwenhoek's observations reveal?

Living things that moved and were smaller than the eye could see

Texas Essential Knowledge and Skills

3D Relate the impact of research on scientific thought and society, including the history of science and contributions of scientists as related to the content.

12E Compare the functions of a cell to the functions of organisms such as waste removal.

12F Recognize that according to cell theory all organisms are composed of cells and cells carry on similar functions such as extracting energy from food to sustain life.

What the Cell Theory Says **Figure 2** highlights people who made key discoveries in the early study of cells. Their work and the work of many others led to the development of the cell theory. **The cell theory states the following:**

- **All living things are composed of cells.**
- **Cells are the basic units of structure in living things and carry on similar functions, such as obtaining energy, removing wastes, and reproducing.**
- **All cells are produced from other cells.**

The cell theory holds true for all living things, no matter how big or how small. Because cells are common to all living things, cells can provide clues about the functions that living things perform. And because all cells come from other cells, scientists can study cells to learn about growth and reproduction.

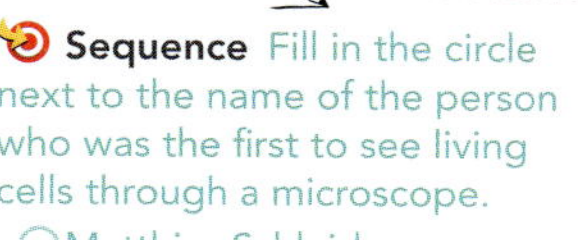

Sequence Fill in the circle next to the name of the person who was the first to see living cells through a microscope.

- ○ Matthias Schleiden
- ○ Robert Hooke
- ● Anton van Leeuwenhoek
- ○ Rudolf Virchow
- ○ Theodor Schwann

Schleiden, Schwann, and Virchow

In 1838, using his own research and the research of others, Matthias Schleiden concluded that all plants are made of cells. A year later, Theodor Schwann reached the same conclusion about animals. In 1855, Rudolf Virchow proposed that new cells are formed only from cells that already exist. "All cells come from cells," wrote Virchow.

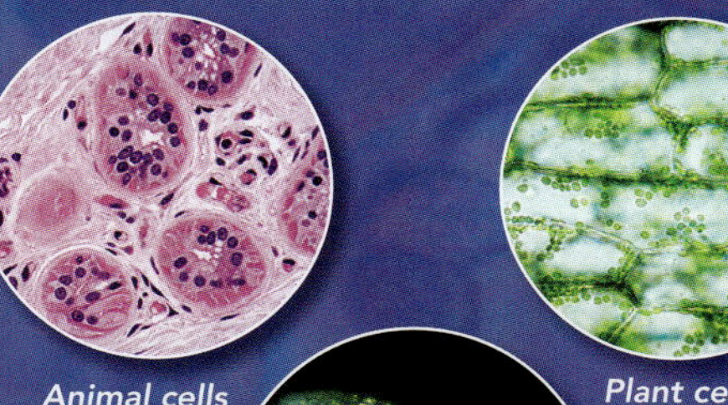

Animal cells

Plant cells

A cell reproducing

To which part of the cell theory did Virchow contribute?

Virchow said all cells come from other cells.

Do the Quick Lab *Observing Cells.* Student Lab Manual, p. 15

Assess Your Understanding

TEKS 3D, 12F

1a. Relate Cause and Effect Why would Hooke's discovery have been impossible without a microscope?

The human eye is not able to see structures as small as most cells.

b. Recognize According to the cell theory, all cells carry on similar functions. Name two functions that all cells perform.

Sample: Obtaining energy and removing wastes

got it?

○ **I get it!** Now I know that the cell theory describes how cells are related to the structure and function of living things.

○ I need extra help with See TE note.

Sequence Students use the skill of sequencing when they organize events in the order they occurred. Tell students to look for dates to determine which discoveries were made first, and which came later.

LESSON 2.2

Elaborate

21st Century Learning

CREATIVITY Direct students' attention to **Figure 2.** Invite them to draw a timeline that shows the order in which the discoveries occurred. You may wish to have students do research to add additional events to the timeline.

Lab Resource: Quick Lab 

L2 OBSERVING CELLS Students will explore living cells under the microscope. This Quick lab can be found in the Student Lab Manual, p. 15, and online.

Evaluate

Assess Your Understanding

After students answer the questions, have them evaluate their understanding by completing the appropriate sentence.

RTI Response to Intervention

1a. If students cannot relate the microscope to the discovery of cells, **then** have them look at a piece of cork with their naked eye and compare it to the image of cork cells in **Figure 2.**

b. If students need help recognizing two cell functions, **then** have them reread the Key Concept statement.

Differentiated Instruction

L1 Cell Size If students have difficulty understanding why they cannot see cells, tell them that one square centimeter of their skin's surface contains more than 100,000 cells.

L3 Discovery of the Cell Cells could not be studied until instruments that let people see them were invented. Have students research the invention and development of early microscopes, comparing Hooke's early compound microscope to Leeuwenhoek's simple (but more powerful) microscope from the same time period.

LESSON 2.2

Explain

Teach Key Concepts

Explain to students that most cells are too small to see unless we use specialized instruments called microscopes to magnify them. Light microscopes magnify cells by focusing reflected light through lenses; electron microscopes use beams of electrons to get even better magnification. Ask: **Why do we use microscopes to look at most cells?** *(Most cells are too small to see with the naked eye.)* **What are two kinds of microscopes?** *(Light and electron microscopes)*

Lead a Discussion

MAGNIFICATION Tell students that microscopes and magnifying glasses both use lenses that bend light in a way that makes objects look bigger. Ask: **What kind of lenses do microscopes use?** *(Convex lenses)* **What is the shape of a convex lens?** *(It is thicker in the center than it is at the edges.)*

Texas Essential Knowledge and Skills

4A Use appropriate tools to collect, record, and analyze information, including life science models, hand lens, stereoscopes, microscopes, beakers, Petri dishes, microscope slides, graduated cylinders, test tubes, meter sticks, metric rulers, metric tape measures, timing devices, hot plates, balances, thermometers, calculators, water test kits, computers, temperature and pH probes, collecting nets, insect traps, globes, digital cameras, journals/notebooks, and other equipment as needed to teach the curriculum.

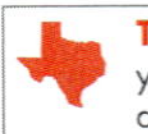

TEKS 4A In this section, you'll explore how microscopes allow people to view objects not visible to the naked eye.

How Do Microscopes Work?

The cell theory could not have been developed without microscopes. **Some microscopes focus light through lenses to produce a magnified image, and other microscopes use beams of electrons.** Both light microscopes and electron microscopes do the same job in different ways. For a microscope to be useful, it must combine two important properties—magnification and resolution.

Vocabulary Prefixes The prefix *magni-* means "great" or "large." Underline all the words in the paragraph at the right that you can find with this prefix.

Magnification and Lenses Have you ever looked at something through spilled drops of water? If so, did the object appear larger? Magnification is the condition of things appearing larger than they are. Looking through a magnifying glass has the same result. A magnifying glass consists of a convex lens, which has a center that is thicker than its edge. When light passes through a convex lens and into your eye, the image you see is magnified. Magnification changes how you can see objects and reveals details you may not have known were there, as shown in **Figure 3.**

1 Leaf; green color and veins

2 Sample: feathers; lines, color pattern

3 Sample: orange; color and dimpled appearance

4 Sample; butterfly wing: color pattern

FIGURE 3

Magnification

The images above have all been magnified, which makes them look unfamiliar. **Infer On the lines, write what you think each photograph shows, and explain your reasoning. (One answer is completed for you.)**

Magnification With a Compound Microscope

Figure 4 shows a microscope that is similar to one you may use in your classroom. This type of instrument, called a compound microscope, magnifies the image using two lenses at once. One lens is fixed in the eyepiece. A second lens is chosen from a group of two or three lenses on the revolving nosepiece. Each of these lenses has a different magnifying power. By turning the nosepiece, you can select the lens you want. A glass slide on the stage holds the object to be viewed.

A compound microscope can magnify an object more than a single lens can. Light from a lamp (or reflecting off a mirror) passes through the object on the slide, the lower lens, and then the lens in the eyepiece. The total magnification of the object equals the magnifications of the two lenses multiplied together. For example, suppose the lower lens magnifies the object 10 times, and the eyepiece lens also magnifies the object 10 times. The total magnification of the microscope is 10 × 10, or 100 times, which is written as "100×."

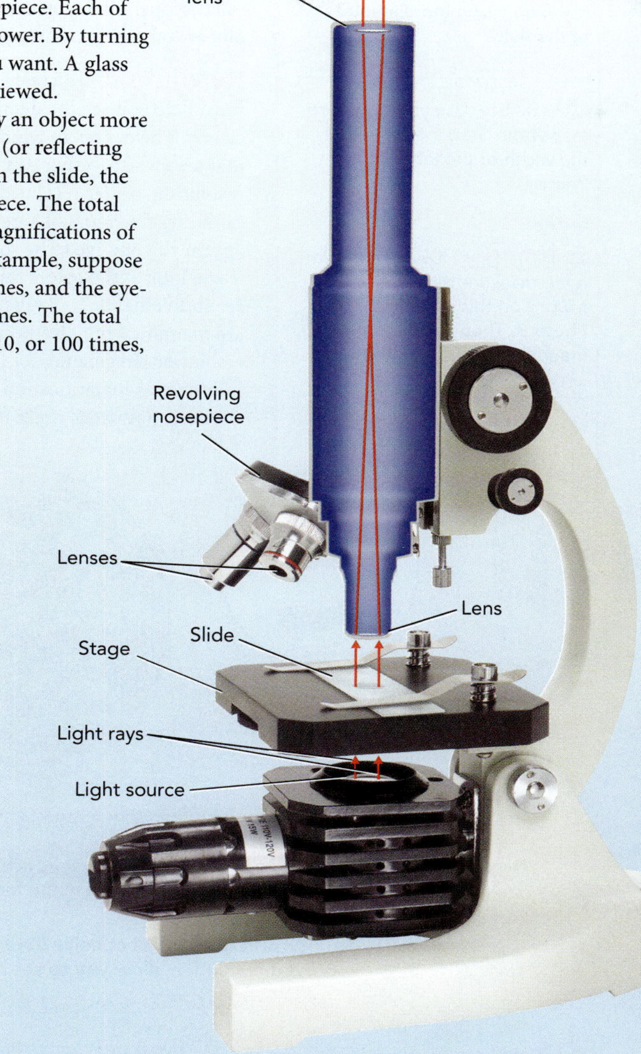

FIGURE 4

A Compound Microscope

This microscope has a 10× lens in the eyepiece. The revolving nosepiece holds three different lenses: 4×, 10×, and 40×.

Complete these tasks.

1. **Calculate** Calculate the three total magnifications possible for this microscope.
   10 × 4 = 40x
   10 × 10 = 100x
   10 × 40 = 400x

2. **Predict** What would happen if the object on the slide were too thick for light to pass through it?
   Light would not reach the lenses, so you would not see anything.

Teach With Visuals

Tell students to look at **Figure 4.** Ask: **Why is this called a compound microscope?** *(It has two lenses.)* **Where is the lens you look through?** *(In the eyepiece)* **Where is the other lens?** *(On the revolving nosepiece)* **Why are there several lenses on the revolving nosepiece?** *(By turning the nosepiece, you can change the total magnification.)* **What is the light source for this microscope?** *(An electric lamp)* **What other light source is used in some microscopes?** *(A mirror reflecting light up through the stage of the microscope)* **How is using a microscope different from using a magnifying glass?** *(Samples: The microscope has a light source, but a magnifying lens does not. You look through the object with the microscope and at the surface with a magnifying lens.)*

Elaborate

Build Inquiry 

L1 LENSES AND MAGNIFICATION

Materials magnifying glasses with varying strengths, small objects

Time 15 minutes

Have students compare the curvature of the lenses. Give students time to examine the objects with the magnifying glasses and estimate the magnification they see with each lens.

Ask: **What shape are these lenses?** *(Convex lenses)* **Which lens provided the greatest magnification?** *(The lens with the greatest curvature)* **What did you have to do to see the image in sharp focus?** *(Move the lens toward and away from the object)*

Differentiated Instruction

L1 Microscopes Have students work in pairs, pointing to a part of a compound microscope and asking their partners to name it and describe its function.

L3 Magnification The amount of magnification a microscope provides determines how much larger objects appear. Have students observe and describe what the same object looks like when observed through three different microscope objectives.

LESSON 2.2

Explain

Lead a Discussion

RESOLUTION OF HEADLIGHTS Students may have experienced changes in resolution when riding in a car at night. From a great distance, the two headlights may look like one bright light. Then, as the car gets closer, it becomes possible to distinguish two separate lights. Ask: **When is the resolution low? When does it improve?** *(It is low when only one light can be detected. It improves when the two lights become distinct.)*

Elaborate

Apply It!

L1 Before beginning the activity, show students a plastic millimeter ruler. Allow them to feel the ridges that indicate millimeters. If such a ruler is not available, use any millimeter ruler and explain that the ruler shown in the activity is a yellow plastic ruler with colored ridges to indicate millimeters. Explain that if students look at a letter e through a microscope, the image would actually appear upside down and backwards because of the way the lenses work. That image is not shown here.

Measure Remind students to start their measurement on the side where a millimeter mark is lined up with the field of the microscope. Make sure students remember that a unit should always be included as part of a measurement. Ask: **What would these images look like if they were viewed at a lower magnification?** *(The e would look smaller, and more millimeter lines would be visible on the ruler.)*

apply it!

1 **Measure** In Photo A, you can see the millimeter markings of a metric ruler in the field of the microscope. What is the approximate diameter of the field?

About 4.4 or 4.5 mm

2 **Estimate** Use your measurement from Step 1 to estimate the width of the letter in Photo B.

About 2.2 mm

3 **CHALLENGE** Using a metric ruler, measure the letter **e** in a word on this page and in Photo B. Then calculate the magnification in the photo.

Magnification is about 10x.

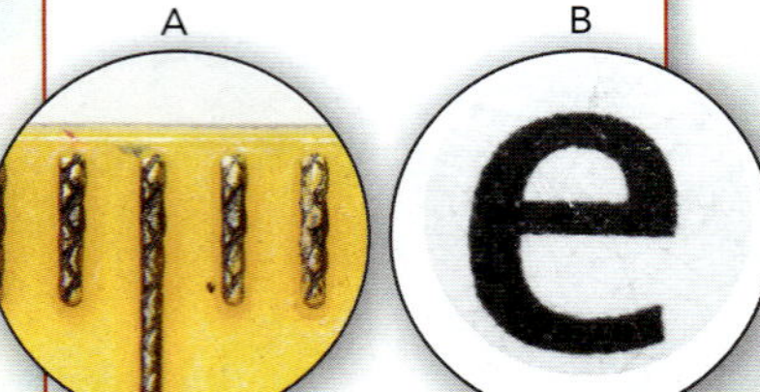

Measuring Microscopic Objects

When you see objects through a microscope, they look larger than they really are. How do you know their true size? One way is to use a metric ruler to measure the size of the circular field in millimeters as you see it through the microscope. Then you can estimate the size of the object you see by comparing it to the width of the field.

Resolution To create a useful image, a microscope must help you see the details of the object's structure clearly. The degree to which two separate structures that are close together can be distinguished is called resolution. Better resolution shows more details. For example, the colors of a newspaper photograph may appear to your eye to be solid patches of color. However, if you look at the colors through a microscope, you will see individual dots. You see the dots not only because they are magnified but also because the microscope improves resolution. In general, for light microscopes, resolution improves as magnification increases. Good resolution, as shown in **Figure 5,** makes it easier to study cells.

FIGURE 5

Resolution

The images in colorful photographs actually consist of only a few ink colors in the form of dots.

Interpret Photos **What color dots does improved resolution allow you to see?**

Sample: Red, yellow, black, blue-green

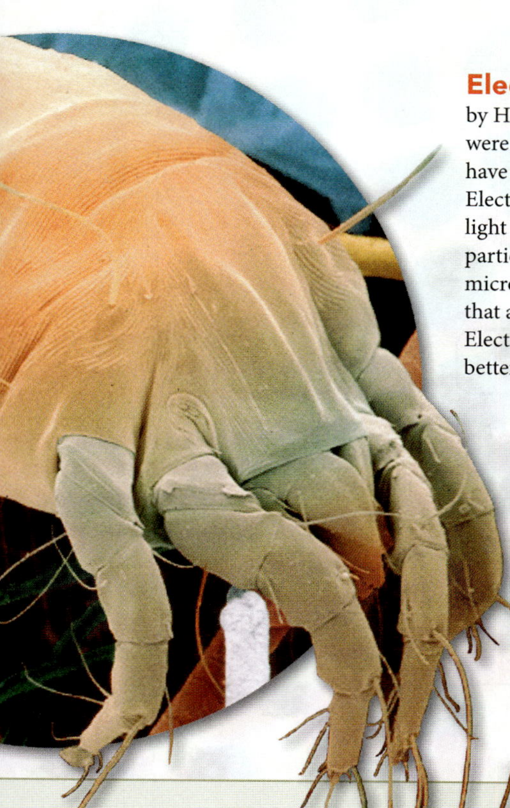

Electron Microscopes The microscopes used by Hooke, Leeuwenhoek, and other early researchers were all light microscopes. Since the 1930s, scientists have developed several types of electron microscopes. Electron microscopes use a beam of electrons instead of light to produce a magnified image. (Electrons are tiny particles that are smaller than atoms.) By using electron microscopes, scientists can obtain pictures of objects that are too small to be seen with light microscopes. Electron microscopes allow higher magnification and better resolution than light microscopes.

FIGURE 6

A Dust Mite

Dust mites live in everyone's homes. A colorized image made with an electron microscope reveals startling details of a mite's body.

Observe List at least three details that you can see in the photo.

Sample: "paddles" at ends of legs, folds in "skin," hairlike tendrils

Do the Lab Investigation *Design and Build a Microscope.* Student Lab Manual, p. 16

Assess Your Understanding

TEKS 4A

2a. Define Magnification makes objects look (smaller/larger) than they really are.

b. Estimate The diameter of a microscope's field of view is estimated to be 0.9 mm. About how wide is an object that fills two thirds of the field? Circle your answer.

1.8 mm 0.6 mm 0.3 mm

c. Compare and Contrast How are magnification and resolution different?

Magnification makes objects look larger. Resolution shows more detail.

d. Explain How do the characteristics of electron microscopes make them useful for studying cells?

Electron microscopes have more powerful magnification and better resolution than light microscopes, so smaller objects can be seen more clearly.

got it?

○ **I get it!** Now I know that light microscopes work by using lenses to focus light, making objects look bigger.

○ I need extra help with See TE note.

73

Differentiated Instruction

L1 Magnification and Resolution If students have difficulty with the distinction between magnification and resolution have them use a graphics program to repeatedly magnify an image on a computer screen. Students can see that as the image gets bigger, the level of detail they can see does not change—and eventually they will see nothing but blurry pixels. Explain that this is the result of increasing magnification without increasing resolution.

L3 Electron Microscopes Have students research the history of the electron microscope and the different types of electron microscopes in use today. Interested students can present their findings to the class.

Teacher Demo 

L2 RESOLUTION OF ELECTRON MICROSCOPES

Materials images of insects taken with both light and electron microscopes

Time 10 minutes

Show students images of the same types of insects photographed with light and electron microscopes.

Ask: **Which photographs are more detailed?** *(The ones from electron microscopes)* Make sure students understand that photographs taken through microscopes may be colored to make details more visible.

Lab Resource: Lab Investigation 

L2 DESIGN AND BUILD A MICROSCOPE Students will explore how microscopes work by designing and building a basic compound microscope. This Lab Investigation can be found in the Student Lab Manual, p. 16, and online.

Evaluate

Assess Your Understanding

After students answer the questions, have them evaluate their understanding by completing the appropriate sentence.

RTI Response to Intervention

2a, b. If students cannot explain magnification, **then** have them look at the same object through first a weak and then a more powerful magnifying glass to see how the apparent size of the object changes.

c. If students have trouble contrasting magnification and resolution, **then** have them reread the paragraph under the red heading *Resolution.*

d. If students need help with explaining the characteristics of an electron microscope, **then** have them reread the paragraph under the red heading *Electron Microscopes.*

Name ____________________ Date __________ Class __________

Assess Your Understanding

Discovering Cells

What Are Cells?

got it?

○ **I get it!** Now I know that a cell is the basic unit of ____________________

____

○ **I need extra help with** ____________________

____

What Is the Cell Theory?

1a. RELATE CAUSE AND EFFECT Why would Hooke's discovery have been impossible without a microscope?

____

____

b. RECOGNIZE Name two functions that all cells perform.

____

____

How Do Microscopes Work?

2a. DEFINE Magnification makes objects look (smaller/larger) than they really are.

b. ESTIMATE The diameter of a microscope's field of view is estimated to be 0.9 mm. About how wide is an object that fills two thirds of the field? Circle your answer.
1.8 mm 0.6 mm 0.3 mm

c. COMPARE AND CONTRAST How are magnification and resolution different?

____

____

d. EXPLAIN How do the characteristics of electron microscopes make them useful for studying cells?

____

____

____

Name ______________________ Date ____________ Class ____________

Enrich

Discovering Cells

Because the beam of electrons used in electron microscopes damages live biological specimens, scientists have developed new kinds of microscopes for use with these kinds of samples. Read the passage and examine the diagram below. Then answer the questions that follow on a separate sheet of paper.

Recent Advances in the Microscope

One new type of microscope is the transmission positron microscope, or TPM. Like the transmission electron microscope, or TEM, the TPM sends a beam of atomic particles through a specimen. However, instead of using a beam of electrons, the TPM uses a beam of positrons, which are positively charged atomic particles that do not harm living specimens as electrons do.

Another new type of microscope is the acoustic microscope. It uses sound waves instead of beams of atomic particles to "see" an object. As shown in the figure, the echoes of sound waves bouncing off the specimen are translated onto a screen as a microscopic image. The sound waves that are used are very high in frequency, but they do no damage to living things. Doctors have used the acoustic microscope to view changes in living cells and to examine living cells for cancer without removing the cells from the body.

How an Acoustic Lens Works

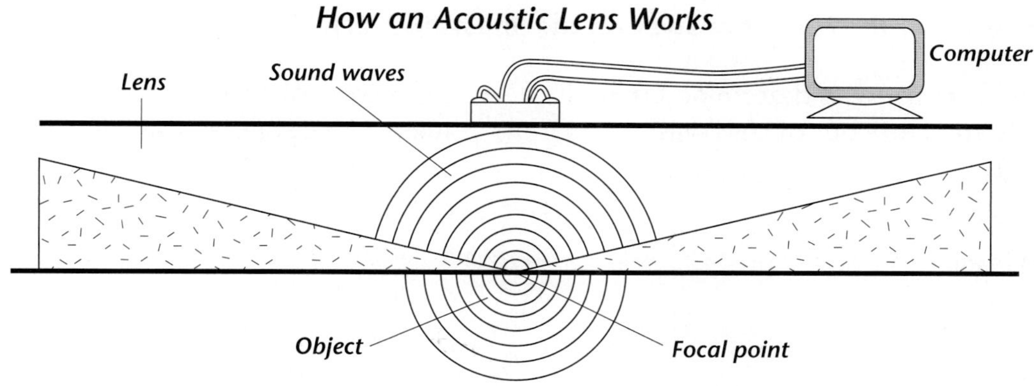

1. Compare and contrast transmission electron microscopes and transmission positron microscopes.
2. Explain how acoustic microscopes work.
3. Why are transmission positron microscopes and acoustic microscopes important tools for understanding how living cells function?

Place the outside corner, the corner away from the dotted line, in the corner of your copy machine to copy onto letter-size paper.

Name ______________________ Date ____________ Class ____________

Lesson Quiz

Discovering Cells

Fill in the blank to complete each statement.

1. A cell's functions can include obtaining food and water and getting rid of waste.

2. Compound microscopes focus light through lenses to produce a magnified image.

3. A large organism is made up of many millions of cells.

4. A(n) convexs lens has a center that is thicker than its edge.

5. The cells theory describes how cells are related to living things.

6. The ability to distinguish between two nearby objects is called resoultions.

Write the letter of the correct answer on the line at the left.

7. ___ The scientist who determined that all animals are made out of cells was
   - A Hooke
   - B Schleiden
   - C Schwann
   - D Virchow

8. ___ A compound microscope with a 10× eyepiece and a 40× objective has a magnification of
   - A 10×
   - B 40×
   - C 50×
   - D 400×

9. ___ Which of the following statements is NOT part of the cell theory?
   - A All cells are produced from other cells.
   - B Cells can absorb food and oxygen.
   - C All living things are composed of cells.
   - D Cells are the basic units of structure and function in living things.

10. ___ The visible field of a microscope is 10 mm wide. How large is an object that takes up $\frac{1}{4}$ of the field?
    - A 1 mm
    - B 2.5 mm
    - C 4 mm
    - D 5 mm

Place the outside corner, the corner away from the dotted line, in the corner of your copy machine to copy onto letter-size paper.

Discovering Cells

Answer Key

Review and Reinforce

Find the worksheet in the Student Workbook.

1. true
2. Microscopes
3. true
4. Electron
5. Magnification
6. Resolution
7. 200×
8. a lens that has a thicker center than edge
9. an instrument that makes small objects look larger by focusing light through two lenses
10. a widely accepted explanation of the relationship between cells and living things
11. Most cells are too small for human eyes to see, and their discovery and the subsequent study of their behavior required instruments – microscopes – that could magnify them and make them visible. Once Anton van Leeuwenhook observed cells through a microscope, scientists like Schleiden, Schwann, and Virchow were able to study cells and develop the cell theory.

Enrich

1. Transmission electron microscopes (TEM) and transmission positron microscopes (TPM) both use beams of atomic particles to view specimens. However, the TPM uses beams of positrons, which do not harm living things, whereas the TEM uses beams of electrons, which living things cannot withstand. Therefore, unlike the TEM, the TPM can be used to view living specimens.
2. Acoustic microscopes bounce high-frequency sound waves off an object. The echoes of the sound waves are then translated onto a screen as a microscopic image.
3. Transmission positron microscopes and acoustic microscopes can be used to view cells that are still alive and functioning. Electron microscopes, on the other hand, can be used to view only cells that are no longer alive.

Lesson Quiz

1. wastes
2. lenses
3. cells
4. convex
5. cell theory
6. resolution
7. C
8. D
9. B
10. B

Looking Inside Cells

 **What structures are found in cells?**

LESSON PACING:
3–4 periods or $1\frac{1}{2}$–2 blocks

Lesson Vocabulary

- cell wall • nucleus • organelle • ribosome • cytoplasm
- mitochondria • endoplasmic reticulum • Golgi apparatus
- vacuole • chloroplast • lysosome • tissue
- organ • organ system

Content Refresher

Evolution of Eukaryotic Cells This chapter focuses on the cell structure of eukaryotes, which include protists, plants, fungi, and animals. Eukaryotic cells contain membrane-bound organelles, and they sequester their DNA inside a nucleus—features that prokaryotes (such as bacteria and viruses) lack.

The endosymbiotic model suggests that eukaryotic cells evolved out of a mutually beneficial relationship between different kinds of prokaryotic cells. Ancient archean cells enveloped bacterial cells and kept them alive inside their cytoplasm instead of digesting them, harvesting the waste products of the bacteria for their own use. The mutualistic relationship evolved over time until the archean and bacterial cells could not survive independently. Evidence supporting the endosymbiotic model includes the fact that mitochondria and chloroplasts contain enzymes, ribosomes, membranes, and DNA that are more similar to those found in bacterial cells than the same structures found in other parts of the eukaryotic cell. Mitochondria and chloroplasts also divide inside the eukaryotic cell in a way that is similar to bacteria reproduction.

Lesson Objectives	TEKS	ELPS
Differentiate between the functions of cell structures and organelles.	3B, 12D, 12F	4.F.2
Recognize how cells are organized in many-celled organisms.	12C	

Texas Essential Knowledge and Skills

3B Use models to represent aspects of the natural world such as human body systems and plant and animal cells.
12C Recognize levels of organization in plants and animals, including cells, tissues, organs, organ systems, and organisms.
12D Differentiate between structure and function in plant and animal cell organelles, including cell membrane, cell wall, nucleus, cytoplasm, mitochondrion, chloroplast, and vacuole.
12F Recognize that according to cell theory all organisms are composed of cells and cells carry on similar functions such as extracting energy from food to sustain life.

English Language Proficiency Standards

ELPS Reading 4.F.2 Use visual and contextual support to enhance and confirm understanding.

DIFFERENTIATED INSTRUCTION KEY
L1 Struggling Students or Special Needs
L2 On-Level Students L3 Advanced Students

LESSON PLANNER 2.3

Investigations and Activities

My Planet Diary, **Student Edition,** p. 74

Inquiry: Inquiry Warm-Up, How Large Are Cells?, **PearsonTexas.com**

Introduce Vocabulary, **Teacher's Edition,** p. 75

Teach Key Concepts, **Teacher's Edition,** p. 75

Lead a Discussion, The Nucleus, **Teacher's Edition,** p. 76

21st Century Learning, Critical Thinking, **Teacher's Edition,** p. 77

Teach With Visuals, **Teacher's Edition,** p. 78

21st Century Learning, Accountability, **Teacher's Edition,** p. 78

Inquiry: Teacher Demo, Compare Slides to Illustrations, **Teacher's Edition,** p. 79

Lead a Discussion, Chloroplasts, **Teacher's Edition,** p. 80

Apply It!, **Student Edition,** p. 80

Inquiry: Teacher Demo, A Function of Plant Vacuoles, **Teacher's Edition,** p. 81

Inquiry: Quick Lab, Gelatin Cell Model, **Lab Manual,** p. 21

Teach Key Concepts, **Teacher's Edition,** p. 82

Lead a Discussion, One Cell or Many? **Teacher's Edition,** p. 82

Teach With Visuals, **Teacher's Edition,** p. 82

Address Misconceptions, Cells Contain the Same Genes, **Teacher's Edition,** p. 83

Differentiated Instruction, **Teacher's Edition,** p. 83

Inquiry: Quick Lab, Tissues, Organs, Organ Systems, **Lab Manual,** p. 22

TEKS Review

Apply the TEKS, Cells in Living Things, **Student Edition,** p. 78

Apply the TEKS, Comparing Elements, **Student Edition,** p. 94

TEKS Practice, **Student Edition,** p. 95

TEKS Practice: Chapter and Cumulative Review, **Student Edition,** p. 98

Lesson 2.3, **TEKS Preparation and Study Guide Workbook,** p. 20

SHORT ON TIME? To do this lesson in approximately half the time, do the Activate Prior Knowledge activity. A discussion of the Key Concepts will familiarize students with the lesson content. Use the Apply the TEKS activity to help students understand scientific inquiry. Have students do the Quick Labs. The rest of the lesson can be completed by students independently.

These editable worksheets are available on **PearsonTexas.com.**
Print versions can be found in the **TEKS Preparation and Study Guide Workbook.**

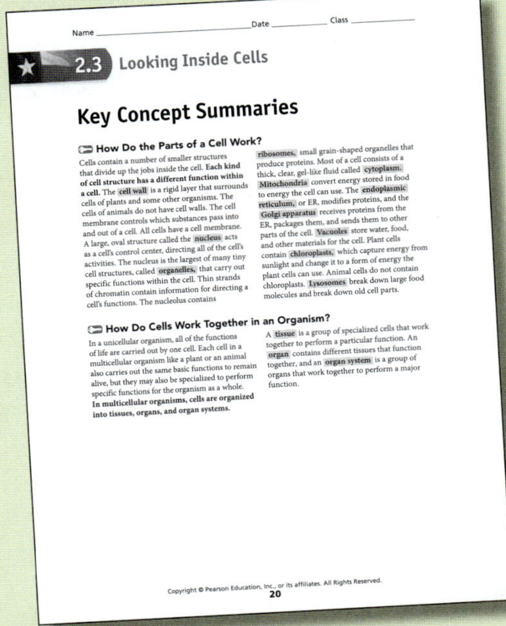

Name ______________ Date ________ Class ________

2.3 Looking Inside Cells

Key Concept Summaries

How Do the Parts of a Cell Work?

Cells contain a number of smaller structures that divide up the jobs inside the cell. **Each kind of cell structure has a different function within a cell.** The **cell wall** is a rigid layer that surrounds cells of plants and some other organisms. The cells of animals do not have cell walls. The cell membrane controls which substances pass into and out of a cell. All cells have a cell membrane. A large, oval structure called the **nucleus** acts as a cell's control center, directing all of the cell's activities. The nucleus is the largest of many tiny cell structures, called **organelles,** that carry out specific functions within the cell. Thin strands of chromatin contain information for directing a cell's functions. The nucleolus contains **ribosomes,** small grain-shaped organelles that produce proteins. Most of a cell consists of a thick, clear, gel-like fluid called **cytoplasm.** **Mitochondria** convert energy stored in food to energy the cell can use. The **endoplasmic reticulum,** or ER, modifies proteins, and the **Golgi apparatus** receives proteins from the ER, packages them, and sends them to other parts of the cell. **Vacuoles** store water, food, and other materials for the cell. Plant cells contain **chloroplasts,** which capture energy from sunlight and change it to a form of energy the plant cells can use. Animal cells do not contain chloroplasts. **Lysosomes** break down large food molecules and break down old cell parts.

How Do Cells Work Together in an Organism?

In a unicellular organism, all of the functions of life are carried out by one cell. Each cell in a multicellular organism like a plant or an animal also carries out the same basic functions to remain alive, but they may also be specialized to perform specific functions for the organism as a whole. **In multicellular organisms, cells are organized into tissues, organs, and organ systems.** A **tissue** is a group of specialized cells that work together to perform a particular function. An **organ** contains different tissues that function together, and an **organ system** is a group of organs that work together to perform a major function.

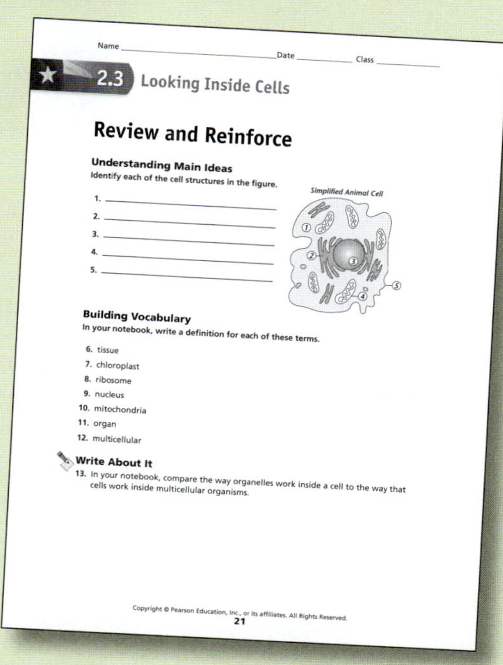

Name ______________ Date ________ Class ________

2.3 Looking Inside Cells

Review and Reinforce

Understanding Main Ideas
Identify each of the cell structures in the figure.

1. ______________
2. ______________
3. ______________
4. ______________
5. ______________

Building Vocabulary
In your notebook, write a definition for each of these terms.

6. tissue
7. chloroplast
8. ribosome
9. nucleus
10. mitochondria
11. organ
12. multicellular

Write About It
13. In your notebook, compare the way organelles work inside a cell to the way that cells work inside multicellular organisms.

Lexile Measure = 960L

LESSON 2.3

Looking Inside Cells

Establish Learning Objectives

After this lesson, students will be able to:

- Differentiate between the functions of cell structures and organelles.
- Recognize how cells are organized in many-celled organisms.

Engage

Activate Prior Knowledge

MY PLANET DIARY Read *Glowing Globs* with the class. Ask students what they think the glowing lines around the nucleus do for the cell. Point out that no one knew those structures existed until scientists found dyes that could stain them. Ask: **How do stains make it easier to study the parts inside a cell?** *(Stains help scientists see structures inside of cells that would otherwise be invisible.)*

Explore

Lab Resource: Inquiry Warm-Up 

L1 **HOW LARGE ARE CELLS?** Students will estimate how tall they are in cells. This Inquiry Warm-Up can be found online.

Texas LESSON 3

Looking Inside Cells

- How Do the Parts of a Cell Work? TEKS 3B, 12D, 12F
- How Do Cells Work Together in an Organism? TEKS 12C

MY PLANET DIARY

Glowing Globs

Do these cells look as if they're glowing? This photograph shows cells that have been stained with dyes that make cell structures easier to see. Scientists view such treated cells through a fluorescent microscope, which uses strong light to activate the dyes and make them glow. Here, each green area is a cell's nucleus, or control center. The yellow "fibers" form a kind of support structure for the cell.

TECHNOLOGY

Communicate Discuss these questions with a partner. Then write your answers below.

1. Why is staining useful when studying cells through a microscope?
   Staining makes a cell's structures easier to see.
2. If you had a microscope, what kinds of things would you like to look at? Why?
   Sample: An eyelash, a grain of sand, a fingernail; I'd like to see how they look.

Lab zone: Do the Inquiry Warm-Up *How Large Are Cells?* Find the lab online.

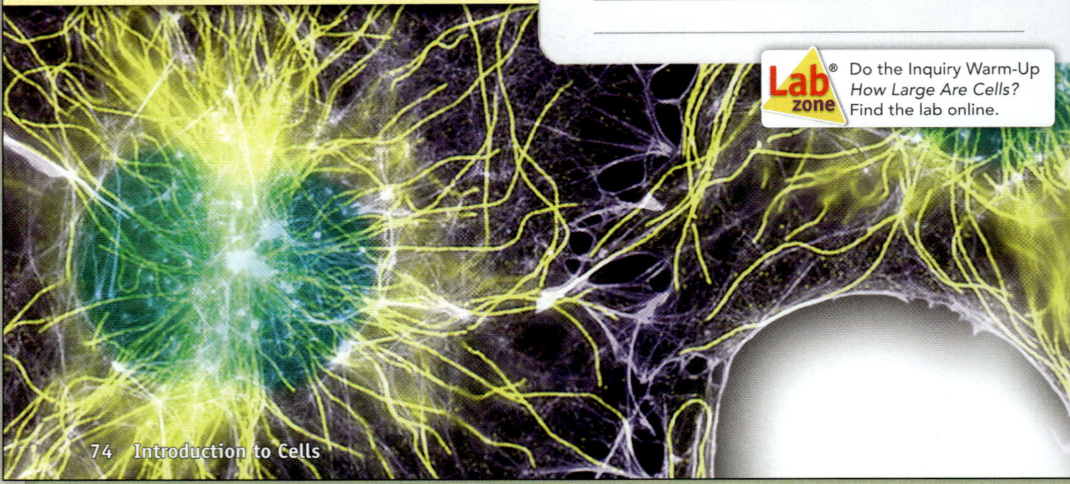

SUPPORT ALL READERS

Lexile Measure = 960L   Lexile Word Count = 1528

Prior Exposure to Content: Many students may have misconceptions on this topic

Academic Vocabulary: *identify, main idea, model*

Science Vocabulary: *nucleus, organelle, mitochondria, vacuole*

Concept Level: Generally appropriate for most students in this grade

Preteach With: My Planet Diary "Glowing Globs" and Figure 1 activity

Vocabulary

- cell wall • nucleus • organelle • ribosome
- cytoplasm • mitochondria • endoplasmic reticulum
- Golgi apparatus • vacuole • chloroplast • lysosome
- tissue • organ • organ system

Skills

Reading: Identify the Main Idea

Inquiry: Make Models

How Do the Parts of a Cell Work?

TEKS 3B, 12D, 12F In this section, you'll use models to learn about the different structures and functions of plant and animal cells.

When you look at a cell through a microscope, you can usually see the outer edge of the cell. Sometimes you can also see smaller structures within the cell.  **Each kind of cell structure has a different function within a cell.** In this lesson, you will read about the structures that plant and animal cells have in common. You will also read about some differences between the cells.

ELPS 4.F.2

Look at the drawing of a cell below. Describe to a partner the structures, or shaped objects, you see within the cell. As you read, label the structures and discuss the functions, or jobs, of each labeled part.

Cell Wall The cell wall is a rigid layer that surrounds the cells of plants and some other organisms. The cells of animals, in contrast, do not have cell walls. A cell wall helps protect and support the cell. A plant's cell wall is made mostly of a strong material called cellulose. Still, many materials, including water and oxygen, can pass through the cell wall easily.

Cell Membrane Think about how a window screen allows air to enter and leave a room but keeps insects out. One of the functions of the cell membrane is something like that of a screen. The cell membrane controls which substances pass into and out of a cell. Everything a cell needs, such as food particles, water, and oxygen, enters through the cell membrane. Waste products leave the same way. In addition, the cell membrane prevents harmful materials from entering the cell.

All cells have cell membranes. In plant cells, the cell membrane is just inside the cell wall. In cells without cell walls, the cell membrane forms the border between the cell and its environment.

Check that students color the cell membrane and key. Verify answer using Figure 3.

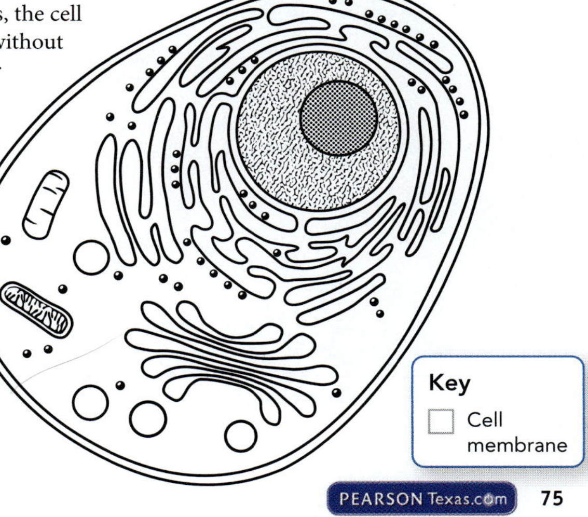

FIGURE 1

A Typical Animal Cell

You will see this diagram of a cell again in this lesson.

 **Identify** **Use a colored pencil to shade the cell membrane and fill in the box in the key.**

Key

☐ Cell membrane

Explain

Introduce Vocabulary

Point out that when Robert Hooke first discovered cells, he saw plant structures called cell walls. Students can remember that the cell wall surrounds the outside of a plant cell like a stone wall surrounds a garden.

Teach Key Concepts

Explain to students that cells contain smaller structures that carry out specific functions inside the cell, just as their bodies contain organs that have specific functions. Ask: **What two structures surround cells?** *(Cell membrane and cell wall)* **Do all cells have these structures?** *(All cells have cell membranes. Cells of plants and some other organisms have cell walls, but animal cells do not have cell walls.)*

Teach With Visuals

Tell students to look at **Figure 1.** Point out that this diagram shows many cell structures, not just the cell membrane. Students do not need to identify all structures at this point. Ask: **Where is the cell membrane?** *(It is the outermost layer of the cell.)* **Without the caption for Figure 1, how would you be able to tell if this was a diagram of an animal cell or a plant cell?** *(It is an animal cell, because it has no cell wall.)*

PEARSON Texas.com

English Language Proficiency Standards

ELPS Reading 4.F.2

Have students label structures in the diagram as they read the lesson.

Beginning Display pages 78–79. Have students repeat each label. Then read aloud page 75 and prompt students: *Is this a diagram of a plant cell or an animal cell? Does it have a cell membrane or a cell wall? Where?* Continue reading aloud and having students identify cell structures.

Intermediate Read the page together. Have students label the cell membrane. Continue through the lesson, having students label each structure.

Advanced Encourage students to elaborate with details as they describe the cell structures.

Advanced High Have partners read the lesson and complete the activity. Encourage them to explain the function of each structure in the cell.

Texas Essential Knowledge and Skills

3B Use models to represent aspects of the natural world such as human body systems and plant and animal cells.

12D Differentiate between structure and function in plant and animal cell organelles, including cell membrane, cell wall, nucleus, cytoplasm, mitochondrion, chloroplast, and vacuole.

12F Recognize that according to cell theory all organisms are composed of cells and cells carry on similar functions such as extracting energy from food to sustain life.

LESSON 2.3

Explain

Lead a Discussion

THE NUCLEUS Ask students to identify the nucleus in the cell shown in **Figure 2.** Then ask them to find the membrane surrounding the nucleus and the dense region inside the nucleus. Ask: **What is the function of the nucleus?** *(It directs the cell's activities.)* **What carries out that function inside the nucleus?** *(The chromatin, which contains the information for directing the cell)* **What is the name of the membrane that surrounds the nucleus?** *(The nuclear envelope)* **What is the dark structure within the nucleus?** *(The nucleolus)* **What does the nucleolus do?** *(It is the region of the nucleus that makes ribosomes.)*

Teach With Visuals

Tell students to look at **Figure 2,** and point out that the images shown are electron micrographs, images from an electron microscope and artists' three-dimensional renderings based on electron micrographs. Have students verify that the nucleus, the mitochondrion, and the endoplasmic recticulum all have outer membranes. Ask: **What function does the membrane have in each of these organelles?** *(It lets the cell keep different things inside the organelle than are in the cytoplasm.)* **How is this similar to the function of the cell membrane?** *(The cell membrane controls which materials leave the cell and which materials stay in the cell.)*

Nucleus A cell doesn't have a brain, but it has something that functions in a similar way. A large oval structure called the **nucleus** (NOO klee us) acts as a cell's control center, directing all of the cell's activities. The nucleus is the largest of many tiny cell structures, called **organelles,** that carry out specific functions within a cell. Notice in **Figure 2** that the nucleus is surrounded by a membrane called the nuclear envelope. Materials pass in and out of the nucleus through pores in the nuclear envelope.

Chromatin You may wonder how the nucleus "knows" how to direct the cell. Chromatin, thin strands of material that fill the nucleus, contains information for directing a cell's functions. For example, the instructions in the chromatin ensure that leaf cells grow and divide to form more leaf cells.

Nucleolus Notice the small, round structure in the nucleus. This structure, the nucleolus, is where ribosomes are made. **Ribosomes** are small grain-shaped organelles that produce proteins. Proteins are important substances in cells.

FIGURE 2

Organelles of a Cell

The structures of a cell look as different as their functions.

Complete each task.

1. **Review** Answer the questions in the boxes.
2. **Relate Text and Visuals** In the diagram on the facing page, use different-colored pencils to color each structure and its matching box in the color key.

Nucleus

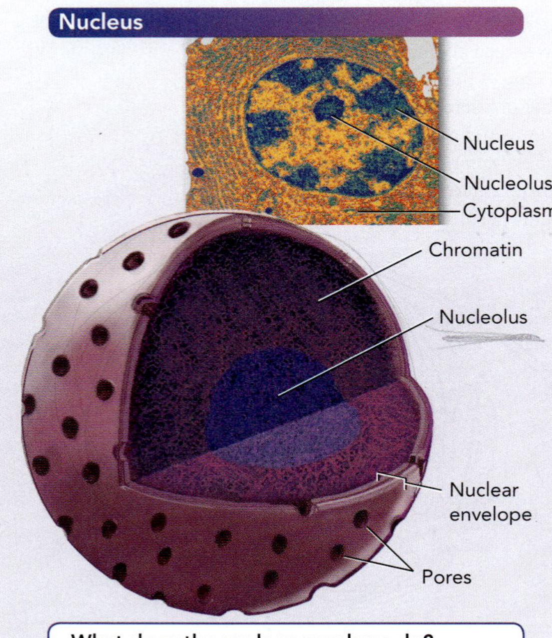

What does the nuclear envelope do?

It surrounds the nucleus.

Mitochondrion

CHALLENGE In what types of cells would you expect to find a lot of mitochondria?

Muscle cells

76 Introduction to Cells

Organelles in the Cytoplasm Most of a cell consists of a thick, clear, gel-like fluid. The **cytoplasm** fills the region between the cell membrane and the nucleus. The fluid of the cytoplasm moves constantly within a cell, carrying along the nucleus and other organelles that have specific jobs.

Mitochondria Floating in the cytoplasm are rod-shaped structures that are nicknamed the "powerhouses" of a cell. Look again at **Figure 2.** **Mitochondria** (myt oh KAHN dree uh; singular *mitochondrion*) convert energy stored in food to energy the cell can use to live and function.

Endoplasmic Reticulum and Ribosomes In **Figure 2,** you can see what looks something like a maze of passageways. The **endoplasmic reticulum** (en doh PLAZ mik rih TIK yuh lum), often called the ER, is an organelle with a network of membranes that produces many substances. Ribosomes dot some parts of the ER, while other ribosomes float in the cytoplasm. The ER helps the attached ribosomes make proteins. These newly made proteins and other substances leave the ER and move to another organelle.

Vocabulary Prefixes The prefix *endo-* is Greek for "within." If the word part *plasm* refers to the "body" of the cell, what does the prefix *endo-* tell you about the endoplasmic reticulum?

The ER is within the body of the cell.

Endoplasmic Reticulum and Ribosomes

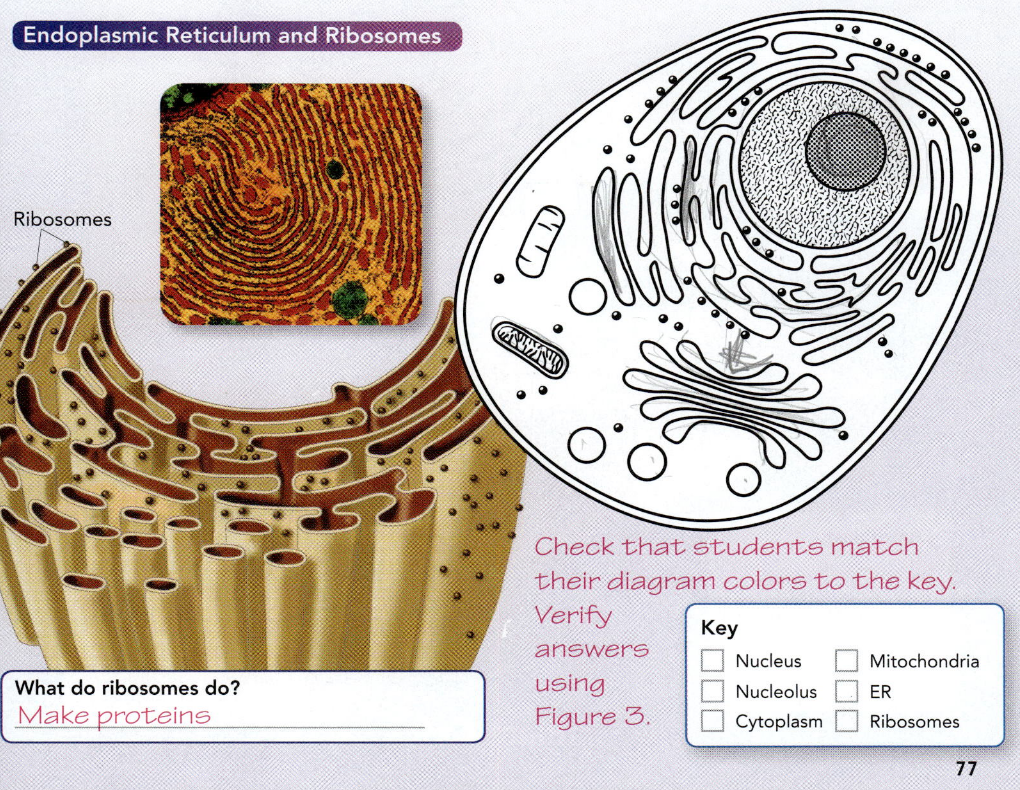

What do ribosomes do?
Make proteins

Check that students match their diagram colors to the key. Verify answers using Figure 3.

Key
- ☐ Nucleus
- ☐ Nucleolus
- ☐ Cytoplasm
- ☐ Mitochondria
- ☐ ER
- ☐ Ribosomes

Address Misconceptions

L1 MITOCHONDRIA IN PLANTS Students may think that plant cells do not contain mitochondria because they think chloroplasts carry out a similar function. Ask what mitochondria do inside a cell. *(Mitochondria convert energy stored in food to a form the cell can use).* Point out that plant cells have other organelles in which food is made, but the cells still have to break down that food for energy. Ask: **Why do plant cells contain mitochondria?** *(To break down the food the cells have made for energy)*

Elaborate

21st Century Learning

CRITICAL THINKING Ribosomes are built inside the nucleolus, but function either on the endoplasmic recticulum or in the cytoplasm. Have students write a list that shows the structures a ribosome would travel through before it enters the cytoplasm. *(From nucleolus through the nucleus, into endoplasmic reticulum, then into cytoplasm)*

Differentiated Instruction

L1 Division of Labor Explain that just as jobs in a town are performed by different people, the tasks inside a cell are performed by different organelles. Have students make a list of jobs in the cell, then write down which organelle performs each task as they read.

L3 Prokaryotic Cells This lesson focuses on the biology of eukaryotic cells. Have students research the cell structure of one group of prokaryotes (either bacteria and archeans) and build a model of their cells.

L3 Endosymbiotic Hypothesis Students may be surprised to learn that mitochondria and chloroplasts seem to have evolved from symbiotic bacteria living inside larger cells. Have them research the evidence that supports this hypothesis.

LESSON 2.3

Explain

Teach With Visuals

Tell students to look at **Figure 3.** Point out that although many structures are found in both kinds of cells, a few are only found in plant cells. Ask: **Which structures are found only in plant cells?** *(Cell wall, chloroplasts)* **Which structure is much larger in a plant cell than in an animal cell?** *(The vacuole)* **Are there any structures shown in the animal cell that are not found in the plant cell?** *(Lysosomes)* If students cannot find the information they need to complete all the labels, tell them that some of the cell structures are identified and explained on the pages that follow. You may wish to discuss those cell structures with the class and then return to **Figure 3** for a review of all structures.

21st Century Learning

L1 ACCOUNTABILITY Students can use **Figure 3** to check their understanding of the names and functions of cell structures. Tell students to use small sticky notes to cover the label boxes, and then to write the names and functions of the structures on the sticky notes.

APPLY THE TEKS 12D, 12F

CELLS IN LIVING THINGS

What structures are found in cells?

FIGURE 3

These illustrations show typical structures found in plant and animal cells. Other living things share many of these structures, too.

Describe **Differentiate the functions of the cell structures in the boxes provided. (Turn the page to find information about the Golgi apparatus, vacuoles, chloroplasts, and lysosomes.)**

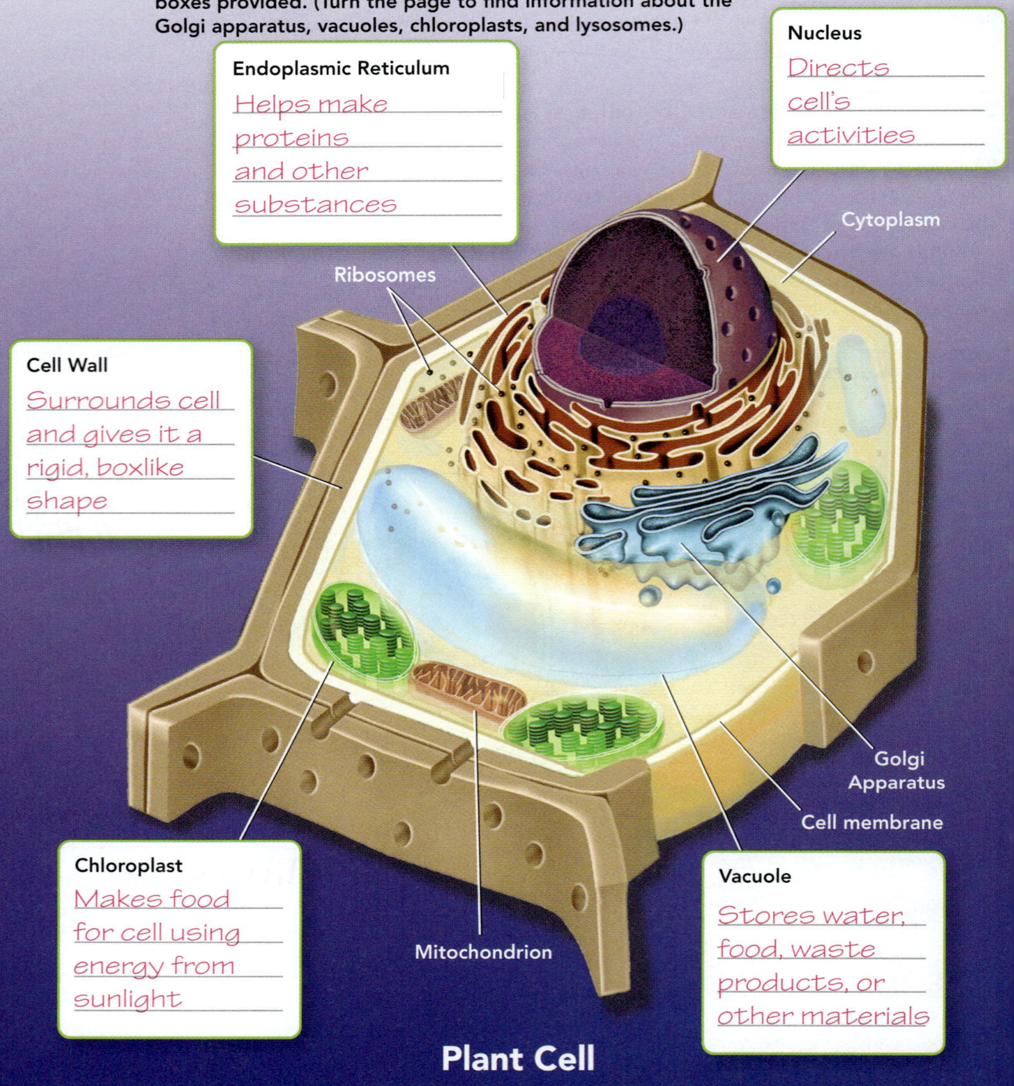

Endoplasmic Reticulum
Helps make proteins and other substances

Nucleus
Directs cell's activities

Cell Wall
Surrounds cell and gives it a rigid, boxlike shape

Chloroplast
Makes food for cell using energy from sunlight

Vacuole
Stores water, food, waste products, or other materials

Ribosomes

Cytoplasm

Golgi Apparatus

Cell membrane

Mitochondrion

Plant Cell

78 Introduction to Cells

Texas Essential Knowledge and Skills

12D Differentiate between structure and function in plant and animal cell organelles, including cell membrane, cell wall, nucleus, cytoplasm, mitochondrion, chloroplast, and vacuole.

12F Recognize that according to cell theory all organisms are composed of cells and cells carry on similar functions such as extracting energy from food to sustain life.

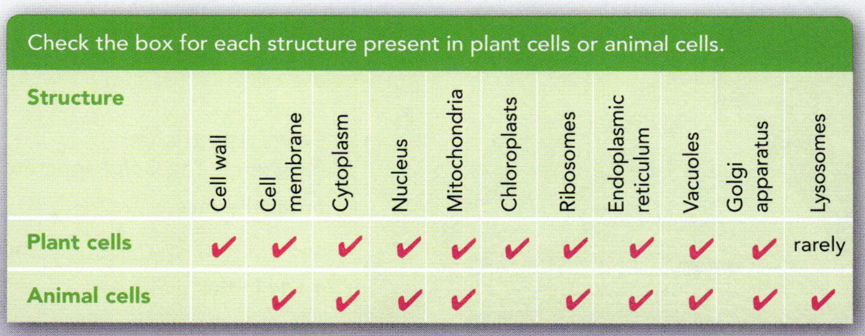

Check the box for each structure present in plant cells or animal cells.

Structure	Cell wall	Cell membrane	Cytoplasm	Nucleus	Mitochondria	Chloroplasts	Ribosomes	Endoplasmic reticulum	Vacuoles	Golgi apparatus	Lysosomes
Plant cells	✔	✔	✔	✔	✔	✔	✔	✔	✔	✔	rarely
Animal cells		✔	✔	✔	✔		✔	✔	✔	✔	✔

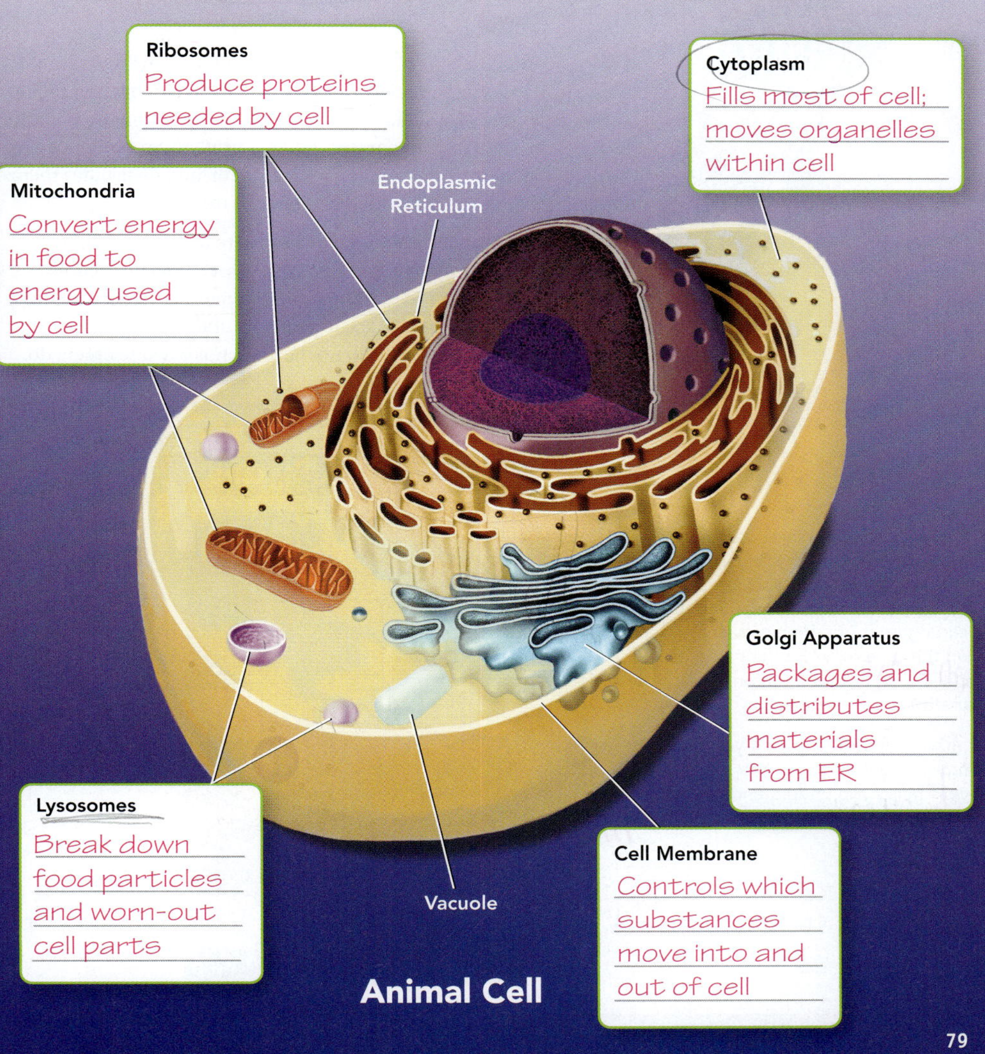

Differentiated Instruction

L1 Compare and Contrast Pair students and have them explain the differences between plant and animal cells to each other. Students can use a two-column chart to help them organize the differences they identify.

L3 Cell Diversity This lesson focuses on the differences between animal and plant cells. Encourage curious students to choose another taxon of eukaryote (fungi, different species of protist) and do research to find out how their cells are organized.

LESSON 2.3

Elaborate

Apply the TEKS

Direct students' attention to the images of the plant and animal cells in **Figure 3.** Point out that both types of cells contain specialized structures with specific functions. Ask: **Which organelles are found in both plant and animal cells?** *(Cell membrane, mitochondria, ribosomes, endoplasmic recticulum, Golgi apparatus, vacuole)* **Does each of these structures perform the same functions in both kinds of cells?** *(Yes)* **Which structures control the movement of materials in the cell?** *(Cell membrane, endoplasmic reticulum, Golgi apparatus)* **Which are involved in making proteins?** *(Nucleus, ribosomes, endoplasmic reticulum, Golgi apparatus)* **Which break down food to release energy?** *(Mitochondria)*

Teacher Demo 

L2 COMPARE SLIDES TO ILLUSTRATIONS

Materials microprojector, prepared slide of plant cells, prepared slide of animal cells, drawings of other types of cells (such as leaf and root cells for plants, and muscle and bone cells for animals)

Time 10 minutes

Point out that the drawings of plant and animal cells shown on these pages are generalized representations of cells. Project the prepared slides, and have students relate the organelles they see to the illustrations. Encourage them to describe how the actual cells vary in shape and structure from the generalized cells in their worktext. Point out that cells have many different shapes and sizes. They also can vary in the specific organelles they contain.

Ask: **Why do you think different cells look so different from one another?** *(Because they have different functions in the organism)*

Texas Essential Knowledge and Skills

12D Differentiate between structure and function in plant and animal cell organelles, including cell membrane, cell wall, nucleus, cytoplasm, mitochondrion, chloroplast, and vacuole.

12F Recognize that according to cell theory all organisms are composed of cells and cells carry on similar functions such as extracting energy from food to sustain life.

LESSON 2.3

Explain

Lead a Discussion

CHLOROPLASTS Tell students that plant cells are green because they contain a colored substance called chlorophyll. Although most leaves look uniformly green, the cells in the leaves do not have chlorophyll spread evenly through them. When cells from a leaf are viewed under a microscope, dots of green are visible. Ask: **What are the green structures in a plant cell?** *(Chloroplasts)* **What is their function?** *(They make food for the plant.)* **Why can't animal cells make their own food?** *(Animal cells do not have chloroplasts.)*

21st Century Learning

L3 **CRITICAL THINKING** Explain that while lysosomes are only shown in the image of the animal cell, they are also occasionally found in plant cells. Ask: **Why are lysosomes more common in animal cells?** *(Lysosomes break down food particles that have been engulfed by the animal cell; plants make their own food inside their cells.)*

Elaborate

Apply It!

L1 Before beginning the activity, review the parts of a cell and their functions.

Make Models Point out to students that models can help them understand things that are too small to observe directly. In this case, they are making a mental model when they apply what they know about a familiar set of relationships to a new situation. Ask: **What other mental models of cell function could you create?** *(Samples: parts of a computer, jobs in a company)*

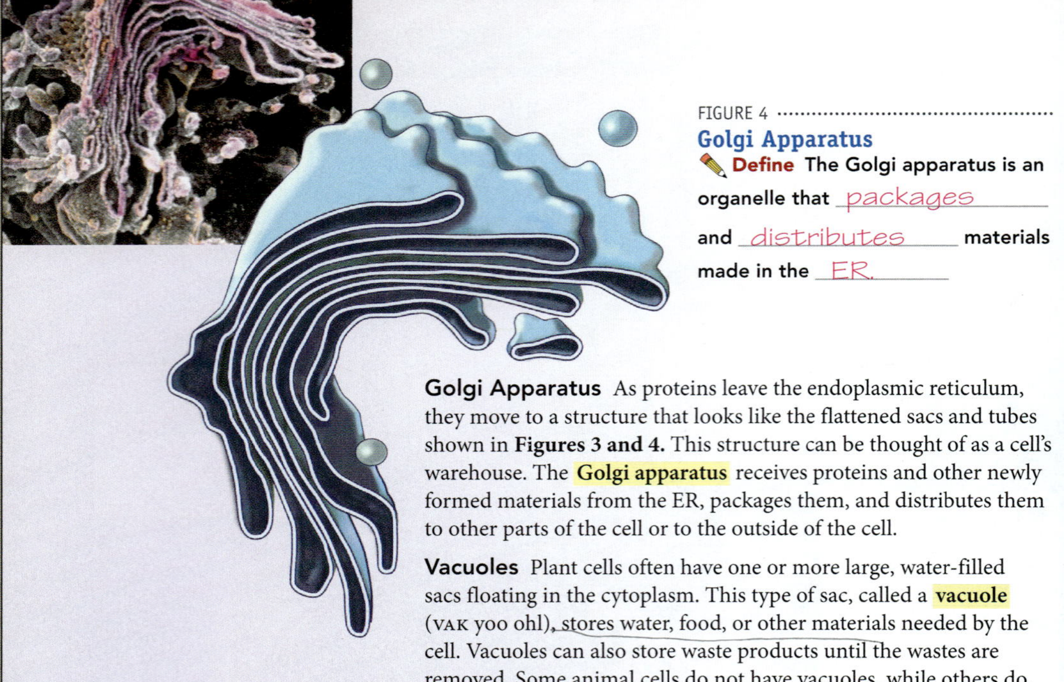

FIGURE 4

Golgi Apparatus

Define The Golgi apparatus is an organelle that packages and distributes materials made in the ER.

Golgi Apparatus As proteins leave the endoplasmic reticulum, they move to a structure that looks like the flattened sacs and tubes shown in **Figures 3 and 4.** This structure can be thought of as a cell's warehouse. The **Golgi apparatus** receives proteins and other newly formed materials from the ER, packages them, and distributes them to other parts of the cell or to the outside of the cell.

Vacuoles Plant cells often have one or more large, water-filled sacs floating in the cytoplasm. This type of sac, called a **vacuole** (VAK yoo ohl), stores water, food, or other materials needed by the cell. Vacuoles can also store waste products until the wastes are removed. Some animal cells do not have vacuoles, while others do.

apply it!

Can a store's building be a model for a cell? If so, how do the parts of a cell function in ways that are similar to the parts of a building? See if you can figure it out. In each blank space on the picture, write the name of a cell structure that functions most like that part of the store.

Make Models How do you think making real-world comparisons with cells helps you understand cell structure and function?

Sample: Real-world comparisons relate what I know to what I don't know.

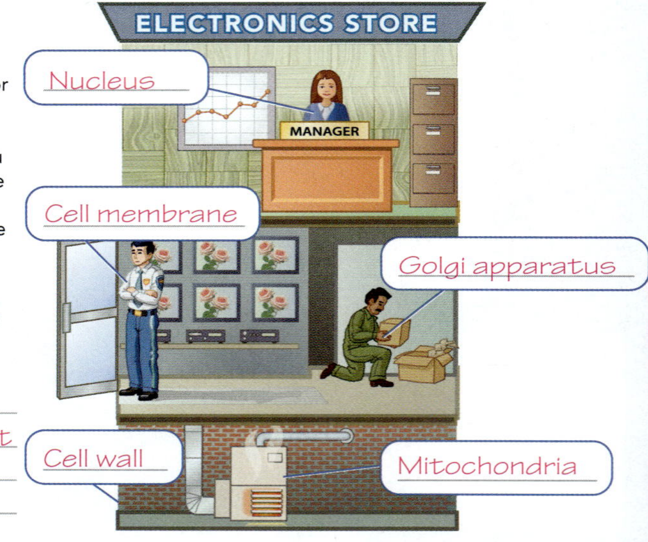

Chloroplasts A typical plant cell contains green structures, called **chloroplasts**, in the cytoplasm. A chloroplast, shown in **Figure 5**, captures energy from sunlight and changes it to a form of energy cells can use in making food. Animal cells don't have chloroplasts, but the cells of plants and some other organisms do. Chloroplasts make leaves green because leaf cells contain many chloroplasts.

Lysosomes Look again at the animal cell in **Figure 3**. Notice the saclike organelles, called **lysosomes** (LY suh sohmz), which contain substances that break down large food particles into smaller ones. Lysosomes also break down old cell parts and release the substances so they can be used again. You can think of lysosomes as a cell's recycling centers. Most plant cells do not have lysosomes, but some do.

FIGURE 5
A Chloroplast
Infer In which part of a plant would you NOT expect to find cells with chloroplasts?
Sample: root, flower, or seed

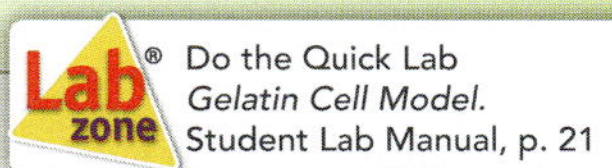

Do the Quick Lab *Gelatin Cell Model.* Student Lab Manual, p. 21

Assess Your Understanding
TEKS 12D, 12F

1a. Differentiate Use the table you completed in **Figure 3** to summarize the differences between a plant cell and an animal cell.
Plant cells have cell walls and chloroplasts. Animal cells have lysosomes, but they are only rarely found in plant cells.

b. Make Generalizations How are the functions of the endoplasmic reticulum and the Golgi apparatus related?
Proteins and other substances made in the ER are stored, packaged, and distributed by the Golgi apparatus.

c. CHALLENGE A solar panel collects sunlight and converts it to heat or electrical energy. How is a solar panel similar to chloroplasts?
Like chloroplasts, a solar panel changes sunlight into another form of usable energy.

d. Recognize What structures and materials make up a cell?
Sample: A cell membrane, a nucleus, cytoplasm, and other organelles

got it?
- O **I get it!** Now I know that different kinds of organelles in a cell have different functions.
- O I need extra help with See TE note.

Differentiated Instruction

L1 Flashcards Have students who are having difficulty remembering the names and functions of cell organelles make flash cards with the name of an organelle on one side and its function on the other. Pair students and have them use the cards to quiz each other.

L3 Making Models Encourage students to develop a model that demonstrates how the nucleus, ribosomes, endoplasmic reticulum, and Golgi apparatus work together to make and modify proteins in the cell.

Teacher Demo 

L1 A FUNCTION OF PLANT VACUOLES

Materials water, wilted houseplant

Time 5 minutes each at the beginning and end of class

Show students the wilted plant at the beginning of class, and remind them that a plant's cell walls can only provide support when the cell is full of water. If the cells lose water, the cell walls lose rigidity and the plant wilts. Water the plant, and have students observe it at intervals throughout the class.

Ask: **Where is water stored inside the plant cell?** *(Inside the vacuole)* **Why has the plant recovered from being wilted?** *(Its vacuoles have filled with water.)*

Lab Resource: Quick Lab 

L2 GELATIN CELL MODEL Students will explore cell structures by making a model with gelatin and other materials. This Quick Lab can be found in the Student Lab Manual, p. 21, and online.

Evaluate

Assess Your Understanding

After students answer the questions, have them evaluate their understanding by completing the appropriate sentence.

RTI Response to Intervention

1. If students cannot explain organelle structures and functions, **then** have them review the *Cells in Living Things* activity.

Explain

Teach Key Concepts

Explain to students that multicellular organisms like plants and animals have specialized cells organized into tissues, organs, and organ systems that perform specific tasks inside their bodies. Ask: **Name an organ inside your body.** *(Sample: heart)* **What kind of tissue is it made up of?** *(Sample: cardiac muscle)* **What organ system is it part of, and what is its function in your body?** *(Sample: It is one part of the circulatory system, which is the body's transport system.)*

Lead a Discussion

ONE CELL OR MANY? Point out to students that some organisms are made of just one cell, while other organisms are made of many cells. Ask: **What is a unicellular organism?** *(An organism made of one cell)* **What is a multicellular organism?** *(An organism made of many cells)* Point out that a unicellular organism is always small, a multicellular organism does not have to be large. Ask: **Which kind of organism shows the most specialization?** *(A multicellular organism)* **What is the term for a group of specialized cells that work together?** *(A tissue)*

Identify the Main Idea Tell students that the biggest idea in a paragraph is its main idea. Other ideas or facts support the main idea. Encourage students to practice the skill of identifying the main idea whenever they read.

Teach With Visuals

Tell students to look at **Figure 6.** Point out that each cell in the figure is specialized for a particular function. Each type of specialized cell tends to have a distinctive appearance. A cell's appearance often reflects its function. Ask: **How do you think each cell's shape helps it do its job in the organism?** *(Red blood cells are small and round, so can squeeze through tiny capillaries; leaf cells contain many food-making chloroplasts for making food; the extensions on the root cells give the cell more area to absorb water; the nerve cell has many extensions to connect and send messages to other cells.)*

Texas Essential Knowledge and Skills

12C Recognize levels of organization in plants and animals, including cells, tissues, organs, organ systems, and organisms.

TEKS 12C In this section, you'll recognize that specialized cells work together to carry out specific functions that benefit the entire organism.

How Do Cells Work Together in an Organism?

Plants and animals (including you) are multicellular, which means "made of many cells." Single-celled organisms are called unicellular. In a multicellular organism, the cells often look quite different from one another. They also perform different functions.

Identify the Main Idea Reread the paragraph about specialized cells. Then underline the phrases or sentences that describe the main ideas about specialized cells.

Specialized Cells All cells in a multicellular organism must carry out key functions, such as getting oxygen, to remain alive. However, cells also may be specialized. That is, they perform specific functions that benefit the entire organism. These specialized cells share what can be called a "division of labor." One type of cell does one kind of job, while other types of cells do other jobs. For example, red blood cells carry oxygen to other cells that may be busy digesting your food. Just as specialized cells differ in function, they also differ in structure. **Figure 6** shows specialized cells from plants and animals. Each type of cell has a distinct shape. For example, a nerve cell has thin, fingerlike extensions that reach toward other cells. These structures help nerve cells transmit information from one part of your body to another. The nerve cell's shape wouldn't be helpful to a red blood cell.

FIGURE 6

The Right Cell for the Job

Many cells in plants and animals carry out specialized functions.

Draw Conclusions Write the number of each kind of cell in the circle of the matching function.

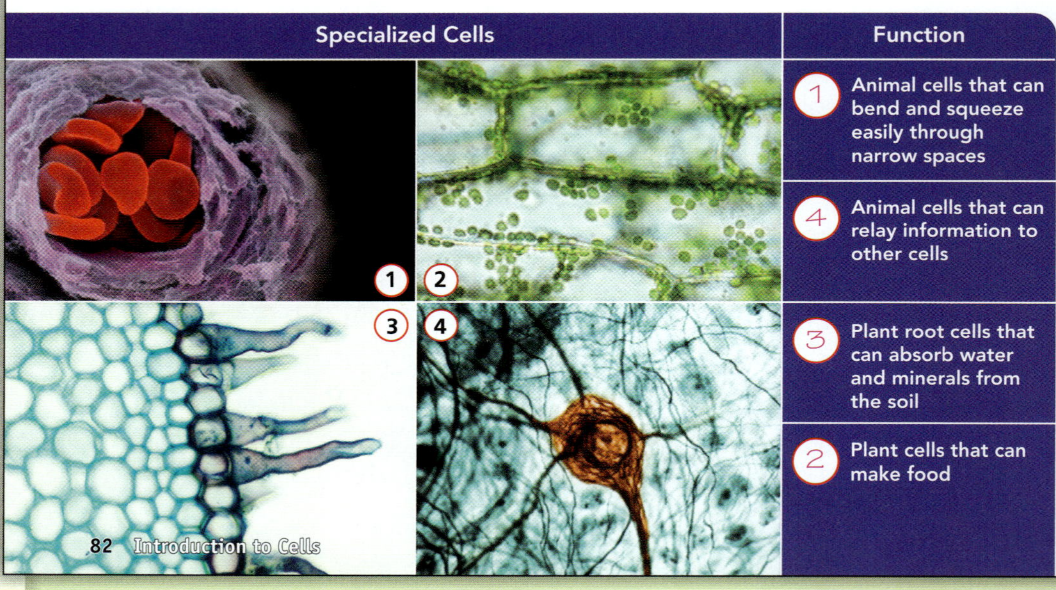

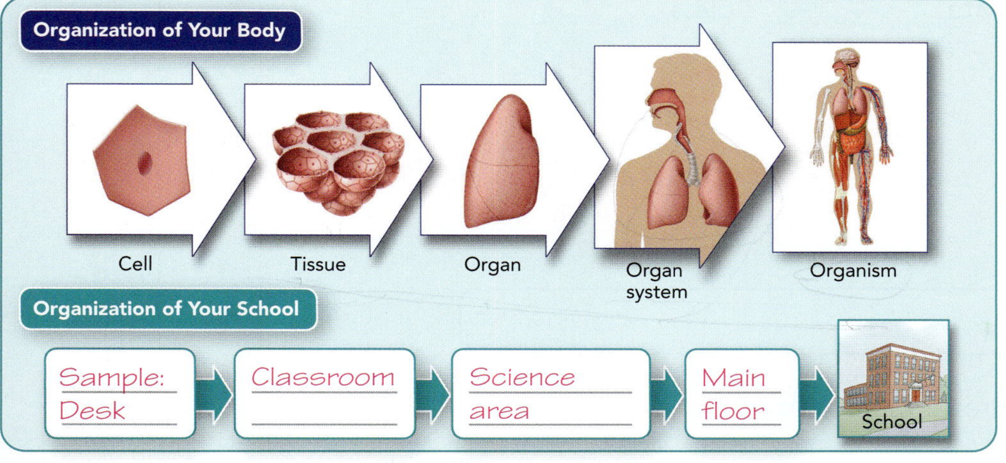

Cells Working Together A division of labor occurs among specialized cells in an organism. It also occurs at other levels of organization. **In multicellular organisms, cells are organized into tissues, organs, and organ systems.** A **tissue** is a group of similar cells that work together to perform a specific function. For example, your brain is made mostly of nerve tissue, which consists of nerve cells that relay information to other parts of your body. An **organ,** such as your brain, is made of different kinds of tissues that function together. For example, the brain also has blood vessels that carry the blood that supplies oxygen to your brain cells. Your brain is part of your nervous system, which directs body activities and processes. An **organ system** is a group of organs that work together to perform a major function. As **Figure 7** shows, the level of organization becomes more complex from cell, to tissue, to organ, to organ systems, to the organism itself.

FIGURE 7

Levels of Organization
Living things are organized in levels of increasing complexity. Many nonliving things, like a school, have levels of organization, too.

Apply Concepts On the lines above, write the levels of organization of your school building, from the simplest level, such as your desk, to the most complex.

Do the Quick Lab *Tissues, Organs, Organ Systems.* Student Lab Manual, p. 22

Assess Your Understanding

TEKS 12C

2a. Describe What does the term *division of labor* mean as it is used in this lesson?

Different cells do different jobs.

b. Recognize Would a tissue or an organ have more kinds of specialized cells? Explain your answer.

An organ, because it's made of different tissues

got it?

O **I get it!** Now I know that the levels of organization in a multicellular organism include cells, tissues, organs, and organ systems.

O I need extra help with See TE note.

83

Address Misconceptions

L1 CELLS CONTAIN THE SAME GENES Students may think that specialized cells in a multicellular organisms contain different genetic material. However, every cell in an individual organism is genetically identical. Specialized cells look and behave differently because different portions of their genetic instructions are active. Ask: **What is the origin of each cell in a multicellular organism?** *(Each cell is the descendant of another genetically identical cell.)*

Elaborate

Lab Resource: Quick Lab 

L1 TISSUES, ORGANS, ORGAN SYSTEMS Students will build a model that shows the organization of a multicellular organism. This Quick Lab can be found in the Student Lab Manual, p. 22, and online.

Evaluate

Assess Your Understanding

After students answer the questions, have them evaluate their understanding by completing the appropriate sentence.

RTI Response to Intervention

2a. If students need help with explaining the term *division of labor,* **then** have them reread the paragraph under the red heading *Cells Working Together.*

b. If students cannot make an inference about specialized cells, **then** have them list four organs in their own body and look up what kinds of tissues each organ contains.

Differentiated Instruction

L1 Prefixes Remind students that a prefix is placed in front of a root word to change its meaning. *Uni-* means "one," and *multi-* means "many."

L3 Cells in Tissues Ask students to do research to find images of cells from a variety of tissues in their bodies and describe how their shapes are related to their functions. If students do their research on the Internet, remind them to follow prescribed guidelines for Internet use.

Name ________________ Date ________ Class ________

Assess Your Understanding

Looking Inside Cells

How Do the Parts of a Cell Work?

1a. **DIFFERENTIATE** Use the table you completed in Figure 3 to summarize the differences between a plant cell and an animal cell.

____

____

b. **MAKE GENERALIZATIONS** How are the functions of the endoplasmic recticulum and the Golgi apparatus related?

____

____

c. **CHALLENGE** A solar panel collects sunlight and converts it to heat or electrical energy. How is a solar panel similar to chloroplasts?

____

____

d. **RECOGNIZE** What structures and materials make up a cell?

____

____

____

How Do Cells Work Together in an Organism?

2a. **DESCRIBE** What does the term *division of labor* mean as it is used in this lesson?

____

b. **RECOGNIZE** Would a tissue or an organ have more kinds of specialized cells? Explain your answer.

____

Place the outside corner, the corner away from the dotted line, in the corner of your copy machine to copy onto letter-size paper.

Name ______________ Date ________ Class ________

Enrich

Looking Inside Cells

The figure below shows a city that is a model for a cell. Study the figure, and answer the questions that follow on a separate sheet of paper.

Modeling Cell Structures

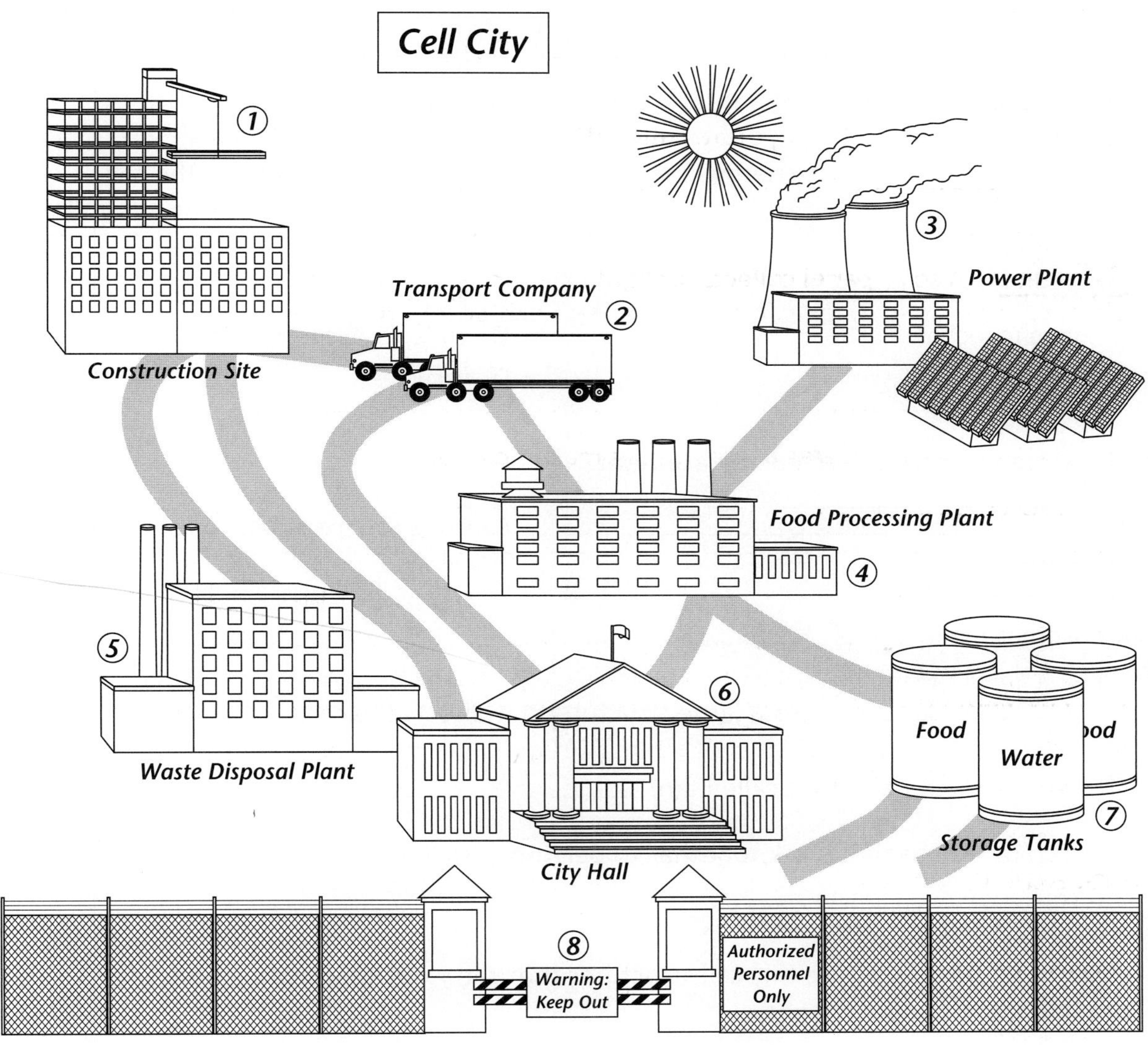

1. State the function performed by each numbered structure in the figure.
2. Now name a cell structure that performs each of these same functions.
3. Does "Cell City" represent a plant cell or an animal cell? Explain your answer.

Place the outside corner, the corner away from the dotted line, in the corner of your copy machine to copy onto letter-size paper.

Name ______________________ Date ____________ Class ____________

Lesson Quiz

Looking Inside Cells

Write the letter of the correct answer on the line at the left.

1. ___ Which of the following organelles are NOT found in animal cells?

A mitochondria

B chloroplasts

C cell membrane

D Golgi apparatus

2. ___ Ribosomes are produced in the

A cell membrane

B vacuoles

C chloroplasts

D nucleolus

3. ___ Mitochondria convert energy in food into

A lysosomes

B energy used by the cell

C more mitochondria

D cytoplasm

4. ___ The region between the cell membrane and the nucleus is filled with

A cytoplasm

B cell wall

C endoplasmic reticulum

D vacuoles

Fill in the blank to complete each statement.

5. The ______________________ controls the materials that enter and leave the cell.

6. Ribosomes make ______________.

7. The ______________ is a large structure that directs the cell's activities.

8. The storage area of a cell is called a(n) ______________.

9. A group of organs that work together to perform a major function is called a(n) ______________________.

10. ______________ are tiny cell structures that carry out specific functions in the cell.

Place the outside corner, the corner away from the dotted line, in the corner of your copy machine to copy onto letter-size paper.

Looking Inside Cells

Answer Key

Review and Reinforce

Find the worksheet in the Student Workbook.

1. cytoplasm
2. endoplasmic reticulum
3. nucleus
4. mitochondrion
5. cell membrane
6. a group of similar cells that work together to perform a function
7. an organelle that uses light to make food in plant cells
8. an organelle that makes proteins
9. the control center of the cell
10. an organelle that breaks down food to make energy
11. a group of different tissues that function together
12. made up of many cells
13. Both the organelles inside a cell and the cells inside a multicellular organism specialize. Specific organelles/cells have specific functions. Organelles are small parts of a cell and cells are small parts of a multicellular organism.

Enrich

1. (1) building new structures (2) moving materials (3) energy production (4) produces food (5) removes wastes (6) control (7) storage (8) controls what enters and leaves city
2. (1) ribosome (2) endoplasmic reticulum (3) chloroplast (4) mitochondria (5) lysosome (6) nucleus (7) vacuole (8) cell membrane
3. Cell City represents a plant cell because it contains chloroplasts.

Lesson Quiz

1. B
2. D
3. B
4. A
5. cell membrane
6. proteins
7. nucleus
8. vacuole
9. organ system
10. Organelles

Chemical Compounds in Cells

 **What structures are found in cells?**

Lesson Objectives	TEKS	ELPS
Define elements and compounds.	6, 6A	4.E
Identify the main compounds that are important in cells.	2E, 6A	

LESSON PACING:
1–2 periods or $\frac{1}{2}$–1 block

Lesson Vocabulary

- element
- compound
- carbohydrate
- lipid
- protein
- enzyme
- nucleic acid
- DNA
- double helix

Content Refresher

Polymers Many of the molecules inside of cells are polymers—long molecules built of many similar units joined together. These units are called *monomers*. Three of the most important classes of biological molecules are polymers: complex carbohydrates, proteins, and nucleic acids.

Complex carbohydrates are built of monomers called sugars. Cellulose, a complex carbohydrate that makes up the cell walls of plants, is made of many identical glucose molecules bound together in a long unbranching strand. Glycogen, a complex carbohydrate that animals use for short-term energy storage, is made of many glucose molecules bound in a branching chain.

Proteins are built of monomers called amino acids. The enormous variety of amino acid sequences in different proteins gives rise to a variety of protein shapes and functions.

Nucleic acids are built of long chains of nucleotides; each nucleotide is made up of a sugar, a nitrogen-containing base, and a phosphate group. Nucleotide units form long chains. In DNA, the bases in one nucleic acid chain form weak bonds with the bases on another, giving the molecule its double-stranded structure.

Texas Essential Knowledge and Skills

2E Analyze data to formulate reasonable explanations, communicate valid conclusions supported by the data, and predict trends.
6 The student knows that matter has physical and chemical properties and can undergo physical and chemical changes.
6A Identify that organic compounds contain carbon and other elements such as hydrogen, oxygen, phosphorus, nitrogen, or sulfur.

English Language Proficiency Standards

ELPS Reading 4.E Read linguistically accommodated content area material with a decreasing need for linguistic accommodations as more English is learned.

DIFFERENTIATED INSTRUCTION KEY
L1 Struggling Students or Special Needs
L2 On-Level Students L3 Advanced Students

LESSON PLANNER 2.4

Investigations and Activities

My Planet Diary, **Student Edition,** p. 84

Inquiry: Inquiry Warm-Up, Detecting Starch, **PearsonTexas.com**

Introduce Vocabulary, **Teacher's Edition,** p. 85

Teach Key Concepts, **Teacher's Edition,** p. 85

Inquiry: Quick Lab, What Is a Compound?, **Lab Manual,** p. 23

Teach Key Concepts, **Teacher's Edition,** p. 86

Lead a Discussion, Carbohydrates, **Teacher's Edition,** p. 86

Lead a Discussion, Structural Proteins, **Teacher's Edition,** p. 87

Address Misconceptions, Lipids as Food, **Teacher's Edition,** p. 87

21st Century Learning, Critical Thinking, **Teacher's Edition,** p. 87

Support the TEKS, Cells Need Proteins and Lipids, **Teacher's Edition,** p. 87

Differentiated Instruction, **Teacher's Edition,** p. 87

Lead a Discussion, Chromatin and DNA, **Teacher's Edition,** p. 88

Do the Math!, **Student Edition,** p. 88

Inquiry: Teacher Demo, How Much Water?, **Teacher's Edition,** p. 89

Differentiated Instruction, **Teacher's Edition,** p. 89

Inquiry: Quick Lab, What's That Taste?, **Lab Manual,** p. 24

TEKS Review

Apply the TEKS, Comparing Elements, **Student Edition,** p. 94

TEKS Practice, **Student Edition,** p. 95

TEKS Practice: Chapter and Cumulative Review, **Student Edition,** p. 98

Lesson 2.4, **TEKS Preparation and Study Guide Workbook,** p. 22

SHORT ON TIME? To do this lesson in approximately half the time, do the Activate Prior Knowledge activity. A discussion of the Key Concepts will familiarize students with the lesson content. Have students do the Quick Labs. The rest of the lesson can be completed by students independently.

These editable worksheets are available on **PearsonTexas.com.**
Print versions can be found in the **TEKS Preparation and Study Guide Workbook.**

Name ____________ Date ______ Class ______

2.4 Chemical Compounds in Cells

Key Concept Summaries

What Are Elements and Compounds?

You have probably heard of carbon, hydrogen, and oxygen. All of these are examples of **elements** found in your body. **An element is any substance that cannot be broken down into simpler substances.** The smallest unit of an element is a particle called an atom. Most elements in living things occur in the form of **compounds. Compounds form when two or more elements combine chemically.** When elements combine, they form units called molecules. Organic compounds contain carbon and other elements such as oxygen, hydrogen, and phosphorus.

What Compounds Do Cells Need?

Some important groups of organic compounds that living things need include carbohydrates, lipids, proteins, and nucleic acids. Water is a necessary inorganic compound.

Carbohydrates (sugars, starches) are energy-rich compounds containing carbon, hydrogen, and oxygen. Carbohydrates are used for energy in living things and are important components in cell walls and on cell membranes. **Lipids** (fats, oils, waxes) are compounds that are made mostly of carbon and hydrogen and some oxygen. Cells use lipids to store energy for later use, and cell membranes are made of lipid.

Proteins are large organic molecules containing carbon, hydrogen, oxygen, and nitrogen (and in some proteins, sulfur). Proteins make up parts of the cell membrane and organelles. Specialized proteins called **enzymes** speed up chemical reactions inside living things.

Nucleic acids are long molecules containing carbon, hydrogen, oxygen, nitrogen, and phosphorus. They include **DNA,** the molecule that stores the genetic information used to control the cell. The shape of a DNA molecule is called a **double helix.**

Water plays many important roles in a cell such as assisting in chemical reactions, giving the cell its shape, regulating cell temperature, and carrying substances in and out of the cell. About two-thirds of every living thing is water.

22

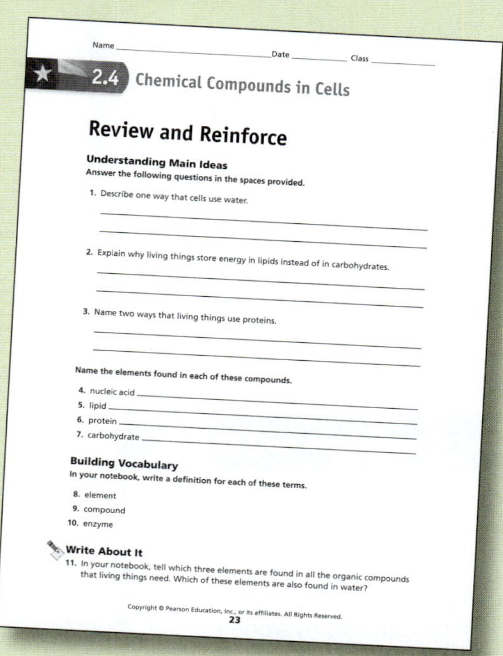

Name ____________ Date ______ Class ______

2.4 Chemical Compounds in Cells

Review and Reinforce

Understanding Main Ideas
Answer the following questions in the spaces provided.

1. Describe one way that cells use water.
2. Explain why living things store energy in lipids instead of in carbohydrates.
3. Name two ways that living things use proteins.

Name the elements found in each of these compounds.

4. nucleic acid ____________
5. lipid ____________
6. protein ____________
7. carbohydrate ____________

Building Vocabulary
In your notebook, write a definition for each of these terms.

8. element
9. compound
10. enzyme

Write About It

11. In your notebook, tell which three elements are found in all the organic compounds that living things need. Which of these elements are also found in water?

23

Lexile Measure = 910L

LESSON 2.4

Chemical Compounds in Cells

Establish Learning Objectives

After this lesson, students will be able to:

- Define elements and compounds.
- Identify the main compounds that are important in cells.

Engage

Activate Prior Knowledge

MY PLANET DIARY Read *Energy Backpacks* with the class. Ask students what foods they eat for energy. Point out that their bodies also store energy as fat, although not in anything as dramatic as a hump. Have students observe the area where this camel lives. Ask: **How do you think having humps has helped camels survive?** *(There isn't a lot of food available, so the stored fat in the humps keeps the animal from starving when conditions are like this.)*

Explore

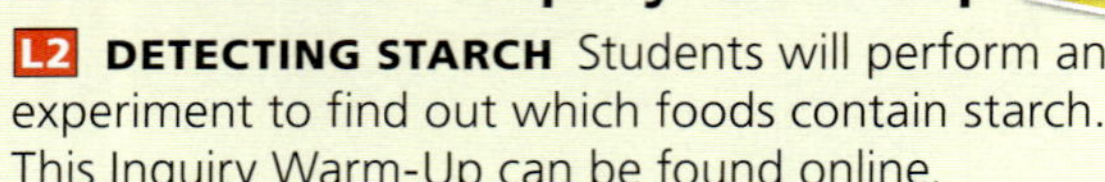

Lab Resource: Inquiry Warm-Up

L2 **DETECTING STARCH** Students will perform an experiment to find out which foods contain starch. This Inquiry Warm-Up can be found online.

Texas Essential Knowledge and Skills

6 The student knows that matter has physical and chemical properties and can undergo physical and chemical changes.

6A Identify that organic compounds contain carbon and other elements such as hydrogen, oxygen, phosphorus, nitrogen, or sulfur.

Texas LESSON 4

Chemical Compounds in Cells

- What Are Elements and Compounds? TEKS 6, 6A
- What Compounds Do Cells Need? TEKS 2E, 6A

my planet Diary

Energy Backpacks

Some people think a camel's humps carry water. Not true! They actually store fat. A hump's fatty tissue supplies energy when the camel doesn't eat. When a camel has enough food, the hump remains hard and round. But when food is scarce, the hump gets smaller and may sag to the side. If the camel then gets more food, the hump can regain its full size and shape in about three or four months.

MISCONCEPTION

Communicate Discuss this question with a group of classmates. Then write your answer below.

How do you think the camel might be affected if it didn't have humps?

Sample: It could run out of energy and be unable to travel as far.

Lab zone Do the Inquiry Warm-Up *Detecting Starch.* Find the lab online.

What Are Elements and Compounds?

TEKS 6, 6A In this section, you'll learn about some properties of elements and compounds, including organic compounds, that are essential to life.

You are made of many substances. These substances supply the raw materials that make up your blood, bones, muscles, and more. They also take part in the processes carried out by your cells.

Elements You have probably heard of carbon, hydrogen, and oxygen. All of these are examples of **elements** found in your body. **An element is any substance that cannot be broken down into simpler substances.** The smallest unit of an element is a particle called an atom. Any single element is made up of only one kind of atom.

SUPPORT ALL READERS

Lexile Measure = 910L Lexile Word Count = 1297

Prior Exposure to Content: Many students may have misconceptions on this topic

Academic Vocabulary: *compare, conclusion, contrast*

Science Vocabulary: *protein, enzyme, nucleic acid, DNA*

Concept Level: Generally appropriate for most students in this grade

Preteach With: My Planet Diary "Energy Backpacks" and Figure 4 activity

Vocabulary
- element
- compound
- carbohydrate
- lipid
- protein
- enzyme
- nucleic acid
- DNA
- double helix

Skills
- Reading: Compare and Contrast
- Inquiry: Draw Conclusions

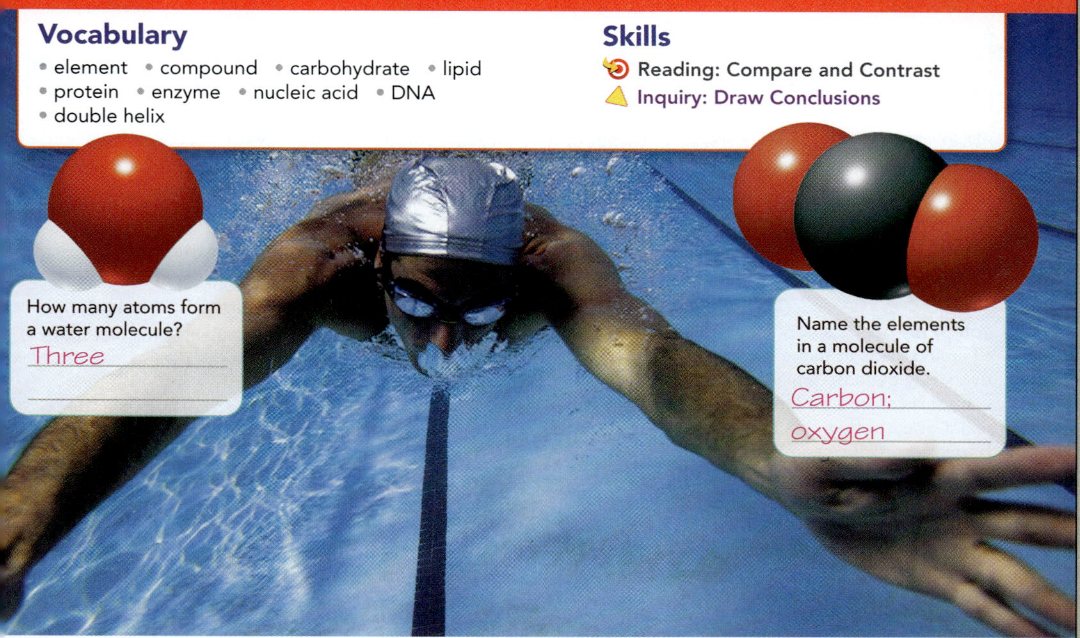

FIGURE 1
Molecules and Compounds
Carbon dioxide, in the air exhaled from the swimmer's lungs, is a compound. Water is also a compound.
Interpret Diagrams **Answer the questions in the boxes provided.**

Compounds

Most elements in living things occur in the form of **compounds. Compounds form when two or more elements combine chemically.** For example, carbon dioxide is a compound made up of the elements carbon and oxygen.

The smallest unit of many compounds is a molecule. A molecule of carbon dioxide consists of one carbon atom and two oxygen atoms. Compare the diagrams of the carbon dioxide molecule and the water molecule in **Figure 1.**

Organic Compounds

With some exceptions, compounds that contain carbon are referred to as organic compounds. Organic compounds are part of the solid matter of every organism on Earth. Carbon atoms act as the backbone or skeleton for these compounds. Organic compounds also contain other elements such as oxygen, hydrogen, or nitrogen. Organic compounds have similar properties. For example, many do not dissolve well in water.

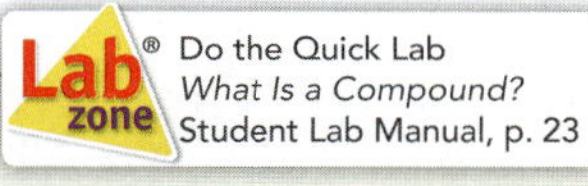
Do the Quick Lab *What Is a Compound?* Student Lab Manual, p. 23

Assess Your Understanding

got it?
- I get it! Now I know that compounds form when two or more elements combine.
- I need extra help with See TE note.

English Language Proficiency Standards

ELPS Reading 4.E

Have students create a written summary of the lesson with the following sentence frames: ________ *are made of* ________. ________ *make up* ________. *They also* ________. Have students use their work to complete the table on page 87.

Beginning Have beginners listen as you read aloud each paragraph. Help students complete sentence frames to summarize the paragraph. Record responses.

Intermediate Have partners read aloud each paragraph and complete one of the sentence frames. Share and record responses.

Advanced Have students work in small groups. Each student writes a summary of one paragraph. Then have students share summaries.

Advanced High Have small groups write lesson summaries.

Explain

Introduce Vocabulary

Students may have heard the term *enzyme* used in advertising. Point out to students that an enzyme is one type of protein, but there are many proteins that are not enzymes.

Teach Key Concepts

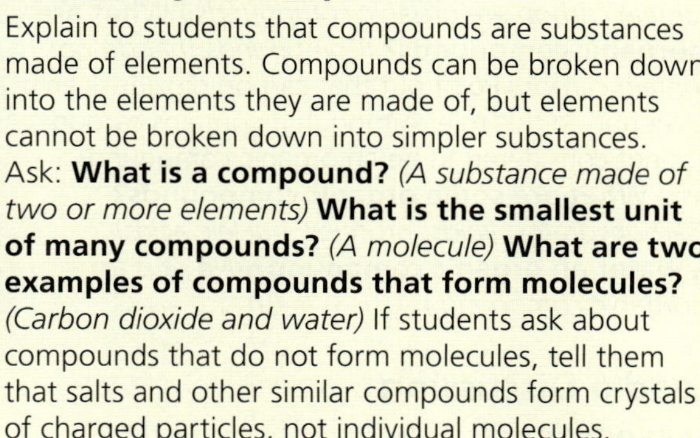

Explain to students that compounds are substances made of elements. Compounds can be broken down into the elements they are made of, but elements cannot be broken down into simpler substances. Ask: **What is a compound?** *(A substance made of two or more elements)* **What is the smallest unit of many compounds?** *(A molecule)* **What are two examples of compounds that form molecules?** *(Carbon dioxide and water)* If students ask about compounds that do not form molecules, tell them that salts and other similar compounds form crystals of charged particles, not individual molecules.

Elaborate

Lab Resource: Quick Lab 

L2 **WHAT IS A COMPOUND?** Students will examine different compounds and learn what elements they are made from. This Quick Lab can be found in the Student Lab Manual, p. 23, and online.

Evaluate

Assess Your Understanding

Have students evaluate their understanding by completing the appropriate sentence.

RTI Response to Intervention

If students cannot explain how compounds form, **then** have them review the first paragraph under the red heading *Compounds*.

PEARSON Texas.com

LESSON 2.4

Explain

Teach Key Concepts

Explain to students that every cell must either make or absorb some compounds to survive. Necessary organic compounds include carbohydrates, lipids, proteins, and nucleic acids. Necessary inorganic compounds include water and salts. Ask: **What are organic compounds?** *(Most compounds that contain carbon are organic compounds.)* **What are inorganic compounds?** *(Compounds that do not contain carbon)* Point out that carbon dioxide is an exception to this rule. Although it contains carbon, it is not considered to be an organic compound. Ask: **What are some organic compounds?** *(Carbohydrates, lipids, proteins, nucleic acids)* **Is water an organic compound?** *(No, it is an inorganic compound.)*

Lead a Discussion

CARBOHYDRATES Have students review the section on carbohydrates. Explain that most of the carbohydrates people eat come from plants. Ask: **What are some foods that contain carbohydrates?** *(Fruit, bread, pasta, potatoes, rice, candy, cookies)* Student may ask about "milk sugar," or lactose. Tell them that lactose is a kind of sugar, so milk does contain carbohydrates. Ask: **What carbohydrate is found in the cell walls of plants?** *(Cellulose)* Tell students that cellulose is a carbohydrate that people cannot digest, so this compound cannot be used by our cells as a source of energy.

Texas Essential Knowledge and Skills

2E Analyze data to formulate reasonable explanations, communicate valid conclusions supported by the data, and predict trends.

6A Identify that organic compounds contain carbon and other elements such as hydrogen, oxygen, phosphorus, nitrogen, or sulfur.

TEKS 2E, 6A In this section, you'll read about the compounds that living things need and analyze data about compounds in cells.

What Compounds Do Cells Need?

Cells require many types of organic compounds and inorganic compounds. Inorganic compounds usually do not contain carbon. Water and table salt are familiar examples of inorganic compounds. Although it does contain carbon, carbon dioxide is also an inorganic compound.

Some important groups of organic compounds that living things need are carbohydrates, lipids, proteins, and nucleic acids. Water is a necessary inorganic compound. Many of these compounds are found in the foods you eat. This fact makes sense because the foods you eat come from living things.

did you know?

Your body needs a new supply of amino acids to make proteins every day because amino acids cannot be stored for later use, as fat or carbohydrates can. Your body can produce some amino acids, while others need to come from food.

Carbohydrates

You have probably heard of sugars and starches. They are examples of **carbohydrates,** energy-rich organic compounds made of the elements carbon, hydrogen, and oxygen.

The food-making process in plants produces sugars. Fruits and some vegetables have a high sugar content. Sugar molecules can combine, forming larger molecules called starches, or complex carbohydrates. Plant cells store excess energy in molecules of starch. Many foods, such as potatoes, pasta, rice, and bread, come from plants and contain starch. When you eat these foods, your body breaks down the starch into glucose, a sugar your cells can use to get energy.

In addition to supplying energy, carbohydrates are important components of some cell parts. For example, the cellulose found in the cell walls of plants is a type of carbohydrate. Carbohydrates are also found on cell membranes.

FIGURE 2
Energy-Rich Compounds
Cooked pasta served with olive oil, spices, and other ingredients makes an energy-packed meal.
Classify **Label each food a starch or a lipid. Next to the label, write another example of a food that contains starch or lipids.**

Lipids Have you ever seen a cook trim fat from a piece of meat before cooking it? The cook is trimming away one kind of lipid. **Lipids** are compounds that are made mostly of carbon and hydrogen and some oxygen. Cell membranes consist mainly of lipids.

Fats, oils, and waxes are all lipids. Gram for gram, fats and oils contain more energy than carbohydrates. Cells store energy from fats and oils for later use. For example, during winter, an inactive bear lives on the energy stored in its fat cells. Foods high in fats include whole milk, ice cream, and fried foods.

Proteins What do a bird's feathers, a spider's web, and a hamburger have in common? They consist mainly of proteins. **Proteins** are large organic molecules made of carbon, hydrogen, oxygen, nitrogen, and, in some cases, sulfur. Foods that are high in protein include meat, dairy products, fish, nuts, and beans.

Much of a cell's structure and function depends on proteins. Proteins form part of a cell's membrane. Proteins also make up parts of the organelles within a cell. A group of proteins known as **enzymes** speed up chemical reactions in living things. Without enzymes, the many chemical reactions that are necessary for life would take too long. For example, an enzyme in your saliva speeds up the digestion of starch. The starch breaks down into sugars while still in your mouth.

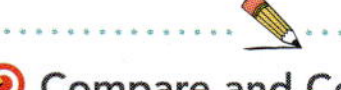

Compare and Contrast
As you read, complete the table below to compare carbohydrates, lipids, and proteins.

Type of Compound	Elements	Functions
Carbohydrate	Carbon, hydrogen, oxygen	Stores energy; makes up cell wall and cell membrane
Lipid	Carbon, hydrogen, oxygen	Makes up most of cell membrane; fats, oils store energy
Protein	Carbon, hydrogen, oxygen, nitrogen, sulfur	Makes up many cell structures; enzymes speed up cell reactions

FIGURE 3
Proteins
A parrot's beak, feathers, and claws are made of proteins.

Apply Concepts **What part of your body most likely consists of proteins similar to those of a parrot's claws?**

Fingernails or toenails

Differentiated Instruction

L1 Model Elements and Compounds Use colored beads to model elements and compounds and help students distinguish between the two. For example, use two beads of one color to represent hydrogen atoms and one bead of another color to represent oxygen atoms. Explain that hydrogen and oxygen are elements, then assemble the atoms into a molecule of water—a compound. If you use glue, explain that the glue represents chemical bonds.

L3 Cholesterol and Steroids are lipids with important health consequences. Have students research these compounds and determine how they are important to organisms and how large concentrations of them in the body can be dangerous.

Lead a Discussion

STRUCTURAL PROTEINS Emphasize that proteins are the building blocks of the human body. Point out to students that all of the tissues they can see and feel—hair, skin, muscle—as well as a large part of all cell membranes are built out of proteins. Ask: **Why are these proteins called structural proteins?** *(They make up much of the structure of the body)* **What is a kind of protein that does not make up part of the body's structure?** *(Enzymes)* **What do enzymes do in the body?** *(They speed up chemical reactions in living things.)*

Address Misconceptions

L1 LIPIDS AS FOOD Students may think that a healthy diet is one that eliminates as many lipids as possible. But lipids play important roles in the body: storing energy and building cell membranes, for example. An extremely low-fat diet could be just as unhealthy as a high-fat one. Ask: **How does your body use lipids?** *(Energy storage, component of cell membranes).* **What are some foods that contain lipids?** *(Dairy products like whole milk, cream, and ice cream; oils; nuts).*

21st Century Learning

CRITICAL THINKING Remind students that four elements make up most of the organic compounds in living things. Ask: **Which four elements are found in all proteins?** *(Carbon, hydrogen, oxygen, and nitrogen)* **Which element is found in some proteins?** *(Sulfur)* **Which three elements make up carbohydrates and lipids?** *(Carbon, hydrogen, and oxygen)* Students may wonder how two groups of compounds can be made of the same element but yield different amounts of energy. Tell them that the difference is in the proportion of the elements in each compound. Lipids have more hydrogen, and the bonds to these hydrogen atoms contain the extra energy.

Compare and Contrast Students compare and contrast information when they find similarities and differences among a set of materials.

Support the TEKS

CELLS NEED PROTEINS AND LIPIDS Ask students to review the information about proteins and lipids. Ask: **Which cell structure is composed mainly of lipids?** *(Cell membrane)* **Which cell structures contain proteins?** *(The cell membrane and other organelles)*

Explain

Lead a Discussion

CHROMATIN AND DNA Remind students that the cell's nucleus is packed with chromatin, which holds the instructions that control cell functions. Explain that the chromatin is made of long strands of DNA wrapped around proteins that help keep them organized. Ask: **Where is the DNA found in the cell?** *(In the nucleus)* **What function does DNA have in the cell?** *(It stores information and directs cell functions.)* **What is the shape of a DNA molecule called?** *(A double helix)* **What does this shape look like?** *(A twisted ladder)* Make sure students understand that the code is in the order of the molecules that make up the rungs of the ladder. *(Additional information about the structure and function of DNA appears in the chapter Genetics: The Science of Heredity.)*

Elaborate

Do the Math!

L1 Review with students how a double bar graph can be used to compare two different things. Point out the key, which tells which bar represents bacteria and which bar represents animals. Ask: **Which bars are tallest?** *(The ones that represent water content)* **What percent of cell weight is water?** *(About 70 percent)*

Draw Conclusions Students draw conclusions by summing up what they have learned from a set of data.

FIGURE 4

DNA

Smaller molecules connect in specific patterns and sequences, forming DNA.

Interpret Diagrams **In the diagram below, identify the pattern of colors. Then color in the ones that are missing.**

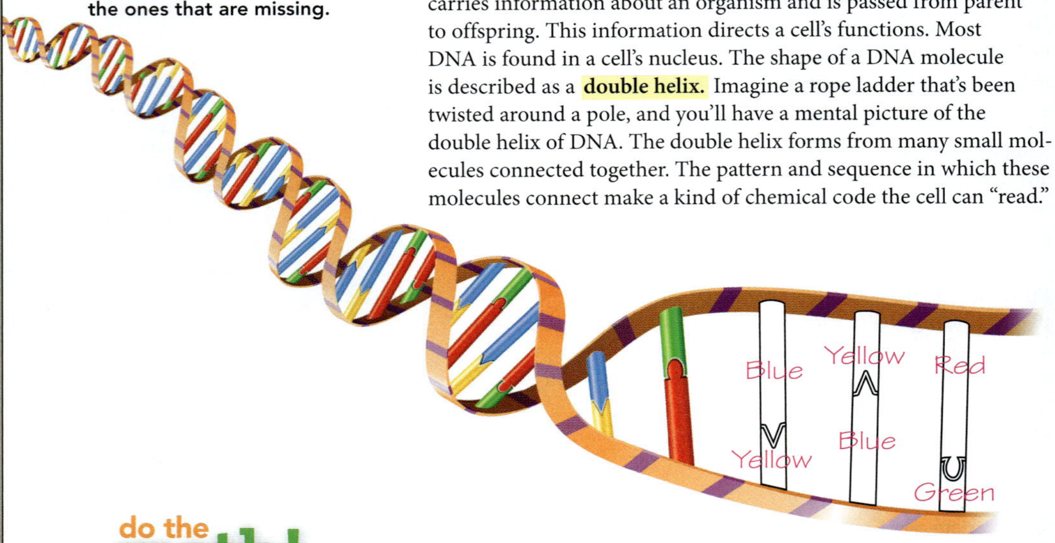

Blue
Yellow
Red
Yellow
Blue
Green

Nucleic Acids

Nucleic acids are very long organic molecules. These molecules consist of carbon, oxygen, hydrogen, nitrogen, and phosphorus. Nucleic acids contain the instructions that cells need to carry out all the functions of life. Foods high in nucleic acids include red meat, shellfish, mushrooms, and peas.

One kind of nucleic acid is deoxyribonucleic acid (dee AHK see RY boh noo KLEE ik), or DNA. **DNA** is the genetic material that carries information about an organism and is passed from parent to offspring. This information directs a cell's functions. Most DNA is found in a cell's nucleus. The shape of a DNA molecule is described as a **double helix.** Imagine a rope ladder that's been twisted around a pole, and you'll have a mental picture of the double helix of DNA. The double helix forms from many small molecules connected together. The pattern and sequence in which these molecules connect make a kind of chemical code the cell can "read."

Most cells contain the same compounds. The graph compares the percentages of some compounds found in a bacterial cell and in an animal cell. Write a title for the graph and answer the questions below.

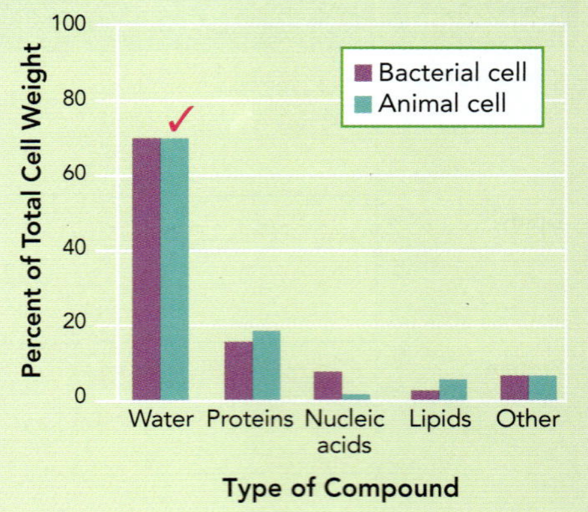

1. **Read Graphs** Put a check above the bar that shows the percentage of water in an animal cell. How does this number compare to the percentage of water in a bacterial cell?
   They are the same.
2. **Analyze Data** (Proteins/Nucleic acids) make up a larger percentage of an animal cell.
3. **Draw Conclusions** In general, how do you think a bacterial cell and an animal cell compare in their chemical composition?
   Similar, except for small differences in nucleic acids, lipids, and proteins

Comparing Compounds in Cells

Percent of Total Cell Weight: 0, 20, 40, 60, 80, 100

Bacterial cell
Animal cell

Water, Proteins, Nucleic acids, Lipids, Other

Type of Compound

FIGURE 5

Mostly Water

About two thirds of the human body is water. But you know you don't really look like a tank of water with a fish! **Graph Complete and label the circle graph to show the percentage of water in your body.**

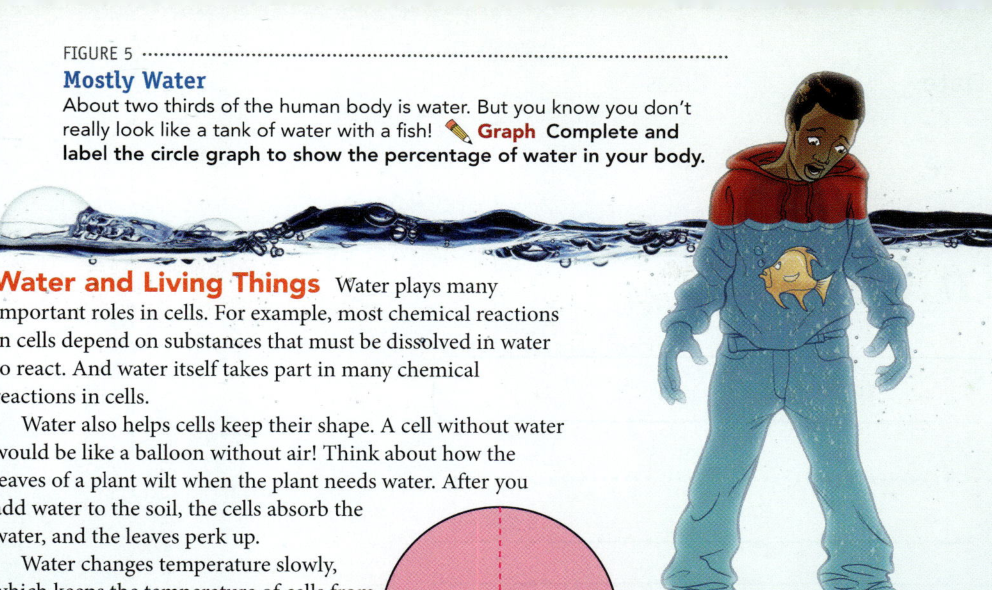

Water and Living Things

Water plays many important roles in cells. For example, most chemical reactions in cells depend on substances that must be dissolved in water to react. And water itself takes part in many chemical reactions in cells.

Water also helps cells keep their shape. A cell without water would be like a balloon without air! Think about how the leaves of a plant wilt when the plant needs water. After you add water to the soil, the cells absorb the water, and the leaves perk up.

Water changes temperature slowly, which keeps the temperature of cells from changing rapidly—a change that can be harmful. Water also plays a key role in carrying substances into and out of cells. Without water, life as we know it would not exist on Earth.

Lab zone: Do the Quick Lab *What's That Taste?* Student Lab Manual, p. 24

Assess Your Understanding

TEKS 6A

1a. Identify A carbohydrate is an organic compound. In addition to carbon, what two other elements are found in this type of organic compound?

Hydrogen and oxygen

b. Classify Which group of organic compounds can speed up the break down of other organic compounds?

proteins

c. Review What is the function of DNA?

DNA directs all of a cell's functions and carries information from parents to offspring.

d. CHALLENGE Describe ways a lack of water could affect cell functions.

Sample: Chemical reactions in a cell could not take place; cells could not keep their shape.

got it?

O **I get it!** Now I know that the important compounds in living things include carbohydrates, lipids, proteins, nucleic acids, and water.

O I need extra help with See TE note.

89

Differentiated Instruction

L1 Compare and Contrast Help students who are having difficulties reading the bar graph in the *Do the Math* activity by reminding them to compare the value for each type of compound from animal cells to the same value from bacterial cells.

L3 Water and Humans People depend on a good supply of fresh water, Have students determine why people cannot drink seawater. *(Drinking seawater is actually dehydrating; the kidneys use more water to remove the extra salts than you get from the water.)*

L3 Properties of Water The physical properties of water have effects on living things. Students can research water's surface tension, heat capacity, and unique change in density at freezing and hypothesize how each of these qualities affects living things.

Teacher Demo 

L1 HOW MUCH WATER?

Materials dry sea sponge, water, electronic scale

Time 10 minutes

Tell students that the dry sponge is the remains of a living animal that has been dried out; there is very little water inside it. Weigh the sponge and show the result to the class. Then drop the sponge in water. Let it absorb water until it is saturated, then weigh the water-filled sponge. Show the result to the class. Explain to the class that the full sponge better represents what the animal would have weighed when it was alive.

Ask: **What can water do for cells?** *(Provide support, play a role in chemical reactions, and move substances in and out of cells.)*

Lab Resource: Quick Lab 

L1 WHAT'S THAT TASTE? Students will explore the effect of saliva enzymes on starch. This Quick Lab can be found in the Student Lab Manual, p. 24, and online.

Evaluate

Assess Your Understanding

After students answer the questions, have them evaluate their understanding by completing the appropriate sentence.

RTI Response to Intervention

1a, b. If students need help with identifying and classifying organic compounds, **then** have them review the Compare and Contrast table.

c. If students cannot explain the function of DNA, **then** have them reread the second paragraph under the red heading *Nucleic Acids*.

d. If students have trouble explaining water's role in cells, **then** have them draw a circle to represent a cell then use arrows to show what water does in the cell.

Name ______________________ Date __________ Class __________

Assess Your Understanding

Chemical Compounds in Cells

What Are Elements and Compounds?

got it?

○ **I get it!** Now I know that compounds form when ______________________

____

○ **I need extra help with** ______________________

____

What Compounds Do Cells Need?

1a. IDENTIFY A carbohydrate is an organic compound. In addition to carbon, what two other elements are found in this type of organic compound?

____

____

b. CLASSIFY Which groups of organic compounds can speed up the breakdown of other organic compounds?

____

c. REVIEW What is the function of DNA?

____

____

d. CHALLENGE Describe ways a lack of water could affect cell functions.

____

____

got it?

○ **I get it!** Now I know that the important compounds in living things include ______________________

____

○ **I need extra help with** ______________________

____

Name ______________________ Date ____________ Class ____________

Enrich

Chemical Compounds in Cells

Read the passage below and complete the table. Then answer the questions that follow the table on a separate sheet of paper.

Amino Acids and Proteins

Though there are only 20 common amino acids, they can be combined in different ways to produce thousands of unique proteins. Proteins that differ in the order or type of amino acids they contain may have different structures and functions. In fact, a change in even a single amino acid can sometimes affect the way a protein works.

Suppose that proteins could consist of just two amino acids. To see how many unique proteins, each composed of just two amino acids, can be formed from five different amino acids, fill in the spaces in the table below. Some of the spaces have been filled in to show you how. Assume that each letter represents a different amino acid.

Amino Acids	*A*	*B*	*C*	*D*	*E*
A	*AA*	*AB*	*AC*		
B	*BA*				
C					
D					
E					

1. What does each letter pair in the table represent?
2. Based on your completed table, how many unique proteins, each composed of just two amino acids, can be formed from five different amino acids?
3. How many unique proteins, each made up of just two amino acids, could be formed from six different amino acids? From 20 different amino acids?
4. Most proteins are made up of not just two, but hundreds or even thousands of amino acids. How does this affect the number of unique proteins that could be formed from just a few amino acids?

Name ______________________ Date __________ Class __________

Lesson Quiz

Chemical Compounds in Cells

Write the letter of the correct answer on the line at the left.

1. ___ A carbohydrate is

A an inorganic compound

B an element found in water

C an energy-rich organic compound

D an element in most organic compounds

2. ___ Water is a necessary

A element

B carbohydrate

C mineral

D inorganic compound

3. ___ Fats, oils, and waxes are all examples of

A enzymes

B inorganic compounds

C lipids

D nucleic acids

4. ___ Which of the following is shaped like a double helix?

A DNA

B a lipid

C water

D a carbohydrate

If the statement is true, write *true*. If the statement is false, change the underlined word or words to make the statement true.

5. ______________ Your body breaks down starch into the sugar glucose.

6. ______________ Proteins are part of cell membranes and store energy.

7. ______________ A(n) enzyme helps speed a chemical reaction.

8. ______________ Carbohydrates direct cell functions.

9. ______________ Water makes up one third of the human body.

10. ______________ Meat, dairy products, fish, nuts, and beans are all foods that are high in protein.

Place the outside corner, the corner away from the dotted line, in the corner of your copy machine to copy onto letter-size paper.

Place the outside corner, the corner away from the dotted line, in the corner of your copy machine to copy onto letter-size paper.

Chemical Compounds in Cells

Answer Key

Review and Reinforce

Find the worksheet in the Student Workbook.

1. Accept one of the following: water's function in chemical reactions, as a carrier of materials, or in keeping the temperature of cells from changing quickly.
2. A given amount of lipid contains more energy than the same amount of carbohydrate.
3. Accept any two of the following: proteins are in cell membranes, organelles, muscle, skin, and are used to speed up chemical reactions in cells.
4. carbon, oxygen, hydrogen, nitrogen, phosphorus
5. carbon, oxygen, hydrogen
6. carbon, oxygen, hydrogen, nitrogen (sometimes sulfur)
7. carbon, oxygen, hydrogen
8. any substance that cannot be broken down into simpler substances
9. the chemical combination of two or more elements
10. a type of protein that speeds up chemical reactions in living things
11. Every organic compound contains three basic elements: carbon, hydrogen, and oxygen. Organic compounds and water both contain hydrogen and oxygen.

Enrich

Completed table should be filled in as follows.
First row: *AA, AB, AC, AD, AE*
Second row: *BA, BB, BC, BD, BE*
Third row: *CA, CB, CC, CD, CE*
Fourth row: *DA, DB, DC, DD, DE*
Fifth row: *EA, EB, EC, ED, EE*

1. Each letter pair represents a unique protein made from two amino acids.
2. 25 (5 × 5)
3. 6 × 6, or 36; 20 × 20, or 400
4. Increasing the number of amino acids each protein contains greatly increases the number of unique proteins that could be formed from just a few amino acids.

Lesson Quiz

1. C
2. D
3. C
4. A
5. true
6. Lipids
7. true
8. Nucleic acids
9. two thirds
10. true

Cells and Homeostasis

 **What structures are found in cells?**

LESSON PACING:
1–2 periods or ½–1 blocks

Lesson Vocabulary

- homeostasis
- diffusion
- active transport
- cell division

Lesson Objectives	TEKS	ELPS
Identify cell processes that maintain homeostasis, such as extracting energy from food, getting rid of waste, and reproducing.	12D, 12E	4.C.3

Content Refresher

Regulation of Body Temperature A mechanical thermostat is a crude imitation of the human body's far more complex and nuanced temperature control system, which involves an array of structures and processes. Maintenance of body temperature involves the hypothalamus region of the brain, as well as organs and tissues ranging from lungs and heart to blood and muscles and skin. Contrary to popular belief, normal body temperature does not stay precisely at 98.6°F. Rather, it fluctuates slightly as the body engages in activity, interacts with various environmental conditions, encounters bacteria and viruses, and (almost immediately) responds to all of these stimuli to maintain homeostasis.

 **Texas Essential Knowledge and Skills**

12D Differentiate between structure and function in plant and animal cell organelles, including cell membrane, cell wall, nucleus, cytoplasm, mitochondrion, chloroplast, and vacuole.
12E Compare the functions of a cell to the functions of organisms such as waste removal.

 **English Language Proficiency Standards**

ELPS Reading 4.C.3 Comprehend English vocabulary used routinely in written classroom materials.

DIFFERENTIATED INSTRUCTION KEY

L1 Struggling Students or Special Needs

L2 On-Level Students L3 Advanced Students

LESSON PLANNER 2.5

Investigations and Activities

My Planet Diary, **Student Edition,** p. 90

Inquiry: Inquiry Warm-Up, Homeostasis, **PearsonTexas.com**

Introduce Vocabulary, **Teacher's Edition,** p. 91

Teach Key Concepts, **Teacher's Edition,** p. 91

21st Century Learning, Interpersonal Skills, **Teacher's Edition,** p. 91

Support the TEKS, Removal of Waste, **Teacher's Edition,** p. 92

Inquiry: Teacher Demo, Modeling Diffusion, **Teacher's Edition,** p. 92

Apply It!, **Student Edition,** p. 92

Make Analogies, In and Out, **Teacher's Edition,** p. 93

Teach Key Concepts, **Teacher's Edition,** p. 93

Differentiated Instruction, **Teacher's Edition,** p. 93

Inquiry: Quick Lab, Effect of Concentration on Diffusion, **Lab Manual,** p. 26

TEKS Review

Apply the TEKS, Comparing Elements, **Student Edition,** p. 94

TEKS Practice, **Student Edition,** p. 95

TEKS Practice: Chapter and Cumulative Review, **Student Edition,** p. 98

Lesson 2.5, **TEKS Preparation and Study Guide Workbook,** p. 24

SHORT ON TIME? To do this lesson in approximately half the time, do the Activate Prior Knowledge activity. A discussion of the Key Concepts will familiarize students with the lesson content. Have students do the Quick Lab. The rest of the lesson can be completed by students independently.

These editable worksheets are available on **PearsonTexas.com.**
Print versions can be found in the **TEKS Preparation and Study Guide Workbook.**

Name ______ Date ______ Class ______

2.5 Cells and Homeostasis

Key Concept Summaries

How Do Cells Maintain Homeostasis?

Homeostasis is the maintenance of internal stable conditions that are necessary for life functions. Entire organisms maintain homeostasis. So do individual cells. **In all cells, the processes that help maintain homeostasis include getting and using energy from food and removing wastes.**

Most organisms get energy from cell processes that break down foods. During cellular respiration, cells break down glucose molecules in the presence of oxygen, releasing energy. Waste products of this process include carbon dioxide and water. Animals and some other organisms get food for energy by eating other organisms.

Unlike animals, plants and some other organisms can make their own food. The process by which cells capture the energy in sunlight and convert it to energy stored in food is called **photosynthesis.** Often, the foods produced in photosynthesis are glucose or other sugars. The raw materials for photosynthesis are carbon dioxide from the air and water. Oxygen leaves the cell as a waste product.

To maintain homeostasis, materials such as food and oxygen must move into cells. At the same time, waste materials must exit. Materials that move in or out of a cell must cross a structure called the cell membrane. The cell membrane surrounds a cell, separates it from the outside environment, and controls which substances enter and leave. One way materials move across the cell membrane involves collisions of molecules. These collisions push molecules away from one another in a process called diffusion. During **diffusion,** molecules move from an area of *higher* concentration to an area of *lower* concentration. Eventually, the molecules spread evenly throughout a space. Some materials, such as calcium and sodium, move in directions opposite to the way they would move during diffusion. **Active transport** is the movement of materials across a cell membrane using cellular energy.

All multicellular organisms grow during some part of their lifetimes. They also have structures that wear out or become injured and must be replaced. Growth and repair can occur because cells reproduce, or make more cells. **Cell division** is a process in which one cell splits into two new cells that are genetically identical to the original cell. Through cell division, your body replaces damaged skin cells and worn out blood cells. **Reproduction of cells is one of the processes that helps multicellular organisms maintain homeostasis.** You grow as your body produces more muscle cells, bone cells, and other kinds of cells. Growth and repair help your body maintain homeostasis.

24

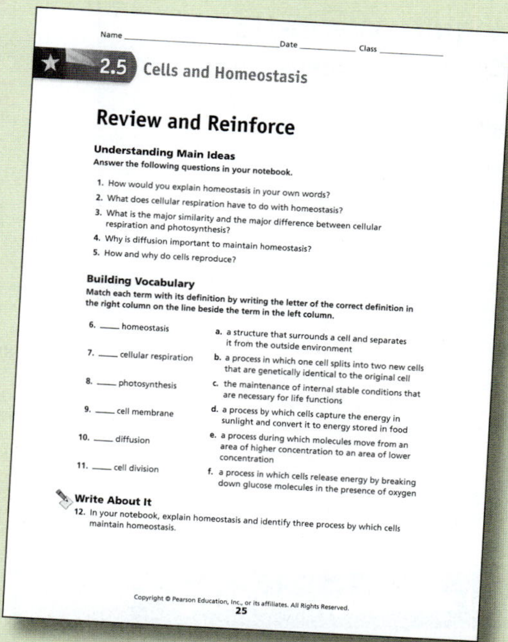

Name ______ Date ______ Class ______

2.5 Cells and Homeostasis

Review and Reinforce

Understanding Main Ideas

Answer the following questions in your notebook.

1. How would you explain homeostasis in your own words?
2. What does cellular respiration have to do with homeostasis?
3. What is the major similarity and the major difference between cellular respiration and photosynthesis?
4. Why is diffusion important to maintain homeostasis?
5. How and why do cells reproduce?

Building Vocabulary

Match each term with its definition by writing the letter of the correct definition in the right column on the line beside the term in the left column.

6. ___ homeostasis
7. ___ cellular respiration
8. ___ photosynthesis
9. ___ cell membrane
10. ___ diffusion
11. ___ cell division

a. a structure that surrounds a cell and separates it from the outside environment
b. a process in which one cell splits into two new cells that are genetically identical to the original cell
c. the maintenance of internal stable conditions that are necessary for life functions
d. a process by which cells capture the energy in sunlight and convert it to energy stored in food
e. a process during which molecules move from an area of higher concentration to an area of lower concentration
f. a process in which cells release energy by breaking down glucose molecules in the presence of oxygen

Write About It

12. In your notebook, explain homeostasis and identify three process by which cells maintain homeostasis.

25

Lexile Measure = 960L

LESSON 2.5

Cells and Homeostasis

Establish Learning Objective

After this lesson, students will be able to:

Identify cell processes that maintain homeostasis, such as extracting energy from food, getting rid of waste, and reproducing.

Engage

Activate Prior Knowledge

MY PLANET DIARY Read *Sleeping Away the Cobwebs* with the class. Invite students to think about their own sleep habits. Ask: **How much sleep do you get during an average night?** *(Sample: Seven or eight hours)* **How much sleep do you need to feel rested?** *(Sample: Eight or nine hours)* **In what ways are you affected by getting too little sleep?** *(Sample: Moodiness, irritability, concentration problems)*

Explore

Lab Resource: Inquiry Warm-Up 

L2 HOMEOSTASIS Students will observe what happens as yeast consumes sugar. This Inquiry Warm-Up can be found online.

 **Texas Essential Knowledge and Skills**

12D Differentiate between structure and function in plant and animal cell organelles, including cell membrane, cell wall, nucleus, cytoplasm, mitochondrion, chloroplast, and vacuole.

12E Compare the functions of a cell to the functions of organisms such as waste removal.

Cells and Homeostasis

How Do Cells Maintain Homeostasis?
TEKS 12D, 12E

MY PLANET DIARY — DISCOVERY

Sleeping Away the Cobwebs

Has your thinking ever felt fuzzy from lack of sleep? New research suggests that brain cells function better when well rested. Tests conducted with fruit flies and rats showed that proteins build up in the brain during wakeful times and decrease during sleep. These proteins take up space between brain cells and use energy. Researchers think sleep helps "clean house," saving energy and space and restoring balance in the brain. In a way, sleep refreshes the brain and prepares it to learn more effectively during the day.

Communicate Complete the tasks.

1. On a separate paper, write a very short story about not getting enough sleep.
2. How is writing a story different from researching the effects of sleep?

Sample: In a story, you can make up ideas. In research, you collect data.

 Do the Inquiry Warm-Up *Homeostasis.* Find the lab online.

TEKS 12D, 12E In this section, you'll explore how cells maintain stable internal conditions in order to function properly.

ELPS 4.C.3

With a partner, do the "Compare and Contrast" on page 91. Notice the words *like* and *unlike* in the text. Discuss other examples where *like* is used to compare or *unlike* is used to contrast.

How Do Cells Maintain Homeostasis?

Have you ever felt sweaty on a very warm day? The evaporation of sweat from your skin helps to cool your body. Keeping your internal body temperature about the same in spite of the temperature outside is just one example of how your body maintains homeostasis. **Homeostasis** (hoh mee oh STAY sis) is the maintenance of stable internal conditions that are necessary for life functions. Just as entire organisms maintain homeostasis, so do individual cells. **In all cells, the processes that help maintain homeostasis include getting and using energy from food and removing wastes.**

SUPPORT ALL READERS

Lexile Measure = 920L **Lexile Word Count = 982**

Prior Exposure to Content: May be the first time students have encountered this topic

Academic Vocabulary: *compare, contrast, predict*

Science Vocabulary: *homeostasis, diffusion, cell division*

Concept Level: May be difficult for students who struggle with abstract ideas

Preteach With: My Planet Diary "Sleeping Away the Cobwebs" and Figure 3

Vocabulary
- homeostasis
- diffusion
- active transport
- cell division

Skills
- Reading: Compare and Contrast
- Inquiry: Predict

Getting and Using Energy

What foods did you eat for breakfast? Have you ever thought about why you should eat breakfast? Food provides the cells in your body with the energy you need to walk to the bus stop, listen in class, and write assignments.

Energy From Food Like you, most organisms get energy from cell processes that break down foods. The most common process involves a sugar called glucose. During cellular respiration, cells break down glucose molecules in the presence of oxygen, releasing energy. Waste products of this process include carbon dioxide and water. Animals and some other organisms get food for energy by eating other organisms. For example, a mouse eats fruits and grains from plants. An owl sometimes eats mice.

Making Food Unlike animals, plants and some other organisms can make their own food. The process by which cells capture the energy in sunlight and convert it to energy stored in food is called photosynthesis (foh toh SIN thuh sis). Often, the foods produced in photosynthesis are glucose or other sugars. Through cellular respiration, these compounds then supply the energy needed for a plant cell's activities. The raw materials for photosynthesis are carbon dioxide from the air and water. Oxygen leaves the cell as a waste product.

Compare and Contrast Underline the sentences that describe how energy processes in organisms are the same and different.

FIGURE 1

Energy From Food

Oranges grow using the energy captured in photosynthesis.

Complete the following tasks.

1. **Recognize** Check the box for each process that applies to the organism.
2. **Apply Concepts** How does energy from the sun become energy the boy can use?

   Photosynthesis changes energy from the sun into energy stored in food. The boy gets this energy from the juice.

Organism	Cellular Respiration	Photosynthesis
Orange tree	✓	✓
Boy	✓	

English Language Proficiency Standards

ELPS Reading 4.C.3

Have students complete the Compare and Contrast activity and these sentence frames: *Plants are* like *animals because* ________. *Plants are* unlike *animals because* ________.

Beginning Display a Venn diagram with the headings Plants, Animals, Both. Read aloud page 91. Together fill in the diagram. Have students point to complete the sentence frames.

Intermediate Have partners underline the sentences with *like* and *unlike*. Model how to complete the sentence frames. Have partners continue on their own.

Advanced Have partners complete the activity. Then have them tell how plants and animals are alike and different.

Advanced High Have students complete the activity and write a paragraph comparing plants and animals.

Explain

Introduce Vocabulary

Explain that the term *homeostasis* comes from two Greek words—*homeo-* ("similar") and *stasis* ("the act of standing"). Point out that when a cell or organism maintains homeostasis, it keeps its conditions stable.

Teach Key Concepts 

Explore with students what it means for a body to be "balanced" or "stable." Encourage them to identify examples from their own experience—such as dizziness or falling down—when their bodies were not stable or in balance. Ask: **What examples can you think of that show stable internal body conditions or processes?** *(Sample: Regular breathing, regular heartbeat, normal body temperature)* **When might one of these processes or conditions change dramatically?** *(Sample: During vigorous exercise or at times of stress)* **What does the body do in response to these changes?** *(A body adjusts to get processes and conditions back to normal again.)*

Compare and Contrast Explain that when you compare and contrast, you examine the similarities and differences between things.

Elaborate

21st Century Learning

INTERPERSONAL SKILLS Have pairs or small groups of students analyze their own dietary habits for a week. Students should record all foods and drinks consumed during meals and snacks during this period. Then have them work together to devise a chart, table, or other graphic organizer to arrange their findings and to analyze their eating habits. Students might examine how their food consumption fits into the six food groups—milk, yogurt, and cheese; vegetables; bread, cereal, rice, and pasta; fruits; meat, fish, beans, eggs, and nuts; and fats, oils, and sweets. Encourage students also to determine which foods in their diet come from plants and which come from animals. Invite students to share their work in oral presentations to the class.

PEARSON Texas.com

LESSON 2.5

Explain

Support the TEKS

REMOVAL OF WASTE Remind students that homeostasis not only involves cells getting and using food, but also removing wastes. Ask: **Why is waste removal so important for cells?** *(Sample: Without it, wastes act as poisons that harm cells and affect their ability to maintain homeostasis.)* **How must the waste material move in order to be removed from the cell?** *(It must move out of the cell via the cell membrane.)* **How does diffusion accomplish this movement of substances?** *(Molecules collide in crowded spaces within a cell, pushing them away from each other. During this process, the molecules travel to less crowded areas on the other side of the cell membrane.)*

Elaborate

Teacher Demo 

L2 MODELING DIFFUSION

Materials clear plastic cup, water, food coloring, plastic dropper

Time 5 minutes

Pour water into the plastic cup and place it on a surface that is easy for students to observe. Invite students to let you know when the water is completely still within the cup. Using the plastic dropper, add one or two drops of food coloring into the water. Encourage students to observe the water as the color diffuses.

Ask: **What begins to happen immediately after the food coloring is added to the water?** *(The color begins to spread through the water.)* **How would you describe the movement of the food coloring in the water?** *(Sample response: It continues to spread out until all of the water is colored.)* **How does this model diffusion in a cell?** *(In a cell, molecules diffuse through a cell membrane, spreading evenly throughout a space; in this demonstration, food coloring diffused, spreading itself evenly throughout the water.)*

Apply It!

L1 Before students begin the activity, have them reread the paragraphs relating to the cell membrane and to diffusion.

Predict Help students understand that when they predict, they make an inference about a future event based on current evidence or past experience.

FIGURE 2

Diffusion

Some materials move across the cell membrane by way of diffusion.

Predict Draw an arrow to show the overall direction that molecules will travel as a result of diffusion.

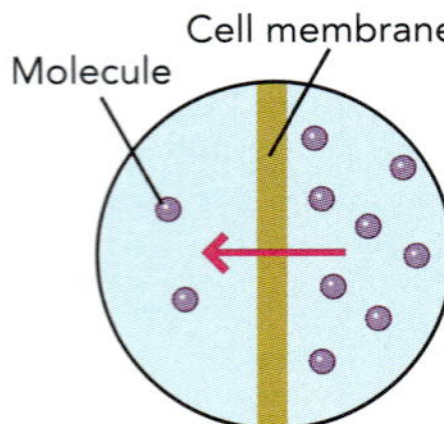

Removing Wastes To maintain homeostasis, an organism must acquire energy and remove harmful wastes. In a similar way, needed materials must move into cells and waste materials must exit. If waste materials build up, they act as poisons that can harm a cell and disrupt homeostasis. The cell membrane controls which substances enter and leave a cell. You can think of the cell membrane much like a gatekeeper who controls the flow of traffic into and out of a parking lot.

Diffusion One way materials move across the cell membrane involves collisions of molecules. Since molecules are always moving, they constantly bump into one another. If they are crowded, or concentrated, in a small space, they collide more often. These collisions push the molecules away from one another in a process called diffusion. During **diffusion** (dih FYOO zhun), molecules move from an area of *higher* concentration (a lot of molecules) to an area of *lower* concentration (fewer molecules). Eventually, the molecules spread evenly throughout a space. Molecules of gas and water move easily across the cell membrane. Look at **Figure 2.**

Active Transport Some materials, such as calcium and sodium, move in directions opposite to the way they would move during diffusion. In these cases, proteins known as transport proteins use energy from the cell to carry molecules across the membrane. **Active transport** is the movement of materials across a cell membrane using cellular energy.

apply it!

Without water, a cell cannot live or function. *Osmosis* is the term used to describe the diffusion of water across a cell membrane. Osmosis happens constantly as water molecules move into and out of a cell. These photos show a plant cell that has been affected by osmosis.

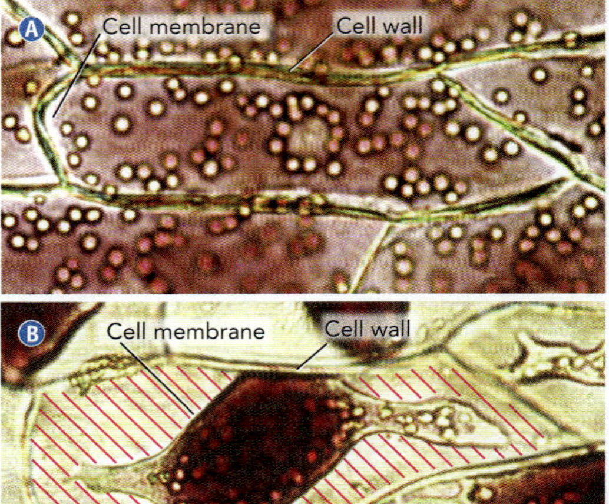

1 **Interpret Photos** Which cell shows the effect of water loss? Explain your reasoning.

B; the cell looks shrunken.

2 **Relate Cause and Effect** How do you think homeostasis would be affected in a cell that loses a lot of water?

The cell would dry up and die.

3 **Predict** Use a colored pencil to show how cell B would change if more water flowed into the cell.

Reproduction of Cells Are you bigger now than when you were a baby? Of course, you are! Have you grown even in the last few months? All multicellular organisms grow during some part of their lifetimes. They also have structures that wear out or become injured and must be replaced. Think about a time when you may have gotten a skinned knee. In just a few days, new skin appeared, and your injury healed.

Just as organisms reproduce and create offspring, cells reproduce to make more cells. **Cell division** is a process in which one cell splits into two new cells that are genetically identical to the original cell. You can see a cell dividing in **Figure 3.**

Reproduction of cells is one of the processes that helps multicellular organisms maintain homeostasis. Through cell division, your body can replace damaged skin cells and worn out blood cells. You grow as your body produces more muscle cells, bone cells, and other kinds of cells. Growth and repair help you maintain stable internal conditions throughout your body.

FIGURE 3

New Cells

This photo shows one cell that is becoming two as a result of cell division.

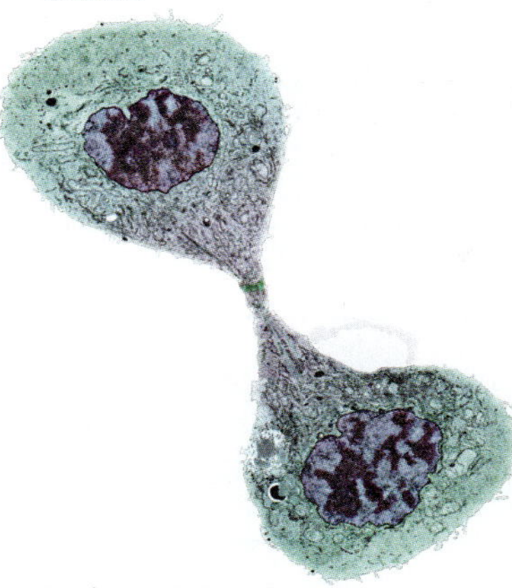

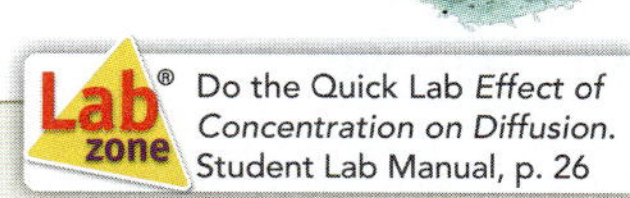

Assess Your Understanding

TEKS 12D, 12E

1a. Define What is homeostasis?

The maintenance of stable internal conditions that are necessary for life functions

b. Compare and Contrast Why must cells and organisms remove harmful wastes?

Both organisms and cells must remove wastes because if harmful wastes build up, it can disrupt homeostasis in the organism or the cell.

c. Relate Cause and Effect Diffusion results in molecules moving from areas of (lower/**higher**) concentration to areas of (**lower**/higher) concentration.

d. CHALLENGE What would happen to a cell if materials could not move across the cell membrane? Justify your answer.

Sample: The cell would die. Oxygen needed for cellular respiration would not be able to get into the cell, so the cell could not get energy from food.

got it?

○ **I get it!** Now I know that maintaining homeostasis in cells and organisms involves getting and using energy from food, removing wastes, and reproduction of cells.

○ I need extra help with See TE note.

Differentiated Instruction

L1 **Sequence of Diffusion** Have students work individually or in pairs to make a series of drawings that illustrate what happens during diffusion in cells. Students might use three or more panels to show molecules colliding, moving away from one another after collision, and seeking less crowded conditions on the other side of a cell membrane.

L3 **Lifetime of a Human Bone** Challenge students to research how bone growth occurs and how it changes during the course of the average human life. Encourage students to focus on changes in the bones of people under the age of 20, in the bones of people between the ages of 20 and 35, and in the bones of people who are 40 and older.

Make Analogies

L1 **IN AND OUT** Point out to students that the basic processes of getting food and oxygen and removing wastes can be seen at the level of the whole organism as well as at the level of cells. Draw students' attention to basic processes in which animals eat food and excrete waste, and inhale oxygen and exhale carbon dioxide. Help students recognize how this parallels the processes that occur on the cellular level.

Explain

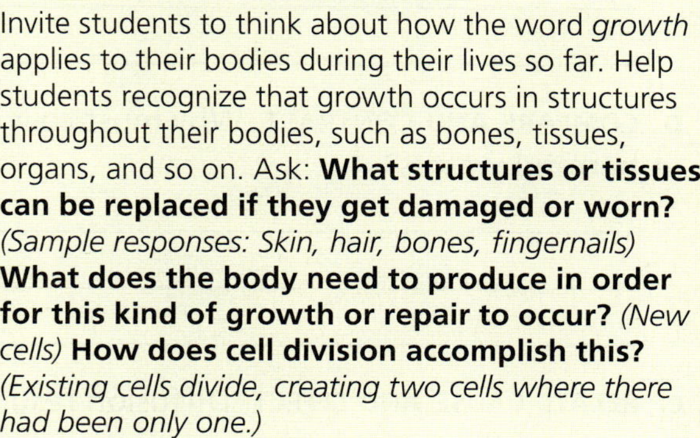

Teach Key Concepts

Invite students to think about how the word *growth* applies to their bodies during their lives so far. Help students recognize that growth occurs in structures throughout their bodies, such as bones, tissues, organs, and so on. Ask: **What structures or tissues can be replaced if they get damaged or worn?** *(Sample responses: Skin, hair, bones, fingernails)* **What does the body need to produce in order for this kind of growth or repair to occur?** *(New cells)* **How does cell division accomplish this?** *(Existing cells divide, creating two cells where there had been only one.)*

Lab Resource: Quick Lab

L2 **EFFECT OF CONCENTRATION ON DIFFUSION** Students will investigate how the concentration of a substance affects its rate of diffusion. This Quick Lab can be found in the Student Lab Manual, p. 26, and online.

Evaluate

Assess Your Understanding

After students answer the questions, have them evaluate their understanding by completing the appropriate sentence.

RTI Response to Intervention

1a. If students need help defining homeostasis, **then** have them review the definition in the lesson's first paragraph.

b. If students have trouble comparing the reasons cells and organisms must remove harmful wastes, **then** have them read the first paragraph after the red head Removing Wastes.

c. If students have trouble understanding the process of diffusion, **then** have them review **Figure 2.**

d. If students need help inferring what would happen to the cell, **then** remind them that materials pass through the cell membrane in both directions to complete processes needed for homeostasis.

Name ______________________ Date __________ Class __________

Assess Your Understanding

Cells and Homeostasis

How Do Cells Maintain Homeostasis?

1a. DEFINE What is homeostasis?

____

____

b. COMPARE AND CONTRAST Why must cells and organisms remove harmful wastes?

____

____

____

c. RELATE CAUSE AND EFFECT Diffusion results in molecules moving from areas of (lower/higher) concentration to areas of (lower/higher) concentration.

d. CHALLENGE What would happen to a cell if materials could not move across the cell membrane? Justify your answer.

____

____

____

got it?

○ **I get it!** Now I know that maintaining homeostasis in cells and organisms involves ________

____

____

○ **I need extra help with** ____

Name ______________________ Date ____________ Class ____________

Enrich

Cells and Homeostasis

Read the passage. Then answer the questions that follow in the spaces provided.

Cell Division

Cell division occurs in unicellular organisms and in multicellular organisms. However, the results of cell division are different depending on how many cells an organism has. Unicellular organisms use cell division to reproduce. In multicellular organisms, most cell division occurs in order to repair or renew old tissue. This renewal process is essentially continuous in some cells, such as those in the skin. In addition, some cell division occurs for the growth of new tissue in multicellular organisms.

Before cell division begins, however, the genetic information it carries in its DNA must be copied exactly and completely. If this duplication of genetic material does not occur, the new cells will not function, and the new tissue will not survive. This copying process is called *DNA replication*.

DNA has a shape called a double helix. Two intertwined strands are each formed by a chain of linked nucleotides. As the replication process begins, enzymes unwind the two strands from one another. Then, other enzymes travel along a molecule in order to "read" the nucleotides in that strand. The copying process proceeds in two directions at once, shortening the time it takes for duplication.

Absolute accuracy is necessary for a new cell to function properly. Fortunately, the body's DNA has a built-in function that checks the accuracy of a new copy. If any mistakes are found, the DNA can erase them and insert correct coding information.

1. What are the functions of cell division?

____

____

2. What must happen before cell division in multicellular organisms, and why?

____

____

____

3. What is the first step of the replication process?

____

4. How are errors in DNA replication handled by the body?

____

Name ____________________ Date __________ Class __________

Lesson Quiz

Cells and Homeostasis

Write the letter of the correct answer on the line at the left.

1. ___ **Which of the following does NOT help your body maintain homeostasis?**

A jogging
B drinking water
C sweating
D eating an apple

2. ___ **Unlike animals, plants can**

A use oxygen as a food source
B make their own food
C get energy by eating other plants
D replace injured cells

3. ___ **The movement of materials across a cell membrane using cellular energy is called**

A diffusion
B photosynthesis
C active transport
D homeostasis

4. ___ **Which process replaces damaged and worn out cells?**

A diffusion
B photosynthesis
C cell division
D active transport

Fill in the blank to complete each statement.

5. ____________ **of cells is one of the processes that helps multicellular organisms maintain homeostasis.**

6. ____________ **is the maintenance of internal stable conditions that are necessary for life functions.**

7. During ____________, **cells break down glucose molecules in the presence of oxygen, releasing energy.**

8. The process by which cells capture the energy in sunlight and convert it to energy stored in food is called ____________.

9. During ____________, **molecules move from an area of higher concentration to an area of lower concentration.**

10. The ____________ **surrounds a cell, separates it from the outside environment, and controls which substances enter and leave.**

Cells and Homeostasis

Answer Key

Review and Reinforce

Find the worksheet in the Student Workbook.

1. Sample: Homeostasis is what the body does to keep everything stable inside itself.
2. Cellular respiration is a process by which organisms get energy from food. Without this energy, homeostasis would be impossible.
3. Sample: Both are cellular processes. Photosynthesis is a process that makes food (sugars) while cellular respiration is a process that breaks down food (sugars).
4. Diffusion allows for the removal of waste products, which is necessary to maintain homeostasis.
5. Cells reproduce to maintain homeostasis, and they do it through cell division.
6. c
7. f
8. d
9. a
10. e
11. b
12. Homeostasis is the process of maintaining stable internal conditions that are necessary for life functions. Homeostasis occurs in cells and in entire organisms. The processes by which cells maintain homeostasis include getting and using energy from food, the removal of wastes from cells, and the reproduction of cells.

Enrich

1. reproduction of unicellular organisms; renewal of tissue and growth of new tissue in multicellular organisms
2. The genetic material carried in the DNA of the cell must be replicated so that it can be transferred to the new cell. If the new cell does not contain DNA that is complete and exactly the same as the original cell, the new cell will not function properly or survive.
3. the unwinding of the two strands of a DNA double helix
4. The DNA has the ability to detect and correct errors in the new copy.

Lesson Quiz

1. A
2. B
3. C
4. C
5. Reproduction
6. Homeostasis
7. cellular respiration
8. photosynthesis
9. diffusion
10. cell membrane

CHAPTER 2

Scientific Investigation and Reasoning

This Apply the TEKS feature will enable students to use the scientific investigation and reasoning skills in TEKS 2E to reinforce the content of TEKS 6A. Students will analyze data about the composition of elements in the human body and Earth's crust to communicate valid conclusions about organic compounds.

Comparing Elements

Have students read about the chemical composition of the human body and Earth's crust. If necessary, explain to students that Earth's crust is the rock that forms our planet's outer layer. The crust is between 5 and 40 kilometers thick.

Make sure students understand the data in the table before they begin answering the questions. Ask: **What do the numbers in each column represent for the given elements?** *(The percentage of mass the element makes up in the human body and in Earth's crust)*

After students answer Question 4, have volunteers share their conclusions and discuss as a class other conclusions that can be drawn from the data in the table.

Comparing ELEMENTS

Elements make up all living and non-living things on Earth. The table below shows the amounts of eight common elements that make up both the human body and Earth's crust. About two-thirds of the human body is composed of the compound H_2O, or water. Earth's crust contains a high percentage of silicate minerals, SiO_2, such as quartz.

Element Name	Approximate Percent of Mass in Human Body	Approximate Percent of Mass in Earth's Crust
Oxygen	65	47
Carbon	18	0.05
Hydrogen	10	0.15
Nitrogen	3	0.005
Calcium	1	4
Phosphorus	1	0.1
Potassium	0.25	3
Silicon	0.002	28

Answer the questions below.

1. **Identify** Which element makes up most of the mass of both the human body and Earth's crust? Why is the percentage of this element so high in both subjects?
   Oxygen; Most of the human body is made up of water, and most of Earth's crust is made up of silicates, both of which contain oxygen.
2. **Interpret Tables** Which element in the table makes up the least mass of the human body? Earth's crust?
   Human body: silicon; Earth's crust: nitrogen
3. **Analyze Data** The elements carbon, hydrogen, and nitrogen together make up less than one percent of the mass of Earth's crust. What percent do they make up in the human body?
   31%
4. **Communicate Valid Conclusions** Remember that organic compounds contain carbon and other elements such as hydrogen, oxygen, and nitrogen. What conclusions about organic compounds can you draw based on these data?
   Sample: The ratio of organic compounds to inorganic compounds is higher in the human body than it is in Earth's crust.

English Language Proficiency Standards

ELPS Listening 2.C.2

Have students complete the following sentence frames to compare the amount of each element in the human body and Earth's crust shown in the table: ________ contains more ________ than ________. ________ contains less ________ than ________.

Beginning Restate the frames as questions. Use students' answers to complete the sentence frames.

Intermediate Model how to complete the sentence frames for the first row or two. Then have partners continue on their own.

Advanced Have partners complete the sentence frames. Then ask them to discuss answers to questions 1–2.

Advanced High Have partners complete the sentence frames. Then ask them to write the answers to questions 1–2.

Texas Essential Knowledge and Skills

2E Analyze data to formulate reasonable explanations, communicate valid conclusions supported by the data, and predict trends.

6A Identify that organic compounds contain carbon and other elements such as hydrogen, oxygen, phosphorus, nitrogen, or sulfur.

CHAPTER 2

TEKS Practice

TEKS 3D, 4A, 6A, 11A, 12C, 12D, 12E, 12F, 13A

LESSON 1 The Characteristics of Life

1. The process by which an organism builds up or breaks down materials through chemical reactions is called
   a. carbohydrates. **b.** metabolism.
   c. multicellular. d. development.

2. The most specific level of classification is species.

3. **Apply Concepts** An owl hears a mouse rustling in some leaves and swoops down to catch it. In this scenario, what is the stimulus? Explain.
   The stimulus is the sound of the mouse in the leaves. A stimulus is a change in an organism's surroundings that causes it to act.

4. **Examine** The scientific name for the American elm tree is *Ulmus americana*. Based on this information what can you infer about an organism called *Ulmus crassifolia*? Explain.
   Sample: Based on the fact that the organism is the same genus as the American elm tree, I can predict the organism is a species of elm tree.

5. **Write About It** Suppose you are searching for new life forms as part of an expedition to the Sahara Desert. At one site, you find 18 reddish-brown objects, each measuring about 50 mm in length. The objects do not appear to have heads or limbs, but you think they might be alive. Explain the steps you would take to determine if the objects are living things.

LESSON 2 Discovering Cells

6. Which tool could help you see a plant cell?
   a. a filter **b.** a microscope
   c. a microwave d. an electromagnet

7. Anton van Leeuwenhoek was the first scientist to observe living cells under a microscope.

8. Because of the work of Schleiden and Schwann, scientists can apply the cell theory to recognize that all living things are made of cells.

9. **Compare and Contrast** How is a light microscope similar to an electron microscope? How do the two types of microscopes differ?
   Both are used to see small objects; one uses light and the other uses electrons.

10. **Estimate** Using a microscope, you see the one-celled organism shown below. The diameter of the microscope's field of view is 0.8 mm. Estimate the cell's length and width, and write your answer in the space provided.

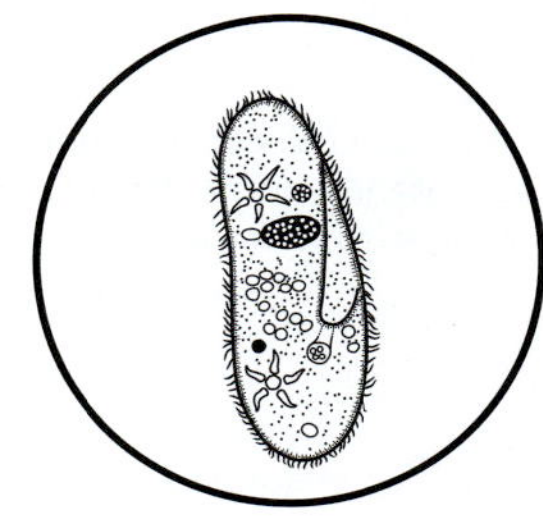

0.6 mm long and 0.2 mm wide

TEKS Practice

Assess Understanding

Have students complete the answers to the TEKS Practice questions. Have a class discussion about what students find confusing. Write Key Concepts on the board to reinforce knowledge.

RTI Response to Intervention

2. If students have trouble identifying the most specific level of classification, **then** have them reread the Key Concept statement about how living things are classified.

8. If students need help recognizing what the cell theory states, **then** have them review the section on the cell theory. Remind them that cells are the basic units of structure and function in living things.

Alternate Assessment

L1 MAKE A MODEL Have students work in groups to create paper diagrams of cells. Each group should create either a plant or animal cell, containing pictures of each organelle and an explanation of its function in the cell. When the cells are complete, plant and animal groups should pair up to compare and contrast their cells.

PEARSON Texas.com

Write About It 5 Assess student's writing using this rubric.

SCORING RUBRIC	SCORE 4	SCORE 3	SCORE 2	SCORE 1
Identify the characteristics of all living things and how objects could be tested	Student identifies all six characteristics of living things and describes how the objects could be tested for each.	Student identifies all six characteristics of living things and describes some ways the objects could be tested.	Student identifies only a few of the six characteristics of living things and does not describe how the objects could be tested.	Students does not identify any characteristics of living things or how objects could be tested.

CHAPTER 2

TEKS Practice, Cont.

RTI Response to Intervention

13. If students cannot identify the nucleus and differentiate its function, **then** have them review **Figure 3.**

18. If students have trouble identifying the element that must be contained in an organic compound, **then** have them reread the paragraph under the red head *Organic Compounds.*

21. If students have difficulty identifying how damaged and worn-out cells are replaced, **then** have them reread the paragraphs under the red head *Reproduction of Cells.*

L3 **WRITING IN SCIENCE** Ask students to write a television interview with a biologist that explains to viewers what cells are made of and how they function.

TEKS Practice

LESSON 3 Looking Inside Cells

11. Which cellular structures are found in plant cells but NOT in animal cells?

- **(a.)** chloroplast and cell wall
- **b.** Golgi apparatus and vacuole
- **c.** mitochondrion and ribosome
- **d.** endoplasmic reticulum and nucleus

12. Mitochondria and chloroplasts are two types of *organelles.*

13. Differentiate What is the function of the cell structure shown in purple in the cell at the right?

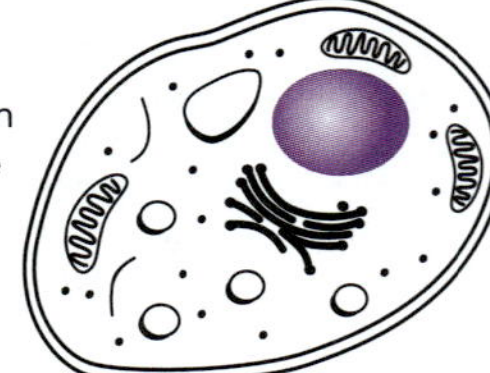

Directs cell activities

14. Sequence Arrange the following, from smallest to largest level of organization: organ system, tissue, organism, cell, organ.

Cell, tissue, organ, organ system, organism

15. Infer A certain cell can no longer package and release materials out of the cell. Which of the cell's organelles is not working?

Golgi apparatus

16. Write About It Imagine you are a tour guide. You and the tour group have shrunk to the size of water molecules. You are now ready to start a tour of the cell! Write a narrative of your tour that you could give a new tour guide to use.

See TE rubric.

LESSON 4 Chemical Compounds in Cells

17. Starch is an example of a

- **a.** lipid.
- **b.** protein.
- **c.** nucleic acid.
- **(d.)** carbohydrate.

18. Identify What element must a compound contain in order to be classified as an organic compound? *carbon.*

19. math! According to the graph, what percentage of the total cell weight is made up of lipids?

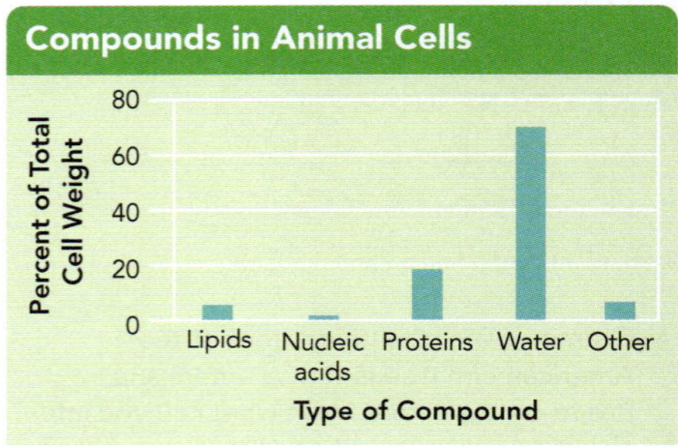

About 5%

LESSON 5 Cells and Homeostasis

20. Cells get energy from food through

- **(a.)** diffusion.
- **b.** homeostasis.
- **c.** cellular respiration.
- **d.** photosynthesis.

21. Replacement of damaged and worn-out cells in an organism can happen as a result of *cell division.*

22. Compare and Contrast In what ways are the functions of a cell and an organism similar?

Both must acquire energy and remove wastes to maintain homeostasis. Both reproduce.

Write About It **16** Assess student's writing using this rubric.

SCORING RUBRIC	SCORE 4	SCORE 3	SCORE 2	SCORE 1
Identifies the structures and functions within a cell.	Student identifies all the major organelles and their functions.	Student identifies several major organelles and their functions.	Student identifies one or two major organelles and their functions.	Student doesn't identify a major organelle or function.

What structures are found in cells?

23. In the space below, sketch a diagram of a simple plant cell. Include the following organelles: cell membrane, cell wall, chloroplast, cytoplasm, mitochondrion, nucleus, and vacuole. On your diagram, differentiate between structure and function by labeling the structures and explaining their functions. Note the differences between plant and animal cells by identifying any structures found in your diagram of a plant cell that would not be found in an animal cell.

Student drawings should resemble the plant cell on page 78. They should indicate the following functions for these structures: cell membrane—controls the substances that pass into and out of a cell cell wall—layer that surrounds a cell and provides protection and support chloroplast—captures energy from the sun and changes it to a form that cells can use to make food cytoplasm—a fluid that fills the region between the cell membrane and nucleus mitochondrion—converts energy stored in food to energy the cell can use nucleus—directs the cell's activities vacuole—stores food, water, and other materials needed by a cell.

Students should indicate that cell walls and chloroplasts are not found in animal cells. Not all animal cells have vacuoles.

REVIEW THE TEKS Interactive Science Chapter 2

12D, 12F

Lesson 1
In Lesson 2.1 you learned that all living things have cellular organization, contain similar chemicals, respond to their surroundings, grow and develop, and reproduce. You also learned that scientists use their observations of characteristics of organisms to organize them into a classification system.
Supporting TEKS: 11A
TEKS: 12E, 13A

Lesson 2
In Lesson 2.2, you learned that cells are the basic unit of structure and function in living things. You also learned that, according to the cell theory, all living things are composed of cells, all cells carry on similar functions, and all cells come from other cells.
Supporting TEKS: 12F
TEKS: 3D, 4A, 12E

Lesson 3
In Lesson 2.3, you learned that each kind of cell structure has a different function within a cell. In multicellular organisms, cells are further organized into tissues, organs, and organ systems.
Supporting TEKS: 12D, 12F
TEKS: 12C

Lesson 4
In Lesson 2.4, you learned that elements contain one type of atom, and compounds form when elements combine. You also learned about organic compounds, which contain carbon and other elements such as hydrogen and oxygen. Important compounds in living things include carbohydrates, lipids, proteins, nucleic acids, and water.
Supporting TEKS: 6A

Lesson 5
In Lesson 2.5, you explored the processes that help maintain homeostasis in cells include getting and using energy from food and removing wastes. You also learned that reproduction of cells is one of the processes that helps organisms maintain homeostasis.
Supporting TEKS: 12D
TEKS: 12E

What structures are found in cells?

Assess student's response using this rubric.

SCORING RUBRIC	SCORE 4	SCORE 3	SCORE 2	SCORE 1
Identifies organelles and explains functions	Correctly identifies and labels cell membrane, nucleus, cytoplasm, and mitochondrion, and explains functions of each organelle	Correctly identifies and labels cell membrane, nucleus, cytoplasm, and mitochondrion, and explains functions of some organelles	Incorrectly identifies organelles, and has trouble explaining function of each one	Cannot identify cell membrane, nucleus, cytoplasm, and mitochondrion, and cannot explain functions
Differentiates between animal and plant cell	Correctly identifies structures only found in plant cells and explains their functions	Identifies structures only found in plant cells and explains some functions	Incorrectly identifies structures only found in plant cells and has trouble explaining functions	Cannot identify structures only found in plant cells or explain functions

What structures are found in cells?

Students should be able to demonstrate understanding of what cells are made of by answering this question. See the scoring rubric below.

Review the TEKS Chapter 2

12D, 12F

PARTNER REVIEW Have partners review the definitions of vocabulary terms and quiz each other. Students can read the Key Concept statements and leave out words for their partners to fill in. They also can change a statement so that it is false and then ask their partners to correct it.

CLASS ACTIVITY: CONCEPT MAP Have students develop a concept map to show how the information in this chapter is related. Have students brainstorm to identify the Key Concepts, vocabulary, details, and examples. Write each response on a sticky note and attach it at random to chart paper or on the board. Tell students that this concept map will be organized in hierarchal order—with Key Concepts at the top. Ask students to use the following questions to help them organize the information on the sticky notes:

- Where do cells come from?
- What organelles are found inside cells?
- What does each organelle do inside a cell?
- What kinds of compounds are important in cells?
- How do cells maintain homeostasis?

SAMPLE CONCEPT MAP:

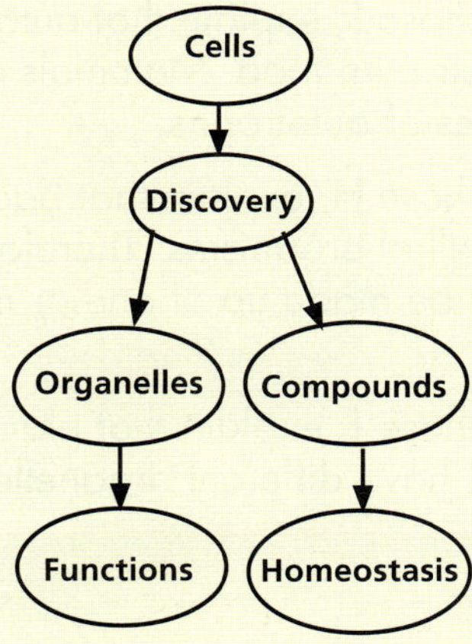

CHAPTER 2

TEKS Practice: Chapter Review

Test-Taking Skills

INTERPRETING DIAGRAMS Tell students that understanding the information in a diagram is the key to answering some questions correctly. When students are shown a diagram, they should examine it carefully. Remind students to look for any important objects or symbols in the diagram and read the labels.

Question 1 TEKS 12D

The correct answer is A. This is the correct label, structure, and function.

If students chose B, explain that the label is pointing correctly to a vacuole, but the function of this structure is to hold materials or waste for the cell.

If students chose C, explain that the label is pointing to the cell wall. In addition, the cell membrane's function is to allow waste and nutrients to move into our out of the cell.

If students chose D, explain that the label is pointing to the nucleus, but the nucleus does not make or transfer proteins.

Question 2 TEKS 12F

The correct answer is G. All living things are made up of one or more cells.

If students chose F, explain that autotrophs produce their own food. Mammals are not examples of autotrophs.

If students chose H, explain that bacteria are single-celled organisms. Therefore, they cannot be made up of one or more tissues.

If students chose J, explain that plant and animal cells have different organelles.

★ TEKS Practice: Chapter Review

Read each question and choose the best answer.

1 Nadia is drawing the diagram shown below of a plant cell.

A Plant Cell

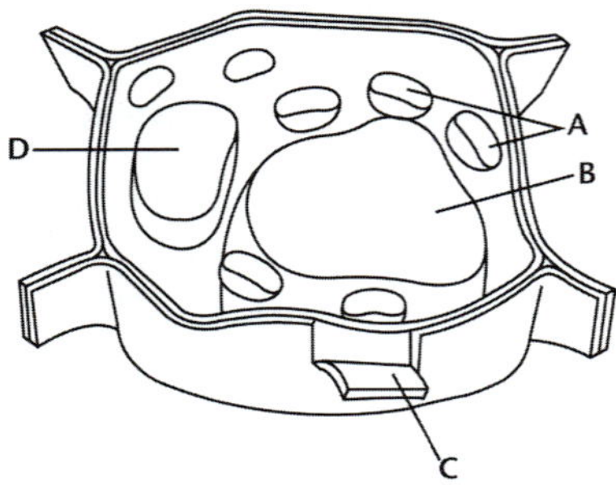

Which of the following labels correctly identifies an organelle and accurately describes its function?

Ⓐ Label A—*Chloroplasts* transform light energy into chemical energy.

B Label B—*Vacuoles* allow waste and nutrients to move into or out of the cell.

C Label C—The *cell membrane* holds materials or waste for the cell.

D Label D—The *nucleus* makes proteins and transports them to other organelles.

2 Organisms come in a variety of forms, such as bacteria, trees, and mammals. Which of the following is an accurate statement about these three examples of living things?

F Each type of organism produces its own food.

Ⓖ Each type of organism is made up of one or more cells.

H Each type of organism is made up of one or more tissues.

J Each type of organism contains cells with identical organelles.

★ TEKS Practice: Cumulative Review

3 The image below shows the setup for an investigation about how plants grow. The hypothesis being tested is that plants grow away from gravity no matter what direction a pot faces.

For this to be a controlled experiment, what must be true of the lamps that shine on each plant?

- **A** The lamps should each give off a different color of light.
- **B** The lamps should each glow at a different level of brightness.
- **C** The lamps should each be turned on for different lengths of time.
- **Ⓓ** The lamps should each be exactly the same and shine for the same length of time.

---

4 You want to find out which color of clothing is most commonly worn by the students in your class. Which of the following is an objective way to investigate this question?

- **F** Ask five randomly selected students to name their favorite color of clothing.
- **Ⓖ** Record the number of students in your class that are wearing each color for two weeks.
- **H** Every day for a month, record the color that the best-dressed student is wearing.
- **J** Record the number of students in your class that are wearing your favorite color.

If You Have Trouble With . . .

Question	1	2	3	4
See Lesson	2.3	2.2	1.3	1.2
TEKS	7.12D	7.12F	7.2B	7.2A

TEKS Practice: Cumulative Review

Additional Assessment Resources

Teacher's Edition: Lesson Quizzes, Texas End-of-Year Test Prep A and B
Online assessments

Question 3 

The correct answer is D. In a controlled experiment, only one variable is manipulated at a time. In this experiment, the manipulated variable is the orientation of the pot.

If students chose A, explain that the experiment is not meant to test how the color of light affects plant growth.

If students chose B, explain that the experiment is not meant to test how the brightness of light affects plant growth.

If students chose C, explain that the experiment is not meant to test how the length of time a plant receives light affects its growth.

Question 4 TEKS 2A

The correct answer is G. This method is objective because it identifies the most commonly worn color of clothing based on available evidence rather than personal feelings.

If students chose F, explain that having someone identify a favorite color is a subjective method because it is based on his or her personal feelings. It also is different than determining the colors that are most commonly worn.

If students chose H, explain that identifying the best-dressed student is a subjective method because it is based on your personal feelings.

If students chose J, explain that identifying a favorite color is a subjective method. It is based on your personal feelings and also introduces personal bias into the investigation.

Science Matters STEM

Technology and History

Have students read *Electron Eyes.* Point out that the first kinds of microscopes that were invented were optical microscopes. Explain that optical microscopes focus light through one or more lenses or mirrors to create an enlarged image. The best optical microscopes can magnify an image up to 1,500 times. Tell students that a Dutch man named Anton van Leeuwenhoek is called the "Father of Microbiology." In 1675, he was the first to observe single-celled organisms, or microorganisms, through a microscope.

Tell students there are many different kinds of microorganisms. Ask them if they can identify any microorganisms. They may name bacteria, fungi, protists, plankton, or amoeba, among others. Explain that many are single-celled, but some are multicellular. Microorganisms' habitats include parts of the biosphere where there is liquid water, high in the atmosphere, and deep inside rocks in Earth's crust. Many of these microorganisms play important roles in their ecosystems by decomposing organic matter and creating nitrogen compounds from inert nitrogen. However, some microorganisms cause diseases. These are called pathogenic microbes.

As students research microscopic images, have them note which images are of microorganisms. Have them group their images by type.

Ask: **How are transmission electron microscopes and scanning electron microscopes different?** *(TEMs create a two-dimensional image to study a section of a specimen. SEMs create a three dimensional image of a specimen.)* **Which human sense does the movement of the probe of a Scanning Tunneling Microscope mimic?** *(Touch)*

SCIENCE MATTERS STEM

Technology and History

Electron EYES

TEKS 3D

▼ Looking through a TEM gives a close-up view of the cells of an onion.

▼ Samples of bread mold as captured by an SEM

The invention of the optical microscope around the year 1600 caused a revolution in science. For the first time, scientists were able to see the cells that make up living things. However, even the most modern optical microscopes that focus light through lenses to produce an enlarged image can magnify an object only about 1,000 times.

Beginning in the early 1930s, new kinds of microscopes have caused new revolutions in science. The electron microscope uses electrons, instead of light, to make very detailed images of specimens. Today, powerful microscopes can magnify images up to 1,000,000 times—enough to enable scientists to see individual atoms!

Scientists use three main types of very powerful microscopes:

Transmission Electron Microscope (TEM) A TEM focuses a beam of electrons so that they pass through a very thinly sliced specimen. They are very useful for studying the interior structures of cells.

Scanning Electron Microscope (SEM) An SEM uses an electron beam to scan the surface of the specimen. The electron beam excites the electrons on the object's surface. The excited electrons are used to make a three-dimensional image of the specimen.

Scanning Tunneling Microscope (STM) An STM works by passing an electrically charged probe very close to the surface of a specimen. As the probe passes over the specimen, the probe moves up and down to keep the current in the probe constant. The path of the probe is recorded and is used to create an image of the specimen's surface.

Write About It How do you think advances in microscope technology have changed the way we understand the world around us? Write a short journal entry in which you explore some of the answers to this question.

Quick Facts

Studying DNA can have many uses. One of these is in DNA forensics. Many times, when a crime is committed, the criminal leaves behind some genetic material—hair, skin, or even blood. Police analyze any DNA found at the scene of a crime to try to determine the identity of a criminal by comparing the evidence to DNA in a database. DNA evidence can also be used to prove a suspect's innocence. Scientists must compare DNA samples at 13 different locations in the sample in order to find enough matches to make a strong connection between the samples. If the samples match in four or five regions, there is a good chance both samples came from the same person. If not, there is a good chance the person is innocent. Ask students to research other applications of DNA forensics.

Texas Essential Knowledge and Skills

3D Relate the impact of research on scientific thought and society, including the history of science and contributions of scientists as related to the content.

SNIFFING OUT DISEASE

Hot Science

TEKS 3D

You may know that trained dogs are used to sniff out bombs and explosives. But did you know that dogs also can detect diseases with their sense of smell?

Volatile organic compounds, or VOCs, are organic compounds that evaporate very easily. Certain VOCs are released from human body cells as part of their basic processes. VOCs enter the blood and are released from the body when a person exhales. VOCs in a person's breath can provide clues about his or her health.

In different studies, dogs have been shown to successfully detect cancers of the skin, breasts, colon, and lung. In some studies, the dogs' results were more accurate than conventional laboratory tests. Researchers are hoping to learn more about the VOCs the dogs are detecting so that they can build an "electronic nose" to test for these chemicals.

Write About It Use print or online sources to research more about the use of dogs to detect the chemical signatures of different cancers. How were the tests and experiments carried out? What were the results? What kinds of chemicals and compounds do scientists think are released by cancer cells? Write a research report in which you summarize your findings.

This border collie is being trained to detect cancer-related VOCs.

CHAPTER 2

Hot Science

Have students read *Sniffing Out Disease*. Explain that dogs have a well-developed sense of smell. Some scientists estimate that a dog's sense of smell is 10,000 to 100,000 times more sensitive than that of a human. Encourage students who own dogs as pets to describe any observations they have witnessed about a dog's acute sense of smell.

Dogs are built to sniff and smell. A dog's nose can contain up to 300 million olfactory receptors. A human's nose, on the other hand, contains about five to six million of these receptors. When humans inhale, air flows through the nose and into the lungs. In dogs, however, inhaled air travels through two different pathways. One travels to the lungs, while the other travels to the dog's olfactory region. When humans breathe out, air leaves the same way it comes in. A dog's nose contains slits on each side, and exhaled air travels out the sides of the dog's nose. At the same time, fresh air travels into the nose, allowing dog's to sniff almost continuously. Dogs, unlike humans, can also move each nostril independently. As a result, dogs can detect the location of a source of an odor.

As students research more about the use of dogs to detect cancer, have them find out about the success rates of standard forms of cancer screening and compare the results to the tests done with dogs.

Ask: **In order to successfully develop an "electronic nose," what must scientists first do?** *(Sample: Scientists must determine how to identify the chemical signatures of the different cancers.)*

Texas Essential Knowledge and Skills

3D Relate the impact of research on scientific thought and society, including the history of science and contributions of scientists as related to the content.

Cell Processes and Energy

Chapter TEKS Overview

This chapter focuses on TEKS 5A, 5C, 7B, and 12F. Students will explore how living things get the energy they need to survive and develop through the processes of photosynthesis and cellular respiration. They will learn how to diagram the flow of energy through living systems, including food chains, food webs, and energy pyramids. Students also will examine how energy is transformed within organisms.

Introduce the TEKS

To pique student interest and introduce TEKS 5A, 7B, and 12F, have students look at the image and read the Focus Question and description. Ask students to develop hypotheses about how the trees get the energy they need to grow. Have volunteers explain their hypotheses. Point out that the giant California redwood trees can grow to such great heights because of the energy they obtain from the sun, which they combine with water and gases from the air to make food. Ask: **What do you think of when you hear the word *energy*?** *(Sample: being able to do things like run fast, play sports)* **Where do you think you get your energy from?** *(Sample: the healthful foods I eat)* **Do you think all living things get energy from eating food?** *(Sample: No, some living things make their own food so they do not have to eat food to get energy.)*

Untamed Science Video

YUM...EATING SOLAR ENERGY Before viewing, invite students to discuss what they know about solar energy. Then play the video. Lead a class discussion and make a list of questions that this video raises. You may wish to have students view the video again after they have completed the chapter to see if their questions have been answered.

HOW DO THESE GIANTS GROW?

How do living things get energy?

Looking straight up into the sky from the ground, you can see the tallest trees on Earth. These giant California redwoods can grow up to 110 meters tall, about the size of a 35-story skyscraper! To grow this big takes energy and raw materials from food. These trees don't eat food as you do. But they do get water through their roots, gases from the air, and lots of sunlight. **Develop Hypotheses How do these trees get the energy they need to grow?**

Sample: The trees get energy from the sun. They use this energy, the water, and the gases to make food.

Watch the **Untamed Science** video to learn more about living things and energy.

UntamedScience

102 Cell Processes and Energy

Professional Development Note

From the Author

Pancakes for breakfast! David and I mixed the flour, sugar, baking soda, and salt in a bowl. Then we added the milk, eggs, and vegetable oil. Our pancake batter was ready. I carefully poured a spoonful of batter onto the skillet to make the first pancake. We waited about two minutes and then used a spatula to flip it over. The pancake batter went right through the spatula! We had forgotten to turn on the burner beneath the skillet. When you learn about photosynthesis, think of the pancakes. Keep in mind how important it is to turn on the burner. For plants to make their own food, they need water and carbon dioxide and energy from the sun to produce sugar and oxygen.

Zipporah Miller

Texas CHAPTER 3

Cell Processes and Energy

Texas Essential Knowledge and Skills

SUPPORTING TEKS: 5C Diagram the flow of energy through living systems. **12F** Recognize that according to cell theory, all cells carry on similar functions such as extracting energy from food to sustain life.

TEKS: 2D Construct graphs, using repeated trials and means, to organize data and identify patterns. **2E** Analyze data to formulate reasonable explanations, communicate valid conclusions supported by the data, and predict trends. **3B** Use models to represent aspects of the natural world. **5A** Recognize that radiant energy from the Sun is transformed into chemical energy through the process of photosynthesis. **7B** Illustrate the transformation of energy within an organism. **12E** Compare the functions of a cell to the functions of organisms such as waste removal.

PEARSON Texas.com 103

Spotlight On Technology

Editable Presentation Each lesson includes an editable presentation to help students understand the energy transformations that occur in cells and cell division.

Flipped Video for Science Show students a Pearson Flipped Video for Science for this chapter so that they can better understand the energy transformations that occur in cells and cell division.

PEARSON Texas.com

Chapter at a Glance

CHAPTER PACING: 6–9 periods or 3–4$\frac{1}{2}$ blocks

INTRODUCE THE CHAPTER: Use the Focus Question and the opening image to get students thinking about cell processes and energy. Activate prior knowledge and preteach vocabulary using the Getting Started pages.

Lesson 1: Photosynthesis

Lesson 2: Cellular Respiration

Lesson 3: Cell Division

ASSESSMENT OPTIONS:
Teacher's Edition: Lesson Quizzes, Texas End-of-Year Test Prep A and B
Online assessments

Texas Essential Knowledge and Skills

2D Construct tables and graphs, using repeated trials and means, to organize data and identify patterns.

2E Analyze data to formulate reasonable explanations, communicate valid conclusions supported by the data, and predict trends.

3B Use models to represent aspects of the natural world such as human body systems and plant and animal cells.

5A Recognize that radiant energy from the Sun is transformed into chemical energy through the process of photosynthesis.

5C Diagram the flow of energy through living systems, including food chains, food webs, and energy pyramids.

7B Illustrate the transformation of energy within an organism such as the transfer from chemical energy to heat and thermal energy in digestion.

12E Compare the functions of a cell to the functions of organisms such as waste removal.

12F Recognize that according to cell theory all organisms are composed of cells and cells carry on similar functions such as extracting energy from food to sustain life.

The following **College and Career Readiness Standards** are covered in this chapter: **I.A.4, VII.H.2, X.B.2**

CHAPTER 3

Getting Started

Check Your Understanding

This activity assesses students' understanding of cellular structures. Plant and animal cells contain a nucleus, which controls and directs the cell's activities and functions. In plant cells, chloroplasts capture energy from sunlight, which the plant uses to help produce food for itself. Mitochondria convert energy in food so that the cell can function.

Preteach Vocabulary Skills

Explain to students that many English words are derived from Greek. Knowing the meaning of Greek word parts can help students determine the meaning of many unfamiliar words. Review the two Greek word parts in the table, and then have students identify other words that contain these word parts.

CHAPTER 3 Getting Started

Check Your Understanding

1. Background Read the paragraph below and then answer the question.

In science class, we looked at both plant and animal cells under the microscope. I could see the **nucleus** in many cells. In plant cells, we could see green-colored **chloroplasts.** Both plant and animal cells have **mitochondria,** but they were too small for us to see with the microscopes we had.

The **nucleus** is the organelle that acts as the cell's control center and directs the cell's activities.

Chloroplasts are organelles that capture energy from sunlight and use it to produce food for the cell.

Mitochondria are organelles that convert energy in food to energy the cell can use to carry out its functions.

- Circle the names of the organelles found only in plant cells. Underline the organelles found in both plant and animal cells.
  nucleus mitochondria chloroplasts

Vocabulary Skill

Greek Word Origins The table below shows English word parts that have Greek origins. Learning the word parts can help you understand some of the vocabulary in this chapter.

Greek Word Part	Meaning	Example
auto-	self	**autotroph,** *n.* an organism that makes its own food; a producer
hetero-	other, different	**heterotroph,** *n.* an organism that cannot make its own food; a consumer

2. Quick Check The word part *-troph* comes from the Greek word *trophe,* which means "food." Circle the word part in two places in the chart above. How does the Greek word relate to the meaning of the terms?

These terms describe whether or not an organism makes its own food.

104 Cell Processes and Energy

English Language Proficiency Standards

ELPS Learning Strategies 1.F

Have students use prior knowledge and vocabulary as they begin the word wall in Preview Vocabulary Terms, page 105.

Beginning Display and read aloud the terms before beginning each lesson. Have students show how familiar they are with each word by signing thumbs up, sideways, or down. Together, discuss clues to meaning in images, labels, captions, and text.

Intermediate Display and read aloud the vocabulary terms. Have students tell what they know about each word. Have small groups preview the text, looking for clues to meaning. Discuss definitions and examples.

Advanced Have partners preview the terms and discuss words or word parts they recognize. Then have them locate the terms in the lesson. Discuss definitions.

Advanced High Have partners preview the terms and discuss words or word parts they recognize. Then have them locate the terms in the glossary. Discuss definitions.

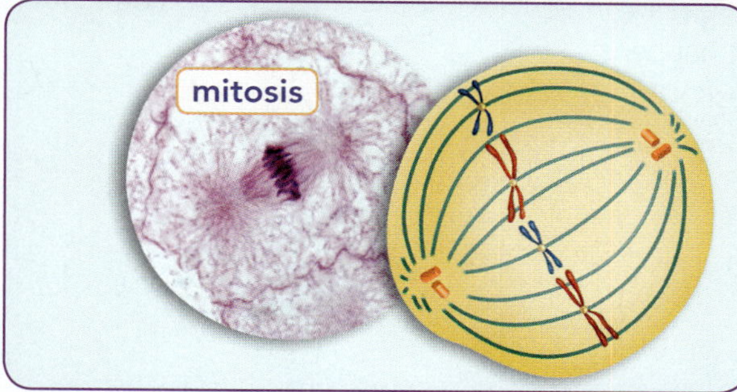

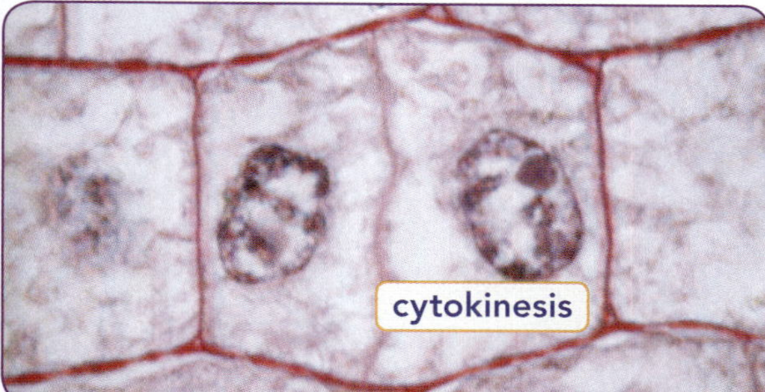

Chapter Preview

LESSON 1

- photosynthesis
- autotroph
- heterotroph
- chlorophyll

Sequence
Classify

LESSON 2

- cellular respiration
- fermentation

Summarize
Control Variables

LESSON 3

- cell cycle
- interphase
- replication
- chromosome
- mitosis
- cytokinesis

Ask Questions
Interpret Data

Preview Vocabulary Terms

Have students work together to create a word wall to display the vocabulary terms for the chapter. Be sure to discuss and analyze each term before posting it on the wall. As the class progresses through the chapter, the words can be sorted and categorized in different ways.

L1 Have students look at the images on this page as you pronounce the vocabulary word. Have students repeat the word after you. Then read the definition below. Use the sample sentence in italics to clarify the meaning of the term.

heterotroph *(HET ur ah trahf)* An organism that cannot make its own food. *Another name for a heterotroph, such as the zebra, is a consumer because it eats other organisms.*

fermentation *(fur mun TAY shun)* An energy-releasing process that does not require oxygen. *The carbon dioxide produced by yeast during alcoholic fermentation causes bread dough to rise.*

mitosis *(my TOH sis)* The second stage of the cell cycle, which is divided into four parts: prophase, metaphase, anaphase, and telophase. *During mitosis, the cell's nucleus divides into two nuclei and one set of DNA is distributed to each daughter cell.*

cytokinesis *(sy toh kih NEE sis)* The final stage of the cell cycle that completes the process of cell division. *During cytokinesis, a plant or animal cell finishes dividing into two daughter cells.*

Academic Vocabulary

Each lesson includes key Academic Vocabulary. See also the Support All Readers box at the start of the lesson.

Lesson 1: classify, sequence

Lesson 2: summarize, variable

Lesson 3: data, interpret, question

Photosynthesis

How do living things get energy?

LESSON PACING:
2–3 periods or 1–1½ blocks

Lesson Vocabulary

- photosynthesis
- autotroph
- heterotroph
- chlorophyll

Content Refresher

Chlorophyll and the Absorption of Light White light, or visible light, is a mixture of different wavelengths, with each wavelength corresponding to a particular color. When white light passes through a prism, it is separated into the visible spectrum: red, orange, yellow, green, blue, and violet.

The pigment chlorophyll absorbs some but not all of the wavelengths of visible light. There are two main types of chlorophyll—chlorophyll a and chlorophyll b, which behave similarly in terms of absorbing different wavelengths. The highest percentage of absorption by chlorophyll occurs in the violet/blue and orange/red areas of the spectrum. In contrast, almost no light is absorbed in the green area of the spectrum.

Chlorophyll makes chloroplasts and leaves look green because it reflects rather than absorbs green light. It is the light reflected by leaves that reaches people's eyes.

Light is a form of energy. When chlorophyll absorbs light, energy is transferred to electrons in atoms of chlorophyll. These "energized" electrons provide the energy for photosynthesis.

Lesson Objectives	TEKS	ELPS
Explain how living things get energy from the sun.	5A, 5C	3.B.2
Describe what happens during photosynthesis.	5A, 7B	

Texas Essential Knowledge and Skills

5A Recognize that radiant energy from the Sun is transformed into chemical energy through the process of photosynthesis.
5C Diagram the flow of energy through living systems, including food chains, food webs, and energy pyramids.
7B Illustrate the transformation of energy within an organism such as the transfer from chemical energy to heat and thermal energy in digestion.

English Language Proficiency Standards

ELPS Speaking 3.B.2 Expand and internalize initial English vocabulary by retelling simple stories and basic information represented or supported by pictures.

STEM Project

To access this project, go to the additional resources for Chapter 3 at PearsonTexas.com. In **Energy Boosters,** students will design and test a method to evaluate how much sugar is in a variety of energy drinks.

DIFFERENTIATED INSTRUCTION KEY

L1 Struggling Students or Special Needs
L2 On-Level Students L3 Advanced Students

LESSON PLANNER 3.1

Investigations and Activities

My Planet Diary, **Student Edition,** p. 106

Inquiry: Inquiry Warm-Up, Where Does the Energy Come From?, **PearsonTexas.com**

Introduce Vocabulary, **Teacher's Edition,** p. 107

Support the TEKS, Consumers, **Teacher's Edition,** p. 107

Lead a Discussion, Plants and Radiant Energy, **Teacher's Edition,** p. 107

Teach Key Concepts, **Teacher's Edition,** p. 108

Lead a Discussion, Producers and Consumers, **Teacher's Edition,** p. 108

Apply It!, **Student Edition,** p. 108

Inquiry: Quick Lab, Energy from the Sun, **Lab Manual,** p. 27

Teach Key Concepts, **Teacher's Edition,** p. 109

Lead a Discussion, Review Organelles, **Teacher's Edition,** p. 109

Lead a Discussion, The Stages of Photosynthesis, **Teacher's Edition,** p. 109

Lead a Discussion, Photosynthesis: Stage 2, **Teacher's Edition,** p. 110

Teach With Visuals, **Teacher's Edition,** p. 110

Inquiry: Teacher Demo, A Leaf's Response to Light, **Teacher's Edition,** p. 110

Lead a Discussion, Chemical Equations, **Teacher's Edition,** p. 111

Differentiated Instruction, **Teacher's Edition,** p. 111

Inquiry: Quick Lab, Looking at Pigments, **PearsonTexas.com**

TEKS Review

Apply the TEKS, Analyzing Metabolic Rates, **Student Edition,** p. 126

TEKS Practice, **Student Edition,** p. 128

TEKS Practice: Chapter and Cumulative Review, **Student Edition,** p. 130

Lesson 3.1, **TEKS Preparation and Study Guide Workbook,** p. 28

SHORT ON TIME? To do this lesson in approximately half the time, do the Activate Prior Knowledge activity. A discussion of the Key Concepts will familiarize students with the lesson content. Have students do the Quick Labs. The rest of the lesson can be completed by students independently.

These editable worksheets are available on **PearsonTexas.com.**
Print versions can be found in the **TEKS Preparation and Study Guide Workbook.**

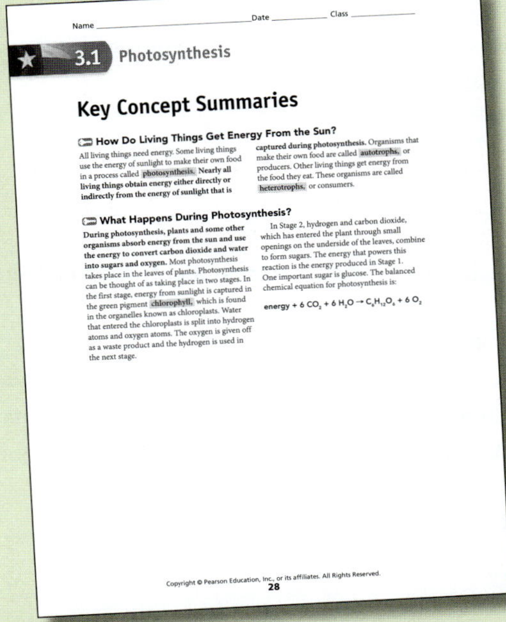

Name ______ Date ______ Class ______

3.1 Photosynthesis

Key Concept Summaries

How Do Living Things Get Energy From the Sun?

All living things need energy. Some living things use the energy of sunlight to make their own food in a process called **photosynthesis. Nearly all living things obtain energy either directly or indirectly from the energy of sunlight that is captured during photosynthesis.** Organisms that make their own food are called **autotrophs,** or producers. Other living things get energy from the food they eat. These organisms are called **heterotrophs,** or consumers.

What Happens During Photosynthesis?

During photosynthesis, plants and some other organisms absorb energy from the sun and use the energy to convert carbon dioxide and water into sugars and oxygen. Most photosynthesis takes place in the leaves of plants. Photosynthesis can be thought of as taking place in two stages. In the first stage, energy from sunlight is captured in the green pigment **chlorophyll,** which is found in the organelles known as chloroplasts. Water that entered the chloroplasts is split into hydrogen atoms and oxygen atoms. The oxygen is given off as a waste product and the hydrogen is used in the next stage.

In Stage 2, hydrogen and carbon dioxide, which has entered the plant through small openings on the underside of the leaves, combine to form sugars. The energy that powers this reaction is the energy produced in Stage 1. One important sugar is glucose. The balanced chemical equation for photosynthesis is:

$$\text{energy} + 6\,CO_2 + 6\,H_2O \rightarrow C_6H_{12}O_6 + 6\,O_2$$

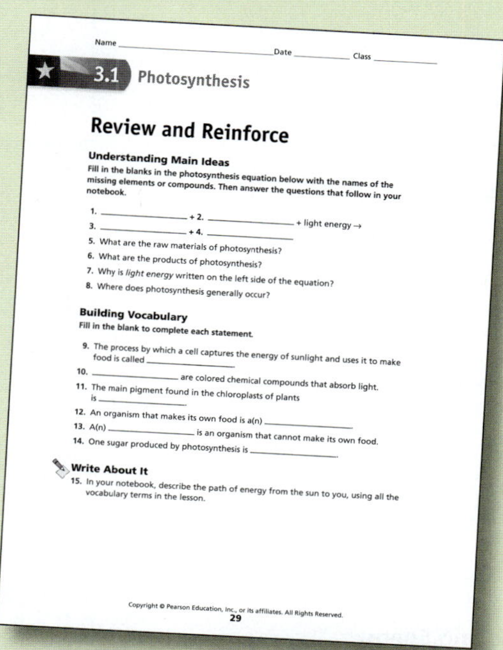

Name ______ Date ______ Class ______

3.1 Photosynthesis

Review and Reinforce

Understanding Main Ideas

Fill in the blanks in the photosynthesis equation below with the names of the missing elements or compounds. Then answer the questions that follow in your notebook.

1. ______ + 2. ______ + light energy →
3. ______ + 4. ______
5. What are the raw materials of photosynthesis?
6. What are the products of photosynthesis?
7. Why is *light energy* written on the left side of the equation?
8. Where does photosynthesis generally occur?

Building Vocabulary

Fill in the blank to complete each statement.

9. The process by which a cell captures the energy of sunlight and uses it to make food is called ______
10. ______ are colored chemical compounds that absorb light.
11. The main pigment found in the chloroplasts of plants is ______.
12. An organism that makes its own food is a(n) ______.
13. A(n) ______ is an organism that cannot make its own food.
14. One sugar produced by photosynthesis is ______.

Write About It

15. In your notebook, describe the path of energy from the sun to you, using all the vocabulary terms in the lesson.

Lexile Measure = 940L

LESSON 3.1

Photosynthesis

Establish Learning Objectives

After this lesson, students will be able to:

- Explain how living things get energy from the sun.
- Describe what happens during photosynthesis.

Engage

Activate Prior Knowledge

MY PLANET DIARY Read *When Is Food Not Food?* with the class. Reinforce the distinction between food and fertilizer by explaining that food is a source of energy and fertilizer is a source of minerals helpful to plant growth. Ask: **Wild plants are not given fertilizer. Where do you think they get the minerals they need?** *(Sample: Most of the minerals a plant needs come from the soil as a result of the decay of dead plant and animal material.)*

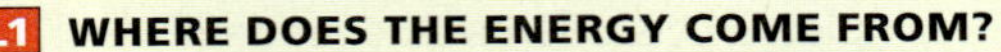

Explore

Lab Resource: Inquiry Warm-Up

L1 WHERE DOES THE ENERGY COME FROM? Students will experiment to discover the energy source for a solar-powered calculator. This Inquiry Warm-Up can be found online.

Photosynthesis

- How Do Living Things Get Energy From the Sun? TEKS 5A, 5C
- What Happens During Photosynthesis? TEKS 5A, 7B

my planet Diary MISCONCEPTION

When Is Food Not Food?

Misconception: Some people think that the plant food they give to house and garden plants is food for the plants. It isn't.

Plants make their own food—in the form of sugars—using water, carbon dioxide, and sunlight. So what is the "food" that people add to plants? It's fertilizer. Fertilizer is a mixture of minerals, such as potassium, calcium, and phosphorus. It helps plants grow but doesn't supply them with energy as food does. Farmers add fertilizer to soil to grow better quality crops. People do the same to grow bigger and healthier plants at home.

Communicate Write your answers to the questions below. Then discuss Question 2 with a partner.

1. What is "plant food"?
   Fertilizer that contains minerals, which help plants grow

2. What do you think would happen if you put a small seedling in complete darkness for a month but kept all other environmental conditions the same?
   Sample: The small seedling would die.

Lab zone: Do the Inquiry Warm-Up *Where Does the Energy Come From?* Find the lab online.

SUPPORT ALL READERS

Lexile Measure = 940L **Lexile Word Count = 1136**

Prior Exposure to Content: Most students have encountered this topic in earlier grades

Academic Vocabulary: *classify, sequence*

Science Vocabulary: *photosynthesis, autotroph, heterotroph, chlorophyll*

Concept Level: Generally appropriate for most students in this grade

Preteach With: My Planet Diary "When Is Food Not Food?" and Figure 2 activity

Vocabulary
- photosynthesis
- autotroph
- heterotroph
- chlorophyll

Skills
- Reading: Sequence
- Inquiry: Classify

How Do Living Things Get Energy From the Sun?

On a plain in Africa, a herd of zebras peacefully eats grass. But watch out! A group of lions is about to attack the herd. The lions will kill one of the zebras and eat it.

Both the zebras and the lion you see in **Figure 1** use the food they eat to obtain energy. Every living thing needs energy. All cells need energy to carry out their functions, such as making proteins and transporting substances into and out of the cell. Like the raw materials used within a cell, energy used by living things comes from their environment. Plant and animal cells obtain their energy differently. Zebra meat supplies the lion's cells with energy. Similarly, grass provides the zebra's cells with energy. But where does the energy in the grass come from? Plants and certain other organisms, such as algae and some bacteria, use the energy in sunlight, sometimes called radiant energy, to make their own food.

TEKS 5A, 5C In this section, you'll explore how radiant energy from the sun is captured by plants. You will also study how this energy flows through living systems.

ELPS 3.B.2 With a partner, read and retell the two paragraphs on this page as a simple story. Use the pictures to help you. Make your story help show how living things get energy from the sun.

FIGURE 1
An Energy Chain
All living things need energy.

Interpret Photos In the boxes, write the direct source of energy for each organism. Which organism shown does not depend on another organism for food?

Grass

Explain

Introduce Vocabulary

To help students understand the terms *autotroph* and *heterotroph*, point out that they share the same root word, *troph*, which means "food." *Auto-* means "self," so autotrophs make food by themselves. *Hetero-* means "other," so heterotrophs get their food from other sources.

Support the TEKS

CONSUMERS Remind students that living things need energy in order to carry out life functions. The larger the organism, the more energy it needs. Ask: **What are some ways animals use energy?** *(Samples: running or walking, chewing food, thinking, growing, reproducing)* **Where do animals get their energy?** *(From the food they eat)* If students have already studied ecology, ask them to suggest another name for **Figure 1.** *(Sample: A Food Chain)*

Lead a Discussion

PLANTS AND RADIANT ENERGY Students may be familiar with the energy needs of plants from growing houseplants at home. Ask: **Where are houseplants usually placed?** *(Near a window or under plant lights)* **Why?** *(So they get enough light, or radiant energy.)* **What happens if a houseplant doesn't get enough light?** *(It loses its leaves, turns yellow, and may die.)* **What does this tell you about why plants need light?** *(Plants need light energy to make food so they can survive and grow.)*

PEARSON Texas.com

English Language Proficiency Standards

ELPS Speaking 3.B.2

Have students complete the ELPS activity using a flow chart to retell the text as a chain of events.

Beginning Point out the grass, lion, and zebras. Ask students to listen for how they are connected. Read aloud the page. Together, fill in a flow chart to show how energy gets from the sun to the lion. Model how to retell based on the chart. Then have partners practice.

Intermediate Together read aloud page 107. Have partners make a flow chart that shows how energy gets from the sun to the lion, then retell the story.

Advanced Have partners read and create a flow chart showing how energy gets from the sun to the lion. Have them tell about events in the story.

Advanced High Have partners read the lesson and create flow charts. Then have them swap charts to retell the story.

Texas Essential Knowledge and Skills

5A Recognize that radiant energy from the Sun is transformed into chemical energy through the process of photosynthesis.

5C Diagram the flow of energy through living systems, including food chains, food webs, and energy pyramids.

LESSON 3.1

Explain

Teach Key Concepts

Explain to students that radiant energy from sunlight is captured during photosynthesis. Ask: **What is photosynthesis?** *(The process by which a cell captures energy from sunlight and uses it to make food)* **Does grass obtain energy directly from sunlight?** *(Yes, it makes its own food.)* **Do people?** *(No)*

Lead a Discussion

PRODUCERS AND CONSUMERS Explain that different terms describe how an organism obtains its energy. Ask: **What does the term *produce* mean to you?** *(Sample: It means "to make.")* **What kinds of organisms make food?** *(Autotrophs)* **What does the term *consume* mean to you?** *(Sample: It means to eat or use something already made.)* **What kinds of organisms eat food?** *(Heterotrophs)*

Elaborate

Apply It!

L1 Review **Figure 1** before beginning the activity. Diagrams should show a path from the sun to the plant to the caterpillar to the spider.

Sequence Tell students that sequencing means organizing information from beginning to end.

Classify Help students understand that when they classify, they group together items that are alike.

Lab Resource: Quick Lab 

L1 **ENERGY FROM THE SUN** Students will look for photosynthetic activity in *Elodea*. This Quick Lab can be found in the Student Lab Manual, p. 27, and online.

Evaluate

Assess Your Understanding

After students answer the questions, have them evaluate their understanding by completing the appropriate sentence.

RTI Response to Intervention

1a. If students cannot classify organisms, **then** have them review the definitions for *autotroph* and *heterotroph*.

b. If students need help with recognizing the kind of energy used in photosynthesis, **then** have them reread the information on the sun as an energy source.

c. If students have trouble applying the concept, **then** have them review the information on the chain of energy and in **Figure 1.**

A spider catches and eats a caterpillar that depends on plant leaves for food.

1 **Sequence** Draw a diagram of your own that illustrates how the sun's energy gets to the spider.

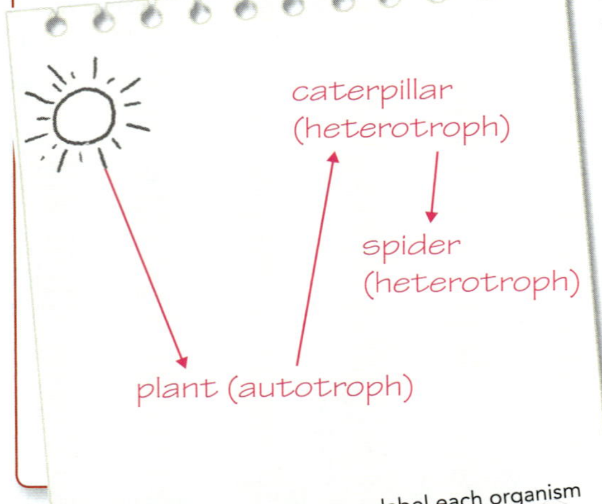

2 **Classify** In your diagram, label each organism as a heterotroph or an autotroph.

The Sun as an Energy Source

The process by which a cell captures energy in sunlight and uses it to make food is called **photosynthesis** (foh toh SIN thuh sis). The term *photosynthesis* comes from the Greek words *photos*, which means "light," and *synthesis*, which means "putting together."

Nearly all living things obtain energy either directly or indirectly from the energy of sunlight that is captured during photosynthesis. Grass obtains energy directly from sunlight and uses it to make its own food during photosynthesis. When the zebra eats grass, it gets energy from the sun that has been stored in the grass as chemical energy. Similarly, the lion obtains energy stored in the zebra. The zebra and lion both obtain the sun's energy indirectly when they consume grass or another organism that consumed grass.

Producers and Consumers

Plants make their own food through the process of photosynthesis. An organism that makes its own food is called a producer, or an **autotroph** (AWT oh trohf). An organism that cannot make its own food, including animals such as the zebra and the lion, is called a consumer, or a **heterotroph** (HET ur oh trohf). Many heterotrophs obtain food by eating other organisms. Some heterotrophs, such as fungi, absorb their food from other organisms.

Do the Quick Lab *Energy From the Sun.* Student Lab Manual, p. 27

Assess Your Understanding

TEKS 5A, 5C

1a. Identify An organism that makes its own food is a(n) (autotroph/heterotroph).

b. Recognize What kind of energy is converted to chemical energy during the process of photosynthesis?
Radiant energy from the sun

c. Apply Concepts Give an example of how energy that first comes from the sun gets into your cells.
Sample: Wheat captures the sun's energy, and I eat bread made from wheat.

got it?

○ **I get it!** Now I know that living things get energy directly from the sun by photosynthesis or indirectly by eating producers or other consumers.

○ I need extra help with See TE note.

What Happens During Photosynthesis?

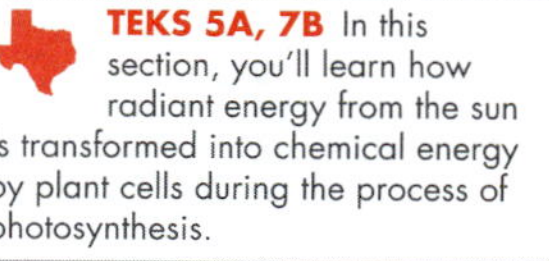

TEKS 5A, 7B In this section, you'll learn how radiant energy from the sun is transformed into chemical energy by plant cells during the process of photosynthesis.

You've just read that plants make their own food. So how do they do that? **During photosynthesis, plants and some other organisms absorb energy from the sun and use the energy to convert carbon dioxide and water into sugars and oxygen.** You can think of photosynthesis as taking place in two stages. First, plants capture the sun's energy. Second, plants produce sugars.

Stage 1: Capturing the Sun's Energy In the first stage of photosynthesis, energy from sunlight is captured. In plants, this process occurs mostly in the leaves. Recall that chloroplasts are green organelles inside plant cells. The green color comes from pigments, colored chemical compounds that absorb light. The main pigment for photosynthesis in chloroplasts is **chlorophyll**.

Chlorophyll functions something like the solar cells in a solar-powered calculator. Solar cells capture the energy in light and transform the energy to power the calculator. Similarly, chlorophyll captures light energy and transforms it into chemical energy that is used in the second stage of photosynthesis.

Also during Stage 1, water in the chloroplasts is split into hydrogen and oxygen, as shown in **Figure 2.** The oxygen is given off as a waste product. The hydrogen is used in Stage 2.

Vocabulary Greek Word Origins The Greek word part *chloros-* means "pale green." Circle two words in the text that begin with this word part. Which word means "a green compound that absorbs light"?

- ○ Chloroplast
- ● Chlorophyll

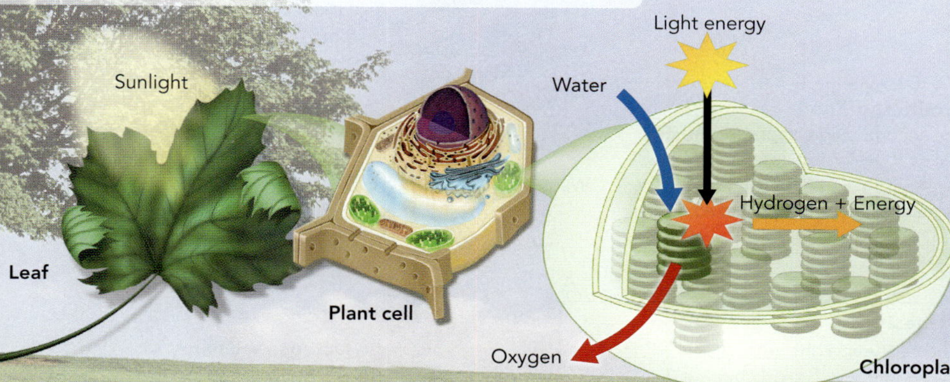

FIGURE 2
First Stage of Photosynthesis
You might say the first stage of photosynthesis powers the "energy engine" of the living world.

Make Generalizations What do you think this sentence means?
Stage 1 starts the chain of energy that moves from producers to consumers.

Differentiated Instruction

L1 Food Consumption Diary To enable students to get a sense of what they consume in terms of its source—autotroph or heterotroph—have them keep a diary of what they eat for a week. Have them indicate whether the food comes from an autotroph or a heterotroph.

L3 Pigments Chlorophyll is a pigment, one of several in a leaf. Have students do research in the library or on the Internet to find out what a pigment is and what other pigments are found in leaves. Have them use this information to explain why leaves change color in many parts of the country in fall.

Explain

Teach Key Concepts

Explain to students that plants and some other organisms are able to carry out photosynthesis. In this process they use the energy of sunlight to combine water and carbon dioxide and produce sugars and oxygen. Ask: **Why can plants perform photosynthesis but animals cannot?** *(Plants have organelles called chloroplasts that contain chlorophyll, which is capable of absorbing the energy of the sun. Animals do not have chloroplasts.)* **Where does photosynthesis take place in a plant?** *(In the leaves)* **Where do the raw materials for photosynthesis come from?** *(Water comes from the soil and enters the plant through the roots. Carbon dioxide comes from the air and enters the plant through the leaves.).*

Lead a Discussion

REVIEW ORGANELLES Ask students to list as many cell parts, or organelles, as they can and identify the function of each. Then have them identify the organelle found in plant cells but not in animal cells. Ask: **What organelle enables a plant to make food?** *(Chloroplast)* **Do animal cells contain this organelle?** *(No)*

Address Misconceptions

L1 A DIFFERENT SORT OF LEAF Because students have learned that photosynthesis takes place in the leaves of plants, they may think that conifers, such as pine trees, do not carry out photosynthesis. Ask students if a pine tree has leaves. Some students may say that it does not. Point out that pine needles are modified leaves. Ask: **How do you think the needles function?** *(They produce food through photosynthesis.)*

Lead a Discussion

THE STAGES OF PHOTOSYNTHESIS Tell students that the process of photosynthesis does not happen all at once. It occurs in two stages. Ask: **What are the two stages of photosynthesis?** *(Capturing the sun's energy and producing sugars)* **What pigment in chloroplasts absorbs light?** *(Chlorophyll)*

Texas Essential Knowledge and Skills

5A Recognize that radiant energy from the Sun is transformed into chemical energy through the process of photosynthesis.

7B Illustrate the transformation of energy within an organism such as the transfer from chemical energy to heat and thermal energy in digestion.

Explain

Lead a Discussion

PHOTOSYNTHESIS: STAGE 2 Remind students that photosynthesis is a two-stage process. The energy captured in Stage 1 converts raw materials into products in Stage 2. Ask: **What happens in the second stage of photosynthesis?** *(Water and carbon dioxide combine chemically to form sugars, and oxygen is released.)* **What substances are the raw materials?** *(Carbon dioxide and water)* **What substances are the products?** *(Sugars and oxygen)* **How are roots important for photosynthesis?** *(Roots take up water.)* **How do plants get carbon dioxide?** *(Through openings on the leaves)*

Sequence Remind students that sequencing means organizing information from beginning to end and helps in understanding a step-by-step process.

Teach With Visuals

Tell students to look at **Figure 3.** Explain that the illustration of the larger chloroplast has been simplified compared to the smaller one to de-emphasize the green, stack-like structures, called grana and focus on Stage 2 of photosynthesis. Ask: **What is the name of the organelle in which photosynthesis takes place?** *(Chloroplast)* **Why is it green?** *(It contains the pigment chlorophyll.)* **What stage of photosynthesis takes place in the stacks of disk-like structures?** *(Stage 1)*

Elaborate

Teacher Demo 

L1 A LEAF'S RESPONSE TO LIGHT

Materials black construction paper, common household plant, paper clip or tape

Time 10 minutes for setup, 1 week before observation

Cut out a small square of black construction paper. Its size should depend on the size of the plant leaf being used. Fold the square over a leaf and use either a paper clip or a small piece of tape to keep it in place. Be sure the leaf is completely shielded from light. Leave the plant in a sunny location for a week, making sure the leaf remains covered for that time. After that time, remove the black square.

Ask: **What is the appearance of the leaf?** *(It is much paler than the other leaves.)* **What do you think that means?** *(There is not as much chlorophyll in the leaf.)* **What does this activity demonstrate?** *(If the leaf stops receiving sunlight, it will stop producing chlorophyll because photosynthesis cannot take place.)*

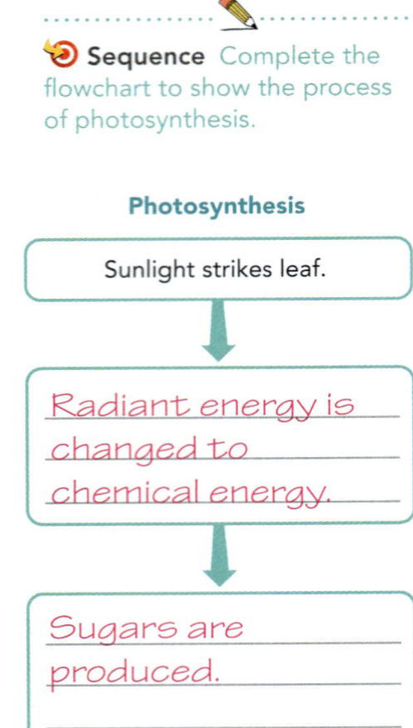

Stage 2: Using Energy to Make Food

In the second stage of photosynthesis, cells produce sugars. As shown in **Figure 3,** cells use hydrogen (H) that came from the splitting of water in Stage 1. Cells also use carbon dioxide (CO_2) from the air. Carbon dioxide enters the plant through small openings on the undersides of the leaves and moves into the chloroplasts.

Powered by the energy from Stage 1, hydrogen and carbon dioxide undergo a series of chemical reactions that result in the formation of sugars. Recall that sugars are a type of carbohydrate. One important sugar produced is glucose. It has the chemical formula $C_6H_{12}O_6$. Cells can use the chemical energy in glucose to carry out vital cell functions.

The other product of photosynthesis is oxygen gas (O_2). Recall that oxygen forms during Stage 1 when water molecules are split apart. Oxygen gas exits a leaf through the openings on its underside. Almost all the oxygen in Earth's atmosphere is produced by living things through the process of photosynthesis.

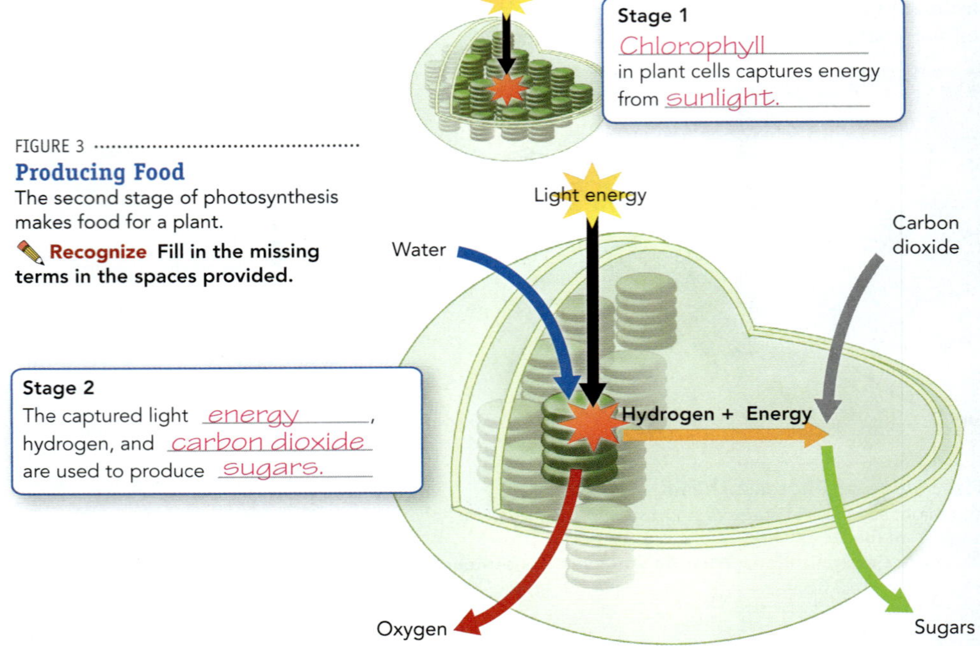

FIGURE 3

Producing Food

The second stage of photosynthesis makes food for a plant.

Recognize Fill in the missing terms in the spaces provided.

The Photosynthesis Equation

The events of photosynthesis that lead to the production of glucose can be summarized by the following chemical equation:

$$\text{light energy} + \underset{\text{carbon dioxide}}{6\,CO_2} + \underset{\text{water}}{6\,H_2O} \longrightarrow \underset{\text{glucose}}{C_6H_{12}O_6} + \underset{\text{oxygen}}{6\,O_2}$$

Notice that six molecules of carbon dioxide and six molecules of water are on the left side of the equation. These compounds are raw materials. One molecule of glucose and six molecules of oxygen are on the right side. These compounds are products. An arrow, meaning "yields," points from the raw materials to the products. Energy is not a raw material, but it is written on the left side of the equation to show that it is used in the reaction.

What happens to the sugars produced in photosynthesis? Plant cells use some of the sugars for food. The cells break down these molecules in a process that releases energy. This energy can then be used to carry out the plant's functions, such as growing and making seeds. Some sugar molecules are made into other compounds, such as cellulose for cell walls. Other sugar molecules may be stored in the plant's cells for later use. When you eat food from plants, such as potatoes or carrots, you are eating the plant's stored energy.

FIGURE 4

From the Sun to You

Carrots store food that is made by the carrot leaf cells.

Explain How are carrots an energy link between you and the sun?

A carrot makes food using the sun's energy. If I eat a carrot, I get some of that energy.

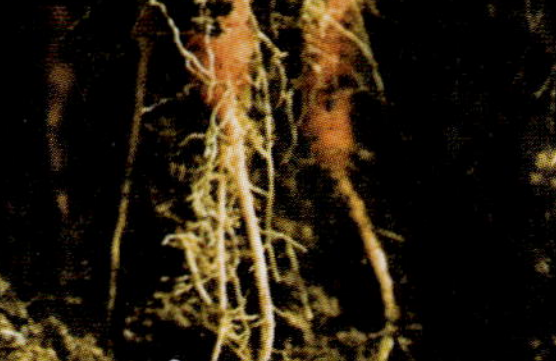

Lab zone® Do the Quick Lab *Looking at Pigments.* Find the lab online.

Assess Your Understanding

TEKS 5A, 7B

2a. Name Circle two products of photosynthesis.
glucose / carbon dioxide / oxygen / chlorophyll
(glucose and oxygen circled)

b. Recognize How is energy transformed within a plant during the process of photosynthesis?

Radiant energy from the sun is transformed into chemical energy.

c. CHALLENGE Would you expect a plant to produce more oxygen on a sunny day or a cloudy day? Explain your answer.

On a sunny day; more sunlight would be available for photosynthesis.

got it?

○ I get it! Now I know that during photosynthesis plants absorb energy from the sun and produce sugar (food) and oxygen.

○ I need extra help with See TE note.

111

Differentiated Instruction

L1 The Photosynthesis Equation Have students work in groups of four to cut out small red, blue, and green construction paper circles as follows: 12 red representing carbon atoms, 36 blue representing oxygen atoms, and 24 green representing hydrogen atoms. Have students arrange the circles to represent the balanced equation for photosynthesis. Have them mount their equation on poster board, and label it.

L3 Earth's Atmosphere Early Earth had an atmosphere that was totally devoid of oxygen. Have students find out how the atmosphere changed over the last 4.6 billion years and the significance of the process of photosynthesis. Encourage students to report their findings in the form of a science fiction travelogue of Earth through time.

Explain

Lead a Discussion

CHEMICAL EQUATIONS Explain to students that a chemical equation is a shorthand way of summarizing a chemical reaction. Elements are represented by chemical symbols, compounds by chemical formulas, and the number of molecules of each by coefficients. Coefficients are used to balance an equation. Any chemical equation must be balanced in order to reflect the law of conservation of matter, which states that matter can neither be created nor destroyed; it is always conserved. Thus, the total number of atoms of each element must be the same on both sides of the equation. Ask: **How many atoms of oxygen are on the left side of the equation? Carbon? Hydrogen?** *(18, 6, 12)* **How many atoms of oxygen are on the right side of the equation? Carbon? Hydrogen?** *(18, 6, 12)* **Has matter been conserved?** *(Yes)*

Lab Resource: Quick Lab 

L2 LOOKING AT PIGMENTS Students will explore the pigments in a leaf using coffee-filter chromatography. This Quick Lab can be found online.

Evaluate

Assess Your Understanding

After students answer the questions, have them evaluate their understanding by completing the appropriate sentence.

RTI Response to Intervention

2a. If students cannot identify two products of photosynthesis, **then** have them review the chemical equation for photosynthesis, focusing on the right side of the equation, which represents the products.

b. If students need help with recognizing how energy is transformed during photosynthesis, **then** have them review **Figure 3.**

c. If students have trouble making an inference about photosynthesis, **then** have them look at the chemical equation for photosynthesis and the placement of light energy in the equation. Have students use that observation to determine the effect of increased and decreased sunlight on the production of oxygen.

Name ______________________ Date __________ Class __________

Assess Your Understanding

Photosynthesis

How Do Living Things Get Energy From the Sun?

1a. IDENTIFY An organism that makes its own food is a(n) (autotroph/heterotroph).

b. RECOGNIZE What kind of energy is converted to chemical energy during the process of photosynthesis?

____

c. APPLY CONCEPTS Give an example of how energy that first comes from the sun gets into your cells. ______________________________

____

got it?

○ **I get it!** Now I know that living things get energy directly from the sun by __________ or indirectly by ______________________________

○ **I need extra help with** ______________________________

____

What Happens During Photosynthesis?

2a. NAME Circle two products of photosynthesis. glucose/carbon dioxide/oxygen/chlorophyll

b. RECOGNIZE How is energy transformed within a plant during the process of photosynthesis?

____

c. CHALLENGE Would you expect a plant to produce more oxygen on a sunny day or a cloudy day? Explain your answer. ______________________________

____

got it?

○ **I get it!** Now I know that during photosynthesis ______________________________

____

○ **I need extra help with** ______________________________

____

Place the outside corner, the corner away from the dotted line, in the corner of your copy machine to copy onto letter-size paper.

Place the outside corner, the corner away from the dotted line, in the corner of your copy machine to copy onto letter-size paper.

Name ______________________ Date ____________ Class ____________

Enrich

Photosynthesis

Read the passage and study the bar graph. Then answer the questions that follow on a separate sheet of paper.

Chlorophyll and the Color of Light

A pigment is a colored chemical compound that absorbs light. You can think of a pigment as a kind of sponge that absorbs light of all colors except the ones that it transmits and reflects. The colors that you see are the colors of light that the pigment reflects. The bar graph below shows the percentages of light of different colors that are reflected by the plant pigment chlorophyll.

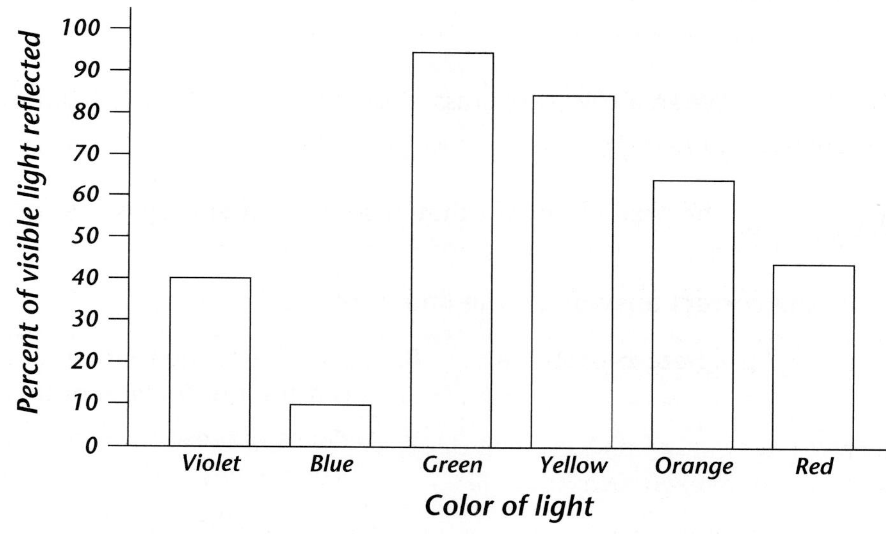

1. Which color of light does chlorophyll reflect most? About what percent of light of this color does chlorophyll reflect?
2. Which color of light does chlorophyll absorb most? About what percent of light of this color does chlorophyll absorb?
3. The colors that are reflected less than 50% contribute very little to what the eye sees. Which colors does your eye respond to when you look at a "green" leaf?
4. Which colors of light do you not see when you look at a "green" leaf?
5. Explain in your own words how chlorophyll makes a leaf look green.

Name ____________________ Date __________ Class __________

Lesson Quiz

Photosynthesis

If the statement is true, write *true*. If the statement is false, change the underlined word or words to make the statement true.

1. ______________ Autotrophs are also known as producers.

2. ______________ The ultimate source of energy for all living things is the leaf.

3. ______________ Plants are able to carry out photosynthesis because they contain the organelle known as a(n) mitochondrion.

4. ______________ One important sugar that results from photosynthesis is cellulose.

5. ______________ When a cow eats grass, it gets energy indirectly from the sugars in the grass.

6. ______________ The green pigment that absorbs light energy is chlorophyll.

Write the letter of the correct answer on the line at the left.

7. ___ **Another name for a heterotroph is a**

A producer
B raw material
C consumer
D plant

8. ___ **Which of the following is NOT true about the products of photosynthesis?**

A Some of the sugar is made into other compounds, such as cellulose.
B Some of the sugar is stored in the plant's cells for later use.
C The waste product carbon dioxide is given off through tiny openings on the underside of the leaves.
D The products are used by both plants and animals for energy.

9. ___ **Which of the following represents the raw materials of photosynthesis?**

A carbon dioxide and oxygen
B carbon dioxide and water
C glucose and oxygen
D water and glucose

10. ___ **The main characteristic of the first stage of photosynthesis is**

A the production of hydrogen and energy
B the production of hydrogen and glucose
C the release of oxygen and carbon dioxide
D the storage of glucose in the plant's cells

Place the outside corner, the corner away from the dotted line, in the corner of your copy machine to copy onto letter-size paper.

Place the outside corner, the corner away from the dotted line, in the corner of your copy machine to copy onto letter-size paper.

Photosynthesis

Answer Key

Review and Reinforce

Find the worksheet in the Student Workbook.

1. carbon dioxide

2. water

3. sugar

4. oxygen

5. carbon dioxide and water

6. oxygen and sugars

7. to show that it is used in the reaction

8. in the chloroplasts of plants and some other organisms, such as algae and some bacteria

9. photosynthesis

10. Pigments

11. chlorophyll

12. autotroph

13. heterotroph

14. glucose

15. Sample: I get energy from the food I eat. Some of the food comes from plants, such as carrots and apples. These plants are *autotrophs*, or producers. They get energy from the sun and use it to make their own food through *photosynthesis*. They can carry out photosynthesis because they contain chloroplasts, which are organelles that contain *chlorophyll*. I also get energy from eating chicken. Chickens are *heterotrophs*, or consumers, that get their energy from eating corn.

Enrich

1. green; about 95 percent

2. blue; about 90 percent

3. You see green, yellow, and orange.

4. You do not see violet, blue, or red.

5. Chlorophyll makes a leaf look green because it reflects green light in the highest percentage.

Lesson Quiz

1. true

2. sun

3. chloroplast

4. glucose

5. sun

6. true

7. C

8. C

9. B

10. A

Cellular Respiration

 **How do living things get energy?**

LESSON PACING:
2–3 periods or 1–1½ blocks

Lesson Vocabulary

- cellular respiration
- fermentation

Content Refresher

A Form of Combustion Respiration is often compared to combustion because both processes involve the breakdown of molecules in the presence of oxygen to produce energy and carbon dioxide. However, respiration is a much slower, more controlled process. The discovery of the nature of cellular respiration is attributed jointly to the French chemist Antoine Laurent Lavoisier and the French physicist, mathematician, and astronomer Pierre Laplace. In 1780, they published the results of their experiments showing that animal respiration is a form of combustion.

Energy and ATP Adenosine triphosphate, or ATP, is one of the main chemical compounds that release energy in organisms—energy that powers processes such as active transport, muscle contraction, and DNA synthesis. During cellular respiration, cells use the energy stored in food molecules to produce ATP. For each molecule of glucose that undergoes the reaction, 36 molecules of ATP are formed. Glycolysis, a stage of respiration that does not take place in a mitochondrion and does not require oxygen, produces a net of only 2 molecules of ATP. The stage of respiration that takes place in a mitochondrion and requires oxygen generates the remaining 34 molecules of ATP.

Lesson Objectives	TEKS	ELPS
Describe the steps in the process of cellular respiration.	5C, 7B, 12E, 12F	4.D
Tell what happens during fermentation.	7B, 12E, 12F	

Texas Essential Knowledge and Skills

5C Diagram the flow of energy through living systems, including food chains, food webs, and energy pyramids.
7B Illustrate the transformation of energy within an organism such as the transfer from chemical energy to heat and thermal energy in digestion.
12E Compare the functions of a cell to the functions of organisms such as waste removal.
12F Recognize that according to cell theory all organisms are composed of cells and cells carry on similar functions such as extracting energy from food to sustain life.

English Language Proficiency Standards

ELPS Reading 4.D Use prereading supports such as graphic organizers, illustrations, and pretaught topic-related vocabulary and other prereading activities to enhance comprehension of written text.

DIFFERENTIATED INSTRUCTION KEY
L1 Struggling Students or Special Needs
L2 On-Level Students L3 Advanced Students

LESSON PLANNER 3.2

Investigations and Activities

My Planet Diary, **Student Edition,** p. 112

Inquiry: Inquiry Warm-Up, Measuring Calories, **Lab Manual,** p. 28

Introduce Vocabulary, **Teacher's Edition,** p. 113

Teach Key Concepts, **Teacher's Edition,** p. 113

Make Analogies, Fuel for Energy, **Teacher's Edition,** p. 113

21st Century Learning, Critical Thinking, **Teacher's Edition,** p. 113

Lead a Discussion, Events During Cellular Respiration, **Teacher's Edition,** p. 114

Address Misconceptions, Cellular Respiration in Plants and Animals, **Teacher's Edition,** p. 114

Teach With Visuals, **Teacher's Edition,** p. 115

21st Century Learning, Critical Thinking, **Teacher's Edition,** p. 115

Differentiated Instruction, **Teacher's Edition,** p. 115

Inquiry: Lab Investigation, Exhaling Carbon Dioxide, **Lab Manual,** p. 30

Teach Key Concepts, **Teacher's Edition,** p. 116

Lead a Discussion, Oxygen Debt, **Teacher's Edition,** p. 116

Apply It!, **Student Edition,** p. 116

Differentiated Instruction, **Teacher's Edition,** p. 117

Inquiry: Quick Lab, Observing Fermentation, **PearsonTexas.com**

TEKS Review

Apply the TEKS, Energy for Life, **Student Edition,** p. 117

Apply the TEKS, Analyzing Metabolic Rates, **Student Edition,** p. 126

TEKS Practice, **Student Edition,** p. 128

TEKS Practice: Chapter and Cumulative Review, **Student Edition,** p. 130

Lesson 3.2, **TEKS Preparation and Study Guide Workbook,** p. 30

SHORT ON TIME? To do this lesson in approximately half the time, do the Activate Prior Knowledge activity. A discussion of the Key Concepts will familiarize students with the lesson content. Use the Apply the TEKS activity to help students understand how living things get energy. Have students do the Quick Lab. The rest of the lesson can be completed by students independently.

These editable worksheets are available on **PearsonTexas.com.**
Print versions can be found in the **TEKS Preparation and Study Guide Workbook.**

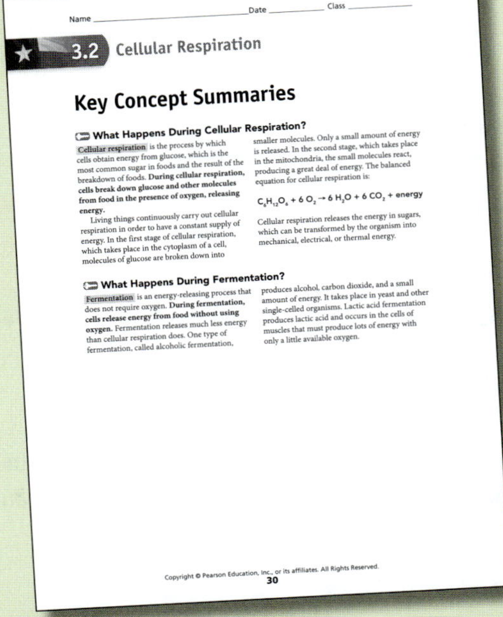

Name ______ Date ______ Class ______

3.2 Cellular Respiration

Key Concept Summaries

What Happens During Cellular Respiration?

Cellular respiration is the process by which cells obtain energy from glucose, which is the most common sugar in foods and the result of the breakdown of foods. **During cellular respiration, cells break down glucose and other molecules from food in the presence of oxygen, releasing energy.**

Living things continuously carry out cellular respiration in order to have a constant supply of energy. In the first stage of cellular respiration, which takes place in the cytoplasm of a cell, molecules of glucose are broken down into smaller molecules. Only a small amount of energy is released. In the second stage, which takes place in the mitochondria, the small molecules react, producing a great deal of energy. The balanced equation for cellular respiration is:

$$C_6H_{12}O_6 + 6\ O_2 \rightarrow 6\ H_2O + 6\ CO_2 + \text{energy}$$

Cellular respiration releases the energy in sugars, which can be transformed by the organism into mechanical, electrical, or thermal energy.

What Happens During Fermentation?

Fermentation is an energy-releasing process that does not require oxygen. **During fermentation, cells release energy from food without using oxygen.** Fermentation releases much less energy than cellular respiration does. One type of fermentation, called alcoholic fermentation, produces alcohol, carbon dioxide, and a small amount of energy. It takes place in yeast and other single-celled organisms. Lactic acid fermentation produces lactic acid and occurs in the cells of muscles that must produce lots of energy with only a little available oxygen.

30

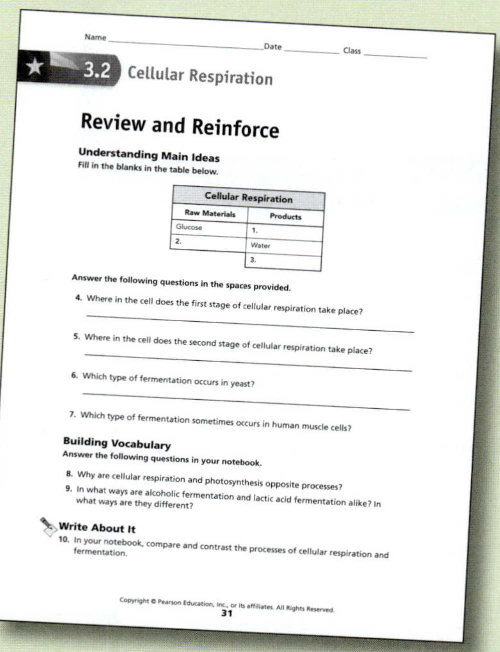

Name ______ Date ______ Class ______

3.2 Cellular Respiration

Review and Reinforce

Understanding Main Ideas
Fill in the blanks in the table below.

Cellular Respiration	
Raw Materials	**Products**
Glucose	1.
2.	Water
	3.

Answer the following questions in the spaces provided.

4. Where in the cell does the first stage of cellular respiration take place?
5. Where in the cell does the second stage of cellular respiration take place?
6. Which type of fermentation occurs in yeast?
7. Which type of fermentation sometimes occurs in human muscle cells?

Building Vocabulary
Answer the following questions in your notebook.

8. Why are cellular respiration and photosynthesis opposite processes?
9. In what ways are alcoholic fermentation and lactic acid fermentation alike? In what ways are they different?

Write About It

10. In your notebook, compare and contrast the processes of cellular respiration and fermentation.

31

Lexile Measure = 930L

LESSON 3.2

Cellular Respiration

Establish Learning Objectives

After this lesson, students will be able to:

- Describe the steps in the process of cellular respiration.
- Tell what happens during fermentation.

Engage

Activate Prior Knowledge

MY PLANET DIARY Read *Going to Extremes* with the class. Tell students that *an extremophile* can survive in places where many other organisms cannot. Ask: **Why do you think scientists study extremophiles?** *(Sample: These organisms manage to survive in ways different from those of most plants and animals. Scientists want to know how extremophiles get energy to survive and how their cells function in very unusual conditions.)*

Explore

Lab Resource: Inquiry Warm-Up 

 **MEASURING CALORIES** Students will calculate the number of Calories in a potato chip. This Inquiry Warm-Up can be found in the Student Lab Manual, p. 28, and online.

Texas Essential Knowledge and Skills

5C Diagram the flow of energy through living systems, including food chains, food webs, and energy pyramids.

7B Illustrate the transformation of energy within an organism such as the transfer from chemical energy to heat and thermal energy in digestion.

12E Compare the functions of a cell to the functions of organisms such as waste removal.

12F Recognize that according to cell theory all organisms are composed of cells and cells carry on similar functions such as extracting energy from food to sustain life.

Cellular Respiration

- What Happens During Cellular Respiration? TEKS 5C, 7B, 12E, 12F
- What Happens During Fermentation? TEKS 7B, 12E, 12F

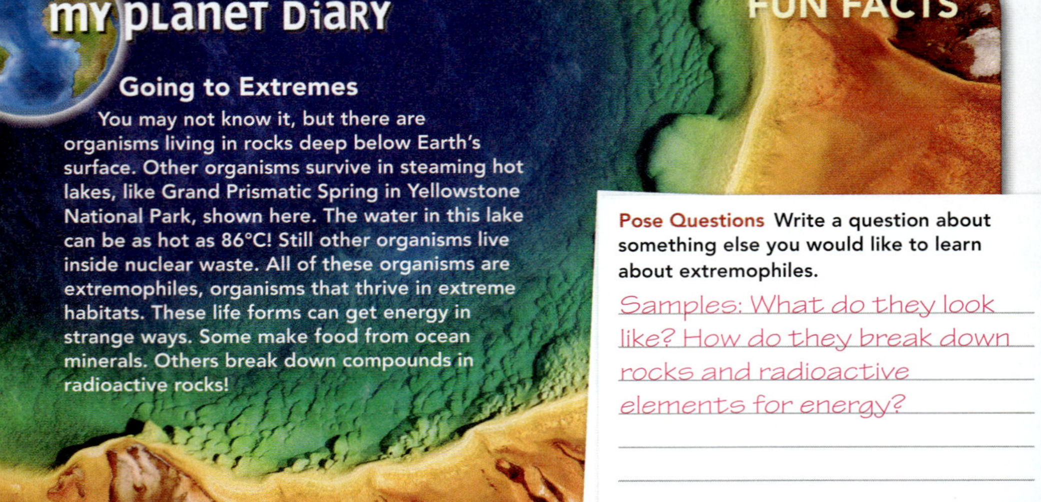

MY PLANET DIARY — FUN FACTS

Going to Extremes

You may not know it, but there are organisms living in rocks deep below Earth's surface. Other organisms survive in steaming hot lakes, like Grand Prismatic Spring in Yellowstone National Park, shown here. The water in this lake can be as hot as 86°C! Still other organisms live inside nuclear waste. All of these organisms are extremophiles, organisms that thrive in extreme habitats. These life forms can get energy in strange ways. Some make food from ocean minerals. Others break down compounds in radioactive rocks!

Pose Questions Write a question about something else you would like to learn about extremophiles.

Samples: What do they look like? How do they break down rocks and radioactive elements for energy?

 Do the Inquiry Warm-Up *Measuring Calories.* Student Lab Manual, p. 28

TEKS 5C, 7B, 12E, 12F In this section, you'll examine how cells extract energy from food during the process of cellular respiration.

What Happens During Cellular Respiration?

You and your friend have been hiking all morning. You look for a flat rock to sit on, so you can eat the lunch you packed. The steepest part of the trail is ahead. You'll need a lot of energy to get to the top of the mountain! That energy will come from food.

SUPPORT ALL READERS

Lexile Measure = 930L   Lexile Word Count = 997

Prior Exposure to Content: May be the first time students have encountered this topic

Academic Vocabulary: *summarize, variable*

Science Vocabulary: *cellular respiration, fermentation*

Concept Level: Generally appropriate for most students in this grade

Preteach With: My Planet Diary "Going to Extremes" and Figure 1 activity

Vocabulary
- cellular respiration
- fermentation

Skills

- Reading: Summarize
- Inquiry: Control Variables

ELPS 4.D Work with a partner, to preview the lesson. Scan the photos and discuss what you see. Then, read the question at the top of page 117 and the "Energy for Life" activity and graphic organizer. Fill in the organizer as you read the lesson.

What Is Cellular Respiration? After you eat a meal, your body breaks down the food into nutrients, such as sugars. The most common sugar in foods is glucose ($C_6H_{12}O_6$). **Cellular respiration** is the process by which cells obtain energy from glucose. **During cellular respiration, cells break down glucose and other molecules from food in the presence of oxygen, releasing energy.** Living things need a constant supply of energy. The cells of living things carry out cellular respiration continuously.

Storing and Releasing Energy Imagine you have money in a savings account. If you want to buy something, you withdraw some money. Your body stores and uses energy in a similar way, as shown in **Figure 1.** When you eat a meal, you add to your body's energy savings account by storing glucose. When cells need energy, they "withdraw" it by breaking down glucose through cellular respiration.

Breathing and Respiration You may have already heard of the word *respiration*. It can mean "breathing"—or moving air in and out of your lungs. Breathing brings oxygen into your lungs, which is then carried to cells for cellular respiration. Breathing also removes the waste products of cellular respiration from your body.

FIGURE 1

Getting Energy

Your body runs on the energy it gets from food.

Complete each task.

1. **Infer** Color in the last three energy scales to show how the hiker's energy changes.
2. **CHALLENGE** How do you think the hiker's breathing rate changes as she climbs?

It is faster as she climbs and slower when she rests.

English Language Proficiency Standards

ELPS Reading 4.D

Have students preview the lesson and complete the organizer on page 117.

Beginning Remind students that plant and animal cells need energy. Prompt students to tell what the images on pages 113, 114, and 115 show about energy: *On page 113, which picture shows the girl with a lot of energy?* Then read aloud the organizer on page 117. Ask students to listen for answers as you read the lesson. Together complete the organizer.

Intermediate Have partners preview images. Then have them read together, pausing after every page to add information to the organizer. Discuss responses before continuing.

Advanced Have partners preview images and page 117. Then have them read the lesson and complete the graphic organizer.

Advanced High Have partners read the lesson and complete the organizer on page 117. Then have then answer the questions at the bottom of the page.

Explain

Introduce Vocabulary

Write the term *cellular respiration* on the board. Point out that *cellular* refers to cells and that cellular respiration occurs in living things. Tell students that it is not a form of breathing. It is a chemical reaction that takes place in cells.

Teach Key Concepts

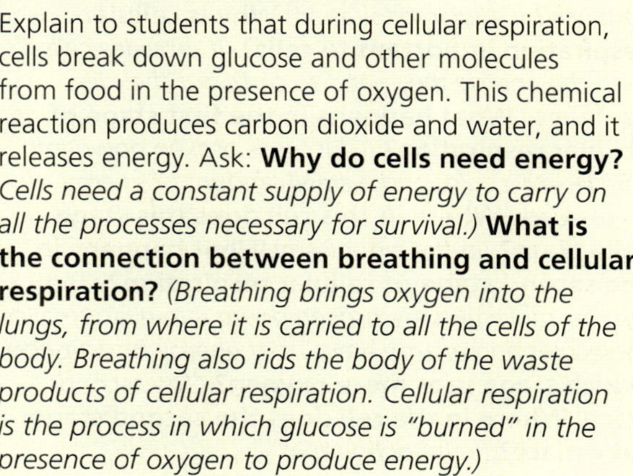

Explain to students that during cellular respiration, cells break down glucose and other molecules from food in the presence of oxygen. This chemical reaction produces carbon dioxide and water, and it releases energy. Ask: **Why do cells need energy?** *Cells need a constant supply of energy to carry on all the processes necessary for survival.)* **What is the connection between breathing and cellular respiration?** *(Breathing brings oxygen into the lungs, from where it is carried to all the cells of the body. Breathing also rids the body of the waste products of cellular respiration. Cellular respiration is the process in which glucose is "burned" in the presence of oxygen to produce energy.)*

Make Analogies

L1 FUEL FOR ENERGY Use another analogy involving filling up a car's gas tank with gasoline. Ask: **Why does a car need gasoline?** *(Gasoline is a fuel that supplies the energy that makes a car work.)* **What happens to the gasoline in the car?** *(It gets burned, or combined with oxygen.)* **What must be done to make sure that the car will keep running?** *(The tank must be refilled when it gets near empty.)*

Elaborate

21st Century Learning

CRITICAL THINKING: Tell students that during winter months, some animals go into a state of hibernation. Hibernating animals do not eat and their bodily functions are greatly reduced. Have students predict what will happen to an animal's rate of cellular respiration during hibernation and explain their prediction. *(Cellular respiration will decrease because the animal's body is not carrying out many of the activities that require energy.)*

PEARSON Texas.com

LESSON 3.2

Explain

Lead a Discussion

EVENTS DURING CELLULAR RESPIRATION Tell students that most glucose from food eaten by humans comes from the digestion of starch or complex sugars such as sucrose, lactose, or maltose. That glucose must then be broken down to release the energy it contains. Ask: **What is the process called that releases energy from glucose?** *(Cellular respiration)* **Why is cellular respiration important to cells?** *(It provides cells with the energy they need to carry out their life functions.)* **What happens in the first stage of cellular respiration?** *(Glucose is broken down into smaller molecules and a small amount of energy is released.)* **Where in this cell does this stage take place?** *(In the cytoplasm)* **What happens in the second stage of cellular respiration?** *(The smaller molecules are broken down into even smaller molecules and a large amount of energy is released.)* **Which stage uses more oxygen?** *(The second stage)* **Where in the cell does the second stage take place?** *(In mitochondria)*

Summarize Explain that summarizing information means giving only the main points. A summary is a brief review of information that has been read or heard in which only the essential ideas are conveyed.

Address Misconceptions

L1 CELLULAR RESPIRATION IN PLANTS AND ANIMALS Because students associate photosynthesis with plants, they might assume that cellular respiration only occurs in animal cells. Reinforce the concept that both plants and animals carry out cellular respiration. Emphasize that plants cannot use energy from the sun directly. They use it to make food and then use food as an energy source for life processes. Ask: **What does cellular respiration provide to the cells of living things?** *(Energy)* **What do living things need to survive?** *(Energy)* **What conclusion can you reach?** *(Living things carry out respiration, which gives them energy.)*

Summarize Complete the concept map below about cellular respiration.

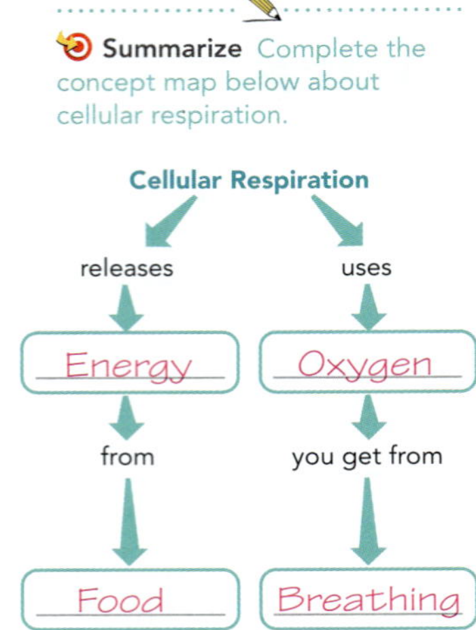

The Two Stages of Cellular Respiration Like photosynthesis, cellular respiration is a two-stage process. See **Figure 2.** The first stage occurs in the cytoplasm of a cell. There, molecules of glucose are broken down into smaller molecules. Oxygen is not involved in this stage, and only a small amount of energy is released.

The second stage takes place in the mitochondria. There, the small molecules are broken down even more. This change requires oxygen and releases a great deal of thermal energy and chemical energy that the cell can use for all its activities. No wonder mitochondria are sometimes called the "powerhouses" of the cell!

The Cellular Respiration Equation Although respiration occurs in a series of complex steps, the overall process can be summarized in the following equation:

$$\underset{\text{glucose}}{C_6H_{12}O_6} + \underset{\text{oxygen}}{6\ O_2} \longrightarrow \underset{\text{carbon dioxide}}{6\ CO_2} + \underset{\text{water}}{6\ H_2O} + \text{energy}$$

Notice that the raw materials for cellular respiration are glucose and oxygen. Animals get glucose from the foods they consume. Plants and other organisms that carry out photosynthesis are able to produce their own glucose. The oxygen needed for cellular respiration is in the air or water surrounding the organism.

FIGURE 2

Releasing Energy

Cellular respiration takes place in two stages.

Recognize Fill in the missing terms in the spaces provided.

Stage 1 In the cytoplasm, glucose is broken down into smaller molecules, releasing a small amount of energy.

Stage 2 In the mitochondria, the smaller molecules react, producing carbon dioxide, water, and large amounts of energy.

Glucose
Energy
Smaller molecules
Mitochondrion
Cytoplasm
Smaller molecules
Oxygen
Carbon dioxide
Water
Energy

Comparing Two Energy Processes

Photosynthesis transforms radiant energy from the sun into chemical energy in sugars. Cellular respiration releases the energy in sugars, which can be transformed by the organism into mechanical, electrical, or thermal energy. If you think the equation for cellular respiration is the opposite of the one for photosynthesis, you're right! Photosynthesis and cellular respiration can be thought of as opposite processes. Together, these two processes form a cycle that keeps the levels of oxygen and carbon dioxide fairly constant in Earth's atmosphere. As you can see from **Figure 3,** living things cycle both gases. Much of the energy released through cellular respiration is lost as thermal energy.

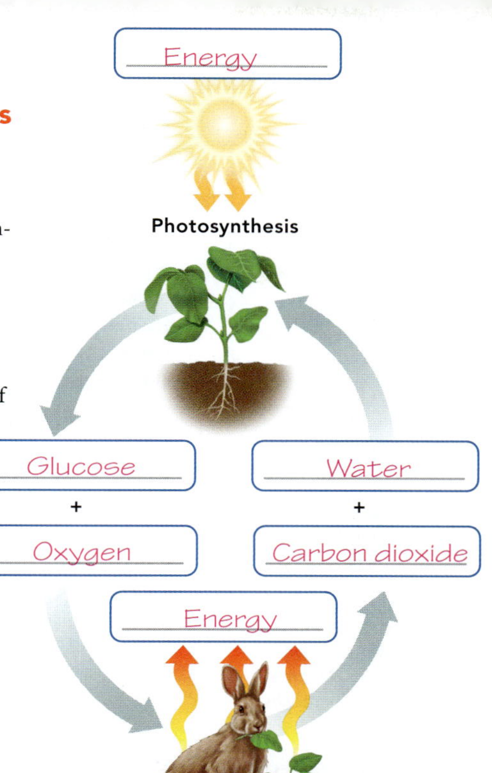

FIGURE 3

Opposite Processes
Producers carry out photosynthesis, but producers and consumers both carry out cellular respiration.

Name Use the following words to fill in the missing terms: Oxygen, Carbon dioxide, Water, Energy, and Glucose. Words can be used more than once.

Lab zone: Do the Lab Investigation *Exhaling Carbon Dioxide.* Student Lab Manual, p. 30

Assess Your Understanding

TEKS 7B, 12F

1a. Recognize Look at **Figure 2** on the facing page. How does Stage 2 of cellular respiration benefit a cell?

A large amount of energy is released, which can be transformed to carry out important processes.

b. Relate Cause and Effect Why does cellular respiration add carbon dioxide to the atmosphere, but photosynthesis does not?

Cellular respiration gives off carbon dioxide; photosynthesis uses it.

got it?

- O **I get it!** Now I know that during cellular respiration, cells break down food, using oxygen and releasing energy.
- O I need extra help with See TE note.

Differentiated Instruction

L1 Comparing Breathing and Cellular Respiration Have students work in pairs and reread the information in *What Is Cellular Respiration?* Have them highlight phrases that will help them compare and contrast the processes. Have partners use the highlighted phrases to explain to each other how breathing and cellular respiration are alike and how they are different.

L3 All About Sugars Glucose, the end product of photosynthesis and the raw material for cellular respiration, is a simple sugar, or monosaccharide. Tell students that there are other types of sugars—disaccharides and polysaccharides. Have students do research to find out more about each type of sugar, and include its definition, characteristics, examples, and uses. Have them share their findings with the class.

Teach With Visuals

Tell students to look at **Figure 3.** Ask: **What are the raw materials for photosynthesis?** *(Water and carbon dioxide)* Explain that the substances that are produced as a result of photosynthesis are called *products.* Ask: **What are water and carbon dioxide in cellular respiration?** *(They are products.)* **What are the products of photosynthesis?** *(Glucose and oxygen)* **What are those two substances in cellular respiration?** *(They are raw materials.)* **What does this indicate about the two processes?** *(Photosynthesis and cellular respiration are opposite processes.)*

Elaborate

21st Century Learning

CRITICAL THINKING: The relationship between the number of mitochondria in a cell and the cell's activity level is an important aspect of cellular respiration. Make sure students understand this relationship by having them explain why muscle cells have many mitochondria. *(Very active cells have many mitochondria because they need more energy. The large number of mitochondria in muscle cells supply a great amount of energy for movement as a result of the second stage of cellular respiration that takes place in them.)*

Lab Resource: Lab Investigation 

L2 EXHALING CARBON DIOXIDE Students will explore the relationship between exercise and the amount of carbon dioxide exhaled. This Lab Investigation can be found in the Student Lab Manual, p. 30, and online.

Evaluate

Assess Your Understanding

After students answer the questions, have them evaluate their understanding by completing the appropriate sentence.

RTI Response to Intervention

1a. If students need help recognizing the benefit of cellular respiration, **then** have them revisit **Figure 2** to determine what is produced other than carbon dioxide and water.

b. If students cannot relate the processes of photosynthesis and cellular respiration to the carbon dioxide cycle, **then** have them review the chemical equations for the two processes, focusing on the products given off into the air.

LESSON 3.2

Explain

Teach Key Concepts

Explain to students that when there is not enough oxygen for cells to carry out cellular respiration, fermentation occurs. In this process, cells release energy from food without using oxygen. Ask: **What are the two types of fermentation?** *(Alcoholic fermentation and lactic acid fermentation)* **What are the products of alcoholic fermentation?** *(Alcohol, carbon dioxide, small amount of energy)* **What are the products of lactic acid fermentation?** *(Lactic acid, small amount of energy)* **How are fermentation and cellular respiration alike and different?** *(Both processes produce energy, but fermentation produces less. Fermentation does not use oxygen, whereas cellular respiration does.)*

Lead a Discussion

OXYGEN DEBT Ask students whether they have had the experience of exercising so hard that they developed cramps and/or soreness in their muscles. Ask: **What was true about the rate of your breathing while you were exercising?** *(It was quite fast.)* **Why do you think this was so?** *(Sample: I was trying to get more oxygen into my body so that more cellular respiration could provide me with energy.)*

Elaborate

Apply It!

 Before beginning the activity review the characteristics of alcoholic fermentation.

Control Variables Help students understand that an experiment should have only one independent variable so that the results can be interpreted in terms of the effect of that variable. By controlling the variable of yeast, they can demonstrate that dough rises as a result of the action of these microorganisms.

Texas Essential Knowledge and Skills

7B Illustrate the transformation of energy within an organism such as the transfer from chemical energy to heat and thermal energy in digestion.

12E Compare the functions of a cell to the functions of organisms such as waste removal.

12F Recognize that according to cell theory all organisms are composed of cells and cells carry on similar functions such as extracting energy from food to sustain life.

TEKS 7B, 12E, 12F In this section, you'll learn how energy is released by cells during the process of fermentation.

What Happens During Fermentation?

Some organisms can live in the presence or absence of oxygen. If not enough oxygen is present to carry out cellular respiration, these organisms switch to another process. **Fermentation** is an energy-releasing process that does not require oxygen. **During fermentation, cells release energy from food without using oxygen.** One drawback to fermentation is that it releases far less energy than cellular respiration does.

Alcoholic Fermentation Did you know that when you eat a slice of bread, you are eating a product of fermentation? Alcoholic fermentation occurs in yeast and other single-celled organisms. This type of fermentation produces alcohol, carbon dioxide, and a small amount of energy. These products are important to bakers and brewers. Carbon dioxide produced by yeast creates gas pockets in bread dough, causing it to rise. Carbon dioxide is also the source of bubbles in alcoholic drinks such as beer and sparkling wine.

Lactic Acid Fermentation Think of a time when you ran as fast and as long as you could. Your leg muscles were pushing hard against the ground, and you were breathing quickly. But, no matter how quickly you breathed, your muscle cells used up the oxygen faster than it could be replaced. Because your muscle cells lacked oxygen, fermentation occurred. Your muscle cells got energy, but they did so by breaking down glucose without using oxygen. One product of this type of fermentation is a compound known as lactic acid. Human muscles cells are not the only cells that can perform this type of fermentation. Bacteria and fungi that perform lactic acid fermentation are used to make foods such as cheese and yogurt.

apply it!

A ball of bread dough mixed with yeast is left in a bowl at room temperature. As time passes, the dough increases in size.

1. **Compare and Contrast** How does fermentation that causes dough to rise differ from fermentation in muscles?
   Fermentation in dough produces carbon dioxide and alcohol. Fermentation in muscles produces lactic acid.

2. **Control Variables** How would you show that yeast was responsible for making the dough rise?
   Mix the dough with all the same ingredients, but leave out the yeast.

5A, 7B, 12F

Energy for Life

How do living things get energy?

FIGURE 4

Energy processes in living things include photosynthesis, cellular respiration, and fermentation.

Review Circle the correct answers and complete the sentences in the spaces provided.

Producers

Plant cells transform the sun's energy into chemical energy during the process of (photosynthesis/fermentation/cellular respiration).

Plants are autotrophs because they make their own food.

Plant cells release energy for cell function by way of (photosynthesis/fermentation/cellular respiration).

Plants get this energy when oxygen reacts with sugars made in photosynthesis.

Consumers

A runner on an easy jog through the woods gets energy by way of (photosynthesis/fermentation/cellular respiration).

The runner is a heterotroph because she gets energy from food that she eats.

If the runner makes a long, fast push to the finish, her muscle cells may get energy by way of (photosynthesis/fermentation/cellular respiration).

This process releases less energy and does not use oxygen.

Lab zone — Do the Quick Lab *Observing Fermentation.* Find the lab online.

Assess Your Understanding

TEKS 12F

2a. Develop Hypotheses When a race ends, why do you think runners continue to breathe quickly and deeply for a few minutes?

To replace oxygen that was used up by the muscle cells

b. Recognize How do living things get energy?

Sample: Living things get energy from food that is made or eaten. Organisms break down food through cellular respiration, releasing energy for cell use.

got it?

○ **I get it!** Now I know fermentation is a way for cells to get energy without using oxygen.

○ I need extra help with See TE note.

Apply the TEKS

5A, 7B, 12F

CELLS AND ENERGY Direct students' attention to the image of a runner against a background of plants. Help students identify the processes that take place in each and what the functions of those processes are. Ask: **What process do producers but not consumers carry out?** *(Photosynthesis)* **What process do both producers and consumers carry out?** *(Cellular respiration)* **How would you define each process?** *(Samples: Photosynthesis is an energy-capturing process in which plants make glucose. Cellular respiration is an energy-releasing process in which glucose combines with oxygen.)* **What process might consumers carry out when there is an absence of oxygen?** *(Fermentation)*

Lab Resource: Quick Lab 

L1 **OBSERVING FERMENTATION** Students will explore the production of carbon dioxide as a result of fermentation when yeast is exposed to sugar. This Quick Lab can be found online.

Evaluate

Assess Your Understanding

After students answer the questions, have them evaluate their understanding by completing the appropriate sentence.

RTI Response to Intervention

2a. If students cannot hypothesize about energy needs and breathing rates, **then** have them reread the information about lactic acid fermentation.

b. If students cannot state how living things get energy, **then** have them reread the Key Concept statements in this chapter.

Differentiated Instruction

L1 **A Variety of Breads** Have students work in pairs to find out about different types of breads. What are the ingredients? Are the breads leavened (made to rise) or not? If so, how? Have students report their finding to the class.

L3 **Measuring Food Energy** The energy contained in food is measured in Calories. A food Calorie is the amount of heat energy needed to raise the temperature of 1,000 grams of water 1 degree Celsius. Have students look at food labels to discover the correlation between ingredients and Calories in a variety of foods. Which ingredients in foods seem to add the most Calories per serving? Why do they think this is the case?

Texas Essential Knowledge and Skills

5A Recognize that radiant energy from the Sun is transformed into chemical energy through the process of photosynthesis.

7B Illustrate the transformation of energy within an organism such as the transfer from chemical energy to heat and thermal energy in digestion.

12F Recognize that according to cell theory all organisms are composed of cells and cells carry on similar functions such as extracting energy from food to sustain life.

Name ______________________ Date ____________ Class ____________

Assess Your Understanding

Cellular Respiration

What Happens During Cellular Respiration?

1a. RECOGNIZE Look at Figure 2 in your textbook. How does Stage 2 of cellular respiration benefit a cell?

____

____

b. RELATE CAUSE AND EFFECT Why does cellular respiration add carbon dioxide to the atmosphere, but photosynthesis does not?

____

____

got it?

○ **I get it!** Now I know that during cellular respiration, cells ____

○ **I need extra help with** ____

What Happens During Fermentation?

2a. DEVELOP HYPOTHESES When a race ends, why do you think runners continue to breathe quickly and deeply for a few minutes?

____

b. RECOGNIZE How do living things get energy?

____

____

____

got it?

○ **I get it!** Now I know fermentation is a way for cells to ____

○ **I need extra help with** ____

Place the outside corner, the corner away from the dotted line, in the corner of your copy machine to copy onto letter-size paper.

Place the outside corner, the corner away from the dotted line, in the corner of your copy machine to copy onto letter-size paper.

Name ______________________ Date ____________ Class ____________

Enrich

Cellular Respiration

Read the passage. Then answer the questions that follow on a separate sheet of paper.

History of Fermentation

In 1854, the French chemist Louis Pasteur determined that fermentation is caused by yeast. His work was influenced by the earlier work of Theodor Schwann, the German scientist who helped develop the cell theory. Around 1840, Schwann concluded that fermentation is the result of processes that occur in living things. In 1907, a German chemist named Eduard Buchner received the Nobel prize for showing that enzymes in yeast cells cause fermentation. About two decades later, two other scientists determined exactly how enzymes cause fermentation. Their names are Arthur Harden and Hans Euler-Chelpin, and they won the Nobel prize for their work in 1929. By the 1940s, technology was developed to use fermentation to produce antibiotics.

Fermentation is a very useful process. Today it is used to produce industrial chemicals, medicines such as antibiotics, and alcoholic beverages, as well as to make bread rise and to preserve many types of food. Some of these uses have been known for thousands of years. For example, the Chinese used fermented soybean curd to treat skin infections 3,000 years ago, and they started using fermented tea to treat a variety of illnesses as early as 220 B.C. The use of fermentation to make bread rise and to produce alcoholic beverages is as old as the development of agriculture itself, which most scholars date to about 8000 B.C.

1. **Use the information provided in the passage above to make a timeline of the history of fermentation.**
2. **What contribution did Louis Pasteur make to the understanding of the process of fermentation?**
3. **What are two of the oldest uses of fermentation?**
4. **How is fermentation used in medicine today?**

Name ______________________ Date __________ Class __________

Lesson Quiz

Cellular Respiration

Write the letter of the correct answer on the line at the left.

1. ___ The process by which cells obtain energy from glucose using oxygen is called

A alcoholic fermentation
B cellular respiration
C lactic acid fermentation
D photosynthesis

2. ___ During Stage 2 of cellular respiration, mitochondria produce energy, carbon dioxide, and

A food
B oxygen
C glucose
D water

3. ___ Alcoholic fermentation occurs when there is a lack of

A glucose
B water
C alcohol
D oxygen

4. ___ Which process occurs because your muscles use up oxygen faster than it can be replaced?

A alcoholic fermentation
B cellular respiration
C lactic acid fermentation
D photosynthesis

Fill in the blank to complete each statement.

5. Oxygen is a raw material in the process of cellular respiration.

6. Pain and weakness in human muscles cells are often the result of the buildup of Latic Acid.

7. Cellular respration releases the energy in sugars which can be transformed by the organism into mechanical, electrical, or thermal energy.

8. The energy-releasing process that does not require oxygen is fermation.

9. Mitco are the powerhouses of the cell because they are the organelles in which the second stage of cellular respiration takes place.

10. The products of photosynthesis are the raw materiaks of cellular respiration.

Place the outside corner, the corner away from the dotted line, in the corner of your copy machine to copy onto letter-size paper.

Cellular Respiration

Answer Key

Review and Reinforce

Find the worksheet in the Student Workbook.

1. carbon dioxide
2. oxygen
3. energy
4. in the cytoplasm
5. in the mitochondria
6. alcoholic fermentation
7. lactic acid fermentation
8. Cellular respiration and photosynthesis are opposite processes because the raw materials of photosynthesis are the products of cellular respiration, and the products of photosynthesis are the raw materials of cellular respiration.
9. Alcoholic fermentation and lactic acid fermentation are both energy-releasing processes that do not require oxygen. Both processes release less energy than cellular respiration. Alcoholic fermentation occurs in yeast and lactic acid fermentation occurs in human muscle cells.
10. Both cellular respiration and fermentation are energy-releasing processes that use food as the source of energy, but fermentation releases less energy. The products of cellular respiration are energy, carbon dioxide, and water. The products of alcoholic fermentation are carbon dioxide and alcohol.

Enrich

1. Students' timelines should include the following events: People used fermatation to make bread rise and produce alcoholic beverages (8000 B.C.); Chinese use fermented soybeans to treat skin infections (1000 B.C.); Chinese use fermented tea to treat several illnesses (220 B.C.); Schwann concludes fermentation is due to living things (1840); Pasteur finds fermentation is caused by yeast (1854); Buchner receives Nobel Prize for showing yeast enzymes cause fermentation (1907); Harden and Euler-Chelpin receive Nobel Prize for determining how enzymes cause fermentation (1929); fermentation produces antibiotics (1940s).
2. Louis Pasteur determined that the process of fermentation is caused by yeast.
3. Two of the oldest uses of fermentation are to make alcoholic beverages and to make bread rise.
4. Fermentation is used in medicine today to produce antibiotics.

Lesson Quiz

1. B
2. D
3. D
4. C
5. cellular respiration
6. lactic acid
7. Cellular respiration
8. fermentation
9. Mitochondria
10. raw materials

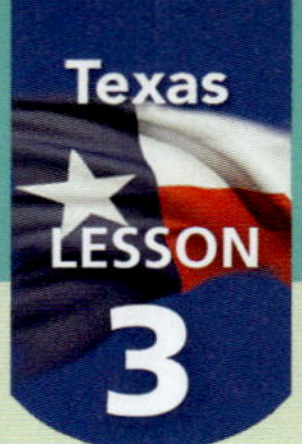

Cell Division

 **How do living things get energy?**

LESSON PACING:
2–3 periods or 1–1½ blocks

Lesson Vocabulary

- cell cycle
- interphase
- replication
- chromosome
- mitosis
- cytokinesis

Content Refresher

DNA Replication In most eukaryotic cells—cells with nuclei and other membrane-bound organelles—the replication of a DNA molecule does not start at one end and finish at the other end. Instead, it occurs simultaneously at many sites along the molecule. At each site, the two DNA strands separate, forming an opening called a bubble. The points at which the strands separate are called replication forks. The process of DNA replication is facilitated by different enzymes as well as other types of protein molecules. Single-strand binding proteins hold the original DNA strands apart. The enzyme DNA polymerase links individual nucleotides in a chain, one by one, to form the new DNA strand. DNA polymerase also "proofreads" each new nucleotide as it is added to the strand, making sure the new nucleotide correctly pairs with the corresponding nucleotide on the original strand. If a nucleotide is incorrect, DNA polymerase removes it.

Lesson Objectives	TEKS	ELPS
Summarize the functions of cell division.	12E, 12F	3.D.1
Identify the events that take place during the three stages of the cell cycle.	2E, 12E, 12F	

 **Texas Essential Knowledge and Skills**

2E Analyze data to formulate reasonable explanations, communicate valid conclusions supported by the data, and predict trends.
12E Compare the functions of a cell to the functions of organisms such as waste removal.
12F Recognize that according to cell theory all organisms are composed of cells and cells carry on similar functions such as extracting energy from food to sustain life.

 **English Language Proficiency Standards**

ELPS Speaking 3.D.1 Speak using grade-level content area vocabulary in context to internalize new English words.

DIFFERENTIATED INSTRUCTION KEY
L1 Struggling Students or Special Needs
L2 On-Level Students L3 Advanced Students

LESSON PLANNER 3.3

Investigations and Activities

My Planet Diary, **Student Edition,** p. 118
Inquiry: Inquiry Warm-Up, What are the Yeast Cells Doing?, **PearsonTexas.com**
Introduce Vocabulary, **Teacher's Edition,** p. 119
Teach Key Concepts, **Teacher's Edition,** p. 119
Teach With Visuals, **Teacher's Edition,** p. 119
Inquiry: Quick Lab, Observing Mitosis, **Lab Manual,** p. 39

Teach Key Concepts, **Teacher's Edition,** p. 120
Lead a Discussion, Interphase and Mitosis, **Teacher's Edition,** p. 120
Teacher to Teacher, **Teacher's Edition,** p. 120
Apply It!, **Student Edition,** p. 121
21st Century Learning, Critical Thinking, **Teacher's Edition,** p. 121
Teach With Visuals, **Teacher's Edition,** p. 122
Inquiry: Build Inquiry, Model the Cell Cycle, **Teacher's Edition,** p. 122
21st Century Learning, Critical Thinking, **Teacher's Edition,** p. 123
Inquiry: Build Inquiry, Chromosome Numbers, **Teacher's Edition,** p. 123
21st Century Learning, Creativity, **Teacher's Edition,** p. 123
Support the TEKS, Cell Division and Energy, **Teacher's Edition,** p. 124
Inquiry: Teacher Demo, Not All Cytokinesis Is the Same, **Teacher's Edition,** p. 124
Do the Math! **Student Edition,** p. 125
Differentiated Instruction, **Teacher's Edition,** p. 125
Inquiry: Quick Lab, Modeling Mitosis, **PearsonTexas.com**

TEKS Review

Apply the TEKS, Analyzing Metabolic Rates, **Student Edition,** p. 126

TEKS Practice, **Student Edition,** p. 128

TEKS Practice: Chapter and Cumulative Review, **Student Edition,** p. 130

Lesson 3.3, **TEKS Preparation and Study Guide Workbook,** p. 32

SHORT ON TIME? To do this lesson in approximately half the time, do the Activate Prior Knowledge activity. A discussion of the Key Concepts will familiarize students with the lesson content. Have students do the Quick Labs. The rest of the lesson can be completed by students independently.

These editable worksheets are available on **PearsonTexas.com.**
Print versions can be found in the **TEKS Preparation and Study Guide Workbook.**

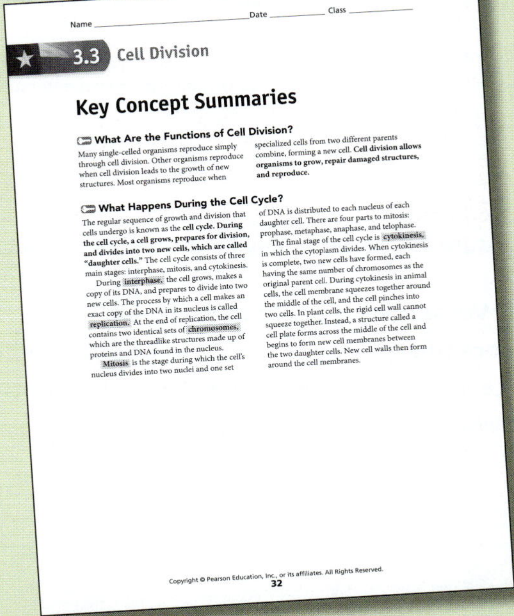

Name ______ Date ______ Class ______

3.3 Cell Division

Key Concept Summaries

What Are the Functions of Cell Division?

Many single-celled organisms reproduce simply through cell division. Other organisms reproduce when cell division leads to the growth of new structures. Most organisms reproduce when specialized cells from two different parents combine, forming a new cell. **Cell division allows organisms to grow, repair damaged structures, and reproduce.**

What Happens During the Cell Cycle?

The regular sequence of growth and division that cells undergo is known as the **cell cycle. During the cell cycle, a cell grows, prepares for division, and divides into two new cells, which are called "daughter cells."** The cell cycle consists of three main stages: interphase, mitosis, and cytokinesis.

During **interphase,** the cell grows, makes a copy of its DNA, and prepares to divide into two new cells. The process by which a cell makes an exact copy of the DNA in its nucleus is called **replication.** At the end of replication, the cell contains two identical sets of **chromosomes,** which are the threadlike structures made up of proteins and DNA found in the nucleus.

Mitosis is the stage during which the cell's nucleus divides into two nuclei and one set of DNA is distributed to each nucleus of each daughter cell. There are four parts to mitosis: prophase, metaphase, anaphase, and telophase.

The final stage of the cell cycle is **cytokinesis,** in which the cytoplasm divides. When cytokinesis is complete, two new cells have formed, each having the same number of chromosomes as the original parent cell. During cytokinesis in animal cells, the cell membrane squeezes together around the middle of the cell, and the cell pinches into two cells. In plant cells, the rigid cell wall cannot squeeze together. Instead, a structure called a cell plate forms across the middle of the cell and begins to form new cell membranes between the two daughter cells. New cell walls then form around the cell membranes.

32

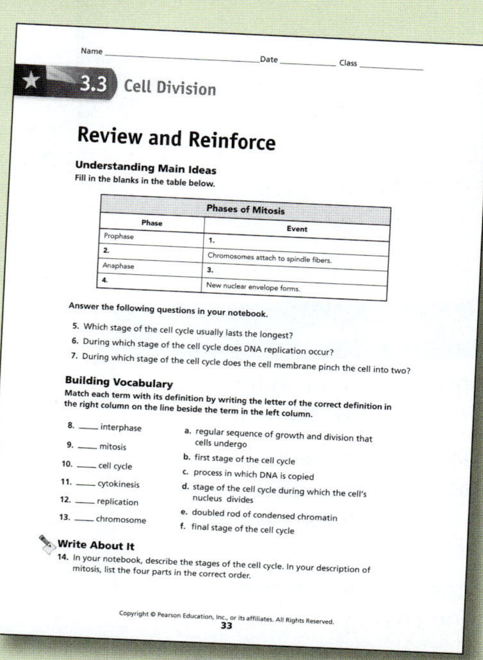

Name ______ Date ______ Class ______

3.3 Cell Division

Review and Reinforce

Understanding Main Ideas
Fill in the blanks in the table below.

Phases of Mitosis	
Phase	Event
Prophase	1.
2.	Chromosomes attach to spindle fibers.
Anaphase	3.
4.	New nuclear envelope forms.

Answer the following questions in your notebook.

5. Which stage of the cell cycle usually lasts the longest?
6. During which stage of the cell cycle does DNA replication occur?
7. During which stage of the cell cycle does the cell membrane pinch the cell into two?

Building Vocabulary
Match each term with its definition by writing the letter of the correct definition in the right column on the line beside the term in the left column.

8. ____ interphase
9. ____ mitosis
10. ____ cell cycle
11. ____ cytokinesis
12. ____ replication
13. ____ chromosome

a. regular sequence of growth and division that cells undergo
b. first stage of the cell cycle
c. process in which DNA is copied
d. stage of the cell cycle during which the cell's nucleus divides
e. doubled rod of condensed chromatin
f. final stage of the cell cycle

Write About It
14. In your notebook, describe the stages of the cell cycle. In your description of mitosis, list the four parts in the correct order.

33

Lexile Measure = 890L

LESSON 3.3

Cell Division

Establish Learning Objectives

After this lesson, students will be able to:

- Summarize the functions of cell division.
- Identify the events that take place during the three stages of the cell cycle.

Engage

Activate Prior Knowledge

MY PLANET DIARY Read *Cycling On* with the class. Ask students to offer a definition of the term *cycle*. Try to elicit that a cycle is a series of events that repeats over and over again in the same order. Ask: **What is an example of a cycle?** *(Sample: the phases of the moon; the daily tides; the months of the year)* **Why is the growth and reproduction of a cell a cycle?** *(They result in two new cells that repeat the processes of growth and reproduction.)*

Explore

Lab Resource: Inquiry Warm-Up

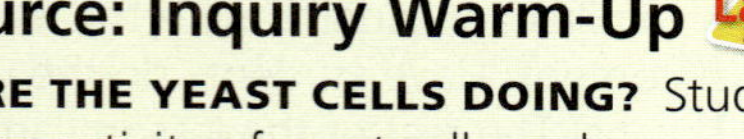

L1 WHAT ARE THE YEAST CELLS DOING? Students will observe the activity of yeast cells under a microscope. This Inquiry Warm-Up can be found online.

Texas Essential Knowledge and Skills

12E Compare the functions of a cell to the functions of organisms such as waste removal.

12F Recognize that according to cell theory all organisms are composed of cells and cells carry on similar functions such as extracting energy from food to sustain life.

Cell Division

- What Are the Functions of Cell Division? TEKS 12E, 12F
- What Happens During the Cell Cycle? TEKS 2E, 12E, 12F

my planet Diary

Cycling On

How long do you think it takes a cell to grow and reproduce, that is, to complete one cell cycle? The answer depends on the type of cell and the organism. Some cells, such as the frog egg cells shown here, divide every 30 minutes, and others take as long as a year! The table below compares the length of different cell cycles.

Comparing Cell Cycles

Frog Egg Cells	Yeast Cells	Fruit Fly Wing Cells	Human Liver Cells
30 minutes	90 minutes	9–10 hours	Over 1 year

SCIENCE STATS

Interpret Data Use the table to help you answer the following questions.

1. Which type of cell completes a cell cycle fastest?
   The frog egg cell
2. With each cell cycle, two cells form from one cell. In three hours, how many cells could form from one frog egg cell?
   64 cells

Do the Inquiry Warm-Up *What Are the Yeast Cells Doing?* Find the lab online.

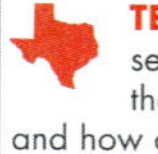

TEKS 12E, 12F In this section, you'll learn about the functions of cell division and how cell division is related to the health of an organism.

ELPS 3.D.1 With a partner, ask the question that starts each paragraph on this page. Answer by summarizing the paragraph in your own words.

What Are the Functions of Cell Division?

How do tiny frog eggs become big frogs? Cell division allows organisms to grow larger. One cell splits into two, two into four, and so on, until a single cell becomes a multicellular organism.

How does a broken bone heal? Cell division produces new healthy bone cells that replace the damaged cells. Similarly, cell division can replace aging cells and those that die from disease.

SUPPORT ALL READERS

Lexile Measure = 890L **Lexile Word Count = 1377**

Prior Exposure to Content: May be the first time students have encountered this topic

Academic Vocabulary: *data, interpret, question*

Science Vocabulary: *cell cycle, replication, chromosome, mitosis*

Concept Level: Generally appropriate for most students in this grade

Preteach With: My Planet Diary "Cycling On" and Figure 2 activity

Vocabulary
- cell cycle
- interphase
- replication
- chromosome
- mitosis
- cytokinesis

Skills
- Reading: Ask Questions
- Inquiry: Interpret Data

Growth and repair are two functions of cell division. A third function is reproduction. Some organisms reproduce simply through cell division. Many single-celled organisms, such as amoebas, reproduce this way. Other organisms can reproduce when cell division leads to the growth of new structures. For example, a cactus can grow new stems and roots. These structures can then break away from the parent plant and become a separate plant.

Most organisms reproduce when specialized cells from two different parents combine, forming a new cell. This cell then undergoes many divisions and grows into a new organism.

The many functions of cell division in living things are shown in **Figure 1.** **Cell division allows organisms to grow, repair damaged structures, and reproduce.**

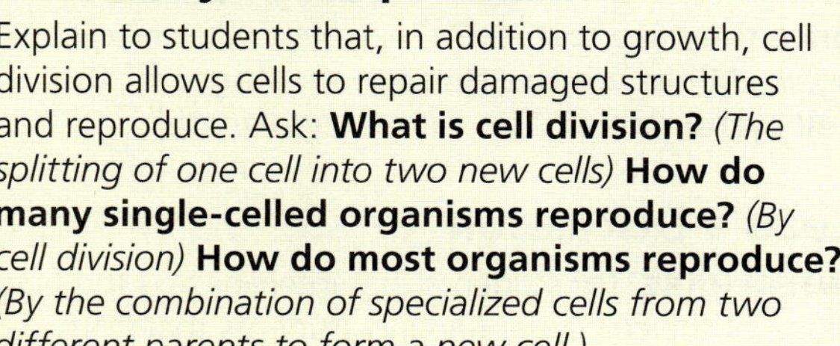

FIGURE 1

Cell Division

Each photo represents a function of cell division.

Answer these questions.

1. **Identify** Label each photo as (A) growth, (B) repair, or (C) reproduction.
2. **CHALLENGE** Which function of cell division is illustrated by a rose bud developing into a bloom? Growth

Do the Quick Lab *Observing Mitosis.* Student Lab Manual, p. 39

Assess Your Understanding

got it?

- O **I get it!** Now I know the functions of cell division are growth, repair, and reproduction.
- O I need extra help with See TE note.

English Language Proficiency Standards

ELPS Speaking 3.D.1

Have students ask and answer questions about cell division.

Beginning Read aloud page 118. Point out the questions at the beginning of each paragraph. Model how to summarize each paragraph to answer the question. Then have partners take turns asking and answering the questions.

Intermediate Together read page 118. Ask partners to find and read aloud the question at the beginning of each paragraph. Then have them cover the page and discuss the answer.

Advanced Have partners read page 118 together. Have one partner ask the questions while the other answers without the help of the text.

Advanced High Have partners read page 118 together, then ask and answer the questions. Encourage students to elaborate on their answers with details.

Explain

Introduce Vocabulary

Students may find the term *cytokinesis* difficult to understand. Remind them that *cyto-* refers to the cell, as in cytoplasm, and *-kinesis* means motion (i.e. kinetic energy).

Teach Key Concepts

Explain to students that, in addition to growth, cell division allows cells to repair damaged structures and reproduce. Ask: **What is cell division?** *(The splitting of one cell into two new cells)* **How do many single-celled organisms reproduce?** *(By cell division)* **How do most organisms reproduce?** *(By the combination of specialized cells from two different parents to form a new cell.)*

Teach With Visuals

Tell students to look at **Figure 1.** Ask: **How would you describe what is happening in the image you labeled A?** *(A seedling is growing from a seed into a new plant as a result of cell division.)* **How would you describe what is happening in the image you labeled B?** *(Damaged skin is being repaired or replaced as a result of cell division.)* **How would you describe what is happening in the image you labeled C?** *(A cell is dividing in order to reproduce.)*

Elaborate

Lab Resource: Quick Lab

L1 OBSERVING MITOSIS Students will observe phases of mitosis in an onion root tip. This Quick Lab can be found in the Student Lab Manual, p. 39, and online.

Evaluate

Assess Your Understanding

Have students evaluate their understanding by completing the appropriate sentence.

RTI Response to Intervention

If students have trouble identifying the three functions of cell division, **then** have them reread the Key Concept statement and review **Figure 1.**

PEARSON Texas.com

LESSON 3.3

Explain

Teach Key Concepts

Explain to students that the cell cycle is the regular sequence of growth and division that cells undergo. Ask: **What happens during the cell cycle?** *(The cell grows, prepares for division, and divides into two new cells.)* **What are the two new cells called?** *(Daughter cells)* **What do the daughter cells do once they are formed?** *(They begin the cell cycle again.)* **What are the three main stages of the cell cycle?** *(Interphase, mitosis, cytokinesis)*

Lead a Discussion

INTERPHASE Tell students that interphase is the first stage of the cell cycle. Remind students that DNA holds all the information that a cell needs in order to carry out its functions. Ask: **What happens during interphase?** *(The cell grows, makes a copy of its DNA, and prepares to divide into two cells.)* **In what process does the cell make a copy of the DNA?** *(Replication)* **What is the result of replication?** *(The cell has two identical sets of chromosomes.)* **How does the cell prepare for cell division?** *(It produces structures that it will use to divide.)*

Lead a Discussion

MITOSIS Review with students that interphase prepares the cell to divide. Remind students that cell growth, replication of DNA, and preparation for division have all taken place. It is now time for the cell's nucleus to divide. Ask: **What happens during mitosis?** *(One copy of the DNA is distributed into each of the two daughter cells.)* **What structures contain the cell's DNA?** *(Chromosomes)* **What is each rod-like part of a chromosome called?** *(Chromatid)* **What are the stages of mitosis?** *(Prophase, metaphase, anaphase, telophase)*

TEKS 2E, 12E, 12F In this section, you'll investigate the stages of the cell cycle, in which a cell divides into two new cells. You will also use a graph to analyze data about cell division in the liver.

What Happens During the Cell Cycle?

The regular sequence of growth and division that cells undergo is known as the **cell cycle.** **During the cell cycle, a cell grows, prepares for division, and divides into two new cells, which are called "daughter cells."** Each of the daughter cells then begins the cell cycle again. The cell cycle consists of three main stages: interphase, mitosis, and cytokinesis.

Stage 1: Interphase

The first stage of the cell cycle is **interphase.** This stage is the period before cell division. During interphase, the cell grows, makes a copy of its DNA, and prepares to divide into two cells.

Growing Early during interphase, a cell grows to its full size and produces the organelles it needs. For example, plant cells make more chloroplasts. And all cells make more ribosomes and mitochondria. Cells also make more enzymes, substances that speed up chemical reactions in living things.

Copying DNA Next, the cell makes an exact copy of the DNA in its nucleus in a process called **replication.** You may know that DNA holds all the information that a cell needs to carry out its functions. Within the nucleus, DNA and proteins form threadlike structures called **chromosomes.** At the end of replication, the cell contains two identical sets of chromosomes.

Preparing for Division Once the DNA has replicated, preparation for cell division begins. The cell produces structures that will help it to divide into two new cells. In animal cells, but not plant cells, a pair of centrioles is duplicated. You can see the centrioles in the cell in **Figure 2.** At the end of interphase, the cell is ready to divide.

FIGURE 2

Interphase: Preparing to Divide

The changes in a cell during interphase prepare the cell for mitosis.

List Make a list of the events that occur during interphase.

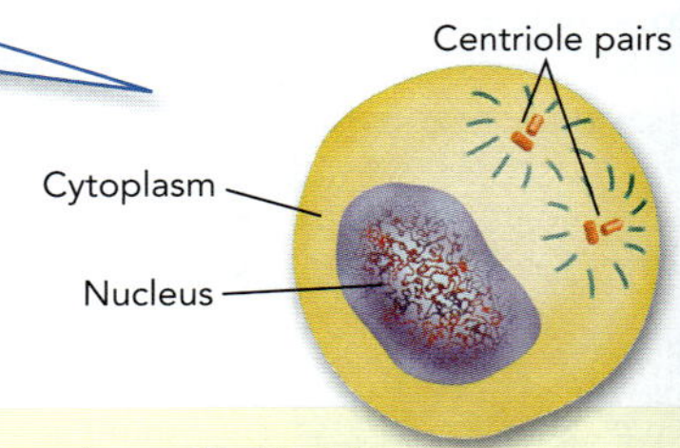

Interphase To-Do List

Sample:

- Grow to full size
- Make more organelles and enzymes
- Copy DNA
- Produce other needed structures

Texas Essential Knowledge and Skills

2E Analyze data to formulate reasonable explanations, communicate valid conclusions supported by the data, and predict trends.

12E Compare the functions of a cell to the functions of organisms such as waste removal.

12F Recognize that according to cell theory all organisms are composed of cells and cells carry on similar functions such as extracting energy from food to sustain life.

Professional Development Note

Teacher to Teacher

Activity Illustrations work well, so make them come to life! Students create a flip-book of the phases of cell division. Layer four sheets of 4 x 11 white card stock, matching the edges carefully. Fold the layered papers at the mid-horizontal point, and staple at what is now the top fold. Using this flip-book, begin with the title and illustrator's name on the front page. Proceed with each cell division phase illustration, one per page, using detail, labels, and color. The key is to encourage the students to illustrate their work in approximately the same part of each consecutive page. When complete, the students flip the pages and "see" cell division in action!

Susan M. Pritchard, Ph.D.
Washington Middle School
La Habra, California

When one cell splits in half during cell division, the result is two new cells. Each of those two cells can divide into two more, and so on.

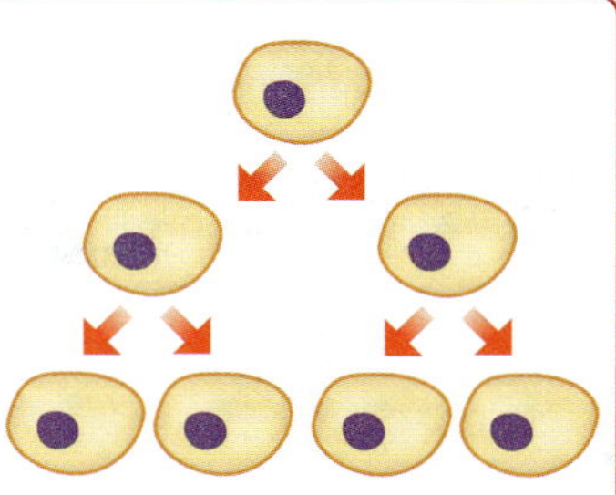

1 Calculate How many cell divisions would it take to produce at least 1,000 cells from one cell?

10 cell divisions

2 Describe What happens to the number of cells after each division?

The number doubles each time.

3 CHALLENGE Do you think all human cells divide at the same rate throughout life? Justify your answer.

Sample: No; cells probably divide at different rates. For example, cells probably divide more often in young children than in adults.

Stage 2: Mitosis

Once interphase ends, the second stage of the cell cycle begins. During **mitosis** (my TOH sis), the cell's nucleus divides into two new nuclei and one set of DNA is distributed into each daughter cell.

Scientists divide mitosis into four parts, or phases: prophase, metaphase, anaphase, and telophase. During prophase, the chromosomes condense into shapes that can be seen under a microscope. In **Figure 3** you can see that a chromosome consists of two rod-like parts, called chromatids. Each chromatid is an exact copy of the other, containing identical DNA. A structure known as a centromere holds the chromatids together until they move apart later in mitosis. One copy of each chromatid will move into each daughter cell during the final phases of mitosis. When the chromatids separate they are called chromosomes again. Each cell then has a complete copy of DNA. **Figure 4** on the next page summarizes the events of mitosis.

FIGURE 3

Mitosis: Prophase

Mitosis begins with prophase, which involves further changes to the cell.

Compare and Contrast How does prophase look different from interphase?

Chromosomes condense; the nucleus is changing.

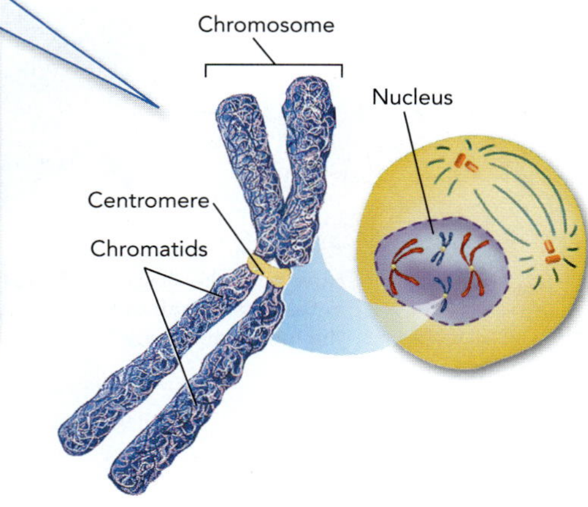

Elaborate

Make Analogies

L1 LIKE MONEY IN THE BANK Use the following activity to illustrate the way in which cells can increase in number. Have students demonstrate visually how much money they would have in the bank by the 15th of the month if they saved one penny on the first day of the month and their money doubled each day for the remaining consecutive days. Ask: **From one penny, how much money would you have saved by the 15th day of the month?** *(16,384 pennies, or $163.84)*

Apply It!

L1 Before beginning the activity, make sure students understand the mathematical operation they are using to solve the problem—doubling the number. Suggest that students can use a calculator or do the math longhand. Encourage students to continue the illustration for the next two or three divisions if it helps them understand the concept.

21st Century Learning

CRITICAL THINKING To help students understand the importance of replication, have them determine the number of chromosomes, and thus the amount of DNA, each daughter cell would have if interphase did not include this important process. *(Half the number of chromosomes and thus half the DNA in the parent cell)*

Differentiated Instruction

L1 Regeneration A sea star can grow a new arm as a result of cell division. A single planaria flatworm can regrow into two separate individuals if split lengthwise or crosswise. This process is called regeneration. Have students find out more about regeneration—what function of cell division it represents and what other organisms undergo the process.

L3 Cell Differentiation When certain cells divide and grow, they also develop specialized shapes and functions. The changes that such cells undergo as they develop are called differentiation. Have students find out more about cell differentiation and its results.

LESSON 3.3

Explain

Teach With Visuals

Tell students to look at **Figure 4.** Ask: **Where is the chromatin in the interphase stage?** *(In the nucleus)* Point out that in prophase, each pair of chromatids consists of the original DNA of the parent cell plus a copy of the DNA made during replication in interphase. Have students trace the movement of chromosomes through the cycle. Ask: **When are the chromosomes completely separated in their own nuclear envelope?** *(During telophase)* **What is the result of cytokinesis?** *(There are two daughter cells, each having an identical set of chromosomes and about half the organelles.)*

Make Analogies

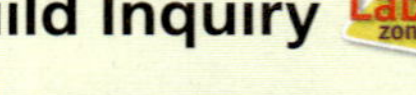

L1 ALL IN THE RIGHT ORDER Making sure students understand that the stages of the cell cycle are always repeated in the same order is essential to their understanding of the process. Ask: **Can you think of something you do almost every day that takes place in the same series of steps?** *(Sample: I wake up, turn off the alarm, wash up, get dressed, have breakfast, and go to school.)* **Is the order of the steps important?** *(Yes)* **What would happen if you did the steps out of order?** *(Sample: I might go to school in my pajamas!)*

Elaborate

Build Inquiry Lab zone

L2 MODEL THE CELL CYCLE

Materials poster board, colored markers, index cards, number cubes, small objects such as colored erasers or game tokens

Time 20 minutes

Challenge students to create a board game that models the cell cycle. Divide the class into groups and supply each group with game materials. Tell students that to get from "start" to "finish" on the game board, players must advance through each stage of the cell cycle by correctly answering questions about that stage.

Ask: **What three stages of the cell cycle must appear on the game board?** *(Interphase, mitosis, cytokinesis)* **What is the "start" point?** *(One cell)* **What is the "finish" point?** *(Two identical daughter cells)*

FIGURE 4

The Cell Cycle

Cells undergo an orderly sequence of events as they grow and divide. The photographs show cells of a developing whitefish.

Interpret Diagrams Answer the questions and draw the missing parts of the stages in the spaces provided.

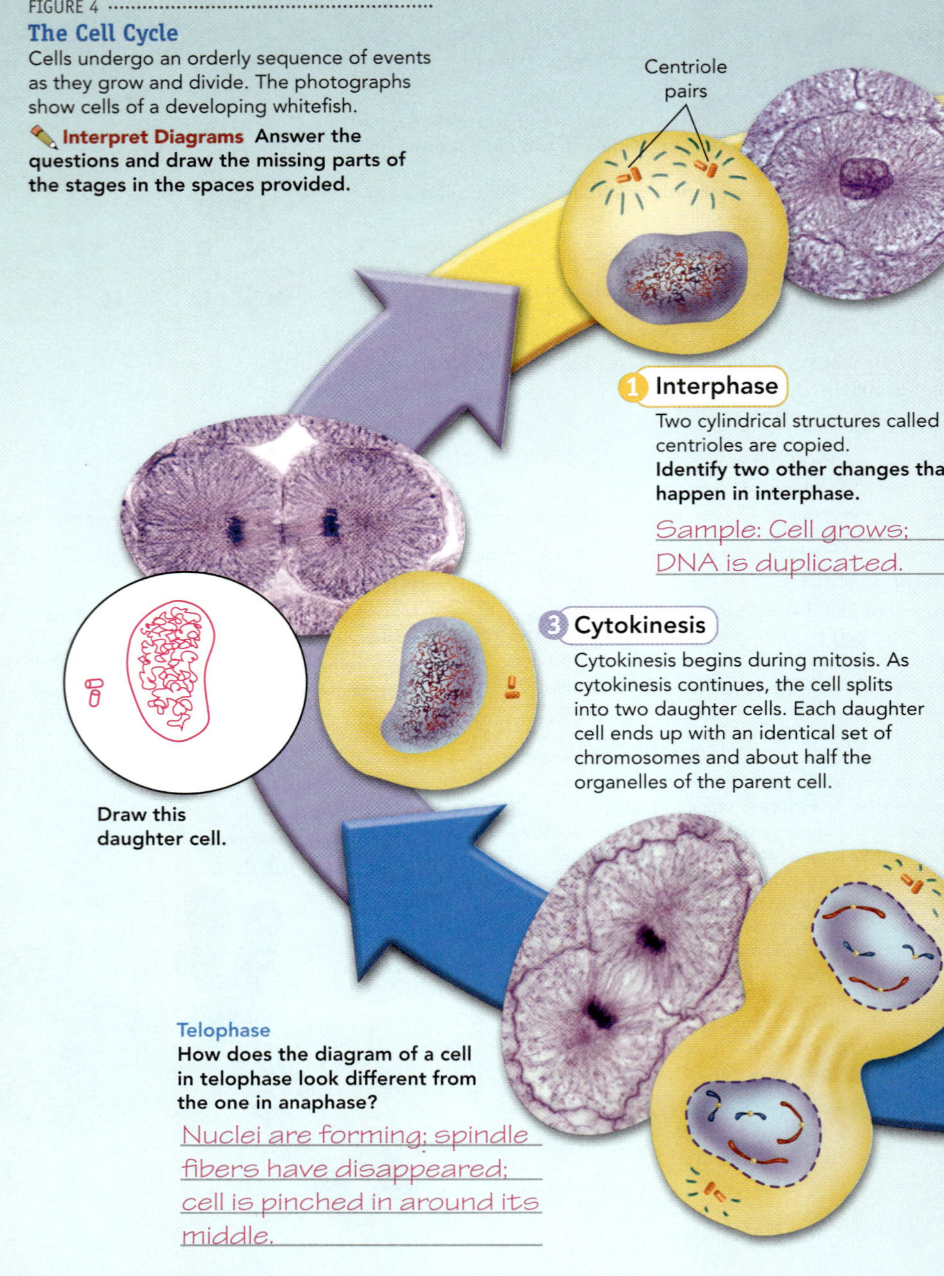

1 Interphase

Two cylindrical structures called centrioles are copied.

Identify two other changes that happen in interphase.

Sample: Cell grows; DNA is duplicated.

3 Cytokinesis

Cytokinesis begins during mitosis. As cytokinesis continues, the cell splits into two daughter cells. Each daughter cell ends up with an identical set of chromosomes and about half the organelles of the parent cell.

Telophase

How does the diagram of a cell in telophase look different from the one in anaphase?

Nuclei are forming; spindle fibers have disappeared; cell is pinched in around its middle.

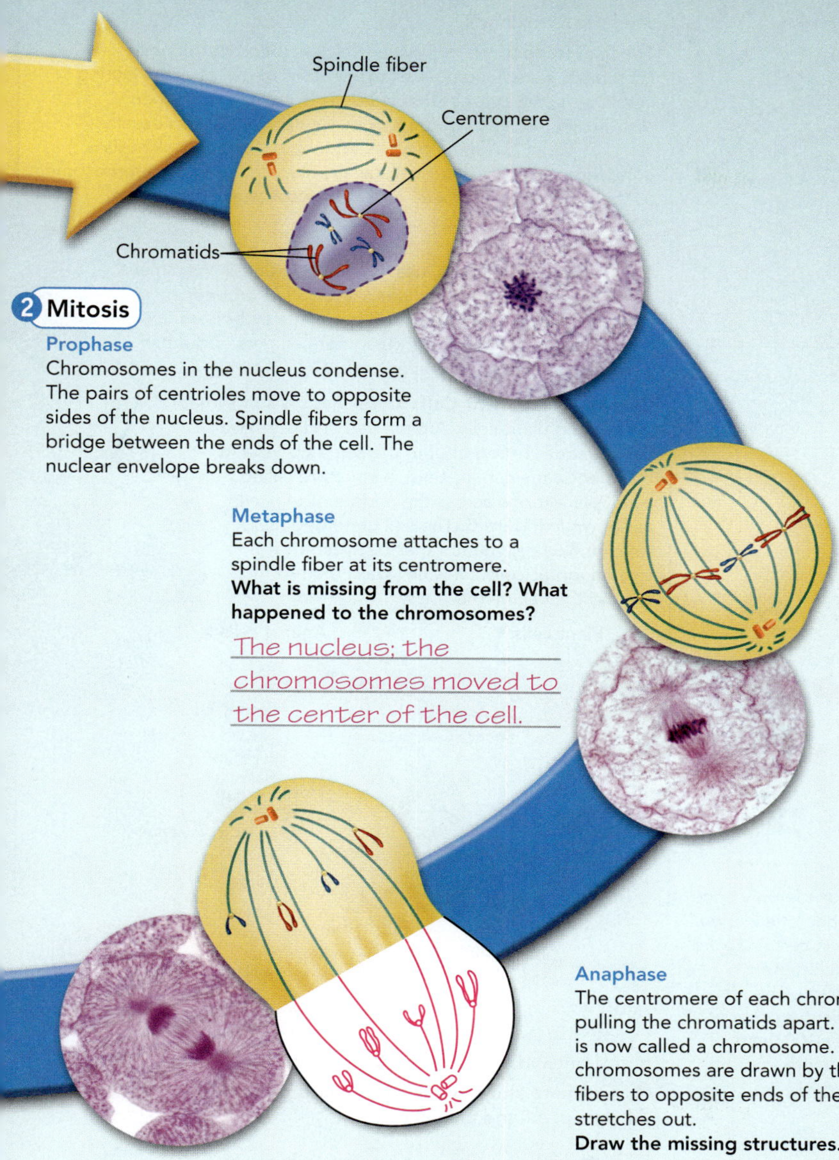

2 Mitosis

Prophase
Chromosomes in the nucleus condense. The pairs of centrioles move to opposite sides of the nucleus. Spindle fibers form a bridge between the ends of the cell. The nuclear envelope breaks down.

Metaphase
Each chromosome attaches to a spindle fiber at its centromere.
What is missing from the cell? What happened to the chromosomes?
The nucleus; the chromosomes moved to the center of the cell.

Anaphase
The centromere of each chromosome splits, pulling the chromatids apart. Each chromatid is now called a chromosome. These chromosomes are drawn by their spindle fibers to opposite ends of the cell. The cell stretches out.
Draw the missing structures.

Differentiated Instruction

L1 Word Analysis Have students work in pairs to identify and remember the stages of mitosis by using the meanings of the prefixes *pro-, meta-, ana-, and telo-*. Then have them explain what relationship interphase has to mitosis by analyzing its prefix.

L3 Cell-Cycle Scramble Have students use **Figure 4** as a guide for creating their own illustrations of the stages of the cell cycle. Have them cut out the drawings and play Cell-Cycle Scramble with a classmate. They should challenge the classmate to put the drawings in the correct sequence and then determine if they did so correctly.

Elaborate

21st Century Learning

L3 CRITICAL THINKING: Tell students that skeletal muscle cells undergo interphase and mitosis, but not cytokinesis. Ask: **How do you think this affects the cells?** *(Sample: The cells grow by getting larger, not increasing in number. Many of these cells will have more than one nucleus.)*

Build Inquiry 

L1 CHROMOSOME NUMBERS

Materials List of the chromosome numbers of selected organisms written on the board

Time 10 minutes

Use the following information to construct a two-column list of the chromosome numbers of selected organisms: amoeba: 50; carrot: 18; cat: 32; lettuce: 18; human: 46; chimpanzee: 48; dog: 78; goldfish: 94; earthworm: 36. Students should organize the data in a way that shows a trend.

Ask: **Do you think the number of chromosomes in the cells of an organism is related to the size of the organism? Explain.** *(No, probably not. Goldfish have the most chromosomes and they are not the largest organism on the list.)* **Do you think the number of chromosomes in the cells of an organism is related to the complexity of the organism? Explain.** *(No, probably not. Humans are the most complex of these organisms, and they do not have the greatest number.)*

21st Century Learning

CREATIVITY Help students remember the stages of mitosis by having them work in pairs to create a poem or song in which they describe each stage. Remind them to keep the stages in the correct sequence. Set aside class time to have a "poem/song fest" in which each student pair presents their material. Have the audience critique the scientific accuracy of each poem/song.

LESSON 3.3

Explain

Lead a Discussion

THE CELL DIVIDES Review with students that during mitosis the nucleus of the cell divides, but the cell is still one cell. Ask: **What happens during cytokinesis?** *(The cytoplasm divides, distributing the organelles into each of the two new cells.)* **What happens during cytokinesis in animal cells?** *(The cell membrane squeezes together around the middle of the cell.)* **What happens during cytokinesis In plant cells?** *(A cell plate forms across the middle of the cell and gradually develops into new cell membranes, and new cell walls form around the cell membranes.)* **Why is the process different in animals and plants?** *(The cell membrane is flexible and can be squeezed; the cell wall is rigid and cannot be squeezed.)*

Ask Questions Discuss with students the value of asking questions: It is an excellent way to focus on and remember new information. Tell students that they can ask questions about material they both read and hear. Remind students that *what* and *how* are probably the most common question words, but they may also ask *who, where, why,* or *when* questions.

Support the TEKS

CELL DIVISION AND ENERGY Discuss with students the fact that cell processes require energy, which is produced during cellular respiration. Cell division is one important process that requires this energy. Have students identify one step in each of the three stages of cell division that require energy to complete *(Sample: centrioles are copied during interphase, chromosomes are drawn to opposite sides of the cell during mitosis, cell splits into two daughter cells during cytokinesis)*

Teacher Demo 

L2 NOT ALL CYTOKINESIS IS THE SAME

Materials Diagram on the board of yeast cell dividing

Time 10 minutes

On the board, draw a diagram of a yeast cell dividing.

Ask: **Does this drawing of cell division look like the photos in Figure 5? Explain.** *(No; the plant cell and animal cell shown dividing in Figure 5 form two new cells of the same size.)* **What is happening with the yeast cell?** *(A small daughter cell is pinching off the parent cell.)* **What conclusion do these images help you reach?** *(Not all cytokinesis is the same. Different organisms may have different patterns of cell division.)*

Ask Questions Before you read details about cytokinesis, write a question that asks something you would like to learn.

Sample: Does cytokinesis happen the same way in plant cells as it does in animal cells?

Stage 3: Cytokinesis

The final stage of the cell cycle, which is called **cytokinesis** (sy toh kih NEE sis), completes the process of cell division. During cytokinesis, the cytoplasm divides. The structures are then distributed into each of the two new cells. Cytokinesis usually starts at about the same time as telophase. When cytokinesis is complete, each daughter cell has the same number of chromosomes as the parent cell. At the end of cytokinesis, each cell enters interphase, and the cycle begins again.

Cytokinesis in Animal Cells During cytokinesis in animal cells, the cell membrane squeezes together around the middle of the cell, as shown here. The cytoplasm pinches into two cells. Each daughter cell gets about half of the organelles of the parent cell.

Cytokinesis in Plant Cells Cytokinesis is somewhat different in plant cells. A plant cell's rigid cell wall cannot squeeze together in the same way that a cell membrane can. Instead, a structure called a cell plate forms across the middle of the cell, as shown in **Figure 5.** The cell plate begins to form new cell membranes between the two daughter cells. New cell walls then form around the cell membranes.

did you know?

Certain bacteria divide only once every 100 years! Bacteria known as *Firmicutes* live in certain rocks that are found 3 kilometers below Earth's surface. The life functions of *Firmicutes* occur so slowly that it takes 100 years or more for them to store enough energy to split in two.

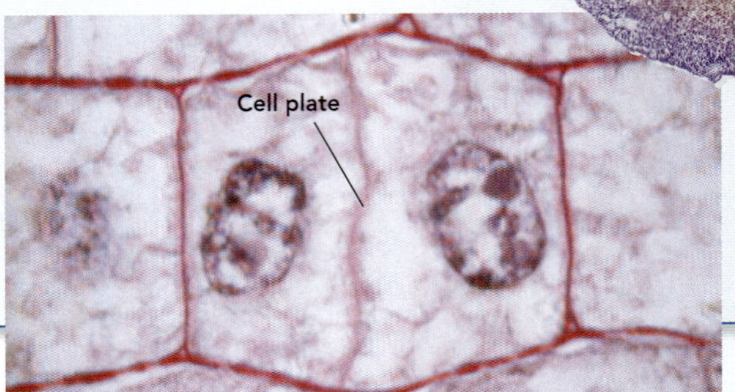

FIGURE 5
Cytokinesis
Both plant and animal cells undergo cytokinesis.

Compare and Contrast How does cytokinesis differ in plant and animal cells?

Plant cells must form a cell plate to divide.

Professional Development Note — Teacher to Teacher

Structure and Function The relationship of structure and function is a powerful tool in the biological sciences. For example, plants that are stationary and somewhat rigid must have something in their cells to give them strength and the ability to hold their shape. When we look at animal cells and plant cells, we see some significant differences. In the plant cell, we see a rigid structure outside the cell membrane, called the *cell wall,* that supports the structure of the plant. It is easy to see that cell walls are necessary in plant cells but not in animal cells.

Joel Palmer, Ed.D.
Mesquite ISD
Mesquite, Texas

Analyzing Data

Length of a liver cell cycle

How long does it take for a cell to go through one cell cycle? It depends on the cell. Human liver cells generally reproduce less than once per year. At other times, they can complete one cell cycle in about 22 hours, as shown in the circle graph. Study the graph and answer the following questions.

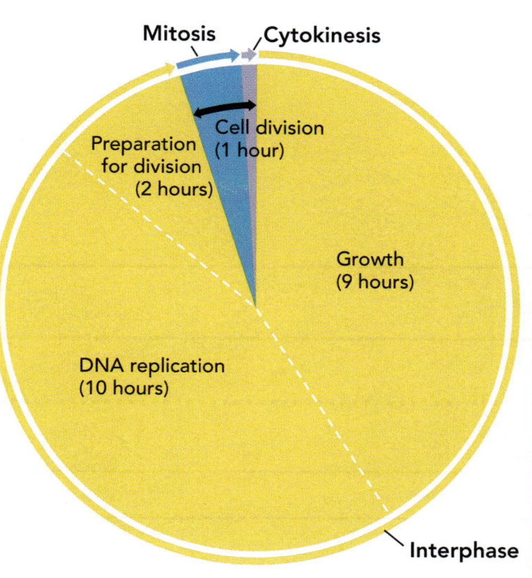

1. **Read Graphs** What do the three curved arrows outside of the circle represent?
   Interphase, mitosis, and cytokinesis
2. **Read Graphs** The wedge representing growth is in which stage of the cell cycle?
   Interphase
3. **Interpret Data** About what percentage of the cell cycle is shown for DNA replication?
   About 45%
4. **Interpret Data** What stage in the cell cycle takes the shortest amount of time? How do you know?
   Cytokinesis; it is the smallest section of the graph.

Do the Quick Lab *Modeling Mitosis.* Find the lab online.

Assess Your Understanding

TEKS 12F

1a. List What are the three stages of the cell cycle?
Interphase, mitosis, cytokinesis

b. Sequence Put the following terms in correct order starting with the first phase of mitosis: anaphase, telophase, metaphase, prophase.
Prophase, metaphase, anaphase, telophase

c. Recognize What do you think would happen if a cell's DNA did not replicate correctly?
The daughter cells would not receive a complete set of DNA and might not be able to carry out the functions needed to live.

got it?

O I get it! Now I know that during the cell cycle the cell grows, duplicates DNA, and divides.

O I need extra help with See TE note.

Elaborate

Do the Math!

L1 Remind students that the sections of a circle graph show portions of a whole. In this case, the whole is the 22-hour cell cycle of mature human liver cells, and the sections are the stages in the cell cycle.

Interpret Data Remind students of the importance of interpreting data, in which they use data to identify trends or patterns in measurements or observations. Ask: **What process in the cell cycle of a mature human liver cell takes the least amount of time?** *(Cytokinesis)*

Lab Resource: Quick Lab 

L1 **MODELING MITOSIS** Students will explore the process of mitosis by constructing a model using pipe cleaners. This Quick Lab can be found online.

Evaluate

Assess Your Understanding

After students answer the questions, have them evaluate their understanding by completing the appropriate sentence.

RTI Response to Intervention

1a. If students have trouble identifying the stages of the cell cycle, **then** have them use the circle graph accompanying the *Do the Math!* activity or **Figure 4** to identify in the correct order the three stages of the cell cycle.

b. If students have trouble sequencing the phases of mitosis, **then** have them carefully trace the phases shown in **Figure 4.**

c. If students cannot predict the outcome of incorrect DNA replication, **then** have them reread the information about copying DNA, the second event of interphase.

LESSON 3.3

Differentiated Instruction

L1 **The Lives of Cells** Not all cells have the same average life span (production as daughter cell to end of cytokinesis). Some last only hours and others more than 100 years. Have students research and compile the interesting data on the life spans of the following human cells and create a graph to represent their findings: red blood cells, skin cells, white blood cells, brain cells.

L3 **A Useful Poison** Colchicine is a poison produced by the flowering plant known as the meadow saffron. However, because of its effect on mitosis, colchicine has important uses in the laboratory and in medicine. Have students find out more about colchicine's effect on mitosis and how it is used by researchers and doctors. Have students report their findings to the class.

Name ______________________ Date __________ Class __________

Assess Your Understanding

Cell Division

What Are the Functions of Cell Division?

got it?

○ **I get it!** Now I know the functions of cell division are ______________________

____

○ **I need extra help with** ______________________

____

What Happens During the Cell Cycle?

1a. LIST What are the three stages of the cell cycle?

____

b. SEQUENCE Put the following terms in correct order starting with the first phase of mitosis: anaphase, telophase, metaphase, prophase.

____

____

c. RECOGNIZE What do you think would happen if a cell's DNA did not replicate correctly?

____

____

____

got it?

○ **I get it!** Now I know that during the cell cycle ______________________

____

○ **I need extra help with** ______________________

____

Name _______________ Date _______ Class _______

Enrich

Cell Division

Read the passage and study the figures below. Then answer the questions that follow on a separate sheet of paper.

Recall that all plant cells have a rigid wall. Because of this rigid cell wall, cytokinesis in plant cells is different from cytokinesis in animal cells. Study the figures below to see how cytokinesis differs in plant cells and animal cells.

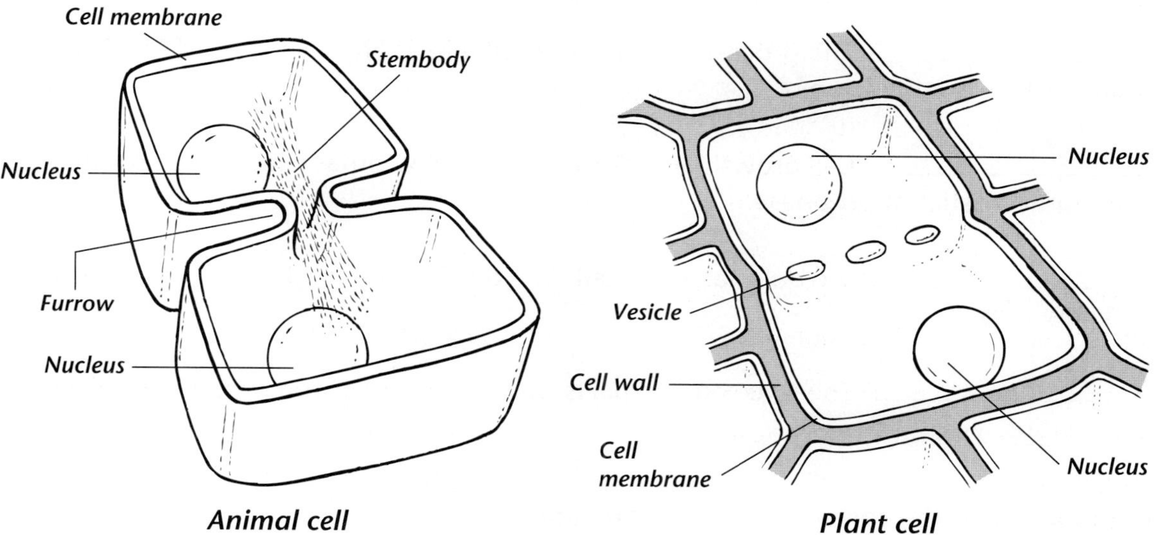

Animal cell Plant cell

In animal cells, as daughter cells pinch into two cells, there is a space between the cells called a furrow. As the furrow gets increasingly narrower, the spindle fibers are pressed into a tight bundle, called a stembody. The stembody eventually is cut in two as the new cell membranes fuse together.

In plant cells, pockets of cell-wall material, called vesicles, line up across the middle of the cell. The vesicles fuse together in two sheets to form new cell walls and cell membranes between the daughter cells.

1. How does the furrow form in an animal cell? What's the furrow's role in cell division?
2. What causes the stembody to form in an animal cell? What happens to the stembody when the cell divides?
3. What are vesicles? Which parts of the plant cell do vesicles develop into?
4. If you observed a cell under a microscope during cytokinesis, how could you tell whether it was a plant cell or an animal cell?

Name ______________________ Date __________ Class __________

Lesson Quiz

Cell Division

If the statement is true, write *true*. If the statement is false, change the underlined word or words to make the statement true.

1. ______________ Cell division allows organisms to grow, repair damaged structures, and <u>produce energy</u>.

2. ______________ <u>Mitosis</u> results in the formation of two daughter cells.

3. ______________ The process in which the cell makes an exact copy of the DNA in its nucleus is <u>replication</u>.

4. ______________ Cell growth and production of new organelles and enzymes are characteristics of <u>prophase</u>.

5. ______________ It would take <u>five</u> cell divisions for one original cell to produce 128 new cells.

6. ______________ The two rod-like parts that make up a chromosome are called <u>chromatids</u>.

Write the letter of the correct answer on the line at the left.

7. ___ The total number of cells in an organism increases as a result of which process?
   A respiration
   B photosynthesis
   C cell division
   D fermentation

8. ___ The formation of a cell plate is a characteristic of
   A cytokinesis in plant cells
   B cytokinesis in animal cells
   C both A and B
   D neither A nor B

9. ___ Chromatids are held together by a
   A spindle fiber
   B centromere
   C cell plate
   D centriole

10. ___ The correct order for the parts of mitosis are
   A prophase, interphase, metaphase, anaphase
   B telophase, anaphase, metaphase, prophase
   C interphase, prophase, metaphase, telophase
   D prophase, metaphase, anaphase, telophase

Place the outside corner, the corner away from the dotted line, in the corner of your copy machine to copy onto letter-size paper.

Cell Division

Answer Key

Review and Reinforce

Find the worksheet in the Student Workbook.

1. Chromatin condenses.
2. Metaphase
3. Chromatids separate.
4. Telophase
5. interphase
6. interphase
7. cytokinesis
8. b
9. d
10. a
11. f
12. c
13. e
14. The cell cycle consists of three main stages: interphase, mitosis, and cytokinesis. During interphase, the cell grows, replicates, and prepares to divide into two new cells. At the end of replication, the cell contains two identical sets of chromosomes. Mitosis is the stage during which the cell's nucleus divides into two nuclei and one set of DNA is distributed to each nucleus of each daughter cell. There are four parts to mitosis: prophase, metaphase, anaphase, and telophase. The final stage of the cell cycle is cytokinesis, in which the cytoplasm divides. When cytokinesis is complete, two new cells have formed, each having the same number of chromosomes as the original parent cell.

Enrich

1. The furrow forms when the cell membrane pinches in around the middle of the cell. It eventually divides the cell in two.
2. The stembody forms when the spindle fibers are pressed together by the furrow. It is cut in two when the cell divides.
3. Vesicles are pockets of cell-wall material. They develop into new cell walls and cell membranes.
4. An animal cell would appear pinched in around the middle by the cell membrane. A plant cell would not be pinched in but would have small structures (vesicles) lined up across the middle of the cell.

Lesson Quiz

1. reproduce
2. Cytokinesis
3. true
4. interphase
5. seven
6. true
7. C
8. A
9. B
10. D

CHAPTER 3

Scientific Investigation and Reasoning

This Apply the TEKS feature will enable students to use the scientific investigation and reasoning skills in TEKS 2D and 2E to reinforce the content of TEKS 7B. Students will construct a scatterplot and analyze the data to draw conclusions about the metabolic rates of different organisms.

Analyzing Metabolic Rates

Have students read about metabolic rates. Make sure students understand how the metabolic rates in the chart are measured. Ask: **How do you read the measurement for metabolic rates in the chart?** *(Cubic millimeters of oxygen consumed per gram of body mass each hour)*

If necessary, explain to students that a scatterplot is a type of graph that shows data plotted as points, like a line graph. Scientists use scatterplots to display data when they are trying to show that two variables might be related to one another. The relationship between the two variables is called their correlation. You may want to show students some examples of scatterplots before they begin answering Question 1.

APPLY THE TEKS 2D, 2E, 7B

ANALYZING METABOLIC RATES

During photosynthesis, plants transform radiant energy from the sun into chemical energy stored in sugars. Cellular respiration releases the energy in sugars. This energy can be transformed into mechanical, electrical, or thermal energy within an organism, allowing it to move, grow, and survive in its environment.

An organism's metabolic rate refers to the amount of energy it uses to carry out basic life processes. Metabolic rates can be measured in several ways. One method involves measuring the amount of oxygen an organism consumes in a certain period of time for every gram of its body mass. By measuring the amount of oxygen used, one can estimate the rate of cellular respiration. The chart below shows the metabolic rates of five land animals.

Land Animal	Average Mass (kg)	Metabolic Rate (mm^3 O_2/g of body mass/hr)
Dog	20	300
Human	85	210
Horse	500	120
Rabbit	10	490
Orangutan	80	230

Texas Essential Knowledge and Skills

2D Construct tables and graphs, using repeated trials and means, to organize data and identify patterns.

2E Analyze data to formulate reasonable explanations, communicate valid conclusions supported by the data, and predict trends.

7B Illustrate the transformation of energy within an organism such as the transfer from chemical energy to heat and thermal energy in digestion.

English Language Proficiency Standards

ELPS Listening 2.C.1

Have students refer to the table on page 126 to complete the following sentence frames: _________ *has a (low/high) metabolic rate.* _________ *has a (lower/higher) metabolic rate.* _________ *has the (lowest/highest) metabolic rate.*

Beginning Use simple drawings to explain the terms in parentheses. Model how to complete the sentence frames. Then have partners complete them independently.

Intermediate Use simple drawings to explain the terms. Model how to complete a sample sentence frame. Have partners continue independently.

Advanced Have partners complete the sentence frames. Then ask them to discuss Question 5 on page 127.

Advanced High Have partners complete the sentence frames. Then ask them to discuss Questions 5–6 on page 127.

Answer the questions below.

1. **Construct Graphs** Draw and label a scatter-plot to represent the data in the table. Label the x-axis *Mass* and the y-axis *Metabolic Rate*. (See page 631 for help with scatterplots.)
2. **Analyze Data** Which animal has the lowest metabolic rate? Which animal has the greatest metabolic rate?

   Lowest: horse Greatest: rabbit
3. **Analyze Data** Which animal will have to eat the most food relative to its body mass in order to survive each day?

   rabbit
4. **Identify Patterns** Examine your scatterplot. What relationship exists between the mass of an animal and its metabolic rate?

   Sample: The less the mass of the animal, the greater its metabolic rate.
5. **Formulate Reasonable Explanations** Cellular respiration is a source of body heat. Smaller animals have larger surface areas for their body size compared to larger animals. They are more likely to lose body heat to the outside environment. Why do you think a rabbit has a higher metabolic rate than a horse?

   Sample: Because a rabbit has a larger surface area for its body size, it loses body heat more easily than a horse. So the rabbit must eat more food to maintain its body temperature.
6. **Predict Trends** The animals in the table all produce their own body heat. Other animals, such as snakes, do not produce body heat. How do you think a snake's metabolic rate would compare with that of an animal of similar mass that produces its own body heat?

   The snake's metabolic rate should be lower because it does not use energy to generate body heat.

Scientific Investigation and Reasoning, Cont.

Before students answer Question 4, explain that the more the points in a scatterplot approximate a line, the stronger the correlation between the two variables. Ask: **Is there a correlation between body mass and metabolic rate?** *(Yes)* **How do you know?** *(The points on the scatterplot fall closely along a single line.)* Instruct students to draw that line on their scatterplots.

To extend the activity, have volunteers identify other land mammals and challenge students to plot where they think the data point for that animal might fall on their scatterplots.

CHAPTER 3

TEKS Practice

Assess Understanding

Have students complete the answers to the TEKS Practice questions. Have a class discussion about what students find confusing. Write Key Concepts on the board to reinforce knowledge.

RTI Response to Intervention

4. If students cannot predict how animals might be affected by the volcanic eruption, **then** reinforce the concept that the sun is the ultimate source of energy. Autotrophs use it directly and heterotrophs use it indirectly.

10. If students need help with summarizing the process of cellular respiration, **then** remind them that both plants and animals carry out cellular respiration, in which they release energy. Invite students to compare and contrast this process with photosynthesis.

Alternate Assessment

L3 **CREATE A STORYBOARD** Have students design a storyboard to illustrate the carbon dioxide-oxygen cycle involving the processes of photosynthesis and cellular respiration/breathing. Remind students to use all of the Key Concepts and vocabulary terms that apply. Tell students to create a "subplot" in which they show how energy is used for the cell cycle as they illustrate the various stages.

PEARSON Texas.com

CHAPTER 3

TEKS Practice

TEKS 5A, 5C, 7B, 12E, 12F

LESSON 1 Photosynthesis

1. Which of the following organisms are autotrophs?
   a. fungi
   b. rabbits
   c. humans
   d. oak trees (circled)

2. Plants are green because of chlorophyll, the main photosynthetic pigment in chloroplasts.

3. **Recognize** Fill in the missing labels in the diagram below.

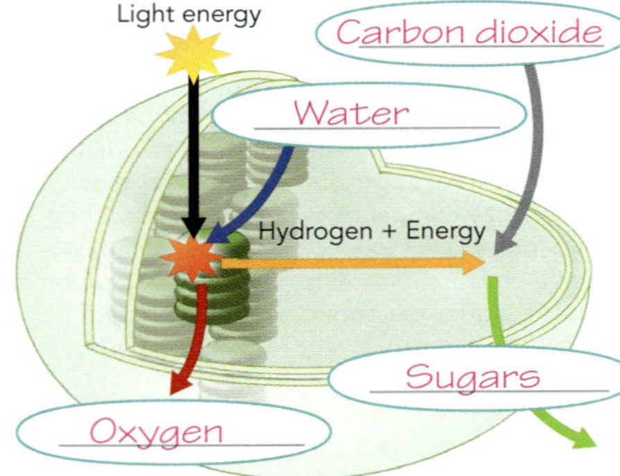

4. **Predict** Suppose a volcano threw so much ash into the air that it blocked much of the sunlight. How might this event affect the ability of animals to obtain energy to live?
   Without the sun, autotrophs could not make food; heterotrophs could not obtain food from autotrophs.

5. **Write About It** How do you get energy? Illustrate the path of energy from the sun to you, using at least two vocabulary terms you learned in this lesson.
   See TE rubric.

LESSON 2 Cellular Respiration

6. In which organelle does cellular respiration take place?
   a. nucleus
   b. chloroplast
   c. chlorophyll
   d. mitochondrion (circled)

7. Fermentation is a process that releases energy in cells without using oxygen.

8. **Identify** What is one common food that is made with the help of fermentation?
   Sample: bread

9. **Explain** Create a flowchart that illustrates how energy is transformed in your body when you eat an apple.
   Flowcharts should illustrate that sugars in the apple are broken down to release chemical energy, which is transformed into other forms of energy in the body.

10. **Recognize** In one or two sentences, summarize what happens during each of the two stages of cellular respiration.
    Stage 1: Glucose is broken down into smaller molecules and some energy is released. Stage 2: Small molecules are broken down further, using oxygen and releasing more energy.

11. **Compare** How is an organism's breathing related to cellular respiration in its cells?
    Breathing brings in oxygen that is used in cellular respiration and removes waste products, such as carbon dioxide.

Write About It 5 Assess student's writing using this rubric.

SCORING RUBRIC	SCORE 4	SCORE 3	SCORE 2	SCORE 1
Illustrate the path of energy from the sun to student	Correctly explains that he/she is a heterotroph that relies on other heterotrophs or autotrophs for energy	Adequately explains that he/she is a heterotroph that relies on other heterotrophs or autotrophs for energy	Knows his/her source of energy is food, but incompletely describes path of energy	Does not know how he/she gets energy

LESSON 3 **Cell Division**

12. During which phase of the cell cycle does DNA replication occur?

a. mitosis
b. division
c. interphase (circled)
d. cytokinesis

13. During mitosis, a cell's nucleus divides into two new nuclei.

14. Recognize Why is cell division a necessary function of living things?

It allows for growth, repair of damaged cells, and reproduction.

15. Relate Cause and Effect Why is replication a necessary step in cell division?

Sample: One copy of DNA is needed for each daughter cell.

16. Sequence Fill in the diagram below with descriptions of each part of the cell cycle.

Interphase
Cell grows, copies DNA, prepares for division.

Mitosis
Nucleus divides in two.

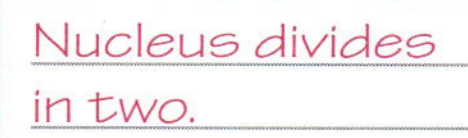

Cytokinesis
Rest of cell divides in two.

How do living things get energy?

17. All living things need energy. Use the terms *autotroph* and *heterotroph* to describe how grass, a rabbit, and a fox obtain energy.

The grass is an autotroph. It uses energy from the sun to make food. The rabbit is a heterotroph. It gets energy from the sun by eating grass. The fox is a heterotroph. It gets energy from the sun by eating the rabbit, which ate the grass. (Foxes also eat grass.)

5A, 7B, 12F

Interactive Science Chapter 3

Lesson 1
In Lesson 3.1 you learned that nearly all living things obtain energy either directly or indirectly from the radiant energy of sunlight. Plants and algae use the energy from sunlight during photosynthesis, which converts carbon dioxide and water to sugars and oxygen.
Supporting TEKS: 5C
TEKS: 5A, 7B

Lesson 2
In Lesson 3.2 you learned that cells use cellular respiration to break down glucose and other large molecules from food in the presence of oxygen. When there is a lack of oxygen, some cells use fermentation to release energy from glucose.
Supporting TEKS: 5C, 12F
TEKS: 7B, 12E

Lesson 3
In Lesson 3.3 you learned that cell division allows organisms to grow, repair damage, and reproduce. You also learned that during the cell cycle, a cell grows, prepares for division, and then divides into two new cells.
Supporting TEKS: 12F
TEKS: 12E

129

TEKS Practice, Cont.

CHAPTER 3

RTI Response to Intervention

16. If students cannot sequence the stages of the cell cycle, **then** have students review **Figure 4.**

L3 **WRITING IN SCIENCE** Ask students to write a blog entry that explains to readers how living things get energy. Then they should describe the processes carried out on the cellular level to use this energy.

How do living things get energy?

Students should be able to demonstrate understanding of how living things get energy by answering this question. See the scoring rubric below.

Review the TEKS Chapter 3

PARTNER REVIEW Have partners review the definitions of vocabulary terms and quiz each other. Students can read the Key Concept statements and leave out words for their partners to fill in. They also can change a statement so that it is false and then ask their partners to correct it.

CLASS ACTIVITY: FLOW CHART Have students develop a flow chart to show how information in this chapter is related. Have students review the Key Concepts, vocabulary, equations, and other details from the chapter. On the board, begin the flow chart with a box for the "sun." Then have students develop the rest of the chart, keeping the following questions in mind:

- What is an autotroph? A heterotroph?
- How do these organisms get energy?
- How are photosynthesis and cellular respiration related?
- What are the functions of cell division?
- What are the stages of the cell cycle?

How do living things get energy?

Assess student's response using this rubric.

SCORING RUBRIC	SCORE 4	SCORE 3	SCORE 2	SCORE 1
Identify organisms as autotrophs or heterotrophs	Identifies organisms correctly and uses terms *photosynthesis, directly,* and *indirectly* in explanation	Identifies organisms correctly but does not use terms *photosynthesis, directly,* and *indirectly* in explanation	Identifies grass as autotroph but cannot identify rabbit and fox	Does not identify organisms in terms of how they obtain energy

CHAPTER 3

TEKS Practice: Chapter Review

Test-Taking Skills

INTERPRETING ILLUSTRATIONS Remind students to review all parts of an illustration carefully, including any labels. They should make sure they understand the meaning of arrows and other symbols. Once they determine exactly what the question is asking, they can eliminate those answer choices that are not supported by the illustration.

Question 1 TEKS 7B

The correct answer is C. During cellular respiration, glucose is broken down in the presence of oxygen to produce carbon dioxide, water, and energy.

If students chose A, explain that oxygen is a product of photosynthesis, but the organelle in the diagram is taking in oxygen.

If students chose B, explain that plants use sunlight to produce glucose during photosynthesis, but the organelle in the diagram is taking in glucose.

If students chose D, explain that fermentation takes place when oxygen is not available, but oxygen is present in the diagram.

Question 2 TEKS 7B

The correct answer is F. A chloroplast uses light energy to produce glucose, which the cell uses as fuel.

If students chose G, explain that a vacuole stores materials and waste in a cell.

If students chose H, explain that mitochondria do not use light energy to provide the cell with fuel.

If students chose J, explain that the nucleus is not directly involved with energy production in the cell.

★ TEKS Practice: Chapter Review

Read each question and choose the best answer.

1 The illustration shows a process that takes place in cells.

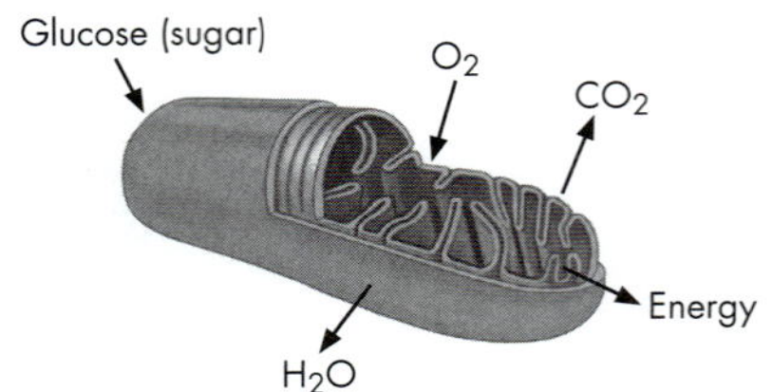

What function does this process serve?

A Using energy to produce oxygen

B Using sunlight to produce glucose

(C) Breaking down glucose to release energy

D Releasing energy when there is no oxygen present

2 A sporting goods store sells a solar-powered battery charger. When you set the charger out in sunlight for several hours, it uses light energy to recharge batteries. Once recharged, the batteries can be used to provide energy for flashlights, radios, and other electronic devices. Which part of a plant cell is most like the solar-powered battery charger?

(F) Chloroplast

G Vacuole

H Mitochondrion

J Nucleus

★ TEKS Practice: Cumulative Review

3 Many of the chemical compounds in living things are known as organic compounds. Which of the following lists a pair of organic compounds that are found in your body?

A Ammonia and water

B Carbon dioxide and oxygen

C Potassium and salt

(D) Carbohydrates and lipids

4 The functions that organelles perform in a cell are comparable to the functions that some organs perform in an organism. Which of the following is the most accurate comparison?

F A mitochondrion is comparable to the kidney in the excretory system.

(G) A nucleus is comparable to the brain in the nervous system.

H A ribosome is comparable to the heart in the circulatory system.

J A vacuole is comparable to the lungs in the respiratory system.

5 The cells in certain tissues of a plant have many more chloroplasts than other cells in the plant do. What function in the plant do these tissues most likely have?

A Absorbing oxygen

(B) Capturing energy

C Supporting the plant

D Taking in water

If You Have Trouble With . . .

Question	1	2	3	4	5
See Lesson	3.2	3.1	2.4	2.5	2.3
TEKS	7.7B	7.7B	7.6A	7.12E	7.12C

TEKS Practice: Cumulative Review

Additional Assessment Resources

Teacher's Edition: Lesson Quizzes, Texas End-of-Year Test Prep A and B
Online assessments

Question 3 TEKS 6A

The correct answer is D. Both of these compounds contain carbon.

If students chose A, explain that both are liquids found in the body, but neither compound is organic.

If students chose B, explain that carbon dioxide contains carbon but is considered inorganic. Oxygen is an element, not a compound.

If students chose C, explain that neither compound is organic. In addition, potassium is an element, not a compound.

Question 4 TEKS 12E

The correct answer is G. The brain directs the operation of the organs in the body just as the nucleus directs the operation of the organelles in a cell.

If students chose F, explain that mitochondria extract energy, which is not the same as excreting wastes.

If students chose H, explain that a ribosome makes proteins, while the circulatory system transports nutrients and wastes through the body.

If students chose J, explain that vacoules store waste, while the stomach breaks down food.

Question 5 TEKS 12C

The correct answer is B. Chloroplasts capture energy from the sun, so these tissues most likely function by capturing energy.

If students chose A, explain that chloroplasts do not absorb oxygen.

If students chose C, explain that the cell wall is responsible for providing support, not the chloroplasts.

If students chose D, explain that chloroplasts capture energy from sunlight and are not involved in taking in water.

CHAPTER 3

Science Matters

Think Like a Scientist

Have students read *Energized!* Explain that there are over 300 species of hummingbird. While these birds come in a variety of shapes and sizes, they are all nectivores and must feed constantly. About 90 percent of their diet consists of sugary flower nectar, and the rest consists of high-protein insects.

Before students begin researching the organs and organ systems of the hummingbird for their diagrams, review the different forms that energy can take. Ask: **What are some forms of energy that you have learned about in science class?** *(Sample: Mechanical, potential, electrical, nuclear, chemical, electromagnetic, and thermal)* **Which of these forms of energy would you expect to observe in a hummingbird?** *(Sample: Mechanical, electrical, chemical, and thermal)*

Have students present their completed diagrams to the class and explain how energy is transformed in the hummingbird's body.

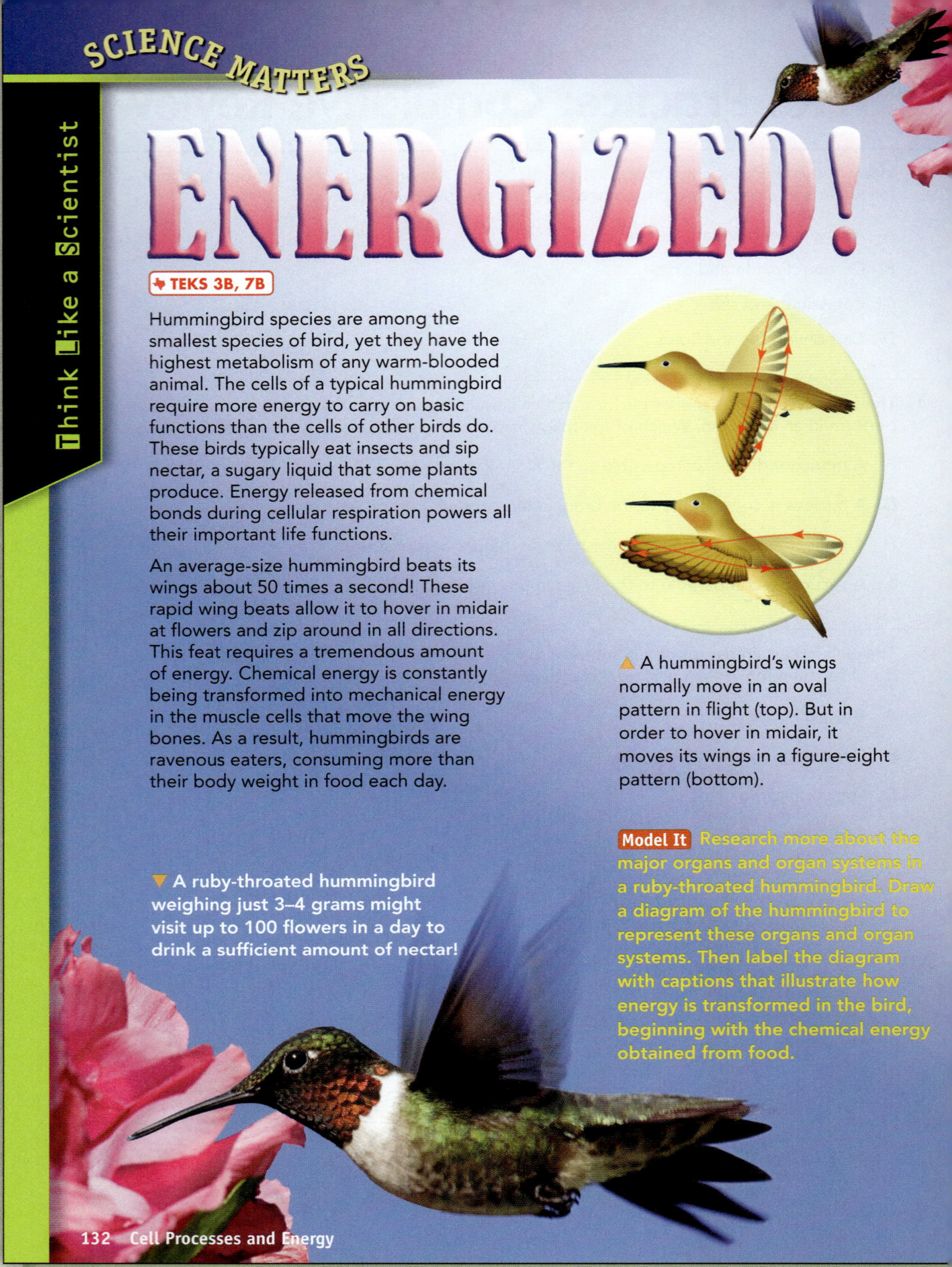

SCIENCE MATTERS

Think Like a Scientist

ENERGIZED!

TEKS 3B, 7B

Hummingbird species are among the smallest species of bird, yet they have the highest metabolism of any warm-blooded animal. The cells of a typical hummingbird require more energy to carry on basic functions than the cells of other birds do. These birds typically eat insects and sip nectar, a sugary liquid that some plants produce. Energy released from chemical bonds during cellular respiration powers all their important life functions.

An average-size hummingbird beats its wings about 50 times a second! These rapid wing beats allow it to hover in midair at flowers and zip around in all directions. This feat requires a tremendous amount of energy. Chemical energy is constantly being transformed into mechanical energy in the muscle cells that move the wing bones. As a result, hummingbirds are ravenous eaters, consuming more than their body weight in food each day.

▲ A hummingbird's wings normally move in an oval pattern in flight (top). But in order to hover in midair, it moves its wings in a figure-eight pattern (bottom).

▼ A ruby-throated hummingbird weighing just 3–4 grams might visit up to 100 flowers in a day to drink a sufficient amount of nectar!

Model It Research more about the major organs and organ systems in a ruby-throated hummingbird. Draw a diagram of the hummingbird to represent these organs and organ systems. Then label the diagram with captions that illustrate how energy is transformed in the bird, beginning with the chemical energy obtained from food.

Quick Facts

There are over 300 species of hummingbirds worldwide, and 17 species can be found in the United States. Hummingbirds have developed a way to conserve energy when they are not eating. At night or during cold periods, hummingbirds go into a sleep-like state called *torpor.* The bird's body temperature and heart rate drop, and its breathing slows to an almost imperceptible level. This deep sleep allows metabolic demands to be reduced and helps the bird survive the night without a constant source of energy. A hummingbird expends about 50 times more energy awake than in torpor.

Texas Essential Knowledge and Skills

3B Use models to represent aspects of the natural world such as human body systems and plant and animal cells.

7B Illustrate the transformation of energy within an organism such as the transfer from chemical energy to heat and thermal energy in digestion.

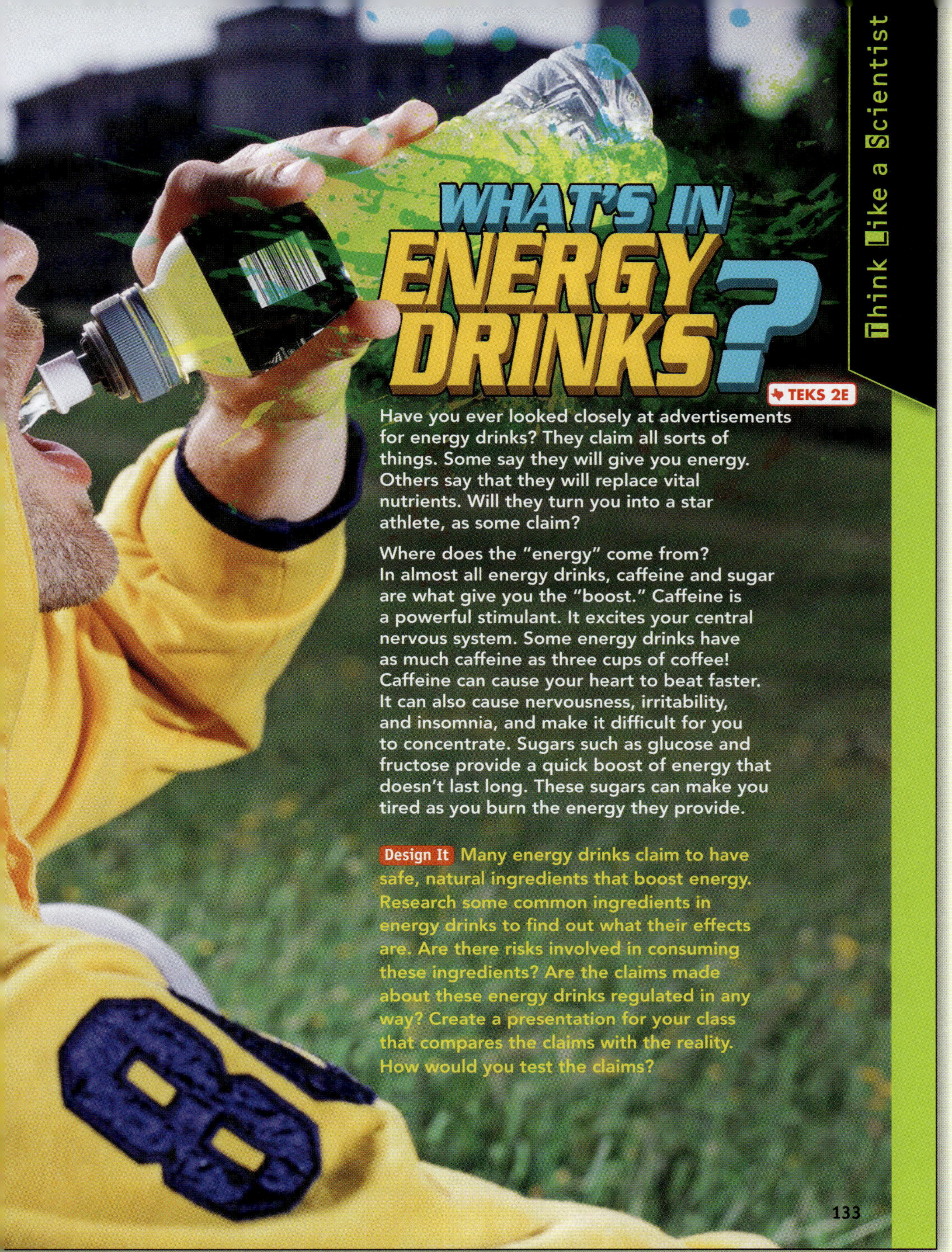

WHAT'S IN ENERGY DRINKS?

TEKS 2E

Have you ever looked closely at advertisements for energy drinks? They claim all sorts of things. Some say they will give you energy. Others say that they will replace vital nutrients. Will they turn you into a star athlete, as some claim?

Where does the "energy" come from? In almost all energy drinks, caffeine and sugar are what give you the "boost." Caffeine is a powerful stimulant. It excites your central nervous system. Some energy drinks have as much caffeine as three cups of coffee! Caffeine can cause your heart to beat faster. It can also cause nervousness, irritability, and insomnia, and make it difficult for you to concentrate. Sugars such as glucose and fructose provide a quick boost of energy that doesn't last long. These sugars can make you tired as you burn the energy they provide.

Design It Many energy drinks claim to have safe, natural ingredients that boost energy. Research some common ingredients in energy drinks to find out what their effects are. Are there risks involved in consuming these ingredients? Are the claims made about these energy drinks regulated in any way? Create a presentation for your class that compares the claims with the reality. How would you test the claims?

Think Like a Scientist

Have students read *What's in Energy Drinks?* Point out that many energy drinks use some ingredients with nutritional value such as green tea, ginseng, or vitamin C in orange juices. By adding these ingredients, people are led to think that a drink is nutritious and healthy. Usually, small amounts of these healthier ingredients are found in energy drinks. They often work as stimulants to strengthen the effect of the caffeine. While energy drinks do achieve the desired energy boost, drinking them in moderation is perhaps the key.

Ask: **How does caffeine give a boost?** *(Sample: Caffeine gives a boost because it stimulates the central nervous system and can cause the heart to beat faster.)* **What are some of the side effects of too much caffeine?** *(Sample: nervousness, irritability, lack of concentration)* **What is a disadvantage of sugar as an energy booster?** *(Sample: The sugar provides a quick boost that does not last long.)*

English Language Proficiency Standards

ELPS Speaking 3.G.1

Have students express opinions about energy drinks.

Beginning Have students listen as you or another student read aloud page 133. Help students express opinions about energy drinks by scaffolding questions, for example *Are energy drinks good for you? Will they help you be an athlete? Will they help you do school work?*

Intermediate Together, read aloud page 133. Have partners express their opinions by completing the following sentence frames: *I think energy drinks are ______ because ______.*

Advanced Have partners read aloud page 133 and express opinions. Tell students to elaborate on their ideas.

Advanced High After students express opinions about energy drinks, have them complete the Design It activity.

Texas Essential Knowledge and Skills

2E Analyze data to formulate reasonable explanations, communicate valid conclusions supported by the data, and predict trends.

CHAPTER 4

Genetics: The Science of Heredity

Chapter TEKS Overview

This chapter focuses on 14A, 14B, and 14C. Students will examine how inherited traits are governed by genes within chromosomes, how genetic information passes from one generation to the next generation, and how the results of sexual and asexual reproduction compare.

Introduce the TEKS

To pique student interest and introduce TEKS 14A, 14B, and 14C, have students look at the image and read the Focus Question and description. Ask the students to describe the joey and its mother. Call on volunteers to share their descriptions while you compile a class list on the board. Point out that not all white animals are albino. Ask: **Does the joey have any pigment in its eyes and nose?** *(No)* **If the pink color is not pigment, what do you think it is?** *(The eyes look pink because there is no pigment to hide the color caused by blood flow through the tissues.)* **Do you think the joey will have albino offspring?** *(Accept all answers at this time. Sample: I think it could happen, if it mates with another albino koala.)*

Untamed Science Video

WHERE'D YOU GET THOSE GENES? Before viewing, invite students to discuss what they know about heredity and genes. Then play the video. Lead a class discussion and make a list of questions that this video raises. You may wish to have students view the video again after they have completed the chapter to see if their questions have been answered.

WHAT MAKES THIS BABY KOALA DIFFERENT?

FOCUS ON TEKS 14A, 14B, 14C

Why don't offspring always look like their parents?

Even though this young koala, or joey, has two fuzzy ears, a long nose, and a body shaped like its mother's, you can see that the two are different. You might expect a young animal to look exactly like its parents, but think about how varied a litter of kittens or puppies can look. This joey is an albino—an animal that lacks the usual coloring in its eyes, fur, and skin.

Observe Describe how this joey looks different from its mother.

Sample: Its eyes and nose are pink, but the mother's eyes are brown. Its fur is white, but the mother's fur is gray.

Watch the **Untamed Science** video to learn more about heredity.

UntamedScience

Professional Development Note

From the Author

A father and son in the D.C. area were successful basketball players and also coached the same university basketball team. In Maryland, we have a father who has light brown eyes, made average grades in school, and was an outstanding athlete; his son also has light brown eyes, makes excellent grades in school, and is not athletically inclined. Traits are characteristics that are passed down from parent to offspring. Based on this example, one may think that getting good grades is not inherited, but playing basketball and eye color may be inherited. This is partially correct. Eye color is inherited; it is passed from parent to offspring. However, playing sports is a behavior that is learned.

Zipporah Miller

Texas
CHAPTER
4

Genetics: The Science of Heredity

Texas Essential Knowledge and Skills

SUPPORTING TEKS: 14B Compare the results of uniform or diverse offspring from sexual reproduction or asexual reproduction. **14C** Recognize that inherited traits of individuals are governed in the genetic material found in the genes within chromosomes in the nucleus.
TEKS: 2C Record data using qualitative means such as graphic organizers. **2E** Analyze data to formulate reasonable explanations, communicate valid conclusions supported by the data, and predict trends. **3B** Use models to represent aspects of the natural world. **3D** Relate the impact of research on scientific thought and society, including the contributions of scientists as related to the content. **14A** Define heredity as the passage of genetic instructions from one generation to the next generation.

PEARSON Texas.com

Spotlight On Technology

Virtual Lab Activity Set the tone for inquiry by having students engage in a quick online, inquiry-based experience to help them explore key questions about chromosomes and inheritance.

Flipped Video for Science Show students a Pearson Flipped Video for Science for this chapter so that they can better understand heredity and the advantages and disadvantages of sexual and asexual reproduction.

PEARSON Texas.com

Chapter at a Glance

CHAPTER PACING: 9–13 periods or $4\frac{1}{2}$–$6\frac{1}{2}$ blocks

INTRODUCE THE CHAPTER: Use the Focus Question and the opening image to get students thinking about heredity. Activate prior knowledge and preteach vocabulary using the Getting Started pages.

Lesson 1: What Is Heredity?

Lesson 2: Probability and Heredity

Lesson 3: Patterns of Inheritance

Lesson 4: Chromosomes and Inheritance

ASSESSMENT OPTIONS:
Teacher's Edition: Lesson Quizzes, Texas End-of-Year Test Prep A and B
Online assessments

Texas Essential Knowledge and Skills

2C Collect and record data using the International System of Units (SI) and qualitative means such as labeled drawings, writing, and graphic organizers.

2E Analyze data to formulate reasonable explanations, communicate valid conclusions supported by the data, and predict trends.

3B Use models to represent aspects of the natural world such as human body systems and plant and animal cells.

3D Relate the impact of research on scientific thought and society, including the history of science and contributions of scientists as related to the content.

14A Define heredity as the passage of genetic instructions from one generation to the next generation.

14B Compare the results of uniform or diverse offspring from sexual reproduction or asexual reproduction.

14C Recognize that inherited traits of individuals are governed in the genetic material found in the genes within chromosomes in the nucleus.

The following **College and Career Readiness Standards** are covered in this chapter: **I.D.3, VI.C.1, VI.C.2, VI.D.1, VI.D.5**

CHAPTER 4

Getting Started

Check Your Understanding

This activity assesses students' understanding of sexual and asexual reproduction. After students have shared their answers, point out that sexual reproduction is carried out by all plants and animals. A few simple animals and some kinds of plants can also reproduce asexually. Many protists and algae can reproduce both sexually and asexually.

Preteach Vocabulary Skills

Explain to students that recognizing suffixes can help them learn new vocabulary words. A suffix often changes a word's part of speech. Point out that the suffix *-ive* can change a noun or verb into an adjective. *Recessive* is an adjective that modifies the noun *allele.* Encourage students to use this suffix to change these verbs into adjectives: *combat, digest, conduct, reflect, elect.*

CHAPTER 4 Getting Started

Check Your Understanding

1. **Background** Read the paragraph below and then answer the question.

> Kent's cat just had six kittens. All six kittens look different from one another—and from their two parents! Kent knows each kitten is unique because cats reproduce through **sexual reproduction,** not **asexual reproduction.** Before long, the kittens will grow bigger and bigger as their cells divide through **mitosis.**

- In what way are the two daughter cells that form by mitosis and cell division identical?

They have the same DNA.

Sexual reproduction involves two parents and combines their genetic material to produce a new organism that differs from both parents.

Asexual reproduction involves only one parent and produces offspring that are identical to the parent.

During **mitosis,** a cell's nucleus divides into two new nuclei, and one copy of DNA is distributed into each daughter cell.

Vocabulary Skill

Suffixes A suffix is a word part that is added to the end of a word to change its meaning. For example, the suffix *-tion* means "process of." If you add the suffix *-tion* to the verb *fertilize,* you get the noun *fertilization. Fertilization* means "the process of fertilizing." The table below lists some other common suffixes and their meanings.

Suffix	Meaning	Example
-ive	performing a particular action	recessive allele, *n.* an allele that is masked when a dominant allele is present
-ance or *-ant*	state, condition of	codominance, *n.* occurs when both alleles are expressed equally

2. **Quick Check** **Fill in the blank with the correct suffix.**

- A domin ant allele can mask a recessive allele.

English Language Proficiency Standards

ELPS Learning Strategies 1.F

Modify the three-column chart activity in Preview Vocabulary Terms to meet students' needs and abilities.

Beginning Work together to create the chart. Have students rate how well they know each term. Have partners Turn and Talk about whether they know the word, or any word parts or cognates. Then add a definition to the chart.

Intermediate Display and read aloud the terms. Have partners work together to create their charts and fill in the first two columns. Then discuss definitions and examples.

Advanced Have partners work together to create their charts. Encourage students to use glossaries or dictionaries to find definitions.

Advanced High Have individual students create the charts. Encourage students to check their definitions using glossaries or dictionaries.

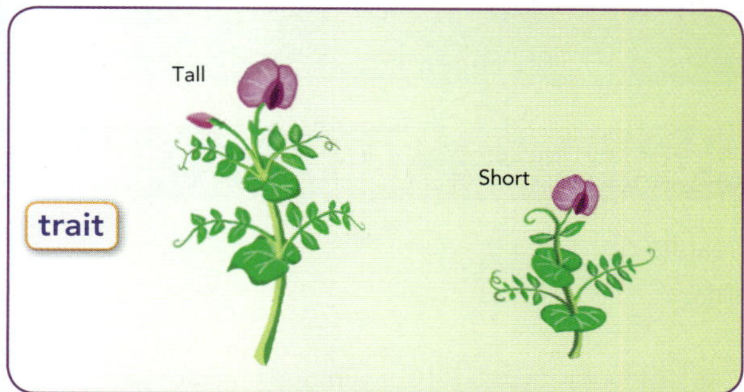

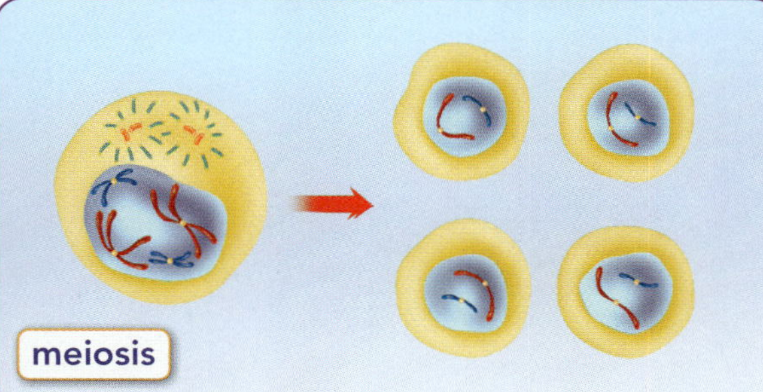

Chapter Preview

LESSON 1
- heredity
- trait
- genetics
- fertilization
- purebred
- gene
- allele
- dominant allele
- recessive allele
- hybrid

Identify Supporting Evidence
Predict

LESSON 2
- probability
- Punnett square
- phenotype
- genotype
- homozygous
- heterozygous

Identify the Main Idea
Draw Conclusions

LESSON 3
- incomplete dominance
- codominance
- multiple alleles
- polygenic inheritance

Compare and Contrast
Interpret Data

LESSON 4
- meiosis

Relate Cause and Effect
Design Experiments

Preview Vocabulary Terms

Have students create a three-column chart to rate their knowledge of the vocabulary terms before they read the chapter. In the first column of the chart, students should list the terms for the chapter. In the second column, students should identify whether they can define and use the word, whether they have heard or seen the word before, or whether they do not know the word. As the class progresses through the chapter, have students write definitions for each term in the last column of the chart.

L1 Have students look at the images on this page as you pronounce the vocabulary word. Have students repeat the word after you. Then read the definition below. Use the sample sentence in italics to clarify the meaning of the term.

trait *(trayt)* A characteristic that can be passed from parents to offspring. *Plant height is a trait that plants pass on to their offspring.*

phenotype *(FEE nuh typ)* An organism's physical appearance. *Black fur is a phenotype of some guinea pigs.*

incomplete dominance *(in kuhm PLEET DAHM uh nuns)* A pattern of inheritance in which one allele is only partially dominant. *Incomplete dominance produces pink snapdragon flowers as a result of a cross between a white flower and a red flower.*

meiosis *(my OH sis)* The process by which the number of chromosomes is reduced by half as sex cells form. *Meiosis results in cells with half the number of chromosomes as the other cells in the organism.*

Academic Vocabulary

Each lesson includes key Academic Vocabulary. See also the Support All Readers box at the start of the lesson.

Lesson 1: evidence, identify, predict

Lesson 2: conclusions, identify, main idea

Lesson 3: compare, contrast, data, interpret

Lesson 4: cause, design, effect, experiment, relate

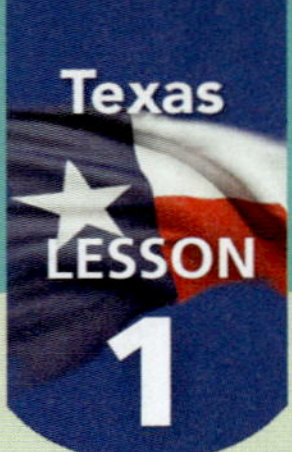

What Is Heredity?

Why don't offspring always look like their parents?

LESSON PACING:
2–3 periods or 1–1½ blocks

Lesson Vocabulary

- heredity
- trait
- genetics
- fertilization
- purebred
- gene
- allele
- dominant allele
- recessive allele
- hybrid

Content Refresher

The Law of Segregation Mendel's First Law can be stated as:

a. Each of the selected garden pea traits is controlled by a pair of alleles.

b. For each trait, an offspring receives one allele from one parent and one allele from the other.

c. The chance of offspring receiving one or the other allele from each pair is equal.

d. The expression of a dominant trait requires only one dominant allele; the expression of a recessive trait requires two recessive alleles.

Lesson Objectives	TEKS	ELPS
Describe the results of Mendel's experiments.	14A	
Identify the role of alleles in controlling the inheritance of traits.	3D, 14A, 14C	2.C.4

Texas Essential Knowledge and Skills

3D Relate the impact of research on scientific thought and society, including the history of science and contributions of scientists as related to the content.
14A Define heredity as the passage of genetic instructions from one generation to the next generation.
14C Recognize that inherited traits of individuals are governed in the genetic material found in the genes within chromosomes in the nucleus.

English Language Proficiency Standards

ELPS Listening 2.C.4 Learn academic vocabulary heard during classroom instruction and interactions.

DIFFERENTIATED INSTRUCTION KEY
L1 Struggling Students or Special Needs
L2 On-Level Students **L3** Advanced Students

LESSON PLANNER 4.1

Investigations and Activities	TEKS Review
My Planet Diary, **Student Edition,** p. 138 **Inquiry:** Inquiry Warm-Up, What Does the Father Look Like?, **PearsonTexas.com** Introduce Vocabulary, **Teacher's Edition,** p. 139 Lead a Discussion, Family Resemblances, **Teacher's Edition,** p. 139 Teach With Visuals, **Teacher's Edition,** p. 139 Lead a Discussion, Purebred Organisms, Teacher's Edition, p. 139 Teach Key Concepts, **Teacher's Edition,** p. 140 Lead a Discussion, Latin Origins, **Teacher's Edition,** p. 140 **Inquiry:** Quick Lab, Observing Pistils and Stamens, **PearsonTexas.com**	Apply the TEKS, Predicting Genotypes, **Student Edition,** p. 166 TEKS Practice: **Student Edition,** p. 167 TEKS Practice: Chapter and Cumulative Review, **Student Edition,** p. 169 Lesson 4.1, **TEKS Preparation and Study Guide Workbook**, p. 36
Teach Key Concepts, **Teacher's Edition,** p. 141 Make Analogies, Traits and Alleles, **Teacher's Edition,** p. 141 Teach With Visuals, **Teacher's Edition,** p. 141 Differentiated Instruction, **Teacher's Edition,** p. 141 Teach With Visuals, **Teacher's Edition,** p. 142 **Inquiry:** Teacher Demo, Observing Crosses in Fruit Flies, **Teacher's Edition,** p. 142 Support the TEKS, No Traits Get Lost, **Teacher's Edition,** p. 143 21st Century Learning, Information Literacy, **Teacher's Edition,** p. 143 Apply It!, **Student Edition,** p. 143 Differentiated Instruction, **Teacher's Edition,** p. 143 **Inquiry:** Quick Lab, Inferring the Parent Generation, **Lab Manual,** p. 40	**SHORT ON TIME?** To do this lesson in approximately half the time, do the Activate Prior Knowledge activity. A discussion of the Key Concepts will familiarize students with the lesson content. Have students do the Quick Labs. The rest of the lesson can be completed by students independently.

These editable worksheets are available on **PearsonTexas.com.**
Print versions can be found in the **TEKS Preparation and Study Guide Workbook.**

Name ______________________ Date __________ Class __________

4.1 What Is Heredity?

Key Concept Summaries

What Did Mendel Observe?

Heredity is the passage of genetic instructions from one generation to the next generation. In the mid-nineteenth century, Gregor Mendel wondered why pea plants had different characteristics. Each characteristic, such as height or seed color, is called a **trait.** Mendel wondered why the forms of the pea plants' traits were often—but not always—similar to their parents. His discoveries form the foundation of **genetics,** the scientific study of heredity.

In a pea plant flower, the pistil produces female sex cells, or eggs. The stamens produce pollen, which contains the male sex cells, or sperm. A new organism begins to form when egg and sperm cells join, a process called **fertilization**. For this to occur in plants, pollen must reach the pistil of a flower, a process called pollination. Pea plants usually self-pollinate: Pollen from a flower lands on the pistil of the same flower.

A **purebred** organism is the offspring of many generations that have the same form of a trait. Mendel cross-pollinated, or "crossed," purebred tall with purebred short plants. Scientists call these the parental, or P, generation. The offspring of the P generation are called the first filial, or F_1, generation. Their offspring are called the second filial, or F_2, generation. **In all of his crosses, Mendel found that only one form of the trait appeared in the F_1 generation. However, in the F_2 generation, the "lost" form of the trait always reappeared in about one fourth of the plants.**

How Do Alleles Affect Inheritance?

Today, scientists use the word **gene** to describe the factors that control a trait. **Alleles** are the different forms of a gene. The gene that controls stem height in peas has one allele for short stems and one allele for long stems. Genes are found within chromosomes in the nucleus of a cell.

An organism's traits are controlled by the alleles it inherits from its parents. Some alleles are dominant, while other alleles are recessive. A **dominant allele** is one whose trait always shows up when the allele is present. A **recessive allele** is hidden whenever the dominant allele is present. A **hybrid** organism has two different alleles for a trait. Geneticists, scientists who study genetics, often use letters to represent alleles. A dominant allele is symbolized by a capital letter. A recessive allele is symbolized by the lowercase version of the *same* letter. Because Mendel's observations eventually influenced scientists' ideas about heredity, he is often called the Father of Genetics.

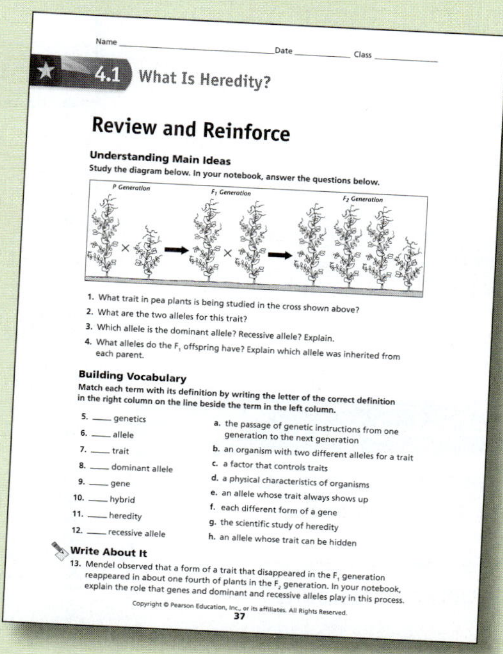
Name ______________________ Date __________ Class __________

4.1 What Is Heredity?

Review and Reinforce

Understanding Main Ideas

Study the diagram below. In your notebook, answer the questions below.

1. What trait in pea plants is being studied in the cross shown above?
2. What are the two alleles for this trait?
3. Which allele is the dominant allele? Recessive allele? Explain.
4. What alleles do the F_1 offspring have? Explain which allele was inherited from each parent.

Building Vocabulary

Match each term with its definition by writing the letter of the correct definition in the right column on the line beside the term in the left column.

5. ____ genetics	a. the passage of genetic instructions from one generation to the next generation
6. ____ allele	b. an organism with two different alleles for a trait
7. ____ trait	c. a factor that controls traits
8. ____ dominant allele	d. a physical characteristics of organisms
9. ____ gene	e. an allele whose trait always shows up
10. ____ hybrid	f. each different form of a gene
11. ____ heredity	g. the scientific study of heredity
12. ____ recessive allele	h. an allele whose trait can be hidden

Write About It

13. Mendel observed that a form of a trait that disappeared in the F_1 generation reappeared in about one fourth of plants in the F_2 generation. In your notebook, explain the role that genes and dominant and recessive alleles play in this process.

LESSON 4.1

Lexile Measure = 950L

What Is Heredity?

Establish Learning Objectives

After this lesson, students will be able to:

Describe the results of Mendel's experiments.

Identify the role of alleles in controlling the inheritance of traits.

Engage

Activate Prior Knowledge

MY PLANET DIARY Read *Almost Forgotten* with the class. Point out that, as a priest in a monastery, Mendel had little contact with scientists. Ironically, his work was rediscovered by three scientists working independently in the same year. Ask: **Why is it important for scientists to do library research before lab research?** *(They could find useful information that will help them plan their experiments or help them realize what tests have been performed in the past.)*

Explore

Lab Resource: Inquiry Warm-Up 

L1 WHAT DOES THE FATHER LOOK LIKE? Students will observe the physical characteristics of a kitten and its mother and make predictions about what the unknown father of the kitten looks like. This Inquiry Warm-Up can be found online.

Texas Essential Knowledge and Skills

14A Define heredity as the passage of genetic instructions from one generation to the next generation.

What Is Heredity?

What Did Mendel Observe? TEKS 14A

How Do Alleles Affect Inheritance? TEKS 3D, 14A, 14C

my planet Diary BIOGRAPHY

Almost Forgotten

When scientists make great discoveries, sometimes their work is praised, criticized, or even forgotten. Gregor Mendel was almost forgotten. He spent eight years studying pea plants, and he discovered patterns in the way characteristics pass from one generation to the next. For almost 40 years, people overlooked Mendel's work. When it was finally rediscovered, it unlocked the key to understanding heredity.

Communicate Discuss the question below with a partner. Then write your answer.

Did you ever rediscover something of yours that you had forgotten? How did you react?

Sample: Yes, I found a softball mitt I bought the end of last season. I was surprised. It was like finding something new and useful.

Do the Inquiry Warm-Up *What Does the Father Look Like?* Find the lab online.

TEKS 14A In this section, you'll learn about the experiments that Gregor Mendel conducted on pea plants, which contributed to the modern study of heredity.

ELPS 2.C.4

With a partner, read the "Symbols for Alleles," paragraph on page 142. Then complete the Figure 4 activity. Present the completed chart to another pair of students. Point out examples of symbols, capital and lowercase letters, and subscript letters.

What Did Mendel Observe?

In the mid-nineteenth century, a priest named Gregor Mendel tended a garden in a central European monastery. Mendel's experiments in that peaceful garden would one day transform the study of heredity. **Heredity** is the passage of genetic instructions from one generation to the next generation.

Mendel wondered why different pea plants had different characteristics. Some pea plants grew tall, while others were short. Some plants produced green seeds, while others had yellow seeds. Each specific characteristic, such as stem height or seed color, is called a **trait.** Mendel observed that the forms of the pea plants' traits were often similar to those of their parents. Sometimes, however, the forms differed.

SUPPORT ALL READERS

Lexile Measure = 950L **Lexile Word Count = 1263**

Prior Exposure to Content: Many students may have misconceptions on this topic

Academic Vocabulary: *evidence, identify, predict*

Science Vocabulary: *heredity, trait, genetics, gene, allele, hybrid*

Concept Level: Generally appropriate for most students in this grade

Preteach With: My Planet Diary "Almost Forgotten" and Figure 2 activity

Vocabulary
- heredity • trait • genetics • fertilization
- purebred • gene • allele • dominant allele
- recessive allele • hybrid

Skills
- Reading: Identify Supporting Evidence
- Inquiry: Predict

Mendel's Experiments Mendel experimented with thousands of pea plants. Today, Mendel's discoveries form the foundation of **genetics,** the scientific study of heredity. **Figure 1** shows the parts of a pea plant's flower. The pistil produces female sex cells, or eggs. The stamens produce pollen, which contains the male sex cells, or sperm. A new organism begins to form when egg and sperm cells join in the process called **fertilization.** Before fertilization can happen in pea plants, pollen must reach the pistil of a pea flower. This process is called pollination.

Pea plants are usually self-pollinating. In self-pollination, pollen from a flower lands on the pistil of the same flower. Mendel developed a method by which he cross-pollinated, or "crossed," pea plants. **Figure 1** shows his method.

Mendel decided to cross plants that had contrasting forms of a trait—for example, tall plants and short plants. He started with purebred plants. A **purebred** organism is the offspring of many generations that have the same form of a trait. For example, purebred tall pea plants always come from tall parent plants.

FIGURE 1

Crossing Pea Plants

Mendel devised a way to cross-pollinate pea plants.

Use the diagram to answer the questions about Mendel's procedure.

1. **Observe** How does flower B differ from flower A?
   Flower B has no stamens.
2. **Infer** Describe how Mendel cross-pollinated pea plants.
   He used a brush to transfer pollen from one plant to another.

English Language Proficiency Standards

ELPS Listening 2.C.4

Have students read the paragraph on page 142 and complete the **Figure 4** activity.

Beginning Read aloud the paragraph on page 142 to students. Before working through the figure with students, review the terms *summarize, cause,* and *effect.*

Intermediate Read aloud the paragraph on page 142 and have students complete the figure in small groups. Before students begin the activity, point out examples of how to summarize, as well as examples of causes and effects.

Advanced Have partners read the paragraph on page 142 and complete the figure. Ask students to explain what it means to *summarize* a text and to define the terms *cause* and *effect.*

Advanced High Have students read the paragraph on page 142 and complete the figure independently. Have students point out examples of a *summary,* a *cause,* and an *effect* in **Figure 4**.

Explain

Introduce Vocabulary

Students have probably heard the word *trait* used in everyday language. Tell them that in science, *trait* has a narrower definition. A trait is an inherited characteristic of an organism.

Lead a Discussion

FAMILY RESEMBLANCES Invite volunteers to share observations they have made about the physical similarities and differences among members of a family, either their own or a family they know. Ask: **Have you ever wondered why some family members look very similar while others look very different?** *(Accept reasonable answers. Many students will have considered this in one way or another.)* Encourage students to share their ideas about the inheritance of traits in families. Be alert for misconceptions students may have, and address these throughout the section. Also be sensitive to the fact that some students may be living with adoptive, foster, or blended families, and may be uncomfortable talking about families and inherited traits.

Teach With Visuals

Tell students to look at **Figure 1.** Point out how the stamens are cut off in step B. Students may have seen lilies with the stamens cut off. Explain that florists do this to lilies such as the Stargazer, which have pollen that stains clothing and other surfaces it lands on. In Mendel's case, cutting off the stamens served a different purpose. Ask: **Why was it necessary to cut off the stamens of the pea flower?** *(To make sure that the pollen that landed on the pistil came from the plant Mendel wanted to use in the cross.)*

Lead a Discussion

PUREBRED ORGANISMS Tell students that every living thing has traits inherited from its parents. Until the work of Mendel, people did not understand how traits were passed from parents to offspring. Ask: **What is a purebred organism?** *(The offspring of many generations that have the same form of a trait.)*

PEARSON Texas.com

LESSON 4.1

Explain

Teach Key Concepts

Explain to students that Mendel began each experiment by crossing plants that differed with respect to one specific trait. Ask: **What did Mendel find when he crossed purebred tall plants with purebred short plants?** *(The offspring all were tall.)* **Was the trait for shortness lost? Explain.** *(No. When Mendel crossed the offspring, or F_1 generation, with one another, some of the offspring were short, so the trait wasn't lost.)* **Are the plants in the F_1 generation purebred plants? Explain.** *(No; they were not all identical to the parent plants.)* Point out to students that Mendel did not get exactly three-fourths and one-fourth in every case for every trait. Although the numbers varied somewhat from the ideal, he was able to recognize a pattern.

Lead a Discussion

LATIN ORIGINS Point out that the terms *F_1 generation* and *F_2 generation* are derived from the Latin words *filius* (son) and *filia* (daughter). Explain to students that in Mendel's time, an educated person would have studied Latin and Greek. In addition, his training as a priest would have included studying Latin. Ask: **What other reason might there be for using a Latin term in science?** *(If scientists in other countries also studied Latin, they would recognize the word.)* **Where else in your study of science have you seen words that are derived from Latin?** *(Sample: names of elements)*

Elaborate

Lab Resource: Quick Lab 

L2 OBSERVING PISTILS AND STAMENS Students will examine the parts of a flower. This Quick Lab can be found online.

Evaluate

Assess Your Understanding

After students answer the questions, have them evaluate their understanding by completing the appropriate sentence.

RTI Response to Intervention

1a. If students have trouble defining *heredity*, **then** have them find the highlighted term and reread its definition.

b. If students need help describing the plants in Mendel's crosses, **then** have them review **Figure 2.**

The F_1 and F_2 Offspring

Mendel crossed purebred tall plants with purebred short plants. Today, scientists call these plants the parental, or P, generation. The resulting offspring are the first filial (FIL ee ul), or F_1, generation. The word *filial* comes from *filia* and *filius*, the Latin words for "daughter" and "son."

Look at **Figure 2** to see the surprise Mendel found in the F_1 generation. All the offspring were tall. The shortness trait seemed to have disappeared!

When these plants were full-grown, Mendel allowed them to self-pollinate. The F_2 (second filial) generation that followed surprised Mendel even more. He counted the plants of the F_2 generation. About three fourths were tall, while one fourth were short.

Experiments With Other Traits

Mendel repeated his experiments, studying other pea-plant traits, such as flower color and seed shape. **In all of his crosses, Mendel found that only one form of the trait appeared in the F_1 generation. However, in the F_2 generation, the "lost" form of the trait always reappeared in about one fourth of the plants.**

FIGURE 2

Results of a Cross

In Mendel's crosses, some forms of a trait were hidden in one generation but reappeared in the next.

Interpret Diagrams **Draw and label the offspring in the F_2 generation.**

Do the Quick Lab *Observing Pistils and Stamens.* Find the lab online.

Assess Your Understanding

TEKS 14A

1a. Define What is heredity?

The passage of genetic instructions from one generation to the next

b. Compare and Contrast In Mendel's cross for stem height, how did the plants in the F_2 generation differ from the F_1 plants?

Some F_2 plants were short.

got it?

O **I get it!** Now I know that Mendel found that one form of a trait could hide another form, but the hidden trait could reappear in the next generation.

O I need extra help with See TE note.

How Do Alleles Affect Inheritance?

TEKS 3D, 14A, 14C In this section, you'll examine how an organism's traits are controlled by the genes it receives from its parents.

Mendel reached several conclusions from his experimental results. He reasoned that individual factors, or sets of genetic instructions, must control the inheritance of traits in peas. The factors that control each trait exist in pairs. The female parent contributes one factor, while the male parent contributes the other factor. Finally, one factor in a pair can mask, or hide, the other factor. The tallness factor, for example, masked the shortness factor.

Genes and Alleles Today, scientists use the word **gene** to describe the factors that control a trait. Genes are found within the chromosomes inside a cell's nucleus. **Alleles** (uh LEELZ) are the different forms of a gene. The gene that controls stem height in peas has one allele for tall stems and one allele for short stems. Each pea plant inherits two alleles—one from the egg and the other from the sperm. A plant may inherit two alleles for tall stems, two alleles for short stems, or one of each.

An organism's traits are controlled by the alleles it inherits from its parents. Some alleles are dominant, while other alleles are recessive. A **dominant allele** is one whose trait always shows up in the organism when the allele is present. A **recessive allele,** on the other hand, is hidden whenever the dominant allele is present. **Figure 3** shows dominant and recessive alleles of the traits in Mendel's crosses.

FIGURE 3

Alleles in Pea Plants

Mendel studied the inheritance of seven different traits in pea plants.

Use the table to answer the questions.

1. **Draw Conclusions** Circle the picture of each dominant form of the trait in the P generation.
2. **Predict** Under what conditions would the recessive form of one of these traits reappear?

In an offspring having two recessive alleles

Inheritance of Pea Plants Studied by Mendel

	Seed Shape	Seed Color	Pod Shape	Pod Color	Flower Color	Flower Position	Stem Height
P	Wrinkled x Round	Yellow x Green	Pinched x Smooth	Green x Yellow	Purple x White	Tip of stem x Side of stem	Tall x Short
F_1	Round	Yellow	Smooth	Green	Purple	Side of stem	Tall

Explain

Teach Key Concepts

Tell students that an organism's alleles come from its parents. Ask: **What are the two kinds of alleles?** *(Dominant and recessive)* **How are they different?** *(The dominant allele is the one whose trait is always seen if the allele is present. The recessive allele is hidden whenever the dominant allele is present.)* **What must be true for the trait controlled by a recessive allele to be seen?** *(There must be no dominant allele. Both alleles must be the recessive allele.)*

Make Analogies

L1 **TRAITS AND ALLELES** Students may find the terms *trait* and *allele* confusing because they seem to refer to the same thing. Explain that *trait* refers to a general characteristic and *allele* refers to the specific forms that characteristic can have. For example, a fiction book has the characteristic or trait of having a cover. The type of cover would be the alleles. One allele is a hardcover, and the other is a paperback.

Teach With Visuals

Tell students to look at **Figure 3**. Remind students that the traits Mendel studied in pea plants have two distinct forms. Direct students' attention to the trait of seed shape. Ask: **What are the two kinds of seed shape?** *(Round or wrinkled)* **Which shape is the result of a dominant allele?** *(Round)* **What combinations of alleles will produce round seeds?** *(Two dominant alleles or one dominant allele and one recessive allele)* Note that high school and college texts generally use the terms *terminal* and *axial* to describe flower position. These textbooks also use terms such as *constricted* and *inflated* to describe pod shape. For easier readability in this text, simpler terms were used.

Predict Tell students that when they predict, they use evidence and prior experience to make an inference about a future event.

LESSON 4.1

Differentiated Instruction

L1 **Dominant Alleles** If students have difficulty identifying the dominant allele for each of Mendel's seven traits, have them look back at **Figure 3** to remind themselves which allele is seen in F_1 plants. Students may find it useful to add the label *Dominant* to the bottom row of the table.

L3 **Mendel's Crosses** Challenge students to select one of the traits, other than stem height, that Mendel studied in peas. Students should create a poster on which they show the crosses from P generation to F_2 generation.

Texas Essential Knowledge and Skills

3D Relate the impact of research on scientific thought and society, including the history of science and contributions of scientists as related to the content.

14A Define heredity as the passage of genetic instructions from one generation to the next generation.

14C Recognize that inherited traits of individuals are governed in the genetic material found in the genes within chromosomes in the nucleus.

LESSON 4.1

Explain

Teach With Visuals

Tell students to look at **Figure 4.** Remind them that the style for writing the alleles is to write the letter for the dominant allele first. Ask: **How would you write the alleles for a plant that is purebred tall?** *(TT)* **How would you write the alleles for the offspring of a cross between a purebred tall pea plant and a purebred short pea plant?** *(Tt)* **What term did Mendel use for an individual with one dominant allele and one recessive allele?** *(Hybrid)*

Elaborate

Teacher Demo 

L2 OBSERVING CROSSES IN FRUIT FLIES

Materials 2 *Drosophila melanogaster* cultures—wild-type and ebony, culture vials and plugs, *Drosophila* media, nonether anesthesia kit, hand lens, paint brush, white index cards, marking pen

Advanced Preparation Set up the parental cross about two weeks in advance by placing 2 to 3 ebony males with 2 to 3 wild-type virgin females in each of three vials. To collect virgin females, remove all adult flies from the culture vial. Then, within 4 to 6 hours, collect the newly emerged females. Females have pointed abdomens with stripes almost to the end. Males have rounded abdomens that are black at the end. Anesthetize flies to sort them and set up the crosses. Place vials on their sides until the flies wake up. Remove parent flies from the vials when the pupae begin to develop. When F_1 adults begin to emerge, remove the flies daily to prevent F_2 offspring from mixing with F_1 offspring. Dispose of unwanted flies in a jar of mineral oil.

Time 20 minutes

Anesthetize the parent flies and place them on index cards for students to examine. *CAUTION: Students should not work with the anesthetic.* Ask: **How do these flies differ?** *(Ebony flies have darker bodies than wild-type flies.)*

Challenge students to predict which trait is controlled by a dominant allele and which is controlled by a recessive allele. Then anesthetize the F_1 flies and place them on index cards for students to count. Ask: **Which trait is controlled by a dominant allele?** *(Lighter body color)* **How do you know?** *(None of the F_1 flies has an ebony body.)* **What color do you predict the F_2 flies will have?** *(Some will have ebony bodies, but most will have lighter bodies.)* Ask students to use symbols for alleles to write out this cross and the cross that would produce an F_2 generation.

FIGURE 4

Dominant and Recessive Alleles

Symbols serve as a shorthand way to identify alleles.

Complete each row of the diagram.

1. **Identify** Fill in the missing allele symbols and descriptions.
2. **Summarize** Use the word bank to complete the statements. (Terms will be used more than once.)
3. **Relate Cause and Effect** Draw the two possible ways the F_2 offspring could look.

Alleles in Mendel's Crosses In Mendel's cross for stem height, the purebred tall plants in the P generation had two alleles for tall stems. The purebred short plants had two alleles for short stems. But each F_1 plant inherited one allele for tall stems and one allele for short stems. The F_1 plants are called hybrids. A **hybrid** (HY brid) organism has two different alleles for a trait. All the F_1 plants are tall because the dominant allele for tall stems masks the recessive allele for short stems.

Symbols for Alleles Geneticists, scientists who study genetics, often use letters to represent alleles. A dominant allele is symbolized by a capital letter. A recessive allele is symbolized by the lowercase version of the same letter. For example, *T* stands for the allele for tall stems, and *t* stands for the allele for short stems. When a plant has two dominant alleles for tall stems, its alleles are written as *TT*. When a plant has two recessive alleles for short stems, its alleles are written as *tt*. These plants are the P generation shown in **Figure 4.** Think about the symbols that would be used for F_1 plants that all inherit one allele for tall stems and one for short stems.

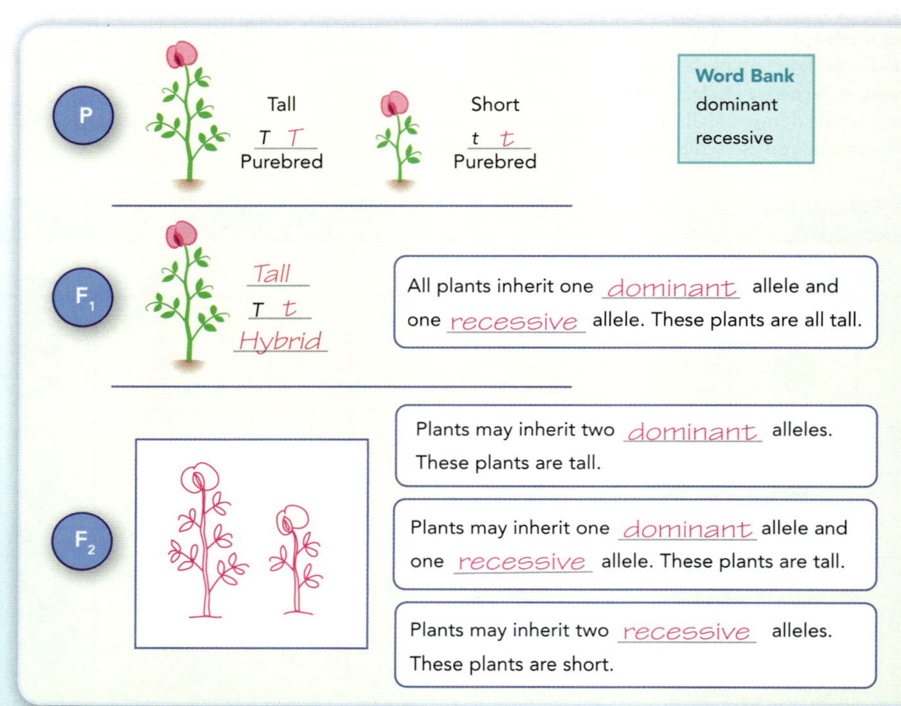

apply it!

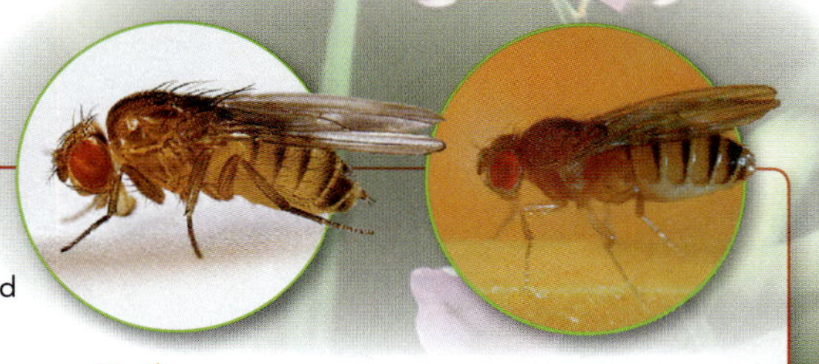

In fruit flies, long wings are dominant over short wings. A scientist crossed a purebred long-winged fruit fly with a purebred short-winged fruit fly.

1. If *W* stands for long wings, write the symbols for the alleles of each parent fly.
   WW and ww
2. **Predict** What will be the wing length of the F_1 offspring?
   All will have long wings; Ww
3. **Predict** If the scientist crosses a hybrid male F_1 fruit fly with a hybrid F_1 female, what will their offspring probably be like?
   About 3/4 will have long wings and 1/4 will have short wings.

Significance of Mendel's Contribution

Mendel's observations eventually influenced scientists' ideas about heredity. Before Mendel, most people thought that the traits of an individual organism were simply a blend of the parents' characteristics. Unfortunately, the value of Mendel's discoveries were not known during his lifetime. But when scientists in the early 1900s rediscovered Mendel's work, they quickly realized its importance. Because of his work, Mendel is often called the Father of Genetics. Mendel's results provided an understanding of the patterns in which traits are passed from one generation to the next. This knowledge can be applied to dog breeding as well as to the study of genetic diseases.

Identify Supporting Evidence What evidence showed Mendel that traits are determined by separate alleles?
The missing form of a trait reappeared in a later generation.

Do the Quick Lab *Inferring the Parent Generation.* Student Lab Manual, p. 40

Assess Your Understanding

TEKS 3D, 14A

2a. Relate Cause and Effect Why is a pea plant that is a hybrid for stem height tall?
The dominant allele for tall stems masks the recessive allele for short stems.

b. Relate How do you think that Mendel's work influenced the agriculture industry?
Sample: Farmers can grow crops with certain traits, such as a particular flavor or resistance to pests.

got it?

- ○ **I get it!** Now I know that an organism's traits are controlled by the alleles it inherits from its parents.
- ○ I need extra help with See TE note.

Differentiated Instruction

L1 Choose a Letter Students may often choose a different letter to represent each allele, such as *T* for tall and *S* for short. Make sure they understand the two alleles must be the same letter, one uppercase and one lower case, as in *T* for tall and *t* for short. The convention is to choose a letter that represents the dominant allele.

L3 Hybrid Seeds Many seeds available in stores, such as foxgloves, lupines, and tomatoes are hybrids. Some gardeners collect and save seeds from plants in their gardens to plant the next year. Ask students if this is a good idea, assuming the gardener wants the same characteristics in his or her plot. *(It is not a good idea. The offspring of a cross between hybrids will not have the same desirable traits. Only 50 percent of the offspring will be hybrids.)*

Explain

Support the TEKS

NO TRAITS GET LOST Remind students that Mendel was the first person to show that traits are inherited as discrete units that do not get "lost" or modified as they are passed from one generation to the next. Ask: **Why do some of the offspring of two hybrid individuals show the recessive form of the trait?** *(The hybrid parents each had one recessive gene. If an offspring receives both of these recessive genes, it will show the recessive form of the trait.)*

21st Century Learning

INFORMATION LITERACY Mendel published his results, but they did not attract attention right away. Although he found patterns in inheritance, he had no way to explain what his "factors" were because so little was known about how cells work in his time. Ask students to suggest the kind of information that would have helped Mendel explain his results more fully. *(Samples: chromosomes, mitosis)*

Identify Supporting Evidence Tell students that a paragraph has one main, or most important idea. Other ideas in the paragraph are details or examples that support or explain the main idea.

Apply It!

L1 Before beginning the activity, remind students that the same principles that apply to genetics of pea plants apply to fruit flies.

Predict Tell students that when they predict, they use evidence and prior experience to make an inference about a future event. A prediction is more than just a guess.

Lab Resource: Quick Lab

L3 INFERRING THE PARENT GENERATION
Students will examine an ear of colored corn to determine the alleles of the parent generation. This Quick Lab can be found in the Student Lab Manual, p. 40, and online.

Evaluate

Assess Your Understanding

After students answer the questions, have them evaluate their understanding by completing the appropriate sentence.

RTI Response to Intervention

2a., b. If students have difficulty explaining the behavior of alleles in a hybrid organism, **then** review with them the information in **Figure 4.**

Name ______________________ Date __________ Class __________

Assess Your Understanding

What Is Heredity?

What Did Mendel Observe?

1a. DEFINE What is heredity? ______________________

____

b. COMPARE AND CONTRAST In Mendel's cross for stem height, how did the plants in the F_2 generation differ from the F_1 plants? ______________________

____

got it?

○ **I get it!** Now I know that Mendel found that one form of a trait ______________________

____

____

○ **I need extra help with** ______________________

____

How Do Alleles Affect Inheritance?

2a. RELATE CAUSE AND EFFECT Why is a pea plant that is a hybrid for stem height tall? ______________________

____

b. RELATE How do you think that Mendel's work influenced the agriculture industry?

____

____

got it?

○ **I get it!** Now I know that an organism's traits are controlled by ______________________

____

○ **I need extra help with** ______________________

____

Place the outside corner, the corner away from the dotted line, in the corner of your copy machine to copy onto letter-size paper.

Place the outside corner, the corner away from the dotted line, in the corner of your copy machine to copy onto letter-size paper.

Name ______________________ Date ____________ Class ____________

Enrich

What Is Heredity?

When an organism has a trait controlled by a dominant allele, it can either be a hybrid or a purebred. To find out which, geneticists use a test cross. Read the passage and study the diagram below. In your notebook, answer the questions that follow the diagram.

The Test Cross

In a test cross, the organism with the trait controlled by a dominant allele is crossed with an organism with a trait controlled by a recessive allele. If all offspring have the trait controlled by the dominant allele, then the parent is probably a purebred. If any offspring has the recessive trait, then the dominant parent is a hybrid.

Test Cross

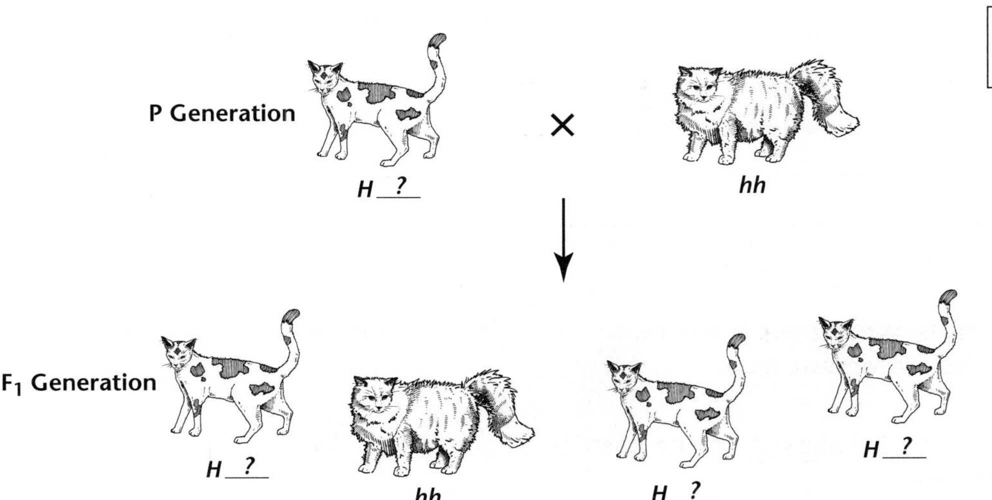

1. Is the long-haired cat in the P generation a hybrid or a purebred? Explain your answer.
2. Is the short-haired cat in the P generation a hybrid or a purebred? Explain your answer.
3. If the short-haired cat in the P generation were purebred, what would you expect the offspring to look like?
4. In horses, the allele for a black coat (*B*) is dominant over the allele for a brown coat (*b*). A cross between a black horse and a brown horse produces a brown foal. Is the black horse a hybrid or a purebred? Explain.
5. In guinea pigs, the allele for a smooth coat (*S*) is dominant over the allele for a rough coat (*s*). Explain how you could find out whether a guinea pig with a smooth coat is a hybrid or a purebred.

Name ______________________ Date ____________ Class ____________

Lesson Quiz

What Is Heredity?

Write the letter of the correct answer on the line at the left.

1. ___ Inherited traits are governed by

A hybrids

B genes

C purebreds

D fertilization

2. ___ When parent plants are crossed, scientists refer to the first generation of offspring as

A P

B F_2

C 1^F

D F_1

3. ___ A seed can be round or wrinkled. Seed shape is

A a trait

B an allele

C a factor

D a gene

4. ___ The alleles for a hybrid tall pea plant are represented as

A *TT*

B *Tt*

C *TS*

D *tt*

If the statement is true, write *true.* If the statement is false, change the underlined word or words to make the statement true.

5. Heredity ______ Fertilization is the passage of genetic instructions from one generation to the next.

6. Purebred ______ A hybrid organism is the offspring of many generations that have the same form of a trait.

7. dominant ______ Capital letters are used to represent recessive alleles.

8. Hybird ______ Mendel called an individual that has one dominant allele and one recessive allele for a trait a purebred.

9. True ______ Mendel said that the factors that control a trait exist in pairs.

10. True ______ Mendel's experiments showed that the traits of an offspring were not a blend of the characteristics of the parents.

What Is Heredity?

Answer Key

Review and Reinforce

Find the worksheet in the Student Workbook.

1. stem height or stem length

2. The two alleles are tall stems and short stems.

3. The dominant allele is tall stems because all F_1 plants are tall. The recessive allele is short stems because about $\frac{1}{4}$ of F_2 plants are short.

4. The F_1 offspring have one allele for tall stems from the tall parent, and one allele for short stems from the short parent.

5. g **6.** f

7. d **8.** e

9. c **10.** b

11. a **12.** h

13. Stem height in pea plants is controlled by one gene with two different alleles. The allele for tall stems (*T*) is dominant, while the allele for short stems (*t*) is recessive. Mendel crossed purebred tall plants (*TT*) with purebred short plants (*tt*). Because one allele is received from each parent, their offspring, the F_1 generation, all had *Tt* alleles. Because dominant genes mask recessive genes, those plants were all tall. But among their offspring, the F_2 generation, were plants with *TT, Tt,* and *tt* alleles. The plants with the double recessive *tt* alleles would be short.

Enrich

1. It is a purebred because it has two recessive alleles for long hair. If it were a hybrid, it would have both the dominant and recessive allele and have short hair because the dominant allele always masks the recessive allele.

2. The short-haired cat is a hybrid because it had offspring with long hair. The offspring receive one allele from each parent, and in order to have long hair, the offspring must receive a recessive allele from each parent.

3. All offspring would have short hair.

4. The black horse is a hybrid because it produced an offspring with the trait controlled by a recessive allele. In order to have the trait controlled by a recessive allele, the offspring must receive a recessive allele from each parent.

5. Cross the smooth-coated guinea pig with a rough-coated guinea pig. If any of the offspring have rough coats, the then smooth-coated parent is a hybrid. If all the offspring have smooth coats, then the smooth-coated parent is a purebred.

Lesson Quiz

1. B **2.** D

3. A **4.** B

5. Heredity **6.** purebred

7. dominant **8.** hybrid

9. true **10.** true

Probability and Heredity

Why don't offspring always look like their parents?

LESSON PACING:
2–3 periods or 1–1½ blocks

Lesson Vocabulary

- probability
- Punnett square
- phenotype
- genotype
- homozygous
- heterozygous

Content Refresher

Reginald C. Punnett English geneticist Reginald C. Punnett devised the Punnett square in the early 1900s. Punnett graduated from Gonville and Caius College at the University of Cambridge in 1898 with a degree in zoology and earned a master's degree from the University of St. Andrews in Scotland. He then returned to the Cambridge zoology department to research worms. While at Cambridge, Punnett worked with biologist William Bateson. Bateson had recognized that the results of his breeding experiments were explained by Mendel's principles. Bateson and Punnett were among the first to show that Mendel's principles applied to animals as well as to plants. Punnett wrote the first textbook, *Mendelism* (originally published in 1905), on the subject of Mendel's principles of genetics. In 1910, he became a professor of biology at Cambridge and eventually taught genetics there. Punnett was awarded the Darwin Medal in 1922.

Lesson Objectives	TEKS	ELPS
Describe how probability helps explain the results of genetic crosses.	2C, 14A	2.I.4
Differentiate phenotype and genotype.	14A	

Texas Essential Knowledge and Skills

2C Collect and record data using the International System of Units (SI) and qualitative means such as labeled drawings, writing, and graphic organizers.
14A Define heredity as the passage of genetic instructions from one generation to the next generation.

English Language Proficiency Standards

ELPS Listening 2.I.4 Demonstrate listening comprehension of increasingly complex spoken English by collaborating with peers commensurate with content and grade-level needs.

DIFFERENTIATED INSTRUCTION KEY
L1 Struggling Students or Special Needs
L2 On-Level Students L3 Advanced Students

LESSON PLANNER 4.2

Investigations and Activities

My Planet Diary for Texas, **Student Edition,** p. 144

Inquiry: Inquiry Warm-Up, What's the Chance?, **Lab Manual, p. 41**

Introduce Vocabulary, **Teacher's Edition,** p. 145

Lead a Discussion, Coin Tosses and Probability, **Teacher's Edition,** p. 145

Do the Math!, **Student Edition,** p. 145

Teach Key Concepts, **Teacher's Edition,** p. 146

Teach With Visuals, **Teacher's Edition,** p. 146

Make Analogies, Fertilization, **Teacher's Edition,** p. 146

Teacher to Teacher, **Teacher's Edition,** p. 146

21st Century Learning, Accountability, **Teacher's Edition,** p. 147

Inquiry: Teacher Demo, Observe Crosses in Tobacco Plants, **Teacher's Edition,** p. 147

Inquiry: Quick Lab, Coin Crosses, **PearsonTexas.com**

Teach Key Concepts, **Teacher's Edition,** p. 148

Support the TEKS, Heterozygous and Homozygous, **Teacher's Edition,** p. 148

Teach With Visuals, **Teacher's Edition,** p. 148

Apply It!, **Student Edition,** p. 149

Differentiated Instruction, **Teacher's Edition,** p. 149

Inquiry: Lab Investigation, Make the Right Call!, **Lab Manual, p. 42**

TEKS Review

Apply the TEKS, Predicting Genotypes, **Student Edition,** p. 166

TEKS Practice, **Student Edition,** p. 167

TEKS Practice: Chapter and Cumulative Review, **Student Edition,** p. 169

Lesson 4.2, **TEKS Preparation and Study Guide Workbook,** p. 38

SHORT ON TIME? To do this lesson in approximately half the time, do the Activate Prior Knowledge activity. A discussion of the Key Concepts will familiarize students with the lesson content. Have students do the Quick Lab. The rest of the lesson can be completed by students independently.

These editable worksheets are available on **PearsonTexas.com.**
Print versions can be found in the **TEKS Preparation and Study Guide Workbook.**

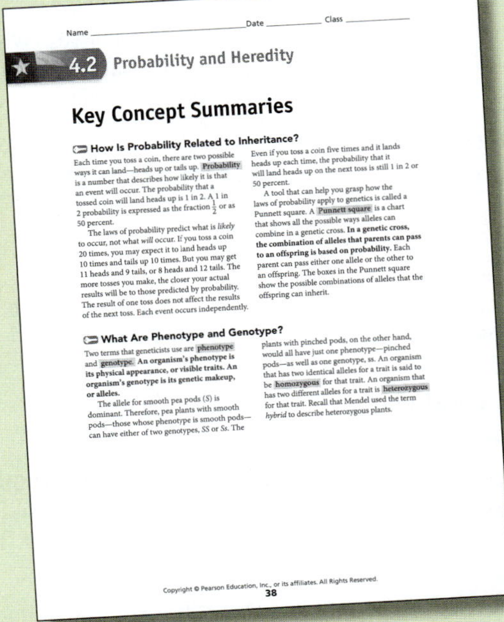

Name ________________ Date ________ Class ________

4.2 Probability and Heredity

Key Concept Summaries

How Is Probability Related to Inheritance?

Each time you toss a coin, there are two possible ways it can land—heads up or tails up. **Probability** is a number that describes how likely it is that an event will occur. The probability that a tossed coin will land heads up is 1 in 2. A 1 in 2 probability is expressed as the fraction $\frac{1}{2}$ or as 50 percent.

The laws of probability predict what is *likely* to occur, not what *will* occur. If you toss a coin 20 times, you may expect it to land heads up 10 times and tails up 10 times. But you may get 11 heads and 9 tails, or 8 heads and 12 tails. The more tosses you make, the closer your actual results will be to those predicted by probability. The result of one toss does not affect the results of the next toss. Each event occurs independently. Even if you toss a coin five times and it lands heads up each time, the probability that it will land heads up on the next toss is still 1 in 2 or 50 percent.

A tool that can help you grasp how the laws of probability apply to genetics is called a Punnett square. A **Punnett square** is a chart that shows all the possible ways alleles can combine in a genetic cross. **In a genetic cross, the combination of alleles that parents can pass to an offspring is based on probability.** Each parent can pass either one allele or the other to an offspring. The boxes in the Punnett square show the possible combinations of alleles that the offspring can inherit.

What Are Phenotype and Genotype?

Two terms that geneticists use are **phenotype** and **genotype. An organism's phenotype is its physical appearance, or visible traits. An organism's genotype is its genetic makeup, or alleles.**

The allele for smooth pea pods (*S*) is dominant. Therefore, pea plants with smooth pods—those whose phenotype is smooth pods—can have either of two genotypes, *SS* or *Ss*. The plants with pinched pods, on the other hand, would all have just one phenotype—pinched pods—as well as one genotype, ss. An organism that has two identical alleles for a trait is said to be **homozygous** for that trait. An organism that has two different alleles for a trait is **heterozygous** for that trait. Recall that Mendel used the term *hybrid* to describe heterozygous plants.

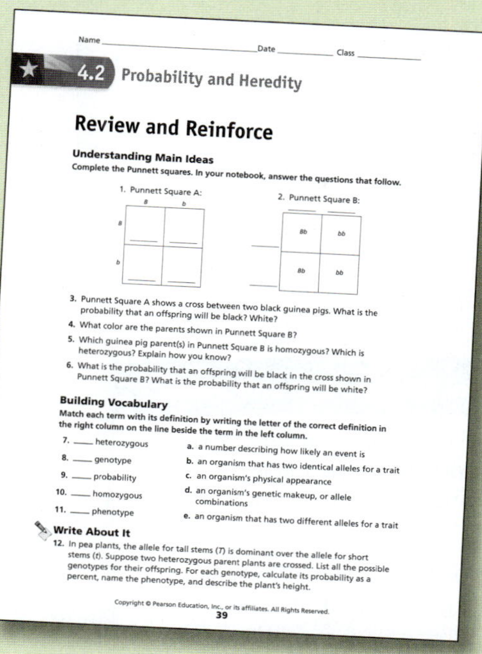

Name ________________ Date ________ Class ________

4.2 Probability and Heredity

Review and Reinforce

Understanding Main Ideas

Complete the Punnett squares. In your notebook, answer the questions that follow.

1. Punnett Square A:

	B	b
B	___	___
b	___	___

2. Punnett Square B:

	___	___
___	Bb	bb
___	Bb	bb

3. Punnett Square A shows a cross between two black guinea pigs. What is the probability that an offspring will be black? White?
4. What color are the parents shown in Punnett Square B?
5. Which guinea pig parent(s) in Punnett Square B is homozygous? Which is heterozygous? Explain how you know?
6. What is the probability that an offspring will be black in the cross shown in Punnett Square B? What is the probability that an offspring will be white?

Building Vocabulary

Match each term with its definition by writing the letter of the correct definition in the right column on the line beside the term in the left column.

7. ____ heterozygous
8. ____ genotype
9. ____ probability
10. ____ homozygous
11. ____ phenotype

a. a number describing how likely an event is
b. an organism that has two identical alleles for a trait
c. an organism's physical appearance
d. an organism's genetic makeup, or allele combinations
e. an organism that has two different alleles for a trait

Write About It

12. In pea plants, the allele for tall stems (*T*) is dominant over the allele for short stems (*t*). Suppose two heterozygous parent plants are crossed. List all the possible genotypes for their offspring. For each genotype, calculate its probability as a percent, name the phenotype, and describe the plant's height.

Lexile Measure = 910L

LESSON 4.2

Probability and Heredity

Establish Learning Objectives

After this lesson, students will be able to:

Describe how probability helps explain the results of genetic crosses.

Differentiate phenotype and genotype.

Engage

Activate Prior Knowledge

MY PLANET DIARY FOR TEXAS Read *Storm on the Way?* with the class. Ask students to share times they have followed a developing storm on the news. Students may say they worried that their family might have to evacuate to avoid a storm. Ask: **Do weather forecasters guess about the weather?** *(No; they rely on data and past experience.)* **What kind of evidence do scientists studying genetics use to make predictions?** *(Experience from crosses they have done)*

Explore

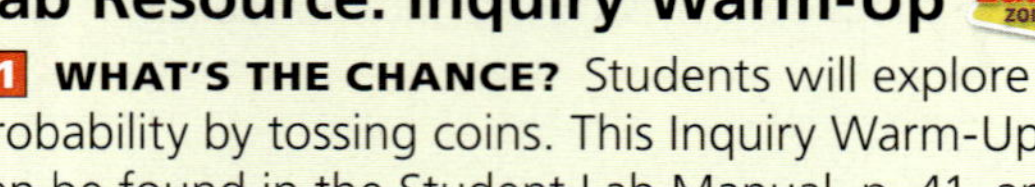

Lab Resource: Inquiry Warm-Up

L1 WHAT'S THE CHANCE? Students will explore probability by tossing coins. This Inquiry Warm-Up can be found in the Student Lab Manual, p. 41, and online.

Texas Essential Knowledge and Skills

2C Collect and record data using the International System of Units (SI) and qualitative means such as labeled drawings, writing, and graphic organizers.

14A Define heredity as the passage of genetic instructions from one generation to the next generation.

Texas LESSON 2

Probability and Heredity

How Is Probability Related to Inheritance? TEKS 2C, 14A

What Are Phenotype and Genotype? TEKS 14A

MY PLANET DIARY for TEXAS — FIELD TRIP

Storm on the Way?

The Southern Region Headquarters (SRH) of the National Weather Service is located in Fort Worth, TX. The SRH oversees weather conditions and issues forecasts for the region extending from Florida and the Caribbean Sea westward to New Mexico. This area sees the most active weather in the country. Researchers study data from aircraft, ocean buoys, and field offices to develop computer models. These models predict the probable path and strength of storms. If the probability of dangerous weather is high, the SRH issues warnings that helps save lives and reduce damage.

Communicate Answer the question below. Then discuss your answer with a partner.

Local weather forecasters often talk about the percent chance for rainfall. What do you think they mean?

Sample: They are describing how likely it is that rain is going to fall in a certain place.

Do the Inquiry Warm-Up *What's the Chance?* Student Lab Manual, p. 41

TEKS 2C, 14A In this section, you'll use Punnett squares to explore how inheritance is based on probability.

How Is Probability Related to Inheritance?

Before the start of a football game, the team captains stand with the referee for a coin toss. The team that wins the toss chooses whether to kick or receive the ball. As the referee tosses the coin, the visiting team captain calls "heads." What is the chance that the visitors will win the toss? To answer this question, you need to understand the principles of probability.

SUPPORT ALL READERS

Lexile Measure = 910L Lexile Word Count = 1302

Prior Exposure to Content: May be the first time students have encountered this topic

Academic Vocabulary: *conclusions, identify, main idea*

Science Vocabulary: *probability, phenotype, genotype*

Concept Level: May be difficult for students who struggle with math

Preteach With: My Planet Diary for Texas "Storm on the Way?" and Figure 1 activity

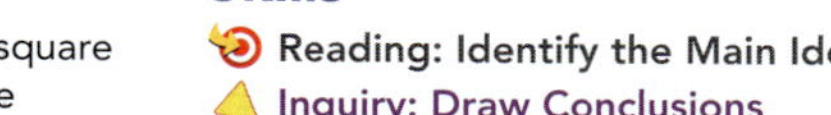

Vocabulary
- probability
- phenotype
- homozygous
- Punnett square
- genotype
- heterozygous

Skills
- Reading: Identify the Main Idea
- Inquiry: Draw Conclusions

What Is Probability?

Each time you toss a coin, there are two possible ways it can land—heads up or tails up. **Probability** is a number that describes how likely it is that an event will occur. In mathematical terms, you can say the probability that a tossed coin will land heads up is 1 in 2. There's also a 1 in 2 probability that the coin will land tails up. A 1 in 2 probability is expressed as the fraction $\frac{1}{2}$ or as 50 percent.

The laws of probability predict what is *likely* to occur, not what *will* occur. If you toss a coin 20 times, you may expect it to land heads up 10 times and tails up 10 times. But you may get 11 heads and 9 tails, or 8 heads and 12 tails. The more tosses you make, the closer your actual results will be to those predicted by probability.

Do you think the result of one toss affects the result of the next toss? Not at all. Each event occurs independently. Suppose you toss a coin five times and it lands heads up each time. What is the probability that it will land heads up on the next toss? If you said the probability is still 1 in 2, or 50 percent, you're right. The results of the first five tosses do not affect the result of the sixth toss.

do the math!

Percentage

One way to express probability is as a percentage. A percentage is a number compared to 100. For example, 50 percent, or 50%, means 50 out of 100. Suppose you want to calculate percentage from the results of a series of basketball free throws in which 3 out of 5 free throws go through the hoop.

 STEP 1 Write the comparison as a fraction.

$$3 \text{ out of } 5 = \frac{3}{5}$$

STEP 2 Calculate the number value of the fraction.

$$3 \div 5 = 0.6$$

STEP 3 Multiply this number by 100%.

$$0.6 \times 100\% = 60\%$$

Practice!

1 **Calculate** Suppose 5 out of 25 free throws go through the hoop. Write this result as a fraction.

$\frac{5}{25}$ or $\frac{1}{5}$

2 **Calculate** Express your answer in Question 1 as a percentage.

$5 \div 25 = 0.2$

$0.2 \times 100\% = 20\%$

English Language Proficiency Standards

ELPS Listening 2.I.4

Have students complete the Do the Math! activity on this page.

Beginning Have beginners listen as you read aloud the activity. Scaffold questions to help students with key concepts: *What is probability? What is a percentage?*

Intermediate Read aloud the activity with students. Have partners orally complete the activity. Provide sentence frames as needed: *5 out of 25 written as a fraction is* ______.

Advanced Have pairs work together to complete the activity. Encourage students to ask each other questions about expressing probabilities as fractions and percentages.

Advanced High Have partners complete the activity. Then have partners ask and answer questions about the probability of other simple events.

Explain

Introduce Vocabulary

Write *homozygous* and *heterozygous* on the board. Tell students that the Greek word part *homo-* means "same" and the Greek word part *hetero-* means "other" or "different." An individual that is homozygous has two identical alleles for a trait. Purebred individuals are homozygous. An individual that is heterozygous has two different alleles for a trait. Hybrid individuals are heterozygous.

Lead a Discussion

COIN TOSSES AND PROBABILITY Invite students to describe situations in which they have used a coin to decide an issue. Ask: **Why did you toss a coin in these situations?** *(Sample: It was the fairest way to make a decision.)* **Why is a coin toss fair?** *(Each person has a 50–50 chance of winning.)* Explain that Mendel used the principles of probability to explain his results. Ask: **What is probability?** *(A number that describes how likely it is that an event will occur)* **Does probability predict what will definitely occur?** *(No; it predicts what is likely to occur.)* **What does probability predict will happen if you toss a coin ten times?** *(The most likely result is that the coin will land heads up 5 times and tails up 5 times.)*

Elaborate

Do the Math!

L1 Show students how to move the decimal point two places to the right to convert a decimal to a percentage.

PEARSON Texas.com

LESSON 4.2

Explain

Teach Key Concepts

Explain to students that the alleles an offspring receives from its parents depends on probability. Ask: **How is this similar to the outcome of a coin toss?** *(The allele that each parent will pass on to its offspring is based on probability and the side of the coin that will land facing up is also based on probability.)* **What tool can be used to predict the results of a cross?** *(A Punnett square)* **What is a Punnett square?** *(A chart that shows all the possible combinations of alleles that can result from a genetic cross)*

Teach With Visuals

Tell students to look at **Figure 1** and have student volunteers read each step. Point out that combinations are pairings of the alleles from a particular row and column. Remind students that, by convention, the letter representing the dominant allele is always written first, regardless of which parent was the source of the allele. Challenge students to make a Punnett square that shows the possible offspring in a cross between a purebred pea plant with purple flowers and a purebred pea plant with white flowers. Ask: **What are the alleles of the plant with purple flowers?** *(PP)* **What are the alleles of the plant with white flowers?** *(pp)* **Where will you write these alleles?** *(Outside the box, with one parent at the top and the other parent at the side.)* Have students complete their Punnett squares. Then ask: **What alleles do the offspring have?** *(All have Pp.)* **What color flowers will these plants have?** *(All will have purple flowers.)*

Make Analogies

L1 **FERTILIZATION** Tell students that a Punnett square represents the process of reproduction. Ask: **What is fertilization?** *(The joining of two sex cells)* Point out that the letters written outside the square represent the alleles in the parents' sex cells. The letters written inside the box represent the joining of two sex cells during fertilization. Remind students that for any trait in an organism, one from each pair of alleles came from each parent.

Probability and Genetics How is probability related to genetics? Think back to Mendel's experiments. He carefully counted the offspring from every cross. When he crossed two plants that were hybrid for stem height (*Tt*), about three fourths of the F_2 plants had tall stems. About one fourth had short stems.

Each time Mendel repeated the cross, he observed similar results. He realized that the principles of probability applied to his work. He found that the probability of a hybrid cross producing a tall plant was 3 in 4. The probability of producing a short plant was 1 in 4. Mendel was the first scientist to recognize that the principles of probability can predict the results of genetic crosses.

Punnett Squares

A tool that can help you grasp how the laws of probability apply to genetics is called a Punnett square. A **Punnett square** is a chart that shows all the possible ways alleles can combine in a genetic cross. Geneticists use Punnett squares to see these combinations and to determine the probability of a particular outcome, or result. **In a genetic cross, the combination of alleles that parents can pass to an offspring is based on probability.**

Figure 1 shows how to make a Punnett square. In this case, the cross is between two hybrid pea plants with round seeds (*Rr*). The allele for round seeds (*R*) is dominant over the allele for wrinkled seeds (*r*). Each parent can pass either one allele or the other to an offspring. The boxes in the Punnett square show the possible combinations of alleles that the offspring can inherit.

FIGURE 1

How to Make a Punnett Square

You can use a Punnett square to find the probabilities of a genetic cross.

Follow the steps in the figure to fill in the Punnett square.

1. **Predict** What is the probability that an offspring will have wrinkled seeds?
   1/4 or 25%

2. **Interpret Tables** What is the probability that an offspring will have round seeds? Explain your answer.
   3/4 or 75%; 1/4 or 25% round seeds RR, plus 1/2 or 50% round seeds Rr

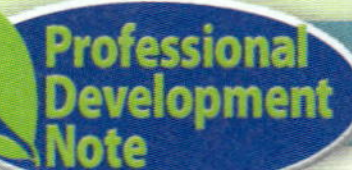

1 Start by drawing a box and dividing it into four squares.

2 The male parent's alleles are written along the top of the square. Fill in the female parent's alleles along the left side.

Professional Development Note

Teacher to Teacher

Activity Genetics was always my favorite area to teach. The students especially had a blast when we did baby simulations. Bunny Babies can be created by flipping coins and using heads or tails to represent male or female gametes and dominant or recessive alleles for specific traits. Have your students flip a different coin for each trait to determine the genotype. For example, heads of a quarter might be dominant *(E)* for long ears, and tails recessive (e) for short ears. Other traits can include oval head/round head and large tail/small tail. Students can use construction paper, glue, and craft sticks to create the bunny babies.

Treva Jeffries
Scott High School
Toledo, Ohio

3 Copy the female parent's alleles into the boxes to their right. The first one is done for you.

	R	r
R	R	R
r	r	r

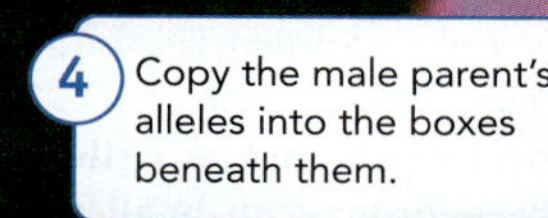

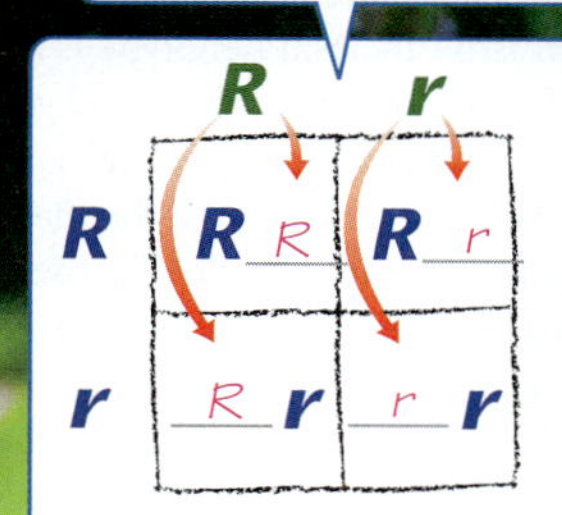

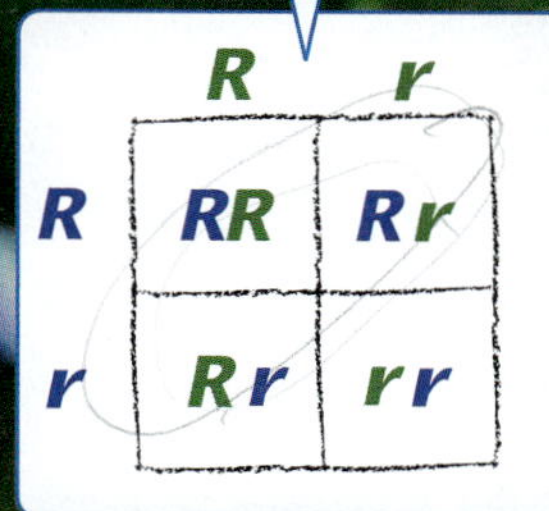

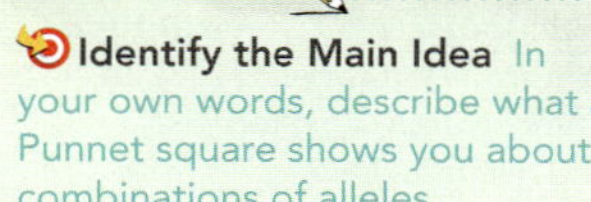

Identify the Main Idea In your own words, describe what a Punnet square shows you about combinations of alleles.

A Punnett square shows all the possible ways the alleles from parents can combine in their offspring.

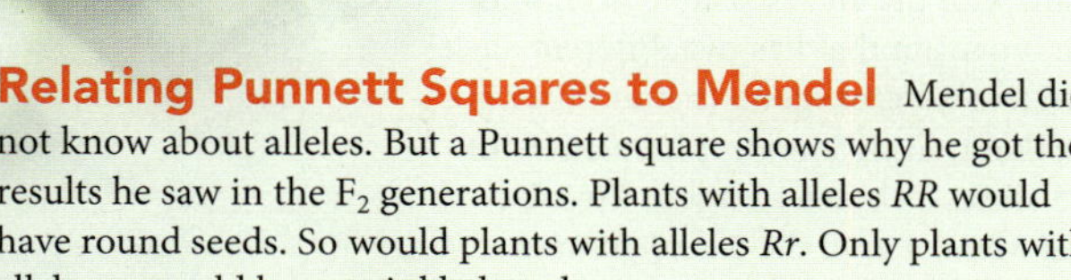

Relating Punnett Squares to Mendel Mendel did not know about alleles. But a Punnett square shows why he got the results he saw in the F_2 generations. Plants with alleles *RR* would have round seeds. So would plants with alleles *Rr*. Only plants with alleles *rr* would have wrinkled seeds.

Do the Quick Lab *Coin Crosses*. Find the lab online.

Assess Your Understanding

TEKS 14A

1a. Review How is probablity related to inheritance?

It predicts how likely a certain trait is to appear in offspring.

b. Compare and Contrast A hybrid pea plant with round seeds is crossed with a plant with wrinkled seeds. Which trait is more likely to appear in offspring—wrinkled or round seeds? Explain.

Each trait is equally likely at 50%.

got it?

- **I get it!** Now I know that the combination of alleles parents can pass to offspring can be determined using probability.
- I need extra help with See TE note.

Differentiated Instruction

L1 Percentages If students have difficulty understanding that 75 percent of the seeds are expected to be round, tell them that each square of the box represents one-fourth, or 25 percent, of the offspring. Because *Rr* appears twice, its 25 percent must be counted twice. So 25 percent *Rr* plus 25 percent *Rr* plus 25 percent *RR* totals 75 percent round seeds.

L3 Reginald Punnett Tell students that the Punnett square was developed by Reginald Punnett, a geneticist. Encourage interested students to research Punnett's work as a teacher and a geneticist. Students can present their findings in short class presentations.

Elaborate

21st Century Learning

ACCOUNTABILITY Have students check their understanding by drawing a Punnett square for a cross between two hybrids (*Aa* × *Aa*). Have students work in pairs to explain the steps in completing a Punnett square and describing the offspring.

Identify the Main Idea Tell students that a paragraph may contain many ideas. The broadest, or most important, is the main idea.

Teacher Demo

L1 OBERVE CROSSES IN TOBACCO PLANTS

Materials F_2 tobacco seeds that produce green (*GG, Gg*) and albino (*gg*) seedlings in a ratio of 3:1, seed starting soil, shallow pan, plastic wrap, water.

Advance Preparation Order the seeds from a science supply house, and plant them 7 to 14 days in advance, or as directed on the seed packet.

Time 15 minutes

Explain that a cross between two hybrid green plants (*Gg, Gg*) produces both green and white offspring. Draw a Punnett square of this cross; 75 percent of the seedlings are likely to be green and 25 percent are likely to be white. Have students count the green and white F_2 seedlings and compare those numbers to predictions.

Ask: **Why might the numbers not match?** *(The Punnett square predicts probability. The actual combinations may be close, but not exact.)*

Lab Resource: Quick Lab

L1 COIN CROSSES Students will explore how alleles combine. This Quick Lab can be found online.

Evaluate

Assess Your Understanding

After students answer the questions, have them evaluate their understanding by completing the appropriate sentence.

RTI Response to Intervention

1a. If students need help explaining how probability is related to heredity, **then** have them reread the Key Concept statement.

b. If students have trouble answering the question, **then** help them complete a Punnett square for the cross.

LESSON 4.2

Explain

Teach Key Concepts

Tell students that an organism can be described by its phenotype or genotype. Emphasize that an organism's phenotype is the result, or expression, of its genotype. Ask: **What is an organism's genotype?** *(Its genetic makeup, or alleles)* **What is an organism's phenotype?** *(Its physical appearance; what it looks like)* **Which of these can you tell by looking at an organism?** *(Its phenotype)*

Support the TEKS

HETEROZYGOUS AND HOMOZYGOUS Review that an individual can have two dominant alleles for a trait, or two recessive alleles for that trait, or one dominant allele and one recessive allele. Ask: **What is the term used to describe an organism whose genotype consists of two identical alleles for a trait?** *(Homozygous)* **What are the two ways an individual can be homozygous for a trait?** *(Homozygous dominant or homozygous recessive)* **What is the term used to describe an organism whose genotype consists of two different alleles for a trait?** *(Heterozygous)* **What term did Mendel use for an individual that was heterozygous?** *(Hybrid)* **Why can you be certain of the genotype of an organism that shows a recessive trait?** *(It must have a homozygous recessive genotype because the recessive allele is not hidden by a dominant allele.)*

Teach With Visuals

Tell students to look at **Figure 2.** Ask: **How do the two smooth pods differ?** *(They don't differ at all.)* **Can you tell the genotypes of the plants that produced these pods?** *(No)* Point out that the table should only show two different phenotypes for shape and then three different genotypes. Students should recognize that they could list *SS* or *Ss* first, since the two smooth pods could be either genotype. Ask: **Why can you be certain of the genotype of the plant that produced the pinched pod?** *(Pinched pod is recessive, so the only way this form of the trait could be seen is if the plant is homozygous,* ss.*)*

Texas Essential Knowledge and Skills

14A Define heredity as the passage of genetic instructions from one generation to the next generation.

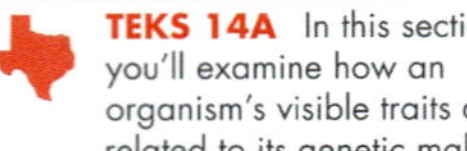

TEKS 14A In this section, you'll examine how an organism's visible traits are related to its genetic makeup.

What Are Phenotype and Genotype?

Two terms that geneticists use are **phenotype** (FEE noh typ) and **genotype** (JEE noh typ). **An organism's phenotype is its physical appearance, or visible traits. An organism's genotype is its genetic makeup, or alleles.** In other words, genotype is an organism's alleles. Phenotype is how a trait looks or is expressed.

To compare phenotype and genotype, look at **Figure 2.** The allele for smooth pea pods (*S*) is dominant over the allele for pinched pea pods (*s*). All the plants with at least one *S* allele have the same phenotype. That is, they all produce smooth pods. However, these plants can have two different genotypes—*SS* or *Ss*. If you were to look at the plants with smooth pods, you would not be able to tell the difference between those that have the genotype *SS* and those with the genotype *Ss*. The plants with pinched pods, on the other hand, would all have the same phenotype—pinched pods—as well as the same genotype—*ss*.

Geneticists use two additional terms to describe an organism's genotype. An organism that has two identical alleles for a trait is said to be **homozygous** (hoh moh ZY gus) for that trait. A smooth-pod plant that has the alleles *SS* and a pinched-pod plant with the alleles *ss* are both homozygous. An organism that has two different alleles for a trait is **heterozygous** (het ur oh ZY gus) for that trait. A smooth-pod plant with the alleles *Ss* is heterozygous. Recall that Mendel used the term *hybrid* to describe heterozygous pea plants.

Vocabulary Suffixes The suffix *-ous* means "having." Circle this suffix in the highlighted terms *homozygous* and *heterozygous* in the paragraph at the right. These terms describe the organism as having identical or different alleles.

FIGURE 2

Describing Inheritance

An organism's phenotype is its physical appearance. Its genotype is its genetic makeup.

Based on what you have read, answer these questions.

1. **Classify** Fill in the missing information in the table.
2. **Interpret Tables** How many genotypes are there for the smooth-pod phenotype? Two

Phenotypes and Genotypes

Phenotype	Genotype	Homozygous or Heterozygous
Smooth pods	SS	Homozygous
Smooth pods	Ss	Heterozygous
Pinched pods	ss	Homozygous

apply it!

Mendel's principles of heredity apply to many other organisms. For example, in guinea pigs, black fur color (*B*) is dominant over white fur color (*b*). Suppose a pair of black guinea pigs produces several litters of pups during their lifetimes. The graph shows the phenotypes of the pups. Write a title for the graph.

1 Read Graphs How many black pups were produced? How many white pups were produced?

About 65 black pups; 15 white pups

2 Infer What are the possible genotypes of the offspring?

Black pups: *BB* or *Bb*; white pups: *bb*

3 Draw Conclusions What can you conclude about the genotypes of the parent guinea pigs? Explain your answer.

Both parents must be *Bb*. To have black fur, each parent must have one *B* allele. Each also must have one *b* allele to produce white pups.

Lab zone Do the Lab Investigation *Make the Right Call!* Student Lab Manual, p. 42

Assess Your Understanding

TEKS 14A

2a. Relate Cause and Effect Explain how two organisms can have the same phenotype but different genotypes.

One organism is heterozygous and one is homozygous for a dominant allele.

b. CHALLENGE In their lifetimes, two guinea pigs produce 40 black pups and 40 white pups. On a separate paper, make a Punnett square and find the likely genotypes of these parents.

One parent is probably *Bb* and the other is probably *bb*.

got it?

○ **I get it!** Now I know that phenotype and genotype are terms that describe an organism's physical appearance and its alleles (or genetic makeup).

○ I need extra help with See TE note.

Elaborate

Apply It!

L1 Before beginning the activity, explain that students need to consider each piece of evidence—the significance of each color of the offspring as well as the relative numbers—before drawing a conclusion about the parents.

Draw Conclusions Explain to students that a conclusion is a statement that sums up what has been learned from an experiment, explanation, diagram, or data. Tell students that they are drawing conclusions when they identify the parents because they are evaluating data about the offspring of these parents.

Lab Resource: Lab Investigation 

L2 MAKE THE RIGHT CALL! Students will predict the results of genetic crosses. This Lab Investigation can be found in the Student Lab Manual, p. 42, and online.

Evaluate

Assess Your Understanding

After students answer the questions, have them evaluate their understanding by completing the appropriate sentence.

RTI Response to Intervention

2a. If students cannot distinguish between genotype and phenotype, **then** have them locate and reread the Key Concept statement for this section.

b. If students have trouble drawing a Punnett square, **then** review the steps for writing alleles outside and then inside the boxes.

Differentiated Instruction

L1 Sample Size Students may notice that the number of guinea pigs in the *Apply It!* activity does not come out to 75 percent black pups and 25 percent white pups. Remind them that in probability, the larger the sample size, the more likely the results will match what is expected. In this case, there are 81 percent black pups and 19 percent white pups, which is close to the expected values.

L3 Sample Problems Challenge students to write their own genetics problems, using fictional plants and animals. Students can work in pairs to solve each other's problems. Students should ensure that their problems can be solved based on the information they provide.

Name ______________________ Date __________ Class __________

Assess Your Understanding

Probability and Heredity

How Is Probability Related to Inheritance?

1a. **REVIEW** How is probability related to inheritance? ______________________

____

b. **COMPARE AND CONTRAST** A hybrid pea plant with round seeds is crossed with a plant with wrinkled seeds. Which trait is more likely to appear in offspring—wrinkled or round seeds? Explain. ______________________

____

got*it*?

○ **I get it!** Now I know that the combination of alleles parents can pass to offspring ______________________

____

○ **I need extra help with** ______________________

____

What Are Phenotype and Genotype?

2a. **RELATE CAUSE AND EFFECT** Explain how two organisms can have the same phenotype but different genotypes. ______________________

____

____

b. **CHALLENGE** In their lifetimes, two guinea pigs produce 40 black pups and 40 white pups. On a separate paper, make a Punnett square and find the likely genotypes of these parents. ______________________

____

got*it*?

○ **I get it!** Now I know that phenotype and genotype are terms that describe ______________________

____

____

○ **I need extra help with** ______________________

____

Place the outside corner, the corner away from the dotted line, in the corner of your copy machine to copy onto letter-size paper.

Name ____________________ Date __________ Class __________

Enrich

Probability and Heredity

Read the passage and study the diagram to its right. Then use a separate sheet of paper to answer the questions that follow.

Genetic Crosses With Two Traits

In his work with garden peas, Mendel also set up crosses in which he studied the inheritance of two traits at one time. For example, he crossed tall plants having green pods (*TTGG*) with short plants having yellow pods (*ttgg*). The F1 offspring showed both traits controlled by dominant alleles, tall and green. Mendel allowed the F_1 offspring to self-pollinate. The F_2 offspring had four different phenotypes: tall plants with green pods, tall plants with yellow pods, short plants with green pods, and short plants with yellow pods. These results led Mendel to formulate the Law of Independent Assortment, which states that alleles of one gene separate or assort independently of alleles of another gene. In other words, the distribution of one gene does not affect the distribution of alleles for another gene.

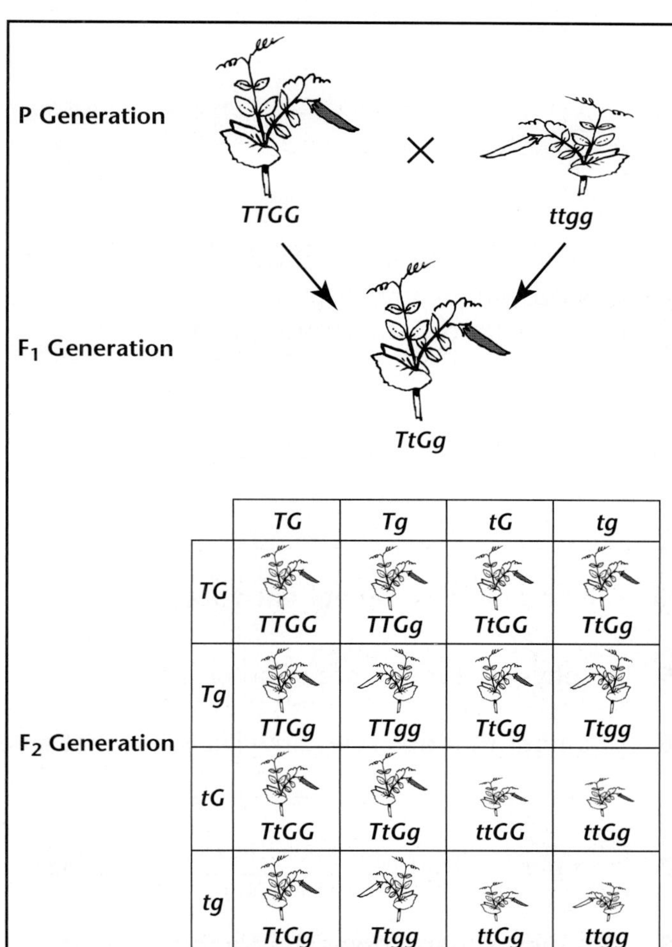

1. What are the possible combinations of alleles that each F_1 parent can pass on to the offspring?
2. What are the possible genotypes of the F_2 offspring? What are the possible phenotypes of the F_2 offspring?
3. What is the probability that an F_2 offspring will be tall with green pods? What is the probability that an F_2 offspring will be short with yellow pods?

Name ______________________ Date ____________ Class ____________

Lesson Quiz

Probability and Heredity

Write the letter of the correct answer on the line at the left.

1. ___ Which of these genotypes is heterozygous?

A *AA*

B *Bb*

C *Cd*

D *ee*

2. ___ Which of these is NOT a phenotype?

A tall

B short

C homozygous

D round

3. ___ In a cross between individuals that are *Aa* × *Aa,* how many boxes of the Punnett square will show an offspring that is *AA?*

A 1

B 2

C 3

D 4

4. ___ Which of these is NOT a way to express probability?

A 1 in 4

B 50 percent

C $\frac{3}{4}$

D 25

Fill in the blank to complete each statement.

5. The physical appearance of an organism is its ____________________.

6. A number that describes how likely it is that an event will occur is the ____________________ of the event.

7. An organism that is ____________________ has two identical alleles for a trait.

8. A Punnett square shows the combination of ____________________ that parents can pass on to offspring.

9. The genetic makeup of an organism is its ____________________.

10. An organism that is ____________________ has two different alleles for a trait.

Place the outside corner, the corner away from the dotted line, in the corner of your copy machine to copy onto letter-size paper.

Probability and Heredity

Answer Key

Review and Reinforce

Find the worksheet in the Student Workbook.

1. first row: *BB, Bb;* second row: *Bb, bb*
2. top: *B, b;* side: *b, b*
3. The probability of an offspring being black is 3 in 4, $\frac{3}{4}$, or 75 percent. The probability of an offspring being white is 1 in 4, $\frac{1}{4}$, or 25 percent.
4. One guinea pig parent is white, and the other is black.
5. The white guinea pig is homozygous (*bb*), and the black guinea pig is heterozygous (*Bb*). If the black guinea pig parent were homozygous (*BB*), then there would be no chance for the offspring to be white; they would all be heterozygous black.
6. The probability of an offspring being black is 2 in 4, which is $\frac{1}{2}$, or 50 percent. The probability of an offspring being white is 2 in 4, which is $\frac{1}{2}$, or 50 percent.
7. e
8. d
9. a
10. b
11. c
12. *TT* genotype: 25%; tall-stem phenotype; tall stemmed
    Tt genotype: 50%; tall-stem phenotype; tall stemmed
    tt genotype: 25%; short-stem phenotype; short stemmed

Enrich

1. All possible allele combinations that each parent can pass on to the offspring are *TG, Tg, tG,* and *tg.*
2. The possible genotypes are *TTGG, TTGg, TtGG, TtGg, TTgg, Ttgg, ttGG, ttGg, or ttgg.* The possible phenotypes are tall with green pods, tall with yellow pods, short with green pods, or short with yellow pods.
3. The probability of tall pants with green pods is 9 out of 16, or about 56 percent. The probability of short plants with yellow pods is 1 out of 16, or about 6 percent.

Lesson Quiz

1. B
2. C
3. A
4. D
5. phenotype
6. probability
7. homozygous
8. alleles
9. genotype
10. heterozygous

Patterns of Inheritance

Why don't offspring always look like their parents?

LESSON PACING:
2–3 periods or 1–1½ blocks

Lesson Vocabulary

- incomplete dominance
- codominance
- multiple alleles
- polygenic inheritance

Content Refresher

Lethal Genotypes Some matings repeatedly produce offspring in a ratio that does not match standard ratios of 3:1 (*Aa* × *Aa*) or 1:1 (*Aa* × *aa*). Fruit flies have a gene that produces curly wings. When a curly-winged fly is crossed with a fly that has normal wings, the offspring are always found in a 1:1 ratio, indicative of parents that are *Cc* × *cc*. When two curly-winged flies are crossed, the offspring are always found in a ratio of two curly-winged flies to one normal-winged fly. A cross between *Cc* × *Cc* is expected to produce a 3:1 ratio, not 2:1. The genotype ratio produced by a cross between *Cc* × *Cc* is 1*CC*:2*Cc*:1*cc*. If the homozygous dominant, curly-winged flies are removed from this ratio, the resulting 2*Cc*:1*cc* ratio matches the phenotype ratio of two curly to one normal. The homozygous dominant genotype is lethal and offspring with this genotype never hatch.

Lesson Objectives	TEKS	ELPS
Describe at least three complex patterns of inheritance.	14A	4.G.3
Discuss how characteristics result from inheritance and environmental factors.	14A	

Texas Essential Knowledge and Skills

14A Define heredity as the passage of genetic instructions from one generation to the next generation.

English Language Proficiency Standards

ELPS Reading 4.G.3 Demonstrate comprehension of increasingly complex English by responding to questions commensurate with content area and grade level needs.

DIFFERENTIATED INSTRUCTION KEY
L1 Struggling Students or Special Needs
L2 On-Level Students L3 Advanced Students

LESSON PLANNER 4.3

Investigations and Activities	TEKS Review
My Planet Diary for Texas, **Student Edition,** p. 150 Inquiry: Inquiry Warm-Up, Observing Traits, **PearsonTexas.com** Introduce Vocabulary, **Teacher's Edition,** p. 151 Teach Key Concepts, **Teacher's Edition,** p. 151 Teach With Visuals, **Teacher's Edition,** p. 151 Lead a Discussion, Multiple Alleles, **Teacher's Edition,** p. 152 Lead a Discussion, Polygenic Inheritance, **Teacher's Edition,** p. 152 Apply It!, **Student Edition,** p. 152 Inquiry: Quick Lab, Patterns of Inheritance, **PearsonTexas.com**	Apply the TEKS, Patterns of Inheritance, **Student Edition,** p. 154 Apply the TEKS, Predicting Genotypes, **Student Edition,** p. 166 TEKS Practice, **Student Edition,** p. 167 TEKS Practice: Chapter and Cumulative Review, **Student Edition,** p. 169 Lesson 4.3, **TEKS Preparation and Study Guide Workbook,** p. 40
Lead a Discussion, Acquired Traits, **Teacher's Edition,** p. 153 21st Century Learning, Interpersonal Skills, **Teacher's Edition,** p. 153 21st Century Learning, Accountability, **Teacher's Edition,** p. 153 Differentiated Instruction, **Teacher's Edition,** p. 153 Teach Key Concepts, **Teacher's Edition,** p. 154 Differentiated Instruction, **Teacher's Edition,** p. 155 Inquiry: Quick Lab, Is It All in the Genes?, **Lab Manual,** p. 47	**SHORT ON TIME?** To do this lesson in approximately half the time, do the Activate Prior Knowledge activity. A discussion of the Key Concepts will familiarize students with the lesson content. Use the Apply the TEKS activity to help students understand why offspring don't always look like their parents. Have students do the Quick Labs. The rest of the lesson can be completed by students independently.

These editable worksheets are available on **PearsonTexas.com.**
Print versions can be found in the **TEKS Preparation and Study Guide Workbook.**

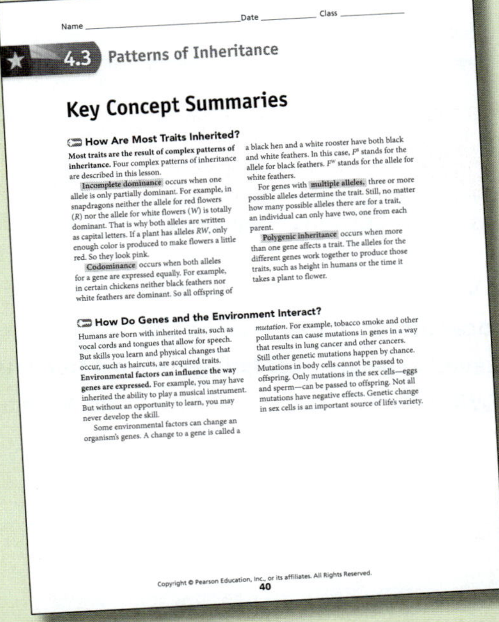

Name ____________ Date ________ Class ________

4.3 Patterns of Inheritance

Key Concept Summaries

How Are Most Traits Inherited?

Most traits are the result of complex patterns of inheritance. Four complex patterns of inheritance are described in this lesson.

Incomplete dominance occurs when one allele is only partially dominant. For example, in snapdragons neither the allele for red flowers (*R*) nor the allele for white flowers (*W*) is totally dominant. That is why both alleles are written as capital letters. If a plant has alleles *RW*, only enough color is produced to make flowers a little red. So they look pink.

Codominance occurs when both alleles for a gene are expressed equally. For example, in certain chickens neither black feathers nor white feathers are dominant. So all offspring of a black hen and a white rooster have both black and white feathers. In this case, F^B stands for the allele for black feathers. F^W stands for the allele for white feathers.

For genes with **multiple alleles,** three or more possible alleles determine the trait. Still, no matter how many possible alleles there are for a trait, an individual can only have two, one from each parent.

Polygenic inheritance occurs when more than one gene affects a trait. The alleles for the different genes work together to produce those traits, such as height in humans or the time it takes a plant to flower.

How Do Genes and the Environment Interact?

Humans are born with inherited traits, such as vocal cords and tongues that allow for speech. But skills you learn and physical changes that occur, such as haircuts, are acquired traits. **Environmental factors can influence the way genes are expressed.** For example, you may have inherited the ability to play a musical instrument. But without an opportunity to learn, you may never develop the skill.

Some environmental factors can change an organism's genes. A change to a gene is called a *mutation*. For example, tobacco smoke and other pollutants can cause mutations in genes in a way that results in lung cancer and other cancers. Still other genetic mutations happen by chance. Mutations in body cells cannot be passed to offspring. Only mutations in the sex cells—eggs and sperm—can be passed to offspring. Not all mutations have negative effects. Genetic change in sex cells is an important source of life's variety.

40

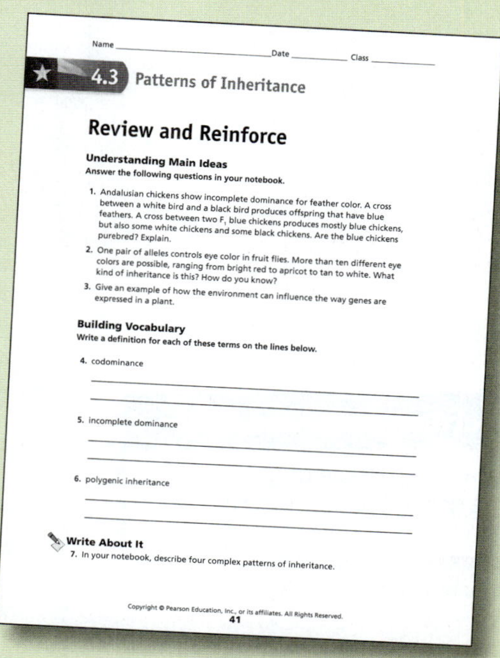

Name ____________ Date ________ Class ________

4.3 Patterns of Inheritance

Review and Reinforce

Understanding Main Ideas
Answer the following questions in your notebook.

1. Andalusian chickens show incomplete dominance for feather color. A cross between a white bird and a black bird produces offspring that have blue feathers. A cross between two F_1 blue chickens produces mostly blue chickens, but also some white chickens and some black chickens. Are the blue chickens purebred? Explain.
2. One pair of alleles controls eye color in fruit flies. More than ten different eye colors are possible, ranging from bright red to apricot to tan to white. What kind of inheritance is this? How do you know?
3. Give an example of how the environment can influence the way genes are expressed in a plant.

Building Vocabulary
Write a definition for each of these terms on the lines below.

4. codominance
5. incomplete dominance
6. polygenic inheritance

Write About It

7. In your notebook, describe four complex patterns of inheritance.

41

Lexile Measure = 850L

LESSON 4.3

Patterns of Inheritance

Establish Learning Objectives

After this lesson, students will be able to:

Describe at least three complex patterns of inheritance.

Discuss how characteristics result from inheritance and environmental factors.

Engage

Activate Prior Knowledge

MY PLANET DIARY FOR TEXAS Read *Cold, With a Chance of Males* with the class. Explain that laws of probability apply to situations that are not affected by anything other than chance. For many animal species, the probability of offspring being male or female is 50 percent because the outcome is a matter of chance. But in turtles and other reptiles, the environment is the deciding factor, not probability.

Explore

Lab Resource: Inquiry Warm-Up

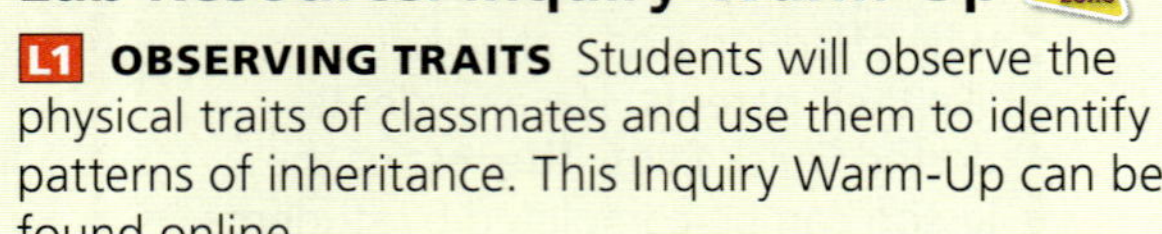

L1 OBSERVING TRAITS Students will observe the physical traits of classmates and use them to identify patterns of inheritance. This Inquiry Warm-Up can be found online.

Texas Essential Knowledge and Skills

14A Define heredity as the passage of genetic instructions from one generation to the next generation.

Texas LESSON 3

Patterns of Inheritance

How Are Most Traits Inherited? TEKS 14A

How Do Genes and the Environment Interact? TEKS 14A

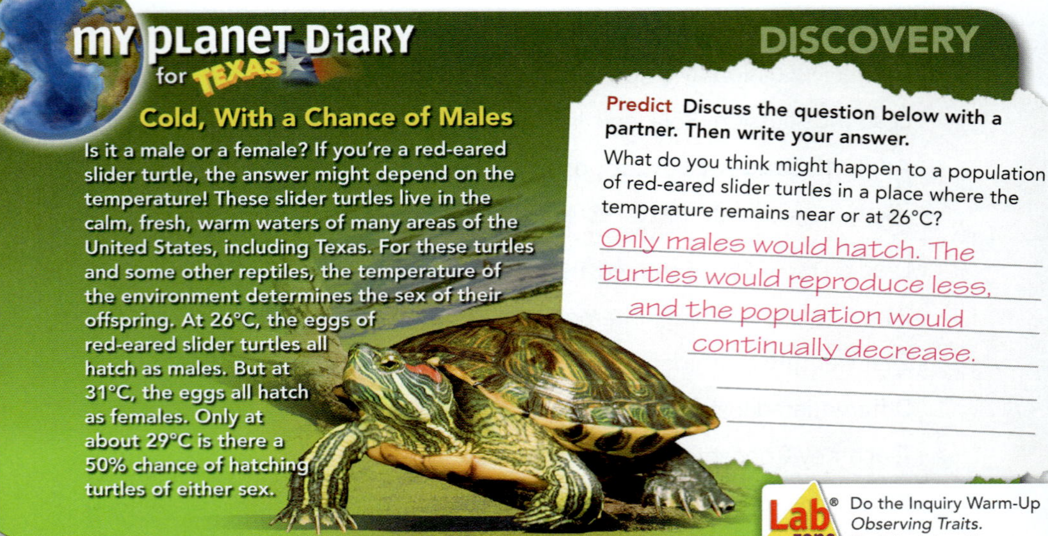

MY PLANET DIARY for TEXAS — DISCOVERY

Cold, With a Chance of Males

Is it a male or a female? If you're a red-eared slider turtle, the answer might depend on the temperature! These slider turtles live in the calm, fresh, warm waters of many areas of the United States, including Texas. For these turtles and some other reptiles, the temperature of the environment determines the sex of their offspring. At 26°C, the eggs of red-eared slider turtles all hatch as males. But at 31°C, the eggs all hatch as females. Only at about 29°C is there a 50% chance of hatching turtles of either sex.

Predict Discuss the question below with a partner. Then write your answer.

What do you think might happen to a population of red-eared slider turtles in a place where the temperature remains near or at 26°C?

Only males would hatch. The turtles would reproduce less, and the population would continually decrease.

Do the Inquiry Warm-Up *Observing Traits.* Find the lab online.

TEKS 14A In this section, you'll learn how complex patterns of inheritance affect the way traits are expressed in offspring.

ELPS 4.G.3

Read with a partner the Key Questions at the top of the page. As you read the lesson together, work out some possible responses to each Key Question and take notes on your ideas. Then, pair up with another group of two. Ask one another the Key Questions. Respond to the questions, using your notes.

How Are Most Traits Inherited?

The traits that Mendel studied are controlled by genes with only two possible alleles. These alleles are either dominant or recessive. Pea flower color is either purple or white. Peas are either yellow or green. Can you imagine if all traits were like this? If people were either short or tall? If cats were either black or yellow?

Studying two-allele traits is a good place to begin learning about genetics. But take a look around at the variety of living things in your surroundings. As you might guess, most traits do not follow such a simple pattern of inheritance. **Most traits are the result of complex patterns of inheritance.** Four complex patterns of inheritance are described in this lesson.

SUPPORT ALL READERS

Lexile Measure = 850L **Lexile Word Count = 1094**

Prior Exposure to Content: May be the first time students have encountered this topic

Academic Vocabulary: *compare, contrast, data, interpret*

Science Vocabulary: *incomplete dominance, codominance*

Concept Level: Generally appropriate for most students in this grade

Preteach With: My Planet Diary for Texas "Cold, With a Chance of Males" and Figure 1 activity

Vocabulary
- incomplete dominance
- codominance
- multiple alleles
- polygenic inheritance

Skills
- Reading: Compare and Contrast
- Inquiry: Interpret Data

Incomplete Dominance Some traits result from a pattern of inheritance known as incomplete dominance. **Incomplete dominance** occurs when one allele is only partially dominant. For example, look at **Figure 1.** The flowers shown are called snapdragons. A cross between a plant with red flowers and one with white flowers produces pink offspring.

Snapdragons with alleles *RR* produce a lot of red color in their flowers. It's no surprise that their flowers are red. A plant with two white alleles (*WW*) produces no red color. Its flowers are white. Both types of alleles are written as capital letters because neither is totally dominant. If a plant has alleles *RW*, only enough color is produced to make the flowers just a little red. So they look pink.

Codominance The chickens in **Figure 1** show a different pattern of inheritance. **Codominance** occurs when both alleles for a gene are expressed equally. In the chickens shown, neither black feathers nor white feathers are dominant. All the offspring of a black hen and a white rooster have both black and white feathers.

Here, F^B stands for the allele for black feathers. F^W stands for the allele for white feathers. The letter *F* tells you the trait is feathers. The superscripts *B* for black and *W* for white tell you the color.

FIGURE 1

Other Patterns of Inheritance

Many crosses do not follow the patterns Mendel discovered.

Apply Concepts **Fill in the missing pairs of alleles.**

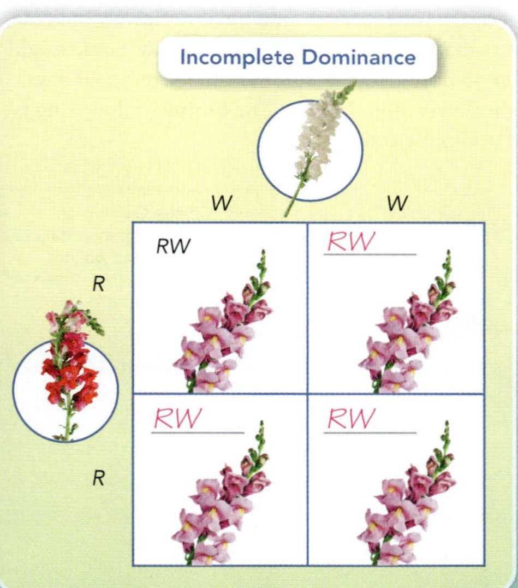

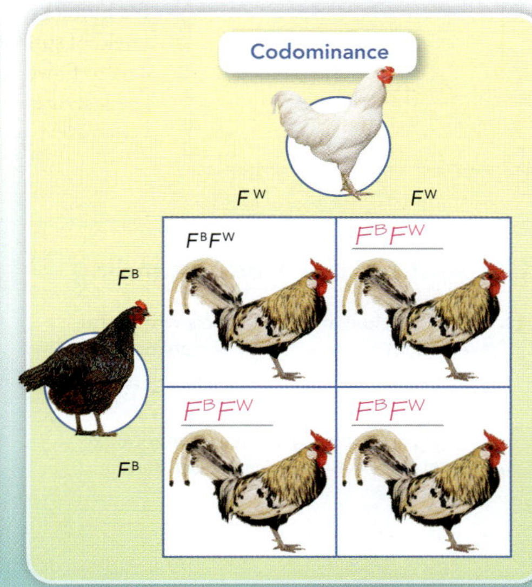

Explain

Introduce Vocabulary

Point out the term *polygenic inheritance.* Students may be familiar with the prefix *poly-*, which means "many." Tell them that polygenic inheritance involves more than one pair of alleles, or "many genes."

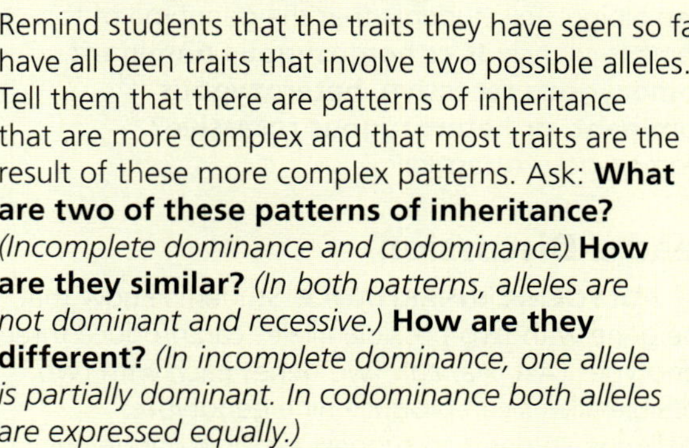

Teach Key Concepts

Remind students that the traits they have seen so far have all been traits that involve two possible alleles. Tell them that there are patterns of inheritance that are more complex and that most traits are the result of these more complex patterns. Ask: **What are two of these patterns of inheritance?** *(Incomplete dominance and codominance)* **How are they similar?** *(In both patterns, alleles are not dominant and recessive.)* **How are they different?** *(In incomplete dominance, one allele is partially dominant. In codominance both alleles are expressed equally.)*

Teach With Visuals

Tell students to look at **Figure 1.** Ask: **Which alleles are expressed in these offspring?** *(Both)* **Which organism shows incomplete dominance?** *(Snapdragons)* **What is incomplete dominance?** *(A pattern of inheritance in which one allele is only partially dominant)* Students may point out that some of the flowers look darker than others. Tell them that it is because the darker flowers are not completely open, which makes them look darker than they really are. **Which organism shows codominance?** *(Chickens)* **What is codominance?** *(A pattern of inheritance in which alleles are neither dominant nor recessive)*

PEARSON Texas.com

LESSON 4.3

English Language Proficiency Standards

ELPS Reading 4.G.3

Have partners confirm understanding by asking and answering the Key Questions on page 150.

Beginning Read aloud the questions. Demonstrate how to use sentence frames to answer the questions: *Traits are inherited through _______. The environment affects genes by _______.*

Intermediate Together, read the questions. Model how to take notes. Then have two sets of partners ask the Key Questions and use their notes to answer. Discuss responses.

Advanced Read aloud a page of the lesson, modeling how to take notes. Then have partners complete the activity. Ask students to share the answers they heard.

Advanced High Have partners complete the activity independently. Ask students to write down the best answers.

LESSON 4.3

Explain

Lead a Discussion

MULTIPLE ALLELES Help students understand that in multiple alleles, any individual has only two alleles for the trait. Each of the rabbits shown has two alleles for coat color. The color of the brown rabbit is called wild type, or agouti. The color of the gray rabbit is called chinchilla. The white rabbit is an albino. Tell students that the genotype of the albino is *cc*. Ask: **Is *cc* homozygous dominant, homozygous recessive, heterozygous dominant, or heterozygous recessive?** *(Homozygous recessive)*

Lead a Discussion

L3 **POLYGENIC INHERITANCE** Students know that one gene with two possible alleles can produce three genotypes (*AA, Aa, aa*). Two genes each with two possible alleles can produce nine genotypes. Challenge students to suggest possible genotypes using the letters *A, a, B,* and *b.* Keep a list on the board, and see if they can figure out all nine. *(AABB, AABb, AAbb, AaBB, AaBb, AAbb, aaBB, aaBb, aabb)*

Elaborate

Apply It!

L1 Before beginning the activity, tell students that not all alleles appear in a genotype. Although there are more than two possible alleles, an individual has only two of the alleles. The more possible alleles, the greater the number of potential genotypes.

Interpret Data When students interpret data, they look for patterns or trends.

Lab Resource: Quick Lab 

L2 **PATTERNS OF INHERITANCE** Students will explore patterns of inheritance. This Quick Lab can be found online.

Evaluate

Assess Your Understanding

After students answer the questions, have them evaluate their understanding by completing the appropriate sentence.

RTI Response to Intervention

1a. If students are not sure how to write the symbols for alleles that show incomplete dominance, **then** have them review the example in **Figure 1.**

b. If students need help with polygenic inheritance, **then** remind them that this is the only inheritance pattern they have seen so far that involves more than one pair of alleles.

apply it!

An imaginary insect called the blingwing has three alleles for wing color: *R* (red), *B* (blue), and *Y* (yellow).

1 List If an organism can inherit only two alleles for a gene, what are the six possible allele pairs for wing color in blingwings? One answer is given.

RB, RY, RR, BB, BY, YY

2 Interpret Data Suppose the three alleles are codominant. What wing color would each pair of alleles produce? One answer is given.

RB: purple

RY: orange

RR: red

BB: blue

BY: green

YY: yellow

Multiple Alleles Some genes have **multiple alleles,** which means that three or more possible alleles determine the trait. Remember that an organism can only inherit two alleles for a gene—one from each parent. Even if there are four, five, or more possible alleles, an individual can only have two. However, more genotypes can occur with multiple alleles than with just two alleles. For example, four alleles control the color of fur in some rabbits. Depending on which two alleles a rabbit inherits, its coat color can range from brownish gray to all white.

Polygenic Inheritance The traits that Mendel studied were each controlled by a single gene. **Polygenic inheritance** occurs when more than one gene affects a trait. The alleles of the different genes work together to produce these traits.

Polygenic inheritance results in a broad range of phenotypes, like human height or the time it takes for a plant to flower. Imagine a field of sunflowers that were all planted the same day. Some might start to flower after 45 days. Most will flower after around 60 days. The last ones might flower after 75 days. The timing of flowering is a characteristic of polygenic traits.

Do the Quick Lab *Patterns of Inheritance.* Find the lab online.

Assess Your Understanding

TEKS 14A

1a. Describe How are the symbols written for alleles that share incomplete dominance?

Both are capital letters.

b. CHALLENGE How is polygenic inheritance different from the patterns described by Mendel?

In polygenic inheritance, more than one gene affects a trait.

got it?

○ **I get it!** Now I know that most traits are produced by complex patterns of inheritance.

○ I need extra help with See TE note.

How Do Genes and the Environment Interact?

TEKS 14A In this section, you'll explore how an organism's environment can affect the way its genes are expressed.

You were not born knowing how to skateboard, but maybe you can skateboard now. Many traits are learned, or acquired. Unlike inherited traits, acquired traits are not carried by genes or passed to offspring. Although inherited traits are determined by genes, they also can be affected by factors in the environment. The phenotypes you observe in an organism result both from genes and from interactions of the organism with its environment.

Inherited and Acquired Traits Humans are born with inherited traits, such as vocal cords and tongues that allow for speech. But humans are not born speaking Spanish, or Mandarin, or English. The languages that a person speaks are acquired traits. Do you have a callus on your finger from writing with your pencil? That is an acquired trait. Skills you learn and physical changes that occur, such as calluses and haircuts, are acquired traits. See if you can tell the inherited traits from the acquired traits in **Figure 2.**

FIGURE 2

Inherited or Acquired?
Which traits shown are carried in genes, and which are not?

Classify Identify each trait shown as inherited or acquired.

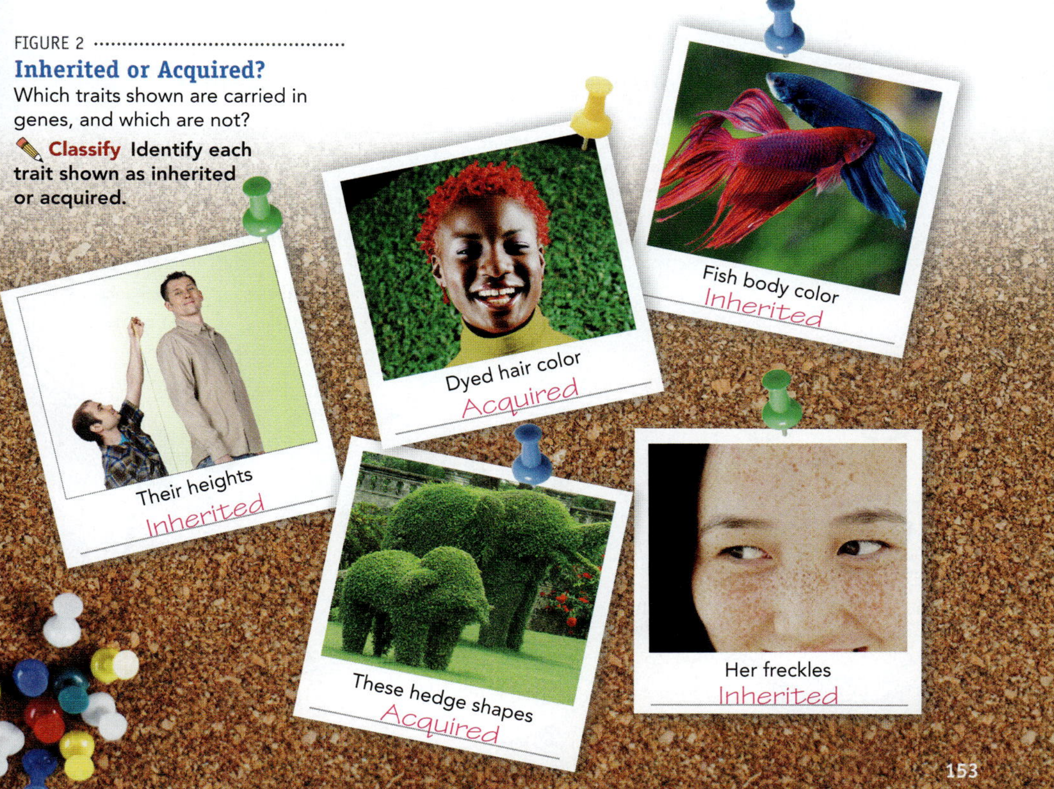

Explain

Lead a Discussion

ACQUIRED TRAITS Ask students to list the subjects they are studying this year. If students mention that a particular subject is difficult, say "nobody is born knowing that." Discuss that, in many ways, a baby is a clean slate. As humans grow up, they learn information and develop skills. Ask students to name famous people who are good at something, such as musicians or athletes. Then, ask if they would expect those people to have children who are good at the same activities. *(Samples: Students will likely mention children of musicians who became musicians. There are many second-generation professional football and baseball players.)* Emphasize that the skills learned by the parents were not inherited by their children. Often, parents teach their skills to their children. These are acquired traits.

21st Century Learning

INTERPERSONAL SKILLS Ask students to use **Figure 2** to make a list of traits. They should then work with a partner to write whether the trait is inherited or acquired. Have students share their responses with the class, and ask if they can think of any other examples of inherited and acquired traits.

21st Century Learning

ACCOUNTABILITY Have students write about what disease or phenomenon they would study if chosen to be a scientist on a DNA sequencing team. Students should explain why their research would be beneficial to society. Ideas can be presented in journal or blog format.

Differentiated Instruction

L1 Alleles Students may not be sure how to combine the alleles in the correct order in the *Apply It!* activity. Tell them that the order in which the alleles is written does not matter in this kind of inheritance. Genotypes *RB* and *BR* are the same. Explain to students that multiple alleles often show codominance or incomplete dominance.

Texas Essential Knowledge and Skills

14A Define heredity as the passage of genetic instructions from one generation to the next generation.

LESSON 4.3

Explain

Teach Key Concepts

Explain to students that an individual's phenotype is the result of the interaction of genes and the environment. Ask students to list some traits that people inherit. *(Samples: height, hair color, skin color, eye color)* Ask: **What inheritance pattern does height show?** *(Polygenic inheritance)* **How might the environment affect how the genotype is expressed?** *(Sample: Someone who has a diet that is low in nutrients might not grow as tall as he or she could have.)* **How do people change the phenotype of their hair?** *(Samples: They color their hair. They straighten curly hair.)* **Do these changes affect the person's genotype? How do you know?** *(No; the body still produces the hair that the genotype says to produce. You can see the natural color of dyed hair at the roots as hair grows. If you straighten your hair with a hair iron, the next time you wash your hair it's curly again.)*

Compare and Contrast Tell students that comparing and contrasting involves examining the similarities and differences between things. When two things are compared, ways in which they are alike are identified. When two things are contrasted, ways in which they are different are identified.

Elaborate

Apply the TEKS

Direct students' attention to **Figure 3.** If they have difficulty deciding on the labels for each image, suggest that they look back through the chapter for previous images. The answer given on the page for height is *Polygenic inheritance,* but some students may add *Environmental factors* because of the potential for environmental influence on that trait. The answer given on the page for insects is *Multiple Alleles,* since the insect was used as an example of multiple alleles in an *Apply It!* activity. However, *Polygenic inheritance* is another possible answer. When so many phenotypes are present, it is impossible to know the inheritance pattern without studying crosses between phenotypes.

Texas Essential Knowledge and Skills

14A Define heredity as the passage of genetic instructions from one generation to the next generation.

14B Compare the results of uniform or diverse offspring from sexual reproduction or asexual reproduction.

14C Recognize that inherited traits of individuals are governed in the genetic material found in the genes within chromosomes in the nucleus.

Compare and Contrast Underline two sentences that tell how changes to genes in body cells differ from changes to genes in egg and sperm cells.

Genes and the Environment Think again about sunflowers. Genes control when the plants flower. But sunlight, temperature, soil nutrients, and water also affect a plant's flowering time. **Environmental factors can influence the way genes are expressed.** For example, you may have inherited the ability to play a musical instrument. But without an opportunity to learn, you may never develop the skill.

Some environmental factors can change an organism's genes. For example, tobacco smoke and other pollutants can affect genes in a person's body cells in a way that may result in lung cancer and other cancers. Still other genetic changes happen by chance. A change to a gene is called a *mutation*.

Mutations in body cells cannot be passed to offspring. Only mutations in the sex cells—eggs and sperm—can be passed to offspring. Not all mutations have negative effects. Genetic change in sex cells is an important source of life's variety.

Patterns of Inheritance

APPLY THE TEKS 14A, 14B, 14C

Why don't offspring always look like their parents?

FIGURE 3 The traits you see in organisms result from their genes and from interactions of genes with the environment.

Define Match the terms in the word bank with the examples shown.

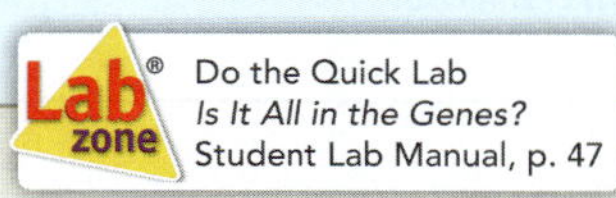

Do the Quick Lab *Is It All in the Genes?* Student Lab Manual, p. 47

Assess Your Understanding

TEKS 14A, 14B

2a. Review Only genetic mutations in (sex cells/ body cells) can be passed to offspring.

b. Describe Give one example of how environmental factors affect gene expression.

Sample: The amounts of sun, water, and nutrients affect the growth of a plant.

c. Compare Why don't offspring always look like their parents?

The way offspring look depends on the combinations of alleles they inherit and on interactions between genes and the environment.

got it?

○ **I get it!** Now I know that the environment can affect the way genes are expressed.

○ I need extra help with See TE note.

155

Differentiated Instruction

L1 Start With Mendel Some students may find the variety of inheritance patterns confusing. Point out that it is more important to understand the differences in how the alleles behave than to be remember the examples. Suggest that they use their understanding of basic Mendelian patterns and compare. Students can make Venn diagrams with Mendelian inheritance in one circle and one of the more complex patterns in the other.

L3 Sex Determination in Reptiles Invite interested students to research the environmental factors that affect the sex of reptiles other than the example of the turtle shown here. Not all species are affected the same way.

Lab Resource: Quick Lab 

L2 IS IT ALL IN THE GENES? Students will explore some inherited and acquired characteristics of an imaginary organism. This Quick Lab can be found in the Student Lab Manual, p. 47, and online.

Evaluate

Assess Your Understanding

After students answer the questions, have them evaluate their understanding by completing the appropriate sentence.

RTI Response to Intervention

2a. If students need help understanding that genetic changes must occur in sex cells to be passed on, **then** have them reread the last paragraph on the previous page.

b. If students have trouble giving examples of environmental affects on gene expression, **then** help them think of ways that individuals are changed by health conditions or by where they live.

c. If students cannot explain why offspring don't always look like their parents, **then** review with them Mendel's observations and patterns of inheritance.

Name ____________________ Date __________ Class __________

Assess Your Understanding

Patterns of Inheritance

How Are Most Traits Inherited?

1a. DESCRIBE How are the symbols written for alleles that share incomplete dominance? ____________________

b. CHALLENGE How is polygenic inheritance different from the patterns described by Mendel? ____________________

got it?

○ **I get it!** Now I know that most traits are produced by ____________________

○ **I need extra help with** ____________________

How Do Genes and the Environment Interact?

2a. REVIEW Only genetic mutations in (sex cells/body cells) can be passed to offspring.

b. DESCRIBE Give one example of how environmental factors affect gene expression. ____________________

c. COMPARE Why don't offspring always look like their parents? ____________________

got it?

○ **I get it!** Now I know that the environment can affect ____________________

○ **I need extra help with** ____________________

Place the outside corner, the corner away from the dotted line, in the corner of your copy machine to copy onto letter-size paper.

Name ______________________ Date ____________ Class ____________

Enrich

Patterns of Inheritance

Many traits are determined by a combination of more than one gene. Read the passage and study the Punnett square below. Then use a separate sheet of paper to answer the questions that follow the Punnett square.

Polygenic Inheritance

In his genetic experiments with pea plants, Gregor Mendel studied traits that had distinct phenotypes: white or purple flowers, smooth or wrinkled skin, and so on. But many traits found in organisms are more complex and diverse. Humans are not either tall or short; instead, there is a wide range of possible heights—an example of polygenic inheritance. With polygenic inheritance, more than one gene contributes to a phenotype.

To help you understand polygenic inheritance, imagine this simplified scenario: Two genes determine the color of a certain plant's flower. The two genes produce a range of colors from white to red depending on the particular combination of alleles. Red plants are homozygous *AABB* and white plants are homozygous *aabb*. When a white plant is crossed with a red plant, the F_1 offspring have pink flowers (*AaBb*). But what happens when two plants with pink flowers are crossed? The results are shown below.

	AB	***Ab***	***aB***	***ab***
AB	*AABB*	*AABb*	*AaBB*	*AaBb*
Ab	*AABb*	*AAbb*	*AaBb*	*Aabb*
aB	*AaBB*	*AaBb*	*aaBB*	*aaBb*
ab	*AaBb*	*Aabb*	*aaBb*	*aabb*

There are five possible phenotypes in the F_2 offspring, corresponding to the number of dominant alleles in the genotype: red (*AABB*), light red (*AABb*, *AaBB*), pink (*AbBb*, *AAbb*, *aaBB*), light pink (*Aabb*, *aaBb*), and white (*aabb*).

1. **In the example above, can you cross two plants with light-red flowers and get offspring with white flowers? Explain.**

2. **Imagine that a certain animal's fur length is determined by three genes. The genotype for long fur is *AABBCC* and *aabbcc* for short fur. A cross between long fur and short fur produces F_1 offspring with medium-length fur (*AaBbCc*). Draw a Punnett square that shows a cross between two F_1 offspring.**

3. **Examine the number of dominant alleles in the genotypes of the F_2 offspring in your Punnett square. How many phenotypes are possible?**

Name ______________________ Date __________ Class __________

Lesson Quiz

Patterns of Inheritance

Write the letter of the correct answer on the line at the left.

1. ___ Height in humans is an example of
- A incomplete dominance
- B codominance
- C polygenic inheritance
- D multiple alleles

2. ___ The pattern of inheritance in which one allele is only partially dominant is
- A incomplete dominance
- B codominance
- C polygenic inheritance
- D multiple alleles

3. ___ The pattern of inheritance in which there are three or more possible alleles for a trait is
- A incomplete dominance
- B codominance
- C polygenic inheritance
- D multiple alleles

4. ___ The pattern of inheritance in which both genes are expressed equally is
- A incomplete dominance
- B codominance
- C polygenic inheritance
- D multiple alleles

Fill in the blank to complete each statement.

5. A cow with a mix of red hairs and white hairs has the genotype H^RH^W. This is an example of Codominace.

6. Having pierced ears is an example of a(n) acquired trait.

7. Four alleles determine if a rabbit is white, brown, or gray. This is an example of Multiple Alleles.

8. The pattern of inheritance in which more than one pair of genes affects a trait is polygentic inheritance.

9. If a plant with red flowers crossed with a plant with white flowers produces a plant with pink flowers, it is an example of incomplete domin.

10. Only mutations in Sex cells can be passed to offspring.

Place the outside corner, the corner away from the dotted line, in the corner of your copy machine to copy onto letter-size paper.

Patterns of Inheritance

Answer Key

Review and Reinforce

Find the worksheet in the Student Workbook.

1. In cases of incomplete dominance, neither allele is dominant. Hybrids have a phenotype that is a blend of the two purebred phenotypes. Blue feathers are caused by one allele for black feathers and one allele for white feathers. There is no allele for blue feathers, so there cannot be a purebred blue chicken.
2. This is an example of multiple alleles. Incomplete dominance and codominance can produce only three different phenotypes. Polygenic inheritance can produce many phenotypes, but in this case, there is only one pair of alleles involved. So it has to be multiple alleles, with a different combination of alleles producing each different color.
3. Sample: If a plant does not get enough water and nutrients, it may not be as tall as it could have been, the leaves might not be as green, or the plant may not be able to produce fruit.
4. both alleles for a gene are expressed equally
5. one allele is only partially dominant
6. more than one pair of alleles affect a trait
7. Incomplete dominance occurs when one allele is only partially dominant, so it may be only partially expressed. Codominance occurs when both alleles for a gene are expressed equally. Some genes have multiple alleles for a trait. Three or more possible alleles determine those traits, although any individual will receive only two, one from each parent. Polygenic inheritance occurs when more than one gene affects a trait. The different alleles for two or more genes work together to produce those traits.

Enrich

1. It is not possible to produce a plant with white flowers by crossing two plants with light-red flowers. Sample: White flowers require no dominant alleles and you cannot get this genotype by crossing two plants with light-red flowers.
2. Check students' Punnett squares.
3. There are seven possible phenotypes.

Lesson Quiz

1. C
2. A
3. D
4. B
5. codominance
6. acquired
7. multiple alleles
8. polygenic inheritance
9. incomplete dominance
10. sex

Chromosomes and Inheritance

Why don't offspring always look like their parents?

LESSON PACING:
3–4 periods or $1\frac{1}{2}$–2 blocks

Lesson Vocabulary

- meiosis

Content Refresher

Crossing Over Before the first division of meiosis occurs, an important event called *crossing over* takes place. During this event, corresponding segments of chromosome pairs are exchanged. In effect, the organism's maternal and paternal chromosomes exchange some alleles and produce some chromosomes that are genetically different from those of either parent of the organism. This process increases the genetic diversity of a species.

Crossing over occurs randomly, but the farther apart genes are located on their chromosomes, the more often crossing over will occur between them. Geneticists use this principle to map the location of genes on chromosomes.

In 1911, research biologist Thomas Hunt Morgan noted crossover during meiosis in the fruit flies he was studying at Columbia University. Based on his studies, he was able to determine that genes are located on chromosomes. His discoveries earned him the Nobel Prize in Physiology or Medicine in 1933, the first in the field of genetics.

Lesson Objectives	TEKS	ELPS
Describe the role chromosomes and genes play in heritance.	14C	4.F.6
Recognize that genes are found within chromosomes and are made up of DNA.	14C	
Identify the events that occur during meiosis and fertilization.	14C	
Compare sexual reproduction requiring meiosis and asexual reproduction involving mitosis.	14B	

Texas Essential Knowledge and Skills

14B Compare the results of uniform or diverse offspring from sexual reproduction or asexual reproduction.
14C Recognize that inherited traits of individuals are governed in the genetic material found in the genes within chromosomes in the nucleus.

English Language Proficiency Standards

ELPS Reading 4.F.6 Use support from peers and teachers to read grade-appropriate content area text.

DIFFERENTIATED INSTRUCTION KEY
L1 Struggling Students or Special Needs
L2 On-Level Students L3 Advanced Students

Investigations and Activities	TEKS Review
My Planet Diary, **Student Edition,** p. 156 **Inquiry:** Inquiry Warm-Up, Which Chromosome Is Which?, **PearsonTexas.com** Introduce Vocabulary, **Teacher's Edition,** p. 157 Apply It!, **Student Edition,** p. 157 Teach Key Concepts, **Teacher's Edition,** p. 158 Support the TEKS, Mixed Pairs, **Teacher's Edition,** p. 159 **Inquiry:** Quick Lab, Chromosomes and Inheritance, **Lab Manual,** p. 48	Apply the TEKS, Predicting Genotypes, **Student Edition,** p. 166 TEKS Practice, **Student Edition,** p. 167 TEKS Practice: Chapter and Cumulative Review, **Student Edition,** p. 169 Lesson 4.4, **TEKS Preparation and Study Guide Workbook,** p. 42
Teach Key Concepts, **Teacher's Edition,** p. 160 **Inquiry:** Teacher Demo, Unique Sequences, **Teacher's Edition,** p. 161 Differentiated Instruction, **Teacher's Edition,** p. 161 **Inquiry:** Quick Lab, Modeling the Genetic Code, **PearsonTexas.com**	
Teach Key Concepts, **Teacher's Edition,** p. 162 **Inquiry:** Teacher Demo, Model Meiosis, **Teacher's Edition,** p. 163 Differentiated Instruction, **Teacher's Edition,** p. 163 **Inquiry:** Quick Lab, Modeling Meiosis, **PearsonTexas.com**	**SHORT ON TIME?** To do this lesson in approximately half the time, do the Activate Prior Knowledge activity. A discussion of the Key Concepts will familiarize students with the lesson content. Have students do the Quick Labs. The rest of the lesson can be completed by students independently.
Teach Key Concepts, **Teacher's Edition,** p. 164 21st Century Learning, Creativity, **Teacher's Edition,** p. 165 Differentiated Instruction, **Teacher's Edition,** p. 165 **Inquiry:** Quick Lab, Comparing Meiosis and Mitosis, **Lab Manual,** p. 49	

These editable worksheets are available on **PearsonTexas.com.**
Print versions can be found in the **TEKS Preparation and Study Guide Workbook.**

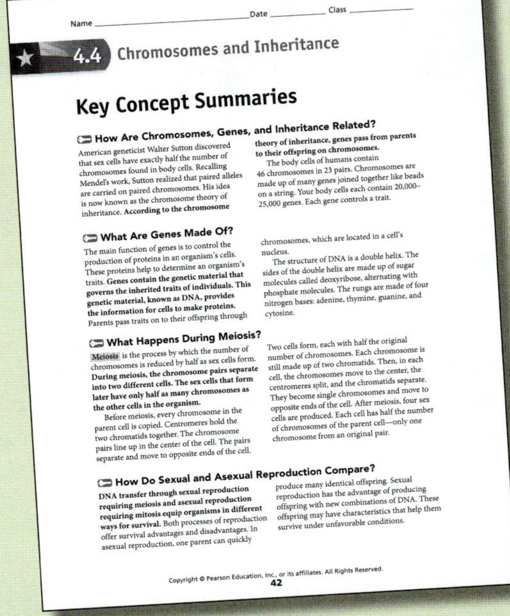
Name ______________ Date ________ Class ________

4.4 Chromosomes and Inheritance

Key Concept Summaries

How Are Chromosomes, Genes, and Inheritance Related?

American geneticist Walter Sutton discovered that sex cells have exactly half the number of chromosomes found in body cells. Recalling Mendel's work, Sutton realized that paired alleles are carried on paired chromosomes. His idea is now known as the chromosome theory of inheritance. **According to the chromosome theory of inheritance, genes pass from parents to their offspring on chromosomes.**

The body cells of humans contain 46 chromosomes in 23 pairs. Chromosomes are made up of many genes joined together like beads on a string. Your body cells each contain 20,000–25,000 genes. Each gene controls a trait.

What Are Genes Made Of?

The main function of genes is to control the production of proteins in an organism's cells. These proteins help to determine an organism's traits. **Genes contain the genetic material that governs the inherited traits of individuals. This genetic material, known as DNA, provides the information for cells to make proteins.** Parents pass traits on to their offspring through chromosomes, which are located in a cell's nucleus.

The structure of DNA is a double helix. The sides of the double helix are made up of sugar molecules called deoxyribose, alternating with phosphate molecules. The rungs are made of four nitrogen bases: adenine, thymine, guanine, and cytosine.

What Happens During Meiosis?

Meiosis is the process by which the number of chromosomes is reduced by half as sex cells form. **During meiosis, the chromosome pairs separate into two different cells. The sex cells that form later have only half as many chromosomes as the other cells in the organism.**

Before meiosis, every chromosome in the parent cell is copied. Centromeres hold the two chromatids together. The chromosome pairs line up in the center of the cell. The pairs separate and move to opposite ends of the cell. Two cells form, each with half the original number of chromosomes. Each chromosome is still made up of two chromatids. Then, in each cell, the chromosomes move to the center, the centromeres split, and the chromatids separate. They become single chromosomes and move to opposite ends of the cell. After meiosis, four sex cells are produced. Each cell has half the number of chromosomes of the parent cell—only one chromosome from an original pair.

How Do Sexual and Asexual Reproduction Compare?

DNA transfer through sexual reproduction requiring meiosis and asexual reproduction requiring mitosis equip organisms in different ways for survival. Both processes of reproduction offer survival advantages and disadvantages. In asexual reproduction, one parent can quickly produce many identical offspring. Sexual reproduction has the advantage of producing offspring with new combinations of DNA. These offspring may have characteristics that help them survive under unfavorable conditions.

42

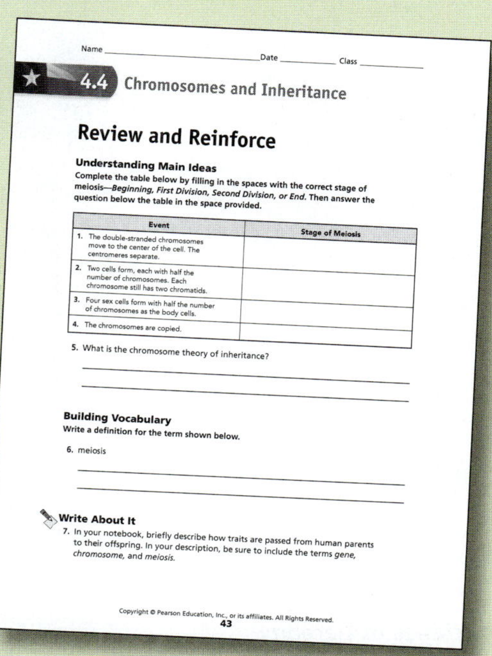
Name ______________ Date ________ Class ________

4.4 Chromosomes and Inheritance

Review and Reinforce

Understanding Main Ideas

Complete the table below by filling in the spaces with the correct stage of meiosis—*Beginning, First Division, Second Division, or End.* Then answer the question below the table in the space provided.

Event	Stage of Meiosis
1. The double-stranded chromosomes move to the center of the cell. The centromeres separate.	
2. Two cells form, each with half the number of chromosomes. Each chromosome still has two chromatids.	
3. Four sex cells form with half the number of chromosomes as the body cells.	
4. The chromosomes are copied.	

5. What is the chromosome theory of inheritance?

Building Vocabulary

Write a definition for the term shown below.

6. meiosis

Write About It

7. In your notebook, briefly describe how traits are passed from human parents to their offspring. In your description, be sure to include the terms *gene, chromosome,* and *meiosis.*

43

Lexile Measure = 920L

LESSON 4.4

Chromosomes and Inheritance

Establish Learning Objectives

After this lesson, students will be able to:

- Describe the role chromosomes and genes play in inheritance.
- Recognize that genes are found within chromosomes and are made up of DNA.
- Identify the events that occur during meiosis and fertilization.
- Compare sexual reproduction requiring meiosis and asexual reproduction involving mitosis.

Engage

Activate Prior Knowledge

MY PLANET DIARY Read *Chromosome Sleuth* with the class. Explain to students that genetic research has yielded information about many genetic disorders. By studying how genes are related to an inherited disorder, scientist can learn how to diagnose the condition early and help the patient reduce its effects. Students may ask why doctors don't just insert a missing gene or fix a defective one. Tell students that scientists are looking for ways to do this, and it may be possible in the future.

Explore

Lab Resource: Inquiry Warm-Up 

L2 **WHICH CHROMOSOME IS WHICH?** Students will explore how chromosomes recombine through meiosis and fertilization. This Inquiry Warm-Up can be found online.

Chromosomes and Inheritance

- How Are Chromosomes, Genes, and Inheritance Related? TEKS 14C
- What Are Genes Made Of? TEKS 14C
- What Happens During Meiosis? TEKS 14C
- How Do Sexual and Asexual Reproduction Compare? TEKS 14B

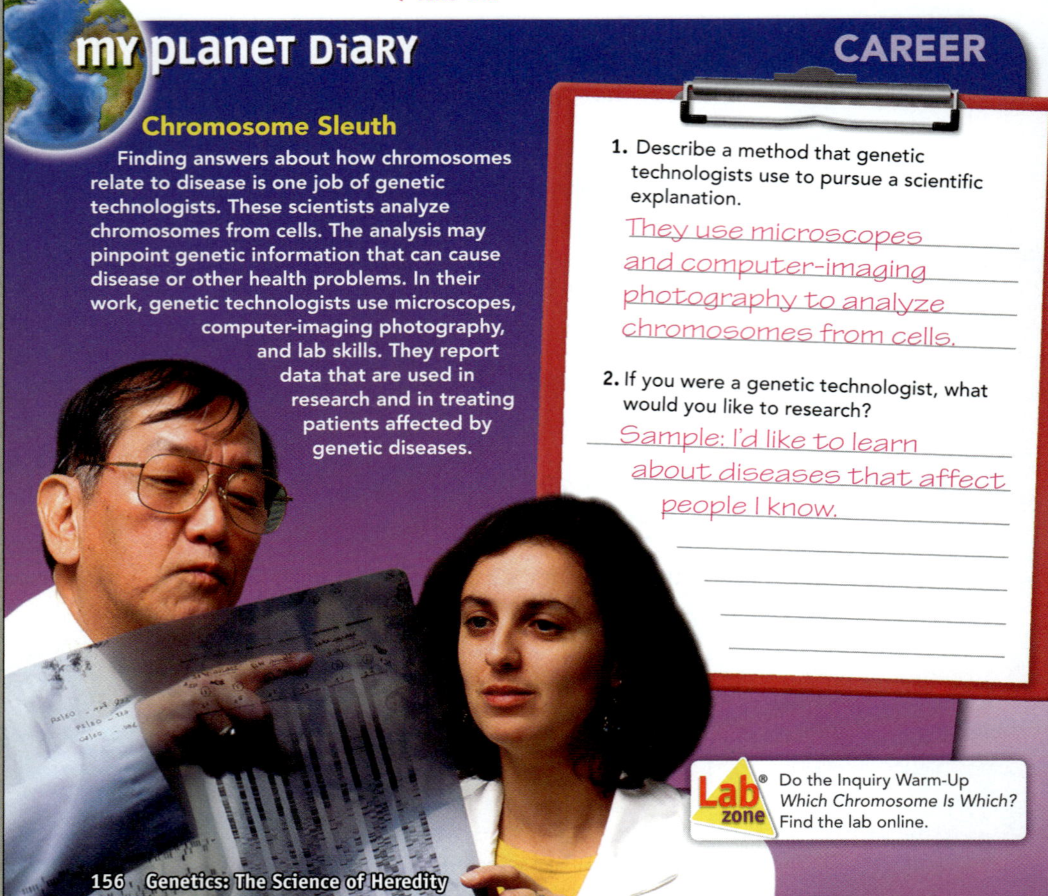

SUPPORT ALL READERS

Lexile Measure = 920L **Lexile Word Count = 1479**

Prior Exposure to Content: May be the first time students have encountered this topic

Academic Vocabulary: *cause, design, effect, experiment, relate*

Science Vocabulary: *meiosis*

Concept Level: Generally appropriate for most students in this grade

Preteach With: My Planet Diary "Chromosome Sleuth" and Figure 1 activity

Vocabulary
- meiosis

Skills
- Reading: Relate Cause and Effect
- Inquiry: Design Experiments

How Are Chromosomes, Genes, and Inheritance Related?

TEKS 14C In this section, you'll learn that genes pass from parents to offspring on chromosomes.

ELPS 4.F.6
Preview the blue and red headings in the lesson. Circle any new words. Discuss with your teacher and classmates your ideas about the words' meanings—use pictures and previous lessons to review. Ask for explanation and pronunciation as you need it. Then read to find out more.

Mendel's work showed that genes exist. (Remember that he called them "factors.") But scientists in the early twentieth century did not know what structures in cells contained genes. The search for the answer was something like a mystery story. The story could be called "The Clue in the Grasshopper's Cells."

At the start of the 1900s, Walter Sutton, an American geneticist, studied the cells of grasshoppers. He wanted to understand how sex cells (sperm and eggs) form. Sutton focused on how the chromosomes moved within cells during the formation of sperm and eggs. He hypothesized that chromosomes are the key to learning how offspring have traits similar to those of their parents.

apply it!

Design Experiments Different types of organisms have different numbers of chromosomes, and some organisms are easier to study than others. Suppose you are a scientist studying chromosomes and you have to pick an organism from those shown below to do your work. Which one would you pick and why?

Sample: I would pick corn because I can buy it in the store and it doesn't look like a bug. Shrimp have too many chromosomes, and skunks smell. Mosquitoes bite, and grasshoppers are too hard to catch.

English Language Proficiency Standards

ELPS Reading 4.F.6

Have students preview the lesson. Provide support to help students think critically about the text.

Beginning Read aloud each heading and have students repeat. Prompt students to connect words to illustrations, for example: *In Figure 2, which is bigger, a gene or a chromosome?*

Intermediate Together, read each lesson heading. Discuss pronunciation and develop meaning as needed. Have partners Turn and Talk to connect the words in the heading to the illustrations on the page. Discuss ideas as a group.

Advanced Have partners take turns reading aloud the headings on each page. Have partners explain the connections between the words and the illustrations. Have students explain their thinking.

Advanced High Have partners read the headings on each page and explain their connections to the illustrations. Have students explain their ideas.

Explain

Introduce Vocabulary

Tell students that *meiosis* is sometimes referred to as "reduction division." This is a form of cell division that reduces the number of chromosomes in the cells that are produced.

Lead a Discussion

PUTTING TOGETHER DISCOVERIES Students may think that one discovery leads directly to another, with scientists adding information like a sequence of beads on a string. In genetics, several discoveries were made independently and were not connected immediately. During the second half of the nineteenth century, Mendel worked out patterns of inheritance when a compound called "nuclein" (later named DNA) was isolated from the nuclei of cells, and chromosomes were observed during cell division. It wasn't until later that this information was put together. Ask: **What was Sutton trying to understand?** *(How sex cells form)* **What aspect of sex cell formation was he studying?** *(How chromosomes move)* **What was his hypothesis?** *(Chromosomes are involved in how traits are passed from parents to offspring.)*

Elaborate

Apply It!

L1 Before beginning the activity, review the organisms shown on the page.

Design Experiments Remind students that in a well-designed experiment, they need to keep all variables except one the same. By choosing one experimental subject, students keep the variable of chromosome number constant.

PEARSON Texas.com

Texas Essential Knowledge and Skills

14C Recognize that inherited traits of individuals are governed in the genetic material found in the genes within chromosomes in the nucleus.

Explain

LESSON 4.4

Teach Key Concepts

Review the definition and location of chromosomes. Make sure students understand that genes are carried on chromosomes. Ask: **What did Sutton observe about the relative numbers of chromosomes in the body cells and sex cells of grasshoppers?** *(The sex cells have half the number of chromosomes as body cells.)* **How many chromosomes does the fertilized egg receive from each parent?** *(The number that is present in the sex cell)* **How are genes passed from parents to offspring?** *(Sex cells contain half of each parent's chromosomes, which include the parent's genes. When the sex cells from each parent join during fertilization, the offspring receives a full set of genes.)* **What is the chromosome theory of inheritance?** *(Genes are carried from parents to their offspring on chromosomes.)* **If human body cells have 46 chromosomes, how many chromosomes do human sex cells have?** *(23)*

Teach With Visuals

Tell students to look at **Figure 1.** Ask: **How many chromosomes are there in the body cells of a grasshopper?** *(24)* **How many chromosomes are there in a grasshopper egg cell?** *(12)* **How many chromosomes are there in a grasshopper sperm cell?** *(12)* **How is the chromosome number returned to normal?** *(When a sperm and egg join in fertilization, their individual 12 chromosomes add up to 24 chromosomes.)*

Chromosomes and Inheritance Sutton needed evidence to support his hypothesis. Look at **Figure 1** to see how he found this evidence in grasshopper cells. To his surprise, he discovered that the nuclei of grasshopper sex cells have exactly half the number of chromosomes found in the nuclei of grasshopper body cells.

Chromosome Pairs Sutton observed what happened when a sperm cell and an egg cell joined. The fertilized egg that formed had 24 chromosomes. It had the same number of chromosomes as each parent. These 24 chromosomes existed as 12 pairs. One chromosome in each pair came from the male parent. The other chromosome came from the female parent.

FIGURE 1

Paired Up

Sutton studied grasshopper cells through a microscope. He concluded that genes are carried on chromosomes.

Relate Text and Visuals Answer the questions in the spaces provided.

1 Body Cell

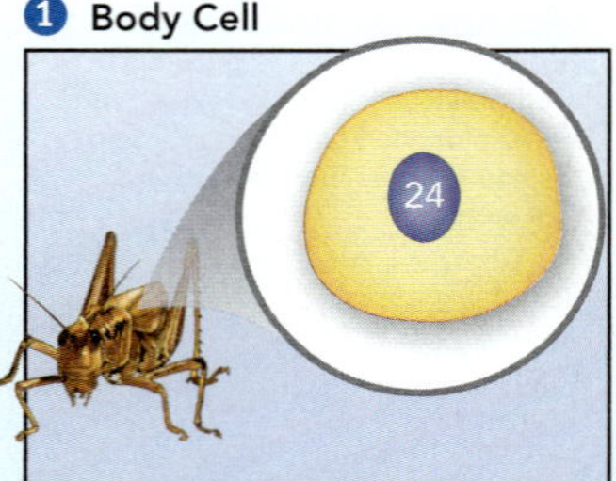

Each grasshopper body cell has 24 chromosomes.

2 Sex Cells

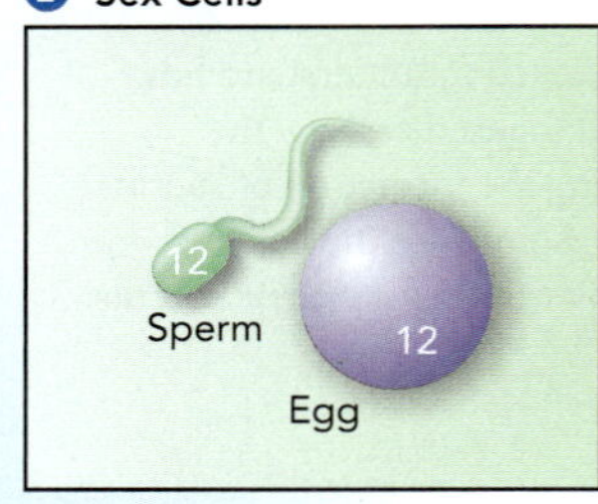

Sutton found that grasshopper sex cells each have 12 chromosomes.

1. How does the number of chromosomes in grasshopper sex cells compare to the number in body cells?

   Sex cells have half the chromosomes of body cells.

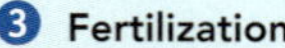

3 Fertilization

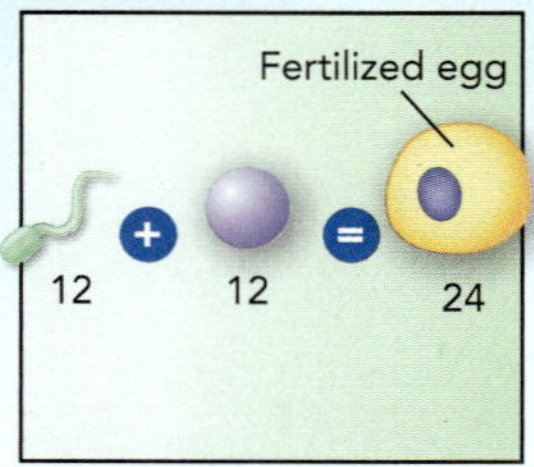

The fertilized egg cell has 24 chromosomes.

4 Grasshopper Offspring

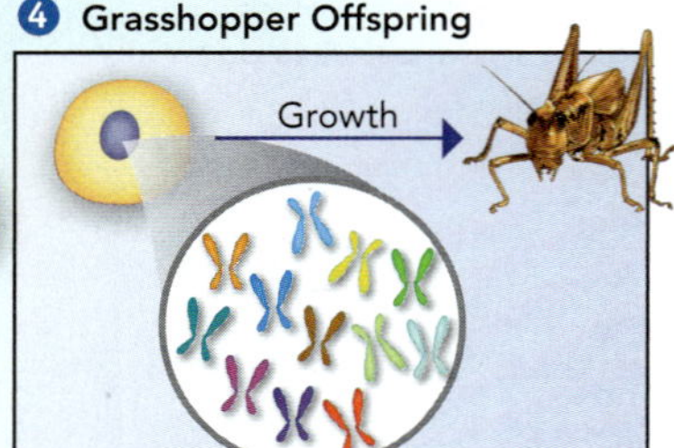

The 24 chromosomes exist as 12 pairs.

2. How is the inheritance of chromosomes similar to what you know about alleles?

   As with alleles, only one chromosome of a pair is inherited from each parent.

Genes on Chromosomes Recall that alleles are different forms of a gene. Because of Mendel's work, Sutton knew that alleles exist in pairs in an organism. One allele comes from the female parent. The other allele comes from the male parent. Sutton realized that paired alleles are carried on paired chromosomes. His idea is now known as the chromosome theory of inheritance. **According to the chromosome theory of inheritance, genes pass from parents to their offspring on chromosomes.**

A Lineup of Genes

The body cells of humans contain 46 chromosomes that form 23 pairs. Chromosomes are made up of many genes joined together like beads on a string. Although you have only 23 pairs of chromosomes, your body cells each contain between 20,000 and 25,000 genes. Genes control traits.

Figure 2 shows a pair of chromosomes from an organism. One chromosome is from the female parent. The other chromosome is from the male parent. Notice that each chromosome has the same genes. The genes are lined up in the same order on both chromosomes. However, the alleles for some of the genes are not identical. For example, one chromosome has allele *A*, and the other chromosome has allele *a*. As you can see, this organism is heterozygous for some traits and homozygous for others.

Relate Cause and Effect Suppose gene A on the left chromosome is damaged and no longer functions. What form of the trait would show? Why?

Recessive; The dominant form of the trait can no longer mask the recessive form of the trait.

FIGURE 2

A Pair of Chromosomes

Chromosomes in a pair may have different alleles for some genes and the same alleles for others.

Interpret Diagrams For each pair of alleles, tell whether the organism is homozygous or heterozygous. The first two answers are shown.

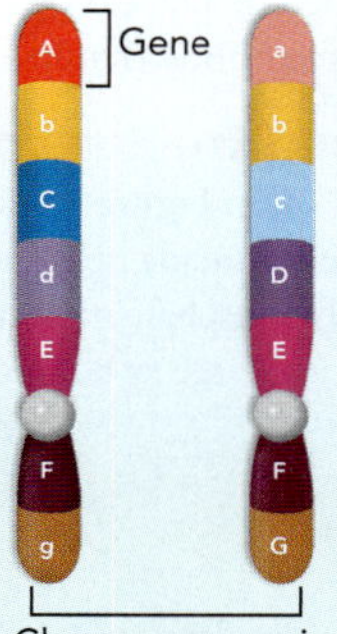

Heterozygous
Homozygous
Heterozygous
Heterozygous
Homozygous
Homozygous
Heterozygous

Do the Quick Lab *Chromosomes and Inheritance.*
Student Lab Manual, p. 48

Assess Your Understanding

TEKS 14C

1a. Describe When two grasshopper sex cells join, the chromosome number in the new cell is (half/double) the number in the sex cells. [double circled]

b. Recognize Describe the arrangement of genes on a pair of chromosomes.

Genes line up like beads on a string in the same order on both chromosomes.

c. Relate Evidence and Explanation How do Sutton's observations support the chromosome theory of inheritance?

He found that pairs of chromosomes followed the same pattern of inheritance as pairs of alleles.

got it?

○ I get it! Now I know that genes are passed from parents to offspring on the chromosomes.

○ I need extra help with See TE note.

Differentiated Instruction

L1 Visualize Chromosomes Show a picture of a cell and point out the chromosomes. Diagram two cells, each with one pair of chromosomes. Work backward to show how one chromosome came from the mother and one from the father. Point out the location of a gene. Show how it can have two alleles.

L3 Model the Function of Meiosis Provide various art materials to students, and challenge them to illustrate what might happen if sex cells did not have half the number of chromosomes as body cells. Have groups present their models to the class and explain why sex cells have half the chromosomes of body cells.

Explain

Support the TEKS

MIXED PAIRS Direct students' attention to **Figure 2.** Help them to understand that the terms heterozygous and homozygous apply to individual pairs of genes, not to chromosomes. Emphasize that an individual can be homozygous for one trait and heterozygous for another trait, which can lead to great variety in offspring. Challenge students to use upper and lower case letters to write a genotype for this individual. *(AabbCcDdEEFFGg)*

Relate Cause and Effect Tell students that a cause makes something happen. An effect is what happens. When students recognize that one event causes another, they are relating cause and effect.

Elaborate

Lab Resource: Quick Lab

L2 CHROMOSOMES AND INHERITANCE Students will investigate genetic crosses in imaginary creatures. This Quick Lab can be found in the Student Lab Manual, p. 48, and online.

Evaluate

Assess Your Understanding

After students answer the questions, have them evaluate their understanding by completing the appropriate sentence.

RTI Response to Intervention

1a. If students need help with chromosome numbers and fertilization, **then** suggest they review **Figure 1.**

b. If students cannot describe the arrangement of genes, **then** have them look at **Figure 2** and note the letters that represent the genes.

c. If students have difficulty explaining the importance of Sutton's observations, **then** review with them the importance of understanding that genes come in pairs.

LESSON 4.4

Explain

Teach Key Concepts

Review **Figure 3** with students. Explain to students that genes are found within chromosomes contained in a cell's nucleus. Genes are made up of DNA, which carries the codes to produce specific proteins. Make sure students understand the relationship between chromosomes, genes, and DNA. Ask: **What is the relationship between a gene and a chromosome?** *(A gene is a segment of a chromosome.)* **How are genes and DNA related?** *(A gene is a section of DNA that codes for a specific protein.)*

Lead a Discussion

DNA BASES Remind students that each person's DNA is unique. However, many sequences are common to all people. These sequences code for proteins that all people have, such as enzymes and the proteins that make up part of the cell membrane. Ask: **How does DNA code for proteins?** *(The sequence of nitrogen bases determines which amino acids will be used to make specific proteins.)* **How does changing the sequence of nitrogen bases affect the protein that is produced?** *(Changing the order of bases changes the amino acids that are used, so the protein is changed.)*

Texas Essential Knowledge and Skills

14C Recognize that inherited traits of individuals are governed in the genetic material found in the genes within chromosomes in the nucleus.

TEKS 14C In this section, you'll examine the structure of genes that are found in a cell's nucleus.

What Are Genes Made Of?

The main function of genes is to control the production of proteins in an organism's cells. Proteins help to determine the size, shape, color, and many other traits of an organism.  **Genes contain the genetic material that govern the inherited traits of individuals. This genetic material, known as DNA, provides the information for cells to make proteins.**

Chromosomes, Genes, and DNA

Parents pass traits to offspring through chromosomes. Chromosomes are made of DNA and proteins and are located in a cell's nucleus. In **Figure 3** you can see the relationship among chromosomes, genes, and DNA. A gene is a section of a DNA molecule that contains the information to code for one specific protein. Each gene is located at a specific place on a chromosome. DNA can be found in every cell in your body that has a nucleus.

The Structure of DNA

The twisted structure of DNA you can see in **Figure 3** is known as a double helix. The sides of the double helix are made up of sugar molecules called deoxyribose, alternating with phosphate molecules. The name DNA, or deoxyribonucleic acid (DEE ahk see ry boh noo klee ik) comes from this structure.

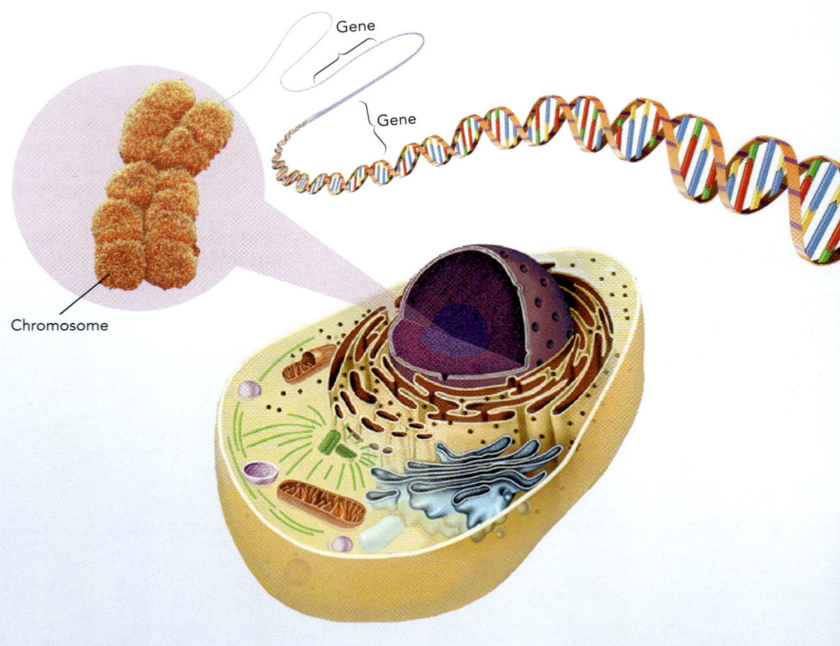

FIGURE 3

Chromosomes and Genes

Humans have between 20,000 and 25,000 genes on their chromosomes. The corals that make up ocean reefs are thought to have as many as 25,000 genes, too!

The rungs of DNA are made of nitrogen bases. Nitrogen bases are molecules that contain nitrogen and other elements. DNA has four kinds of nitrogen bases: adenine, thymine, guanine, and cytosine. The capital letters A, T, G, and C are used to represent the bases. A single gene on a chromosome may contain anywhere from several hundred to a million or more of these bases. Because there are so many possible combinations of bases and genes, each individual organism has a unique set of DNA.

Order of the Bases The order of the nitrogen bases along a gene forms a genetic code that specifies the protein that will be produced. Remember that proteins are long-chain molecules made of individual amino acids. In the genetic code, a group of three DNA bases codes for one specific amino acid. For example, the three-base sequence CGT (cytosine-guanine-thymine) always codes for the amino acid alanine. The order of the three-base code units determines the order in which amino acids are put together to form a protein.

Do the Quick Lab *Modelling the Genetic Code.* Find the lab online.

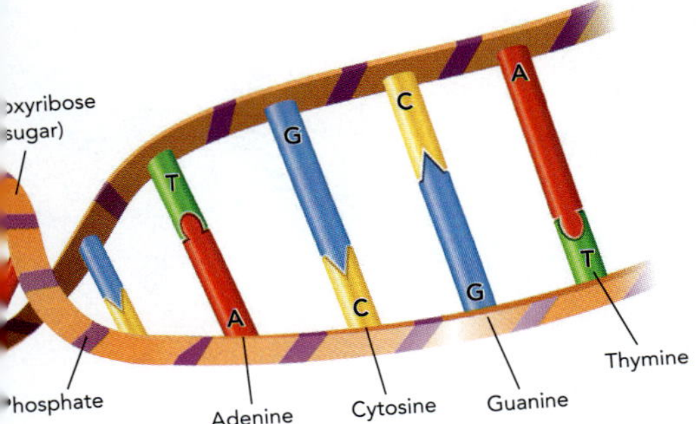

FIGURE 4

DNA Bases

Notice the pattern in the DNA bases.

Interpret Diagrams **Which base always pairs with cytosine?**

Guanine

Assess Your Understanding

TEKS 14C

2a. Sequence List the following terms from the smallest level of organization to the largest level of organization: chromosome, DNA, gene nitrogen base, nucleus.

nitrogen base, DNA, gene, chromosome, nucleus

b. Identify These letters represent the nitrogen bases on one strand of DNA: GGCTATCCA. What letters would form the other strand of the helix?

CCGATAGGT

got it?

○ I get it! Now I know that genes contain DNA which provides the information for cells to make proteins.

○ I need extra help with See TE note.

161

Elaborate

Teacher Demo

L1 UNIQUE SEQUENCES

Materials paper and pencil

Time 10 minutes

Write a sequence of DNA bases 20 letters long. Do not show it to the class. Ask each student to write a sequence of bases 20 letters long. Tell students to raise their hands if the first letter in your sequence matches theirs, and to keep their hands raised until you read a letter that does not match theirs. Read your sequence slowly, and stop when no more hands are raised. Discuss with students how long it took before the best match ceased to match yours.

Ask: **We could not get a match for a sequence of only 20 letters. How do you think the probability of finding two identical DNA sequences changes as the length of the sequence increases?** *(The probability decreases.)* **What does this tell you about DNA?** *(Each person's DNA is unique.)*

Lab Resource: Quick Lab

L2 MODELING THE GENETIC CODE Students will construct a model of DNA. This Quick Lab can be found online.

Evaluate

Assess Your Understanding

After students answer the questions, have them evaluate their understanding by completing the appropriate sentence.

RTI Response to Intervention

2a. If students have trouble organizing the terms, **then** have them review the information in **Figure 3**.

b. If students need help determining the pattern in the DNA bases, **then** point out the relationships between the two strands in **Figure 4**.

LESSON 4.4

Differentiated Instruction

L1 Compare DNA Segments Some students, especially less proficient readers, may have trouble understanding the DNA sequences shown in **Figure 4**. Work with students to cut out paper pieces to match the bases show in the figure. Demonstrate how only certain bases pair together. As an extension, provide students with a three-base sequence and challenge them to use their paper pieces to model the sequence.

L3 How Long Does It Take? Television detectives get the results of DNA tests almost immediately. But, in reality, it takes longer than that. Have interested students research DNA testing to find out how large the sample must be and how long the process takes. If students do their research on the Internet, remind them to follow prescribed guidelines for Internet use.

LESSON 4.4

Explain

Teach Key Concepts

Explain to students that meiosis is a kind of cell division that produces cells with only half the chromosomes as the other cells in the organism. Ask: **What kind of cells are produced in meiosis?** *(Sex cells)* **How many divisions take place during meiosis?** *(Two)* **What happens in the first division?** *(Chromosome pairs line up together and then move apart.)* **What happens in the second division?** *(Chromosomes split into identical halves, which move to new cells.)* **In which division is the number of chromosomes reduced by half?** *(In the first division)*

Teach With Visuals

Tell students to look at **Figure 5.** Ask: **How would you describe the shape of a chromosome?** *(It looks like an X.)* **What are the two sides of the X?** *(The chromatids)* **What holds the chromatids together?** *(The centromere)* **How do the chromatids of a chromosome compare?** *(They are identical.)* **When do the chromatids separate?** *(During the second division)*

Students may need help understanding the difference between the starting cell, which has replicated pairs of chromosomes, and the final cells that result, which have unreplicated, separated pairs. Walk them through the steps, emphasizing the changes in each one.

TEKS 14C In this section, you'll learn how sex cells form during the process of meiosis.

What Happens During Meiosis?

Body cells contain a complete set of all chromosomes with genes that code for every protein an organism needs to survive. How do sex cells end up with half the number of chromosomes as body cells? The answer to this question is a form of cell division called meiosis. **Meiosis** (my OH sis) is the process by which the number of chromosomes is reduced by half as sex cells form. You can trace the events of meiosis in **Figure 5.** Here, the parent cell has four chromosomes arranged in two pairs. **During meiosis, the chromosome pairs separate into two different cells. The sex cells that form later have only half as many chromosomes as the other cells in the organism.**

FIGURE 5

Meiosis

During meiosis, a cell produces sex cells with half the number of chromosomes.

Interpret Diagrams **Fill in the missing terms in the spaces provided, and complete the diagram.**

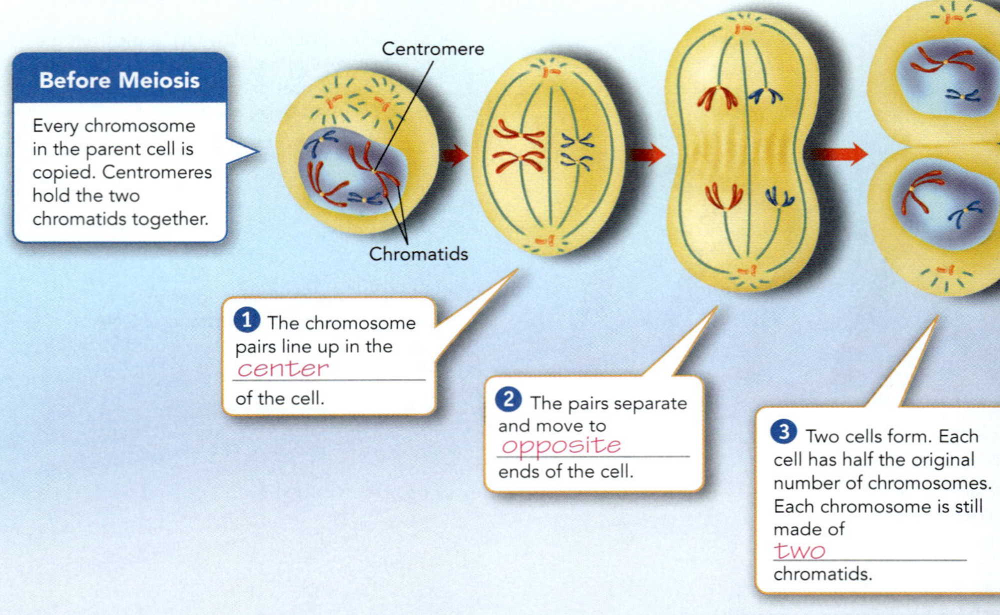

Texas Essential Knowledge and Skills

14C Recognize that inherited traits of individuals are governed in the genetic material found in the genes within chromosomes in the nucleus.

During meiosis, a cell divides into two cells. Then each of these cells divides again, forming a total of four cells. The chromosomes duplicate only before the first cell division.

Each of the four sex cells shown below receives two chromosomes—one chromosome from each pair in the original cell. When two sex cells join at fertilization, the new cell that forms has the full number of chromosomes. In this case, the number is four. The organism that grows from this cell got two of its chromosomes from one parent and two from the other parent.

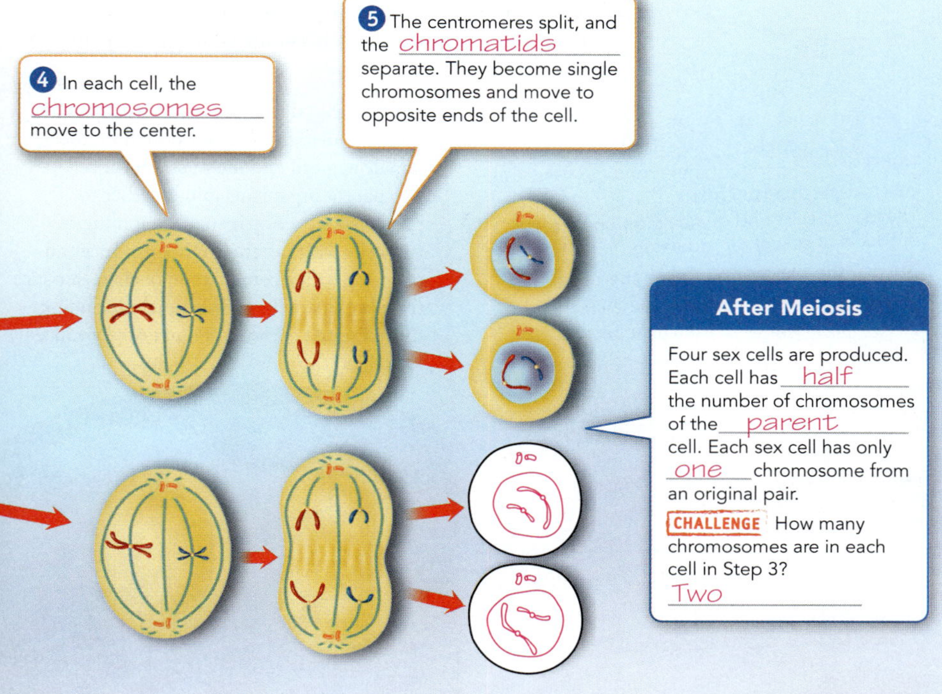

After Meiosis

Four sex cells are produced. Each cell has half the number of chromosomes of the parent cell. Each sex cell has only one chromosome from an original pair.

CHALLENGE How many chromosomes are in each cell in Step 3? Two

Lab zone® Do the Quick Lab *Modeling Meiosis.* Find the lab online.

Assess Your Understanding

got it?

○ I get it! Now I know that during meiosis, the number of chromosomes becomes half of what the number was in the body cells.

○ I need extra help with See TE note.

Differentiated Instruction

L1 Chromosome Pairs Make sure students understand that chromosome pairs do not normally stay together in the nucleus of the cell. Explain that this meeting occurs only during meiosis. The rest of the time, it does not matter where the chromosomes of a pair are in relation to each other.

L3 Organisms and Their Chromosomes Encourage interested students to research the number of chromosomes different organisms have. Students should also find out what the terms *diploid* and *haploid* mean.

Elaborate

Teacher Demo 

L2 MODEL MEIOSIS

Materials two decks of playing cards

Time 10 minutes

Use all the cards of one suit from both decks. Fan out the cards so that students can view the cards and see that there are two of each card. Shuffle the cards and count out 13 cards for one pile and 13 for a second pile. Invite volunteers to examine each pile and determine if there are any duplicates in the piles.

Ask: **Are the two piles the same?** *(No)* **If these cards were chromosomes in sex cells, how could this be a problem for the organism that results from one of these sex cells?** *(The new organism would be missing some genes and have extras of others.)* **How could you make sure the two piles of cards will be the same?** *Match up pairs before you separate them into piles.*

Allow students to guide you in lining up the cards in pairs, and then gathering them into two piles.

Ask: **How do the two piles compare?** *(They are the same.)* **Would sex cells that contain these groups of chromosomes have all the genes they should have?** *(Yes)* **How is lining up the cards like what happens in meiosis?** *(Pairs of chromosomes line up together so that one from each pair moves into each new cell.)*

Lab Resource: Quick Lab 

L1 MODELING MEIOSIS Students will model the movement of chromosomes during meiosis. This Quick Lab can be found online.

Evaluate

Assess Your Understanding

Have students evaluate their understanding by completing the appropriate sentence.

RTI Response to Intervention

If students have trouble explaining what happens in meiosis, **then** help them review **Figure 5.**

LESSON 4.4

Explain

Teach Key Concepts

Explain to students that the processes and results of sexual and asexual reproduction differ. Ask: **How does sexual reproduction differ from asexual reproduction?** *(Sexual reproduction occurs through meiosis, the process by which the number of chromosomes is reduced by half. Offspring receive chromosomes and DNA from both parents. Asexual reproduction occurs through mitosis. A parent's cell divides into two cells.)*

You may want students to make a T-chart with the headings *Sexual Reproduction* and *Asexual Reproduction* and note characteristics of each.

Teach With Visuals

Use **Figure 6** to discuss the differences in pairs and offspring in sexual reproduction. Ask: **Which characteristics differ and which characteristics seem similar?** *(Differences: fur color, size; Similarities: eye shape, body shape, presence of fur)* **Do you think some of these differences will change as the offspring grows?** *(Yes, size will increase and fur may change color.)*

Texas Essential Knowledge and Skills

14B Compare the results of uniform or diverse offspring from sexual reproduction or asexual reproduction.

TEKS 14B In this section, you'll compare how genetic material is transferred through sexual and asexual reproduction. You'll also learn how those differences affect offspring.

How Do Sexual and Asexual Reproduction Compare?

You now know that sexual reproduction starts with the joining of two sex cells. As a result, offspring receive DNA within chromosomes from both parents. Recall that some organisms can reproduce asexually. In many of these organisms, asexual reproduction takes place through mitosis. During mitosis, a parent cell divides into two new cells. No new genetic material is introduced during mitosis. **DNA transfer through sexual reproduction requiring meiosis and asexual reproduction requiring mitosis equip organisms in different ways for survival.**

Sexual Reproduction Like many animals, you developed after two sex cells joined. During sexual reproduction, the female egg cell and the male sperm cell of two parent organisms join together to produce a new organism. The joining of two cells with different DNA produces an offspring with a combination of characteristics from both parents. Most animals, including the mammals shown in **Figure 6**, reproduce sexually.

FIGURE 6

Sexual Reproduction

These wolf cubs and guinea pig pups are products of sexual reproduction.

Use the photos to answer the questions.

1. **Compare and Contrast** How do the offspring in each photo differ from their parent?
   The fur colors are different and the offspring are smaller.

2. **Explain** Why do the parent and the offspring look different?
   Because the offspring have a combination of physical characteristics from two parents.

Asexual Reproduction

During asexual reproduction, one parent produces a new organism identical to itself. This new organism receives an exact copy of the parent's DNA. Many plants reproduce asexually. Some animals, including sponges, jellyfish, and worms, reproduce asexually. The hydra, shown in **Figure 7**, reproduces asexually through budding. In budding, a new animal grows out of the parent and breaks off. Some animals reproduce asexually by dividing in two.

Comparing Asexual and Sexual Reproduction

Both sexual and asexual reproduction offer survival advantages and disadvantages. An advantage of asexual reproduction is that one parent can quickly produce many identical offspring. But a major disadvantage is that there is uniformity among offspring because they have the same DNA as the parent. With no variation from the parent, the offspring may not survive changes in the environment. In contrast, sexual reproduction has the advantage of producing offspring with new combinations of DNA. These offspring are more diverse and may have characteristics that help them survive under unfavorable conditions. However, sexual reproduction also has disadvantages. Sexual reproduction requires finding a mate. Also, development of offspring takes longer than it does during asexual reproduction.

FIGURE 7

A Chip Off the Old Block

Budding is the most common form of asexual reproduction for this hydra, a type of cnidarian.

Relate Text to Visuals How does this photo show asexual reproduction?

The offspring is growing out of the parent and looks just like the parent, except smaller.

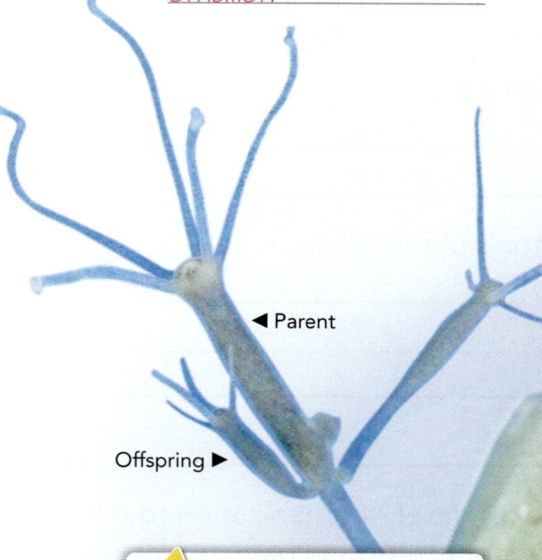

Lab zone — Do the Quick Lab *Comparing Meiosis and Mitosis.* Student Lab Manual, p. 49

Assess Your Understanding

TEKS 14B

3a. Define (Sexual/Asexual) reproduction involves the joining of sperm and egg. [circled: Sexual]

b. Compare and Contrast The offspring of (sexual/asexual) reproduction have a better chance of surviving changes in the environment than the offspring of (sexual/asexual) reproduction. [circled: sexual; asexual]

got it?

○ **I get it!** Now I know that organisms are equipped for survival in different ways as a result of sexual reproduction through meiosis and asexual reproduction through mitosis.

○ I need extra help with See TE note.

Elaborate

21st Century Learning

CREATIVITY Have students invent an offspring produced by asexual reproduction. Have them tell how it is like and unlike the parent, and the advantages and disadvantages to its type of reproduction. Have students do the same for an offspring produced by sexual reproduction. Encourage students to make their offspring interesting and memorable.

Teach With Visuals

Direct students' attention to **Figure 7.** Ask: **How does budding differ from asexual reproduction by division in two?** *(Budding is when a part breaks off the parent rather than the parent dividing in two.)*

Lab Resource: Quick Lab 

L2 **COMPARING MEIOSIS AND MITOSIS** Students will observe the differences between the two types of cell division. This Quick Lab can be found in the Student Lab Manual, p. 49, and online.

Evaluate

Assess Your Understanding

After students answer the questions, have them evaluate their understanding by completing the appropriate sentence.

RTI Response to Intervention

3a. If students have trouble differentiating between sexual and asexual reproduction, **then** have them review the definition of each term.

b. If students do not know whether the offspring of sexual or asexual reproduction have a better chance of surviving environmental change, **then** have them review the section *Comparing Asexual and Sexual Reproduction.*

Differentiated Instruction

L1 **Make a Venn Diagram** Work with students to make a Venn diagram that shows the similarities and differences for sexual and asexual reproduction. *(Similarities might include the following: produces offspring, related to parent, has advantages and disadvantages; differences might include the following: offspring produced by asexual reproduction do not have new DNA, offspring produced by sexual reproduction include new DNA)*

Name ______________________ Date __________ Class __________

Assess Your Understanding

Chromosomes and Inheritance

How Are Chromosomes, Genes, and Inheritance Related?

1a. **DESCRIBE** When two grasshopper sex cells join, the chromosome number in the new cell is (half/double) the number in the sex cells.

b. **RECOGNIZE** Describe the arrangement of genes on a pair of chromosomes. ______________________

____

c. **RELATE EVIDENCE AND EXPLANATION** How do Sutton's observations support the chromosome theory of inheritance? ______________________

____

What Are Genes Made Of?

2a. **SEQUENCE** List the following terms from the smallest level of organization to the largest level of organization: chromosome, DNA, gene, nitrogen base, nucleus.

____

b. **IDENTIFY** These letters represent the nitrogen bases on one strand of DNA: GGCTATCCA. What letters would form the other strand of the helix?

____

____

What Happens During Meiosis?

got it?

○ **I get it!** Now I know that during meiosis, the number of chromosomes ______________________

____

○ **I need extra help with** ______________________

How Do Sexual and Asexual Reproduction Compare?

3a. **DEFINE** (Sexual/Asexual) reproduction involves the joining of sperm and egg.

b. **COMPARE AND CONTRAST** The offspring of (sexual/asexual) reproduction have a better chance of surviving changes in the environment than the offspring of (sexual/asexual) reproduction.

Place the outside corner, the corner away from the dotted line, in the corner of your copy machine to copy onto letter-size paper.

Name ______________________ Date ____________ Class ____________

Enrich

Chromosomes and Inheritance

Follow the procedure below to make a model of meiosis. Then use a separate sheet of paper to answer the questions that follow.

A Model of Meiosis

Materials

different colors of pipe cleaners or yarn

beads

macaroni

string

glue

scissors

marker

construction paper or poster board

Procedure

1. Study the diagram of the stages of meiosis in your textbook.
2. Decide how you can use the materials listed above, or other materials of your choice, to make a model of meiosis. Your model should include the beginning of meiosis, the first division, the second division, and the end of meiosis.
3. Create your model. Begin with at least six copied pairs of chromosomes. Label the stages of meiosis and all the important structures. On a separate sheet of paper, write a description in your own words of what happens in each stage of meiosis.

Analyze and Conclude

1. What is meiosis?
2. What must happen before meiosis can begin?
3. What happens to chromosomes during the first division?
4. What happens to chromosomes during the second division?
5. Compare the sex cells produced by meiosis to the parent cell. Why is the difference between the sex cells and parent cell important?
6. Why are chromosomes important to heredity?

Name ______________________________ Date ____________ Class ____________

Lesson Quiz

Chromosomes and Inheritance

Write the letter of the correct answer on the line at the left.

1. ___ The process that produces sex cells is called

A DNA

B meiosis

C inheritance

D reproduction

2. ___ Each chromosome contains two identical

A genes

B nuclei

C chromatids

D cells

3. ___ The section of DNA that codes for a protein is a(n)

A trait

B gene

C meiosis

D chromosome

4. ___ Uniformity among identical offspring is a result of

A asexual reproduction

B centromeres

C meiosis

D sexual reproduction

If the statement is true, write *true*. If the statement is false, change the underlined word or words to make the statement true.

5. ________________ DNA provides the information for cells to make proteins.

6. ________________ Body cells of humans have 46 pairs of chromosomes.

7. ________________ Sex cells have twice the number of chromosomes as body cells.

8. ________________ Genes pass from parents to offspring on chromosomes.

9. ________________ The two chromosomes in a pair have the same genes lined up in the same order.

10. ________________ A fertilized egg has twice the number of chromosomes as the body cells of the parent.

Chromosomes and Inheritance

Answer Key

Review and Reinforce

Find the worksheet in the Student Workbook.

1. Second Division

2. First Division

3. End

4. Beginning

5. Genes are carried from parents to their offspring on chromosomes.

6. Meiosis is the process by which the number of chromosomes is reduced by half to form sex cells.

7. The body cells of humans contain 46 chromosomes in 23 pairs. On each chromosome are many, many genes joined together like beads on a string. Each gene controls a trait. During meiosis, the chromosome pairs separate into two different cells. The sex cells that form later have only 23 chromosomes, instead of 23 pairs. When two sex cells—an egg cell from the mother and a sperm cell from the father—join at fertilization, the new cell that forms again has 23 pairs of chromosomes. One chromosome of each pair is from the mother, and one is from the father. The paired genes on these chromosomes determine which traits the offspring will have.

Enrich

1. Meiosis is the process by which the number of chromosomes is reduced by half to form sex cells—sperm and eggs.

2. Every chromosome in the cell must be copied.

3. The chromosome pairs line up next to each other, and then they separate and move to opposite ends of the cell. The cell divides, and each new cell contains one double-stranded chromosome from each pair.

4. The double-stranded chromosomes line up in the center of the cell and split apart. The chromatids move to opposite ends of the cell.

5. Each sex cell has only half the number of chromosomes that the parent cell had. This is important because when sex cells combine to produce offspring, each sex cell contributes half the normal number of chromosomes, not double the number.

6. Genes are carried from parents to their offspring on chromosomes.

Lesson Quiz

1. B

2. C

3. B

4. A

5. true

6. 23

7. half

8. true

9. true

10. the same

CHAPTER 4

Scientific Investigation and Reasoning

This Apply the TEKS feature will enable students to use the scientific investigation and reasoning skills in TEKS 2C, 2E, and 3B to reinforce the content of TEKS 14A. Students will record and analyze data to predict the genotypes of the parent generation.

Analyzing Data From a Punnett Square

Before students answer Questions 1–4, review basic information about genetics and how a Punnett square can be used to determine the possible genotypes produced from a genetic cross. Ask: **What is an allele?** *(An allele is one of two or more genes that contain specific inherited characteristics.)* **What is the difference between a phenotype and a genotype?** *(Phenotype is a physical description of a genetic trait and genotype is the combination of genes for a trait.)* **What is a Punnett square?** *(A Punnett square is a diagram that is used to predict the outcome of a particular genetic cross.)*

After students have completed Questions 1–4, ask: **If the rooster in this Punnett square had a genotype of YY instead of Yy, how many chicks would you expect to have white legs? Why?** *(None of the chicks would have white legs because the genotypes of the chicks would be either YY or Yy and both of those genotypes produces chickens with yellow legs.)* **Would it be possible for a white-legged hen and a white-legged rooster to produce yellow-legged chicks? Why?** *(No, because neither parent would have the Y allele necessary for yellow legs to contribute to the offspring.)*

APPLY THE TEKS 2C, 2E, 3B, 14A

Predicting GENOTYPES

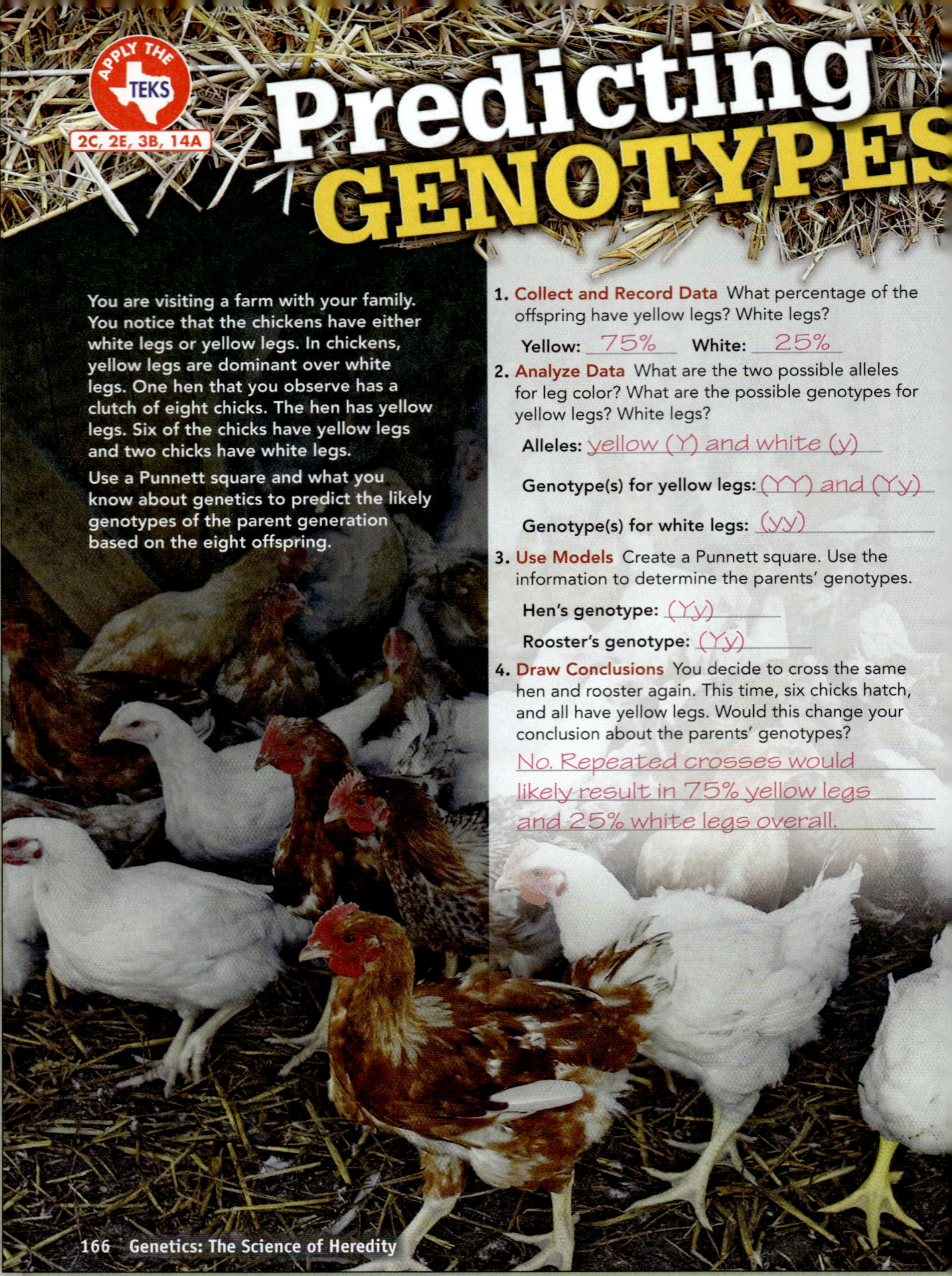

You are visiting a farm with your family. You notice that the chickens have either white legs or yellow legs. In chickens, yellow legs are dominant over white legs. One hen that you observe has a clutch of eight chicks. The hen has yellow legs. Six of the chicks have yellow legs and two chicks have white legs.

Use a Punnett square and what you know about genetics to predict the likely genotypes of the parent generation based on the eight offspring.

1. **Collect and Record Data** What percentage of the offspring have yellow legs? White legs?

   Yellow: 75% **White:** 25%

2. **Analyze Data** What are the two possible alleles for leg color? What are the possible genotypes for yellow legs? White legs?

   Alleles: yellow (Y) and white (y)

   Genotype(s) for yellow legs: (YY) and (Yy)

   Genotype(s) for white legs: (yy)

3. **Use Models** Create a Punnett square. Use the information to determine the parents' genotypes.

   Hen's genotype: (Yy)

   Rooster's genotype: (Yy)

4. **Draw Conclusions** You decide to cross the same hen and rooster again. This time, six chicks hatch, and all have yellow legs. Would this change your conclusion about the parents' genotypes?

   No. Repeated crosses would likely result in 75% yellow legs and 25% white legs overall.

Texas Essential Knowledge and Skills

2C Collect and record data using the International System of Units (SI) and qualitative means such as labeled drawings, writing, and graphic organizers.

2E Analyze data to formulate reasonable explanations, communicate valid conclusions supported by the data, and predict trends.

3B Use models to represent aspects of the natural world such as human body systems and plant and animal cells.

14A Define heredity as the passage of genetic instructions from one generation to the next generation.

English Language Proficiency Standards

ELPS Reading 4.F.6

Have students preview the chapter and provide support to help students think critically about the lesson.

Beginning Read aloud each heading and have students repeat. Prompt students to connect words to illustrations, for example: *In Figure 2, which is bigger, a gene or a chromosome?*

Intermediate Together, read each lesson heading. Discuss pronunciation and develop meaning as needed. Have partners Turn and Talk to connect the words in the heading to the illustrations on the page. Discuss ideas as a group.

Advanced Have partners take turns reading aloud the headings on each page. Have partners explain the connections between the words and the illustrations. Have students explain their thinking.

Advanced High Have partners read the headings on each page and explain their connections to the illustrations. Have students explain their ideas.

CHAPTER 4

TEKS Practice

TEKS 3D, 14A, 14B, 14C

LESSON 1 What Is Heredity?

1. Different forms of a gene are called
   - (a.) alleles.
   - b. hybrids.
   - c. genotypes.
   - d. chromosomes.

2. Heredity is the passage of genetic instructions from one generation to the next generation.

3. **Explain** Mendel crossed two pea plants: one with green pods and one with yellow pods. The F_1 generation all had green pods. What color pods did the F_2 generation have? Explain your answer.
   Most F_2 plants had green pods, but some had yellow pods because they inherited two recessive alleles.

4. **Predict** The plant below is purebred for height (tall). Write the alleles of this plant. In any cross for height, what kind of offspring will this plant produce? Why?

TT

This plant can produce only tall offspring because it has only tall alleles and tall is dominant over short.

5. **Compare and Contrast** How do dominant alleles and recessive alleles differ?
   Dominant alleles have an effect even if only one is present. Recessive alleles have an effect only if two are present.

6. **Write About It** Write a paragraph that explains the impact Mendel had on scientific thought.
   See TE rubric.

LESSON 2 Probability and Heredity

7. Which of the following represents a heterozygous genotype?
   - a. *YY*
   - b. *yy*
   - (c.) *Yy*
   - d. Y^HY^H

8. An organism's phenotype is the way its genotype is expressed.

9. **Make Models** Fill in the Punnett square below to show a cross between two guinea pigs that are heterozygous for coat color. *B* is for black coat color, and *b* is for white coat color.

	B	b
B	BB	Bb
b	Bb	bb

10. **Interpret Tables** What is the probability that an offspring from the cross above has each of the following genotypes?
    - *BB* 1/4 or 25%
    - *Bb* 2/4 or 1/2 or 50%
    - *bb* 1/4 or 25%

11. **Apply Concepts** What kind of cross might tell you if a black guinea pig is *BB* or *Bb*? Why?
    A cross with a white guinea pig (bb); if a white pup is born, the black guinea pig had to be Bb.

12. **math!** A garden has 80 pea plants. Of this total, 20 plants have short stems and 60 plants have tall stems. What percentage of the plants have short stems? What percentage have tall stems?
    See TE note.

Write About It 6 Assess student's writing using this rubric.

SCORING RUBRIC	SCORE 4	SCORE 3	SCORE 2	SCORE 1
Explain one impact of Mendel's work	Chooses appropriate aspect of Mendel's work and fully discusses impact on society	Chooses appropriate aspect of Mendel's work and adequately discusses impact on society	Chooses appropriate aspect of Mendel's work but cannot discuss impact on society	Cannot choose appropriate aspect of Mendel's work or discuss his impact on society

TEKS Practice

Assess Understanding

Have students complete the answers to the TEKS Practice questions. Have a class discussion about what students find confusing. Write Key Concepts on the board to reinforce knowledge.

RTI Response to Intervention

5. If students cannot distinguish between dominant and recessive alleles, **then** have them review the definitions of these terms and the table that shows Mendel's traits in pea plants.

9. If students have difficulty completing the Punnett square, **then** remind them how to work across and down the table to fill in the boxes.

12. math!

SHORT STEMS

20 out of 80 = $\frac{20}{80}$

20 ÷ 80 = 0.25

0.25 × 100% = 25%

TALL STEMS

60 out of 80 = $\frac{60}{80}$

60 ÷ 80 = 0.75

0.75 × 100% = 75%

Alternate Assessment

L1 DESIGN A GAME Have students work in pairs or small groups to design a game about heredity. Students can design a board game that requires players to answer trivia questions in order to advance. Remind students to create rules, spinners, game pieces, and questions for their games. The questions for the game should include vocabulary terms and Key Concepts from the chapter. Students can exchange games with other groups or pairs or play their own game.

PEARSON Texas.com

CHAPTER 4

CHAPTER 4

TEKS Practice, Cont.

RTI Response to Intervention

16. If students do not identify polygenic inheritance, **then** help them review the complex patterns of inheritance.

20. If students have trouble explaining how meiosis affects chromosome number, **then** have them review the steps of meiosis and note what happens to the chromosomes during each division.

L3 **WRITING IN SCIENCE** Ask students to write an article for a science magazine that explains to readers how traits are passed from one generation to the next.

TEKS Practice

LESSON 3 Patterns of Inheritance

13. Which of the following terms describes a pattern of inheritance in which one allele is only partially dominant?

a. codominance
b. acquired traits
c. multiple alleles
d. incomplete dominance (circled)

14. Traits that have three or more phenotypes may be the result of multiple alleles.

15. Compare and Contrast How is codominance different from incomplete dominance?

Codominance happens when two alleles are expressed equally. Incomplete dominance happens when one allele is only partially dominant.

16. Relate Cause and Effect Human height is a trait with a very broad range of phenotypes. Which pattern of inheritance could account for human height? Explain your answer.

Polygenic inheritance; multiple alleles could account for several phenotypes, but not for the very wide range of human heights.

17. Identify Faulty Reasoning Neither of Josie's parents plays a musical instrument. Josie thinks that she won't be able to play an instrument because her parents can't. Is she right? Why or why not?

No; Josie's parents may not have learned to play an instrument, but Josie may find out she has the ability to play if she has the chance to learn.

LESSON 4 Chromosomes and Inheritance

18. Genes are carried from parents to offspring on structures called

a. alleles.
b. chromosomes. (circled)
c. phenotypes.
d. genotypes.

19. The process of meiosis results in the formation of sex cells.

20. Summarize If an organism's body cells have 12 chromosomes, how many chromosomes will the sex cells have? Explain your answer.

Six; the process of meiosis forms sex cells with half the number of chromosomes of the body cells.

21. Compare and Contrast In terms of offspring, how are the results of sexual and asexual reproduction similar? How are the results different?

In both cases, genetic material is passed from parent to offspring. In sexual reproduction, offspring get genetic material from two parents, which leads to variation. In asexual reproduction, offspring get all genetic information from one parent, which means there is no variation.

22. Recognize How are inherited traits in individuals governed by DNA within chromosomes in the nucleus?

Sample: DNA controls the production of proteins in cells. These proteins help to determine an organism's traits.

168 Genetics: The Science of Heredity

Why don't offspring always look like their parents?

23. A species of butterfly has three alleles for wing color: blue, orange, and pale yellow. A blue butterfly mates with an orange butterfly. The following offspring result: about 25% are blue and 25% are orange. However, another 25% are speckled blue and orange, and 25% are yellow. Explain how these results could occur. Then provide an example of a situation in which these offspring could benefit from the diversity.

Offspring of a blue butterfly and an orange butterfly

Yellow color is recessive to blue and orange, but blue and orange have codominance. The diversity in the offspring might help them survive changes in their environment. For example, a certain type of animal that eats butterflies might only be attracted to certain colors. Having offspring with variations in coloring might help some of them avoid getting eaten.

Review the TEKS — Interactive Science Chapter 4

14A, 14B, 14C

Lesson 1
In Lesson 4.1, you learned that heredity is the passage of genetic instructions from one generation to the next generation. You also learned that an organism's traits are controlled by the alleles it receives from its parents. Some alleles are dominant, while other alleles are recessive.
Supporting TEKS: 14C
TEKS: 14A

Lesson 2
In Lesson 4.2, you learned that, in a genetic cross, the combination of alleles that parents can pass to an offspring is based on probability. You also learned that an organism's phenotype is its physical appearance, or visible traits. An organism's genotype is its genetic makeup, or alleles.
TEKS: 14A

Lesson 3
In Lesson 4.3, you learned that most traits are the result of complex patterns of inheritance. Environmental factors can influence the way genes are expressed.
TEKS: 14A

Lesson 4
In Lesson 4.4, you learned that the chromosome theory of inheritance states that genes pass from parents to their offspring on chromosomes. Chromosomes are contained in the nuclei of cells. You also learned that the transfer of genetic material through meiosis and sexual reproduction or through mitosis and asexual reproduction provides different advantages and disadvantages for offspring.
Supporting TEKS: 14B, 14C
TEKS: 3D

Why don't offspring always look like their parents?

Assess student's response using this rubric.

SCORING RUBRIC	SCORE 4	SCORE 3	SCORE 2	SCORE 1
Identify the patterns of inheritance shown by four butterflies	Student identifies dominant and recessive alleles, as well as alleles that show incomplete dominance.	Student identifies dominant and recessive alleles, but does not identify alleles that show incomplete dominance.	Student identifies the dominant allele, but cannot identify the nature of the other alleles.	Student cannot explain the relationship among any of the alleles.

TEKS Practice, Cont.

Why don't offspring always look like their parents?

Students should be able to demonstrate understanding of heredity. See the scoring rubric below.

Review the TEKS Chapter 4

PARTNER REVIEW Have partners review definitions of vocabulary terms and quiz each other. Students can read the Key Concept statements and leave out words for their partners to fill in. They also can change a statement so that it is false and then ask their partner to correct it.

CLASS ACTIVITY: CONCEPT MAP Have students develop a concept map to show how the information in this chapter is related. Have students brainstorm to identify Key Concepts, vocabulary, details, and examples. Then write each on a self-sticking note and attach it at random on chart paper or on the board. Explain that the concept map will begin at the top with Key Concepts. As students to use the following questions to help them organize the information on the notes:

- What did Mendel observe?
- How do alleles affect inheritance?
- How is probability related to inheritance?
- What are some complex patterns of inheritance?
- How do genes and the environment interact?
- How are chromosomes, genes, and DNA related?
- What happens during meiosis?
- How do sexual and asexual reproduction compare?

SAMPLE CONCEPT MAP:

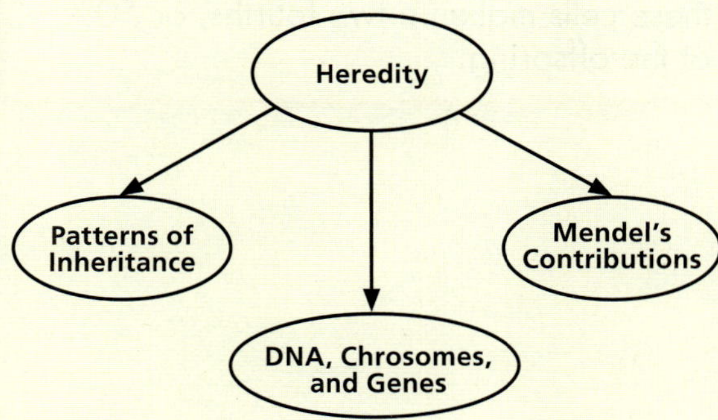

CHAPTER 4

TEKS Practice: Chapter Review

Test-Taking Skills

INTERPRETING TABLES Explain to students that some questions may require them to interpret data in a table. Tell students that they should take a few seconds to scan the table and describe the data to themselves. For example, the table in Question 1 contains information about the pod color and number of offspring for a cross of pea plants.

Question 1 TEKS 14A

The correct answer is A. Because yellow pods must be a recessive trait, both parents must have an allele for yellow pods.

If students chose B, explain that all genes must come from the parents, so the parents must have provided the alleles for pod color even though their colors are different.

If students chose C, explain that the data suggest that yellow pods are a recessive trait because there are more offspring with green pods than with yellow pods.

If students chose D, explain that traits are a result of genes from both parents, not just one parent.

Question 2 TEKS 14A

The correct answer is 50. The two *Rr* cells represent offspring with round seeds, and these cells make up two fourths, or 50%, of the offspring.

★ TEKS Practice: Chapter Review

Read each question and choose the best answer.

1 In one of Gregor Mendel's experiments, he produced a new generation of pea plants by cross-pollinating pea plants that had green pods. However, as the data table shows, he found that not all offspring had green pods.

New Generation of Pea Plants	
Pod Color of Offspring	**Number of Offspring**
green	428
yellow	152

Which statement provides the best explanation for these results?

Ⓐ Both parents provided the alleles for the yellow pod color.

B Neither parent provided alleles to the offspring for pod color.

C One parent provided alleles for green pods, while the other parent provided alleles for yellow pods.

D One parent provided alleles for the yellow pods, while the other parent did not provide any alleles for pod color.

2 In pea plants, the allele for round seeds (*R*) is dominant to the allele for oval seeds (*r*). The Punnett square below shows a cross between a parent with a *Rr* genotype and a parent with a *rr* genotype.

	R	**r**
r	*Rr*	*rr*
r	*Rr*	*rr*

What percent of the offspring of these parents will likely have round seeds?

		5	0	.		
0	0	0	●		0	0
1	1	1	1		1	1
2	2	2	2		2	2
3	3	3	3		3	3
4	4	4	4		4	4
5	5	●	5		5	5
6	6	6	6		6	6
7	7	7	7		7	7
8	8	8	8		8	8
9	9	9	9		9	9

★ TEKS Practice: Cumulative Review

3 Which part of a cell divides during mitosis?

- (A) Nucleus
- B Cell membrane
- C Vacuole
- D All of these

4 A student is comparing a cell to a city. She wants to label cell structures in the image below with the parts of a city that best represent them. In a city, a power plant burns fuel to generate electrical energy needed by the city. Cells also use "fuel" in the form of glucose to gather the energy they need.

Cell Structures

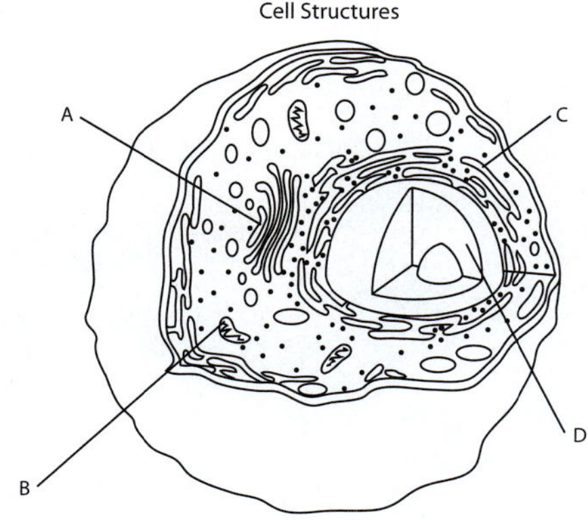

Which structure in the cell should the student label as the power plant?

- F Structure A
- (G) Structure B
- H Structure C
- J Structure D

If You Have Trouble With . . .				
Question	1	2	3	4
See Lesson	4.1	4.2	3.3	2.3
TEKS	7.14A	7.14A	7.12F	7.12D

TEKS Practice: Cumulative Review

Additional Assessment Resources

Teacher's Edition: Lesson Quizzes, Texas End-of-Year Test Prep A and B
Online assessments

Question 3 TEKS 12F

The correct answer is A. During mitosis, the nucleus of a cell divides into two.

If students chose B, explain that the cell membrane pinches together around the middle of a cell during cytokinesis.

If students chose C, explain that vacuoles do not divide during cell division.

If students chose D, explain that only the nucleus divides during mitosis.

Question 4 TEKS 12D

The correct answer is G. Mitochondria carry out cellular respiration, which breaks down glucose to produce energy for the cell.

If students chose F, explain that the Golgi apparatus does not produce energy for the cell. It packages and distributes proteins.

If students chose H, explain that the endoplasmic reticulum does not produce energy for the cell. It helps ribosomes make proteins.

If students chose J, explain that the nucleus does not produce energy for the cell. It regulates cell activity.

CHAPTER 4

Science Matters STEM

Science and Society

Have students read *Fruit That Won't Freeze?* Discuss the fact that in the late 1960s Arthur DeVries discovered that the resistance to freezing of some Antarctic fish was due to blood serum glycoproteins. Since then there has been much research on these proteins commonly called AFPs. The same proteins that keep fish from freezing in cold weather can prevent ice from melting at warmer temperatures. To keep fish from freezing, the protein stops the formation of ice crystals in bodily fluids. Superheated ice crystals can also be stabilized above the melting point by AFP. And the protein's ability to stop ice growth is far higher than its ability to arrest melting.

Before having students begin the research activity, have them look at look at weather records to discover when in the last five years the temperatures dropped below freezing in the crop-growing regions of Texas.

Ask: **How would climate change impact the problem?** *(Sample: If temperatures rise, the freezing of crops will not be such a big problem. However, there may be other problems.)* **Do you think the flavor or texture of fruits would be compromised by AFPs?** *(Accept all reasonable responses.)*

SCIENCE MATTERS STEM

Science and Society

Fruit That Won't Freeze?

TEKS 3D

I scream, you scream, we all scream for...fish genes? When the temperature drops below freezing, some fish and other animals and insects do something interesting...they don't freeze! The genes of these organisms contain antifreezing proteins, or AFPs, that help prevent the development of ice crystals inside of cells. Fish AFPs have been used in the medical field and in ice cream manufacturing. Some ice cream makers use AFPs to keep ice crystals small.

In 2010, many farmers lost strawberry crops to frost damage.

Freezing temperatures threaten many fruit crops, which can be damaged when temperatures fall. Plants make AFPs, but the AFPs of animals work at colder temperatures than those of plants. So what if these plants could become more frost-tolerant? In the 1990s, researchers tried inserting a fish gene for AFP into strawberries and other crops. Although the genes were expressed in the plants, the proteins did not provide much, if any, protection from freezing. Still, scientists continue to work to improve frost tolerance of crop plants.

Research It **Use reference materials to research what citrus farmers currently do to help their crops tolerate cold temperatures. Write a letter to the farmers explaining how scientific understanding of other organisms' frost tolerance methods might help their crops.**

Quick Facts

There is little agreement about the origin of the Dalmatian breed, although its name suggests it has roots in Dalmatia (a historical region of Croatia). The only spotted breed of dog, the Dalmatian probably is best known as a firehouse mascot. However, the breed also has been used as a rat catcher, a shepherd, a retriever, and a war dog. Dalmatians have a natural affinity for horses, and they commonly were used to guard horse-drawn carriages and coaches. The American Kennel Club (AKC) first recognized this breed in 1888, in the Non-Sporting Group.

Texas Essential Knowledge and Skills

3D Relate the impact of research on scientific thought and society, including the history of science and contributions of scientists as related to the content.

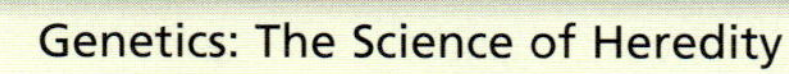

Seeing Spots

TEKS 2C, 14A

Everyday Science

You would probably recognize a Dalmatian if you saw one—Dalmatians typically have white coats with distinctive black or brown spots. Spots are a defining characteristic of the Dalmatian breed. These spots can be large or small, but all Dalmatians have them.

In Dalmatians, spots are a dominant trait. When two Dalmatians breed, each parent contributes a gene for spots. The trait for spots is controlled by one set of genes with only two possible alleles that control color. No matter how many puppies are in a litter, they will all develop spots.

But what if a Dalmatian breeds with another dog that isn't a Dalmatian? While the puppies won't develop the distinctive Dalmatian pattern, they will have spots, because the allele for spots is dominant. Some puppies will have many tiny spots and some will have large patches! Dalmatians, like leopards, cannot change their spots.

Newborn Dalmatian puppies are white—their spots develop when the puppies are about a week old. ▼

Predict It! Dalmatians' spots may be black or liver (brown), but never both on the same dog. Liver is a recessive allele. Use a Punnett square to predict the color of the spots on the offspring of a liver Dalmatian and a black Dalmatian with a recessive liver allele. Display your prediction on a poster.

Everyday Science

Have students read *Seeing Spots.* Point out that genes determine genetic characteristics. Explain that each characteristic is determined by a pair of alleles, or coding sequences, on that gene. Alleles may be dominant or recessive. When an offspring is created, each parent contributes a pair of alleles for each trait. The offspring receives one allele from each parent. If one or both of the alleles it receives are dominant, the offspring will have the dominant trait. Only if it receives two recessive alleles, will the offspring have the recessive trait. For example, a Dalmatian that has liver spots must have received the recessive allele for spot color from both parents.

Ask students to raise their hand if the bottoms of their earlobes are directly attached to the sides of their head. Record the number of hands. Then do the same for free earlobes. Ask students to predict which trait is dominant or recessive. Explain that free earlobes are dominant. You may want to tell students to observe their parents' earlobes to tell where the trait comes from.

As students create their Punnett squares for Dalmatians, remind them that since black spots are dominant, the parent dog with black spots may either have one or two dominant spot color alleles. Either of two different Punnett squares could show the correct combinations.

Ask: **How many alleles for liver spots does a Dalmatian with liver spots have?** *(Two)* **When an offspring receives a dominant trait, must they have received a dominant allele from each parent?** *(No, the dominant trait only requires one dominant allele come from one parent.)*

Texas Essential Knowledge and Skills

2C Collect and record data using the International System of Units (SI) and qualitative means such as labeled drawings, writing, and graphic organizers.

14A Define heredity as the passage of genetic instructions from one generation to the next generation.

Plant Structure, Function, and Response

Chapter TEKS Overview

This chapter focuses on TEKS 7C, 12A, 12C, 13A, and 13B. Students will explore how the structures of plants help them to function and survive, and how plants respond to internal stimuli and external stimuli in their environments.

Introduce the TEKS

Have students look at the image and read the Focus Question and description. Ask students to suggest explanations for the unusual shape of the baobab tree, especially its wide trunk and short, topmost branches. Have volunteers share their explanations. Tell students that baobab trees can store as much as 120,000 liters of water in their swollen trunks. Ask: **Considering where these trees live, why is water storage essential?** *(It prevents the trees from dying during the harsh drought conditions of the dry season.)* Tell students that almost every part of a baobab tree, which is sometimes called the "Tree of Life," is used by native peoples. Leaves and roots are used for medicinal purposes. The bark is used for cloth and rope. The fruit, which is called "monkey bread," is a good source of vitamin C and is eaten, as are the seeds. The trunk and roots can be tapped for water during dry periods. And because mature trees are frequently hollow, they can provide living space for many organisms, including humans. **The baobab tree is indeed unusual. Are there other trees that you might consider unusual or odd looking?** *(Samples: Giant sequoias, giant redwoods, banyans, mangroves, cacti, and the Divi-Divi tree of the Caribbean islands, such as Aruba)*

Untamed Science Video

AMAZING PLANT DEFENSES Before viewing, invite students to suggest ways in which plants defend and protect themselves. Then play the video. Lead a class discussion and make a list of questions that this video raises. You may wish to have students view the video again after they have completed the chapter to see if their questions have been answered.

WHAT'S UNUSUAL ABOUT THESE TREES?

FOCUS ON TEKS 7C, 13A

How do plants respond to their environment?

With its wide trunk, stubby branches, and just a few tiny leaves, a baobab tree looks like a sweet potato or an upside-down tree. Seen for miles across the hot, dry environment of the African savannah, the baobab can live for over 1,000 years and can grow to over 23 meters high and 27 meters around the trunk!

Draw Conclusions **Why do you think the baobab tree has such a wide trunk and short branches only at the very top?**

The baobab trunk could be thick to retain water and the branches might be short and only at the top so animals cannot reach them.

Watch the **Untamed Science** video to learn more about plants.

UntamedScience

From the Author

Humans are now one of the most significant agents of geologic change on Earth's surface. However, they have a long way to go before their impact equals that of plants. Oxygen in Earth's atmosphere is the result of the evolution of photosynthesizing bacteria—the precursors of modern plants—2.5 billion years ago. For 4 billion years the continents were huge expanses of barren rock devoid of life. Then plants evolved vascular tissue that allowed them to spread across the land, and animal life followed right behind. Even now, plant life controls the rate of land erosion and soil development. So next time you see a plant, acknowledge its contribution and show it a bit of respect.

Michael Wysession

Plant Structure, Function, and Response

Texas CHAPTER 5

Texas Essential Knowledge and Skills

SUPPORTING TEKS: 11A Examine organisms or their structures such as leaves.

TEKS: 2A Plan comparative investigations by making observations, asking well-defined questions, and using appropriate equipment and technology. **2C** Collect and record data using qualitative means. **2E** Analyze data to communicate valid conclusions supported by the data and predict trends. **7C** Demonstrate forces that affect motion in everyday life such as turgor pressure and geotropism. **12A** Investigate and explain how internal structures of organisms have adaptations that allow specific functions such as xylem in plants. **12C** Recognize levels of organization in plants, including cells, tissues, organs, organ systems, and organisms. **13A** Investigate how organisms respond to external stimuli found in the environment such as phototropism. **13B** Describe and relate responses in organisms that may result from internal stimuli such as wilting in plants that allow them to maintain balance.

PEARSON Texas.com

Spotlight On Technology

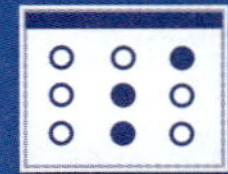

Online Chapter Test Assess student understanding of key standards with rigorous test items in an online format that hones in on student strengths and weaknesses to provide remediation opportunities.

Flipped Video for Science Show students a Pearson Flipped Video for Science for this chapter so that they can better understand how the structure and function of plants and how plants respond to their environment.

PEARSON Texas.com

Chapter at a Glance

CHAPTER PACING: 6–9 periods or 3–4$\frac{1}{2}$ blocks

INTRODUCE THE CHAPTER: Use the Focus Question and the opening image to get students thinking about plants. Activate prior knowledge and preteach vocabulary using the Getting Started pages.

Lesson 1: What Is a Plant?

Lesson 2: Plant Structures

Lesson 3: Plant Responses and Growth

ASSESSMENT OPTIONS:
Teacher's Edition: Lesson Quizzes, Texas End-of-Year Test Prep A and B
Online assessments

PEARSON Texas.com

Texas Essential Knowledge and Skills

2A Plan and implement comparative and descriptive investigations by making observations, asking well-defined questions, and using appropriate equipment and technology.

2C Collect and record data using the International System of Units (SI) and qualitative means such as labeled drawings, writing, and graphic organizers.

2E Analyze data to formulate reasonable explanations, communicate valid conclusions supported by the data, and predict trends.

7C Demonstrate and illustrate forces that affect motion in everyday life such as emergence of seedlings, turgor pressure, and geotropism.

11A Examine organisms or their structures such as insects or leaves and use dichotomous keys for identification.

12A Investigate and explain how internal structures of organisms have adaptations that allow specific functions such as gills in fish, hollow bones in birds, or xylem in plants.

12C Recognize levels of organization in plants and animals, including cells, tissues, organs, organ systems, and organisms.

13A Investigate how organisms respond to external stimuli found in the environment such as phototropism and fight or flight.

13B Describe and relate responses in organisms that may result from internal stimuli such as wilting in plants and fever or vomiting in animals that allow them to maintain balance.

The following **College and Career Readiness Standards** are covered in this chapter: **I.A.3, I.A.4, I.B.1**

CHAPTER 5

Getting Started

Check Your Understanding

This activity assesses students' understanding of the classification of plants as autotrophs because they make their own food through the process of photosynthesis. After students have shared their answers, point out that the other classification of organisms based on how they obtain food is heterotrophs, which are living things that eat autotrophs or other heterotrophs to survive.

Preteach Vocabulary Skills

Draw students' attention to the table of Greek word parts and their meanings. Review with students the concept of a word part as having a specific meaning that alone or in combination with another word part forms many of the terms common to science. Tell students that learning the meaning of word parts can help them remember and define new and unfamiliar vocabulary terms. Point out the vocabulary term *stoma* in the Chapter Preview. Tell students this term is derived from the Greek word for "mouth." As students work through the chapter, have them identify other vocabulary terms that are derived from Greek.

CHAPTER 5

Getting Started

Check Your Understanding

1. **Background** Read the paragraph below and then answer the question.

> Rahim and Malika were in the park after school. "Plants are such cool **organisms,**" said Rahim. "Can you imagine if humans had green **pigment** in their skin?" "Yeah," said Malika. "If we were **autotrophs,** I'd never have to get up early to pack my lunch!"

An **organism** is a living thing.

A **pigment** is a colored chemical compound that absorbs light.

An **autotroph** is an organism that makes its own food.

- Give an example of an autotrophic organism that has green pigment.

Sample: a tree

Vocabulary Skill

Greek Word Origins Many science words come to English from ancient Greek. Learning the Greek word parts can help you understand some of the vocabulary in this chapter.

Greek Word Part	Meaning	Example Word
chloros	pale green	chloroplast, *n.* green cellular structure in which photosynthesis occurs
petalon	leaf	petal, *n.* colorful, leaflike flower structure

2. **Quick Check** ***Chlorophyll*** **is a pigment found in plants. Which part of the word** ***chlorophyll*** **tells you that it is a green pigment?**

Chloro

English Language Proficiency Standards

ELPS Learning Strategies 1.F

Modify the three-column chart activity in Preview Vocabulary Terms to meet students' needs and abilities.

Beginning Work with students to create the chart. Have students rate how well they know each term. Have partners turn and talk about whether they know the word, or any word parts or cognates. Then add a definition to the chart.

Intermediate Display and read aloud the terms. Have partners work together to create their charts and fill in the first and second columns. Then discuss definitions and examples.

Advanced Have partners work together to create their charts. Encourage students to use glossaries or dictionaries to help them find definitions.

Advanced High Have individual students create the charts. Then have students use a dictionary or glossary to confirm their definitions.

Chapter Preview

LESSON 1

- vascular tissue • xylem
- phloem • cuticle
- gymnosperm • angiosperm

Compare and Contrast
Predict

LESSON 2

- root cap • cambium • stoma
- transpiration • embryo • fruit
- germination • flower
- pollination • sepal • petal
- stamen • pistil • ovary

Relate Cause and Effect
Observe

LESSON 3

- tropism • hormone
- auxin • photoperiodism
- critical night length
- short-day plant • long-day plant
- day-neutral plant • dormancy

Relate Text and Visuals
Draw Conclusions

Preview Vocabulary Terms

Have students create a three-column chart to rate their knowledge of the vocabulary terms before they read the chapter. In the first column of the chart, students should list the terms for the chapter. In the second column, students should identify whether they can define and use the word, whether they have heard or seen the word before, or whether they do not know the word. As the class progresses through the chapter, have students write definitions for each term in the last column of the chart.

L1 Have students look at the images on this page as you pronounce the vocabulary word. Have students repeat the word after you. Then read the definition below. Use the sample sentence in italics to clarify the meaning of the term.

gymnosperm *(JIM nuh spurm)* A vascular plant that uses seeds to reproduce but does not produce flowers. *A pine tree is a perfect example of a gymnosperm.*

flower *(FLOW ur)* The reproductive organ of an angiosperm plant. *A flower can give off a strong odor.*

sepal *(SEE pul)* Leaf-like structure that encloses a flower when it is a bud. *A sepal, often green in color, protects the developing flower and folds back as the flower blooms.*

tropism *(TROH pizm)* A plant's growth response toward or away from a stimulus. *A plant growing toward sunlight is an example of a tropism.*

Academic Vocabulary

Each lesson includes key Academic Vocabulary. See also the Support All Readers box at the start of the lesson.

Lesson 1: compare, contrast, predict

Lesson 2: cause, effect, observe, relate

Lesson 3: conclusions, relate

What Is a Plant?

How do plants respond to their environment?

LESSON PACING:
2–3 periods or 1–1½ blocks

Lesson Vocabulary

- vascular tissue
- xylem
- phloem
- cuticle
- gymnosperm
- angiosperm

Lesson Objectives	TEKS	ELPS
Identify the characteristics that all plants share.	11A, 12C	2.D.1
Name all the things that a plant needs to live successfully on land.	2E, 12A	2.D.1
Explain how plants are classified.	11A, 12A	2.D.1

Content Refresher

Plant Traits and Diversity Plants are defined by a combination of traits. Plants are multicellular eukaryotes, with cell walls that contain cellulose. The great majority of plants are autotrophs: They capture the sun's energy and photosynthesize, thus making their own food. However, some plant species are entirely or partly heterotrophic.

Land plants in particular have developed successful adaptations that enable them to live in a wide variety of environments. One important adaptation of land plants is the production of lignin, a chemical that stiffens cell walls. Lignin provides strength and support, allowing a plant to grow large and tall. Plants play critical roles in the ecosystems in which they are found. They are primary producers and form the foundation of many food webs. Plants also release oxygen gas, a product of photosynthesis essential for other organisms.

Texas Essential Knowledge and Skills

2E Analyze data to formulate reasonable explanations, communicate valid conclusions supported by the data, and predict trends.
11A Examine organisms or their structures such as insects or leaves and use dichotomous keys for identification.
12A Investigate and explain how internal structures of organisms have adaptations that allow specific functions such as gills in fish, hollow bones in birds, or xylem in plants.
12C Recognize levels of organization in plants and animals, including cells, tissues, organs, organ systems, and organisms.

English Language Proficiency Standards

ELPS Listening 2.D.1 Monitor understanding of spoken language during classroom instruction and interactions.

DIFFERENTIATED INSTRUCTION KEY
L1 Struggling Students or Special Needs
L2 On-Level Students L3 Advanced Students

LESSON PLANNER 5.1

Investigations and Activities

My Planet Diary, **Student Edition,** p. 178

Inquiry: Inquiry Warm-Up, What Do Leaves Reveal About Plants?, **PearsonTexas.com**

Teach Key Concepts, **Teacher's Edition,** p. 179

Support the TEKS, Plant Characteristics, **Teacher's Edition,** p. 180

Inquiry: Quick Lab, Algae and Other Plants, **PearsonTexas.com**

Teach Key Concepts, **Teacher's Edition,** p. 181

Differentiated Instruction, **Teacher's Edition,** p. 181

Make Analogies, Vascular Systems, **Teacher's Edition,** p. 182

21st Century Learning, Critical Thinking, **Teacher's Edition,** p. 182

Apply It!, **Student Edition,** p. 182

Inquiry: Teacher Demo, Vascular Tissue, **Teacher's Edition,** p. 183

Differentiated Instruction, **Teacher's Edition,** p. 183

Inquiry: Quick Lab, Local Plant Diversity, **Lab Manual,** p. 50

Teach Key Concepts, **Teacher's Edition,** p. 184

21st Century Learning, Communication, **Teacher's Edition,** p. 185

Differentiated Instruction, **Teacher's Edition,** p. 185

Inquiry: Quick Lab, Will Mosses Absorb Water?, **Lab Manual,** p. 51

TEKS Review

Apply the TEKS, Exploring Geotropism in Plants, **Student Edition,** p. 202

TEKS Practice, **Student Edition,** p. 204

TEKS Practice: Chapter and Cumulative Review, **Student Edition,** p. 206

Lesson 5.1, **TEKS Preparation and Study Guide Workbook,** p. 46

SHORT ON TIME? To do this lesson in approximately half the time, do the Activate Prior Knowledge activity. A discussion of the Key Concepts will familiarize students with the lesson content. Have students do the Quick Labs. The rest of the lesson can be completed by students independently.

These editable worksheets are available on **PearsonTexas.com.**
Print versions can be found in the **TEKS Preparation and Study Guide Workbook.**

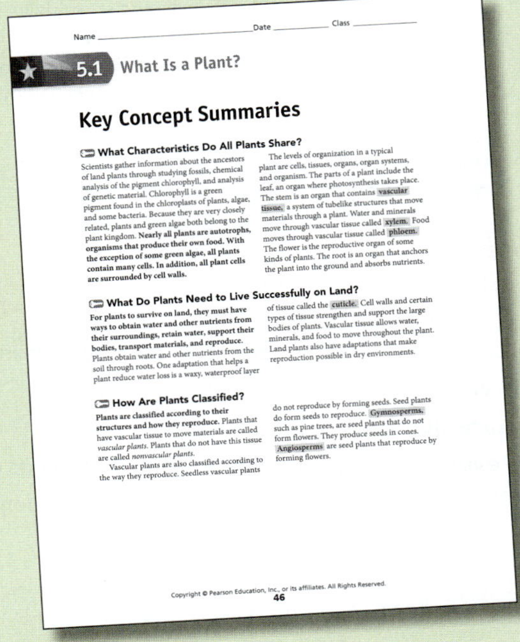

Name ____________ Date ________ Class ________

5.1 What Is a Plant?

Key Concept Summaries

What Characteristics Do All Plants Share?

Scientists gather information about the ancestors of land plants through studying fossils, chemical analysis of the pigment chlorophyll, and analysis of genetic material. Chlorophyll is a green pigment found in the chloroplasts of plants, algae, and some bacteria. Because they are very closely related, plants and green algae both belong to the plant kingdom. **Nearly all plants are autotrophs, organisms that produce their own food. With the exception of some green algae, all plants contain many cells. In addition, all plant cells are surrounded by cell walls.**

The levels of organization in a typical plant are cells, tissues, organs, organ systems, and organism. The parts of a plant include the leaf, an organ where photosynthesis takes place. The stem is an organ that contains **vascular tissue,** a system of tubelike structures that move materials through a plant. Water and minerals move through vascular tissue called **xylem.** Food moves through vascular tissue called **phloem.** The flower is the reproductive organ of some kinds of plants. The root is an organ that anchors the plant into the ground and absorbs nutrients.

What Do Plants Need to Live Successfully on Land?

For plants to survive on land, they must have ways to obtain water and other nutrients from their surroundings, retain water, support their bodies, transport materials, and reproduce. Plants obtain water and other nutrients from the soil through roots. One adaptation that helps a plant reduce water loss is a waxy, waterproof layer of tissue called the **cuticle.** Cell walls and certain types of tissue strengthen and support the large bodies of plants. Vascular tissue allows water, minerals, and food to move throughout the plant. Land plants also have adaptations that make reproduction possible in dry environments.

How Are Plants Classified?

Plants are classified according to their structures and how they reproduce. Plants that have vascular tissue to move materials are called *vascular plants.* Plants that do not have this tissue are called *nonvascular plants.*

Vascular plants are also classified according to the way they reproduce. Seedless vascular plants do not reproduce by forming seeds. Seed plants do form seeds to reproduce. **Gymnosperms,** such as pine trees, are seed plants that do not form flowers. They produce seeds in cones. **Angiosperms** are seed plants that reproduce by forming flowers.

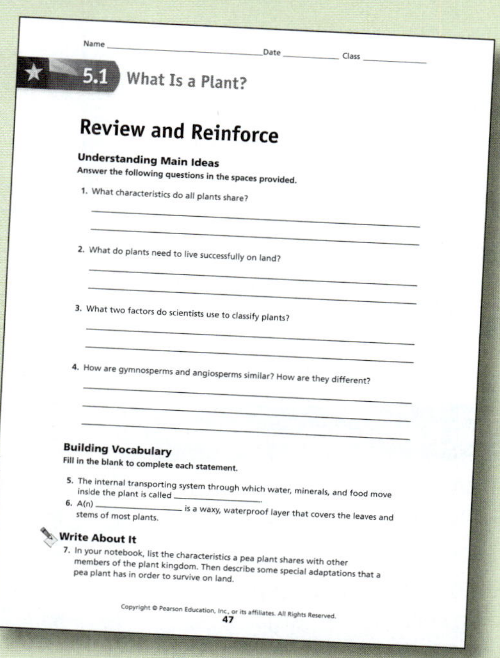

Name ____________ Date ________ Class ________

5.1 What Is a Plant?

Review and Reinforce

Understanding Main Ideas
Answer the following questions in the spaces provided.

1. What characteristics do all plants share?
2. What do plants need to live successfully on land?
3. What two factors do scientists use to classify plants?
4. How are gymnosperms and angiosperms similar? How are they different?

Building Vocabulary
Fill in the blank to complete each statement.

5. The internal transporting system through which water, minerals, and food move inside the plant is called ____________.
6. A(n) ____________ is a waxy, waterproof layer that covers the leaves and stems of most plants.

Write About It

7. In your notebook, list the characteristics a pea plant shares with other members of the plant kingdom. Then describe some special adaptations that a pea plant has in order to survive on land.

Lexile Measure = 920L

LESSON 5.1

What Is a Plant?

Establish Learning Objectives

After this lesson, students will be able to:

- Identify the characteristics that all plants share.
- Name all the things that a plant needs to live successfully on land.
- Explain how plants are classified.

Engage

Activate Prior Knowledge

MY PLANET DIARY Read *How Does Your Garden Grow?* with the class. Explain that native plants are those that evolved naturally in a particular area. Emphasize that different areas have different native plants adapted to the conditions there. Ask: **What are some conditions that determine the types of native plants that grow in an area?** *(Temperature, moisture, type of soil, altitude)*

Explore

Lab Resource: Inquiry Warm-Up 

L1 **WHAT DO LEAVES REVEAL ABOUT PLANTS?** Students will observe leaves from plants growing in two different environments and infer the growing place of each plant based on leaf thickness, texture, and size. This Inquiry Warm-Up can be found online.

Texas Essential Knowledge and Skills

11A Examine organisms or their structures such as insects or leaves and use dichotomous keys for identification.

12C Recognize levels of organization in plants and animals, including cells, tissues, organs, organ systems, and organisms.

What Is a Plant?

- What Characteristics Do All Plants Share? TEKS 11A, 12C
- What Do Plants Need to Live Successfully on Land? TEKS 2E, 12A
- How Are Plants Classified? TEKS 11A, 12A

MY PLANET DIARY — PROFILE

How Does Your Garden Grow?

Students at The Hilldale School in Daly City, California, get to play in the dirt during class. The students planted and maintain a garden filled with native species. Native plants, or plants that have been in an area for a long time, can struggle to survive if new plants are introduced. This creates problems for the insects, animals, and other organisms that rely on the native plants. The students spent three months removing nonnative plants before creating a garden that will help local organisms right outside their school.

Communicate Discuss the activity with a group of classmates. Write your answer below.

Describe a plant project you would like to do at your school.

Sample: We could grow tomato plants in the science classroom and have the cafeteria staff use them to prepare a meal.

 Do the Inquiry Warm-Up *What Do Leaves Reveal About Plants?* Find the lab online.

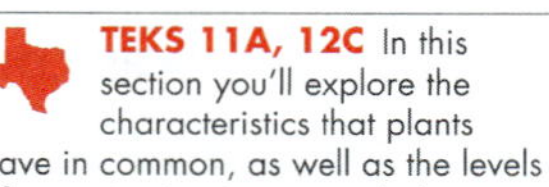

What Characteristics Do All Plants Share?

Which organisms were the ancestors of today's plants? In search of answers, biologists study fossils, the traces of ancient life forms preserved in rock and other substances. The oldest discovered plant fossils are about 400 million years old. These fossils show that even at that early date, plants had many adaptations for life on land.

SUPPORT ALL READERS

Lexile Measure = 920L **Lexile Word Count = 1521**

Prior Exposure to Content: Many students may have misconceptions on this topic

Academic Vocabulary: *compare, contrast, predict*

Science Vocabulary: *vascular tissue, cuticle*

Concept Level: Generally appropriate for most students in this grade

Preteach With: My Planet Diary "How Does Your Garden Grow?" and Figure 1 activity

Vocabulary

- vascular tissue
- xylem
- phloem
- cuticle
- gymnosperm
- angiosperm

Skills

- Reading: Compare and Contrast
- Inquiry: Predict

Better clues to the origin of plants came from comparing the chemicals in modern plants to those in other organisms. Biologists studied chlorophyll. Recall that chlorophyll is a green pigment found in the chloroplasts of plants, algae, and some bacteria. Land plants and green algae contain the same forms of chlorophyll. Further comparisons of genetic material clearly showed that plants and green algae are very closely related. Today, green algae are classified as plants.

Members of the plant kingdom share several characteristics. **Nearly all plants are autotrophs, organisms that produce their own food. With the exception of some green algae, all plants contain many cells. In addition, all plant cells are surrounded by cell walls.**

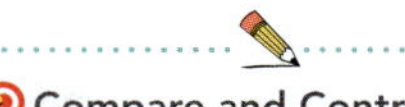

Compare and Contrast According to the text, how are land plants and green algae similar?

Land plants and green algae have the same forms of chlorophyll.

Plants Are Autotrophs

You can think of a typical plant as a sun-powered, food-making factory. Sunlight provides the energy for this food-making process, called photosynthesis. During photosynthesis, a plant uses carbon dioxide gas and water to make food. Oxygen is released during the process.

Plants Are Multicellular

Except for some green algae, all plants are made of many cells. No matter how large or small a plant is, its cells are organized into tissues. Tissues are groups of similar cells that perform a specific function in an organism.

Plant Cells

Unlike animal cells, plant cells are enclosed by a cell wall. See **Figure 1.** The cell wall surrounds the cell membrane. Plant cell walls contain cellulose, a material that makes the walls rigid. Cell walls provide support and make apples and carrots crunchy. Chloroplasts, which look like green jelly beans, are the structures in which food is made. A vacuole is a storage sac that can expand and shrink. Vacuoles store many substances, including water, wastes, and food. A plant wilts when too much water has left its vacuoles.

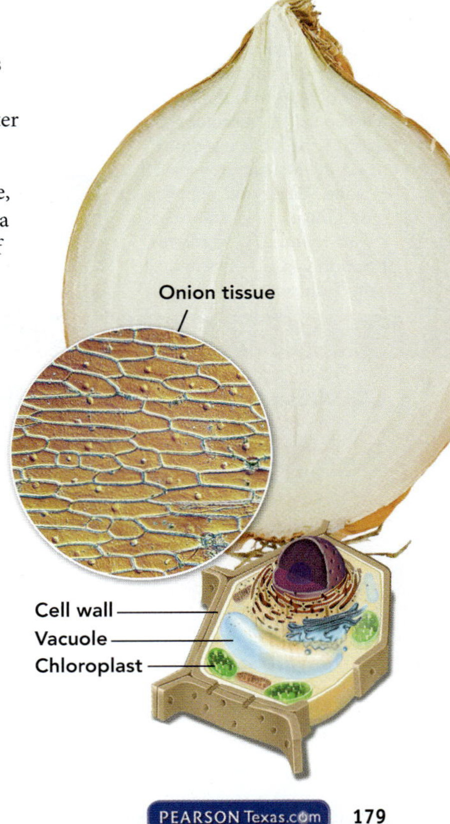

FIGURE 1

Plant Cells

Plant cells are different from animal cells. Cell walls make onions crunchy.

Infer How does having cell walls affect a plant's ability to grow tall?

Sample: The rigid cell walls provide support for the plant to grow tall.

Explain

Introduce Vocabulary

Point out to students the vocabulary term *cuticle.* The cuticle is the waxy outer layer of a plant that helps to keep the plant from drying out. The outer layer of human skin and hair is also called the cuticle. Our cuticle keeps moisture and germs outside our bodies. People also call the hardened, dead skin around the toenails and fingernails the cuticle.

Teach Key Concepts 

Explain to students that members of the plant kingdom share certain characteristics: almost all are autotrophs; all are multicellular; and all have cells surrounded by cell walls. Discuss each of these characteristics by first explaining that autotrophs convert energy from sunlight into the chemical energy of the food they make, which plants use to live. Ask: **What is the name of the food-making process in plants?** *(Photosynthesis)* **What are the raw materials and products of photosynthesis?** *(Raw materials are carbon dioxide and water. Products are food and oxygen.)* **In what structures of the plant cell does photosynthesis take place?** *(Chloroplasts)* **What pigment in the chloroplasts is able to capture the energy of sunlight?** *(Chlorophyll)* **Are animals autotrophs? How do you know?** *(No, they eat plants or other animals.)*

Compare and Contrast Tell students that comparing and contrasting involves examining the similarities and differences between things. By comparing and contrasting information, students can determine how concepts, facts, and events are alike and different.

Teach With Visuals

Tell students to look at **Figure 1.** Ask: **Which structures are found only in plant cells?** *(Cell wall, chloroplasts, one large vacuole)* **Which structure makes plant parts stiff?** *(Cell wall)* **In which structure is food made?** *(Chloroplast)*

PEARSON Texas.com

English Language Proficiency Standards

ELPS Listening 2.D.1

Have students listen to the lesson and complete the following sentence frames: *All plants ______. Plants can have ______. Plants need ______. The types of plants are ______.*

Beginning Have students listen as you read aloud each section. Encourage students to ask you to repeat or explain things they do not understand. Model how to complete the sentence frames. Have students repeat.

Intermediate Assign small groups to sections of the text. Ask students to listen as you or another student read aloud the section. Have groups complete sentence frames and share responses.

Advanced Assign small groups to each section of the lesson. Have one student in each group read aloud. Then have groups summarize the section, confirm answers, and share their responses.

Advanced High Have partners take turns reading aloud, pausing after each section to summarize.

LESSON 5.1

Explain

Support the TEKS

PLANT CHARACTERISTICS Review with students the meaning of the term *multicellular,* and point out that all plants are multicellular except for some green algae. Explain to students that similar cells that perform a specific function are organized into tissues. Groups of different types of tissues are arranged together to form organs. Finally, organs are grouped into organ systems. Ask: **What are the major organ systems of plants?** *(Root system and shoot system)* **What are the four main plant organs?** *(Roots, stems, leaves, flowers)* **Which of the organs are part of the shoot system?** *(Stems, flowers, leaves)* **What is the purpose of the root system?** *(Anchors the plant to the ground, stores food, absorbs water and minerals)*

Teach With Visuals

Tell students to look at the illustration *The Organs of a Flowering Plant.* Ask: **What tissue makes up most of the stem?** *(Vascular tissue)* **What is the function of vascular tissue?** *(Water, minerals, and food move through it)* **How are plants anchored to the ground?** *(Roots)*

Elaborate

Lab Resource: Quick Lab 

L2 **ALGAE AND OTHER PLANTS** Students will observe algae and plant leaves to determine how they are alike and what evidence supports the idea that plants are descended from green algae. This Quick Lab can be found online.

Evaluate

Assess Your Understanding

After students answer the questions, have them evaluate their understanding by completing the appropriate sentence.

RTI Response to Intervention

1a. If students need help identifying the site of food production in a plant, **then** have them review the paragraph under the red heading *Plant Cells.*

b. If students cannot recognize the major organ systems of plants, **then** review plant organization.

c. If students have trouble understanding the role of the vacuole as a storage sac for water, **then** review the functions of this cell part.

The Organs of a Flowering Plant
The tissues of flowering plants are organized into four main organs: flowers, leaves, stems, and roots.

Flower The flower is the reproductive organ of some kinds of plants. These can be made up of male or female reproductive parts—or both.

Leaf The leaf is a plant organ that makes food for the plant. Except for a thin layer of tissue that covers the leaf, most cells in the leaf contain chloroplasts.

Stem The stem is a plant organ that holds leaves up so that they can absorb sunlight. Most of a stem is made up of a tissue called **vascular tissue**. Vascular tissue is a system of tubelike structures inside a plant. Water and minerals move through vascular tissue called **xylem**. Food moves through vascular tissue called **phloem**.

Root The root is a plant organ that anchors the organism in the ground. A root contains tissues that absorb water and minerals. In some plants, such as carrot and radish, the roots contain tissue that stores food made in the leaves.

Plant Organization Most of the plants that you see around you are flowering plants. The levels of organization in a typical plant include cells, tissues, organs, organ systems, and organism. There are two main organ systems in plants: the root system and the shoot system. The root system is the part of the plant below the ground and the shoot system is the part above ground.

Do the Quick Lab *Algae and Other Plants.* Find the lab online.

Assess Your Understanding

TEKS 12C

1a. Review (Flowers/Chloroplasts) are the site where a plant produces its food.

b. Recognize What are the two organ systems of a typical flowering plant?

The root system and the shoot system

c. Infer What do you think happens to a plant cell if the plant is given too much water?

Sample: The vacuoles will not be able to hold all of the water.

got it?

O I get it! Now I know that nearly all plants are autotrophs, are multicellular, and have cell walls.

O I need extra help with See TE note.

What Do Plants Need to Live Successfully on Land?

TEKS 2E, 12A In this section you'll find out how plants obtain water and other nutrients from their environments, as well as how they transport these materials.

Imagine algae floating in the ocean. The algae obtain water and other materials directly from the water around them. They are held up toward the sunlight by the water. Now imagine plants living on land. What adaptations would help them meet their needs without water all around them? **For plants to survive on land, they must have ways to obtain water and other nutrients from their surroundings, retain water, support their bodies, transport materials, and reproduce.**

Obtaining Water and Other Nutrients

Recall that all organisms need water to survive. Obtaining water is easy for algae because water surrounds them. To live on land, plants need adaptations for obtaining water from the soil. One adaptation is the way a plant's roots grow, as shown in **Figure 2.** Roots also obtain other nutrients from the soil, such as nitrogen and phosphorus. Plant cells use these nutrients to build proteins and for other processes. Some plants, such as tulips, onions, and garlic, can store food in a bulb. The bulb is a modified stem that grows underground, helping the plant to survive dry or cold seasons.

FIGURE 2

Getting Water in the Desert

The saguaro cactus and the acacia tree both survive with limited water in deserts. Saguaro roots spread out horizontally. When it rains, the roots quickly absorb water over a wide area. Acacia trees in the Negev Desert of Israel get their water from deep underground instead of at the surface.

Interpret Diagrams Draw the roots of the acacia tree. Then describe how the growth of the roots differs between the plants.

Sample: The cactus roots grow horizontally, while the acacia roots grow vertically.

Explain

Teach Key Concepts

Tell students that the evolution of plants from water-dwellers to land-dwellers required adaptations that would help plants meet their needs without water all around them. Discuss with students the five requirements of plants to live successfully on land. Ask: **What do plants need to survive on land?** *(Plants must have ways to obtain water and other nutrients from their surroundings, retain water, support their bodies, transport materials to all plant parts, and reproduce.)* **Where do plants get water and nutrients?** *(From the soil)* **What plant adaptation helps a plant get water from the soil?** *(Roots)* **Are all roots alike? Explain.** *(No, plant roots differ depending on the environment and the soil conditions.)*

Address Misconceptions

L1 PLANTS OR ANIMALS FIRST? When discussing the history of life on land, students often think of some exotic animal emerging from the sea, breathing air, and starting the long line of land-dwelling organisms to come. Students do not always consider whether plants beat animals to it. Ask: **What did animals need to survive on land?** *(A source of food)* **Where did this food come from?** *(Plants that were already thriving on land)*

Differentiated Instruction

L1 Plant Cell Structure Have student pairs draw a large illustration of a cell. Then have them make several sticky labels for cell features such as the nucleus, cytoplasm, cell wall, vacuole, and chloroplasts. Have students take turns placing a sticky note in the appropriate part of the illustration as they describe the cell part. For example, "A plant cell has many chloroplasts."

L1 Plant Adaptations Have students use old magazines or newspapers to find a photograph of a land plant. Have them label the plant with the five adaptations that a plant needs to survive on land.

L3 Botany The scientific study of plants is known as *botany.* Have interested students find out about the various areas of specialization in botany as well as the numerous career opportunities the subject offers.

Texas Essential Knowledge and Skills

2E Analyze data to formulate reasonable explanations, communicate valid conclusions supported by the data, and predict trends.

12A Investigate and explain how internal structures of organisms have adaptations that allow specific functions such as gills in fish, hollow bones in birds, or xylem in plants.

LESSON 5.1

Explain

Lead a Discussion

PLANT ADAPTATIONS Continue to discuss the plant adaptations that make plants successful at survival on land. Ask: **What plant adaptation reduces water loss?** *(Cuticle)* **Why does a plant tend to lose water?** *(When the amount of water in plant cells is greater than the amount of water in the air, water moves out of the cells and into the air.)* Discuss with students the need for land plants to support their body and to transport essential materials throughout the body. Ask: **Why must land plants support their body?** *(They must position those parts of their body that make food in such a way to get maximum sunlight.)* **How do they accomplish this?** *(Cell walls and vascular tissue strengthen and support plant bodies.)* **What is vascular tissue?** *(A system of tubelike structures for transporting water, minerals, and food)*

Make Analogies

VASCULAR SYSTEMS Discuss with students the role of the vascular system in plants in transporting materials from one part of the plant body to another. Then remind students of the cardiovascular system in humans. Ask: **What does the term *vascular* refer to?** *(A system of vessels that have certain functions in getting material to and from cells)* **How are the two systems similar?** *(They both transport vital materials to and from all parts of the body.)*

Elaborate

21st Century Learning

CRITICAL THINKING Tell students that some of the earliest land plants to appear on Earth were very tiny plants that lived in moist environments. These plants were sometimes described as being only partly adapted to life on land. Ask: **Why is this description appropriate?** *(Because they were small, they did not have to have an extensive vascular system to transport materials throughout the plant or support the plant. Because they grew in moist places, water and nutrients were available and plentiful. Water for reproduction was also available.)*

Apply It!

L1 Before beginning the activity, remind students that a graph shows the relationship between two variables—in this case the time of day and the amount of water loss.

Predict Remind students that when they predict, they make an inference about a future event or situation based on what they already know.

FIGURE 3

Waterproof Leaves

The waxy cuticle of many leaves, like the one below, looks shiny under light.

Retaining Water When there is more water in plant cells than in the air, the water leaves the plant and enters the air. The plant could dry out if it cannot hold onto water. One adaptation that helps a plant reduce water loss is a waxy, waterproof layer of tissue called the **cuticle.** You can see the cuticle on the leaf in **Figure 3.**

Support A plant on land must support its own body. It's easier for small, low-growing plants to support themselves. In larger plants, the food-making parts must be exposed to as much sunlight as possible. Cell walls and certain types of tissues strengthen and support the large bodies of these plants.

Transporting Materials A plant needs to transport water, minerals, food, and other materials from one part of its body to another. In general, water and minerals are taken up by the root system, while food is made in the shoot system. But all of the plant's cells need water, minerals, and food.

In small plants, materials can simply move from one cell to the next. Larger plants need a more efficient way to transport materials from one part of the plant to another. These plants have vascular tissue. Water, minerals, and food move through vascular tissue. See vascular tissue in action in **Figure 4.**

apply it!

This graph shows how much water a plant loses during the day. Give the graph a title.

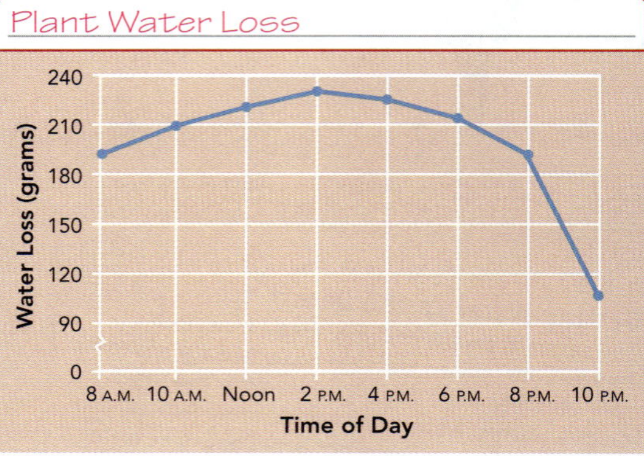

1 **Interpret Graphs** During what part of the day did the plant lose the most water?

Around 2 P.M.

2 **Predict** How might the line in the graph look from 10 P.M. to 8 A.M.? Why?

Sample: The line would continue to decrease through the night until morning and then increase. There is less water loss when there is no sun.

3 CHALLENGE Do you think this graph would be the same for plants all around the world? Why?

Sample: No; plants in different parts of the world are exposed to different amounts of sunlight.

Reproduction For algae and some other plants, reproduction can only occur if there is water in the environment. This is because the sperm cells of these plants swim through the water to the egg cells. Most land plants have adaptations that make reproduction possible in dry environments.

FIGURE 4

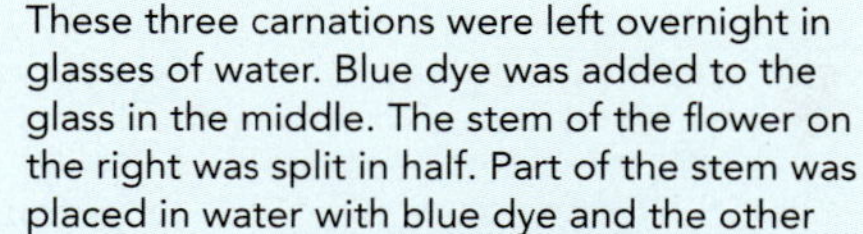

Colorful Carnations

These three carnations were left overnight in glasses of water. Blue dye was added to the glass in the middle. The stem of the flower on the right was split in half. Part of the stem was placed in water with blue dye and the other part was placed in water with red dye.

Draw Conclusions Why did the flowers in the glasses with dye change color?

Sample: In both cases, the dye was transported through vascular tissue in the stem into the flower, changing the flower's color.

Lab zone Do the Quick Lab *Local Plant Diversity.* Student Lab Manual, p. 50

Assess Your Understanding

TEKS 12A

2a. **Define** What is a cuticle?

A cuticle is a waxy, waterproof layer on a leaf that helps the plant reduce water loss.

b. **Explain** Describe the pros and cons of being a tall land plant.

Sample: Tall plants have to support themselves more than shorter plants. But, they get more sunlight if they are taller than the plants around them.

got it?

○ **I get it!** Now I know that to live on land, plants need to obtain and retain water and nutrients, support themselves, transport materials, and reproduce.

○ I need extra help with See TE note.

Differentiated Instruction

L1 Growing Tall To ensure that students understand the importance of vascular tissue as an adaptation of land plants, have them explain why different plants are able to grow to many different heights

L1 Problem/Solution As a summarizing activity, have students make a two-column chart with the headings *Problem* and *Solution*. In the *Problem* column, have them identify the five problems plants had to overcome in order to successfully live on land. In the *Solution* column, have students identify the plant adaptation that solved each problem.

L3 Plants in Moist Places Have students do the Problem/Solution activity above and add a third column with the heading *Small Plants Growing in Moist Places*. Have students explain how these land plants are adapted to life on land.

Teacher Demo 

L1 VASCULAR TISSUE

Materials two celery stalks, medium-sized jar, dark food coloring, water, knife

Time 10 minutes and then observation 24 hours later

Fill the jar one-quarter full with water and color the water with a few drops of food coloring. Place the stalks leaf-side up in the jar of colored water. Have students observe the appearance of the stalks. Leave the stalks undisturbed for 24 hours. Have students again observe the stalks. Remove the stalks from the water. Using the knife, cut off the part of the stalk that was submerged and discard it. Then cut the stalks into cross-sectional pieces and provide students with pieces to observe.

Ask: **What did you observe about the stalk and leaf color?** *(The stalk and leaf started to show streaks of the food coloring.)* **What did you observe about the pieces of stalk?** *(There were dots the color of the food coloring in the cross-sectional pieces.)* **What do you think those dots are?** *(The structures that make up the vascular tissue that carried the colored water up the stalk to the leaves)*

Lab Resource: Quick Lab 

L2 LOCAL PLANT DIVERSITY Students will go on a plant hunt around the school describing and drawing as many different species as they can find. This Quick Lab can be found in the Student Lab Manual, p. 50, and online.

Evaluate

Assess Your Understanding

After students answer the questions, have them evaluate their understanding by completing the appropriate sentence.

RTI Response to Intervention

2a. If students have trouble describing a cuticle, **then** remind them that a cuticle is an adaptation of a plant to living on land and aids in water retention.

b. If students need help with identifying the advantages and disadvantages of being a tall land plant, **then** tell them to think about size in terms of support and of getting the energy required for photosynthesis.

Teach Key Concepts

Explain to students that the earliest plants resembled green algae, which are nonvascular, seedless plants. Over millions of years, vascular plants evolved. Seeds and flowers were later developments that helped plants to survive and expand into new environments. Discuss with students the two major characteristics used to classify plants. Ask: **What is the major difference between vascular and nonvascular plants?** *(Only vascular plants have tube-like structures, or vascular tissues, that help them carry food, water, and other materials from one part of the plant to another.)* **What advantages do vascular plants have over nonvascular plants?** *(Vascular plants can grow larger because they can transport materials more efficiently from one part of the plant to another.)* **What are the three major types of vascular plants? What is an example of each?** *(Seedless vascular plants such as ferns, angiosperms such as apple trees, and gymnosperms such as pine trees)* **What is the major difference between gymnosperms and angiosperms?** *(Gymnosperms produce seeds without flowers, while angiosperms reproduce by using flowers to form seeds.)*

Teach With Visuals

Tell students to look at **Figure 5.** Ask: **Why are liverworts and mosses such small, low-lying plants?** *(They are nonvascular plants. They have no way to transport materials across a large area.)* **What kind of plant makes its seeds in cones?** *(Gymnosperm)* **Which kind of plant has no roots?** *(Nonvascular plant)*

Address Misconceptions

L1 GYMNOSPERM DIVERSITY Students may think that all gymnosperms look like pine trees and that they produce pinecones. Explain that gingko trees are also gymnosperms. Male gingko trees produce a small pollen cone, while female gingkos make a seed with a fleshy covering that looks like a small apricot. Cycads are gymnosperms that look similar to palm trees. Ask: **How could you tell whether a plant is an angiosperm or a gymnosperm?** *(Only angiosperms produce flowers.)*

TEKS 11A, 12A In this section you'll learn about the major features scientists use to classify plants.

How Are Plants Classified?

There are hundreds of thousands of plant species on Earth. They range from tiny mosses carpeting the ground to giant sequoia trees towering toward the sky. All plants have cells with cell walls and chloroplasts. But plants differ in other ways. **Plants are classified according to their structures and how they reproduce.**

Transporting Materials One difference in structure that is used to classify plants is related to how materials are transported in the plant. Plants that have vascular tissue are called *vascular plants*. Plants that do not have vascular tissue are called *nonvascular plants*.

Reproduction Vascular plants are also grouped according to the way they reproduce. Seedless vascular plants do not reproduce by forming seeds. Other vascular plants, called seed plants, do form seeds. One group of seed plants produces flowers. The other group does not.

FIGURE 5

Classifying Plants

All plants share basic characteristics, such as the ability to make their own food. But plant structures can vary greatly.

Identify What three plant structures do scientists use to classify plants?

Scientists determine whether or not there is vascular tissue, whether or not the plant forms seeds, and whether or not the plant produces flowers.

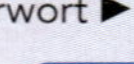

Liverwort ▶

Nonvascular Plants

Mosses and liverworts are two examples of nonvascular plants. They do not have vascular tissue or roots. Water and other nutrients travel through individual cells. Because of this, these plants are small. They usually live in moist places.

◀ Horsetail

Seedless Vascular Plants

Ferns and horsetails are two examples of seedless vascular plants. They have roots, stems, and leaves, but they do not produce seeds. Instead, they reproduce by forming spores.

English Language Proficiency Standards

ELPS Listening 2.I.4

Have students listen to pages 184–185 and work with peers to create a classification tree for plants.

Beginning Display a blank classification tree. Explain that this is a way of showing how plants are like or unlike each other. Start the tree with the word *Plants* at the bottom. Then read aloud the pages. After reading about each class of plants, have students show where to add to the chart.

Intermediate Distribute blank classification trees to small groups. Have students read aloud the pages. Pause after reading about each class of plants and have students add it to their charts.

Advanced Have students in small groups take turns reading aloud. Have them create a classification tree to show types of plants.

Advanced High Have student partners read the pages and create a classification tree to show how plants are related. Encourage them to tell details about each class of plant.

Texas Essential Knowledge and Skills

11A Examine organisms or their structures such as insects or leaves and use dichotomous keys for identification.

12A Investigate and explain how internal structures of organisms have adaptations that allow specific functions such as gills in fish, hollow bones in birds, or xylem in plants.

◀ Ponderosa pinecone

Gymnosperms

Gymnosperms are vascular plants that use seeds to reproduce but do not form flowers. Most gymnosperms produce seeds in structures called cones. These gymnosperms include pine trees and are called conifers. Other gymnosperms include cycads and gingko trees.

Texas bluebonnet

Angiosperms

Angiosperms are vascular plants that reproduce by forming flowers that produce seeds. Some flowers are large and colorful, like the Texas bluebonnet. Some flowers are small and harder to see, like those of an oak tree.

Lab zone Do the Quick Lab *Will Mosses Absorb Water?* Student Lab Manual, p. 51

Assess Your Understanding

got it?

- O **I get it!** Now I know that plants are classified according to their structure and how they reproduce.
- O I need extra help with See TE note.

185

Differentiated Instruction

L1 Vascular vs. Nonvascular Plants To help students distinguish between vascular and nonvascular plants, have them create Venn diagrams. The overlapping area in the center of the diagram should include features shared by all plants. The separate areas on each side of the diagram should include words and phrases unique to vascular or nonvascular plants.

L3 Dichotomous Key Have student pairs prepare dichotomous keys for classifying plants by their structures and how they reproduce. Then ask the students to trade keys with another group. Next, have the students go outside to try out their keys on some of the plants in the schoolyard or neighborhood. If students discover any problems with their keys, they should correct the mistakes. Finally, have each pair produce a final, illustrated draft of their key.

Elaborate

21st Century Learning

COMMUNICATION Have students work in pairs to find images of plants in newspapers, magazines, or online. For each image, students work together to classify each plant based on their structures and how they reproduce. Encourage students to include photos or drawings of local plants.

Lab Resource: Quick Lab 

L2 WILL MOSSES ABSORB WATER? Students will test whether mosses or sand can better absorb water. This Quick Lab can be found in the Student Lab Manual, p. 51, and online.

Evaluate

Assess Your Understanding

Have students evaluate their understanding by completing the appropriate sentence.

RTI Response to Intervention

If students have trouble explaining plant classification, **then** remind them that plants are classified by their structures and how they reproduce.

Name ______________________ Date ____________ Class ____________

Assess Your Understanding

What Is a Plant?

What Characteristics Do All Plants Share?

1a. **REVIEW** (Flowers/Chloroplasts) are the site where a plant produces its food.

b. **RECOGNIZE** What are the two organ systems of a typical flowering plant? ______________________

____

c. **INFER** What do you think happens to a plant cell if the plant is given too much water? ______________________

____

What Do Plants Need to Live Successfully on Land?

2a. **DEFINE** What is a cuticle? ______________________

____

b. **EXPLAIN** Describe the pros and cons of being a tall land plant. ______________________

____

____

How Are Plants Classified?

got it?

○ **I get it!** Now I know that plants are classified according to ______________________

____

____

○ **I need extra help with** ______________________

____

Place the outside corner, the corner away from the dotted line, in the corner of your copy machine to copy onto letter-size paper.

Place the outside corner, the corner away from the dotted line, in the corner of your copy machine to copy onto letter-size paper.

Name ________________ Date __________ Class __________

Enrich

What Is a Plant?

Desert plants have special survival needs. Read the passage and study the diagram below. Then use a separate sheet of paper to answer the questions that follow.

Desert Survival

To obtain water, some desert plants have very deep root systems that can absorb moisture far underground. Others have shallow, horizontal root systems that can quickly absorb a large amount of water when it rains.

The aboveground surfaces of many desert plants are covered with spines. These spines help to shade the plant from the sun and keep it from getting too hot. They also help to reduce water loss from the plant by shielding it from dry winds. Some plants in the desert have thick, fleshy stems that can store water for long periods of time.

Many plants, such as the one shown below, survive dry periods by becoming *dormant,* or inactive. When a plant is dormant, it needs very little water.

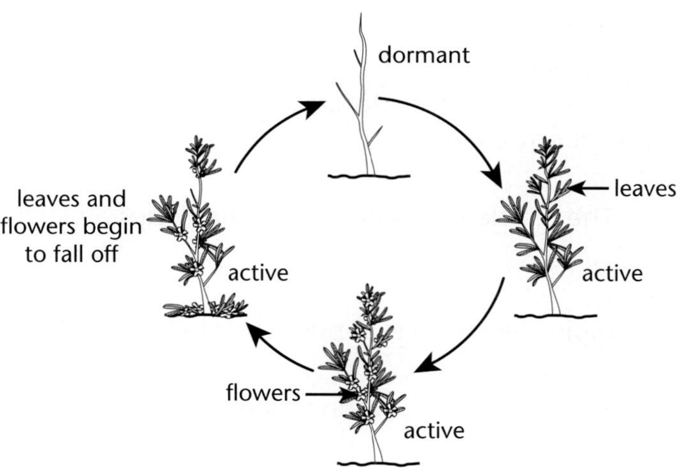

1. What do you think causes the plant in the figure above to come out of dormancy and become active?
2. Why do you think the plant loses its leaves and flowers when it becomes dormant?
3. Do you think the plant shown above is more likely to have deep roots or shallow horizontal roots? Explain.
4. A desert plant called the creosote bush has a double root system: It has both shallow horizontal roots and deep vertical roots. Why would this type of root system be an advantage to a desert plant?
5. Do you think a desert plant is more likely to have a thin or thick cuticle?

Name ______________________ Date __________ Class __________

Lesson Quiz

What Is a Plant?

Write the letter of the correct answer on the line at the left.

1. ___ In a typical plant, vascular tissue makes up the

A roots

B flowers

C stem

D leaves

2. ___ Which plant structure is responsible for obtaining nutrients from the soil?

A leaves

B flowers

C cuticles

D roots

3. ___ Which level of organization is a plant leaf?

A cell

B tissue

C organ

D organ system

4. ___ Vascular plants that do not form flowers to reproduce are called

A gymnosperms

B angiosperms

C mosses

D spores

If the statement is true, write *true*. If the statement is false, change the underlined word or words to make the statement true.

5. ______________ The <u>cuticle</u> is an adaptation that helps a plant reduce water loss.

6. ______________ Rigid <u>chloroplasts</u> in plant cells help provide support.

7. ______________ The tissues of flowering plants are organized into four main <u>organ systems</u>.

8. ______________ The system of tubelike structures inside a plant through which water, minerals, and food move is called <u>root</u> tissue.

9. ______________ Nearly all plants are <u>unicellular</u>.

10. ______________ Flowers are structures that help a plant <u>support itself</u>.

Place the outside corner, the corner away from the dotted line, in the corner of your copy machine to copy onto letter-size paper.

What Is a Plant?

Answer Key

Review and Reinforce

Find the worksheet in the Student Workbook.

1. Nearly all plants are autotrophs. All are multicellular. All plant cells are surrounded by cell walls.
2. To live successfully on land, plants need ways to obtain water and other nutrients from their surroundings, retain water, support their bodies, transport materials, and reproduce.
3. Plants are classified according to their structures and how they reproduce.
4. Both are vascular plants that use seeds to reproduce. Gymnosperms, however, do not form flowers, which angiosperms do.
5. vascular tissue
6. cuticle
7. Like other plants, a pea plant is made up of many cells that are surrounded by cell walls, and it produces its own food via photosynthesis. In order to survive on land, a pea plant has roots that enable it to obtain water and other nutrients from the soil. A waxy, waterproof cuticle on its leaves and stems helps it retain that water. Cell walls and tissue strengthen and support a pea plant, enabling it to lift its food-making leaves toward the light of the sun. Water, minerals, and food are transported to all the parts of a pea plant through vascular tissues.

Enrich

1. The plant comes out of dormancy after a rain shower.
2. It loses its leaves and flowers to help conserve water.
3. Because the plant grows leaves and flowers after a rain shower, it probably absorbs a large amount of rainwater. This suggests that the plant has shallow horizontal roots.
4. It could absorb large amounts of water after a rain with its roots near the surface and could get water from deep in the earth during dry periods with its vertical roots.
5. Because desert plants must survive with very little water, they need to reduce water loss as much as possible and they are more likely to have thick cuticles.

Lesson Quiz

1. C
2. D
3. C
4. A
5. true
6. cell walls
7. organs
8. vascular
9. multicellular
10. reproduce

Plant Structures

How do plants respond to their environment?

LESSON PACING:
3–4 periods or $1\frac{1}{2}$–2 blocks

Lesson Vocabulary

- root cap • cambium • stoma • transpiration • embryo • fruit
- germination • flower • pollination • sepal • petal • stamen
- pistil • ovary

Content Refresher

Structures of Angiosperms Angiosperms, the flowering plants, make up more than 88 percent of all living plant species and can be found almost anywhere on Earth. In addition to being classified according to the number of cotyledons present in the plant embryo, angiosperms can also be classified according to the characteristics of their stems. Woody plants have stems that are hard and rigid. Their cells have thick cell walls that support the plant body. In contrast, herbaceous plants have stems that are generally soft and smooth. Angiosperms have unique characteristics—flowers and fruits—that unify the group. Flowers enclose the ovary in which seeds develop; fruits are the mature ovaries.

Lesson Objectives	TEKS	ELPS
Describe the functions of roots, stems, and leaves.	7C, 11A, 12A, 13B	4.F.7
Explain how seeds become new plants.	7C, 13A	4.F.7
Describe the structures of a flower.	11A, 12C	4.F.7

Texas Essential Knowledge and Skills

7C Demonstrate and illustrate forces that affect motion in everyday life such as emergence of seedlings, turgor pressure, and geotropism.
11A Examine organisms or their structures such as insects or leaves and use dichotomous keys for identification.
12A Investigate and explain how internal structures of organisms have adaptations that allow specific functions such as gills in fish, hollow bones in birds, or xylem in plants.
12C Recognize levels of organization in plants and animals, including cells, tissues, organs, organ systems, and organisms.
13A Investigate how organisms respond to external stimuli found in the environment such as phototropism and fight or flight.
13B Describe and relate responses in organisms that may result from internal stimuli such as wilting in plants and fever or vomiting in animals that allow them to maintain balance.

English Language Proficiency Standards

ELPS Reading 4.F.7 Use support from peers and teachers to enhance and confirm understanding.

DIFFERENTIATED INSTRUCTION KEY
L1 Struggling Students or Special Needs
L2 On-Level Students **L3** Advanced Students

LESSON PLANNER 5.2

Investigations and Activities

Inquiry: Inquiry Warm-Up, Which Plant Part Is It?, **PearsonTexas.com**
Teach Key Concepts, **Teacher's Edition,** p. 187
Teach Key Concepts, **Teacher's Edition,** p. 188
21st Century Learning, Critical Thinking, **Teacher's Edition,** p. 188
Apply It!, **Student Edition,** p. 189
Inquiry: Teacher Demo, Observing Tree Rings, **Teacher's Edition,** p. 189
Differentiated Instruction, **Teacher's Edition,** p. 189
Teach Key Concepts, **Teacher's Edition,** p. 190
Support the TEKS, Leaf Structures, **Teacher's Edition,** p. 190
Teacher to Teacher, **Teacher's Edition,** p. 190
21st Century Learning, Critical Thinking, **Teacher's Edition,** p. 191
Inquiry: Lab Investigation, Investigating Stomata, **Lab Manual,** p. 52

Teach Key Concepts, **Teacher's Edition,** p. 192
Inquiry: Build Inquiry, Modeling Seed Dispersal, **Teacher's Edition,** p. 193
Differentiated Instruction, **Teacher's Edition,** p. 193
Inquiry: Lab Investigation, Fruit Basket, **Lab Manual,** p. 57

Teach Key Concepts, **Teacher's Edition,** p. 194
Inquiry: Build Inquiry, Observing the Structure of a Flower, **Teacher's Edition,** p. 195
Differentiated Instruction, **Teacher's Edition,** p. 195
Inquiry: Quick Lab, Modeling Flowers, **Lab Manual,** p. 63

TEKS Review

Apply the TEKS, Exploring Geotropism in Plants, **Student Edition,** p. 202

TEKS Practice, **Student Edition,** p. 204

TEKS Practice: Chapter and Cumulative Review, **Student Edition,** p. 206

Lesson 5.2, **TEKS Preparation and Study Guide Workbook,** p. 48

SHORT ON TIME? To do this lesson in approximately half the time, do the Activate Prior Knowledge activity. A discussion of the Key Concepts will familiarize students with the lesson content. Have students do the Quick Lab. The rest of the lesson can be completed by students independently.

These editable worksheets are available on **PearsonTexas.com.**
Print versions can be found in the **TEKS Preparation and Study Guide Workbook.**

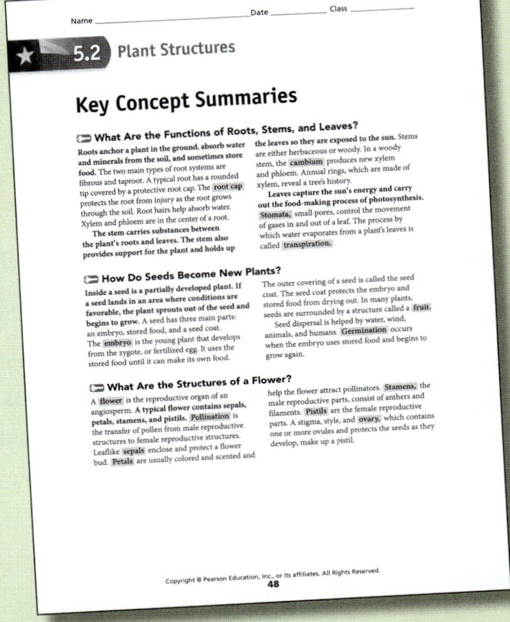

Name ______ Date ______ Class ______

5.2 Plant Structures

Key Concept Summaries

What Are the Functions of Roots, Stems, and Leaves?

Roots anchor a plant in the ground, absorb water and minerals from the soil, and sometimes store food. The two main types of root systems are fibrous and taproot. A typical root has a rounded tip covered by a protective root cap. The **root cap** protects the root from injury as the root grows through the soil. Root hairs help absorb water. Xylem and phloem are in the center of a root.

The stem carries substances between the plant's roots and leaves. The stem also provides support for the plant and holds up the leaves so they are exposed to the sun. Stems are either herbaceous or woody. In a woody stem, the **cambium** produces new xylem and phloem. Annual rings, which are made of xylem, reveal a tree's history.

Leaves capture the sun's energy and carry out the food-making process of photosynthesis. **Stomata,** small pores, control the movement of gases in and out of a leaf. The process by which water evaporates from a plant's leaves is called **transpiration.**

How Do Seeds Become New Plants?

Inside a seed is a partially developed plant. If a seed lands in an area where conditions are favorable, the plant sprouts out of the seed and begins to grow. A seed has three main parts: an embryo, stored food, and a seed coat. The **embryo** is the young plant that develops from the zygote, or fertilized egg. It uses the stored food until it can make its own food. The outer covering of a seed is called the seed coat. The seed coat protects the embryo and stored food from drying out. In many plants, seeds are surrounded by a structure called a **fruit.**

Seed dispersal is helped by water, wind, animals, and humans. **Germination** occurs when the embryo uses stored food and begins to grow again.

What Are the Structures of a Flower?

A **flower** is the reproductive organ of an angiosperm. **A typical flower contains sepals, petals, stamens, and pistils.** **Pollination** is the transfer of pollen from male reproductive structures to female reproductive structures. Leaflike **sepals** enclose and protect a flower bud. **Petals** are usually colored and scented and help the flower attract pollinators. **Stamens,** the male reproductive parts, consist of anthers and filaments. **Pistils** are the female reproductive parts. A stigma, style, and **ovary,** which contains one or more ovules and protects the seeds as they develop, make up a pistil.

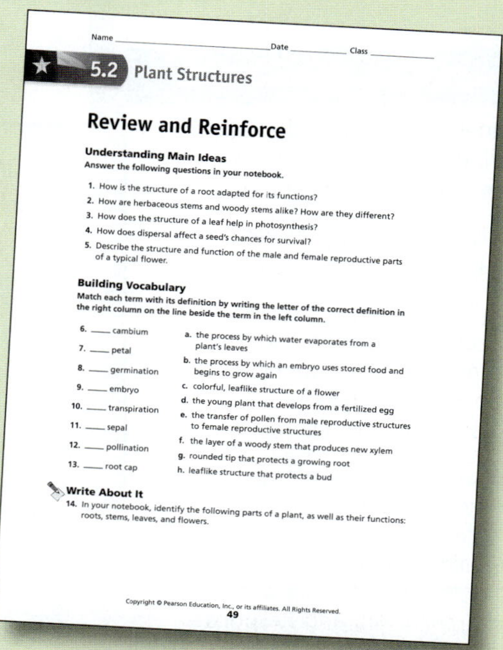

Name ______ Date ______ Class ______

5.2 Plant Structures

Review and Reinforce

Understanding Main Ideas
Answer the following questions in your notebook.

1. How is the structure of a root adapted for its functions?
2. How are herbaceous stems and woody stems alike? How are they different?
3. How does the structure of a leaf help in photosynthesis?
4. How does dispersal affect a seed's chances for survival?
5. Describe the structure and function of the male and female reproductive parts of a typical flower.

Building Vocabulary
Match each term with its definition by writing the letter of the correct definition in the right column on the line beside the term in the left column.

6. ____ cambium	a. the process by which water evaporates from a plant's leaves
7. ____ petal	b. the process by which an embryo uses stored food and begins to grow again
8. ____ germination	c. colorful, leaflike structure of a flower
9. ____ embryo	d. the young plant that develops from a fertilized egg
10. ____ transpiration	e. the transfer of pollen from male reproductive structures to female reproductive structures
11. ____ sepal	f. the layer of a woody stem that produces new xylem
12. ____ pollination	g. rounded tip that protects a growing root
13. ____ root cap	h. leaflike structure that protects a bud

Write About It
14. In your notebook, identify the following parts of a plant, as well as their functions: roots, stems, leaves, and flowers.

LESSON 5.2

Lexile Measure = 880L

Plant Structures

Establish Learning Objectives

After this lesson, students will be able to:

- Describe the functions of roots, stems, and leaves.
- Explain how seeds become new plants.
- Describe the structures of a flower.

Engage

Activate Prior Knowledge

MY PLANET DIARY Read *Plant Giants* with the class. Explain that the aroid leaf is almost 3 meters long, which is the height of a basketball hoop off the ground. Tell students that the smallest flower and fruit belong to duckweed plants (genus *Wolffia*) which are about 0.6 mm long, 0.3 mm wide, and 120–150 μg in mass, or the mass of two salt grains. Ask: **What are four important parts of an angiosperm?** *(Roots, stems, leaves, and flowers)*

Explore

Lab Resource: Inquiry Warm-Up 

L1 **WHICH PLANT PART IS IT?** Students will observe common food items and identify them as roots, stems, or leaves. This Inquiry Warm-Up can be found online.

Texas Essential Knowledge and Skills

7C Demonstrate and illustrate forces that affect motion in everyday life such as emergence of seedlings, turgot pressure, and geotropism.

11A Examine organisms or their structures such as insects or leaves and use dichotomous keys for identification.

12A Investigate and explain how internal structures of organisms have adaptations that allow specific functions such as gills in fish, hollow bones in birds, or xylem in plants.

13B Describe and relate responses in organisms that may result from internal stimuli such as wilting in plants and fever or vomiting in animals that allow them to maintain balance.

Plant Structures

- What Are the Functions of Roots, Stems, and Leaves?
  TEKS 7C, 11A ,12A, 13B
- How Do Seeds Become New Plants?
  TEKS 7C,13A
- What Are the Structures of a Flower?
  TEKS 11A, 12C

MY PLANET DIARY — SCIENCE STATS

Plant Giants

- The aroid plant (as shown here) on the island of Borneo in Asia has leaves that can grow three meters long! These are the largest undivided leaves on Earth!
- The rafflesia flower can grow up to one meter wide and weigh seven kilograms.
- The jackfruit can weigh up to 36 kilograms. That's the world's largest fruit that grows on trees!

Write your answer below.

Why do you think the aroid plant has such big leaves?

Sample: so the plant can make more food

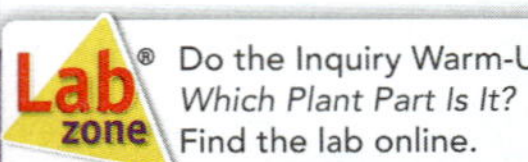

Do the Inquiry Warm-Up *Which Plant Part Is It?* Find the lab online.

TEKS 7C, 11A, 12A, 13B In this section you'll discover how structures inside plants allow them to perform everyday functions needed to survive.

ELPS 4.F.7

As you read about the plant structures in this lesson, confirm and enhance your understanding by discussing with a partner or group examples of each plant structure on different plants. Your teacher may show you examples on plants in or near your classroom.

What Are the Functions of Roots, Stems, and Leaves?

Each part of a plant plays an important role in its structure and function. Roots, stems, and leaves are just three structures we will look into further.

Roots Have you ever tried to pull a dandelion out of the soil? It's not easy, is it? That is because most roots are good anchors. Roots have three main functions. **Roots anchor a plant in the ground, absorb water and minerals from the soil, and sometimes store food.** The more root area a plant has, the more water and minerals it can absorb.

SUPPORT ALL READERS

Lexile Measure = 880L **Lexile Word Count = 2144**

Prior Exposure to Content: Many students may have misconceptions on this topic

Academic Vocabulary: *cause, effect, observe, relate*

Science Vocabulary: *stoma, transpiration, pollination, stamen, pistil*

Concept Level: Generally appropriate for most students in this grade

Preteach With: My Planet Diary "Plant Giants" and Figure 8 activity

Vocabulary

- root cap • cambium • stoma • transpiration
- embryo • fruit • germination • flower • pollination
- sepal • petal • stamen • pistil • ovary

Skills

- Reading: Relate Cause and Effect
- Inquiry: Observe

Types of Roots The two main types of root systems are shown in **Figure 1.** A fibrous root system consists of many similarly sized roots that form a dense, tangled mass. Plants with fibrous roots take a lot of soil with them when you pull them out of the ground. Lawn grass, corn, and onions have fibrous root systems. In contrast, a taproot system has one long, thick main root. Many smaller roots branch off the main root. A plant with a taproot system is hard to pull out of the ground. Carrots, dandelions, and cacti have taproots.

FIGURE 1

Root Systems and Structure

There are two main root systems with many structures.

Interpret Photos Label the taproot *T* and the fibrous roots *F*.

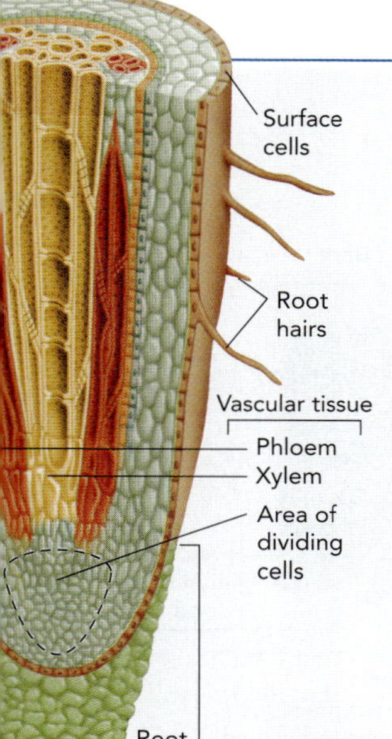

Root Structure

In **Figure 2,** you can see the structure of a typical root. The tip of the root is rounded and is covered by the root cap. The **root cap** protects the root from injury as the root grows through the soil. Behind the root cap are the cells that divide to form new root cells.

Root hairs grow out of the root's surface. These tiny hairs can enter the spaces between soil particles, where they absorb water and minerals. The root hairs also help to anchor the plant in the soil.

Locate the vascular tissue in the center of the root. The water and nutrients that are absorbed from the soil quickly move into the xylem. From there, these substances are transported upward to the plant's stems and leaves. Phloem transports food manufactured in the leaves to the root. The root tissues then use the food for growth or store it for future use by the plant.

FIGURE 2

Root Structure

Roots have many structures.

Define What is the function of the root cap?

It protects the root as it grows through the soil.

English Language Proficiency Standards

ELPS Reading 4.F.7

Have students find and discuss examples of different plant structures. Provide support to confirm understanding.

Beginning Review the diagram on page 180. Then read aloud the lesson. Have small groups find two or three examples of each structure.

Intermediate Form small groups. Together, read aloud the lesson, pausing after the description of each plant structure. Ask each student to find and name an example of the structure, and tell his or her group what the structure does.

Advanced Have partners read the lesson, find two examples of each plant structure, and tell what each structure does. Call on individuals to share responses.

Advanced High Have partners read the lesson, find three examples of each plant structure, and tell why it is important. Call on individuals to share responses.

Explain

Introduce Vocabulary

Point out to students the three vocabulary terms that contain the suffix *-tion*: *transpiration, germination,* and *pollination*. Tell students that this suffix is used to form nouns from verbs.

Teach Key Concepts

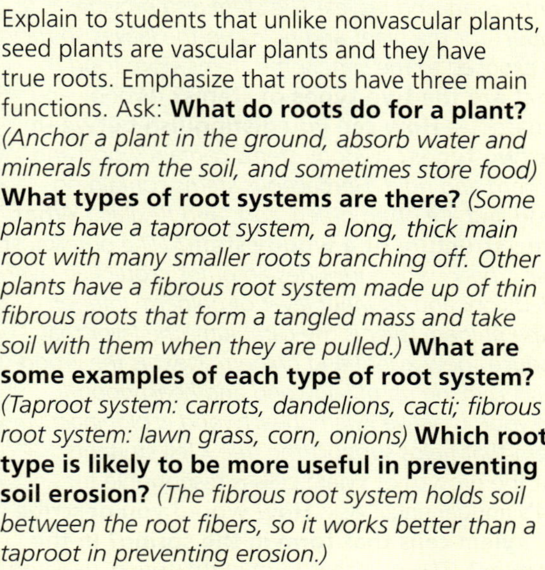

Explain to students that unlike nonvascular plants, seed plants are vascular plants and they have true roots. Emphasize that roots have three main functions. Ask: **What do roots do for a plant?** *(Anchor a plant in the ground, absorb water and minerals from the soil, and sometimes store food)* **What types of root systems are there?** *(Some plants have a taproot system, a long, thick main root with many smaller roots branching off. Other plants have a fibrous root system made up of thin fibrous roots that form a tangled mass and take soil with them when they are pulled.)* **What are some examples of each type of root system?** *(Taproot system: carrots, dandelions, cacti; fibrous root system: lawn grass, corn, onions)* **Which root type is likely to be more useful in preventing soil erosion?** *(The fibrous root system holds soil between the root fibers, so it works better than a taproot in preventing erosion.)*

Teach With Visuals

Tell students to look at **Figure 2.** Ask: **Why is the root cap important?** *(Behind the root cap are the cells that divide to form new root cells. They need to be protected. The root cap protects the root as the root grows through the soil.)* **What are the functions of the root hairs?** *(They enter the spaces between soil particles and absorb water and minerals. They also help anchor the plant.)* **What do xylem and phloem do?** *(Xylem transports water and minerals to the stems and leaves. Phloem transports food manufactured in the leaves to the roots.)* **What does the root tissue do with the food?** *(It uses the food for growth or stores it for later use by the plant.)*

PEARSON Texas.com

LESSON 5.2

Explain

Teach Key Concepts

Remind students that stems are unique to vascular plants and are an adaptation that allowed vascular plants to survive successfully on land. Explain to students that the stem of a plant has two main functions. Ask: **What are the functions of a stem?** *(The stem carries substances between the plant's roots and leaves. The stem also provides support for the plant and holds up the leaves so they are exposed to the sun. Some stems store food.)* **What are the two types of stems?** *(Woody and herbaceous)* **How can you distinguish between the two types?** *(Woody stems contain wood and are hard and rigid. Herbaceous stems contain no wood and are often green, soft, and flexible.)* **What is the structure of a woody stem?** *(The outermost layer is bark, which includes an outer protective layer called cork and an inner layer of phloem. Interior to the phloem layer is the cambium, the layer that produces new xylem and phloem. Next is the wood, which is made up of sapwood and heartwood. Sapwood is active xylem; heartwood is old, inactive xylem that helps support the tree.)* Tell students that the growth of xylem cells is responsible for a tree's annual rings. Ask: **How would you describe the xylem cells that form in the spring? In the summer?** *(The cells that form in the spring are large and have thin walls because they grow rapidly. They produce a wide, light brown ring. The cells that grow in the summer grow slowly, so they are small and have thick walls. They produce a thin, dark ring.)* **What represents one year's growth?** *(One pair of light and dark rings)*

Address Misconceptions

L1 GROW UP OR GROW WIDE? Some students think that trees simply "grow up" the same way that grasses grow. Tell students to look at **Figure 3.** Point out that the heartwood is old xylem and is at the center of the trunk. Ask: **Where would you expect to find new growth?** *(Near the outer portion of the trunk)* **How do trees grow wider?** *(By adding layers to the outside of the stem)*

21st Century Learning

CRITICAL THINKING Reinforce the fact that trees grow by adding layers to the outside of the stem. Tell students that to collect sap from maple trees, farmers cut a "V" shape into the bark of the tree a few feet above ground. Ask: **If a farmer comes back to the same tree in two years, where will the "V" cut be? Explain.** *(At about the same height; as the tree grows wider, its upward growth takes place mainly at the top of the tree.)*

Stems The stem of a plant has two main functions. **The stem carries substances between the plant's roots and leaves. The stem also provides support for the plant and holds up the leaves so they are exposed to the sun.** In addition, some stems, such as those of asparagus, store food.

The Structure of a Stem Stems can be either woody or herbaceous (hur BAY shus). Woody stems are hard and rigid, such as in maple trees. Herbaceous stems contain no wood and are often soft. Plants with herbaceous stems include daisies, ivy, and asparagus (pictured left).

Herbaceous and woody stems consist of phloem and xylem tissue as well as many other supporting cells. As you can see in **Figure 3,** a woody stem contains many layers of tissue. The outermost layer is bark. Bark includes an outer protective layer and an inner layer of living phloem, which transports food through the stem. Next is a layer of cells called the **cambium** (KAM bee um), which divides to produce new phloem and xylem. It is xylem that makes up most of what you call "wood." Sapwood is active xylem that transports water and minerals through the stem. The older, darker, heartwood is inactive but provides support.

FIGURE 3

Stem Structure

The woody stem of a tree contains many different structures.

Interpret Diagrams **Label the active xylem and phloem on the tree trunk below.**

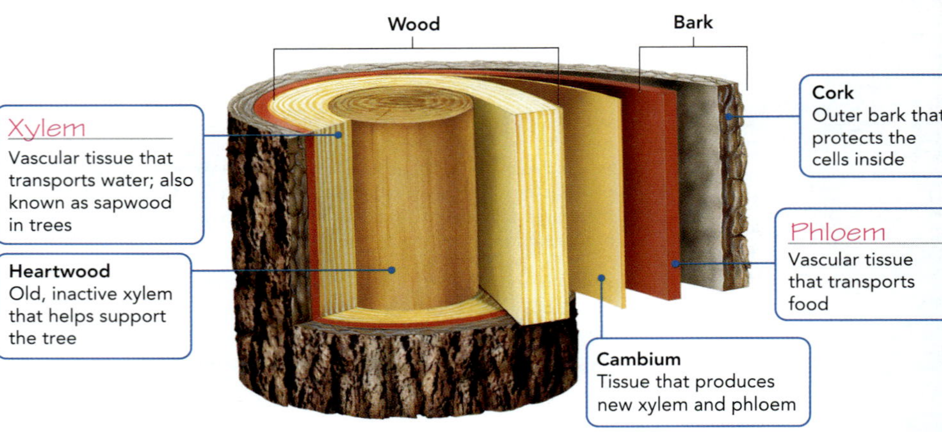

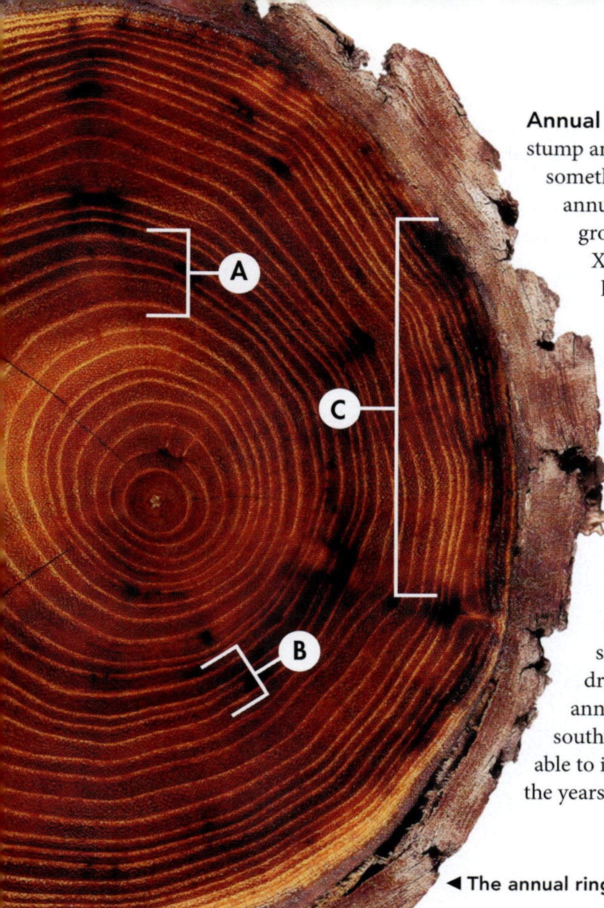

Annual Rings Have you ever looked at a tree stump and seen a pattern of circles that looks something like a target? These circles are called annual rings. They represent a tree's yearly growth. Annual rings are made of xylem. Xylem cells that form in the spring are large and have thin walls because they grow rapidly. They produce a wide, light brown ring. Xylem cells that form in the summer grow slowly and, therefore, are small and have thick walls. They produce a thin, dark ring. One pair of light and dark rings represents one year's growth. You can estimate a tree's age by counting its annual rings.

The width of a tree's annual rings can provide important clues about past weather conditions, such as rainfall. In rainy years, more xylem is produced, so the tree's annual rings are wide. In dry years, rings are narrow. By examining annual rings from some trees in the southwestern United States, scientists were able to infer that severe droughts occurred in the years 840, 1067, 1379, and 1632.

◀ **The annual rings in a tree reveal the tree's history.**

apply it!

1 **Calculate** How old was the tree when it was cut down?

About 25 years old

2 **Observe** The area at Area C is blackened from a fire that affected one side of the tree. Describe how the tree grew after the fire.

Sample: The tree grew more on the side that was not damaged by fire.

3 CHALLENGE Areas A and B both represent four years of growth. What might account for their difference in size?

Sample: Area A was formed when conditions such as rainfall were better for the tree than when Area B formed.

Elaborate

LESSON 5.2

Apply It!

L1 Before beginning the activity, review the idea of annual rings. Also discuss the fact that any damage to the bark of a tree affects the phloem in that area. Such damage could prevent the cells in that area from getting food and growing.

Observe Tell students that when they observe, they use one or more of their five senses to gather information about the world. In this case, they are gathering information about the growth of the tree by looking at the series of annual rings in Area C.

Teacher Demo 

L2 **OBSERVING TREE RINGS**

Materials cross-sectional slice of a tree trunk (you can obtain by slicing through a log for firewood), hand lens

Time 15 minutes

Show students the slice of tree trunk. Have students arrange themselves in groups of five or six. Pass the slice and the hand lens around the class, allowing each group about two or three minutes to examine the slice. Tell students to find three pairs of dark and light rings of different widths.

Ask: **Which pair represents the year of heaviest rainfall?** *(The ring pair with the greatest width grew in the year of heaviest rainfall.)* **Which pair represents the year of lightest rainfall?** *(The ring pair with the smallest width grew in the year of the lightest rainfall.)* **What are annual rings made of?** *(Xylem)*

Differentiated Instruction

L1 **Modeling a Tree Stem** Have student pairs roll up several sheets of newspaper and place them inside the core of an empty roll of paper toweling or bathroom tissue. Tell students that the core and the paper in it represent the center of a woody stem. Next, have students label five additional sheets of paper *old xylem, active xylem, cambium, active phloem,* and *outer bark*. Direct students to wrap these sheets around their model in the correct order. Finally, have partners take turns unwrapping the model, naming and explaining the function of each layer.

L3 **Specialized Stems** Have students research tubers, bulbs, and rhizomes and find examples of each.

LESSON 5.2

Explain

Teach Key Concepts

Remind students that plants are autotrophs—they make their own food. Discuss the fact that leaves come in all different shapes and sizes, but regardless of shape and size, they all play the same important role in a plant. Ask: **What is the main function of leaves?** *(They capture the sun's energy and carry out photosynthesis.)* **In what structures does photosynthesis take place?** *(Chloroplasts)* **What pigment do chloroplasts contain? What is the function of this pigment?** *(Chloroplasts contain chlorophyll, which traps the sun's energy.)* **Why is the structure of a leaf ideal for carrying out photosynthesis?** *(The cells that contain the most chloroplasts are located near the leaf's upper surface, where they get the most light.)* **Where are the stomata mostly found?** *(On a leaf's underside)* **What do they do?** *(They open to let in carbon dioxide and allow water vapor and oxygen to leave. They close to conserve water.)* **If the temperature is not very hot, when would stomata generally be open and closed?** *(Open during the daytime when sunlight is available and photosynthesis is active; closed at night when open stomata would only lead to water loss)* **What is the process by which water evaporates from a plant's leaves?** *(Transpiration)*

Support the TEKS

LEAF STRUCTURES Tell students to look at **Figure 4.** Ask: **Where is the leaf cuticle and what does it do?** *(It covers the leaf's top and bottom surfaces and prevents water loss.)* **What structure contains the xylem and phloem?** *(Vein)* **What are stomata? What do they do?** *(Small pores in the surface layers that open and close to control when gases enter and leave the leaf)* **What is the benefit of upper leaf cells that are densely packed? Of lower leaf cells that are loosely packed?** *(Densely packed cells can collect more energy for photosynthesis. Loosely packed cells allow carbon dioxide to reach cells and oxygen to escape into the air.)*

Lead a Discussion

STOMATA AND GUARD CELLS Explain to students how stomata, the tiny pores in a leaf's surface layers, open and close. Tell students that each stoma is formed by two slightly curved cells, called guard cells. When the guard cells fill up with water, they arch and curve away from each other, opening the stoma. When the guard cells have less water, they straighten out and come together, closing the stoma. Ask: **In a period of little water, what would the guard cells look like? The stoma?** *(The guard cells would be straight and close together because there is little water in them. The stoma would be closed.)*

Vocabulary Greek Word Origins The Greek word *stoma* means "mouth." How are the stomata of a plant like mouths?

Sample: Stomata open and close to let matter in and out of plants. Mouths open and close to let matter in and out of organisms.

Leaves Leaves vary greatly in size and shape. Pine trees have needle-shaped leaves. Birch trees have small rounded leaves with jagged edges. Regardless of their shape, leaves play an important role in a plant. **Leaves capture the sun's energy and carry out the food-making process of photosynthesis.**

The Structure of a Leaf If you were to cut through a leaf and look at the edge under a microscope, you would see the structures in **Figure 4.** The leaf's top and bottom surface layers protect the cells inside. Between the layers of cells are veins that contain xylem and phloem.

The surface layers of the leaf have small openings, or pores, called **stomata** (stoh MAH tuh; *singular* stoma). The stomata open and close to control when gases enter and leave the leaf. When the stomata are open, carbon dioxide enters the leaf, and oxygen and water vapor exit.

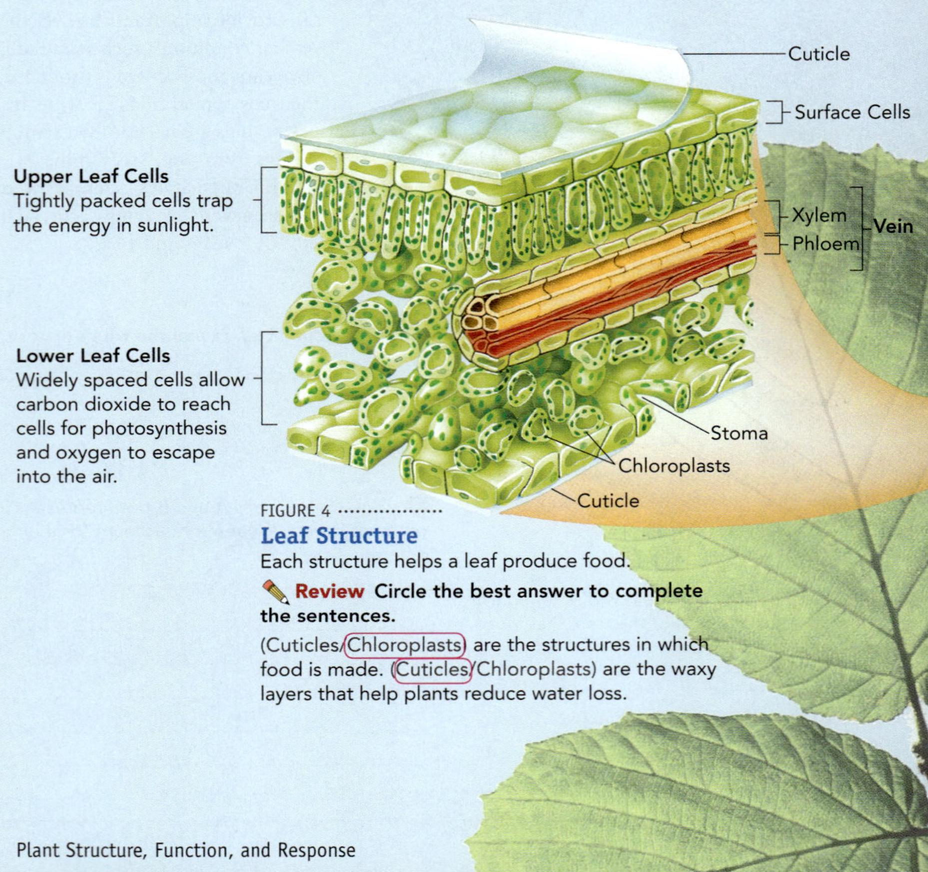

FIGURE 4 **Leaf Structure**
Each structure helps a leaf produce food.

Review Circle the best answer to complete the sentences.
(Cuticles/Chloroplasts) are the structures in which food is made. (Cuticles/Chloroplasts) are the waxy layers that help plants reduce water loss.

Professional Development Note

Teacher to Teacher

Misconception A common misconception that students have about plants is that if all plants died, humans would suffocate from lack of oxygen. In fact, there is enough atmospheric oxygen to breathe for many years. The accumulation of CO_2 would poison humans before they ran out of oxygen, but that could take over 1,000 years. What students do not appreciate is that if all plants died, humans would starve to death long before they suffocated. It has been estimated that if all plants died today, there would be massive starvation within a month, with no food available anywhere within a year.

Joel Palmer, Ed.D.
Mesquite ISD
Mesquite, Texas

The Leaf and Photosynthesis The structure of a leaf is ideal for carrying out photosynthesis. The cells with the most chloroplasts are located near the leaf's surface, where they get the most light. The chlorophyll in the chloroplasts traps the sun's energy.

Carbon dioxide enters the leaf through open stomata. Water, which is absorbed by the plant's roots, travels up the stem to the leaf through the xylem. During photosynthesis, sugar and oxygen are produced from the carbon dioxide and water. Oxygen passes out of the leaf through the open stomata. The sugar enters the phloem and then travels throughout the plant.

Controlling Water Loss Because a leaf is exposed to the air, water can quickly evaporate from it. The process by which water evaporates from a plant's leaves is called **transpiration.** A plant can lose a lot of water through transpiration. Without a way to slow down the process of transpiration, a plant would wilt and die.

Fortunately, plants have ways to slow down transpiration. The cells around a stoma are called guard cells. When the cells contain a lot of water, the water exerts force on their cell walls. This force is called turgor pressure. The pressure causes the guard cells to bulge outward and open a stoma. At night or during dry conditions, the cells lose water and collapse. This closes the stoma and prevents transpiration.

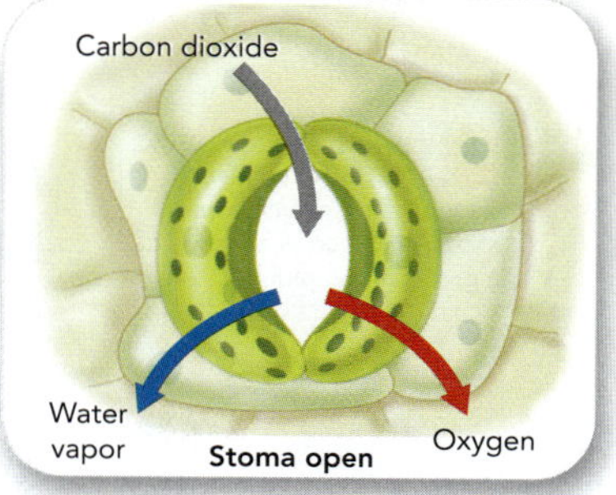

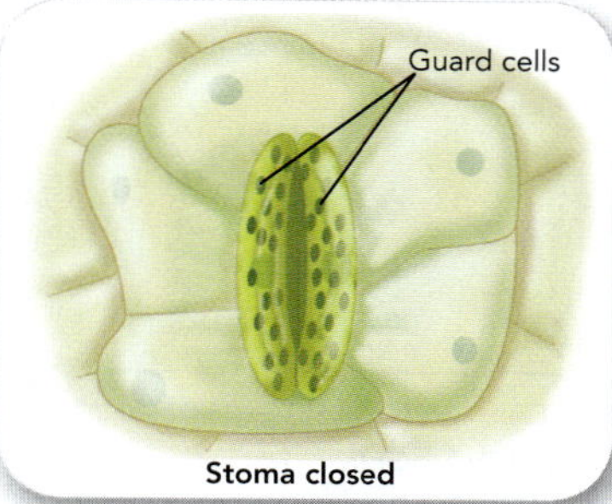

FIGURE 5

Stomata

Stomata can slow water loss.

Name What three substances enter and leave a plant through stomata?

Carbon dioxide, oxygen, and water vapor

Do the Lab Investigation *Investigating Stomata.* Student Lab Manual, p. 52

Assess Your Understanding

TEKS 12A, 13B

1a. Explain What are the functions of a stem?

To carry substances between roots and leaves, provide support, and hold up leaves for sunlight exposure

b. Relate If you forget to water a houseplant for a few days, would its stomata be open or closed? Why?

The stomata would be closed to prevent water loss.

got it?

O **I get it!** Now I know that roots, stems, and leaves perform functions like anchoring the plant, absorbing and transporting nutrients, and making and storing food.

O I need extra help with See TE note.

191

Differentiated Instruction

L1 The Parts of a Plant Have students create a visual representation of the parts of a plant and their functions. Tell students to make a poster of a generic plant, one that has roots, stem, and leaves. Have students label the parts and then use callouts to describe the functions of each part. Tell students that you will display the posters, so they should be as artistic and accurate as possible.

L3 Specialized Leaves Plants have specialized leaves. Have students research the following plants and describe their leaves: Venus' flytrap, pitcher plant, pea plant, kalanchoe, sundew, cactus, and poinsettia.

21st Century Learning

CRITICAL THINKING Review with students the functions of a leaf and the fact that leaves come in many shapes and sizes. Tell students that the shapes and sizes represent adaptations of the plants to their environments. Ask: **Why do you think plants in shady forests often have large leaves?** *(To capture as much sunlight for photosynthesis as possible)* **Why do you think plants in dry areas have small, thick, wax-coated leaves?** *(To prevent as much water loss as possible)* **Why do you think plants in very rainy areas often have long, pointy leaves?** *(To help shed water and thereby prevent bacteria and fungi that cause disease from growing and damaging the leaves)*

Make Analogies

L1 FOOD FACTORY Explain to students that a plant leaf is often compared to a factory. Ask: **What is true about any type of factory?** *(Sample: It has a process that turns raw materials into products under specific conditions.)* **Using your answer, why is a leaf like a food factory?** *(In a leaf, the raw materials water and carbon dioxide are combined in the presence of light energy in a process called photosynthesis to produce sugar and oxygen.)*

Elaborate

Lab Resource: Lab Investigation

L2 INVESTIGATING STOMATA Students will examine stomata in a land plant and in a floating water plant. This Lab Investigation can be found in the Student Lab Manual, p. 52, and online.

Evaluate

Assess Your Understanding

After students answer the questions, have them evaluate their understanding by completing the appropriate sentence.

RTI Response to Intervention

1a. If students need help identifying the functions of a stem, **then** show them a picture of a plant with an herbaceous stem and have them look at the picture as a volunteer reads aloud the Key Concept about stems.

b. If students have trouble relating the condition of the stomata to environmental conditions, **then** first review the function of stomata and then analyze the environmental condition in terms of water availability.

LESSON 5.2

Explain

Teach Key Concepts

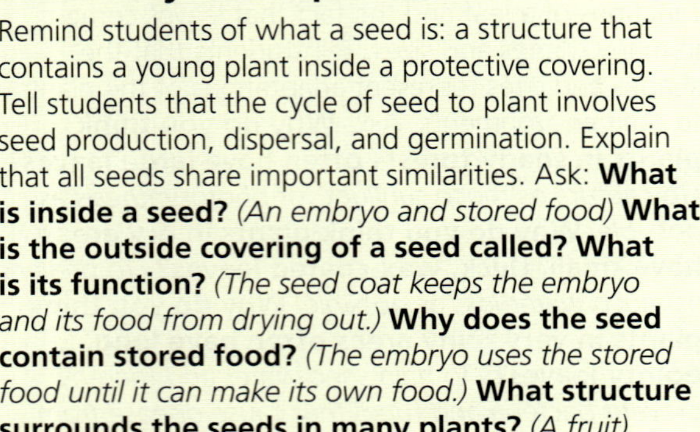

Remind students of what a seed is: a structure that contains a young plant inside a protective covering. Tell students that the cycle of seed to plant involves seed production, dispersal, and germination. Explain that all seeds share important similarities. Ask: **What is inside a seed?** *(An embryo and stored food)* **What is the outside covering of a seed called? What is its function?** *(The seed coat keeps the embryo and its food from drying out.)* **Why does the seed contain stored food?** *(The embryo uses the stored food until it can make its own food.)* **What structure surrounds the seeds in many plants?** *(A fruit)*

Teach With Visuals

Tell students to look at **Figure 6.** Ask: **What structures are common to all seeds?** *(Seed coat, embryo, stored food, cotyledon)* **What plant characteristics can be seen in the embryo?** *(The beginnings of roots, stems, leaves)* **Why do you think a seed germinates once it has absorbed water?** *(When a seed absorbs water, the food-storing tissues swell. This cracks open the seed coat.)*

Lead a Discussion

SEED DISPERSAL AND GERMINATION Tell students that seed dispersal, or the scattering of seeds, can happen in various ways. Ask: **What are some ways in which seeds are dispersed?** *(By wind, water, ejection, animals, humans)* Discuss seed germination, or the process by which an embryo begins to grow again and pushes out of the seed. Ask: **What must the seed do in order to germinate?** *(Absorb water)* **How does the seed germinate?** *(After absorbing water, the embryo uses the stored food to start growing. The roots first grow downward and then the stem and leaves grow upward.)* **Why is it an advantage for seeds to be able to remain inactive and not germinate immediately after the embryo forms?** *(This allows for seeds to be dispersed but not germinate until there are ideal growing conditions.)*

Relate Cause and Effect Tell students that science involves many cause-and-effect relationships. A cause makes something happen. An effect is what happens as a result.

Texas Essential Knowledge and Skills

7C Demonstrate and illustrate forces that affect motion in everyday life such as emergence of seedlings, turgot pressure, and geotropism.

13A Investigate how organisms respond to external stimuli in the environment such as phototropism and fight or flight.

TEKS 7C, 13A In this section you'll examine the changes a seed undergoes as it grows into a plant.

How Do Seeds Become New Plants?

Many plants begin their life cycle as a seed. You can follow the cycle from seed to plant in **Figure 6.** All seeds share important similarities. **Inside a seed is a partially developed plant. If a seed lands in an area where conditions are favorable, the plant sprouts out of the seed and begins to grow.**

Seed Structure A seed has three main parts—an embryo, stored food, and a seed coat. The young plant that develops from the zygote, or fertilized egg, is called the **embryo.** The embryo already has the beginnings of roots, stems, and leaves. In the seeds of most plants, the embryo stops growing when it is quite small. When the embryo begins to grow again, it uses the food stored in the seed until it can make its own food by photosynthesis. In all seeds, the embryo has one or more seed leaves, or cotyledons. In some seeds, food is stored in the cotyledons. In others, food is stored outside the embryo.

The outer covering of a seed is called the seed coat. The seed coat acts like plastic wrap, protecting the embryo and its food from drying out. This allows a seed to remain inactive for a long time. In many plants, the seeds are surrounded by a structure called a **fruit.**

FIGURE 6

Story of a Seed

Read the text on this page and the next page. Then complete the activities about seeds becoming new plants.

Complete each task.

1. **Review** On the diagram, label the seed's embryo, cotyledons, and seed coat.

Seed Dispersal After seeds form, they are usually scattered. The scattering of seeds is called seed dispersal. Seeds can be dispersed in many different ways. When animals eat fruit, the seeds inside the fruit pass through the animal's digestive system and are deposited in new areas. Other seeds are enclosed in barblike structures that hook onto fur or clothing. The seeds fall off in a new area. Water also disperses seeds that fall into oceans and rivers. Wind disperses lightweight seeds, such as those of dandelions and maple trees. Some plants eject their seeds. The force scatters the seeds in many directions. A seed that is dispersed far from competition has a better chance of survival.

Relate Cause and Effect Underline a cause of seed dispersal and circle its effect in the text on this page.

Germination After a seed is dispersed, it may remain inactive for a while. This period of inactivity lasts until the seed absorbs water and germination begins. **Germination** (jur muh NAY shun) occurs when the embryo uses stored food and begins to grow again. The force of the growing embryo causes it to push out of the seed. First the roots grow downward. Then the stem and leaves grow upward.

2. Explain Give two reasons why this seed can be successfully dispersed by wind.

Sample: It is lightweight and has soft structures to catch the wind.

B

A

3. CHALLENGE Which young plant, A or B, is more likely to grow into an adult plant? Why?

Plant B; it does not have to compete for resources with as many plants.

Lab zone Do the Lab Investigation *Fruit Basket.* Student Lab Manual, p. 57

Assess Your Understanding

got it?

○ **I get it!** Now I know that a seed becomes a new plant when it lands in an area with favorable conditions and a new plant sprouts.

○ I need extra help with See TE note.

Differentiated Instruction

L1 Structure of a Seed Have students draw a seed and label the embryo, stored food, and seed coat. Have students describe each part.

L1 All About Seed Plants Have students work in groups of four to write 10 questions about the characteristics of seed plants. Suggest they compose a variety of short-answer, fill-in-the-blank, and matching questions. Remind them to create answer keys as well. Have groups exchange questions and complete them. Students can check their answers using the answer keys.

L3 Germination Have student pairs design an experiment that illustrates seed germination. Remind students of the necessary conditions for germination. Also remind them that their design should include a control. You may wish to have students perform the experiment for the class.

Elaborate

Build Inquiry 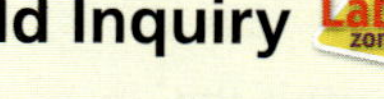

L2 MODELING SEED DISPERSAL

Materials tissue paper, modeling clay, plastic foam balls, plastic spoons, table tennis balls, hook-and-loop fastener strips, and other arts and crafts materials

Time 25 minutes

Tell students to review the various means of seed dispersal. Explain that the fruit (the structure that encloses the seeds) helps the seeds disperse. Have students work together to build model seeds that can be dispersed by wind, water, or by sticking to clothes or animal fur. Encourage students to predict how far their seeds will travel and then to test their predictions. Students can present their models to the class, identifying how their model seeds are similar to real seeds and how their models could be improved.

Ask: **What characteristics are most important for each method of dispersal?** *(Samples: Animal fur—seed must stick to fur with barbs or similar structures; water—seed must float; wind—seed has structures to catch the wind and seed is lightweight; ejection—seed is small, dense, and compact)* **A coconut produces one of the largest seeds in the plant kingdom. The seed is encased in a tough husk made of strong fibers that have air spaces between them. How do you think coconut seeds are dispersed? Explain.** *(On ocean currents; the fruit can float and the husk prevents the seeds from getting too wet before they reach land.)*

Lab Resource: Lab Investigation 

FRUIT BASKET Students will plan and implement an investigation about the internal structures of fruits. This Lab Investigation can be found in the Student Lab Manual, p. 57, and online.

Evaluate

Assess Your Understanding

Have students evaluate their understanding by completing the appropriate sentence.

RTI Response to Intervention

If students need help describing how seeds become new plants, **then** have them work with a partner to talk through the processes of seed dispersal and seed germination.

LESSON 5.2

Explain

Teach Key Concepts

Explain to students that the flower is the reproductive structure in angiosperms. Discuss with students that although flowers come in all sizes, shapes, and colors, they all have the same function—reproduction. Tell students that a typical flower contains four main parts. Ask: **What are the four main parts of a flower?** *(Sepals, petals, stamens, and pistils)* **What are sepals? What function do they perform?** *(Sepals are leaflike structures that protect the developing flower while it is still a bud. Sepals are often green in color.)* **What are petals? What function do they perform?** *(Petals are colorful, leaflike structures that attract pollinators with their color and scent.)* **What is the male part of a flower called? What structures make it up?** *(The stamen consists of filaments and anthers.)* **What is produced in the anther?** *(Pollen)* **What is the female part of a flower called? What structures does it include?** *(The pistil consists of the stigma, style, and ovary.)* **What does the ovary contain?** *(One or more ovules)* Discuss the process of pollination and the role of pollinators. Ask: **What is pollination?** *(The transfer of pollen from male reproductive structures to female reproductive structures)* **What are examples of pollinators?** *(Birds, bats, bees, and flies)*

Teach With Visuals

Tell students to look at **Figure 7.** Ask: **The plants shown here have brightly colored petals or strong scents. Why do you think this is so?** *(Flowers with bright colors or strong scents have a better chance of attracting birds, bats, bees, and flies, which they must do in order to ensure pollination.)* Explain to students that many plants produce flowers without brightly colored petals or strong scents. These plants are usually pollinated by wind. Plants that are pollinated by wind produce much more pollen than plants that are pollinated by animals. Ask: **Why do you think wind-pollinated flowers produce so much pollen?** *(Pollen carried by animals has a better chance of reaching another plant of the same species than does pollen carried by wind.)*

Texas Essential Knowledge and Skills

11A Examine organisms or their structures such as insects or leaves and use dichotomous keys for identification.

12C Recognize levels of organization in plants and animals, including cells, tissues, organs, organ systems, and organisms.

TEKS 11A, 12C In this section you'll learn about the structures of a flower and their functions.

What Are the Structures of a Flower?

Flowers come in all sorts of shapes, sizes, and colors. But, despite their differences, all flowers have the same function—reproduction. A **flower** is the reproductive organ of an angiosperm. **A typical flower contains sepals, petals, stamens, and pistils.**

The colors and shapes of most flower structures and the scents produced by most flowers attract insects and other animals. These organisms ensure that pollination occurs. **Pollination** is the transfer of pollen from male reproductive structures to female reproductive structures. Pollinators, such as those shown in **Figure 7,** include birds, bats, and insects such as bees and flies. As you read, keep in mind that some flowers lack one or more of the parts. For example, some flowers have only male reproductive parts, and some flowers do not have petals.

Sepals and Petals When a flower is still a bud, it is enclosed by leaflike structures called **sepals** (SEE pulz). Sepals protect the developing flower and are often green in color. When the sepals fold back, they reveal the flower's colorful, leaflike **petals.** The petals are generally the most colorful parts of a flower. The shapes, sizes, and number of petals vary greatly between flowers.

Stamens Within the petals are the flower's male and female reproductive parts. The **stamens** (STAY munz) are the male reproductive parts. Locate the stamens inside the flower in **Figure 8.** The thin stalk of the stamen is called the filament. Pollen is made in the anther, at the top of the filament.

FIGURE 7

Pollinator Matchup

Some pollinators are well adapted to the plants they pollinate. For example, the long tongue of the nectar bat helps the bat reach inside the agave plant, as shown below.

Apply Concepts Write the letter of the pollinator on the plant it is adapted to pollinate.

Pistils The female parts, or **pistils** (PIS tulz), are found in the center of most flowers, as shown in **Figure 8.** Some flowers have two or more pistils; others have only one. The sticky tip of the pistil is called the stigma. A slender tube, called a style, connects the stigma to a hollow structure at the base of the flower. This hollow structure is the **ovary,** which protects the seeds as they develop. An ovary contains one or more ovules.

FIGURE 8

Structures of a Typical Flower

Flowers have many structures.

Relate Text and Visuals
Use the word bank to fill in the missing labels.

Sepals are the small, leaflike parts of a flower. They protect the developing flower.

Petals are usually the most colorful parts of a flower. Pollinators are attracted by their color and scent.

Stamens are the male reproductive parts of a flower. Pollen is produced in the anther, at the top of the stalklike filament.

Pistils are the female reproductive parts of a flower. They consist of a sticky stigma, a slender tube called the style, and a hollow structure called the ovary at the base.

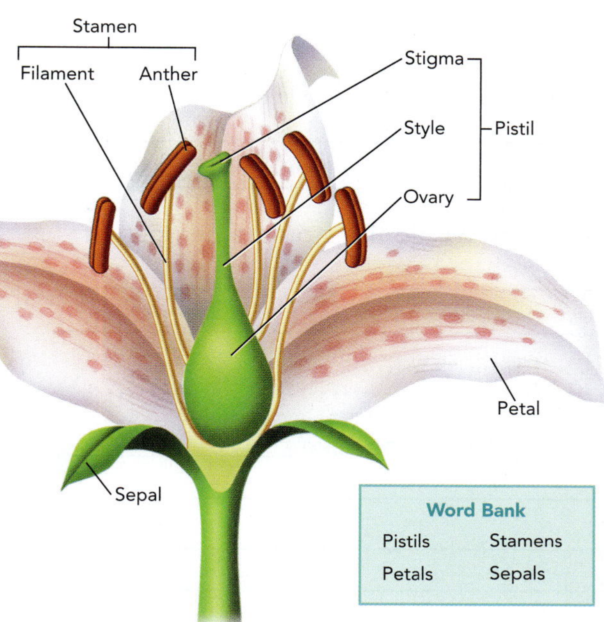

Word Bank	
Pistils	Stamens
Petals	Sepals

Do the Quick Lab *Modeling Flowers.* Student Lab Manual, p. 63

Assess Your Understanding

got it?

- I get it! Now I know that the structures of a flower include sepals, petals, stamens, and pistils.
- I need extra help with See TE note.

Differentiated Instruction

L1 Structure of a Flower Have students make a compare/contrast table listing the male and female parts of a flower.

L1 Flower Parts and Functions Have students summarize what they have learned about flower parts and their functions by drawing a generic flower with sepals, petals, stamen, and pistil. Have them label each part and describe its functions. Have them exchange their drawings with a classmate for evaluation.

L3 Life Cycle Ask students to write the life story of a plant, beginning with the formation of a seed in a flower ovary and continuing until the plant produces its own seeds.

Elaborate

Build Inquiry

L1 OBSERVING THE STRUCTURE OF A FLOWER

Materials gladiolus flower, paper towel, hand lens, metric ruler, lens paper, safety goggles

Time 25 minutes

Instruct students to wear safety goggles. Provide student pairs with a flower and tell them to observe its parts with the hand lens. Have them draw a diagram of the flower. Have them examine and describe the sepals, petals, stamen, and pistil. Have students observe the structures that make up the stamen and the pistil. Tell students to use the ruler to measure the height of a stamen and a pistil. Have them gently place a small piece of lens paper on the stigma and then remove it.

Ask: **What are the functions of the sepals and petals?** *(The sepals protect the flower bud and the petals attract pollinators with their color or scent.)* **How does the structure of a flower enable it to perform its reproductive function?** *(The stamen produces pollen, which contains sperm cells. The pistil produces eggs in the ovules of the ovary. Reproduction occurs when a sperm fertilizes an egg as a result of pollination.)* **Why might it be an advantage to have a longer stamen? A longer pistil?** *(A longer stamen would ensure that an insect picks up pollen on the anther and transfers to another plant, or that it falls from the anther to the stigma of the same plant. A longer pistil would make it easier for pollinators to land on it.)* **What did you discover about the stigma using the lens paper?** *(It is sticky.)*

Lab Resource: Quick Lab 

L2 MODELING FLOWERS Students will make models of flowers with different arrangements of reproductive structures to present to the class and to compare and contrast. This Quick Lab can be found in the Student Lab Manual, p. 63, and online.

Evaluate

Assess Your Understanding

Have students evaluate their understanding by completing the appropriate sentence.

RTI Response to Intervention

If students need help identifying the structures of a flower, **then** have them review the highlighted vocabulary terms and **Figure 8.**

Name ______________________ Date ____________ Class ____________

Assess Your Understanding

Plant Structures

What Are the Functions of Roots, Stems, and Leaves?

1a. EXPLAIN What are the functions of a stem? ______________________

____

____

b. RELATE If you forgot to water a houseplant for a few days, would the stomata be open or closed? Why? ______________________

____

got*it*?

○ **I get it!** Now I know that roots, stems, and leaves perform functions like ______________________

____

○ **I need extra help with** ______________________

____

How Do Seeds Become New Plants?

got*it*?

○ **I get it!** Now I know that a seed becomes a new plant when ______________________

____

○ **I need extra help with** ______________________

____

What Are the Structures of a Flower?

got*it*?

○ **I get it!** Now I know that the structures of a flower include ______________________

____

○ **I need extra help with** ______________________

____

Place the outside corner, the corner away from the dotted line, in the corner of your copy machine to copy onto letter-size paper.

Name ______________________ Date ____________ Class ____________

Enrich

Plant Structures

Read the passage and study the diagrams below. Then use a separate sheet of paper to answer the questions that follow.

Bubbling Leaves

Carbon dioxide enters leaves through stomata. Oxygen, produced during photosynthesis, passes out of leaves through stomata. April designed an experiment to find out more about these tiny pores on a leaf. She picked a few fresh leaves from the trees near her house. Then, while pinching the stalk of one of the leaves, she dipped the leaf in a glass of hot water. After observing what happened, she did the same thing with the rest of the leaves, one at a time. The diagram below shows what April saw when she dipped two different leaves into the glass of hot water.

1. What did April observe coming out of the stomata of each leaf?
2. In the figure, bubbles are coming out of both sides of one leaf, while bubbles are coming out of only one side of the other leaf. What does this tell you about the location of the stomata on these two leaves?
3. What do you think would happen if April did not pinch the stalk of the leaf before dipping it into the hot water?
4. In most plants, most of the stomata are located on the lower surface of the leaves. Explain how this adaptation helps control water loss.
5. Would you expect to find the stomata on a lily pad on the top or bottom? Explain your answer.

Name ______________________ Date __________ Class __________

Lesson Quiz

Plant Structures

Write the letter of the correct answer on the line at the left.

1. ___ Animals are helpful to plants in the process of

A germination

B pollination

C transpiration

D fertilization

2. ___ Which of the following is NOT part of a flower's pistil?

A stigma

B ovary

C style

D anther

3. ___ Which part of a plant is responsible for absorbing water and minerals and anchoring the plant?

A roots

B stems

C anthers

D filaments

4. ___ The three parts of a seed are

A stored food, embryo, cambium

B embryo, seed coat, ovary

C cotyledon, seed coat, ovule

D embryo, stored food, seed coat

Fill in the blank to complete each statement.

5. Seed ______________ is the scattering of seeds.

6. A flower bud is protected by leaflike structures called ______________.

7. The ______________________ protects the root as it grows through the soil.

8. A tree has 24 light rings and 24 dark rings. The tree is ______________ years old.

9. ______________ on the surface of a leaf control the movement of gases into and out of the leaf.

10. The hollow structure at the base of a pistil that protects seeds as they develop is the ______________.

Place the outside corner, the corner away from the dotted line, in the corner of your copy machine to copy onto letter-size paper.

Place the outside corner, the corner away from the dotted line, in the corner of your copy machine to copy onto letter-size paper.

Plant Structures

Answer Key

Review and Reinforce

Find the worksheet in the Student Workbook.

1. Root hairs growing out of the root's surface anchor the plant and absorb water and minerals. Vascular tissue in the root's center moves water and minerals away from the root and moves food made in the leaves to it.

2. Both woody stems and herbaceous stems carry substances between a plant's roots and leaves, provide support, and hold leaves up to the sun. Woody stems are hard and rigid; herbaceous stems are soft and flexible.

3. Cells that contain the most sunlight-absorbing chloroplasts are tightly packed into the top of a leaf, so they get the most light. Openings in the leaf surface called stomata allow carbon dioxide to enter and oxygen to leave. Widely spaced cells in the bottom of the leaf allow these gases to circulate. A waxy cuticle on top and bottom prevents excessive water loss.

4. The farther a seed is dispersed the better its chance of survival because it will not compete with its parent for light, water, and nutrients.

5. In the male reproductive part, the stamen, a filament holds up an anther, in which pollen is formed. The female reproductive part, the pistil, includes a sticky tip called a stigma that traps pollen, an ovary that protects the seeds as they develop, and a style that connects both parts.

6. f **7.** c **8.** b **9.** d

10. a **11.** h **12.** e **13.** g

14. Roots anchor a plant in the ground, absorb water and minerals from the soil, and sometimes store food. Vascular tissues in the stem carry substances between the plant's roots and leaves. Xylem carries water and minerals from the roots to the leaves. Phloem carries food made in the leaves down to the roots. The stem also provides support for the plant and holds up the leaves so they are exposed to the sun. Leaves capture the sun's energy and carry out the food-making process of photosynthesis. Flowers are the reproductive structures of an angiosperm.

Enrich

1. Accept either air bubbles or gas bubbles.

2. One leaf has stomata on one side only, while the other leaf has stomata on both sides.

3. Air inside the leaf might escape.

4. Water would be more readily lost from the upper surface of the leaves, because that is the side more exposed to sun and wind.

5. on the top—the surface that is exposed to air

Lesson Quiz

1. B **2.** D

3. A **4.** D

5. dispersal **6.** sepals

7. root cap **8.** 24

9. Stomata **10.** ovary

Plant Responses and Growth

How do plants respond to their environment?

LESSON PACING:
1–2 periods or $\frac{1}{2}$–1 block

Lesson Vocabulary

- tropism
- hormone
- auxin
- photoperiodism
- critical night length
- short-day plant
- long-day plant
- day-neutral plant
- dormancy

Content Refresher

Rapid Responses All plants respond to light and gravity: Plant shoots move upward toward light and against gravity, whereas roots move downward away from light and with gravity. The response center for gravity appears to be in the root cap and may be governed by starch molecules. Response to light is governed by a plant hormone called *auxin*.

Some plant responses do not involve growth. If the leaves of *Mimosa pudica*, appropriately called the "sensitive plant," are touched, the leaflets fold together completely within only two or three seconds. The leaflets are held apart due to osmotic pressure at the base of the leaflets, where they join. When a leaf is touched, cells near the center of the leaflets pump out ions and lose water due to osmosis. Pressure from cells on the underside of the leaf, which do not lose water, force the leaflets together.

Lesson Objectives	TEKS	ELPS
Identify three stimuli that produce plant responses.	7C, 13A	3.D.2
Describe how plants respond to seasonal changes.	13A	

Texas Essential Knowledge and Skills

7C Demonstrate and illustrate forces that affect motion in everyday life such as emergence of seedlings, turgor pressure, and geotropism.
13A Investigate how organisms respond to external stimuli found in the environment such as phototropism and fight or flight.

English Language Proficiency Standards

ELPS Speaking 3.D.2 Speak using grade-level content area vocabulary in context to build academic language proficiency.

DIFFERENTIATED INSTRUCTION KEY
L1 Struggling Students or Special Needs
L2 On-Level Students L3 Advanced Students

LESSON PLANNER 5.3

Investigations and Activities

My Planet Diary, **Student Edition,** p. 196

Inquiry: Inquiry Warm-Up, Can a Plant Respond to Touch?, **Lab Manual, p. 64**

Introduce Vocabulary, **Teacher's Edition,** p. 197

Teach Key Concepts, **Teacher's Edition,** p. 197

Lead a Discussion, Plant Hormones, **Teacher's Edition,** p. 198

Inquiry: Teacher Demo, Modeling Plant Response, **Teacher's Edition, p. 198**

Inquiry: Quick Lab, Watching Roots Grow, **Lab Manual, p. 65**

Teach Key Concepts, **Teacher's Edition,** p. 199

Teach With Visuals, **Teacher's Edition,** p. 199

21st Century Learning, Critical Thinking, **Teacher's Edition,** p. 199

Teach Key Concepts, **Teacher's Edition,** p. 200

Make Analogies, Dormancy, **Teacher's Edition,** p. 200

Apply It!, **Student Edition,** p. 200

Differentiated Instruction, **Teacher's Edition,** p. 201

Inquiry: Quick Lab, Seasonal Changes, **Lab Manual, p. 67**

TEKS Review

Apply the TEKS, Plant Responses, **Student Edition,** p. 201

Apply the TEKS, Exploring Geotropism in Plants, **Student Edition,** p. 202

TEKS Practice, **Student Edition,** p. 204

TEKS Practice: Chapter and Cumulative Review, **Student Edition,** p. 206

Lesson 5.3, **TEKS Preparation and Study Guide Workbook,** p. 50

SHORT ON TIME? To do this lesson in approximately half the time, do the Activate Prior Knowledge activity. A discussion of the Key Concepts will familiarize students with the lesson content. Use the Apply the TEKS activity to help students understand how plants respond to their environment. Have students do the Quick Labs. The rest of the lesson can be completed by students independently.

These editable worksheets are available on **PearsonTexas.com.**
Print versions can be found in the **TEKS Preparation and Study Guide Workbook.**

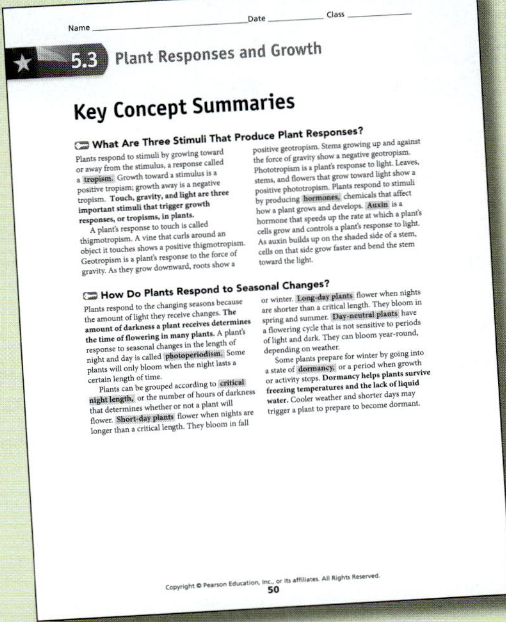

Name ____________ Date ________ Class ________

5.3 Plant Responses and Growth

Key Concept Summaries

What Are Three Stimuli That Produce Plant Responses?

Plants respond to stimuli by growing toward or away from the stimulus, a response called a **tropism.** Growth toward a stimulus is a positive tropism; growth away is a negative tropism. **Touch, gravity, and light are three important stimuli that trigger growth responses, or tropisms, in plants.**

A plant's response to touch is called thigmotropism. A vine that curls around an object it touches shows a positive thigmotropism. Geotropism is a plant's response to the force of gravity. As they grow downward, roots show a positive geotropism. Stems growing up and against the force of gravity show a negative geotropism. Phototropism is a plant's response to light. Leaves, stems, and flowers that grow toward light show a positive phototropism. Plants respond to stimuli by producing **hormones,** chemicals that affect how a plant grows and develops. **Auxin** is a hormone that speeds up the rate at which a plant's cells grow and controls a plant's response to light. As auxin builds up on the shaded side of a stem, cells on that side grow faster and bend the stem toward the light.

How Do Plants Respond to Seasonal Changes?

Plants respond to the changing seasons because the amount of light they receive changes. **The amount of darkness a plant receives determines the time of flowering in many plants.** A plant's response to seasonal changes in the length of night and day is called **photoperiodism.** Some plants will only bloom when the night lasts a certain length of time.

Plants can be grouped according to **critical night length,** or the number of hours of darkness that determines whether or not a plant will flower. **Short-day plants** flower when nights are longer than a critical length. They bloom in fall or winter. **Long-day plants** flower when nights are shorter than a critical length. They bloom in spring and summer. **Day-neutral plants** have a flowering cycle that is not sensitive to periods of light and dark. They can bloom year-round, depending on weather.

Some plants prepare for winter by going into a state of **dormancy,** or a period when growth or activity stops. **Dormancy helps plants survive freezing temperatures and the lack of liquid water.** Cooler weather and shorter days may trigger a plant to prepare to become dormant.

50

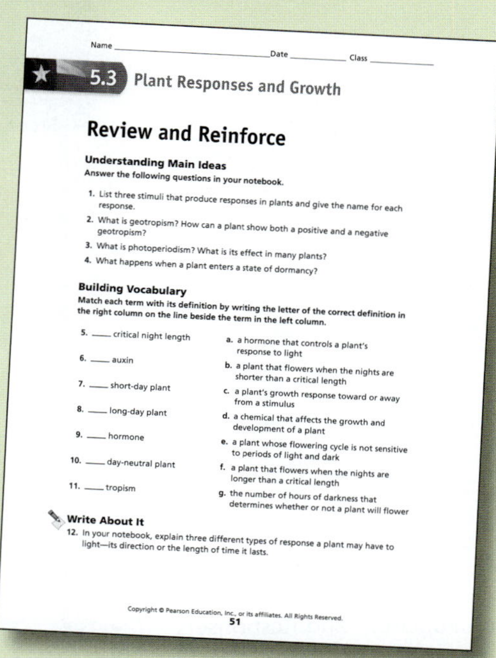

Name ____________ Date ________ Class ________

5.3 Plant Responses and Growth

Review and Reinforce

Understanding Main Ideas
Answer the following questions in your notebook.

1. List three stimuli that produce responses in plants and give the name for each response.
2. What is geotropism? How can a plant show both a positive and a negative geotropism?
3. What is photoperiodism? What is its effect in many plants?
4. What happens when a plant enters a state of dormancy?

Building Vocabulary
Match each term with its definition by writing the letter of the correct definition in the right column on the line beside the term in the left column.

5. ____ critical night length	a. a hormone that controls a plant's response to light
6. ____ auxin	b. a plant that flowers when the nights are shorter than a critical length
7. ____ short-day plant	c. a plant's growth response toward or away from a stimulus
8. ____ long-day plant	d. a chemical that affects the growth and development of a plant
9. ____ hormone	e. a plant whose flowering cycle is not sensitive to periods of light and dark
10. ____ day-neutral plant	f. a plant that flowers when the nights are longer than a critical length
11. ____ tropism	g. the number of hours of darkness that determines whether or not a plant will flower

Write About It

12. In your notebook, explain three different types of response a plant may have to light—its direction or the length of time it lasts.

51

LESSON 5.3

Lexile Measure = 860L

Plant Responses and Growth

Establish Learning Objectives

After this lesson, students will be able to:

Identify three stimuli that produce plant responses.

Describe how plants respond to seasonal changes.

Engage

Activate Prior Knowledge

MY PLANET DIARY Read *Flower Power* with the class. Initiate a discussion about global climate change. Elicit from students the difference between global warming *(rise in average temperature over surface of Earth)* and global climate change *(changes in regional climate characteristics around the world).* Ask: **How are they related?** *(As temperatures increase, some areas will become drier, others wetter, and severe-weather events will increase.)*

Explore

Lab Resource: Inquiry Warm-Up 

L1 CAN A PLANT RESPOND TO TOUCH? Students will determine the response to touch in various plants and infer an advantage to the plant. This Inquiry Warm-Up can be found in the Student Lab Manual, p. 64, and online.

Plant Responses and Growth

What Are Three Stimuli That Produce Plant Responses?
TEKS 7C, 13A

How Do Plants Respond to Seasonal Changes?
TEKS 13A

my planet Diary DISCOVERY

Flower Power

What makes a plant flower? Plants detect the amount of light each day. When there is just enough light, the plant sends a signal to the flower. But what is this signal? For almost 80 years, the answer remained a mystery. In 2008, scientists discovered the protein that was responsible. They linked the protein they thought controlled flowering to a fluorescent, or glowing, protein they obtained from a jellyfish. Then they watched the bright green protein travel with the flowering protein through the stem to make the plant bloom. Why does this experiment matter?

Global climate change is starting to hurt crops. Some places near the equator are becoming too warm to farm. Areas closer to Earth's poles may be needed to grow more crops as they warm. These areas, however, do not get as much sunlight. Scientists could use the flowering protein to encourage plants to flower without direct sunlight.

Communicate Discuss the question with a group of classmates. Then write your answer below.

In addition to getting the plants to flower with no light, what other challenges might scientists have to overcome when trying to get plants to succeed in a new area?

Accept all reasonable answers. Sample: The soil conditions might not be good for the plants. Animals could eat the plants.

 Do the Inquiry Warm-Up *Can a Plant Respond to Touch?* Student Lab Manual, p. 64

The green you see in these plant cells is from a fluorescent protein like the one used in the flowering experiment.

SUPPORT ALL READERS

Lexile Measure = 860L **Lexile Word Count = 1184**

Prior Exposure to Content: May be the first time students have encountered this topic

Academic Vocabulary: *conclusions, relate*

Science Vocabulary: *tropism, hormone, photoperiodism, dormancy*

Concept Level: Generally appropriate for most students in this grade

Preteach With: My Planet Diary "Flower Power" and Figure 1 activity

Vocabulary
- tropism
- hormone
- auxin
- photoperiodism
- critical night length
- short-day plant
- long-day plant
- day-neutral plant
- dormancy

Skills
- Reading: Relate Text and Visuals
- Inquiry: Draw Conclusions

What Are Three Stimuli That Produce Plant Responses?

TEKS 7C, 13A In this section you'll learn how plants respond to various external stimuli such as touch, the force of gravity, and light.

You may be one of those people who close their window shades at night because the morning light wakes you up. People respond to many stimuli each day. Did you know plants also respond to some of the same stimuli, including light?

ELPS 3.D.2

With a partner, read aloud the "My Planet Diary" on page 196. As you discuss the question, describe the experiment that was conducted. Identify different elements of the experiment. What was the question, the hypothesis, and the conclusion?

Tropisms Most animals can respond to stimuli by moving. Because plants are rooted in the ground, they usually respond by growing toward or away from a stimulus. A plant's growth response toward or away from a stimulus is called a **tropism** (TROH piz um). If a plant grows toward the stimulus, it is said to show a positive tropism. If a plant grows away from a stimulus, it shows a negative tropism. **Touch, gravity, and light are three important stimuli that trigger growth responses, or tropisms, in plants.**

Touch

Some plants show a response to touch called thigmotropism. The prefix *thigmo-* comes from a Greek word that means "touch." The stems of many vines, such as morning glories, sweet peas, and grapes, show positive thigmotropism. As the vines grow, they coil around any object they touch.

FIGURE 1

Plant Responses to Stimuli

The stimuli in space are not always the same as those on Earth.

Develop Hypotheses How might the roots of a plant grow in space without the influence of gravity?

Sample: The roots would grow in all directions.

Gravity

Plants can respond to the force of gravity. This response is called geotropism. Roots show positive geotropism if they grow downward. Stems, on the other hand, show negative geotropism. Stems grow upward against the force of gravity.

English Language Proficiency Standards

ELPS Speaking 3.D.2

Have students create a flow chart for My Planet Diary. Have them use the chart to identify the question, hypothesis, and conclusion in the experiment.

Beginning Review the terms *question, hypothesis,* and *conclusion.* Read aloud My Planet Diary. Create a flow chart to show the steps. Ask students to find in the chart the question scientists wanted to answer, the hypothesis, and the conclusion.

Intermediate Review the terms *question, hypothesis,* and *conclusion.* Together, read My Planet Diary. Have partners create a flow chart to show the steps of the experiment, including the question, the hypothesis, and the conclusion.

Advanced Have partners read My Planet Diary, chart the experiment, and identify the question, hypothesis, and conclusion.

Advanced High Have partners complete the My Planet Diary activity. Then have them create a flow chart that shows the question, hypothesis, and conclusion.

Explain

Introduce Vocabulary

Have students look at the vocabulary term *photoperiodism*. Tell them that *photo* refers to light and *periodism* to a recurring cycle of given length. The term refers to a plant's response to a recurring cycle of light and dark periods.

Teach Key Concepts

Explain to students that although plants cannot move from place to place, they can move by growing toward or away from a stimulus. Review with students the meanings of the terms *stimulus* and *response*. Ask: **What is a plant's growth response toward or away from a stimulus called?** *(A tropism)* **In terms of direction of growth, what are the two types of tropisms?** *(A positive tropism is growth toward a stimulus. A negative tropism is growth away from a stimulus.)* **What are three important stimuli to which plants show tropisms?** *(Touch, light, and gravity)* **What is a response to touch called?** *(Thigmotropism)* **How do vines show thigmotropism?** *(They show a positive response by wrapping around objects that they touch.)* **What are two other tropisms plants show?** *(Geotropism, or response to gravity, and phototropism, or response to light)* **What kind of geotropism do a plant's roots show?** *(Roots show a positive geotropism, because they grow downward toward the pull of gravity.)* **Which kind of geotropism do a plant's stems show?** *(Stems show a negative geotropism, because they grow upward away from the pull of gravity.)* **How do plants show phototropism?** *(The stems and leaves of a plant grow toward a light source.)*

PEARSON Texas.com

Texas Essential Knowledge and Skills

7C Demonstrate and illustrate forces that affect motion in everyday life such as emergence of seedlings, turgor pressure, and geotropism.

13A Investigate how organisms respond to external stimuli found in the environment such as phototropism and fight or flight.

LESSON 5.3

Explain

Lead a Discussion

PLANT HORMONES Introduce the idea that plants are able to respond to stimuli because they produce hormones. Ask: **What is a hormone?** *(A chemical that affects a plant's growth and development.)* **What hormone controls a plant's response to light? How does it work?** *(Auxin speeds up the rate at which plant cells grow. When light shines on one side of a plant's stem, auxins build up in the cells on the shaded side. Cells on the shaded side grow faster and become longer. The stem bends toward the light.)*

Relate Text and Visuals Tell students that when they relate text and visuals, they synthesize information in the text and in any accompanying visuals or diagrams.

Elaborate

Teacher Demo 

L1 MODELING PLANT RESPONSE

Materials coiled spring toy

Time 10 minutes

Have students decide which side of a coiled spring toy will represent the shaded side of a plant. Hold the toy and cause the shaded side to elongate, or "grow." Have students observe how the toy bends.

Ask: **Which way does the toy bend?** *(It bends away from the coils.)* **How does this model a plant's response?** *(A plant bends toward the light.)* **What causes this?** *(Auxin makes the cells on the shaded side grow faster. They become longer.)*

Lab Resource: Quick Lab 

L2 WATCHING ROOTS GROW Students will explore the effects of gravity on a sprouting seed. This Quick Lab can be found in the Student Lab Manual, p. 65, and online.

Evaluate

Assess Your Understanding

After students answer the questions, have them evaluate their understanding by completing the appropriate sentence.

RTI Response to Intervention

1a. If students cannot define *tropism,* **then** have them reread the definition.

b. If students have trouble identifying the functions of plant hormones, **then** remind them that hormones are chemicals that affect plant growth.

Relate Text and Visuals Use what you have read to label the side of the plant with more auxin and the side with less auxin.

Light

All plants exhibit a response to light called phototropism. The leaves, stems, and flowers of plants show positive phototropism because they grow toward light. A plant receives more energy for photosynthesis by growing toward the light.

Plants are able to respond to stimuli because they produce hormones. A **hormone** produced by a plant is a chemical that affects how the plant grows and develops. One important plant hormone is named **auxin** (AWK sin). Auxin speeds up the rate at which a plant's cells grow and controls a plant's response to light. When light shines on one side of a plant's stem, auxin builds up in the shaded side of the stem. The cells on the shaded side begin to grow faster. The cells on the stem's shaded side become longer than those on its sunny side. The stem bends toward the light.

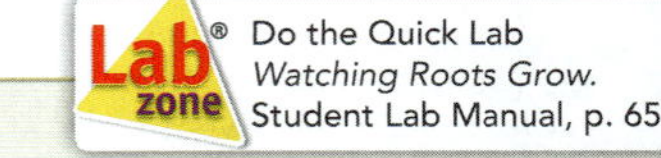

Do the Quick Lab *Watching Roots Grow.* Student Lab Manual, p. 65

Assess Your Understanding

TEKS 7C, 13A

1a. Define What is a tropism? Provide an example.

A tropism is a plant's response toward or away from a stimulus. Positive phototropism, or a plant's growth toward sunlight, is an example of a tropism.

b. Describe What do you think would happen if a plant did not create enough of the hormone that controlled flower formation?

Sample: The flowers may not form correctly or may not form at all.

got it?

○ **I get it!** Now I know that plants respond to touch, gravity, and light.

○ I need extra help with See TE note.

How Do Plants Respond to Seasonal Changes?

TEKS 13A In this section you'll find out more about the external stimuli that cause plants to flower in the spring or summer and lose their leaves in the autumn.

People have long observed that plants respond to the changing seasons. Some plants bloom in early spring, while others don't bloom until summer. The leaves on some trees change color in autumn and then fall off by winter.

Photoperiodism What kinds of external stimuli can trigger a plant to flower? **The amount of darkness a plant receives determines the time of flowering in many plants.** A plant's response to seasonal changes in the length of night and day is called **photoperiodism.**

Plants respond differently to the length of nights. Some plants will only bloom when the nights last a certain length of time. This length, called the **critical night length,** is the number of hours of darkness that determines whether or not a plant will flower. Some plants can flower when darkness lasts longer than their critical night length, and some plants can flower when darkness lasts less time than their critical night length. You can read more on how different plants respond to night length in **Figure 2.**

Photoperiodism

Plants and Night Length		Examples
Short-day plants flower when the nights are longer than a critical length. They bloom in fall or winter.	Midnight / Noon	Chrysanthemums, poinsettias
Long-day plants flower when nights are shorter than a critical length. They bloom in spring or summer.	Midnight / Noon	Irises, lettuce
Day-neutral plants have a flowering cycle that is not sensitive to periods of light and dark. They can bloom year-round depending on weather.	Midnight / Noon; Midnight / Noon	Dandelions, rice, tomatoes

FIGURE 2
Photoperiodism
Flowering plants can be grouped as short-day plants, long-day plants, and day-neutral plants.

Infer **Suppose you are a farmer in a climate that supports plant growth all year-round but night length varies. Based on the categories in the chart, would you plant mostly one type of plant or a mixture of all three? Explain.**

Sample: I would plant all three so that there would be plants in bloom all year no matter how long the nights were.

Differentiated Instruction

L1 Types of Photoperiodism Have students make a compare/contrast table of the types of photoperiodism in plants.

L1 Chrysanthemum Flowers Chrysanthemums are short-day plants. Have students describe the appearance of these plants in the early summer and in the fall. Tell students to explain why the plants appear as they do. If students are unfamiliar with this plant, show them a photograph of one in full bloom. Encourage students to draw pictures for their explanations.

L3 Flower Induction Invite students to research how greenhouse managers bring flowers, such as poinsettias, to bloom for specific seasons, and how they induce seasonal plants, such as chrysanthemums, to bloom all year.

Explain

Teach Key Concepts

Have students think about what happens when the seasons change. Elicit from students the fact that the length of days and nights changes with the change of seasons. Explain to students that the amount of darkness a plant receives determines the time of flowering in many plants. Ask: **Why don't some plants bloom in winter in locations that have seasonal changes?** *(Temperatures are too low and days are too short.)* **What environmental factor triggers plants to flower?** *(The amount of darkness a plant receives)* **What is photoperiodism?** *(A plant's response to hours of light and darkness)* **When does a short-day plant flower?** *(When nights are longer than its critical night length)* **When does a long-day plant?** *(When nights are shorter than its critical night length)*

Teach With Visuals

Tell students to look at **Figure 2.** Ask: **What do the clock faces represent?** *(The number of hours of darkness and light)* **Which is more important to a plant's flowering, the number of hours of darkness or the number of hours of light?** *(The number of hours of darkness)* **Based on your answer, what would be more appropriate names for the plants?** *(Long-night and short-night)* Point out to students that the terms *short-day plant* and *long-day plant* were used before it was understood that it is the amount of time the plant is in the dark, not the light, that is important to flowering.

21st Century Learning

CRITICAL THINKING Have students think about photoperiodism in plants. Ask: **What is the advantage of different plants flowering at different times of the year?** *(Samples: The plant's pollinators may only be active only during certain times of the year. Plants have adapted to the climate—for example, a particular plant may not be able to flower during the summer.)* **How might gardeners get plants to bloom out of season?** *(By controlling the temperature and the amount of light the plants receive)*

Texas Essential Knowledge and Skills

13A Investigate how organisms respond to external stimuli found in the environment such as phototropism and fight or flight.

LESSON 5.3

Explain

Teach Key Concepts

Explain to students that dormancy is another way in which plants respond to seasonal changes. Ask: **What is dormancy?** *(A period when an organism's growth or activity stops)* **How is dormancy beneficial to a plant?** *(It helps plants survive freezing temperatures and the lack of liquid water.)* **What are the changes a tree undergoes as it becomes dormant?** *(The leaves begin to turn color. Sugar and water are transported out of the leaves. The leaves fall off.)* Discuss the changes in leaf color with students. Emphasize that chlorophyll is not changing color; it is breaking down, and the other leaf pigments are now more visible. **Which pigments are visible with the breakdown of chlorophyll?** *(Yellow and orange)* **Which pigment is newly produced?** *(Red)*

Make Analogies

L1 **DORMANCY** Discuss with students other examples of dormancy. Ask: **To what other event in nature have you heard the term *dormancy or dormant* applied?** *(Sample: volcanoes that have not erupted for a time)* **What do some animals do in winter that is similar to trees becoming dormant?** *(They hibernate.)*

Elaborate

Apply It!

L1 Before beginning the activity, remind students that a graph shows how two variables are related. This graph shows the relationship between the number of germinated seeds and each day of growth for two temperatures at which the seeds are kept. The two sets of data are plotted for each temperature and the relationship between the growing conditions can be seen in a comparison of the slopes of the lines.

Draw Conclusions Tell students that when they draw conclusions, they use what they have learned or observed to summarize or explain related information, or answer related questions. In this case, they are using the slopes of the two lines to explain how temperature affects seed germination.

Winter Dormancy Some plants prepare differently than others for certain seasons. As winter draws near, many plants prepare to go into a state of **dormancy.** Dormancy is a period when an organism's growth or activity stops. **Dormancy helps plants survive freezing temperatures and the lack of liquid water.**

With many trees, the first visible change is that the leaves begin to turn color. Cooler weather and shorter days cause the leaves to stop making chlorophyll. As chlorophyll breaks down, yellow and orange pigments become visible. In addition, the plant begins to produce new red pigments. This causes the brilliant colors of autumn leaves. Over the next few weeks, sugar and water are transported out of the tree's leaves. When the leaves fall to the ground, the tree is ready for winter.

apply it!

One hundred radish seeds were planted in two identical trays of soil. One tray was kept at 10°C. The other tray was kept at 20°C. The trays received equal amounts of sun and water. The graph shows how many seeds germinated over time at each temperature.

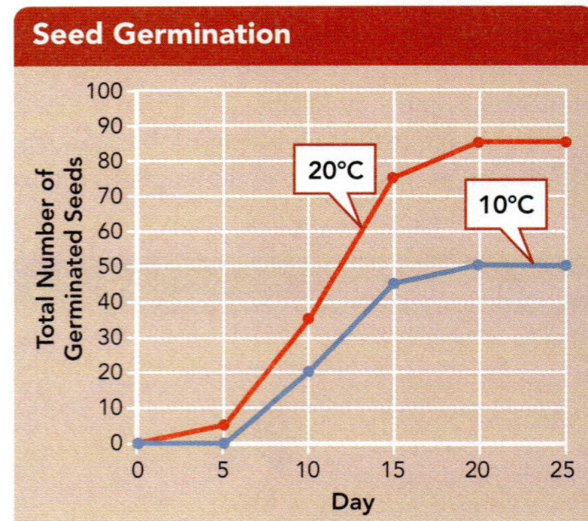

1 **Read Graphs** About how many seeds in the 20°C tray germinated on Day 13?

About 58 seeds
(accept plus/minus 2)

2 **Draw Conclusions** Based on the graph, what can you conclude about the relationship between the two temperatures and germination?

The number of germinating seeds increased with the higher temperature compared to the lower temperature.

3 **CHALLENGE** After the experiment, a fellow scientist concludes that more seeds will *always* germinate at higher temperatures. Is the scientist right? Why?

Sample: No; this experiment does not test temperatures above 20°C. There may be a point at which the temperature is too high for seeds to survive. More experiments with higher temperatures are needed to support that conclusion.

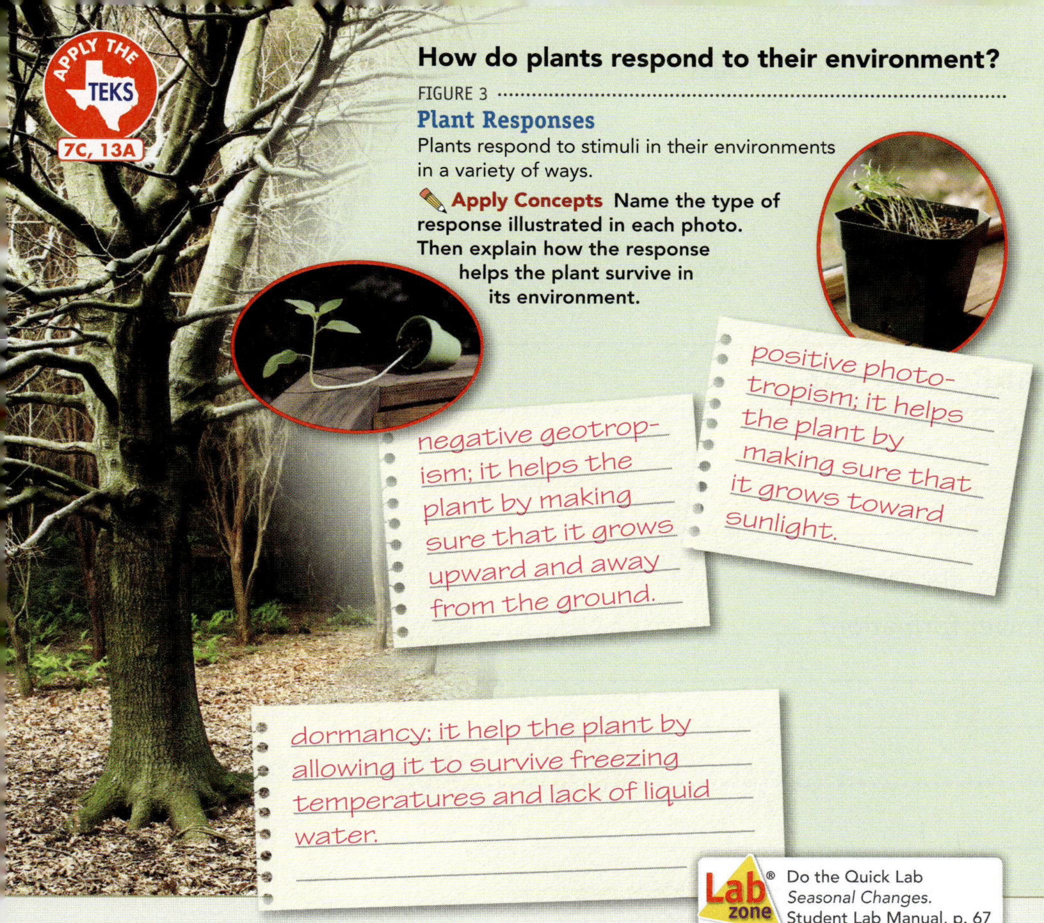

How do plants respond to their environment?

FIGURE 3

Plant Responses

Plants respond to stimuli in their environments in a variety of ways.

Apply Concepts Name the type of response illustrated in each photo. Then explain how the response helps the plant survive in its environment.

Do the Quick Lab *Seasonal Changes.* Student Lab Manual, p. 67

Assess Your Understanding

TEKS 13A

2a. **Review** (Short-day/Long-day) plants flower when nights are shorter than a critical length.

b. **Explain** Why do the leaves of some trees change color in autumn?

The chlorophyll in the leaves of the trees breaks down.

c. **Describe** How do plants respond to their environment?

Sample: Plants respond by growing toward sunlight, growing roots down into the soil, flowering at certain times, and going dormant in the winter.

got it?

○ I get it! Now I know that plants respond to seasonal changes because the amount of darkness determines the flowering times of many plants.

○ I need extra help with See TE note.

201

Differentiated Instruction

L1 Winter Dormancy Have students sketch the steps showing the changes a tree undergoes as winter approaches. Have them label each step and write in their own words what happens.

L1 Plant Response Have students summarize plant response by creating a concept map that includes the following terms: *tropism, stimulus, hormone, auxin, geotropism, phototropism, thigmotropism.*

L3 Autumn Leaves The colors that leaves exhibit in autumn are often characteristic of a particular type of tree. Have students find out which common trees display the following colors: yellow; purple; light brown or dark brown; bright red and orange; dark red. Then have them use the information to draw or paint a vista of trees in autumn color, labeling each type of tree.

Apply the TEKS

APPLY THE TEKS 7C, 13A

Emphasize to students that plants are not passive. Instead, they respond to their environments in ways that help them to survive, make food, and reproduce successfully. Remind students that plants cannot get up and move to a new location when conditions are difficult. A plant is anchored by its root system to the location where its seed originally germinated. Direct students' attention to **Figure 3**. Ask: **What examples have you seen around your home or school of plants responding to stimuli?** *(Samples: plants changing color or losing leaves in autumn, growing toward a window, or climbing up a wall)*

Lab Resource: Quick Lab 

L2 SEASONAL CHANGES Students will explore why the leaves of some trees change color when the seasons change. This Quick Lab can be found in the Student Lab Manual, p. 67, and online.

Evaluate

Assess Your Understanding

After students answer the questions, have them evaluate their understanding by completing the appropriate sentence.

RTI Response to Intervention

2a. If students need help with photoperiodism, **then** remind them that it is the length of darkness that determines a plant's flowering time. Have them think of the two choices in terms of "night" rather than "day."

b. If students have trouble explaining the color change in leaves in autumn, **then** remind them of the connection between chlorophyll and food-making and the fact that plants are getting ready to be dormant, a state in which growth or activity stops.

c. If students cannot explain how plants respond to their environment, **then** have them examine **Figure 3.**

Texas Essential Knowledge and Skills

7C Demonstrate and illustrate forces that affect motion in everyday life such as emergence of seedlings, turgor pressure, and geotropism.

13A Investigate how organisms respond to external stimuli found in the environment such as phototropism and fight or flight.

Name ______________________ Date __________ Class __________

Assess Your Understanding

Plant Responses and Growth

What Are Three Stimuli That Produce Plant Responses?

1a. DEFINE What is a tropism? Provide an example. ______________________

____

b. DESCRIBE What do you think would happen if a plant did not create enough of the hormone that controlled flower formation? ______________________

____

____

got it?

○ **I get it!** Now I know that plants respond to ______________________

____

○ **I need extra help with** ______________________

____

How Do Plants Respond to Seasonal Changes?

2a. REVIEW (Short-day/Long-day) plants flower when nights are shorter than a critical length.

b. EXPLAIN Why do the leaves of some trees change color in autumn? ______________________

____

c. DESCRIBE How do plants respond to their environment? ______________________

____

____

got it?

○ **I get it!** Now I know that plants respond to seasonal changes because ______________________

____

○ **I need extra help with** ______________________

____

Place the outside corner, the corner away from the dotted line, in the corner of your copy machine to copy onto letter-size paper.

Name ______________________ Date __________ Class __________

Enrich

Plant Responses and Growth

Read the passage and study the diagrams below. In your notebook, answer the questions that follow.

Carnivorous Plants

The Venus' flytrap is an example of the almost 400 species of carnivorous plants. A *carnivorous plant* is a plant that traps and then digests insects and other small animals, obtaining their nitrogen. Most carnivorous plants grow in marshy areas such as swamps and bogs where the soil is low in nitrogen. Because carnivorous plants do not have to rely on nitrogen absorbed from the soil by their roots, they are well suited to their environments.

Carnivorous plants respond to the stimulus of touch to trap their prey. For example, an insect touching a hair on the leaf of a Venus' flytrap triggers a specific response. Water moves from cells on the inside of the flytrap to cells on the outside of the flytrap. This causes the leaf of the flytrap to snap shut, quickly, catching the insect.

Carnivorous plants called sundews use another method to trap their prey. Sundews have small leaves that produce a sweet, sticky liquid at their tips. Insects fly into the stalks and stick to them. This triggers a different response. Other leaves begin to curl inward toward a trapped insect by using cell growth. The cells on one side of the stalks grow faster than the cells on the other side. This causes the leaves to bend. The leaves then produce a chemical that digests the insect, so it can be used to nourish the plant.

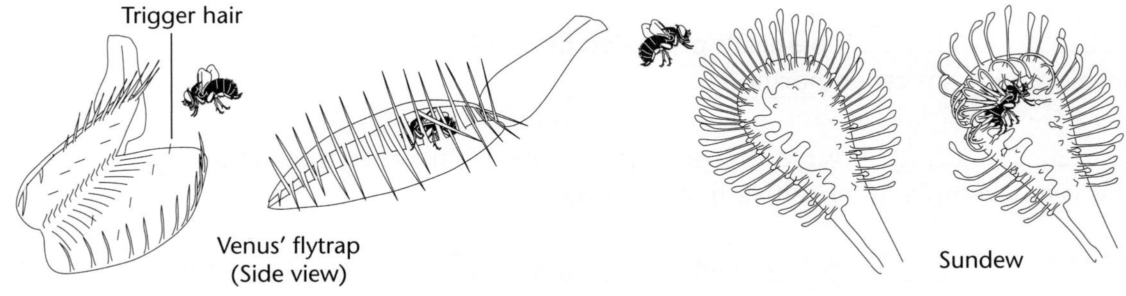

1. How do carnivorous plants get the nitrogen they need?
2. Why is it helpful to a sundew to produce the sweet, sticky liquid?
3. Do you think carnivorous plants also produce food by photosynthesis? Explain your answer.
4. What is one advantage that quick movement gives to a Venus' flytrap?
5. Why doesn't the sundew have to move as quickly as the Venus' flytrap to imprison an insect?

Name ______________________ Date __________ Class __________

Lesson Quiz

Plant Responses and Growth

Write the letter of the correct answer on the line at the left.

1. ___ A plant stem that grows toward light is an example of a

A negative thigmotropism

B negative geotropism

C positive phototropism

D positive geotropism

2. ___ To which three stimuli do plants respond?

A water, light, and touch

B light, gravity, and soil conditions

C touch, gravity, and chlorophyll

D touch, gravity, and light

3. ___ When chlorophyll in leaves breaks down,

A photosynthesis begins

B red pigments disappear

C yellow and orange pigments become visible

D the plant becomes very active

4. ___ Irises and lettuce flower when nights are shorter than a critical length. They can be described as

A short-day plants

B long-day plants

C dormant plants

D day-neutral plants

If the statement is true, write *true*. If the statement is false, change the underlined word or words to make the statement true.

5. ______________ Chemicals produced by a plant that control its growth and development are called hormones.

6. ______________ A plant's roots grow away from a rock they hit in the soil. This is an example of a positive thigmotropism.

7. ______________ Chlorophyll speeds up that rate at which a plant's cells grow and controls a plant's response to light.

8. ______________ The critical night length for a certain plant is 10 hours. This plant will flower only when nights are shorter than 10 hours.

9. ______________ A plant adaptation that helps it survive freezing temperatures and lack of liquid water is dormancy.

10. ______________ The blooming of poinsettias in winter and irises in spring is an example of phototropism.

Plant Responses and Growth

Answer Key

Review and Reinforce

Find the worksheet in the Student Workbook.

1. touch: thigmotropism; gravity: geotropism; light: phototropism
2. When a plant's roots grow downward, it shows a positive geotropism. When its stems grow upward against gravity it shows a negative geotropism.
3. Photoperiodism is a plant's response to seasonal changes in the length of day and night. It determines the time of flowering in many plants.
4. During dormancy a plant's growth or activity stops.
5. g
6. a
7. f
8. b
9. d
10. e
11. c
12. One plant response to light is phototropism. Leaves, stems, and flowers that grow toward light show a positive phototropism. Another type of plant response to light is photoperiodism, a response to seasonal changes in the length of night and day. Some plants will only bloom when the night lasts a certain length of time. A third type of plant response to light is to enter a state of dormancy, a period when growth or activity stops. Both shorter days and colder weather may trigger dormancy, which can help plants survive freezing temperatures and a lack of liquid water.

Enrich

1. by digesting insects and other animals that they trap
2. The sweet sticky liquid acts as bait and trap for insects.
3. Yes the plants still need to make food even though they digest insects to obtain nitrogen.
4. Answers will vary. Sample: Speed allows the Venus' flytrap to trap an insect before the insect can fly away.
5. The sticky liquid traps the insect. Therefore, the plant can move relatively slowly to imprison and digest the insect.

Lesson Quiz

1. C
2. D
3. C
4. B
5. true
6. negative
7. Auxin
8. longer
9. true
10. photoperiodism

CHAPTER 5

Scientific Investigation and Reasoning

This Apply the TEKS feature will enable students to use the scientific investigation and reasoning skills in TEKS 2A and 2C to reinforce the content of TEKS 7C. Students will plan a comparative investigation to explore geotropism in germinating bean seeds.

Planning an Investigation on Geotropism in Plants

Before students answer Questions 1–3, have them review the basic needs of plants. Ask: **What is photosynthesis?** *(The process by which plants make their own food)* **What do plants need for photosynthesis to occur?** *(Water and sunlight)*

Remind students that a tropism is a plant movement triggered by an outside stimulus force. Ask: **What is geotropism?** *(When a plant's roots grow down in the soil as a response to gravity)* **How does geotropism affect the survival of a plant?** *(Roots need to grow down in the soil to absorb water necessary for photosynthesis)*

Explain to students that geotropism was first studied in the early 1800s and many scientists believed that root tips were pulled downward because of their weight. New theories suggest that small structures in the root cells respond to gravity, causing the roots to grow in a certain direction. Have students do research on other tropisms.

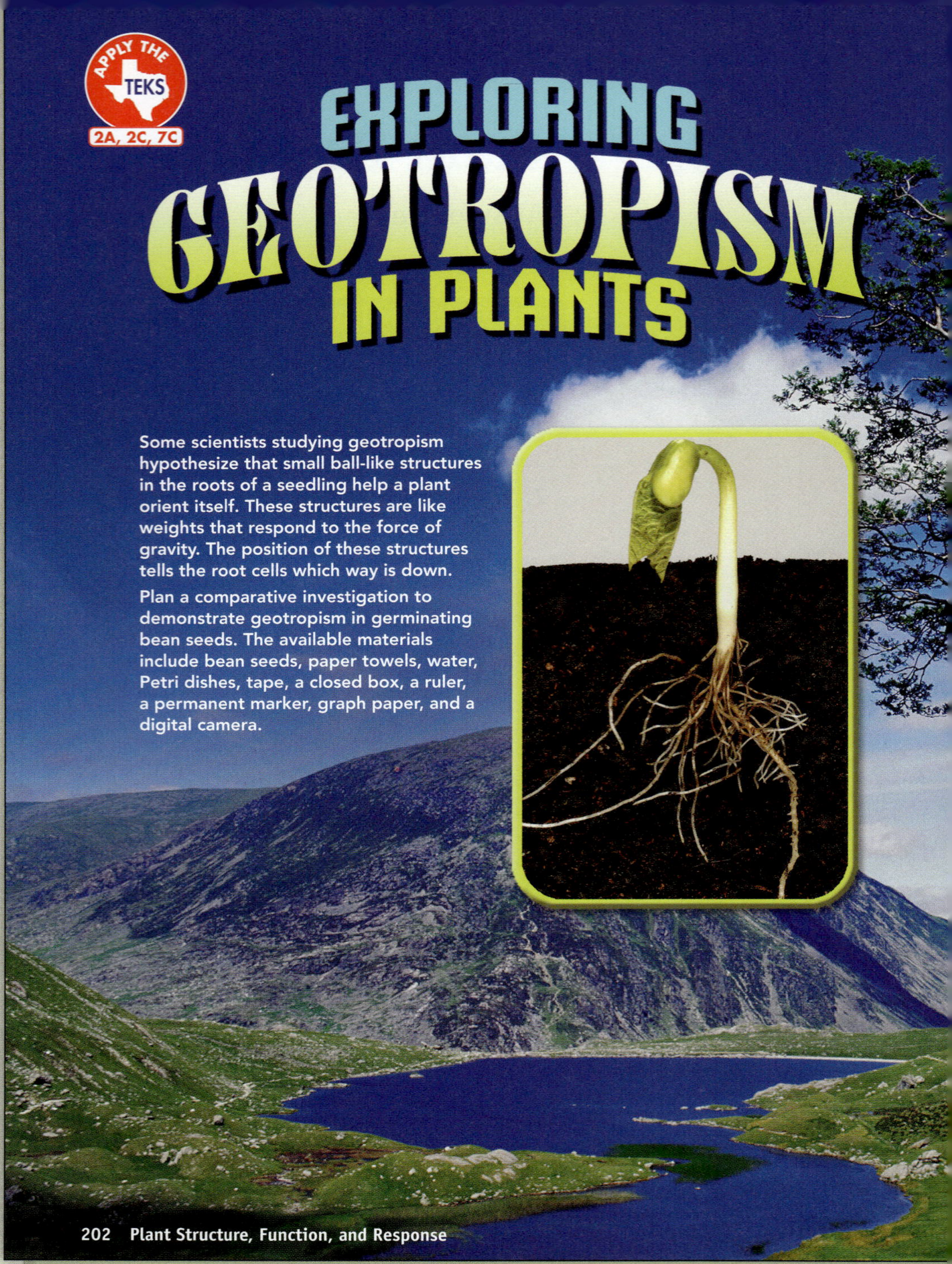

APPLY THE TEKS 2A, 2C, 7C

EXPLORING GEOTROPISM IN PLANTS

Some scientists studying geotropism hypothesize that small ball-like structures in the roots of a seedling help a plant orient itself. These structures are like weights that respond to the force of gravity. The position of these structures tells the root cells which way is down.

Plan a comparative investigation to demonstrate geotropism in germinating bean seeds. The available materials include bean seeds, paper towels, water, Petri dishes, tape, a closed box, a ruler, a permanent marker, graph paper, and a digital camera.

Texas Essential Knowledge and Skills

2A Plan and implement comparative and descriptive investigations by making observations, asking well-defined questions, and using appropriate equipment and technology.

2C Collect and record data using the International System of Units (SI) and qualitative means such as labeled drawings, writing, and graphic organizers.

7C Demonstrate and illustrate forces that affect motion in everyday life such as emergence of seedlings, turgor pressure, and geotropism.

Answer the questions below.

1. **Ask Questions** How will you test the seeds' geotropism ability? Write your challenge in the form of a question.

   Sample: How do plants adjust to being turned over at different frequencies?

2. **Plan an Investigation** Describe how you can implement your comparative investigation. Explain how you will use appropriate equipment and technology. Remember to include a control group. Then construct a chart or other graphic organizer that you can use to record your results.

   I will divide the beans into four different groups. Each group will get the same amount of water and sunlight. The first group will be left alone to grow under normal circumstances on a wet paper towel. The other three groups will be turned over at different intervals each week. I will use the camera to document the growth of each group. I can use a chart to record and compare my observations.

3. **Observe** What are the possible outcomes of your investigation?

   Sample: The plants turned over the most frequently will either die or be deformed. All the plants may survive, but some may be deformed.

203

Scientific Investigation and Reasoning, Cont.

After students have designed their investigations and answered Questions 1–3, have them share their plans with the rest of the class. Students should be able to identify the dependent and independent variables and the controls in their experiments. Ask: **Why is it important to include a control group in the investigation?** *(Sample: The control group shows what happens to a plant under normal circumstances. It allows you to evaluate how changing the situation affects the seeds' geotropism abilities.)*

You may wish to have students find out about the plant experiments that NASA has carried out in space. Some recent experiments have shown that sprouting seedlings begin to grow out in every direction, but they adapt to the conditions of microgravity and start to grow in a more stable direction. Ask: **How might the understanding of successful plant growth in space be important to future space missions?** *(Sample: Astronauts might need to use plants as food sources for longer missions. Plants could also be used to absorb carbon dioxide and provide water during the mission.)* **How might plant experiments conducted in space be different than those conducted on Earth?** *(Sample: Since space is a near-weightless environment, the plants would have to be grown in a contained area so water, nutrients, and soil would not be able to freely move throughout the spacecraft.)*

CHAPTER 5

TEKS Practice

Assess Understanding

Have students complete the answers to the TEKS Practice questions. Have a class discussion about what students find confusing. Write Key Concepts on the board to reinforce knowledge.

RTI Response to Intervention

4. If students need help describing plant adaptations for life on land, **then** review the functions water serves for water-dwelling plants.

7. If students need help relating the condition of stomata to the process of transpiration, **then** review the structure and function of stomata and how the process of transpiration takes place.

Alternate Assessment

L1 DESIGN A GAME Have students work in small groups to design a game about plants. Students can design a board game that requires players to answer questions in order to advance. Remind students to create rules, spinners, game pieces, and questions for their games. The questions for the game should include vocabulary terms and Key Concepts from the chapter. Students can exchange games or play their own game against other groups.

CHAPTER 5

TEKS Practice

TEKS 7C, 11A, 12A, 12C, 13A, 13B

LESSON 1 What Is a Plant?

1. In which cellular structure do plants store water and other substances?
   - a. cuticle
   - (b.) vacuole
   - c. cell wall
   - d. chloroplast

2. The pigment chlorophyll is found in chloroplasts.

3. **Recognize** What are the two main organ systems in a typical plant?
   The root system and the shoot system

4. **Describe** Complete the table to explain internal and external plant adaptations for life on land.

Structure	Function
Roots	Help obtain water and nutrients
Cuticle	Helps prevent water loss
Vascular tissue	Helps move water and food, gives support

5. **Recognize** Identify the levels of organization in a typical plant.

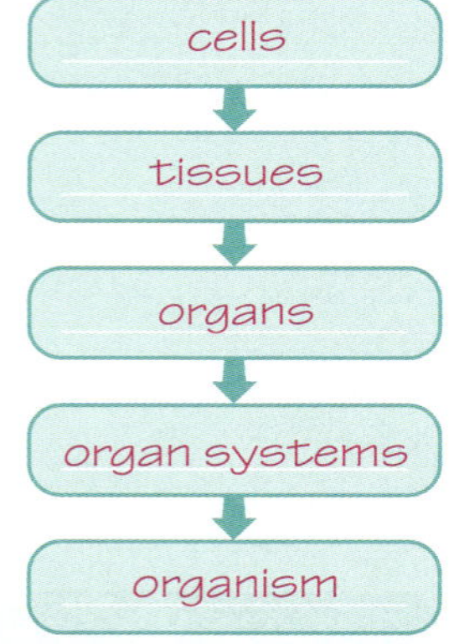

LESSON 2 Plant Structures

6. A plant absorbs water and minerals through
   - (a.) roots.
   - b. stems.
   - c. leaves.
   - d. stomata.

7. Transpiration slows down when stomata are closed.

8. **Relate Cause and Effect** When a strip of bark is removed all the way around the trunk of a tree, the tree dies. Explain why.
   Sample: Cutting the bark cuts off the phloem. The tree would not be able to transport food.

9. **Explain** Describe the internal response that occurs in plants if water loss leads to low turgor pressure in guard cells.
   When turgor pressure is low, the guard cells collapse. This causes the stomata to close, which prevents more water loss from transpiration.

10. **Identify** What is the main role of the leaves in a plant?
    Sample: Leaves use the energy in sunlight to make food through photosynthesis.

11. **Write About It** While all plants share some basic structures, these structures do not look the same among all plants. For example, some plants have short leaves and others have long leaves. Explain why you think there is so much variation among plant structures.
    See TE rubric.

204 Plant Structure, Function, and Response

Write About It **11** Assess student's writing using this rubric.

SCORING RUBRIC	SCORE 4	SCORE 3	SCORE 2	SCORE 1
Explain the variation in plant parts among all plants.	Explanation is an accurate presentation of plant adaptations to different environments that result in variations in structures.	Explanation includes idea of plants' adaptations to their environments but provides no details.	Explanation includes idea that plants grow in different places but does not refer to adaptation to environment.	Explanation includes inaccurate information and is incomplete.

LESSON 3 Plant Responses and Growth

12. A plant's response to gravity is an example of a

a. dormancy. **b.** hormone.
(c.) tropism. **d.** critical night length.

13. Relate Plant stems may bend toward sunlight. Relate how this is also a response to an internal stimulus.

Plant cells grow longer in response to the hormone auxin.

14. Predict A particular short-day plant has a critical night length of 15 hours. Fill in the chart below to predict when this plant would flower.

Day Length	Night Length	Will It Flower?
9 h	15 h	*Yes*
10 h	14 h	*No*
7.5 h	16.5 h	*Yes*

15. Apply Concepts In the space below, illustrate the affect that the force of gravity has on a plant. Use arrows to indicate the direction of the force. Describe your illustration.

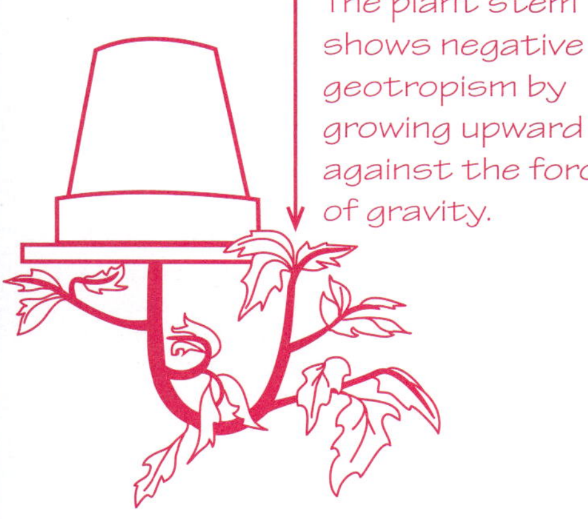

The plant stem shows negative geotropism by growing upward against the force of gravity.

How do plants respond to their environment?

16. Plants are all around us. Describe a plant that you see often and then explain how it responds to its environment.

Sample: A big tree grows outside of my school. The tree's stem grows opposite the force of gravity and toward sunlight. It responds to things like seasons by dropping its leaves before winter and growing new leaves in the spring. See TE rubric.

Interactive Science Chapter 5

Lesson 1
In Lesson 5.1 you learned about the characteristics that plants share and the levels of organization in a plant. You also learned how plants live successfully on land, and how they are classified.
Supporting TEKS: 11A
TEKS:12A, 12C

Lesson 2
In Lesson 5.2 you learned about the basic structures of plants, including roots, stems, and leaves, and their functions. You also learned about the structures of flowers and how seeds develop into new plants.
Supporting TEKS: 11A
TEKS: 7C, 12A, 12C, 13A, 13B

Lesson 3
In Lesson 5.3 you learned about how plants respond to external stimuli, such as light, touch, the force of gravity, and changes in the seasons.
TEKS: 7C, 13A

How do plants respond to their environment?

Assess student's response using this rubric.

SCORING RUBRIC	SCORE 4	SCORE 3	SCORE 2	SCORE 1
Explain how a familiar plant responds to its environment.	Student can identify two or more environmental stimuli and explain how the plant responds to each stimulus.	Student can identify one environmental stimulus and how the plant responds to it.	Student can identify an environmental stimulus, but cannot explain how the plant responds to it.	Student can identify a familiar plant but cannot explain how it responds to its environment.

TEKS Practice, Cont.

RTI Response to Intervention

14. If students need help predicting when a short-day plant will flower based on its critical night length, **then** remind them that the length is the minimum, or least, amount of darkness required.

L3 WRITING IN SCIENCE Ask students to write a television interview with a botanist that explains to viewers what makes a plant a plant. The interview should cover how plants are classified, plant structures, plant reproduction, how plants grow and respond to their environments, and the uses of plants in everyday life.

How do plants respond to their environment?

Student answers should describe a familiar plant, the stimuli it receives from the environment, and how it responds to those stimuli. See the scoring rubric below.

Review the TEKS Chapter 5

REVIEW THE TEKS 7C, 13A

PARTNER REVIEW Have partners review the definitions of vocabulary terms and quiz each other. Students can read the Key Concept statements and leave out words for their partners to fill in. They also can change a statement so that it is false and then ask their partners to correct it.

CLASS ACTIVITY: MURAL Divide the class into two groups and instruct each group to organize the information they have learned about plants in a mural. Provide large sheets of paper to accommodate the murals. Encourage students to make their murals colorful and visual, using drawings, photos, and actual plant parts whenever appropriate. Have students keep the following questions in mind as they develop their murals:

- What are the characteristics of nearly all plants?
- What do plants need to survive on land?
- How are plants classified?
- What are the functions of plant roots, stems, leaves, seeds, and flowers?
- What are the stages in a plant's life cycle?
- How do plants respond to conditions in their environments?

PEARSON Texas.com

CHAPTER 5

TEKS Practice: Chapter Review

Test-Taking Skills

ELIMINATING INCORRECT ANSWERS Explain to students that when they answer a multiple-choice question, they can often eliminate some of the answer choices because they are clearly incorrect. By doing this, they increase the odds of choosing the correct answer.

Question 1 TEKS 13A

The correct answer is D. The plant is demonstrating positive phototropism.

If students chose A, explain that the window does not cut off the sunlight.

If students chose B, explain that a lack of carbon dioxide would not cause the plant to wilt.

If students chose C, explain that gravity pulls straight down and would not cause this plant to lean toward the window.

Question 2 TEKS 11A

The correct answer is F. Epiphytes can get sunlight that might not pass through the canopy below.

If students chose G, explain that all plants need nutrients.

If students chose H, explain that plants are not considered prey.

If students chose J, explain that carbon dioxide is available above and below the canopy.

★ TEKS Practice: Chapter Review

Read each question and choose the best answer.

1 Maria placed a houseplant on a table next to a window. She left the plant in the same position for a couple of days. Then, she examined the plant and drew the following sketch.

Which statement is the best explanation for why the plant is leaning toward the window?

A It is wilting due to a lack of sunlight.

B It is wilting due to a lack of carbon dioxide.

C It is growing in the direction opposite the force of gravity.

(D) It is growing in the direction of the sunlight coming through the window.

2 Many types of plants live in rain forests. Trees in a rain forest form a thick canopy. Plants called epiphytes use their roots to anchor to the top of the tree canopy. What advantage does this give epiphytes in their ecosystem?

(F) They are able to get enough sunlight.

G They are able to live without nutrients.

H They are able to escape from predators.

J They are able to get more carbon dioxide.

★ TEKS Practice: Cumulative Review

3 Which situation would result in the most genetically diverse offspring?

Ⓐ Bees bring pollen from the flower of one plant to the flower of another.
B An amoeba in a freshwater pond reproduces through cell division.
C An orchid pistil receives pollen from a stamen of the same flower.
D A piece of potato containing a bud is removed and planted.

4 The art below shows the steps of cell division out of order. Which list places them in the correct order?

Cell Division

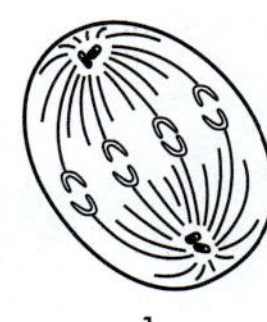
1

2

3

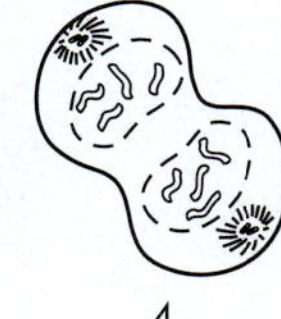
4

5

F 2, 3, 5, 4, 1
G 2, 5, 3, 4, 1
Ⓗ 3, 2, 5, 1, 4
J 5, 2, 3, 1, 4

If You Have Trouble With . . .				
Question	1	2	3	4
See Lesson	5.3	5.1	4.4	3.3
TEKS	7.13A	7.11A	7.14B	7.12F

207

TEKS Practice: Cumulative Review

Additional Assessment Resources

Teacher's Edition: Lesson Quizzes, Texas End-of-Year Test Prep A and B
Online assessments

Question 3 TEKS 14B

The correct answer is A. This is the only example of sexual reproduction, which results in the most genetically diverse offspring.

If students chose B, explain that mitosis results in identical cells. Offspring produced in this way are not diverse.

If students chose C, explain that self-pollination is an example of asexual reproduction because there is only one parent.

If students chose D, explain that a plant growing from a bud is an example of asexual reproduction because there is only one parent.

Question 4 TEKS 12F

The correct answer is H. This sequence places the steps of mitosis in the proper order.

If students chose F, explain that cytokinesis is the last stage of cell division.

If students chose G, explain that the chromosomes are not condensed at the beginning of the process.

If students chose J, explain that chromosomes do not line up in the first stage of cell division.

Science Matters

Focus on Texas

Have students read *Got Water?* Review the concept of transpiration and emphasize its importance in the water cycle. Explain that it is the process by which water is carried through plants from roots to small openings called stomata on the underside of their leaves. The water is released as vapor through the stomata and into the atmosphere. Ask: **A cactus has spines instead of the large broad leaves of a live oak tree. How does this adaptation help the cactus conserve water?** *(Sample: Cactus spines return less water back into the atmosphere than leaves do.)*

Explain to students that the name *cactus* comes from the Greek work *kaktos,* which means "spiny plant." In 1995, the Texas Legislature named the prickly pear cactus as the state plant. Different species of prickly pear cactus are found throughout the western two-thirds of Texas. The prickly pear cactus can survive in areas with large temperature ranges and very little water. Small mammals and birds feed on its fruit and find protection in the areas between the cactus spines. The prickly pear cactus has long been a food source to Native Americans and other cultures living in the southwestern part of the United States.

TEXAS SCIENCE MATTERS

Focus on Texas

TEKS 11A

GOT WATER?

Did you ever wonder why live oaks grow near the coast and cacti grow in the Chihuahuan desert?

Live oaks thrive in areas that are wet all year. With proper drainage, they can survive where other trees would drown or die from contact with salt water. Live oak roots can creep along the surface of the ground to stay above the water table. The trees grow broad leaves that collect a lot of sunlight for photosynthesis. Broad leaves also cause the trees to lose a lot of water to transpiration.

In contrast, desert cacti have several features that help them survive dry conditions. A cactus has a thick, green, and waxy trunk. The trunk is green because this is where chlorophyll is located and where photosynthesis occurs. The waxy surface traps water inside. The ribbed shape of some cacti channels precious rainwater to the thick roots.

Devilshead cactus (*Echinocactus horizonthalon*)

Investigate It Find a list of native plants that live near you. How is each of these plants especially suited to the environment in your area? Make a diagram of one of the plants, pointing out the features that help it survive. Present your diagram to your class.

Southern live oak (*Quercus virginiana*)

208 Plant Structure, Function, and Response

Quick Facts

The wood of live oak trees has played an important role in United States naval history. The *USS Constitution* was one of America's first naval vessels and was launched on October 21, 1797 from Boston Harbor. The hull of the *Constitution* was built from the wood of live oaks and white oaks, which both have very high densities. The density of the wood in the *Constitution's* hull allowed it to repel or absorb enemy cannon shots, preventing serious damage to the ship or great loss of life. The *Constitution* was nicknamed "Old Ironsides" because of its strong, metal-like hull and was victorious in over 30 battles from 1798 to 1854. The *Constitution* is now a floating museum docked once again in Boston Harbor.

Texas Essential Knowledge and Skills

11A Examine organisms or their structures such as insects or leaves and use dichotomous keys for identification.

PLANTING ROOTS IN OUTER SPACE

STEM

Frontiers of Technology

TEKS 2A, 13A

Far from farms and greenhouses on Earth, future space explorers will need to grow their own food, and recycle and purify their air and water. Astronauts from the National Aeronautics and Space Administration (NASA) have been experimenting with plants in space for many years.

Which Way Is Up?

On Earth, plant roots grow downward and outward in response to Earth's gravity, while plant shoots grow upward. In space, where there is no clear up or down, roots and shoots both grow toward the light! In order to grow with the roots at the bottom and the stems at the top, plants need gravity. So space stations need special plant chambers that rotate continuously to create artificial gravity for plants.

Tomatoes From Outer Space

To study whether radiation in space will affect the ability of seeds to grow, NASA scientists placed 12.5 million tomato seeds in a satellite that orbited Earth for six years! Students around the world then planted the seeds, which grew normally and produced normal tomatoes. So scientists now know that seeds will survive for a long time in orbit.

Design It Scientists are still learning about how to grow plants to support space travel. Find out about current NASA research on plants in space. Identify one question you have about plant growth in space. Plan an investigation that might help you answer the question. Identify any appropriate equipment or technology that your investigation would require.

A researcher holds tiny *Arabidopsis* seedlings. *Arabidopsis* plants are related to the cabbage plant, and are often used as model plants in research projects.

209

STEM

Frontiers of Technology

Have students read *Planting Roots in Outer Space.* Explain that one reason these NASA experiments are conducted is so that astronauts can find ways to sustain themselves on extended space voyages. Astronauts bring food with them when they travel into space, but those supplies are limited. Astronauts will be able to stay in space longer and travel farther from Earth if they can grow their own food and do not need to be re-supplied periodically.

Ask students what plants need to grow. Then ask them how they think the plants get those things in space. Tell students that plants are supplied with water that is recycled from other uses aboard a space shuttle or the International Space Station (ISS). Scientists have also built energy-efficient LED lights. For space greenhouses, scientists only supply light in frequencies that plants use to help them grow. No energy is wasted by using these lights. Explain that human waste can be broken down by microbes in bioreactors and then given to the plants to supply valuable nutrients.

Tell students that the best plants to grow in space will be compact, grow well under low light, and have a strong chance of providing a healthy crop with as many edible parts as possible. Scientists are working to identify types of wheat, rice, lettuce, potatoes, and other plants that best meet these criteria.

Ask: **Why do plants need artificial gravity?** *(to get the most growth from the shoots)* **Why would it be helpful to know whether seeds will survive long periods in orbit?** *(Astronauts may need to travel great distances to reach a planet where plants can grow.)*

English Language Proficiency Standards

ELPS Speaking 3.G.2

Have students name some of the problems with growing plants in space. Then have them tell possible solutions.

Beginning Have students listen as you or another student read aloud page 209. Help students express ideas by scaffolding questions: *How do shoots and roots grow on Earth? In space, what do roots and shoots grow toward?*

Intermediate Together, read aloud page 209. Have partners express their ideas by completing the following sentence frames: *One problem with growing plants in space is _____. Scientists think _____.*

Advanced Have partners read page 209. Than have them tell what problems and solutions scientists have found.

Advanced High Have partners read page 209. Than have them tell what problems and solutions scientists have found. Have each student tell one other problem scientists might have growing plants in space, and how they might solve it.

Texas Essential Knowledge and Skills

2A Plan and implement comparative and descriptive investigations by making observations, asking well-defined questions, and using appropriate equipment and technology.

13A Investigate how organisms respond to external stimuli found in the environment such as phototropism and fight or flight.

CHAPTER 6

Introduction to the Human Body

Chapter TEKS Overview

This chapter focuses on TEKS 12B and 12C. Students will explore the levels of organization in the human body and how the major human body systems work together. They also will examine the main functions of the skeletal, muscular, and integumentary systems.

Introduce the TEKS

To pique student interest and introduce TEKS 12B, have students look at the image and read the Focus Question and description. Ask students to make an inference about what jobs bones, muscles, blood vessels, the brain, and nerves do. Point out that the work of keeping a person alive is a group effort: It requires that the different parts of the body perform their specific tasks in "cooperation" with one another. Ask: **Why is it important for blood to reach all parts of the body?** *(Blood brings oxygen to the cells and removes carbon dioxide from the cells.)* **Why might the human body be called a "well-tuned machine"?** *(Sample: A machine performs one or more jobs by having its various parts work together correctly and smoothly. The human body performs its functions by having its parts do the same.)*

Untamed Science Video

FEELING JUST SPINE, THANK YOU Before viewing, invite students to discuss what they know about the ways their muscles and bones interact to make their bodies move. Then play the video. Lead a class discussion and make a list of questions that this video raises. You may wish to have students view the video again after they have completed the chapter to see if their questions have been answered.

Chapter at a Glance

CHAPTER PACING: 11–17 periods or $5\frac{1}{2}$–$8\frac{1}{2}$ blocks

INTRODUCE THE CHAPTER: Use the Focus Question and the opening image to get students thinking about the human body. Activate prior knowledge and preteach vocabulary using the Getting Started pages.

Lesson 1: Body Organization

Lesson 2: System Interactions

Lesson 3: Homeostasis

Lesson 4: The Skeletal System

Lesson 5: The Muscular System

Lesson 6: The Integumentary System

ASSESSMENT OPTIONS:
Teacher's Edition: Lesson Quizzes, Texas End-of-Year Test Prep A and B
Online assessments

WHAT CAN THESE BODY PARTS DO?

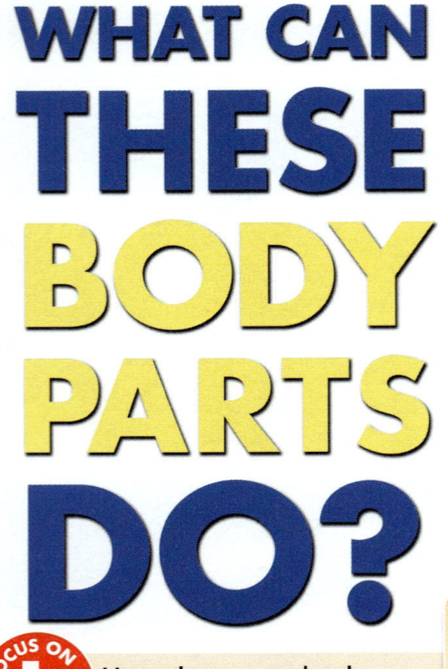

How does your body work?

Your body is an amazingly complex mass of trillions of cells. These cells work together, doing all the functions that keep you alive. On average, an adult has 206 bones, 96,500 kilometers of blood vessels, and a brain with a mass of 1.4 kilograms. Your nerves can send signals at speeds of up to 120 meters per second. Your smallest muscle is in your ear.

Infer What jobs do some of these body parts do?

Sample: My brain tells my body what to do by sending messages through my nerves. My bones hold me up and help me move. My blood vessels carry blood to all parts of my body.

Watch the **Untamed Science** video to learn more about body systems.

UntamedScience

From the Author

I just moved to a new state where the state police do not tolerate speeding. I am also on the road quite a bit on my job. When in the car, I listen to the radio or play audiobooks. Sometimes when I listen to a particularly exciting passage or new story, I find I speed up too much. So I have learned to use my cruise control. My cruise control is a good example of how some homeostatic systems work. I set my control for a certain speed, much like the human body sets its temperature control for 37°C. The hypothalamus acts as the control center, like cruise control, and provides feedback to control body temperature. So far—no tickets.

Michael Padilla

Introduction to the Human Body

Texas CHAPTER 6

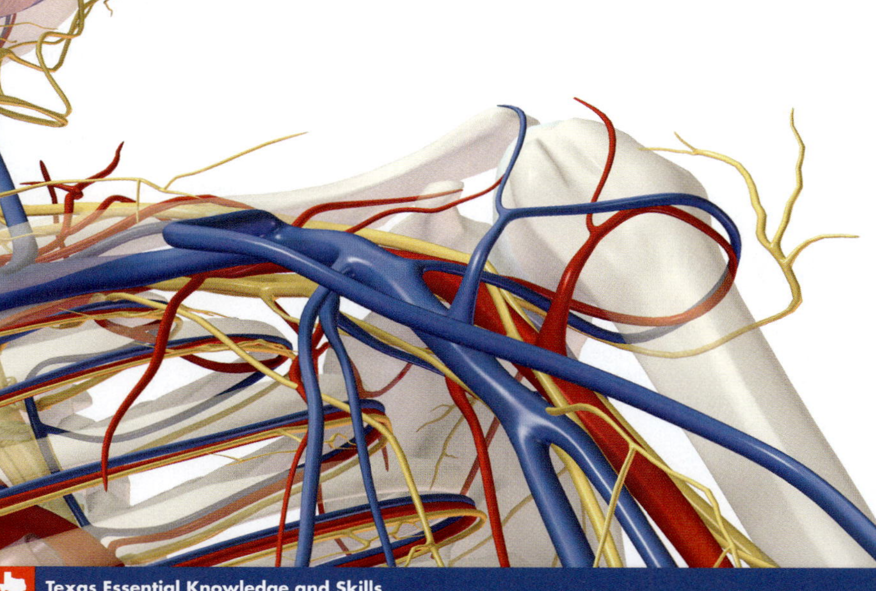

Texas Essential Knowledge and Skills

SUPPORTING TEKS: 7A Contrast situations where work is done with different amounts of force to situations where no work is done. **12B** Identify the main functions of the systems of the human organism, including the skeletal, muscular, and integumentary systems.

TEKS: 2C Collect and record data using the International System of Units (SI). **2D** Construct tables, using repeated trials and means, to organize data and identify patterns. **2E** Analyze data to formulate reasonable explanations, communicate valid conclusions supported by the data, and predict trends. **3B** Use models to represent aspects of the natural world such as human body systems. **3D** Relate the impact of research on scientific thought and society. **12C** Recognize levels of organization in animals, including cells, tissues, organs, organ systems, and organisms. **13A** Investigate how organisms respond to external stimuli found in the environment such as fight or flight. **13B** Describe and relate responses in organisms that may result from internal stimuli such as fever in animals that allows them to maintain balance.

PEARSON Texas.com

Spotlight On Technology

Interactivity Make images come to life to help illustrate the main functions of human body systems with an interactivity that asks students to engage with the content in an active, visual way.

Flipped Video for Science Show students a Pearson Flipped Video for Science for this chapter so that they can better understand the organization of the human body; homeostasis; and the main functions of the skeletal, muscular, and integumentary systems.

PEARSON Texas.com

Texas Essential Knowledge and Skills

2C Collect and record data using the International System of Units (SI) and qualitative means such as labeled drawings, writing, and graphic organizers.

2D Construct tables and graphs, using repeated trials and means, to organize data and identify patterns.

2E Analyze data to formulate reasonable explanations, communicate valid conclusions supported by the data, and predict trends.

3B Use models to represent aspects of the natural world such as human body systems and plant and animal cells.

3D Relate the impact of research on scientific thought and society, including the history of science and contributions of scientists as related to the content.

7A Contrast situations where work is done with different amounts of force to situations where no work is done.

12B Identify the main functions of the systems of the human organism, including the circulatory, respiratory, skeletal, muscular, digestive, excretory, reproductive, integumentary, nervous, and endocrine systems.

12C Recognize levels of organization in plants and animals, including cells, tissues, organs, organ systems, and organisms.

13A Investigate how organisms respond to external stimuli found in the environment such as phototropism and fight or flight.

13B Describe and relate responses in organisms that may result from internal stimuli such as wilting in plants and fever or vomiting in animals that allow them to maintain balance.

The following **College and Career Readiness Standards** are covered in this chapter: **I.A.4, I.B.1, I.E.1, II.B.2, III.D.2, VIII.D.3**

CHAPTER 6

Getting Started

Check Your Understanding

This activity assesses students' understanding of the relationship between an object's structure and its function. After students have shared their answers, point out that all of the other structures listed have specific functions for the fish. Emphasize the fact that the structures interact to keep the fish alive.

Preteach Vocabulary Skills

Draw students' attention to the table of suffixes. Review with students the definition of a suffix as a word part with a specific meaning that is added to the end of a word to change its meaning. Tell students that many science terms can be difficult to remember. Learning the meaning of suffixes can help in understanding new vocabulary terms. Point out the vocabulary term *skeletal muscle.* Ask students to define the word *skeletal* based on what they learned about the suffix *-al.* You might want to continue the discussion by asking students to name other suffixes with which they are familiar. Make a list on the board.

Getting Started

Check Your Understanding

1. **Background** Read the paragraph below and then answer the question.

> Fara is learning how to build and fix bicycles. First, she learned to change a tire using a tire lever. This tool's **structure**—a lever with a curved end—matches its **function**—to pry a tire off a metal rim. She's also learning about the **interactions** between the different bike parts—how the chain, gears, wheels, and brakes all work together as a **system.**

An object's **structure** is its shape or form.

An object's **function** is the action it performs or the role it plays.

An **interaction** occurs when two or more things work together or affect one another.

A **system** is a group of parts that work together to perform a function or produce a result.

- Circle the structure below that best matches the function of helping a fish to swim.
  scales gills (fins) eyes

Vocabulary Skill

Suffixes A suffix is a word part that is added to the end of a word to change its meaning. For example, the suffix *-tion* means "process of." If you add the suffix *-tion* to the verb *digest*, you get the noun *digestion*. *Digestion* means "the process of digesting." The table below lists some other common suffixes and their meanings.

Suffix	Meaning	Example
-al	of, like, or suitable for	epithelial, *adj.* describes a tissue that covers inner and outer surfaces of the body
-ive	of, relating to, belonging to, having the nature or quality of	connective, *adj.* describes a tissue that provides support for the body and connects all of its parts

2. **Quick Check** Circle the suffix in each of the terms below.
   skelet(al) diges(tive) intern(al)

English Language Proficiency Standards

ELPS Learning Strategies 1.F

Help students assimilate the chapter's vocabulary.

Beginning To preview vocabulary terms, display and read aloud the terms listed at the beginning of each lesson. Together, page through the lesson looking for clues to meaning in images, labels, and captions. Discuss clues, including cognates. Then provide definitions and examples.

Intermediate For each lesson, have small groups complete the three-column chart in Preview Vocabulary Terms on page 213. To complete the second column, have students preview the text, look for clues to meaning, and add drawings, phrases, associations, or other clues to the chart. Have groups share their work.

Advanced Have partners preview the text and discuss terms as they complete of the three-column chart in Preview Vocabulary Terms on page 213.

Advanced High Have students independently complete the first two columns of the chart in Preview Vocabulary Terms on page 213. Then have pairs share answers.

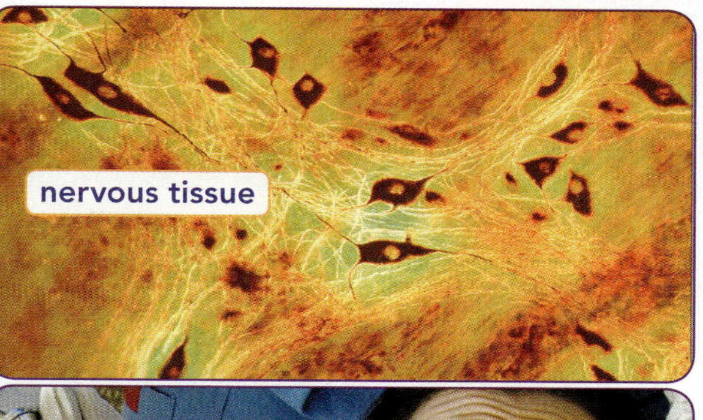

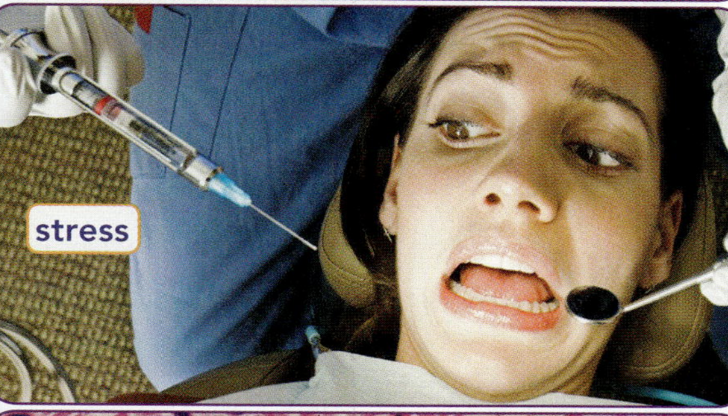

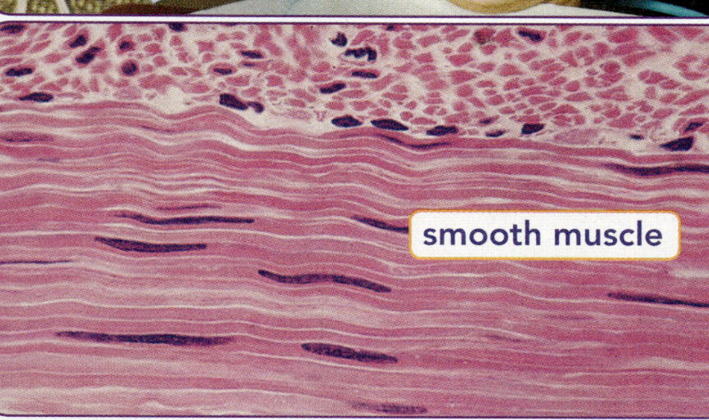

Chapter Preview

LESSON 1
- muscle tissue • nervous tissue
- connective tissue
- epithelial tissue
- organ • organ system

Identify the Main Idea
Make Models

LESSON 2
- skeleton • skeletal muscle • joint
- nutrient • absorption • gland
- stimulus • response • hormone

Summarize
Develop Hypotheses

LESSON 3
- stress

Relate Cause and Effect
Communicate

LESSON 4
- vertebrae • ligament
- compact bone • spongy bone
- marrow • cartilage
- osteoporosis

Summarize
Classify

LESSON 5
- involuntary muscle
- voluntary muscle • tendon
- smooth muscle • cardiac muscle
- striated muscle

Compare and Contrast
Infer

LESSON 6
- epidermis • melanin • dermis
- pore • follicle

Relate Cause and Effect
Observe

CHAPTER 6

Preview Vocabulary Terms

Have students create a three-column chart to rate their knowledge of the vocabulary terms before they read the chapter. In the first column of the chart, students should list the terms for the chapter. In the second column, students should identify whether they can define and use the word, whether they have heard or seen the word before, or whether they do not know the word. As the class progresses through the chapter, have students write definitions for each term in the last column of the chart.

L1 Have students look at the images on this page as you pronounce the vocabulary word. Have students repeat the word after you. Then read the definition below. Use the sample sentence in italics to clarify the meaning of the term.

nervous tissue *(NUR vus TISH oo)* A type of tissue that carries electrical messages to and from the brain and other parts of the body. *While muscle tissue carries out movement, nervous tissue directs and controls the process.*

stress *(STREHS)* A reaction of the body to possible threats, challenges, and uncomfortable events. *Many people feel stress when they go to the dentist for a checkup.*

smooth muscle *(smooth MUS ul)* Tissue found on the inside of many body organs that works to control certain movements inside the body. *This magnified image of smooth muscle shows cells that do not have a striped or banded appearance.*

epidermis *(ep uh DUR mis)* The outer layer of the skin. *It is important to protect the epidermis from too much exposure to the sun's ultraviolet rays.*

Academic Vocabulary

Each lesson includes key Academic Vocabulary. See also the Support All Readers box at the start of the lesson.

Lesson 1: identify, main idea, model
Lesson 2: develop, hypotheses, summarize
Lesson 3: cause, communicate, effect, relate
Lesson 4: classify, summarize
Lesson 5: compare, contrast, infer
Lesson 6: cause, effect, observe, relate

Body Organization

How does your body work?

LESSON PACING:
2–3 periods or 1–1$\frac{1}{2}$ blocks

Lesson Vocabulary

- muscle tissue
- nervous tissue
- connective tissue
- epithelial tissue
- organ
- organ system

Lesson Objectives	TEKS	ELPS
Recognize the levels of organization in the body.	3B, 12B, 12C	3.F.2

Content Refresher

Professional Development Note

A Division of Labor The cell is the basic unit of structure and function of all living things. Although the shape and function of individual cells may vary, cells are remarkably similar in terms of internal makeup and processes.

In multicellular organisms, cells are specialized and organized to perform specific tasks. This allows cells to better meet the needs of the organism. Cells are organized into tissues, tissues into organs, organs into organ systems, and organ systems into organisms. These levels of organization—cells, tissues, organs, organ systems, and organisms—represent a division of labor in a living thing.

The work of keeping a living thing alive and healthy is divided among the different parts of the body. Each part performs its specific function while working together with other parts to ensure the health and survival of the living thing.

Texas Essential Knowledge and Skills

3B Use models to represent aspects of the natural world such as human body systems and plant and animal cells.
12B Identify the main functions of the systems of the human organism, including the circulatory, respiratory, skeletal, muscular, digestive, excretory, reproductive, integumentary, nervous, and endocrine systems.
12C Recognize levels of organization in plants and animals, including cells, tissues, organs, organ systems, and organisms.

English Language Proficiency Standards

ELPS Speaking 3.F.2 Give information ranging from using a very limited bank of high-frequency, high-need concrete vocabulary, including key words and expressions needed for basic communication in academic and social contexts, to using abstract and content-based vocabulary during extended speaking assignments.

DIFFERENTIATED INSTRUCTION KEY
L1 Struggling Students or Special Needs
L2 On-Level Students L3 Advanced Students

LESSON PLANNER 6.1

Investigations and Activities	TEKS Review
My Planet Diary, **Student Edition,** p. 214 Inquiry: Inquiry Warm-Up, How Is Your Body Organized?, **Lab Manual,** p. 68 Introduce Vocabulary, **Teacher's Edition,** p. 215 Teach Key Concepts, **Teacher's Edition,** p. 215 Lead a Discussion, Cells, **Teacher's Edition,** p. 215 Make Analogies, A Well Constructed Machine, **Teacher's Edition,** p. 215 Support the TEKS, Types of Tissue, **Teacher's Edition,** p. 216 21st Century Learning, Critical Thinking, **Teacher's Edition,** p. 216 Address Misconceptions, Connective Tissue, **Teacher's Edition,** p. 216 Differentiated Instruction, **Teacher's Edition,** p. 217 Inquiry: Build Inquiry, More About Muscle Tissue, **Teacher's Edition,** p. 217 Apply It!, **Student Edition,** p. 217 Lead a Discussion, Organs and Organ Systems Work Together, **Teacher's Edition,** p. 218 21st Century Learning, Critical Thinking, **Teacher's Edition,** p. 218 Teach With Visuals, **Teacher's Edition,** 218 Inquiry: Teacher Demo, All Systems Go, **Teacher's Edition,** p. 218 Inquiry: Quick Lab, Observing Cells and Tissues, **Lab Manual,** p. 70	Apply the TEKS, Doing Work, **Student Edition,** p. 254 TEKS Practice, **Student Edition,** p. 256 TEKS Practice: Chapter and Cumulative Review, **Student Edition,** p. 260 Lesson 6.1, **TEKS Preparation and Study Guide Workbook,** p. 54 **SHORT ON TIME?** To do this lesson in approximately half the time, do the Activate Prior Knowledge activity. A discussion of the Key Concepts will familiarize students with the lesson content. Have students do the Quick Lab. The rest of the lesson can be completed by students independently.

These editable worksheets are available on **PearsonTexas.com.**
Print versions can be found in the **TEKS Preparation and Study Guide Workbook.**

Name ______ Date ______ Class ______

6.1 Body Organization

Key Concept Summary

How Is Your Body Organized?

Every minute of the day, no matter what a person is doing, the body is busy at work. Each part of the body has a specific job to do. All the parts work together to keep the body functioning smoothly and effectively. The organization of the body is in part responsible for this smooth functioning.

The levels of organization in the human body consist of cells, tissues, organs, and organ systems. A cell is the basic unit of structure and function in a living thing, or organism. Almost all cells in the human body have the same basic parts. The cell membrane forms the outside border of a cell. The nucleus directs the cell's activities and holds information that controls a cell's function. The cytoplasm, which forms the rest of the cell, is a clear, jellylike substance that contains many cell structures, each of which has a specific job to perform.

A group of similar cells that perform the same function is called a tissue. Muscle tissue is made up of muscle cells. **Muscle tissue** contracts and thus makes body parts move. **Nervous tissue,** made up of nerve cells, carries electrical messages to and from the brain and spinal cord, thereby directing and controlling body processes. **Connective tissue** provides support for the body and connects all its parts. Connective tissue includes bone tissue and fat tissue. **Epithelial tissue** covers both the internal (inside) and external (outside) surfaces of the body.

A group of different types of tissue performing a specific function is called an **organ.** Each type of tissue in an organ does its specific job and in that way contributes to the organ's function. Each organ is part of an **organ system,** or a group of organs that work together to perform a major function. Organ systems also work together, forming the next level of organization, the organism.

54

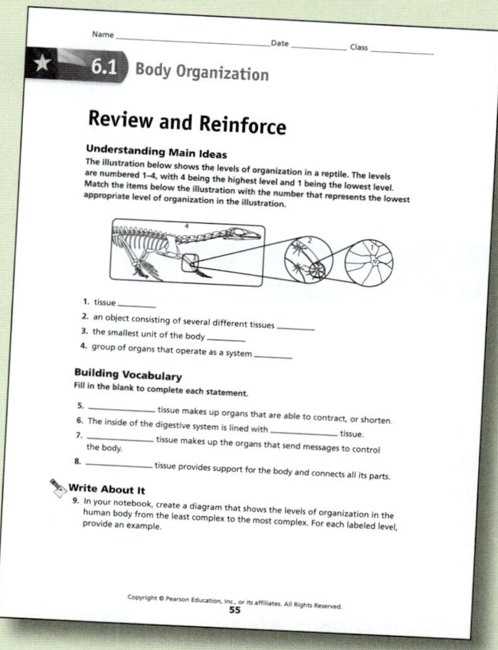
Name ______ Date ______ Class ______

6.1 Body Organization

Review and Reinforce

Understanding Main Ideas

The illustration below shows the levels of organization in a reptile. The levels are numbered 1–4, with 4 being the highest level and 1 being the lowest level. Match the items below the illustration with the number that represents the lowest appropriate level of organization in the illustration.

1. tissue ______
2. an object consisting of several different tissues ______
3. the smallest unit of the body ______
4. group of organs that operate as a system ______

Building Vocabulary

Fill in the blank to complete each statement.

5. ______ tissue makes up organs that are able to contract, or shorten.
6. The inside of the digestive system is lined with ______ tissue.
7. ______ tissue makes up the organs that send messages to control the body.
8. ______ tissue provides support for the body and connects all its parts.

Write About It

9. In your notebook, create a diagram that shows the levels of organization in the human body from the least complex to the most complex. For each labeled level, provide an example.

55

Lexile Measure = 900L

LESSON 6.1

Body Organization

Establish Learning Objective

After this lesson, students will be able to:

Recognize the levels of organization in the body.

Engage

Activate Prior Knowledge

MY PLANET DIARY Read *Medical Illustrator* with the class. Tell students that some medical illustrators specialize in certain areas of work, such as rendering hands, the heart, eyes, and so on. Explain that most medical illustrators have college degrees. Ask: **Have you heard the saying "A picture is worth a thousand words"? What do you think it means?** *(Sample: A picture can provide information just as well as lots of descriptive text.)*

Explore

Lab Resource: Inquiry Warm-Up

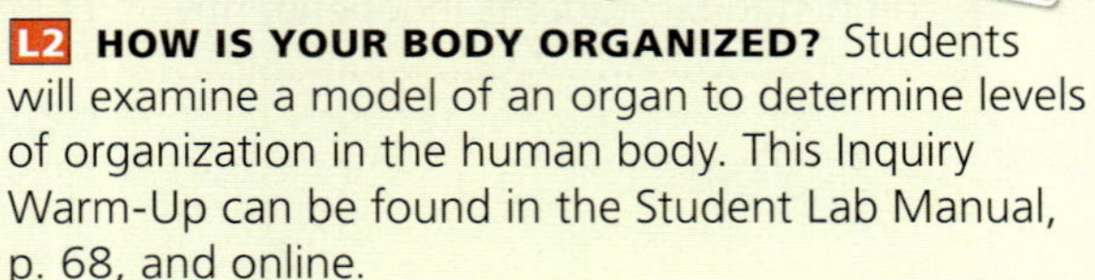

L2 **HOW IS YOUR BODY ORGANIZED?** Students will examine a model of an organ to determine levels of organization in the human body. This Inquiry Warm-Up can be found in the Student Lab Manual, p. 68, and online.

Texas Essential Knowledge and Skills

3B Use models to represent aspects of the natural world such as human body systems and plant and animal cells.

12B Identify the main functions of the systems of the human organism, including the circulatory, respiratory, skeletal, muscular, digestive, excretory, reproductive, integumentary, nervous, and endocrine systems.

12C Recognize levels of organization in animals, including cells, tissues, organs, organ systems, and organisms.

Texas LESSON 1

Body Organization

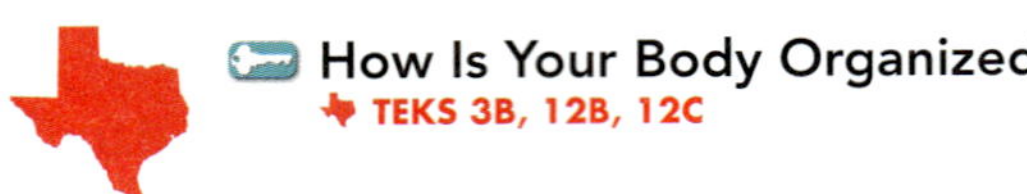

How Is Your Body Organized?

TEKS 3B, 12B, 12C

MY PLANET DIARY

CAREER

Medical Illustrator

Who made the colorful drawings of human body structures in this book? The drawings are the work of specialized artists called medical illustrators. These artists use their drawing skills and knowledge of human biology to make detailed images of body structures. Many artists draw images, such as the one on this page, using 3-D computer graphics. The work of medical illustrators appears in textbooks, journals, magazines, videos, computer learning programs, and many other places.

Communicate Answer the question below. Then discuss your answer with a partner.

Why do you think medical illustrations are important to the study of human biology?

Sample: A picture helps explain things that are hard to understand.

Lab zone® Do the Inquiry Warm-Up *How Is Your Body Organized?* Student Lab Manual, p. 68

TEKS 3B, 12B, 12C In this section, you'll use models to learn to recognize the levels of organization in the human body and identify the main functions of body systems.

How Is Your Body Organized?

The bell rings—lunchtime! You hurry to the cafeteria, fill your tray, and pay the cashier. You look around the cafeteria for your friends. Then you walk to the table, sit down, and begin to eat.

Think about how many parts of your body were involved in the simple act of getting and eating your lunch. Every minute of the day, whether you are eating, studying, walking, or even sleeping, your body is busily working. Each part of the body has a specific job to do. And all these different parts usually work together so smoothly that you don't even notice them.

SUPPORT ALL READERS

Lexile Measure = 900L **Lexile Word Count = 1068**

Prior Exposure to Content: Most students have encountered this topic in earlier grades

Academic Vocabulary: *identify, main idea, model*

Science Vocabulary: *organ, organ system*

Concept Level: Generally appropriate for most students in this grade

Preteach With: My Planet Diary "Medical Illustrator" and Figure 1 activity

Vocabulary
- muscle tissue
- nervous tissue
- connective tissue
- epithelial tissue
- organ
- organ system

Skills
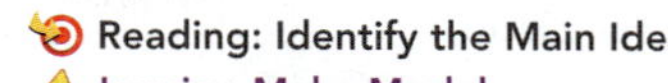

Reading: Identify the Main Idea

Inquiry: Make Models

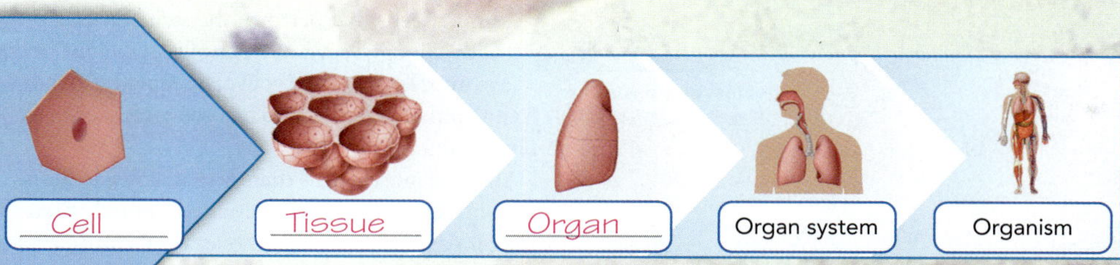

FIGURE 1

Body Organization

You will see this diagram three more times in this lesson. It will help you track the levels of organization in the body.

Name **Fill in the missing terms in the diagram above.**

The smooth functioning of your body is due partly to how the body is organized. **The levels of organization in the human body consist of cells, tissues, organs, and organ systems.** The smallest unit of organization is a cell. The next largest unit is a tissue, then an organ. Finally, an organ system is the largest unit of organization in an organism. **Figure 1** shows body organization.

Cells

A cell is the basic unit of structure and function in a living thing. Complex organisms are made up of many cells in the same way that your school is made up of many rooms. The human body contains about 100 trillion tiny cells. Most cells cannot be seen without a microscope.

Structures of Cells Almost all cells in the human body have the same basic parts, as shown in **Figure 2.** The cell membrane forms the outside border of a cell. The nucleus holds information that directs the cell's activities and controls a cell's function. The rest of the cell, called the cytoplasm (SYT oh plaz um), is made of a clear, jellylike substance that contains many cell structures. Each of these structures has a specific job, or function.

Functions of Cells Cells carry on the processes that keep organisms alive. Inside cells, for example, molecules from digested food undergo changes that release energy that the cells can use. Cells also grow, reproduce, and get rid of the waste products that result from these activities.

Cell membrane

Nucleus

Cytoplasm

FIGURE 2

Cell Structure

A microscope reveals some of the parts of a human cheek cell.

ELPS 3.F.2

Work with a small group, to preview the body systems shown on pages 218–219. Choose one system to explain to your group. Tell what the system does. Then use your book to give some examples of organs, tissues, or cells in the system.

English Language Proficiency Standards

ELPS Speaking 3.F.2

Help students use vocabulary as they preview the body systems on pages 218–219.

Beginning Read aloud the body systems on pages 218–219. Name activities such as running, eating, or breathing. Have students name or point to the system(s) used for each activity.

Intermediate Have groups read and discuss the body systems on pages 218–219. Have students take turns describing one of the systems to the group.

Advanced Have group members choose one body system and preview the corresponding lesson. Have students take turns giving examples of organs, cells, and tissues in the systems they chose.

Advanced High Have groups play a guessing game. Each group member chooses one body system, previews the corresponding lesson, and describes the system's functions. Others name the system based on the clues.

Explain

Introduce Vocabulary

Point out the four types of tissue on page 216. Tell students that the word *tissue* comes from a Latin word meaning "to weave." Tell students to think of the cells that make up tissues as "woven together." Then explain that the difference in the four types of tissue is related to the structure and function of the cells that are woven together.

Teach Key Concepts

Tell students that the human body works as well as it does because its organization allows different parts to perform their specific functions while working together to meet the needs of the organism. On the board, write the terms *cell, tissue, organ, organ system,* and *organism,* and tell students that these are the five levels of organization. Help students understand the levels by telling them to think of their town. Ask: **What are some of the jobs different people in the town perform?** *(Sample: Delivering mail, collecting garbage, maintaining law and order)* **How do they keep the town running?** *(They each perform their specific job but also work together for the benefit of the town.)* Tell students that the human body has a similar organization. Individual workers are like cells; the town is like an organism.

Lead a Discussion

CELLS Tell students that the smallest unit of organization is the cell. Cells are the basic unit of structure and function in all living things. Discuss the fact that most cells in the human body have the same basic parts. Ask: **What is the function of the cell membrane?** *(It is a thin layer that holds in the contents of the cell.)* **Why is the nucleus the control center?** *(It directs the cell's activities.)* **Where do you think most processes occur?** *(In the cytoplasm)*

Make Analogies

L1 **A WELL CONSTRUCTED MACHINE** Explain to students that the human body is like a machine. In a machine, each part has its specific function, and all parts work together to achieve a goal. Have students think about a car. Ask: **What are some of a car's parts and their functions?** *(Sample: engine provides power; tires roll smoothly over road surface; frame provides structure)* **What happens if one of the parts does not work?** *(The car will function poorly or not at all.)*

PEARSON Texas.com

LESSON 6.1

Explain

Support the TEKS

TYPES OF TISSUE Tell students that cells are specialized to perform different functions and that groups of cells that are similar in structure and function are joined to form tissues. For example, bone cells form bone tissue, which gives strength and support to the body. Blood cells form blood tissue, which carries vital substances throughout the body. Tissues are the second level of organization. Like cells, tissues work for themselves as well as for the benefit of the entire organism. Refer students to the photographs of the four types of tissue. As you discuss them, make sure students observe the appearance of each. Ask: **What is the main function of muscle tissue?** *(Allow movement)* **Nervous tissue?** *(Carry messages throughout the body)* **Connective tissue** *(Support and connect parts)* **Epithelial tissue?** *(Protect delicate structures)* Tell students that the micrographs of tissues they are looking at are striated skeletal muscle, a group of neurons that are part of the network that controls peristalsis in the small intestine (The dark, red-brown structures are the cell bodies), bone tissue (connective tissue), and epithelial tissue from the skin.

Identify the Main Idea Tell students that identifying the main idea means determining the most important, or biggest, idea in a paragraph or section. The other information in the paragraph or section supports or further explains the main idea.

21st Century Learning

CRITICAL THINKING Remind students that a tissue is a group of cells that work together to perform a specific function. Ask: **Which type of tissue are tendons that attach muscle to bone?** *(Connective)* **Which type of tissue is the spinal cord made of?** *(Nervous)* **Which type of tissue makes up the lining of the mouth?** *(Epithelial)* **Which type of tissue makes up the stomach wall that moves to mix food?** *(Muscle)*

Address Misconceptions

L1 CONNECTIVE TISSUE Because students have learned that connective tissue provides support for the body and connects all its parts, they may think that all connective tissue is solid. Tell them that blood and lymph are liquid connective tissue. Ask: **What is the role of blood in the body?** *(Blood carries food, oxygen, and wastes throughout the body.)* **How does it act as connective tissue?** *(Sample: Blood joins and nourishes all the organs of the body.)*

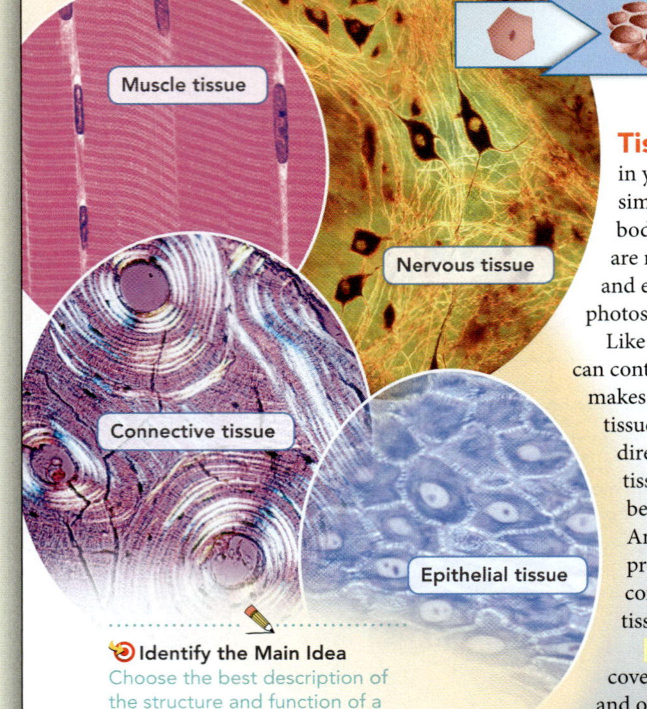

Tissues The next largest unit of organization in your body is a tissue. A tissue is a group of similar cells that perform the same function. Your body contains several types of tissue. Four of these are muscle tissue, nervous tissue, connective tissue, and epithelial tissue. You can see examples in the photos at left.

Like the muscle cells that form it, **muscle tissue** can contract, or shorten. By doing so, muscle tissue makes parts of your body move. While muscle tissue carries out movement, **nervous tissue** directs and controls the process. Nervous tissue carries electrical messages back and forth between the brain and other parts of the body. Another type of tissue, **connective tissue,** provides support for your body and connects all its parts. Bone tissue and fat tissue are examples of connective tissue. **Epithelial tissue** (ep uh THEE lee ul) covers the surfaces of your body, inside and out. Some epithelial tissue, such as your skin, protects the delicate structures that lie beneath it. The lining of your digestive system consists of epithelial tissue that allows you to digest and absorb the nutrients in your food.

Identify the Main Idea
Choose the best description of the structure and function of a tissue.

- ○ A group of different cells that have the same function
- ○ A group of similar cells that have different functions
- ● A group of similar cells that have the same function

FIGURE 3

The Heart
The heart, like your other organs, is made of different kinds of tissues that have different functions.

Answer the following questions.

1. **Relate Text and Visuals** In each box, fill in the kind of tissue that matches the function described.
2. CHALLENGE Pick one type of tissue shown and describe how the heart would be affected if the tissue did not function properly.

Sample: Muscle tissue; the heart would not be able to pump blood effectively.

Connective tissue provides strength and flexible support for muscle tissue and other structures inside and outside the heart.

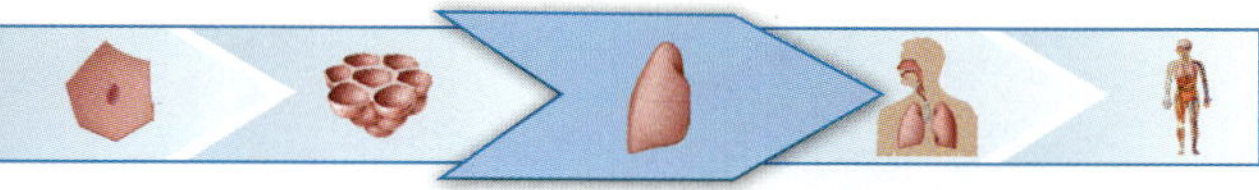

Organs Your stomach, heart, brain, and lungs are all organs. An **organ** is a structure that is made up of different kinds of tissue. Like a tissue, an organ performs a specific job. The job of an organ, however, is usually more complex than that of a tissue. For example, the heart pumps blood through your body over and over again. The heart contains muscle, connective, and epithelial tissues. In addition, nervous tissue connects to the heart and helps control heart function. **Figure 3** shows a diagram of a human heart and describes how some of the heart's tissues work. Each type of tissue contributes in a different way to the organ's job of pumping blood.

Epithelial tissue covers the inside surfaces of the heart and of the blood vessels that lead into and out of the heart.

Nervous tissue carries electrical messages from the brain to the heart but is not shown in this diagram.

Muscle tissue contracts, squeezing the heart so blood moves through the heart's chambers and then into blood vessels that lead to the body.

apply it!

Books are a nonliving model of levels of organization. Find out how a book is organized.

STEP 1 **Observe** Examine this book to see how its chapters, lessons, and other parts are related.

STEP 2 **Make Models** Next, compare levels of organization in this book to those in the human body. Draw lines to show which part of this book best models a level in the body.

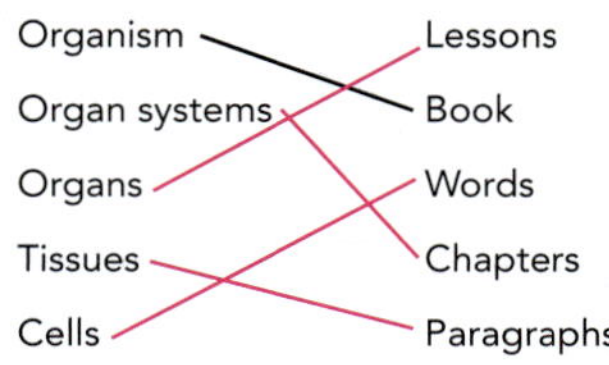

STEP 3 **Make Models** Where in the book model do you think this Apply It fits? What level of organization in the body does the Apply It represent?

Sample: The Apply It is one part of the lesson, so the Apply It is like a tissue.

Lead a Discussion

ORGANS Tell students that an organ performs a specific function. An organ is composed of two or more types of tissue working together. Emphasize that one type of tissue alone would not be able to do all of the things that several tissue types working together can do. Ask: **What are some examples of organs?** *(Samples: eyes, ears, liver, pancreas, kidneys, small intestine, skin)* **What level of organization of the human body do organs represent?** *(Third level)* **Do you know what the largest organ in the human body is?** *(The skin)* Tell students that the epithelial tissue that lines the inside surface of blood vessels and the heart is commonly referred to as *endothelium.* It is made of epithelial tissue that is thin and slick and reduces friction. The term *endothelium* reflects its specific location and function as an inner coating.

Elaborate

Build Inquiry 

L1 MORE ABOUT MUSCLE TISSUE

Materials rubber balls, bowls of ice water (Have students take turns doing the activity so that you need to supply only a minimum of materials.)

Time 15 minutes

Remind students that muscle tissue contracts, and by so doing enables parts of the body to move. Have each student count the number of times he/she can squeeze the ball in 30 seconds for three trials in a row and record the results. Have students rest for one minute and then immerse the squeezing hand in the bowl of ice water for 10 to 15 seconds. Have students perform one trial of squeezing the ball for 30 seconds and record that result.

Ask: **How did the number of squeezes change from your first trial to your third trial?** *(The number decreased.)* **What was the effect of immersing your hand in cold water?** *(It decreased the number of squeezes.)* **What does this activity tell you about muscle tissue?** *(Muscle tissue gets tired and also reacts to temperature.)*

Apply It!

L1 Before beginning the activity, review with students the concept of levels of organization in the human body. Emphasize how the smallest unit builds up to the largest and that the largest is the sum of its parts.

Make Models Remind students that some models are mental models that describe how things work through analogies. For example, the organization of a book from words to chapters to an entire book is similar to the organization of the human body from cells to tissues to organs to body system.

Differentiated Instruction

L1 Levels of Organization To ensure that students understand the levels of organization of the human body, have them create a diagram that correctly identifies the levels and rank from least complex to most complex. Encourage students to be creative.

L1 Writing to Learn One of the most effective ways to help students understand new information is to have them teach it to someone else. Have students write a brief lesson for a third or fourth grade textbook on the organization of the human body. Tell them to identify each level, provide one or two examples, and explain the relationships.

L3 Word Meanings The suffix *-ology* means "study of." Have students find out the tissues, organs, or organ systems studied in cardiology, dermatology, and endocrinology.

LESSON 6.1

Explain

Lead a Discussion

ORGANS AND ORGAN SYSTEMS WORK TOGETHER Tell students that organs seldom work alone. They work with other organs to perform a specific function. A group of organs working together to perform a major function is an organ system. Emphasize that the failure of any part of an organ system can affect the entire system. Ask: **Which level of organization do organ systems represent?** *(Fourth)* **Can an organ be part of more than one system? Explain and give an example.** *(Yes; if the organ's functions are vital to more than one system, it will belong to both systems. The skin is an example because it belongs to the integumentary system and the immune system.)* Tell students that although an organ system has a specific function to perform, no system acts alone. Each system contributes to the work of other systems in keeping the organism alive and well. Ask: **What is the fifth level of organization?** *(Organism)*

21st Century Learning

CRITICAL THINKING Have students develop their own metaphors to describe the human body and the roles played by the body systems. Encourage students to create visual displays based on their metaphors.

Teach With Visuals

Tell students to look at **Figure 4.** Ask: **What is the function of the skeletal system?** *(Supports body, protects organs, allows movement, stores minerals, produces blood cells)* **Which system chemically controls many body processes?** *(Endocrine system)* **What are the organs of the nervous system?** *(Brain, spinal cord, nerves)* **What is the relationship between organs and organ systems?** *(Each organ performs one function that is related to the function of the other organs in the system. The organs work together to carry out one major function in the organ system.)*

21st Century Learning

CRITICAL THINKING If students have previously studied ecology, point out that just as there are levels of organization in the human body, there are also levels of organization in the environment. Ask: **Starting with organisms, what levels of organization exist in the environment?** *(Organisms, populations, communities, ecosystems, and biomes)*

BODY SYSTEM	Skeletal System	Integumentary System	Muscular System	Circulatory System	Respiratory System
STRUCTURES	Bones, cartilage, ligaments, tendons	Skin, hair, nails, sweat glands, oil glands	Skeletal muscle, smooth muscle, cardiac muscle	Heart, blood vessels	Nose, pharynx, larynx, trachea, bronchi, lungs
FUNCTIONS	Supports body; protects internal organs; allows movement; stores minerals; produces blood cells	Guards against infection and injury; helps regulate body temperature	With skeletal system, produces movement; helps circulate blood and move food through the digestive system	Transports oxygen, nutrients, and wastes; fights infection; helps regulate body temperature	Brings in oxygen needed by cells; removes carbon dioxide from body

FIGURE 4

Body Systems

Apply Concepts Describe the levels of organization in a complex system. Write about a sports team, a supermarket, a digital audio player, or an orchestra. Or choose your own example.

Sample: An orchestra is made up of instruments that are grouped with others of the same kind, like the violins, and then grouped into sections of instruments that are similar, like the strings, brass, and percussion. Together with the conductor, they make music.

Systems Each organ in your body is part of an **organ system,** which is a group of organs that work together, carrying out major functions. For example, your heart is part of your circulatory system, which carries oxygen and other materials throughout your body. The circulatory system also includes blood vessels and blood. **Figure 4** shows most of the organ systems in the human body.

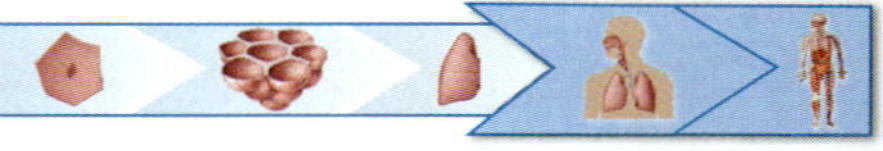

Organisms After an organ system, the next and most complex level of organization is an organism. You, as an organism, consist of many organ systems working together. And all organisms are part of levels of organization within their environment.

Professional Development Note

Teacher to Teacher

Human Body Systems Creating analogies between various human body systems and familiar systems can be an effective tool to facilitate learning. The circulatory system can be compared to the road system in our country. The skeletal system can be compared to the frame of a house. This technique can be enhanced if students can be trained to develop these analogies themselves. An analogy is not just a comparison; it is a comparison with reasons. The circulatory system is like the road system in the US because it carries materials from one place to another and has large paths (interstate highways), smaller paths (state highways), and small paths (city streets) like the circulatory systems have various sized veins and arteries.

Joel Palmer, Ed.D.
Mesquite ISD
Mesquite, Texas

BODY SYSTEM	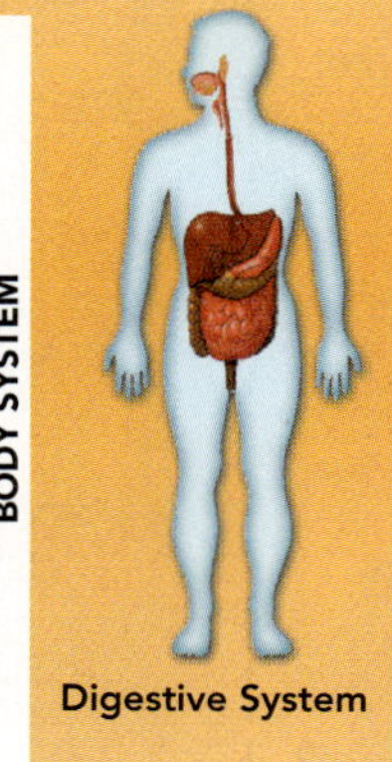 Digestive System	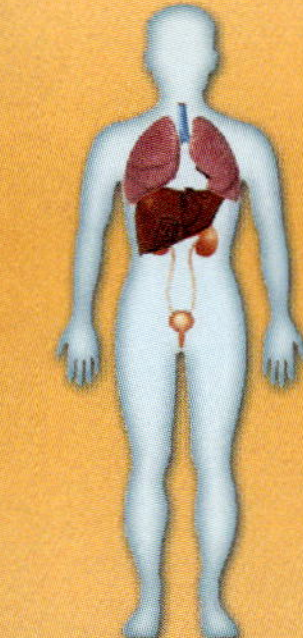 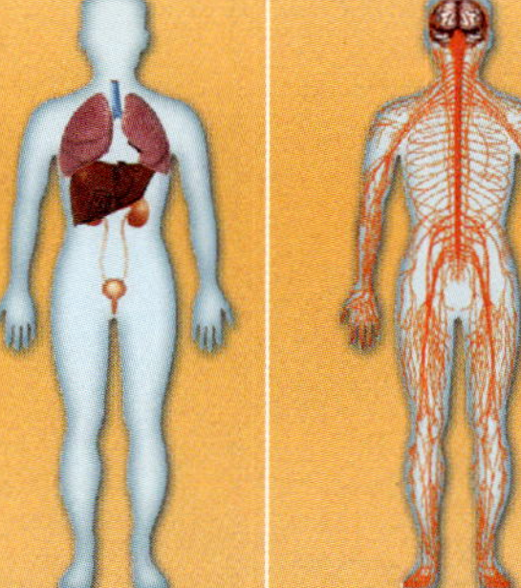 Excretory System	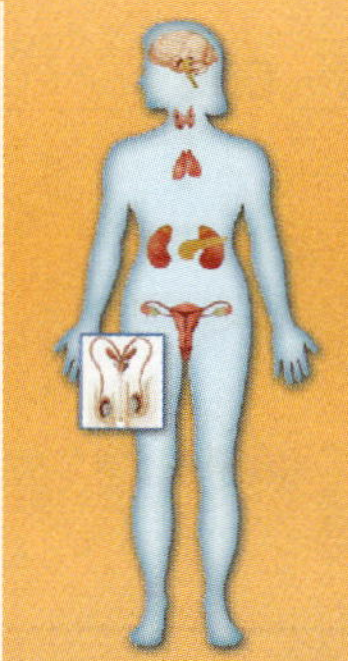 Nervous System	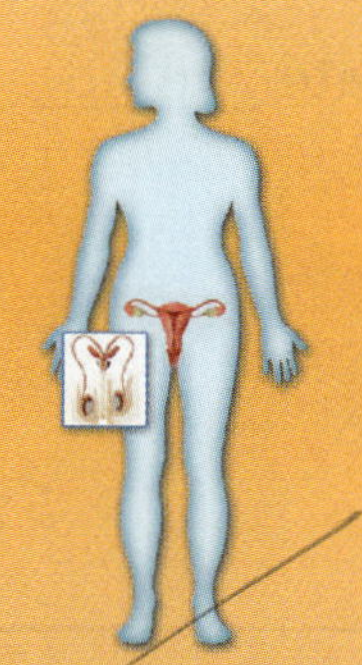 Endocrine System	Reproductive System
STRUCTURES	Mouth, esophagus, stomach, small intestine, liver, pancreas, large intestine, rectum	Skin, lungs, liver, kidneys, urinary bladder, urethra	Brain, spinal cord, nerves	Glands, such as the thyroid, pancreas, adrenals, ovaries, testes, and others	In males: testes, ducts, urethra, penis; in females: ovaries, ducts, uterus, vagina
FUNCTIONS	Breaks down food; absorbs nutrients; removes food wastes	Removes waste products from the body	Controls the body's responses to changes within the body and outside it	Controls growth, development, and energy processes; helps maintain homeostasis	Produces and delivers sex cells; in females, nurtures and protects developing embryo

Do the Quick Lab *Observing Cells and Tissues.* Student Lab Manual, p. 70

Assess Your Understanding

TEKS 12B, 12C

1a. Recognize What are the five levels of organization in the human body from the most basic to the most complex?

Cells, tissues, organs, organ systems, and organism

b. Identify What are the main functions of the muscular system?

It helps produce movement and helps with blood circulation and digestion

c. Apply Concepts What level of organization is the human eye? To what system does it belong? Explain.

Sample: The eye is an organ. It is part of the nervous system because it is involved in perceiving the outside world.

got it?

O **I get it!** Now I know that the body's levels of organization, from least complex to most complex, are cells, tissues, organs, and organ systems.

O I need extra help with See TE note.

219

Differentiated Instruction

L1 Concept Map Review the terms *cell, tissue, organ,* and *organ system* with students. Using **Figure 4** as a guide, construct a concept map. Write the name of one body system in the center and connect the structures and functions of that body system to the center by lines. Then choose another body system, and have students repeat the activity.

L3 Body Systems Have student choose an activity they perform regularly, such as eating a meal, riding a bike, playing with a pet, and so on. Students should identify the systems of the body that are involved in performing the activity, explaining how the structures of these body systems function to allow them to do the activity.

Elaborate

Teacher Demo 

L1 ALL SYSTEMS GO

Materials Large, stop-action photo of diver or trapeze artist doing a triple or quadruple somersault

Time 10 minutes

Show the photo to the class and tell them to think about the action the diver/trapeze artist is performing.

Ask: **Which body systems do you think are working together to enable the performer to do this feat?** *(Samples: skeletal, muscular, nervous, endocrine, circulatory, respiratory)* **Which system is most responsible for the strength it takes to do the somersaults?** *(Muscular)* **Which system coordinates all of the systems involved in the action?** *(Nervous)* **Although you may not perform a feat such as this, what action do you do that involves the same systems?** *(Samples: running, riding a bike, playing a sport, reading a scary book)*

Lab Resource: Quick Lab 

L3 OBSERVING CELLS AND TISSUES Students will observe prepared slides of different types of tissues and explain how the structure of each type is related to its function. This Quick Lab can be found in the Student Lab Manual, p. 70, and online.

Evaluate

Assess Your Understanding

After students answer the questions, have them evaluate their understanding by completing the appropriate sentence.

RTI Response to Intervention

1a. If students cannot recognize the levels of organization in the human body, **then** have them explain what cells, tissues, organs, and organ systems are.

b. If students have trouble identifying the main functions of the muscular system, **then** help them review **Figure 4.**

c. If students need help identifying the level of organization of the eye and explaining what system it belongs to, then have them review **Figure 1** and **Figure 4.**

Name ______________________ Date __________ Class __________

Assess Your Understanding

Body Organization

How Is Your Body Organized?

1a. RECOGNIZE What are the five levels of organization in the human body from the most basic to the most complex?

____

____

b. IDENTIFY What are the main functions of the muscular system?

____

____

c. APPLY CONCEPTS What level of organization is the human eye? To what system does it belong? Explain.

____

____

____

____

got it?

○ **I get it!** Now I know that the body's levels of organization, from least complex to most complex, are ______________________________

____

○ **I need extra help with** ______________________________

____

Name ______________________ Date ____________ Class ____________

Enrich

Body Organization

Read the passage and study the illustration. Then use a separate sheet of paper to answer the questions that follow.

Organ and Tissue Transplants

When a doctor performs a transplant operation, he or she replaces a diseased or damaged organ or tissue. Sometimes a tissue is moved from one place to another on the same person. This procedure is called an *autograft*. (*Auto*- means "self," and *-graft* means "transplant.") A burn victim may have an autograft in which a section of his or her healthy skin is transplanted to cover the burn.

Sometimes a person receives an organ or tissue from another person. This is called an *allograft*. (*Allo*- means "different.") An example of an allograft is the transplantation of a kidney from the body of one person into that of another person. One problem with allografts is *rejection*. Rejection occurs when the patient's body recognizes the transplanted organ or tissues as foreign, similar to the way in which a mother cat recognizes a kitten from another litter as not belonging to her. Rejection is a serious problem because the body begins to attack the transplanted organ or tissue. One way of preventing rejection is by giving the patient certain drugs.

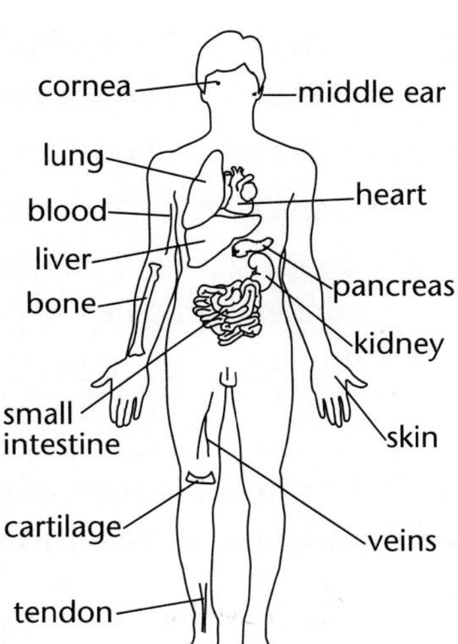

Transplants are performed to save a patient's life or to correct a serious medical condition. For example, a person with severe liver disease might need a new liver in order to survive. Transplanting a part of the eye called the cornea can help some blind people to see. The illustration shows some of the many organs and tissues that doctors can transplant.

1. **Autografts are never rejected. Why do you think this is true?**
2. **Why do you think doctors try to use autografts rather than allografts on burn patients?**
3. **A patient's body is less likely to reject an allograft if it comes from a close relative. Why do you think this is true?**
4. **Do you think doctors would have more difficulty transplanting an organ system than transplanting an organ? Think of an example to explain your answer.**

Name ______________________ Date __________ Class __________

Lesson Quiz

Body Organization

Write the letter of the correct answer on the line at the left.

1. ___ The levels of organization in the human body are

A organ, organ system, tissue, cell, organism
B cell, tissue, organ, organ system, organism
C cell, organ system, organ, tissue, organism
D organism, tissue, organ, cell, organ system

2. ___ Skin, ears, and kidneys are examples of

A organs
B tissues
C organ systems
D cells

3. ___ Which of the following is NOT true about connective tissue?

A It provides support for the body.
B It connects all of the body's parts.
C Bone tissue and fat tissue are examples of connective tissue.
D It makes parts of the body move.

4. ___ A tissue that has the ability to contract is

A nerve tissue
B epithelial tissue
C muscle tissue
D connective tissue

If the statement is true, write *true*. If the statement is false, change the underlined word or words to make the statement true.

5. ______________ The skin is made up of nervous tissue. epithelial tissue

6. Excretory ______________ The endocrine system removes waste products from the body.

7. True ______________ The least complex level of organization of the human body is a(n) cell.

8. tissue ______________ A group of similar cells performing the same function is a(n) organ.

9. True ______________ Each organ in the body is part of a(n) organ system performing a major function.

10. More ______________ As one moves from tissues to organs, the levels become less complex.

Place the outside corner, the corner away from the dotted line, in the corner of your copy machine to copy onto letter-size paper.

Place the outside corner, the corner away from the dotted line, in the corner of your copy machine to copy onto letter-size paper.

Body Organization

Answer Key

Review and Reinforce

Find the worksheet in the Student Workbook.

1. 2

2. 3

3. 1

4. 4

5. Muscle

6. epithelial

7. Nervous

8. Connective

9. The diagram should show, from least complex to most complex: cell, tissue, organ, organ system, and organism. Examples of each level include: muscle cell; nervous/epithelial/muscle/connective tissue; heart/lungs/eyes/liver/stomach/skin; digestive/circulatory/skeletal/nervous system; human being.

Enrich

1. Because autografts come from the patient's own body, the patient's immune system does not recognize them as foreign.

2. If doctors use an autograft, tissue probably won't be rejected.

3. Organs or tissues from a close relative are more likely to be similar to the patient's own organs and tissues and so they are less likely to be recognized as foreign by the patient's immune system.

4. Answers will vary. Sample: It would be more difficult to transplant an organ system than an organ because an organ system is made of more than one organ. Transplanting a skeleton, for example, might be impossible, because there are more than 200 bones in the human body.

Lesson Quiz

1. B

2. A

3. D

4. C

5. epithelial

6. excretory

7. true

8. tissue

9. true

10. more

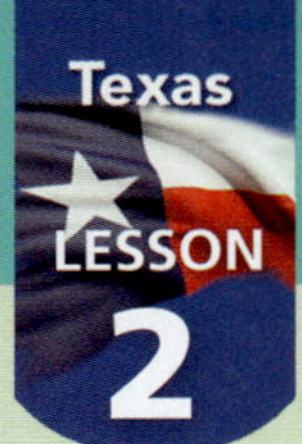

System Interactions

How does your body work?

Lesson Objectives	TEKS	ELPS
Explain how the skeletal and muscular systems work together.	12B	2.I.4
Identify which body systems work together to obtain and transport materials.	12B, 12C	
Identify which body systems control communication and regulation.	12B, 12C	

LESSON PACING:
2–3 periods or 1–1$\frac{1}{2}$ blocks

Lesson Vocabulary

- skeleton
- skeletal muscle
- joint
- nutrient
- absorption
- gland
- stimulus
- response
- hormone

Content Refresher

Bones and Muscles Bone is living, growing tissue. The outermost layer of bone is called *periosteum*. This fibrous membrane contains blood vessels that nourish the bone. Beneath the periosteum is a dense layer of compact bone, which consists of blood vessels, nerve cells, and osteocytes held by a framework of minerals such as calcium and phosphorus. Beneath the compact bone is a layer of cancellous, or spongy, bone—a lightweight, honeycomb-like bone that provides strength and support. Both compact and cancellous bone contain osteoblasts and osteoclasts, which are responsible for bone growth and repair.

Three types of muscle tissue make up the muscles of the body. Skeletal muscles attach to bone by connective tissue called *tendons*. Smooth muscles are found in internal organs. Cardiac muscles are found only in the heart. The actions of smooth and cardiac muscles are involuntary; skeletal muscle can be voluntary or involuntary.

Texas Essential Knowledge and Skills

12B Identify the main functions of the systems of the human organism, including the circulatory, respiratory, skeletal, muscular, digestive, excretory, reproductive, integumentary, nervous, and endocrine systems.
12C Recognize levels of organization in plants and animals, including cells, tissues, organs, organ systems, and organisms.

English Language Proficiency Standards

ELPS Listening 2.I.4 Demonstrate listening comprehension of increasingly complex spoken English by collaborating with peers commensurate with content and grade-level needs.

DIFFERENTIATED INSTRUCTION KEY
L1 Struggling Students or Special Needs
L2 On-Level Students L3 Advanced Students

Investigations and Activities	TEKS Review
My Planet Diary, **Student Edition,** p. 220 Introduce Vocabulary, **Teacher's Edition,** p. 221 Inquiry: Inquiry Warm-Up, How Does Your Body Respond?, **PearsonTexas.com** Teach Key Concepts, **Teacher's Edition,** p. 221 Teach With Visuals, **Teacher's Edition,** p. 221 Lead a Discussion, Move the Body, **Teacher's Edition,** p. 222 Apply It!, **Student Edition,** p. 222 Differentiated Instruction, **Teacher's Edition,** p. 222 Inquiry: Lab Investigation, A Look Beneath the Skin, **Lab Manual,** p. 71	Apply the TEKS, Doing Work, **Student Edition,** p. 254 TEKS Practice, **Student Edition,** p. 256 TEKS Practice: Chapter and Cumulative Review, **Student Edition,** p. 260 Lesson 6.2, **TEKS Preparation and Study Guide Workbook,** p. 56
Teach Key Concepts, **Teacher's Edition,** p. 223 Address Misconceptions, The Color of Blood, **Teacher's Edition,** p. 223 21st Century Learning, Creativity, **Teacher's Edition,** p. 223 Support the TEKS, Food In and Wastes Out, **Teacher's Edition,** p. 224 Teach With Visuals, **Teacher's Edition,** p. 224 Inquiry: Quick Lab, Working Together, Act I, **Lab Manual,** p. 76	**SHORT ON TIME?** To do this lesson in approximately half the time, do the Activate Prior Knowledge activity. A discussion of the Key Concepts will familiarize students with the lesson content. Have students do the Quick Labs. The rest of the lesson can be completed by students independently.
Teach Key Concepts, **Teacher's Edition,** p. 226 Differentiated Instruction, **Teacher's Edition,** p. 227 Apply It!, **Student Edition,** p. 227 Inquiry: Quick Lab, Working Together, Act II, **Lab Manual,** p. 77	

These editable worksheets are available on **PearsonTexas.com.**
Print versions can be found in the **TEKS Preparation and Study Guide Workbook.**

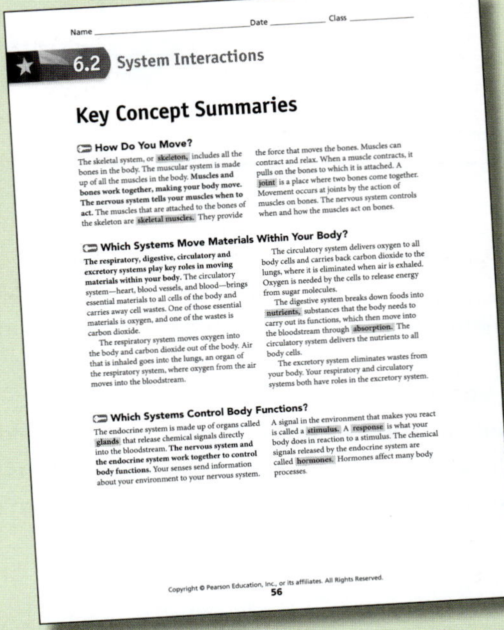
Name ______________ Date ________ Class ________

6.2 System Interactions

Key Concept Summaries

How Do You Move?

The skeletal system, or **skeleton,** includes all the bones in the body. The muscular system is made up of all the muscles in the body. **Muscles and bones work together, making your body move. The nervous system tells your muscles when to act.** The muscles that are attached to the bones of the skeleton are **skeletal muscles.** They provide the force that moves the bones. Muscles can contract and relax. When a muscle contracts, it pulls on the bones to which it is attached. A **joint** is a place where two bones come together. Movement occurs at joints by the action of muscles on bones. The nervous system controls when and how the muscles act on bones.

Which Systems Move Materials Within Your Body?

The respiratory, digestive, circulatory and excretory systems play key roles in moving materials within your body. The circulatory system—heart, blood vessels, and blood—brings essential materials to all cells of the body and carries away cell wastes. One of those essential materials is oxygen, and one of the wastes is carbon dioxide.

The respiratory system moves oxygen into the body and carbon dioxide out of the body. Air that is inhaled goes into the lungs, an organ of the respiratory system, where oxygen from the air moves into the bloodstream.

The circulatory system delivers oxygen to all body cells and carries back carbon dioxide to the lungs, where it is eliminated when air is exhaled. Oxygen is needed by the cells to release energy from sugar molecules.

The digestive system breaks down foods into **nutrients,** substances that the body needs to carry out its functions, which then move into the bloodstream through **absorption.** The circulatory system delivers the nutrients to all body cells.

The excretory system eliminates wastes from your body. Your respiratory and circulatory systems both have roles in the excretory system.

Which Systems Control Body Functions?

The endocrine system is made up of organs called **glands** that release chemical signals directly into the bloodstream. **The nervous system and the endocrine system work together to control body functions.** Your senses send information about your environment to your nervous system. A signal in the environment that makes you react is called a **stimulus.** A **response** is what your body does in reaction to a stimulus. The chemical signals released by the endocrine system are called **hormones.** Hormones affect many body processes.

56

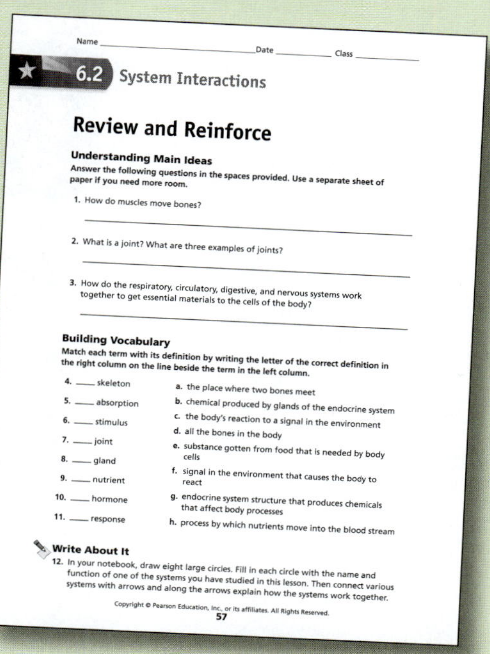
Name ______________ Date ________ Class ________

6.2 System Interactions

Review and Reinforce

Understanding Main Ideas

Answer the following questions in the spaces provided. Use a separate sheet of paper if you need more room.

1. How do muscles move bones?
2. What is a joint? What are three examples of joints?
3. How do the respiratory, circulatory, digestive, and nervous systems work together to get essential materials to the cells of the body?

Building Vocabulary

Match each term with its definition by writing the letter of the correct definition in the right column on the line beside the term in the left column.

4. ____ skeleton	a. the place where two bones meet
5. ____ absorption	b. chemical produced by glands of the endocrine system
6. ____ stimulus	c. the body's reaction to a signal in the environment
7. ____ joint	d. all the bones in the body
8. ____ gland	e. substance gotten from food that is needed by body cells
9. ____ nutrient	f. signal in the environment that causes the body to react
10. ____ hormone	g. endocrine system structure that produces chemicals that affect body processes
11. ____ response	h. process by which nutrients move into the blood stream

Write About It

12. In your notebook, draw eight large circles. Fill in each circle with the name and function of one of the systems you have studied in this lesson. Then connect various systems with arrows and along the arrows explain how the systems work together.

57

LESSON 6.2

Lexile Measure = 920L

System Interactions

Establish Learning Objectives

After this lesson, students will be able to:

Explain how the skeletal and muscular systems work together.

Identify which body systems work together to obtain and transport materials.

Identify which body systems control communication and regulation.

Engage

Activate Prior Knowledge

MY PLANET DIARY Read *Do you hear in color?* with the class. Ask: **Which body system do you think is involved in synesthesia? Explain.** *(Synesthesia involves the sense organs, which are part of the nervous system.)* Explain that the term *synesthesia* comes from two Greek words: *syn* meaning "together" and *aesthesis* meaning "perception." People with synesthesia literally have "linked perception." How these perceptions link is unique to each synesthete.

Explore

Lab Resource: Inquiry Warm-Up 

L1 HOW DOES YOUR BODY RESPOND? Students will investigate how body parts work together and how muscles become fatigued. This Inquiry Warm-Up can be found online.

System Interactions

How Do You Move?
TEKS 12B

Which Systems Move Materials in Your Body?
TEKS 12B, 12C

Which Systems Control Body Functions?
TEKS 12B, 12C

MY PLANET DIARY — FUN FACTS

Do you hear in color?

What color is the letter *b* or the roar of a tiger? You might not see colors when you hear sounds, but some people do. In people with synesthesia (sin us THEE zhuh), their senses overlap. Some people with synesthesia may taste a shape or hear music in colors. Others may hear a sound when they see motion. Even people without synesthesia experience some connections between their senses. You can explore how your own senses overlap in the first question on this page.

Communicate Answer the questions and then discuss your answers with a partner.

1. Look at the shapes below. One of them is called kiki and the other bouba. Which name do you think matches each shape?

A B

Sample: I think A is bouba and B is kiki.

2. Most people call the rounded shape bouba and the pointed shape kiki. Why do you think that is?

Sample: Kiki is a sharp noise, while bouba is a softer and fuller noise.

Lab zone: Do the Inquiry Warm-Up *How Does Your Body Respond?* Find the lab online.

SUPPORT ALL READERS

Lexile Measure = 920L **Lexile Word Count = 1548**

Prior Exposure to Content: Most students have encountered this topic in earlier grades

Academic Vocabulary: *develop, hypotheses, summarize*

Science Vocabulary: *skeleton, joint, gland, stimulus, hormone*

Concept Level: Generally appropriate for most students in this grade

Preteach With: My Planet Diary "Do you hear in color?" and Figure 2 activity

Vocabulary
- skeleton
- skeletal muscle
- joint
- nutrient
- absorption
- gland
- stimulus
- response
- hormone

Skills
- Reading: Summarize
- Inquiry: Develop Hypotheses

How Do You Move?

TEKS 12B In this section, you'll explore how the muscular, skeletal, and nervous systems work together to make the human body move.

Carefully coordinated movements let you thread a needle, ride a bicycle, brush your teeth, and dance. These movements—and all of your body's other movements—happen as a result of the interactions between body systems. Your muscular system is made up of all the muscles in your body. Your skeletal system, or **skeleton,** includes all the bones in your body. **Muscles and bones work together, making your body move. The nervous system tells your muscles when to act.**

Muscles and Bones

Skeletal muscles are attached to the bones of your skeleton and provide the force that moves your bones. Muscles contract and relax. When a muscle contracts, it shortens and pulls on the bones to which it is attached, as shown in **Figure 1.**

FIGURE 1

Muscles Moving Bones

As this dancer's muscles pull on his leg bones, he can make rapid, skillful moves.

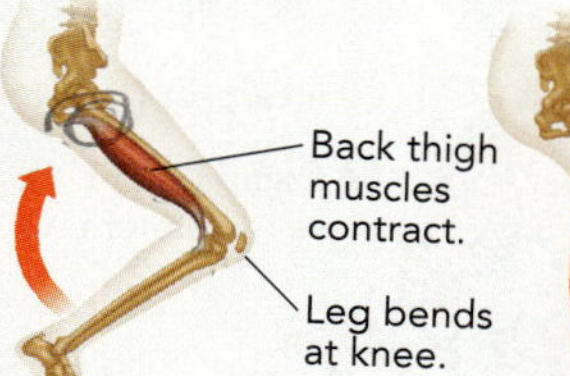

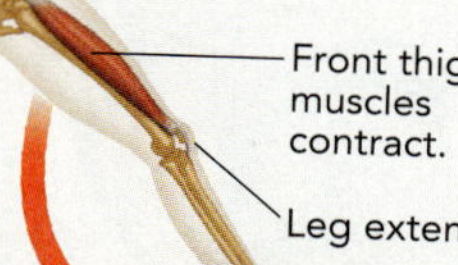

Develop Hypotheses An octopus has no bones. Explain how you think it moves.

Sample: The octopus uses muscles to squeeze water out of its body like a jet stream, which propels the octopus through the water.

PEARSON Texas.com

Explain

LESSON 6.2

Introduce Vocabulary

Have students look at the vocabulary terms *stimulus* and *response.* Tell students that these terms are related in a cause-and-effect way. A stimulus is the cause and the response is the effect.

Teach Key Concepts 

Explain to students that three of the body's organ systems work together to allow for all of the body's movements. Review the ideas that an organ system is a group of organs working together to perform a major function, and that two or more organ systems work together to enable an organism to carry out its life functions and survive. Ask: **What is another name for the skeletal system?** *(Skeleton)* **What is it made up of?** *(All the bones of the body)* **With what other systems does it work to produce movement?** *(The muscular system and the nervous system)* **How do muscles move bones?** *(Muscles contract and relax. Skeletal muscles provide the force that moves bones. When a muscle contracts, it shortens and pulls on the bone it is attached to.)* **What is the role of the nervous system?** *(It tells the muscles when to act.)*

Teach With Visuals

Tell students to look at **Figure 1.** Remind students that skeletal muscles only work by contracting, or shortening. Tell students that the muscles at the back of the thigh are the hamstring group and those at the front are the quadriceps group. Ask: **As the leg is bent at the knee, what happens to the hamstring muscles? The quadriceps muscles?** *(The hamstring muscles contract and the quadriceps muscles relax.)* **As the leg straightens out, what happens to the hamstring muscles? The quadriceps muscles?** *(The hamstring muscles relax and the quadriceps muscles contract.)*

Develop Hypotheses Tell students that a hypothesis is a possible explanation for a set of observations.

PEARSON Texas.com

English Language Proficiency Standards

ELPS Listening 2.1.4

Read aloud page 221 and **Figure 1.**

Beginning Scaffold questions that help students develop hypotheses: *Does an octopus use bones to move? Does it use muscles? Show how it uses muscles to move.* Have pairs share their hypotheses.

Intermediate Discuss **Figure 1.** Have partners orally complete the activity. Provide sentence frames as needed: *An octopus uses _____ to move. It moves by _____.* Have students switch partners to share their hypotheses.

Advanced Have pairs work together to complete the activity in **Figure 1.** Encourage students to ask each other questions about and elaborate on the hypotheses.

Advanced High Have students complete the activity in **Figure 1.** Then have partners ask and answer questions about the hypotheses.

Texas Essential Knowledge and Skills

12B Identify the main functions of the systems of the human organism, including the circulatory, respiratory, skeletal, muscular, digestive, excretory, reproductive, integumentary, nervous, and endocrine systems.

LESSON 6.2

Explain

Lead a Discussion

BONES AND JOINTS Review the definition of *joint.* Tell students human hands and feet have more joints than other parts of the body Ask: **Why are joints important?** *(Joints allow movement of bones.)* **Why do hands and feet have so many joints?** *(They perform complex movements that need great skill. Many joints allow these movements.)*

Lead a Discussion

MOVE THE BODY Discuss with students how muscles work in pairs. You may have students remain in their seats and bend their arms at the elbow. Students will likely know that the biceps muscle bends the elbow. Tell them that the triceps, on the back of the arm, straightens the elbow. Ask: **Why did you bend your arm just now instead of your leg?** *(Sample: My brain told my arm muscles to move. It did not tell my leg to move.)*

Summarize Tell students that when they summarize, they briefly restate the main ideas of what they have read or heard in their own words.

Elaborate

Apply It!

L1 Before beginning the activity, have students think about places in the body where bones come together and movement takes place.

Lab Resource: Lab Investigation

L2 **A LOOK BENEATH THE SKIN** Students will observe characteristics of skeletal muscles and identify how skeletal muscles work. This Lab Investigation can be found in the Student Lab Manual, p. 71, and online.

Evaluate

Assess Your Understanding

Have students evaluate their understanding by completing the appropriate sentence.

RTI Response to Intervention

If students need help explaining how the body moves, **then** remind them that skeletal muscles pull on the bones to which they are attached.

Summarize In your own words, describe which of your systems work together when you write in this book.

Sample: My nervous system tells my muscles to move the bones in my arm, hand, and fingers.

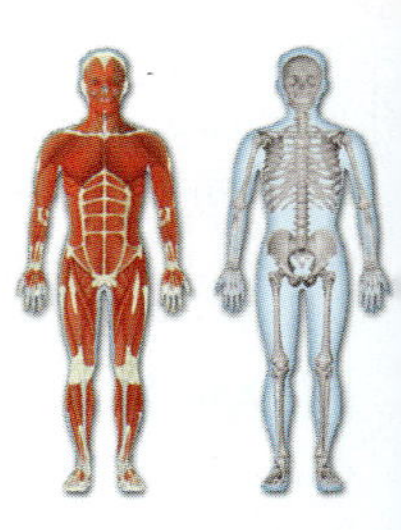

Bones and Joints What happens when you wiggle your fingers or touch your toes? Even though your bones are rigid, your body can bend in many ways. Your skeleton bends at its joints. A **joint** is a place in the body where two bones come together. For example, your elbow and your shoulder are two joints that move when you raise your hand.

Making Movement Happen Muscles make bones move at their joints. Try standing on one leg and bending the other leg at the knee. Hold that position. You can feel that you are using the muscles at the back of your thigh. Now straighten your leg. You can feel the muscles in the back of your leg relax, but the muscles in the front of your leg are at work. Your nervous system controls when and how your muscles act on your bones. You will read more about the nervous system later in this lesson.

apply it!

1 **Interpret Diagrams** Circle three of the football player's joints.

2 **Compare and Contrast** Describe how your shoulder and elbow move in different ways.

My elbow only bends back and forth, but my arm can move in a whole circle at my shoulder.

3 **CHALLENGE** From a standing position, bend down and grab your ankles. List six places or joints where your skeleton bends.

Samples: spine (waist), hip, knee, shoulder, neck, wrist, fingers, ankles

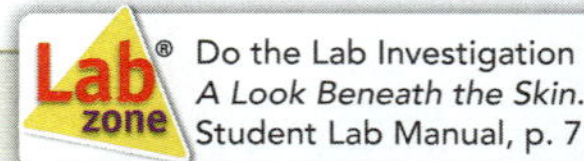

Do the Lab Investigation *A Look Beneath the Skin.* Student Lab Manual, p. 71

Assess Your Understanding

got it?

○ **I get it!** Now I know that muscles and bones work together to make the body move.

○ I need extra help with See TE note.

222 Introduction to the Human Body

Professional Development Note

Teacher to Teacher

Activity After studying muscular structure, types of tissue, and some common muscle names, I like to have the students make clay models of themselves. I make sure the clay is soft enough to be molded but hard enough to hold its shape. I then offer some wood dowels to give the model muscles some support, just like the skeletal system. After the models are complete, I get out a cow's knee joint, which you can get from a butcher, and show how the two systems work closely together.

Mrs. Anne Rice
Woodland Middle School
Gurnee, Illinois

Which Systems Move Materials in Your Body?

TEKS 12B, 12C In this section, you'll learn about the different systems that work together to transport materials throughout the human body.

The trillions of cells that make up your body need materials to function. Cells also produce wastes that must be removed. If the processes of moving these materials were made into a movie, your nervous system would be the director. The movie set would include the muscular and skeletal systems. And the main characters would be some of your other systems. **The circulatory, respiratory, digestive, and excretory systems play key roles in moving materials in your body.**

Transporting Materials Your circulatory system includes your heart, blood vessels, and blood. Blood vessels are found throughout your body. Blood that flows through these vessels carries materials such as water, oxygen, and food to every cell, as shown in **Figure 2.** Materials that your cells must get rid of, such as carbon dioxide and other cell wastes, are also moved through the body in the blood.

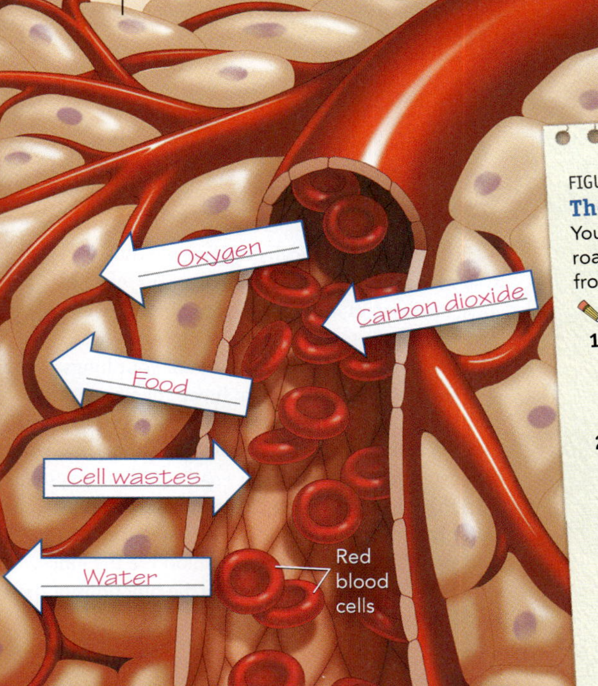

Word Bank
- Carbon dioxide
- Cell wastes
- Food
- Oxygen
- Water

FIGURE 2

The Body's Highway

Your circulatory system is like a set of roadways that carry materials to and from cells.

Answer the following questions.

1. **Identify** Use the word bank to identify the materials that move between cells and the blood. Write the words on the arrows.
2. **Predict** How do you think a blocked blood vessel would affect cells?

   If flow is blocked, oxygen and food could not get to cells, and wastes would build up, so cells could die.

223

Differentiated Instruction

L1 Working in Pairs Tell students that the muscle in the front of their upper arm is the biceps, and the muscle in the back is the triceps. Tell students to raise their lower arm at the elbow and observe what happens to these two muscles. Have them do the same as they straighten the lower arm. Have students explain what is happening in each movement.

L3 Body Levers Skeletal muscles, tendons, and joints work together like levers to move bones. A lever is a simple machine. Have students research the three classes of levers, defining the terms *fulcrum, effort,* and *load.* For each class, have students identify bones that act like that lever. Have students present their findings as an illustrated poster.

Explain

LESSON 6.2

Teach Key Concepts

Explain to students that in order for cells to function, they must obtain certain essential materials. They must also get rid of certain materials that are the waste products of cellular processes. Ask: **What materials are essential to the survival of cells?** *(Oxygen, water, food)* **What are the materials that a cell must get rid of?** *(Carbon dioxide and other cell wastes)* **What systems of the body move these materials to and from all the cells?** *(Circulatory system, respiratory system, digestive system, and excretory system)*

Address Misconceptions

L1 THE COLOR OF BLOOD Have students look at their wrists or the underside of their forearm. Ask: **What color does your blood appear to be?** *(Blue)* Explain to students that although deoxygenated blood looks blue in the vessels and despite the fact that textbook illustrations often show it that way, blood carrying little oxygen and lots of waste materials is dark red in color. Ask: **What color is blood carrying lots of oxygen and little waste?** *(Bright red)* Tell students the reason deoxygenated blood looks blue is a yellow pigment in the skin.

Make Analogies

L1 A SENSE OF SIZE To give students a sense of the size of the network of blood vessels in the human body, tell them that if the vessels were unraveled, they would encircle Earth more than two times. Ask: **What is the circumference of Earth?** *(About 40,000 km)* **How long does that make all the blood vessels in the body?** *(More than 80,000 km)*

21st Century Learning

CREATIVITY Have students write a motivational speech for a marathon team. Students should include tips on hydration, proper layering, injury prevention and how to prepare for "the wall." Students may want to research information about sports nutrition and running. As an extension, students can record themselves delivering their speeches.

Texas Essential Knowledge and Skills

12B Identify the main functions of the systems of the human organism, including the circulatory, respiratory, skeletal, muscular, digestive, excretory, reproductive, integumentary, nervous, and endocrine systems.

12C Recognize levels of organization in plants and animals, including cells, tissues, organs, organ systems, and organisms.

LESSON 6.2

Explain

Support the TEKS

FOOD IN AND WASTES OUT Remind students that oxygen is essential to the survival of cells and gets to the cells as a result of the respiratory and circulatory systems working together. Introduce the digestive system. Ask: **How is oxygen used by cells?** *(Oxygen is used to release energy from sugar molecules.)* **Where do the sugar molecules come from?** *(Foods that are eaten)* **Which system is responsible for breaking down foods into sugars and other nutrient molecules the body can use?** *(The digestive system)* Explain that all the foods people eat must be broken down into substances that can be absorbed into the bloodstream and delivered to all parts of the body. Ask: **Which two systems work together to do this?** *(The digestive and circulatory systems)* Introduce the excretory system by explaining that cells produce wastes that must be eliminated from the body. Ask: **Which system gets rid of body wastes?** *(The excretory system)*

Teach With Visuals

Tell students to look at the small human body illustrations that accompany the discussion of each system. The organs are not labeled, but students may be able to identify some of them from prior studies. Ask: **What structures make up the circulatory system?** *(Heart, blood vessels)* **The respiratory system?** *(Nose, pharynx, larynx, trachea, bronchi, lungs)* **The digestive system?** *(Mouth, esophagus, stomach, small intestine, liver, pancreas, large intestine, rectum)* **The excretory system?** *(Skin, lungs, liver, kidneys, urinary bladder, urethra)* **Which organ is part of both the respiratory system and the excretory system?** *(Lungs)*

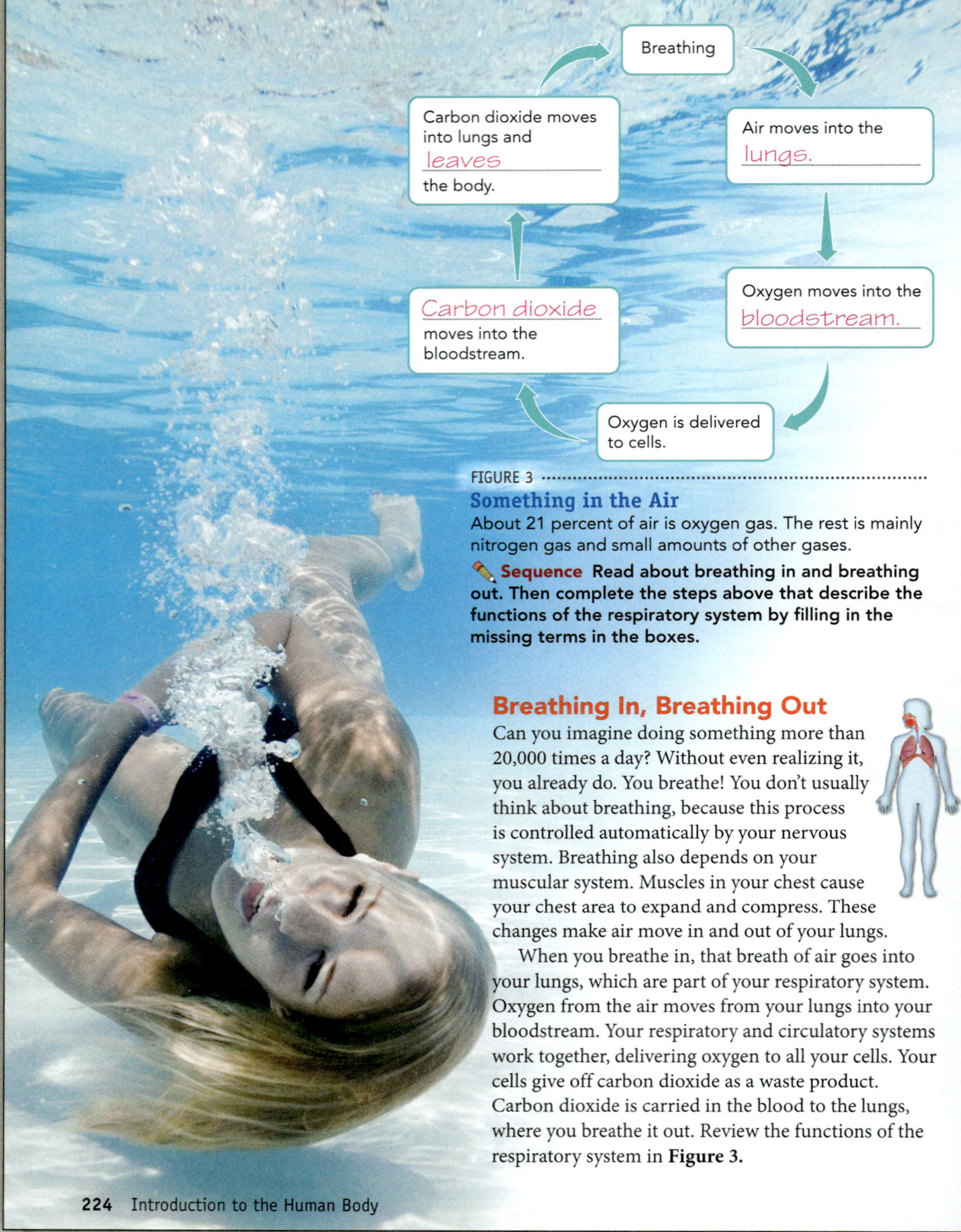

FIGURE 3

Something in the Air

About 21 percent of air is oxygen gas. The rest is mainly nitrogen gas and small amounts of other gases.

Sequence **Read about breathing in and breathing out. Then complete the steps above that describe the functions of the respiratory system by filling in the missing terms in the boxes.**

Breathing In, Breathing Out

Can you imagine doing something more than 20,000 times a day? Without even realizing it, you already do. You breathe! You don't usually think about breathing, because this process is controlled automatically by your nervous system. Breathing also depends on your muscular system. Muscles in your chest cause your chest area to expand and compress. These changes make air move in and out of your lungs.

When you breathe in, that breath of air goes into your lungs, which are part of your respiratory system. Oxygen from the air moves from your lungs into your bloodstream. Your respiratory and circulatory systems work together, delivering oxygen to all your cells. Your cells give off carbon dioxide as a waste product. Carbon dioxide is carried in the blood to the lungs, where you breathe it out. Review the functions of the respiratory system in **Figure 3.**

Getting Food Your respiratory system takes in oxygen, and your circulatory system delivers it to your cells. Oxygen is used in cells to release energy from sugar molecules that come from the food you eat. But how do sugar molecules get to your cells? Your digestive system helps to break down foods into sugars and other nutrient molecules that your body can use. A **nutrient** is a substance that you get from food and that your body needs to carry out processes, such as contracting muscles. Through a process called **absorption,** nutrients move from the digestive system into the bloodstream. The circulatory system then delivers the nutrients to all the cells in your body. In this way, your digestive system and circulatory system work together to get food to your cells.

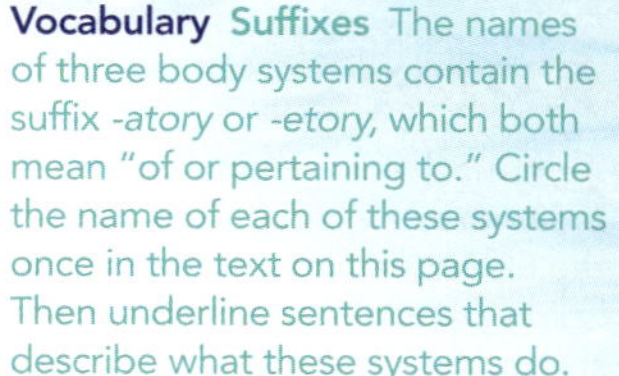

Moving Wastes The excretory system eliminates wastes from your body. Your respiratory and circulatory systems both have roles in the excretory system. You already read that carbon dioxide passes from the circulatory system into the respiratory system and leaves the body when you exhale. Other cellular wastes also pass into the blood. These wastes are altered by the liver and filtered out of the blood by the kidneys. This process produces urine, which then carries the wastes out of your body.

Vocabulary Suffixes The names of three body systems contain the suffix *-atory* or *-etory,* which both mean "of or pertaining to." Circle the name of each of these systems once in the text on this page. Then underline sentences that describe what these systems do.

Sample answers are shown.

Lab zone Do the Quick Lab *Working Together, Act I.* Student Lab Manual, p. 76

Assess Your Understanding

TEKS 12B

1a. Identify Name four body systems whose main functions involve getting oxygen to your cells.

Nervous, muscular, respiratory, and circulatory systems

b. Explain How is absorption an important function of the digestive system?

Absorption moves nutrients into the blood, which carries them to the cells.

c. Draw Conclusions How does the circulatory system help other systems function?

Carries nutrients absorbed by the digestive system and oxygen absorbed by the respiratory system to cells; takes wastes to the lungs and kidneys for excretion by the respiratory and excretory systems

got it?

○ **I get it!** Now I know that materials are moved within my body by the circulatory, respiratory, digestive, and excretory systems.

○ I need extra help with See TE note.

225

Differentiated Instruction

L1 All Systems Go Have students write a paragraph that describes how their circulatory, respiratory, digestive, and excretory systems work together as the students prepare for and run a race.

L1 A Balanced Diet Have students use the concept of the body's need for nutrients to explain the value of a balanced diet.

L3 Blood Types Have students research blood types including the following: 1) major blood types and how they must be considered in blood transfusions, 2) the blood type that is the universal donor and the one that is the universal recipient, and 3) the percentage of the human population with each blood type. Have students report their findings.

Elaborate

Build Inquiry 

L1 MECHANICAL DIGESTION

Materials 2 small beakers, water, 2 sugar cubes, small plastic bag

Time 10 minutes and then observation one hour later

Tell students to fill both beakers with the same amount of cold tap water. Tell them to place one sugar cube in the plastic bag and crush it using a book. Have them place the crushed sugar in one beaker and the sugar cube in the other. Tell them to observe both beakers at this point and then again an hour later.

Ask: **What is different about the sugars?** *(The crushed sugar is mostly or completely dissolved, while the cube is not.)* **Which step in digestion is represented by crushing the sugar cube?** *(Chewing)* **What does this activity suggest about the importance of this step of digestion?** *(Food can be more easily and quickly digested if it is first broken down into smaller pieces.)*

Lab Resource: Quick Lab 

L2 WORKING TOGETHER, ACT I Students will model the ways in which body systems move nutrients and wastes through the body. This Quick Lab can be found in the Student Lab Manual, p. 76, and online.

Evaluate

Assess Your Understanding

After students answer the questions, have them evaluate their understanding by completing the appropriate sentence.

RTI Response to Intervention

1a. If students need help identifying the body systems, **then** discuss the sequence of steps involved in bringing air into the body and getting it to the cells, identifying each system at each step.

b. If students have trouble explaining the importance of absorption, **then** have them review what cells need, what process gets the substances into the body, and how those substances get to the cells.

c. If students need help identifying how the circulatory system helps other systems, **then** have them review the functions of the circulatory system.

LESSON 6.2

Explain

Teach Key Concepts

Explain to students that in order for an organism to function and survive, it must be able to gather and interpret information about its external and internal environments and then communicate that information to all parts of the body. Each part of the body communicates with other parts under the control of the nervous system. The nervous system acts like command central, collecting, processing, and communicating information. Ask: **What are the parts of the nervous system?** *(Brain, spinal cord, and nerves)* **Which organs gather information about the external world for the nervous system?** *(Eyes, ears, nose, taste buds, and skin)* **Which is a stimulus? A response?** *(A stimulus is a signal in the environment that causes a person to react. A response is a reaction of the body to the stimulus.)* Tell students that information travels through nerve cells in the form of electric impulses at an average speed of about 100 meters/second. Ask: **Why is speed so important?** *(Information must be communicated as quickly as possible.)* Introduce the endocrine system by telling students that the nervous system is not the only system that communicates with other body systems. Emphasize that the nervous system and the endocrine system work together to control body functions. Ask: **What is the endocrine system?** *(A system made up of glands that produce hormones, which regulate certain body processes)* **How do hormones reach all parts of the body?** *(Hormones are released directly into the bloodstream and transported by the circulatory system.)*

Make Analogies

L1 **MESSENGER SYSTEMS** Tell students that although the nervous system and the endocrine system are both messenger systems, they work in different ways. Have students imagine that they have important information they need to share with other people via email. Ask: **What is one way you could share the information?** *(Send each person an individual email)* **What is a more efficient way?** *(Send a mass email)* **Which way represents the nervous system? The endocrine system?** *(Individual emails represent the nervous system; a mass email represents the endocrine system.)*

Texas Essential Knowledge and Skills

12B Identify the main functions of the systems of the human organism, including the circulatory, respiratory, skeletal, muscular, digestive, excretory, reproductive, integumentary, nervous, and endocrine systems.

12C Recognize levels of organization in plants and animals, including cells, tissues, organs, organ systems, and organisms.

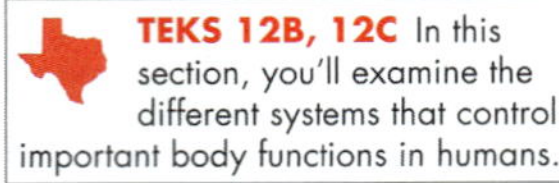

TEKS 12B, 12C In this section, you'll examine the different systems that control important body functions in humans.

did you know?

An optical illusion fools your brain about what you see. Are the horizontal lines in the picture below parallel to one another or slanted?

The lines are parallel.

Which Systems Control Body Functions?

To function properly, each part of your body must be able to communicate with other parts of your body. For example, if you hear a phone ring, that message must be sent to your brain. Your brain then directs your muscles to move your bones so you can answer the phone. These actions are controlled by the nervous system, which is made up of the brain, spinal cord, and nerves. In your nervous system, information travels through nerve cells.

Other messages are sent by chemical signals that are produced by the endocrine system. The endocrine system is made up of organs called **glands** that release chemical signals directly into the bloodstream. For example, when you exercise, your endocrine system sends signals that make you perspire, or sweat. As sweat evaporates, it helps you cool down. **The nervous system and the endocrine system work together to control body functions.**

Nervous System Your eyes, ears, skin, nose, and taste buds send information about your environment to your nervous system. Your senses let you react to bright light, hot objects, and freshly baked cookies. A signal in the environment that makes you react is called a **stimulus** (plural *stimuli*). A **response** is what your body does in reaction to a stimulus. Responses are directed by your nervous system but often involve other body systems. For example, your muscular and skeletal systems help you reach for a cookie. And your digestive system releases saliva before the cookie even reaches your mouth.

FIGURE 4

Stimulus and Response

Have you ever been startled by something unexpected?

Use the pictures to complete these tasks.

1. **Sequence** Use numbers 1, 2, and 3 to put the pictures in order.
2. **Explain** Use the terms *stimulus* and *response* to explain what happened.

The spider is the stimulus that scared the girl. Her response was to jump up.

apply it!

Among the drugs that affect the nervous system, caffeine is one of the most commonly used worldwide. Caffeine is found in coffee, tea, soda, other beverages, and even in chocolate.

1 Explain How does caffeine reach the brain after someone drinks a cup of coffee or tea? In your answer, be sure to identify the systems involved.

It is absorbed through the digestive system into the blood and carried to the brain by the circulatory system.

2 Infer Caffeine is addictive, which means that the body can become physically dependent on the drug. Which body system do you think would be most involved in an addiction? Explain your answer.

The nervous system; it controls body functions.

Endocrine System The chemical signals released by the endocrine system are called **hormones.** Hormones are transported through your body by the circulatory system. These chemicals affect many body processes. For example, one hormone interacts with the excretory system and the circulatory system to control the amount of water in the bloodstream. Another hormone interacts with the digestive system and the circulatory system to control the amount of sugar in the bloodstream. Hormones also affect the reproductive systems of both males and females.

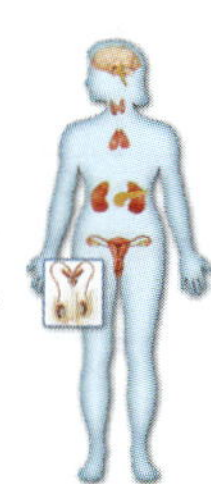

Do the Quick Lab *Working Together, Act II.* Student Lab Manual, p. 77

Assess Your Understanding

TEKS 12B

2a. Identify How are the functions of the nervous system and the endocrine system different?

The nervous system carries information through nerve cells. The endocrine system releases chemical signals into the blood.

b. Apply Concepts Describe an example of a stimulus and response that involves your sense of hearing.

Sample: I hear a song and decide to sing along. The song is the stimulus. Deciding to sing is my response.

got it?

○ I get it! Now I know that the nervous system and endocrine system work together to control body functions.

○ I need extra help with See TE note.

227

Differentiated Instruction

L1 Control Systems Have students create a Venn diagram to compare and contrast the nervous system and the endocrine system. Tell students to be sure to include the terms *stimulus, response, glands, hormones, senses,* and *nerve cells.*

L3 Optical Illusions Optical illusions use shapes, colors, and line distortions to trick the eye. In an optical illusion, what one perceives visually is different from what is actually real. Have students gather examples of optical illusions and present them to the class.

Elaborate

LESSON 6.2

21st Century Learning

CRITICAL THINKING Review with students the function of the nervous system. Tell students that there are three types of neurons, or nerve cells: sensory neurons, interneurons, and motor neurons. Ask: **Knowing what the job of the nervous system is, what do you think is the function of each type of neuron?** *(Sensory neurons carry messages from the surroundings to the brain and spinal cord. Interneurons connect sensory neurons to motor neurons. Motor neurons carry messages from the brain and spinal cord to muscles and glands.)*

21st Century Learning

CREATIVITY To ensure that students understand the relationship between a stimulus and a response, tell them to create a drawing similar to that in **Figure 4** using another example. Students may wish to add "balloons" to show what the person says or thinks. Encourage students to be creative and artistic.

Apply It!

L1 Before beginning the activity, review the functions of all the systems that students have learned about. Remind them that systems work together to allow the body to function.

Lab Resource: Quick Lab

L2 WORKING TOGETHER, ACT II Students will model how the changing needs of cells affect the interactions among body systems. This Quick Lab can be found in the Student Lab Manual, p. 77, and online.

Evaluate

Assess Your Understanding

After students answer the questions, have them evaluate their understanding by completing the appropriate sentence.

RTI Response to Intervention

2a. If students have trouble contrasting the nervous and endocrine systems, **then** have them identify how each system sends messages.

b. If students need help describing a stimulus-response example involving hearing, **then** review the definitions of the two terms.

Name ______________________ Date ____________ Class ____________

Assess Your Understanding

System Interactions

How Do You Move?

got it?

○ **I get it!** Now I know that ________________ and ________________ work together to make the body move.

○ **I need extra help with** ____

____

Which Systems Move Materials in Your Body?

1a. IDENTIFY Name four body systems whose main functions involve getting oxygen to your cells.

____

____

b. EXPLAIN How is absorption an important function of the digestive system?

____

____

c. DRAW CONCLUSIONS How does the circulatory system help other systems function?

____

____

Which Systems Control Body Functions?

2a. IDENTIFY How are the functions of the nervous system and the endocrine system different?

____

____

b. APPLY CONCEPTS Describe an example of a stimulus and response that involves your sense of hearing.

____

____

Place the outside corner, the corner away from the dotted line, in the corner of your copy machine to copy onto letter-size paper.

Name ______________________ Date ____________ Class ____________

Enrich

System Interactions

You know that the nervous system and endocrine system work together to control body functions. All of the body's systems interact with the nervous system in some way. Read the passage and study the table below. Then answer the questions that follow on a separate sheet of paper.

Nervous System Interactions

The nervous system, containing the brain, spinal cord, and nerves, controls the body's reactions to its environment. To do this, the nervous system must send information to and receive information from other systems in the body. The table below explains how the nervous system interacts with some of the other systems in the body.

System	Interactions with Nervous System
Skeletal	• Bones provide calcium that helps nervous system function • Bones protect nervous system organs from injury
Cardiovascular	• Brain regulates heart rate and blood pressure • Endothelial cells prevent materials in the blood from entering the brain
Muscular	• Brain controls contraction of skeletal muscle
Respiratory	• Brain monitors volume of lungs and blood gas levels • Brain regulates respiratory rate
Digestive	• Brain controls drinking and feeding • Brain controls muscles for eating and elimination

1. Which systems provide a benefit to the nervous system? Explain.
2. Which systems receive a benefit from the nervous system? Explain.
3. Choose a body system from the table and name an interaction that is not listed.
4. Name a system not listed in the table and explain how it interacts with the nervous system.

Name ____________________ Date __________ Class __________

Lesson Quiz

System Interactions

Write the letter of the correct answer on the line at the left.

1. ___ The two systems that control body functions are the

A digestive and circulatory systems
B excretory and nervous systems
C nervous and endocrine systems
D endocrine and respiratory systems

2. ___ Which of the following is NOT a stimulus?

A hearing a loud noise
B sneezing
C touching a hot object
D tasting a lemon

3. ___ The gas cells need in order to release energy from sugar molecules is

A carbon dioxide
B water vapor
C nitrogen
D oxygen

4. ___ The muscles attached to bones that provide the force to move the bones are

A striated muscles
B skeletal muscles
C smooth muscles
D connective muscles

Fill in the blank to complete each statement.

5. The circulatory system works with the ____________ system to get nutrients to all body cells.

6. Chemical substances produced by glands that affect many body processes are called ____________.

7. ____________ is the process by which nutrients move from the digestive system into the bloodstream.

8. Chemical substances needed by body cells that result from the process of digestion are called ____________.

9. The ____________ system eliminates wastes from the body.

10. The elbow and shoulder are examples of ____________.

System Interactions

Answer Key

Review and Reinforce

Find the worksheet in the Student Workbook.

1. Skeletal muscles attached to bones can contract and relax. When a muscle contracts, it shortens and pulls on the bones. The pull is the force that moves the bones.
2. A joint is where two bones come together in the body. Examples: knee, elbow, shoulder, hip, ankle.
3. Body cells need nutrients, water, and oxygen. Nutrients and water from foods are broken down by the *digestive system*. Oxygen from air is taken in by the *respiratory system*. The *circulatory system* transports the nutrients and oxygen in the blood to all body cells. All three systems are controlled by the *nervous system*.

4. d **5.** h **6.** f **7.** a

8. g **9.** e **10.** b **11.** c

12. Circles should be filled in as follows: Skeletal system/Supports and protects body, allows movement; Muscular system/Produces movement; Circulatory system/Transports essential materials to the cells and waste materials away from the cells; Respiratory system/Brings oxygen into the body and eliminates carbon dioxide from the body; Digestive system/Breaks down foods into nutrients for absorption; Excretory system/Eliminates wastes from the body; Nervous system/Controls body functions through stimulus-response nerve-cell messages; Endocrine system/Controls body process through hormones produced by glands.

Enrich

1. The skeletal system provides nourishment and protection to the nervous system. The cardiovascular system provides protection to the brain.
2. The muscular, respiratory, and digestive systems are regulated and controlled by the nervous system.
3. Sample: The muscular system sends information about movement and body position to the nervous system.
4. Sample: Receptors in skin in the integumentary system send information to the nervous system.

Lesson Quiz

1. C **2.** B

3. D **4.** B

5. digestive **6.** hormones

7. Absorption **8.** nutrients

9. excretory **10.** joints

Homeostasis

 **How does your body work?**

LESSON PACING:
2–3 periods or 1–1½ blocks

Lesson Vocabulary

- stress

Lesson Objectives	TEKS	ELPS
Define homeostasis and explain how body systems interact to maintain homeostasis.	12B, 13A, 13B	1.A.2

Content Refresher

Homeostasis and Equilibrium Homeostasis is a state of equilibrium, or balance, in the body. It is the physiological tendency of an organism to keep its internal equilibrium constant despite changes in its external environment. The term *homeostasis,* formed from two Greek words, means literally "standing still." It was coined around 1930 by American physiologist Walter B. Cannon, but the concept was described decades earlier by French physiologist Claude Bernard. Various mechanisms within the body—for example, processes that regulate body temperature or contents in the blood—maintain the balance that is essential for survival. When a condition in the body's internal or external environment changes, the body reacts with a series of its own changes—changes that are equal in magnitude and opposite in direction to the environmental change. That is the nature of equilibrium: the opposition to change and the maintenance of internal balance. Thus, homeostasis is an indicator of the proper functioning of the body.

Texas Essential Knowledge and Skills

12B Identify the main functions of the systems of the human organism, including the circulatory, respiratory, skeletal, muscular, digestive, excretory, reproductive, integumentary, nervous, and endocrine systems.
13A Investigate how organisms respond to external stimuli found in the environment such as phototropism and fight or flight.
13B Describe and relate responses in organisms that may result from internal stimuli such as wilting in plants and fever or vomiting in animals that allow them to maintain balance.

English Language Proficiency Standards

ELPS Learning Strategies 1.A.2 Use prior experiences to understand meanings in English.

DIFFERENTIATED INSTRUCTION KEY
L1 Struggling Students or Special Needs
L2 On-Level Students L3 Advanced Students

LESSON PLANNER 6.3

Investigations and Activities

My Planet Diary, **Student Edition,** p. 228

Inquiry: Inquiry Warm-Up, Out of Balance, **PearsonTexas.com**

Introduce Vocabulary, **Teacher's Edition,** p. 229

Teach Key Concepts, **Teacher's Edition,** p. 229

Lead a Discussion, The Mechanism Behind Homeostasis, **Teacher's Edition,** p. 230

Make Analogies, Homeostasis in Your Home, **Teacher's Edition,** p. 230

Differentiated Instruction, **Teacher's Edition,** p. 231

Address Misconceptions, Endotherms and Ectotherms, **Teacher's Edition,** p. 231

Inquiry: Build Inquiry, Evaporation as a Cooling Process, **Teacher's Edition,** p. 231

Lead a Discussion, Stress and Homeostasis, **Teacher's Edition,** p. 232

Teach With Visuals, **Teacher's Edition,** p. 232

Inquiry: Teacher Demo, Getting Stressed, **Teacher's Edition,** p. 232

21st Century Learning, Critical Thinking, **Teacher's Edition,** p. 233

Lead a Discussion, The Immune System, **Teacher's Edition,** p. 233

Apply It!, **Student Edition,** p. 233

Teach With Visuals, **Teacher's Edition,** p. 234

Differentiated Instruction, **Teacher's Edition,** p. 235

Inquiry: Quick Lab, Working to Maintain Balance, **Lab Manual,** p. 78

TEKS Review

Apply the TEKS, **Student Edition,** p. 234

Apply the TEKS, Doing Work, **Student Edition,** p. 254

TEKS Practice, **Student Edition,** p. 256

TEKS Practice: Chapter and Cumulative Review, **Student Edition,** p. 260

Lesson 6.3, **TEKS Preparation and Study Guide Workbook,** p. 58

SHORT ON TIME? To do this lesson in approximately half the time, do the Activate Prior Knowledge activity. A discussion of the Key Concepts will familiarize students with the lesson content. Use the Apply the TEKS activity to help students understand how the body works. Have students do the Quick Lab. The rest of the lesson can be completed by students independently.

These editable worksheets are available on **PearsonTexas.com.**
Print versions can be found in the **TEKS Preparation and Study Guide Workbook.**

Name ______________ Date ________ Class ________

6.3 Homeostasis

Key Concept Summary

How Does Your Body Stay in Balance?

Although conditions outside the human body may change, conditions inside the body stay stable. Such conditions include chemical makeup of the cells, their water content, and body temperature. The condition in which an organism's internal environment is kept stable in spite of changes in the outside environment is called homeostasis. Homeostasis is necessary for an organism's proper functioning and survival.

All of your body systems working together maintain homeostasis and keep the body in balance. Body responses that maintain homeostasis in the face of changes in external conditions include shivering, sweating, being hungry, and being thirsty. In each of these cases, the nervous and endocrine systems respond to a change in the body's internal environment and control the responses. They also signal other body systems to play a role in the response.

Homeostasis is never the responsibility of only one system; it relies on the interaction of many body systems. Maintaining body balance in terms of position involves structures in the inner ear that sense the position of the head and send messages to the brain.

Stress is the reaction of the body to possibly threatening, challenging, or uncomfortable events. Some stress is normal and healthy, and once the stress is over, the body returns to a healthier condition. However, too much negative stress can be unhealthy. Homeostasis can be disrupted by ongoing stress. Thus managing stress is important to having a healthy lifestyle. When homeostasis is maintained, a person is healthy. Bacteria and viruses can upset homeostasis and make a person sick. The body's immune system helps fight disease.

58

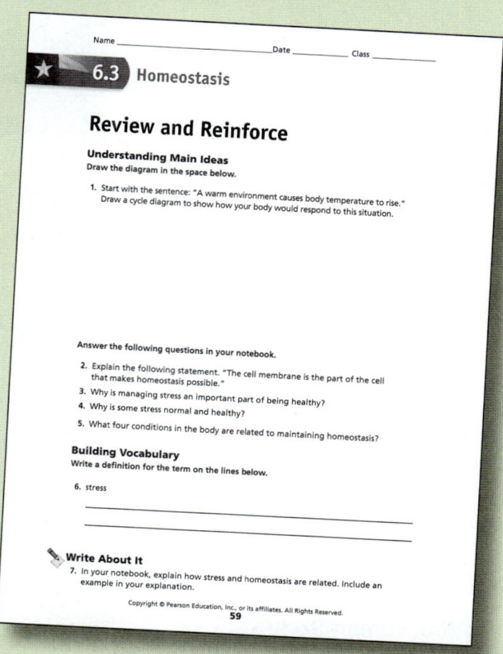

Name ______________ Date ________ Class ________

6.3 Homeostasis

Review and Reinforce

Understanding Main Ideas
Draw the diagram in the space below.

1. Start with the sentence: "A warm environment causes body temperature to rise." Draw a cycle diagram to show how your body would respond to this situation.

Answer the following questions in your notebook.

2. Explain the following statement. "The cell membrane is the part of the cell that makes homeostasis possible."
3. Why is managing stress an important part of being healthy?
4. Why is some stress normal and healthy?
5. What four conditions in the body are related to maintaining homeostasis?

Building Vocabulary
Write a definition for the term on the lines below.

6. stress

____

____

Write About It

7. In your notebook, explain how stress and homeostasis are related. Include an example in your explanation.

59

Lexile Measure = 890L

LESSON 6.3

Homeostasis

Establish Learning Objective

After this lesson, students will be able to:

Define homeostasis and explain how body systems interact to maintain homeostasis.

Engage

Activate Prior Knowledge

MY PLANET DIARY Read *Worried Sick—Not Just an Expression* with the class. Discuss the idea of stress and what students think it means. Ask: **What are some situations that you consider stressful?** *(Sample: taking a test, giving a speech, attending a new school, having family problems, performing in a concert or play, completing a too-busy schedule)*

Explore

Lab Resource: Inquiry Warm-Up 

L1 OUT OF BALANCE Students will apply an activity to the way balance is maintained in the human body. This Inquiry Warm-Up can be found online.

Homeostasis

How Does Your Body Stay in Balance?

TEKS 12B, 13A, 13B

my planet Diary

SCIENCE STATS

Worried Sick—Not Just an Expression

Starting in the 1980s, scientists began to gather evidence that stress can affect the immune system. For example, the graph below shows the relationship between the length of time a person is stressed and the risk of catching a cold when exposed to a virus. Today, scientists know that high levels of stress and long periods of stress can increase a person's risk for many diseases. Therefore, managing stress is an important part of a healthy lifestyle. Many activities, including hanging out with friends, getting enough sleep, and exercising moderately, can help lower stress levels.

Read Graphs Use the graph to answer the questions.

1. Summarize the information given in the graph.
   The longer someone feels stressed, the greater the risk is that the person will get a cold.

2. What do you do to manage stress?
   Sample: Take a walk, play with my dog, listen to music

Stress and Catching a Cold

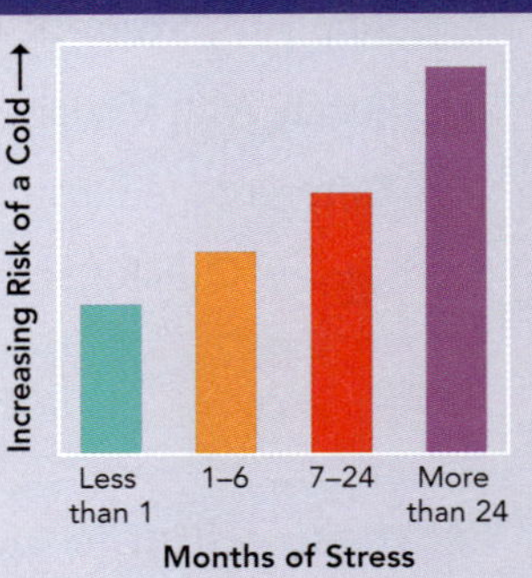

Lab zone: Do the Inquiry Warm-Up *Out of Balance.* Find the lab online.

SUPPORT ALL READERS

Lexile Measure = 890L   Lexile Word Count = 1201

Prior Exposure to Content: Many students may have misconceptions on this topic

Academic Vocabulary: *cause, communicate, effect, relate*

Science Vocabulary: *stress*

Concept Level: Generally appropriate for most students in this grade

Preteach With: My Planet Diary "Worried Sick" and Figure 2 activity

Vocabulary
- stress

Skills
 Reading: Relate Cause and Effect
Inquiry: Communicate

How Does Your Body Stay in Balance?

It may be summer or winter. You may be indoors or outdoors. You may be running or sitting still. Regardless, your internal body temperature is almost exactly 37°C. The external conditions may change. But the conditions inside your body stay stable, or steady. Most of these conditions, including the chemical makeup of your cells, their water content, and your body temperature, stay about the same.

Homeostasis The condition in which an organism's internal environment is kept stable in spite of changes in the external environment is called homeostasis. Keeping this balance is necessary for an organism to function properly and survive. **All of your body systems working together maintain homeostasis and keep the body in balance.**

TEKS 12B, 13A, 13B In this section, you'll investigate how different systems in the human body work together to maintain stable internal conditions.

ELPS 1.A.2
Work with a partner, to complete the "Keeping Warm in the Cold" activity on page 229. Think about your own experience being "really hot or really cold." Describe the example. Then write about it.

FIGURE 1

Keeping Warm in the Cold

The clothes on this snowboarder help keep him warm. But his body is working hard, too. His nervous, circulatory, and muscular systems work together, keeping his body warm.

Describe Think of a time when you were really hot or really cold. Describe the changes you felt as your body adjusted to that condition.

Sample: I played baseball on a really hot day last summer. I was sweating all over, and my face was red. I was so thirsty that I drank a big bottle of water between innings.

Explain

Introduce Vocabulary

Write the term *homeostasis* on the board. Tell students that the term comes from two Greek words: *homoios,* meaning "same," and *stasis,* meaning "act or condition of standing." Have students use the word meanings to define the term.

Teach Key Concepts

Explain to students that homeostasis is the condition in which the body's internal environment is kept stable in spite of changes in the outside environment. Homeostasis is necessary for the functioning and survival of an organism. Emphasize that all of the body systems students have learned about play a role in maintaining homeostasis and keeping the body in balance. Ask: **What conditions must your body keep stable?** *(Body temperature, amount of water in body cells, amount of sugar in blood, amount of salt in body cells, chemical makeup of cells)* **Why?** *(Sample: The processes that keep an organism alive involve many kinds of chemical reactions that can take place only in specifically balanced environments.)* Have students think about running a race. Ask: **What happens to your breathing when you run?** *(You breathe faster and deeper.)* **Why do you think this happens?** *(Breathing changes to get more air into the lungs and more oxygen to the cells, which need it to release energy from nutrient molecules.)* **Why is this an example of homeostasis?** *(In homeostasis, the organ systems work together to keep conditions inside the body balanced. If your body needs more energy because it is expending lots of energy, the respiratory system will speed up the rate and amount of oxygen intake.)* **How does your body help maintain homeostasis when you are cold? When you are hot?** *(The body shivers when it is cold and sweats when it is hot.)*

PEARSON Texas.com

LESSON 6.3

English Language Proficiency Standards

ELPS Learning Strategies 1.A.2

Read aloud **Figure 1** and have students complete the "Keeping Warm in the Cold" activity.

Beginning Read aloud the "Describe" activity. Together, list activities that would be really hot or really cold. Model how to describe a few of them. Then have partners complete the activity orally.

Intermediate Have students read aloud the activity with you. Have partners take turns telling about an experience. Provide sentence frames as needed: *Last _____, I _____. I was so _____ that I _____.*

Advanced As students work in pairs, encourage them to elaborate with details as they describe their bodily changes.

Advanced High Have partners read the text and complete the activity. Encourage them to tell how each of their body systems was affected.

Texas Essential Knowledge and Skills

12B Identify the main functions of the systems of the human organism, including the circulatory, respiratory, skeletal, muscular, digestive, excretory, reproductive, integumentary, nervous, and endocrine systems.

13A Investigate how organisms respond to external stimuli found in the environment such as phototropism and fight or flight.

13B Describe and relate responses in organisms that may result from internal stimuli such as wilting in plants and fever or vomiting in animals that allow them to maintain balance.

LESSON 6.3

Explain

Lead a Discussion

THE MECHANISM BEHIND HOMEOSTASIS Remind students that the cells of the body can function only within a narrow range of conditions in terms of temperature, nutrient availability, and water content. Ask: **What happens if any of these internal conditions change and the balance is shifted?** *(The appropriate body systems react to counter the change and restore the balance.)* **In the case of regulating body temperature, what systems are working together?** *(The nervous, muscular, endocrine, circulatory, and excretory systems)* Introduce the negative-feedback mechanism as the usual means of maintaining homeostasis. Tell students that it is called a negative-feedback system because once the body senses an internal or external change, it activates processes that reverse, or negate, the change. Use the example of temperature regulation. Ask: **If the external change is a decrease in temperature, what happens internally?** *(The body temperature drops.)* **According to the negative-feedback system, what does the body do?** *(It sends messages to the appropriate body systems to initiate actions to reverse the change.)* **What are some of the actions the body takes?** *(The muscles shiver, thereby creating heat and warming the body. Blood vessels supplying the skin constrict, thereby keeping warm blood deeper in the body and minimizing heat lost from the surface.)* Tell students that this mechanism is behind all of the actions the body takes to maintain homeostasis.

Make Analogies

L1 HOMEOSTASIS IN YOUR HOME Explain to students that the negative-feedback system for maintaining homeostasis is similar to the heating and cooling system in a home. Ask: **What device controls when a furnace or air conditioner goes on and off?** *(A thermostat)* **How does it work?** *(A desired room temperature is set on the thermostat. If the room temperature falls below the set temperature, the furnace goes on, producing heat. When the room temperature returns to the set temperature, the furnace goes off. In the case of an air conditioner, it switches on when the room temperature rises above the set temperature and stays on until the temperature returns to the set temperature.)*

Maintaining Homeostasis You experience homeostasis in action when you shiver, sweat, or feel hungry, full, or thirsty. Your nervous and endocrine systems control these responses. Other systems, including the digestive, respiratory, circulatory, and muscular systems, also play roles in your body's responses.

Regulating Temperature When you are cold, your nervous system signals your muscles to make you shiver. Shivering produces heat that helps keep you warm. As explained in the diagram below, when you warm up, shivering stops. When you are too warm, your endocrine system releases hormones that make you perspire. As the sweat evaporates, your body cools. The circulatory system and skin also help regulate temperature. Changes in the amount of blood flow in the skin can help prevent heat loss or carry heat away. In this way, your body temperature stays steady.

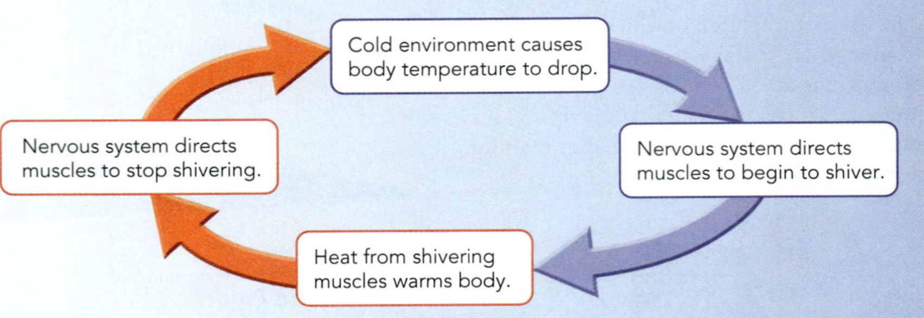

Meeting Energy Needs If your body needs more energy, hormones from the endocrine system signal the nervous system to make you feel hungry. After you eat, other hormones tell your brain to make you feel full. Other body systems are also involved. For example, your muscular system helps move food through your digestive system. Your respiratory system takes in the oxygen that is used in cells to release energy from food.

FIGURE 2

Hungry or Not?

Signals between your nervous system and your digestive system control your feelings of hunger.

Sequence Fill in the missing steps in the cycle diagram.

I feel hungry.

I start to eat.

I feel full.

I stop eating.

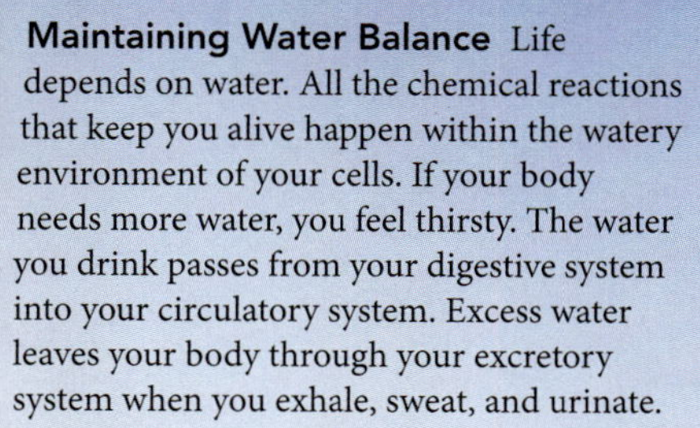

Maintaining Water Balance Life depends on water. All the chemical reactions that keep you alive happen within the watery environment of your cells. If your body needs more water, you feel thirsty. The water you drink passes from your digestive system into your circulatory system. Excess water leaves your body through your excretory system when you exhale, sweat, and urinate.

Keeping Your Balance You know that you hear with your ears. But did you know your ears also help you keep your balance? Structures in your inner ear sense the position of your head. They send this information to your brain, which interprets the signals. If your brain senses that you are losing your balance, it sends messages to your muscles to move in ways that help you stay steady, as in **Figure 3.**

Relate Cause and Effect Complete the cause-and-effect table below to help you organize what you have learned about homeostasis.

Cause	Effect
Body gets cold.	Shivers
Body gets overheated.	Sweat
Body needs more energy.	Hunger
Body needs more water.	Thirst

FIGURE 3

Balancing Act

Signals from this diver's ears to her brain lead to movements that help her balance on the edge of the diving board.

Describe What systems of the diver's body play a role in keeping her balanced on her toes?

The nervous system, muscular system, and skeletal system

Differentiated Instruction

L1 Balancing Acts To ensure that students understand the concept of how the body maintains homeostasis, have them use the information in the cause-and-effect table to write four sentences that identify stimuli and responses. Tell students to be as complete as possible in identifying the systems involved.

L3 Body Temperature Ask students to measure and record their temperature several times in one day. Have them record what they were doing just before they measured it. Students will discover that body temperature varies slightly during a day. Ask students to graph and explain their results using references.

21st Century Learning

CRITICAL THINKING Have students think about how body systems respond to stimuli. Ask: **What is the relationship between the ability of an organism to respond to stimuli and homeostasis?** *(The ability of an organism to respond to stimuli keeps its body functioning well and improves its chances of survival. Homeostasis is the process by which an organism responds to stimuli in ways that allow it to maintain stable internal conditions that ensure its survival.)*

Relate Cause and Effect Tell students that relating cause and effect involves looking at two events to see if one caused the other. A cause makes something happen. The effect is the result.

Address Misconceptions

L1 ENDOTHERMS AND ECTOTHERMS Students may be familiar with the concepts of warm-blooded and cold-blooded animals. Tell students that the correct terms are *endotherms* and *ectotherms,* respectively. Endotherms can generate their own heat, and ectotherms rely on environmental sources of heat. Ask: **Does the body temperature of an endotherm change with changes in the external environment? Explain.** *(No, because of homeostasis, endotherms maintain a stable body temperature.)* Tell students that the blood of ectotherms is not really cold. Rather, these animals must use behavior to help maintain homeostasis.

Elaborate

Build Inquiry 

L1 EVAPORATION AS A COOLING PROCESS

Materials small amount of hand sanitizer

Time 5 minutes

Have students pour a small amount of hand sanitizer on their hands and rub their hands together as if they were washing them. Have students observe how their hands feel at that moment and then again a few seconds later. After one minute, repeat the activity, with students waving their hands.

Ask: **How did the sanitizer make your hands feel?** *(First wet and then cool)* **What happened to the liquid?** *(It evaporated.)* **What does a liquid need to evaporate?** *(Heat)* **Where did the heat come from?** *(The body heat in the hands)* **How did the sanitizer make your hands feel after you waved them in the air? Explain.** *(Even cooler, because waving made the sanitizer evaporate faster, and that used more heat from the hands)* **How does this activity illustrate homeostasis?** *(When the body gets too warm, it perspires. As sweat evaporates, it cools the body.)*

LESSON 6.3

Explain

Lead a Discussion

STRESS AND HOMEOSTASIS Engage students in a discussion about stress and how the body responds to it. Ask: **What is stress?** *(Stress is the reaction of the body to possibly threatening, challenging, or uncomfortable events.)* **Is stress always negative? Explain.** *(Stress is a normal physical response to upsetting, challenging, or threatening events. Some stress is helpful. It prepares the body to meet a particular situation with focus, heightened alertness, strength, and stamina.)* **When is stress unhealthy?** *(When it lasts too long and disrupts homeostasis)* Tell students that the events that bring on stress are called *stressors.* Stressors can be anything from taking a test to outright danger. The body's response to stressors involves the nervous and endocrine systems. The hormones adrenaline and cortisol are released into the bloodstream. Ask: **What do you think are some of the effects of these hormones?** *(Increased breathing rate, heart rate, blood pressure, energy output; dilation of blood vessels and pupils; onset of sweating)* **What happens when the stressor is removed?** *(When the stressor is removed, the body returns to its normal condition.)*

Teach With Visuals

Tell students to look at **Figure 4.** Have students replace the four photos in the figure with descriptions of four activities that rank from least stressful (1) to most stressful (4) in their lives. Encourage students to draw pictures to supplement their descriptions. **Is an event that is very stressful to you necessarily the same to your classmates? Explain.** *(No, people perceive stress differently.)*

Elaborate

Teacher Demo

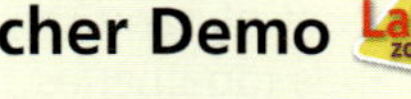

L1 GETTING STRESSED

Materials none

Time 5 minutes

Suddenly announce to students that they should take out a sheet of paper, put their name on it, and number from 1 to 20. Act as if you are about to administer a quiz.

Ask: **What was your immediate reaction?** *(Samples: My heart started to beat faster; my breathing rate increased; I started to sweat.)* **Are those reactions voluntary or involuntary?** *(Involuntary)* **What accounts for those reactions?** *(The body senses a challenging or threatening situation and the body systems respond to the stress.)*

2

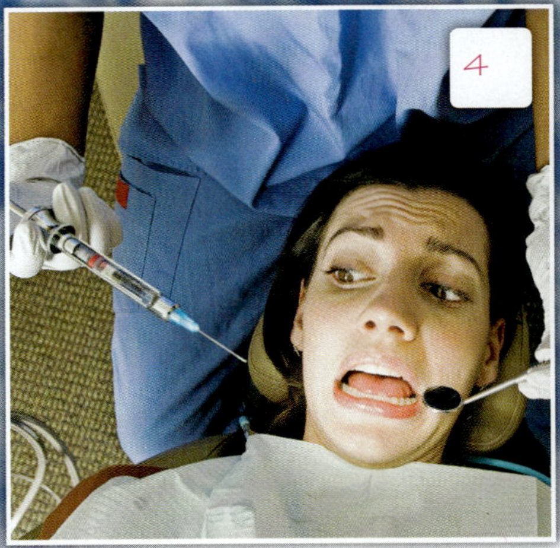

4

Sample rankings are shown.

FIGURE 4

Stressed Out?

Different people view stress differently.

Use the photos to complete these tasks.

1. **Interpret Photos** Using numbers 1–4, rank these activities, in your opinion, from least stressful (1) to most stressful (4).
2. **CHALLENGE** Which activity could be very stressful or not stressful at all? Explain.

Sample: Rock climbing could be stressful if you're afraid of heights, but not stressful if you like to do it.

Responding to Stress Imagine you are out for a walk. Suddenly, a big, snarling dog jumps in front of you! In a process called the fight-or-flight response, your endocrine system instantly pumps the hormone adrenaline into your bloodstream, making your heart beat faster and your breathing rate increase. These changes prepare your body to face or run away from the threat. You may feel these same changes when you start a race or get ready to make a speech in class. In general, **stress** is the reaction of your body to possibly threatening, challenging, or uncomfortable events.

Some stress is normal and healthy. If stress is over quickly, your body returns to a healthier condition. However, ongoing stress can disrupt homeostasis. For example, it can disrupt your body's ability to fight disease. It also can cause depression, headaches, digestion problems, heart problems, and other health issues. Managing stress is an important part of a healthy lifestyle.

apply it!

Communicate A soccer game, a music recital, a class presentation, and many other events can cause stress. Think of an event from your life when you felt stress. Describe how your body responded during the event, and then after the event or when the stress went away.

Sample: I had to give a speech in English class. My heart beat so hard and fast I could feel it. I started to sweat and had trouble breathing, and my hands were shaking. When I sat down, the shaking soon stopped and my breathing and heartbeat felt normal again.

Fighting Disease When your body systems are in balance, you are healthy. However, bacteria and viruses that cause disease can disrupt homeostasis and make you sick. Think about the last time you had a cold or influenza (the flu). You may have had a fever and less energy. You also may have slept more than usual. Over a few days, your immune system probably fought off the disease.

The immune system includes specialized cells that can attack and destroy viruses. When you are sick, these cells temporarily increase in number. Fighting infection sometimes causes your body temperature to go up. It also uses extra energy. As you get well, your fever goes away and your energy comes back. If you are sick for more than a few days, you may need medical attention to help your body fight the infection and become healthy again.

Differentiated Instruction

L1 Types of Stress Have students create a Venn diagram to compare and contrast negative stress and normal/helpful stress.

L3 The Body's Defenses The body has two main defenses against disease: the nonspecific defense system and the specific defense system, or the immune system. Have students find out about both systems: their parts, roles, and actions. Tell students to include the following terms in their report: *infectious disease, noninfectious disease, pathogen, symptom, inflammation, antigen, antibody, active immunity, passive immunity, vaccine.*

Explain

21st Century Learning

CRITICAL THINKING Tell students that the response to stress is sometimes called the "fight-or-flight reaction." Ask: **Why is this name appropriate?** *(The physical changes in the body prepare a person to either engage the stressor or flee from it—in other words, to fight or take flight.)*

Lead a Discussion

THE IMMUNE SYSTEM Explain to students that the immune system is a group of cells, tissues, and organs whose job is to defend the body against disease-causing agents. Ask: **What happens to homeostasis when bacteria and viruses invade the body?** *(It is disrupted.)* **What system goes into action when a person is sick?** *(The immune system)* **How does stress affect the immune system?** *(It can weaken the immune system.)*

21st Century Learning

CREATIVITY Discuss with students that negative stress that lasts for a long time can be unhealthy. Have students work in small groups to do research about stress and healthful ways to manage it. Ask them to prepare and present a poster and a skit or role-play of a stressful situation and healthful ways to manage the stress. Examples include declining grades, social conflicts, and increased responsibilities. Ask: **What are some signs of stress?** *(Samples: feeling irritable, frustrated, restless; being unable to concentrate, easily confused, and forgetful; feeling listless, upset, indecisive; experiencing muscle tension, low back pain, nervous tics, indigestion, sweaty palms, shortness of breath, sleeplessness, heart palpitations)* **What are some ways to manage stress?** *(Develop strong relationships, resilience, problem solving ability, inner calmness, optimism, good health habits, a sense of humor, organizational skills, self-forgiveness, etc.)*

Apply It!

L1 Before beginning the activity, review that stress is the body's reaction to possibly threatening, challenging, or uncomfortable events. Tell students to concentrate on those descriptors as they select an experience and to be as descriptive as possible when they explain their response.

Communicate Remind students that when they communicate, they share ideas and information. Tell students that they should be clear, concise, and thorough in their writing.

LESSON 6.3

Explain

Teach With Visuals

Tell students to look at **Figure 5.** Review the meanings of the terms *homeostasis* and *stress* and the relationship between the two. Ask: **Is this situation an example of stress? Explain.** *(Yes, running this race is a challenging event because the runner wants to push herself to excel.)* **What is the effect of the stress on homeostasis?** *(Homeostasis is maintained as the body systems react in their specific ways. The runner performs at her peak, and when the stress is over, the body systems return to their normal conditions.)*

Elaborate

Apply the TEKS

Direct students' attention to the description of the body systems in action as the runner jumps over the hurdles. Work with students to identify all the activities taking place in the runner's body that allow her to perform at her best. Remind students that the systems that perform those activities work together to ensure proper functioning of the body. Emphasize the fact that although a function seems to be the job of one specific system, that system does not work alone. Systems work together for any specific function. Be aware that many students may identify only one or two systems for each description, so encourage students to analyze every aspect of the function being described and identify the corresponding system. Ask: **What are the obvious systems at work when the runner lifts her legs over the hurdle?** *(Muscular and skeletal systems)* **What other systems must be at work to enable the runner to do this?** *(Nervous, circulatory, respiratory, and digestive systems)* **How do those systems function in her action?** *(The nervous system directs the muscles that allow the runner to lift her legs. The circulatory system absorbs nutrients from the digestive system and oxygen from the respiratory system and delivers them to the cells for energy. The circulatory and respiratory systems then work together to get rid of carbon dioxide and other cell wastes.)*

Texas Essential Knowledge and Skills

12B Identify the main functions of the systems of the human organism, including the circulatory, respiratory, skeletal, muscular, digestive, excretory, reproductive, integumentary, nervous, and endocrine systems.

Systems in Action

APPLY THE TEKS 12B

How does your body work?

FIGURE 5

The body systems of this runner work together as she pushes herself to excel.

Apply Concepts Read the descriptions of functions happening in the runner's body. Then identify the main systems involved.

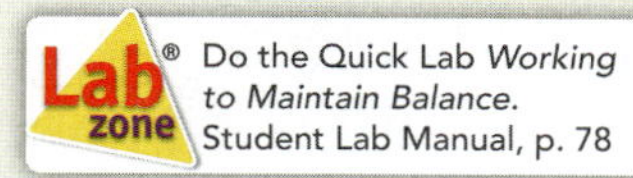

Do the Quick Lab *Working to Maintain Balance.* Student Lab Manual, p. 78

The runner's breathing rate and heart rate increase, supplying more oxygen to her muscle cells.

Respiratory and circulatory systems

The runner's legs lift her off the ground and over the hurdle.

Muscular and skeletal systems, with nervous system

Assess Your Understanding

TEKS 12B, 13A, 13B

1a. Define What is homeostasis?

The condition in which an organism's internal environment is kept stable in spite of changes in the outside environment

b. Describe Give four examples of conditions in your body that are related to maintaining homeostasis.

Sample: temperature, energy (food), water, and physical balance

c. Relate Cause and Effect Give an example of how the body might respond to internal stimuli caused by stress.

Sample: A person who is nervous about singing for an audience may sweat or shake.

d. Identify How does your body work? Use what you have learned about how your body systems function to write your answer.

Sample: The systems in my body work together, helping me get food, oxygen, and energy, and do things like move and respond to stress.

got it?

○ **I get it!** Now I know that maintaining homeostasis depends on my body systems working together.

○ I need extra help with See TE note.

235

Lab Resource: Quick Lab 

L2 WORKING TO MAINTAIN BALANCE Students will monitor a partner's heart rate to determine the effect of relaxed breathing. This Quick Lab can be found in the Student Lab Manual, p. 78, and online.

Evaluate

Assess Your Understanding

After students answer the questions, have them evaluate their understanding by completing the appropriate sentence.

RTI Response to Intervention

1a. If students need help defining *homeostasis,* **then** have them recall the meanings of the word parts and reread the definition of the term.

b. If students have trouble identifying conditions in the body that are related to maintaining homeostasis, **then** review the concept that cell functions can take place only at specific temperatures, within specific ranges of water and nutrient content, and with physical balance.

c. If students cannot provide an example of how the body might respond to internal stimuli caused by stress, **then** help them imagine a stressful situation and the responses of the body to it.

d. If students have difficulty explaining how the body works, **then** have them review the example in **Figure 5.**

Differentiated Instruction

L1 Systems in Action Have students work in pairs to develop another scenario in which all the body systems work together to perform a task. Remind students that the task need not be physically strenuous. Suggest that students consider activities such as doing homework, taking a walk, watching a scary movie, and even sleeping. Have them create a visual similar to that in **Figure 5.**

L3 Peak Performance Just how hard can the body be pushed to perform? Have interested students find out current world records for swimming events, track-and-field events, and marathons. Have them share this information with the class and lead a discussion of how the body must work to achieve such performances.

Name ______________________ Date __________ Class __________

Assess Your Understanding

Homeostasis

How Does Your Body Stay in Balance?

1a. DEFINE What is homeostasis?

____

____

____

b. DESCRIBE Give four examples of conditions in your body that are related to maintaining homeostasis.

____

____

c. RELATE CAUSE AND EFFECT Give an example of how the body might respond to internal stimuli caused by stress.

____

____

d. IDENTIFY How does your body work? Use what you have learned about how your body systems function to write your answer.

____

____

____

____

got it?

○ **I get it!** Now I know that maintaining homeostasis depends on ______________________

____

○ **I need extra help with** ______________________

____

Place the outside corner, the corner away from the dotted line, in the corner of your copy machine to copy onto letter-size paper.

Place the outside corner, the corner away from the dotted line, in the corner of your copy machine to copy onto letter-size paper.

Name ______________________ Date __________ Class __________

Enrich

Homeostasis

It is very important that your blood pH remains constant. Read the passage and study the table below. Then use a separate sheet of paper to answer the questions that follow.

Blood and pH

In blood, pH is maintained by the interaction of carbonic acid (H_2CO_3), bicarbonate ion (HCO_3^-), hydrogen ion (H^+), and carbon dioxide (CO_2). The table below describes what happens when blood pH becomes too low or too high.

What Happens When Blood pH is Low	What Happens When Blood pH is High
1. Hydrogen ions react with bicarbonate ions to produce carbonic acid. $H^+ + HCO_3^- \rightarrow H_2CO_3$	**1.** The breathing rate decreases, so less carbon dioxide is exhaled from the body.
2. Carbonic acid decomposes to produce carbon dioxide and water. $H_2CO_3 \rightarrow CO_2 + H_2O$	**2.** The concentration of carbon dioxide in the blood increases.
3. The concentration of carbon dioxide in the blood increases.	**3.** Carbon dioxide and water in the blood react to produce carbonic acid, $CO_2 + H_2O \rightarrow H_2CO_3$
4. Increased carbon dioxide in the blood causes the breathing rate to increase. More carbon dioxide is exhaled from the body.	**4.** Carbonic acid dissolves to produce hydrogen ions and bicarbonate ions. $H_2CO_3 \rightarrow H^+ + HCO_3^-$

1. How do the reactions on the left side of the table compare with the reactions on the right side of the table?

2. Which reaction removes hydrogen ions from the blood? Which reaction adds hydrogen ions?

3. When molecules in food react in cells to produce energy, carbon dioxide forms as a waste product. The carbon dioxide then enters the blood. How could this addition of carbon dioxide lead to a decrease in blood pH?

4. The reaction of carbon dioxide and water to produce carbonic acid occurs very slowly unless a specific enzyme is present. How would a person who lacked this enzyme be affected?

Name ______________________ Date ____________ Class ____________

Lesson Quiz

Homeostasis

Write the letter of the correct answer on the line at the left.

1. ___ The condition in which the body's internal environment is kept stable is called

A homeopathy

B homeostasis

C metabolism

D equilibrium

2. ___ The reaction of your body to possible threatening, challenging, or uncomfortable events is called

A homeostasis

B hunger

C metabolism

D stress

3. ___ What is the body's response to the stimulus of getting overheated?

A sweating and thirst

B shivering and hunger

C sweating and shivering

D shivering and thirst

4. ___ Which of the following statements about homeostasis is **NOT** true?

A Maintaining homeostasis requires that all of the body systems work together.

B Long periods of stress can disrupt homeostasis.

C Body temperature is a factor of homeostasis.

D Only the nervous and endocrine systems are involved in maintaining homeostasis.

If the statement is true, write *true*. If the statement is false, change the underlined word or words to make the statement true.

5. ____________ The nose helps the body keep its balance.

6. ____________ The endocrine system includes specialized cells that help fight bacteria and viruses.

7. ____________ High levels and long periods of stress can increase a person's risk for many diseases.

8. ____________ Thirst is the body's response to the need for energy.

9. ____________ Regardless of external conditions or activities, the body's internal temperature is almost exactly 37°C.

10. ____________ The condensation of sweat from body surfaces cools the body.

Place the outside corner, the corner away from the dotted line, in the corner of your copy machine to copy onto letter-size paper.

Homeostasis

Answer Key

Review and Reinforce

Find the worksheet in the Student Workbook.

1. The given statement should be in the top box. Right box: Endocrine system releases hormones that make a person perspire. Bottom box: Evaporation of sweat cools the body. Left box: Endocrine system stops releasing hormones and person stops perspiring.
2. By controlling the amounts of nutrients, oxygen, and water that enter the cell and the carbon dioxide and other wastes that leave the cell, the cell membrane keeps the cell in balance, or maintain homeostasis.
3. Stress can disrupt homeostasis. Keeping the body's internal environment stable is a sign of health. By managing stress, a person can eliminate or minimize the negative effects of stress on homeostasis.
4. Some stress is healthy because it better prepares the body's systems to work together to achieve high or peak performance.
5. temperature, energy, water, physical balance
6. the reaction of the body to threatening, challenging, or uncomfortable events
7. Low levels or short periods of stress are normal and healthy. But high levels and long periods of stress can disrupt the body's balance, or homeostasis, by disrupting the immune system and/or causing depression, headaches, digestive problems, heart problems, and other health issues. These conditions seriously affect a person's overall health.

Enrich

1. The reactions on the left side of the table are the reverse of those on the right side.
2. The reaction $H^+ + HCO_3^- \rightarrow H_2CO_3$ removes hydrogen ions. The reaction $H_2CO_3 \rightarrow H^+ + HCO_3^-$ adds hydrogen ions.
3. The carbon dioxide could react with water to produce carbonic acid which could decompose to produce bicarbonate and hydrogen ions which would cause a decrease in pH.
4. Without the enzyme, less carbonic acid and fewer hydrogen ions would be produced so that blood pH would be high.

Lesson Quiz

1. B
2. D
3. A
4. D
5. ears
6. immune
7. true
8. Hunger
9. true
10. evaporation

The Skeletal System

How does your body work?

LESSON PACING:
2–3 periods or 1–1½ blocks

Lesson Vocabulary

- vertebrae
- ligament
- compact bone
- spongy bone
- marrow
- cartilage
- osteoporosis

Content Refresher

Building and Rebuilding Bone Bones do not become inactive after the body stops growing. Bone is constantly being remodeled. Cells called osteoblasts form new bone tissue; cells called osteoclasts continually break down some bone tissue.

Weight-bearing activities, such as playing soccer or tennis, promote the formation of new bone tissue. The ability to build stronger, denser bones in response to exercise may help prevent osteoporosis. A good calcium intake throughout life is needed to form the mineral salts that give bone tissue its strength.

The ability of bone cells to respond to environmental stress is important in wellness and illness. When a person fractures a bone, all of the osteoblasts in the region of the break begin to form new bone tissue.

Orthopedic surgeons can apply metal pieces of apparatus to build a frame that holds the broken ends of bones together. Stress on the bone ends caused by the apparatus speeds up the osteoblasts' task of building new tissue to heal the fractures.

Lesson Objectives	TEKS	ELPS
Identify the main functions of the skeletal system.	12B	4.F.3
Explain the role that joints play in the body.	12B, 12C	
Describe the characteristics of bones and how to keep bones strong and healthy.	12B, 12C	

Texas Essential Knowledge and Skills

12B Identify the main functions of the systems of the human organism, including the circulatory, respiratory, skeletal, muscular, digestive, excretory, reproductive, integumentary, nervous, and endocrine systems.
12C Recognize levels of organization in plants and animals, including cells, tissues, organs, organ systems, and organisms.

English Language Proficiency Standards

ELPS Reading 4.F.3 Use visual and contextual support to develop vocabulary needed to comprehend increasingly challenging language.

DIFFERENTIATED INSTRUCTION KEY
L1 Struggling Students or Special Needs
L2 On-Level Students L3 Advanced Students

LESSON PLANNER 6.4

Investigations and Activities

My Planet Diary, **Student Edition,** p. 236
Inquiry: Inquiry Warm-Up, Hard as a Rock?, **PearsonTexas.com**
Teach Key Concepts, **Teacher's Edition,** p. 237
Support the TEKS, Functions of the Skeleton, **Teacher's Edition,** p. 238
Lead a Discussion, Production and Storage, **Teacher's Edition,** p. 238
Inquiry: Quick Lab, The Skeleton, **Lab Manual, p. 79**

Teach Key Concepts, **Teacher's Edition,** p. 239
Teach With Visuals, **Teacher's Edition,** p. 239
Differentiated Instruction, **Teacher's Edition,** p. 239
Apply It!, **Student Edition,** p. 240
Inquiry: Quick Lab, Observing Joints, **Lab Manual, p. 80**

Teach Key Concepts, **Teacher's Edition,** p. 241
Inquiry: Build Inquiry, Observing Bone Structure, **Teacher's Edition, p. 241**
Address Misconceptions, Living Bones, **Teacher's Edition,** p. 242
Lead a Discussion, Characteristics of Bones, **Teacher's Edition,** p. 242
21st Century Learning, Information Literacy, **Teacher's Edition,** p. 242
Differentiated Instruction, **Teacher's Edition,** p. 243
Inquiry: Teacher Demo, Flexible Cartilage, **Teacher's Edition, p. 243**
Inquiry: Quick Lab, Soft Bones?, **PearsonTexas.com**

TEKS Review

Apply the TEKS, Doing Work, **Student Edition,** p. 254

TEKS Practice, **Student Edition,** p. 256

TEKS Practice: Chapter and Cumulative Review, **Student Edition,** p. 260

Lesson 6.4, **TEKS Preparation and Study Guide Workbook,** p. 60

SHORT ON TIME? To do this lesson in approximately half the time, do the Activate Prior Knowledge activity. A discussion of the Key Concepts will familiarize students with the lesson content. Have students do the Quick Labs. The rest of the lesson can be completed by students independently.

These editable worksheets are available on **PearsonTexas.com.**
Print versions can be found in the **TEKS Preparation and Study Guide Workbook.**

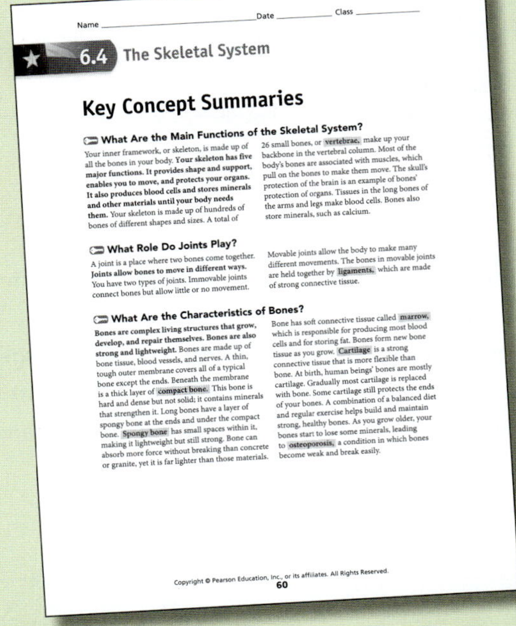

Name ______________ Date ________ Class ________

6.4 The Skeletal System

Key Concept Summaries

What Are the Main Functions of the Skeletal System?

Your inner framework, or skeleton, is made up of all the bones in your body. **Your skeleton has five major functions. It provides shape and support, enables you to move, and protects your organs. It also produces blood cells and stores minerals and other materials until your body needs them.** Your skeleton is made up of hundreds of bones of different shapes and sizes. A total of 26 small bones, or **vertebrae,** make up your backbone in the vertebral column. Most of the body's bones are associated with muscles, which pull on the bones to make them move. The skull's protection of the brain is an example of bones' protection of organs. Tissues in the long bones of the arms and legs make blood cells. Bones also store minerals, such as calcium.

What Role Do Joints Play?

A joint is a place where two bones come together. **Joints allow bones to move in different ways.** You have two types of joints. Immovable joints connect bones but allow little or no movement. Movable joints allow the body to make many different movements. The bones in movable joints are held together by **ligaments,** which are made of strong connective tissue.

What Are the Characteristics of Bones?

Bones are complex living structures that grow, develop, and repair themselves. Bones are also strong and lightweight. Bones are made up of bone tissue, blood vessels, and nerves. A thin, tough outer membrane covers all of a typical bone except the ends. Beneath the membrane is a thick layer of **compact bone.** This bone is hard and dense but not solid; it contains minerals that strengthen it. Long bones have a layer of spongy bone at the ends and under the compact bone. **Spongy bone** has small spaces within it, making it lightweight but still strong. Bone can absorb more force without breaking than concrete or granite, yet it is far lighter than those materials. Bone has soft connective tissue called **marrow,** which is responsible for producing most blood cells and for storing fat. Bones form new bone tissue as you grow. **Cartilage** is a strong connective tissue that is more flexible than bone. At birth, human beings' bones are mostly cartilage. Gradually most cartilage is replaced with bone. Some cartilage still protects the ends of your bones. A combination of a balanced diet and regular exercise helps build and maintain strong, healthy bones. As you grow older, your bones start to lose some minerals, leading to **osteoporosis,** a condition in which bones become weak and break easily.

60

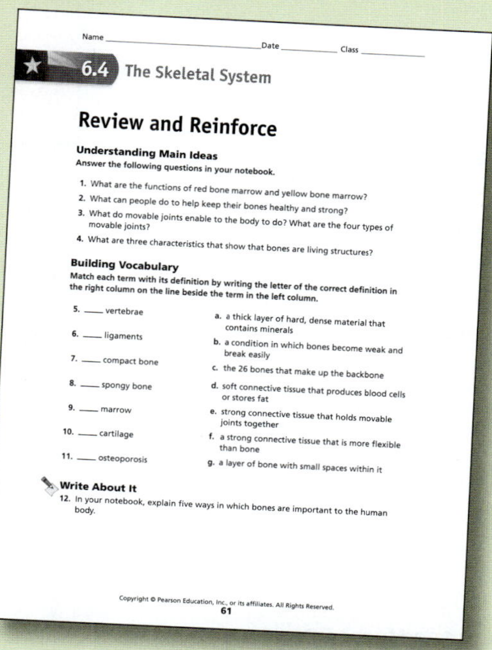

Name ______________ Date ________ Class ________

6.4 The Skeletal System

Review and Reinforce

Understanding Main Ideas
Answer the following questions in your notebook.

1. What are the functions of red bone marrow and yellow bone marrow?
2. What can people do to help keep their bones healthy and strong?
3. What do movable joints enable to the body to do? What are the four types of movable joints?
4. What are three characteristics that show that bones are living structures?

Building Vocabulary
Match each term with its definition by writing the letter of the correct definition in the right column on the line beside the term in the left column.

5. ____ vertebrae
6. ____ ligaments
7. ____ compact bone
8. ____ spongy bone
9. ____ marrow
10. ____ cartilage
11. ____ osteoporosis

a. a thick layer of hard, dense material that contains minerals
b. a condition in which bones become weak and break easily
c. the 26 bones that make up the backbone
d. soft connective tissue that produces blood cells or stores fat
e. strong connective tissue that holds movable joints together
f. a strong connective tissue that is more flexible than bone
g. a layer of bone with small spaces within it

Write About It

12. In your notebook, explain five ways in which bones are important to the human body.

61

LESSON 6.4

Lexile Measure = 830L

The Skeletal System

Establish Learning Objectives

After this lesson, students will be able to:

- Identify the main functions of the skeletal system.
- Explain the role that joints play in the body.
- Describe the characteristics of bones and how to keep bones strong and healthy.

Engage

Activate Prior Knowledge

MY PLANET DIARY Read *Know Your Bones!* with the class. Point out that the hands and feet contain more than 50 percent of the bones in the body. Encourage students to think about the number, sizes, and shapes of bones in various parts of their bodies. Ask: **Which parts of your body might have long bones?** *(Leg, arm, back)* **Which parts of the body are likely spots for breaks or fractures?** *(Arms, wrists, legs, ankles)*

Explore

Lab Resource: Inquiry Warm-Up 

L1 **HARD AS A ROCK?** Students will compare and contrast several characteristics of a rock and an animal bone. This Inquiry Warm-Up can be found online.

The Skeletal System

- What Are the Main Functions of the Skeletal System? TEKS 12B
- What Role Do Joints Play? TEKS 12B, 12C
- What Are the Characteristics of Bones? TEKS 12B, 12C

my planet Diary

FUN FACTS

Know Your Bones!

Here are some fascinating facts you may not know about your bones.

- You have the same number of bones in your neck as a giraffe. However, a single bone in the neck of a giraffe can be as long as 25 centimeters.
- You have 27 bones in each hand and 26 bones in each foot. They account for 106 of the 206 bones in your body.
- You do not have a funny bone. You have a sensitive spot on your elbow where a nerve passes close to the skin. If you hit this spot, the area feels funny.
- No one is truly "double-jointed." People who are able to twist in weird directions have very flexible joints.

Communicate Discuss the question with a partner. Then write your answer below.

Why do you think it is helpful for your hand to have 27 bones?

Sample: It is helpful because my hand is able to bend in many ways.

Lab zone® Do the Inquiry Warm-Up *Hard as a Rock?* Find the lab online.

SUPPORT ALL READERS

Lexile Measure = 830L **Lexile Word Count = 1386**

Prior Exposure to Content: Many students may have misconceptions on this topic

Academic Vocabulary: *classify, summarize*

Science Vocabulary: *vertebrae, ligament, marrow, cartilage*

Concept Level: Generally appropriate for most students in this grade

Preteach With: My Planet Diary "Know Your Bones!" and Figure 2 activity

Vocabulary
- vertebrae
- ligament
- compact bone
- spongy bone
- marrow
- cartilage
- osteoporosis

Skills

Reading: Summarize
Inquiry: Classify

What Are the Main Functions of the Skeletal System?

TEKS 12B In this section, you'll identify the important roles of the skeletal system.

ELPS 4.F.3
With a partner, read or listen to the vocabulary terms, listed on page 237. Look through the lesson to find a visual—photo or diagram—that shows each term.

If you have ever visited a construction site, you have seen workers assemble steel pieces into a rigid frame for a building. Once the building is finished, this framework is invisible.

Like a building, you have an inner framework. Your framework, or skeleton, is made up of all the bones in your body. Just as a building would fall without its frame, you would collapse without your skeleton. **Your skeleton has five major functions. It provides shape and support, enables you to move, and protects your organs. It also produces blood cells and stores minerals and other materials until your body needs them.**

Shape and Support Your skeleton shapes and supports your body. It is made up of about 206 bones of different shapes and sizes. Your backbone, or vertebral column, is the center of your skeleton. A total of 26 small bones, or **vertebrae** (VUR tuh bray) (singular *vertebra*), make up your backbone. **Figure 1** shows how vertebrae connect to form the vertebral column.

FIGURE 1

The Vertebral Column

Just like a flexible necklace of beads, your vertebrae move against each other, allowing you to bend and twist.

Use the photo to answer the questions about your vertebrae.

1. **Interpret Photos** Which body parts does the vertebral column support?
Sample: The vertebral column supports the weight of the head and arms.

2. **CHALLENGE** What is the advantage of having large vertebrae at the base of the vertebral column?
Sample: The large vertebrae can support the weight of the upper body.

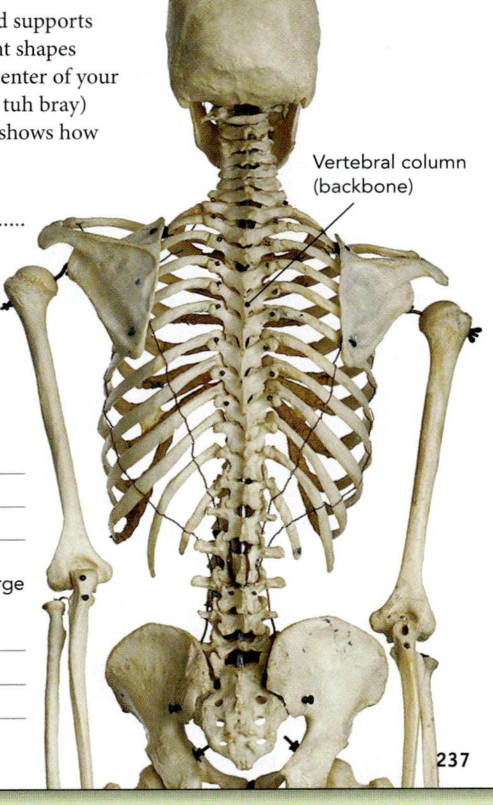

LESSON 6.4

Explain

Introduce Vocabulary

To help students understand the term *osteoporosis,* write the word on the board. Underline *osteo* and circle *porosis*. Explain that *osteo* is derived from the Greek word for "bone" and *porosis* from the Greek word for "pore" or "passage." Have volunteers explain how these word parts are related to the term's meaning.

Teach Key Concepts

Explain to students that the human skeleton's main functions are providing shape and support, allowing movement, protecting organs, producing blood cells, and storing minerals and other materials. Ask: **What would your body be like if you did not have a skeleton?** *(Samples: It would not have a shape; you would not be able to stand; you would not be able to move.)* **Why is the vertebral column so important to your skeleton?** *(It is the "center" of the skeleton and supports the upper body.)*

Lead a Discussion

IMPRESSIONS OF BONES Ask students to feel the bones beneath the skin in one of their hands. Then have them draw a picture of what they think the bones look like. Have students brainstorm a list of facts about bones or characteristics of bones on the board. Revisit items on this list as you teach the section. Ask: **In what parts of your body are bones most obvious to you?** *(Samples: Fingers, toes, hands, feet, legs, arms, shoulders)*

English Language Proficiency Standards

ELPS Reading 4.F.3

Have students find visuals in the lesson for each vocabulary term.

Beginning Have beginners listen as you read aloud the ELPS activity on student book page 237. One at a time, display and read aloud the vocabulary terms. Together, look through the lesson for visuals. Have partners record each term and draw a corresponding image.

Intermediate Together, read the ELPS activity. Have partners page through the lesson looking for photos or diagrams of each term. Have pairs share their findings.

Advanced As partners complete the ELPS activity, have them tell what each visual shows about the term.

Advanced High As partners complete the ELPS activity, have them tell what questions they have about each term. Discuss answers.

Texas Essential Knowledge and Skills

12B Identify the main functions of the systems of the human organism, including the circulatory, respiratory, skeletal, muscular, digestive, excretory, reproductive, integumentary, nervous, and endocrine systems.

LESSON 6.4

Explain

Support the TEKS

FUNCTIONS OF THE SKELETON Explain that a bone's structure relates to its function. Tell students to refer to **Figure 2** as they answer the following questions. Ask: **What is the primary function of the sternum?** *(Protection)* **What is the primary function of the carpals?** *(Movement)* **What are the primary functions of the femur?** *(Support and movement)* **What is the primary function of the ribs?** *(Protection)* **What are the primary functions of the vertebral column?** *(Protection, support, and movement)*

Lead a Discussion

PRODUCTION AND STORAGE Support, movement, and protection are fairly obvious functions of the skeleton. Students may be surprised to learn that the skeletal system has the additional functions of storage and production of substances that the body needs. Ask: **What mineral is stored in bones?** *(Calcium)* **What kinds of cells are produced in some bones?** *(Some kinds of blood cells)*

Elaborate

Lab Resource: Quick Lab 

L1 THE SKELETON Students will use models to determine whether hollow or solid bones are stronger. This Quick Lab can be found in the Student Lab Manual, p. 79, and online.

Evaluate

Assess Your Understanding

Have students evaluate their understanding by completing the appropriate sentence.

RTI Response to Intervention

If students have trouble listing functions of the skeleton, **then** remind them that the skeleton has five major functions, and call on volunteers to each name one function.

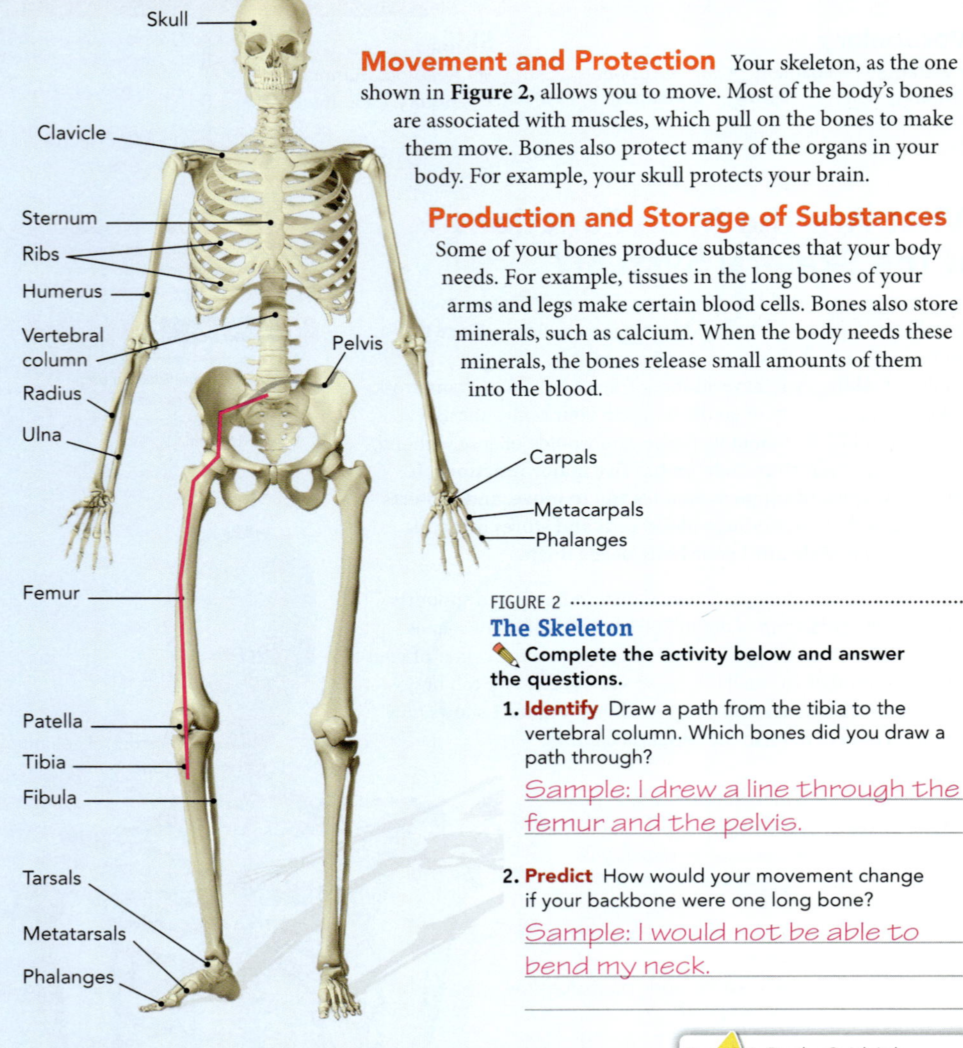

Movement and Protection Your skeleton, as the one shown in **Figure 2,** allows you to move. Most of the body's bones are associated with muscles, which pull on the bones to make them move. Bones also protect many of the organs in your body. For example, your skull protects your brain.

Production and Storage of Substances Some of your bones produce substances that your body needs. For example, tissues in the long bones of your arms and legs make certain blood cells. Bones also store minerals, such as calcium. When the body needs these minerals, the bones release small amounts of them into the blood.

FIGURE 2

The Skeleton

Complete the activity below and answer the questions.

1. **Identify** Draw a path from the tibia to the vertebral column. Which bones did you draw a path through?
   Sample: I drew a line through the femur and the pelvis.
2. **Predict** How would your movement change if your backbone were one long bone?
   Sample: I would not be able to bend my neck.

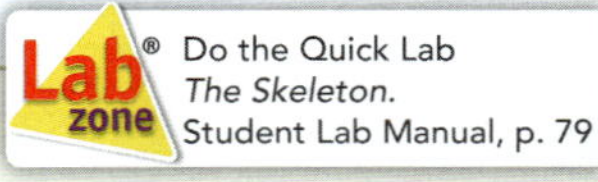
Do the Quick Lab *The Skeleton.* Student Lab Manual, p. 79

Assess Your Understanding

got it?

- I get it! Now I know that my skeleton *provides shape and support, enables movement, protects organs, makes blood cells, and stores materials.*
- I need extra help with *See TE note.*

What Role Do Joints Play?

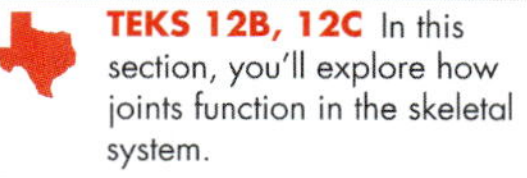

TEKS 12B, 12C In this section, you'll explore how joints function in the skeletal system.

If your leg had only one long bone, how would you get out of bed? Luckily, legs have many bones so they can move easily. A joint is a place where two bones come together. **Joints allow bones to move in different ways.** You have two kinds of joints: immovable and movable.

Immovable Joints Immovable joints connect bones but allow little or no movement. The bones of the skull are held together by immovable joints.

Movable Joints Most joints are movable. They allow the body to make many different movements such as those shown in **Figure 3.** The bones in movable joints are held together by **ligaments,** which are made of strong connective tissue.

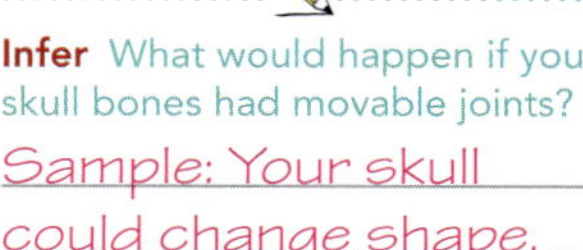

Infer What would happen if your skull bones had movable joints?

Sample: Your skull could change shape.

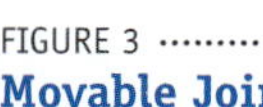

FIGURE 3

Movable Joints

Movable joints allow you to move in different ways.

Classify Write the name of another joint of each type on the line in each box.

Differentiated Instruction

L1 Names of Bones To help students become more familiar with various parts of the skeleton, have pairs use the image and the labels in **Figure 2** to practice identifying the names of bones. Encourage students to point to a particular bone in the illustration, point to the area on his or her own body where that bone is located, and name it.

L3 Backbones Invite students to do research to compare and contrast the backbones of several animals. Encourage students to include photographs or illustrations that show the skeletons of animals such as a cat, dog, mouse, and deer, in addition to a human skeleton. Students should label each image and include two or three important characteristics of each backbone.

Explain

Teach Key Concepts

Explain to students that joints allow bones to move. Ask students to imagine what it would be like to have a skeleton in which all the bones were solidly connected. *(Students will say that the couldn't do anything because they would be unable to move.)* Point out to students that there are several ways that bones can move, including *gliding, pivoting, turning,* and *rotating.* Ask: **Why are joints important?** *(They allow the body to make many different movements.)* **Where are joints located?** *(Where two bones come together)* Explain that joints hold bones together. Not all joints are movable; two bones can come together and not move—for example, in the skull.

Teach With Visuals

Tell students to look at **Figure 3.** Ask: **What types of movements does the ball-and-socket joint in the hip allow the leg to make?** *(Up and down, backward and forward, in a circle)* **What types of movements does the hinge joint in the knee allow the leg to make?** *(Forward and backward, or open and closed)* **What types of movements does the gliding joint in the wrist allow the hand to make?** *(Up and down, side to side)*

Texas Essential Knowledge and Skills

12B Identify the main functions of the systems of the human organism, including the circulatory, respiratory, skeletal, muscular, digestive, excretory, reproductive, integumentary, nervous, and endocrine systems.

12C Recognize levels of organization in plants and animals, including cells, tissues, organs, organ systems, and organisms.

LESSON 6.4

Elaborate

Apply It!

L1 Before beginning the activity, review the descriptions and illustrations of movable joints in **Figure 3.** Point out that similar kinds of movable joints can be found in some inanimate objects, such as hinges on a door and a ball-and-socket joint in a single-lever faucet handle.

Classify Remind students that when they classify, they group items that are alike in some way. Point out that in this chart students are asked to group types of joints.

Lab Resource: Quick Lab 

L1 **OBSERVING JOINTS** Students will manipulate the bones of a model of the human skeleton to explore how the joints move. This Quick Lab can be found in the Student Lab Manual, p. 80, and online.

Evaluate

Assess Your Understanding

After students answer the questions, have them evaluate their understanding by completing the appropriate sentence.

RTI Response to Intervention

1a. If students have trouble explaining the why both types of joints are necessary, **then** have them review the descriptions of how each joint works.

b. If students need help identifying the effect of a ball-and-socket joint on the knee, **then** suggest that students compare and contrast ball-and-socket joints with hinge joints.

Without movable joints, your body would be as stiff as a board.

1 Observe Perform each activity below. Write the type of joint you use.

Move your arm from the shoulder in a circle. Ball-and-socket

Move your wrist to wave. Gliding

Turn your head from side to side. Pivot

2 Classify In the chart, write the name of the type of joint each object has.

Object	Type of Joint
Book	Hinge
Sliding Door	Gliding
Steering Wheel	Pivot

3 Apply Concepts What type of joint do you have in your toes? Explain your answer.

Sample: I have hinge joints in my toes because they move forward and backward.

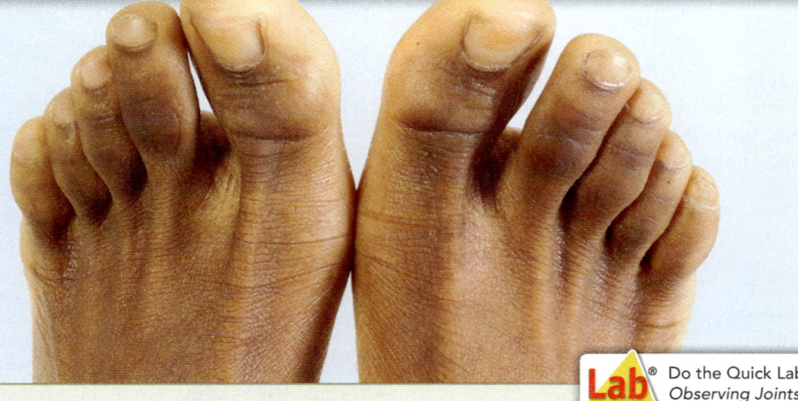

Lab zone® Do the Quick Lab *Observing Joints.* Student Lab Manual, p. 80

Assess Your Understanding

TEKS 12B, 12C

1a. Explain Why does the skeletal system contain both immovable and movable joints?

Immovable joints connect my bones; movable joints allow my bones to move.

b. Relate Cause and Effect How would your legs move if your knees were ball-and-socket joints?

Sample: My legs would swing in all directions.

got it?

O **I get it!** Now I know that joints allow bones to move in different ways.

O I need extra help with See TE note.

What Are the Characteristics of Bones?

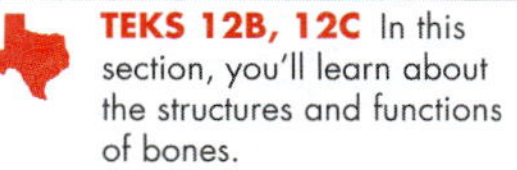

TEKS 12B, 12C In this section, you'll learn about the structures and functions of bones.

The word *skeleton* comes from the Greek words meaning "a dried body." This suggests that a skeleton is dead, but bones are not dead at all. **Bones are complex living structures that grow, develop, and repair themselves. Bones are also strong and lightweight.**

These organs are made up of bone tissue, blood vessels, and nerves. A thin, tough outer membrane covers all of a typical bone except the ends. Beneath the membrane is a thick layer of **compact bone,** which is hard and dense but not solid. Compact bone contains minerals that give bones strength. Small canals in the compact bone carry blood vessels and nerves from the bone's surface to its living cells.

Long bones, such as the femur in **Figure 4,** have a layer of spongy bone at the ends and under the compact bone. The small spaces within **spongy bone** make it lightweight but still strong. Bone also has two types of soft connective tissue called **marrow.** Red bone marrow fills the spaces in some of your spongy bone. It produces most of your blood cells. Yellow bone marrow is found in a space in the middle of the bone. It stores fat.

FIGURE 4

Bone Structure

Many tissues make up the femur, the body's longest bone.

Relate Text and Visuals Write notes to describe each part of the bone and what it does.

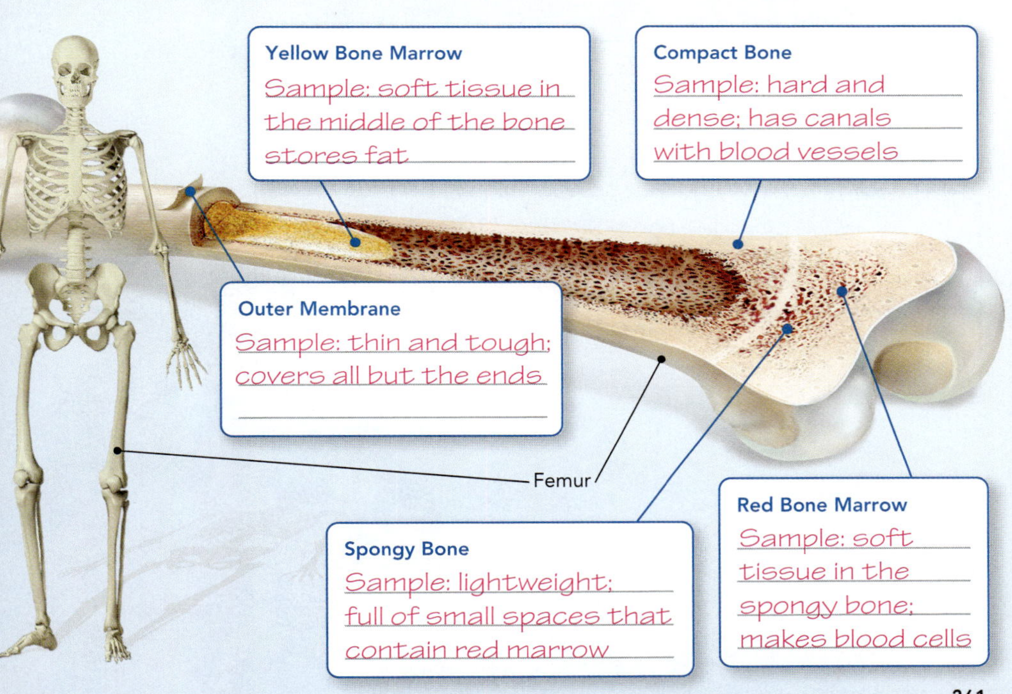

241

Differentiated Instruction

L1 Four Parts of Bones To help students understand and remember what parts of the bone are and do, invite pairs of students to stand opposite one another. One student can point to the yellow bone marrow in **Figure 4** as he or she explains its function, and the other student can repeat the process with red bone marrow. In the same way, the two students can distinguish compact and spongy bone.

L3 Red and Yellow Marrow Have students find out how marrow in the human body changes from birth to adulthood. Students can prepare an oral or written report on the reason why all human marrow is red until approximately age seven. At that time, fat tissue gradually replaces much of the red marrow.

LESSON 6.4

Explain

Teach Key Concepts 

Explain to students that bones are living structures that can grow, develop, and repair themselves. Ask: **What function of the skeleton gives you a clue that bones are alive?** *(Sample: Bones make blood cells.)* Have them refer to **Figure 4.** Ask: **What covers the bone except for the ends?** *(A thin, tough membrane called the* outer membrane*)* **What is just beneath this membrane?** *(Compact bone)* **How do blood and other materials get to the living cells inside the bone?** *(Blood vessels run through canals in the compact bone.)* **How does the structure of spongy bone relate to its function?** *(Spongy bone has many small spaces that make it lightweight but strong.)*

Elaborate

Build Inquiry Lab zone

L2 OBSERVING BONE STRUCTURE

Materials cleaned leg bones from thoroughly cooked chickens or turkeys, dissecting trays, gloves, hand lens. Before class, use a small kitchen saw to slice one leg bone crosswise and one lengthwise for each group. Place the bones in the trays.

Time 15 minutes setup; 15 minutes in class

Remind students that bones are organs that are composed of different types of tissue. Have groups use **Figure 4** to identify the various parts of the bones. Students should write down and sketch their observations.

CAUTION: *After students examine the chicken bones, make sure they wash their hands after removing the gloves.*

Ask: **Where are the largest concentrations of compact bone and spongy bone?** *(Compact bone in the shaft, spongy bone at the ends)*

Explain that short bones, such as those in the fingers, are mostly spongy bone. Ask students to infer why long bones have more compact bone. *(Long bones, such as those in the legs, must have harder, denser compact bone to support the body.)*

Texas Essential Knowledge and Skills

12B Identify the main functions of the systems of the human organism, including the circulatory, respiratory, skeletal, muscular, digestive, excretory, reproductive, integumentary, nervous, and endocrine systems.

12C Recognize levels of organization in plants and animals, including cells, tissues, organs, organ systems, and organisms.

LESSON 6.4

Explain

Address Misconceptions

L1 LIVING BONES Some students might think that bone is dead. Explain that some people may think of bones as lifeless because they have seen skeletons in movies or the skeletons of dead animals. Bone does contain nonliving material, such as calcium. However, bone is composed of living tissue. Point out that bones are living because they are made of cells, which compose all living things. Ask: **Would a bone bleed if it were cut?** *(Yes, because bones have blood vessels.)*

Lead a Discussion

CHARACTERISTICS OF BONES Help students summarize the structural characteristics of bones. Ask: **How does the structure of bones allow the bones to grow and become strong?** *(Sample: The outer membrane contains blood vessels that run through canals in the compact bone and deliver materials that the bone needs to function. Spongy bone has spaces within it that allow it to absorb large amounts of force, leading to the growth of new bone tissue.)* **How have your bones developed since you were a baby?** *(Most of a baby's bones are cartilage. As a person grows, most of that cartilage is replaced with bone.)*

Summarize Tell students that when they summarize, they briefly restate the main ideas of what they have read or heard in their own words.

21st Century Learning

INFORMATION LITERACY Have students research osteoporosis, a bone disease primarily affecting older women that can lead to an increased risk of bone breaks. Students should find information about symptoms of the disease, medical issues associated with the disease, and forms of treatment. Students can summarize their findings in short multimedia presentations.

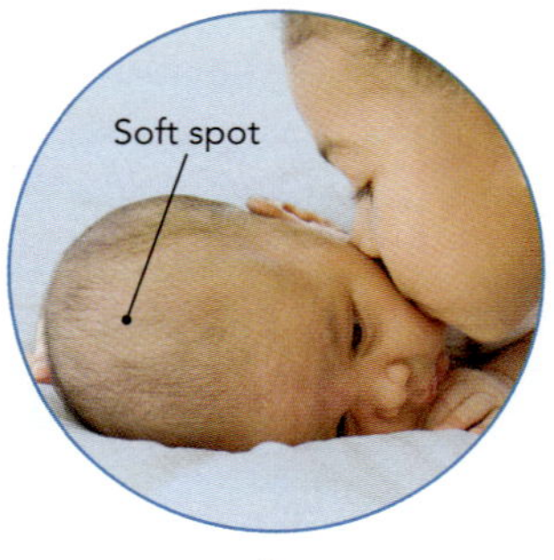

Bone Strength Bone is both strong and lightweight. Bones can absorb more force without breaking than concrete or granite rock can. Yet bones weigh much less than those materials. In fact, only about 20 percent of an average adult's body weight is bone. Bone feels as hard as a rock because it is made of tightly packed minerals—mainly phosphorus and calcium.

Bone Growth Because bones are alive, they form new bone tissue as you grow. Your bones are growing longer now, making you taller. Even after you are fully grown, bone tissue continues to form. For example, every time you play soccer or basketball, some of your bones absorb the force of your weight. They respond by making new bone tissue. New bone tissue also forms when a bone breaks.

Bone Development When you were born, most of your bones were cartilage. **Cartilage** is a strong connective tissue that is more flexible than bone. As you grew, most of that cartilage was replaced with bone. Some cartilage still protects the ends of your bones. You also have cartilage in your ears and at the tip of your nose.

Infer Infants are born with soft spots in their skull made out of cartilage. What do you think happens to soft spots over time?

Sample: Bone replaces the cartilage.

Summarize Write a summary about the characteristics of bones.

Bone Strength	Bones are strong and hard but not heavy.
Bone Growth	My bones grow, making me taller. Bone tissue forms when I run or break a bone.
Bone Development	Most of my cartilage was replaced with bone as I grew.

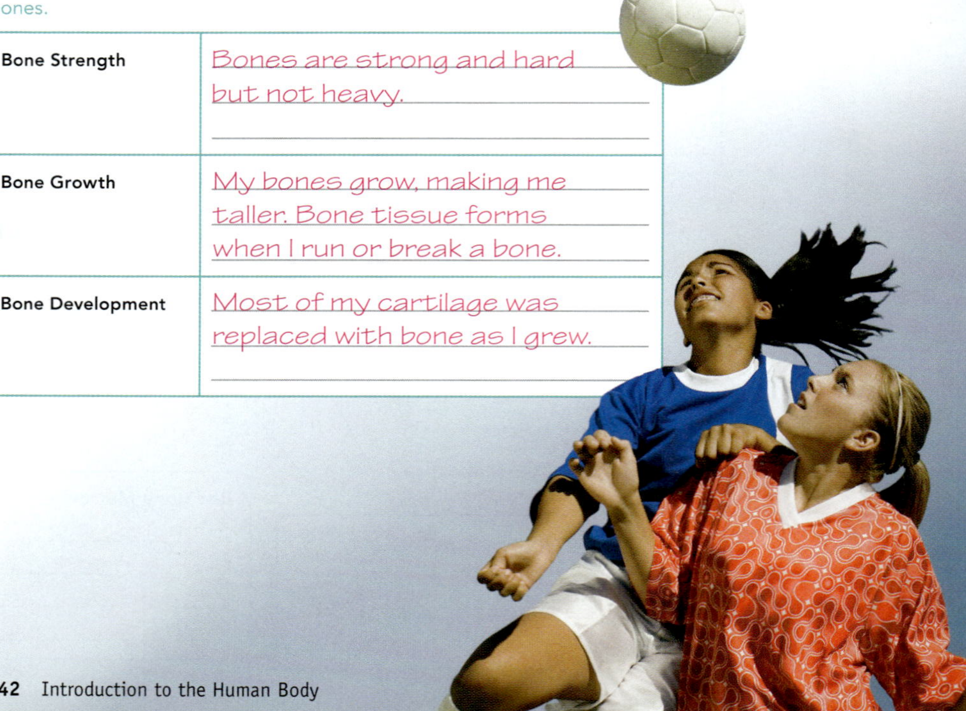

Healthy Bones A combination of a balanced diet and regular exercise are important for healthy bones. A balanced diet includes foods that contain enough calcium and phosphorus to keep your bones strong while they are growing. You should eat dairy products; meats; whole grains; and green, leafy vegetables.

Exercise helps build and maintain strong bones. During activities such as running and dancing, your bones support the weight of your entire body. These weight-bearing activities help your bones grow stronger. However, to prevent injury, always wear appropriate safety equipment when exercising.

As you age, your bones start to lose some minerals. This mineral loss can lead to **osteoporosis** (ahs tee oh puh ROH sis), a condition in which bones become weak and break easily. You can see how osteoporosis causes the spaces in a bone to become larger, reducing its density and strength in **Figure 5**.

FIGURE 5

Osteoporosis

Regular exercise and a diet rich in calcium with vitamin D may help prevent osteoporosis later in life.

Compare and Contrast The photos show two bones. Label the healthy bone and the bone with osteoporosis. Then explain your choices.

Sample: Healthy bone has more bone tissue than bone with osteoporosis.

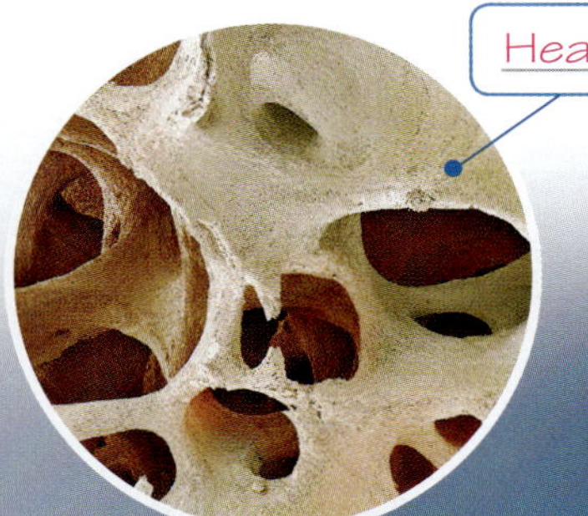

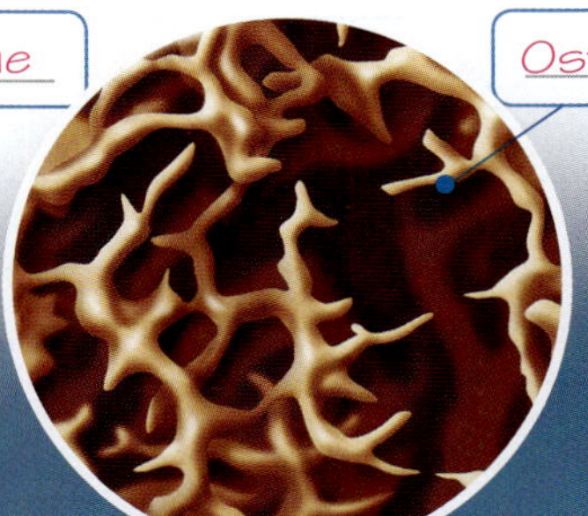

Do the Quick Lab *Soft Bones?* Find the lab online.

Assess Your Understanding

TEKS 12B, 12C

2a. Explain How do eating a balanced diet and exercising regularly benefit your skeletal system?

Diet provides minerals, and exercise keeps bone strong.

b. Recognize How do you know that bone is made up of living tissue?

I know it is alive because it grows and changes.

got it?

○ **I get it!** Now I know that my bones are living structures that grow, develop, and repair themselves. Bones are also strong and lightweight.

○ I need extra help with See TE note.

243

Differentiated Instruction

L1 Weight-Bearing vs. Non-Weight-Bearing Encourage small groups of students to create lists of activities that are weight-bearing and a list of those that are not. Students' lists of weight-bearing activities may include jogging, aerobics, walking, jumping rope, weight lifting, skating, dancing, soccer, and basketball. Students' lists of non-weight-bearing activities may include cycling and swimming.

L3 The Effects of Microgravity Tell students that weight-bearing exercise promotes bone growth and prevents bone loss. Exercises performed against the force of gravity stimulate osteoblasts, the cells that make bone. Ask students to find out the effects of microgravity on the bones of astronauts and how scientists are using this information to help people with spinal cord injuries and osteoporosis.

Lead a Discussion

HABITS FOR STRONG, HEALTHY BONES Remind students that calcium is a mineral that helps make bones hard. Ask: **How can you be sure to get enough calcium?** *(Eat a well-balanced diet that includes good sources of calcium, such as milk and other dairy products.)* **What kind of exercise helps your bones grow stronger and denser?** *(Activities in which your bones support the weight of your body)* **How can you reduce the risk of osteoporosis?** *(Eat calcium-rich foods and get plenty of exercise.)*

Elaborate

Teacher Demo

L1 FLEXIBLE CARTILAGE

Materials none **Time** 5 minutes

Demonstrate the flexibility of cartilage. Ask students to bend their outer ear gently, then let it go.

Ask: **Could you do this if your ear were made of bone?** *(No)* **How is cartilage different from bone?** *(Cartilage is flexible and bone is hard.)* **How would your body be different if your skeleton were made completely of cartilage?** *(It might not be able to hold itself up because the cartilage would bend under the weight of the body.)* Students may know that the skeleton of a shark is made of cartilage. If students ask why this is not a problem, remind them that a shark lives in water where the buoyant force of water supports the fish's body.

Lab Resource: Quick Lab

L2 SOFT BONES? Students will explore the importance of calcium in bones. This Quick Lab can be found online.

Evaluate

Assess Your Understanding

After students answer the questions, have them evaluate their understanding by completing the appropriate sentence.

RTI Response to Intervention

2a. If students have trouble explaining the relationship between diet and healthy bones, **then** remind them that a balanced diet includes foods with calcium and phosphorous to keep bones strong.

b. If students need help explaining how a bone is living tissue, **then** have them review the descriptions of bone make up.

Name ______________________ Date __________ Class __________

Assess Your Understanding

The Skeletal System

What Are the Main Functions of the Skeletal System?

got it?

○ **I get it!** Now I know that my skeleton ______________________

○ **I need extra help with** ______________________

What Role Do Joints Play?

1a. EXPLAIN Why does the skeletal system contain both immovable and movable joints?

b. RELATE CAUSE AND EFFECT How would your legs move if your knees were ball-and-socket joints?

What Are the Characteristics of Bones?

2a. EXPLAIN How do eating a balanced diet and exercising regularly benefit your skeletal system?

b. RECOGNIZE How do you know that bone is made up of living tissue?

Place the outside corner, the corner away from the dotted line, in the corner of your copy machine to copy onto letter-size paper.

Name ______________________ Date __________ Class __________

Enrich

The Skeletal System

While most animals support their backs with four legs, humans use only two. Some scientists think that this is one reason why humans are likely to suffer from back pain. Read the passage below and then answer the questions that follow on a separate sheet of paper.

A Pain in the Back

Back pain can be caused by such things as incorrect posture when sitting or standing, lifting heavy objects incorrectly, sleeping on a mattress that does not provide enough support for the neck and back, or being overweight. Also, working on a computer that is not correctly positioned can be responsible for back pain. Even stress can cause a painful back. The figures below illustrate the correct way to sit while working at a computer and to lift heavy objects.

Sitting at a Computer Don't slouch. Sit with your lower back against the back of the chair. The keyboard should be directly in front of you. The screen should be centered in front of you and the top of the screen should be at eye level.

Lifting Heavy Objects First, make sure you can lift the object without straining. Then lift by bending at your knees, not at your back. Keep the object as close to your body as possible. Don't twist your body. It is better to push a heavy object than to pull it.

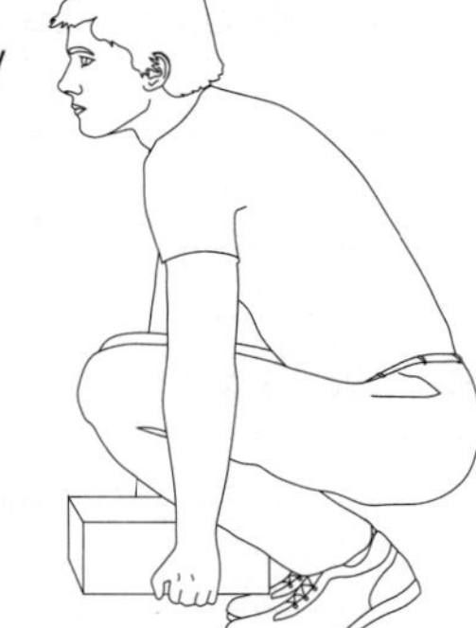

1. Why do you think the computer keyboard and screen should be centered directly in front of the person seated at the computer?
2. You are helping your next-door neighbor move some boxes of books from your house to his. How should you move the boxes to avoid injuring your back?
3. What type of shoes do you think would be likely to cause back pain? Why?
4. A friend tells you she has pain in her lower back. What are some back-care tips you could give her?

Name ______________________ Date __________ Class __________

Lesson Quiz

The Skeletal System

1. Which of the following is NOT a main function of the skeletal system?
   A produces blood cells
   B maintains homeostasis
   C provides shape and support
   D protects organs

2. Bones are made up of bone tissue, blood vessels, and
   A skin
   B nerves
   C joints
   D vertebrae

3. Joints that allow little or no movement are called
   A immovable joints
   B ball-and-socket joints
   C hinge joints
   D movable joints

4. Which of the following can bones store for later use by the body?
   A blood cells
   B cartilage
   C minerals
   D marrow

Fill in the blank to complete each statement.

5. One important function of bones is to produce ______________________.

6. Twenty-six small bones make up the ______________.

7. The bones in movable joints are held together by strong connective tissue called ______________.

8. ______________ is a condition in which bones become weak and break easily because they have lost some minerals.

9. ______________ bone contains minerals that give bones strength.

10. ______________ bone marrow is found in the middle of bone and stores fat.

Place the outside corner, the corner away from the dotted line, in the corner of your copy machine to copy onto letter-size paper.

The Skeletal System

Answer Key

Review and Reinforce

Find the worksheet in the Student Workbook.

1. Red bone marrow produces blood cells and yellow bone marrow stores fat.
2. People can eat a balanced diet and get regular exercise.
3. Moveable joints allow the body to make many different movements. The four types are hinge joints, ball-and-socket joints, gliding joints, and pivot joints.
4. Bones grow, develop, and repair themselves.

5. c **6.** e **7.** a

8. g **9.** d **10.** f

11. b

12. Bones give shape and support to the human body. They allow us to move, and they protect our internal organs. Bones produce blood cells, and store minerals and other materials until the body needs them.

Enrich

1. The keyboard and screen should be centered directly in front of the person to avoid the necessity of twisting the back to type or to look at the screen.
2. First, see if you can pick up a box without straining. Then, bend your knees, keeping your back straight, and lift the box. Keep your back straight as you straighten your knees until you are standing up. Keep the box as close to your body as possible. If the boxes are too heavy to be lifted without strain, push them along the ground.
3. Shoes that do not allow you to stand with your feet flat on the floor, such as shoes with high heels, would be more likely to cause back pain.
4. Samples: Try to reduce the amount of stress she feels. Make sure her mattress is firm enough to provide proper support for the back and neck. When she sits, make sure her lower back is against the chair back for support. Make sure she lifts heavy objects correctly.

Lesson Quiz

1. B **2.** B

3. A **4.** C

5. blood cells **6.** backbone

7. ligaments **8.** Osteoporosis

9. Compact **10.** Yellow

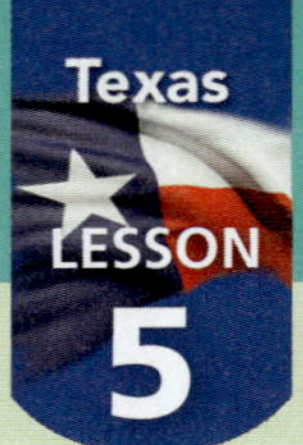

The Muscular System

 **How does your body work?**

LESSON PACING:
2–3 periods or 1–1½ blocks

Lesson Vocabulary

- involuntary muscle
- voluntary muscle
- tendon
- smooth muscle
- cardiac muscle
- striated muscle

Content Refresher

Structure of a Skeletal Muscle Skeletal muscles are composed of bundles of cylindrical muscle fibers. Each muscle fiber is actually one muscle cell. Some muscle fibers are only 1 millimeter long, whereas others can be as long as about 30 centimeters. Many muscle fibers extend the length of the muscle; connective tissues hold the fibers together.

Muscle fibers are composed of mitochondria, an extensive smooth endoplasmic reticulum, many nuclei, and individual units called *myofibrils*. Each skeletal muscle cell (fiber) contains hundreds to thousands of myofibrils. Each myofibril, in turn, is made up of threadlike filaments. Some filaments are thick and contain a protein called *myosin*. Other filaments are thin and are composed mainly of a protein called *actin*. These thin filaments also contain trace amounts of troponin and tropomyosin. Myosin and actin filaments cause muscle fibers to contract. When the thin filaments slide over the thick filaments, a muscle fiber contracts.

Lesson Objectives	TEKS	ELPS
Identify the types of muscles found in the body.	12B, 12C	2.C.1
Explain how skeletal muscles work in pairs.	7A, 12B, 12C	

Texas Essential Knowledge and Skills

7A Contrast situations where work is done with different amounts of force to situations where no work is done such as moving a box with a ramp and without a ramp, or standing still.
12B Identify the main functions of the systems of the human organism, including the circulatory, respiratory, skeletal, muscular, digestive, excretory, reproductive, integumentary, nervous, and endocrine systems.
12C Recognize levels of organization in plants and animals, including cells, tissues, organs, organ systems, and organisms.

English Language Proficiency Standards

ELPS Listening 2.C.1 Learn new language structures heard during classroom instruction and interactions.

DIFFERENTIATED INSTRUCTION KEY
L1 Struggling Students or Special Needs
L2 On-Level Students L3 Advanced Students

LESSON PLANNER 6.5

Investigations and Activities	TEKS Review
My Planet Diary, **Student Edition,** p. 244 Inquiry: Inquiry Warm-Up, How Do Muscles Work?, **Lab Manual, p. 81** Introduce Vocabulary, **Teacher's Edition,** p. 245 Lead a Discussion, How Muscles Move, **Teacher's Edition,** p. 245 21st Century Learning, Critical Thinking, **Teacher's Edition,** p. 245 Teach Key Concepts, **Teacher's Edition,** p. 246 Teach With Visuals, **Teacher's Edition,** p. 246 Support the TEKS, Connective Tissue, **Teacher's Edition,** p. 247 Differentiated Instruction, **Teacher's Edition,** p. 247 Inquiry: Quick Lab, Observing Muscle Tissue, **Lab Manual, p. 82**	Apply the TEKS, Doing Work, **Student Edition,** p. 254 TEKS Practice, **Student Edition,** p. 256 TEKS Practice: Chapter and Cumulative Review, **Student Edition,** p. 260 Lesson 6.5, **TEKS Preparation and Study Guide Workbook,** p. 62
Teach Key Concepts, **Teacher's Edition,** p. 248 Lead a Discussion, Doing Work, **Teacher's Edition,** p. 248 Apply It!, **Student Edition,** p. 248 21st Century Learning, Accountability, **Teacher's Edition,** p. 249 Differentiated Instruction, **Teacher's Edition,** p. 249 Inquiry: Quick Lab, Modeling How Skeletal Muscles Work, **PearsonTexas.com**	**SHORT ON TIME?** To do this lesson in approximately half the time, do the Activate Prior Knowledge activity. A discussion of the Key Concepts will familiarize students with the lesson content. Have students do the Quick Labs. The rest of the lesson can be completed by students independently.

These editable worksheets are available on **PearsonTexas.com.**
Print versions can be found in the **TEKS Preparation and Study Guide Workbook.**

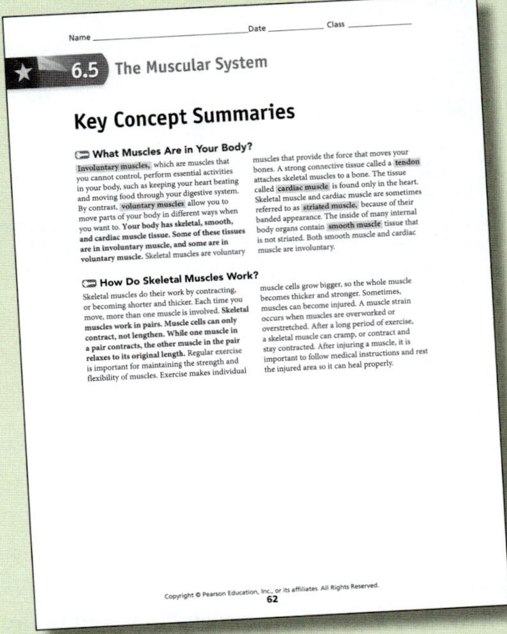
Name ______ Date ______ Class ______

6.5 The Muscular System

Key Concept Summaries

What Muscles Are in Your Body?

Involuntary muscles, which are muscles that you cannot control, perform essential activities in your body, such as keeping your heart beating and moving food through your digestive system. By contrast, **voluntary muscles** allow you to move parts of your body in different ways when you want to. **Your body has skeletal, smooth, and cardiac muscle tissue. Some of these tissues are in involuntary muscle, and some are in voluntary muscle.** Skeletal muscles are voluntary muscles that provide the force that moves your bones. A strong connective tissue called a **tendon** attaches skeletal muscles to a bone. The tissue called **cardiac muscle** is found only in the heart. Skeletal muscle and cardiac muscle are sometimes referred to as **striated muscle,** because of their banded appearance. The inside of many internal body organs contain **smooth muscle** tissue that is not striated. Both smooth muscle and cardiac muscle are involuntary.

How Do Skeletal Muscles Work?

Skeletal muscles do their work by contracting, or becoming shorter and thicker. Each time you move, more than one muscle is involved. **Skeletal muscles work in pairs. Muscle cells can only contract, not lengthen. While one muscle in a pair contracts, the other muscle in the pair relaxes to its original length.** Regular exercise is important for maintaining the strength and flexibility of muscles. Exercise makes individual muscle cells grow bigger, so the whole muscle becomes thicker and stronger. Sometimes, muscles can become injured. A muscle strain occurs when muscles are overworked or overstretched. After a long period of exercise, a skeletal muscle can cramp, or contract and stay contracted. After injuring a muscle, it is important to follow medical instructions and rest the injured area so it can heal properly.

62

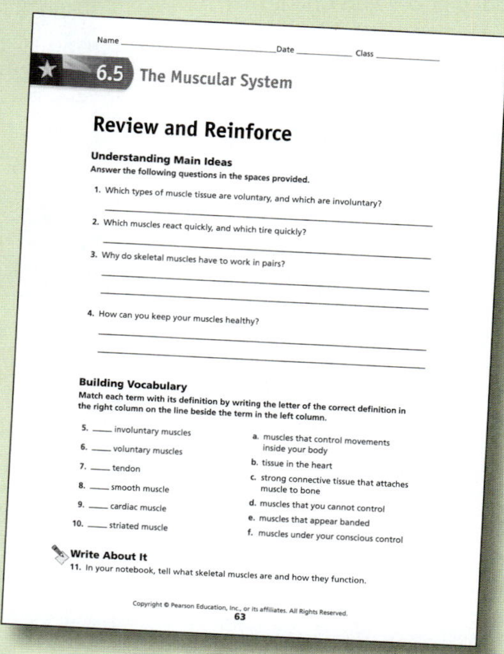
Name ______ Date ______ Class ______

6.5 The Muscular System

Review and Reinforce

Understanding Main Ideas
Answer the following questions in the spaces provided.

1. Which types of muscle tissue are voluntary, and which are involuntary?
2. Which muscles react quickly, and which tire quickly?
3. Why do skeletal muscles have to work in pairs?
4. How can you keep your muscles healthy?

Building Vocabulary
Match each term with its definition by writing the letter of the correct definition in the right column on the line beside the term in the left column.

5. ___ involuntary muscles
6. ___ voluntary muscles
7. ___ tendon
8. ___ smooth muscle
9. ___ cardiac muscle
10. ___ striated muscle

a. muscles that control movements inside your body
b. tissue in the heart
c. strong connective tissue that attaches muscle to bone
d. muscles that you cannot control
e. muscles that appear banded
f. muscles under your conscious control

Write About It
11. In your notebook, tell what skeletal muscles are and how they function.

63

Lexile Measure = 870L

LESSON 6.5

The Muscular System

Establish Learning Objectives

After this lesson, students will be able to:

- Identify the types of muscles found in the body.
- Explain how skeletal muscles work in pairs.

Engage

Activate Prior Knowledge

MY PLANET DIARY Read *Will's Blog* with the class. Point out that minor muscle injuries are a common occurrence among athletes and young people because of their high level of physical activity. Ask: **Why do you think the pain in Will's shoulder eventually went away?** *(Sample: With the aid of ibuprofen, Will's body healed the bruised muscle.)* **What other injuries can affect muscles?** *(Samples: strains, sprains, and tears)*

Explore

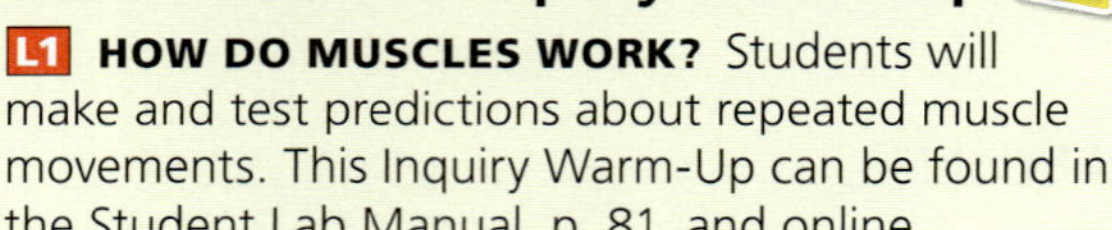

Lab Resource: Inquiry Warm-Up

L1 HOW DO MUSCLES WORK? Students will make and test predictions about repeated muscle movements. This Inquiry Warm-Up can be found in the Student Lab Manual, p. 81, and online.

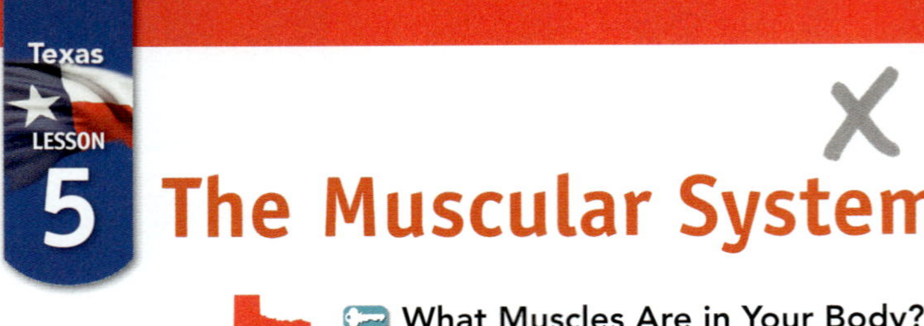

Texas LESSON 5

The Muscular System

- What Muscles Are in Your Body? TEKS 12B, 12C
- How Do Skeletal Muscles Work? TEKS 7A, 12B, 12C

my planet Diary BLOG

Posted by: Will
Location: Moore, Oklahoma

I hurt my shoulder while participating in tackling drills during football practice. The doctor said I had a deep muscle contusion, which is a bruise deep in a muscle. I was unable to lift my right arm for more than a week because of the injury. I had to take three tablets of ibuprofen every day for two weeks because it helped the swelling go down. I missed playing in only one game, and the pain eventually went away.

Answer the questions below.

1. What are two things Will had to do because of his injury?
   Sample: Will had to take ibuprofen tablets and missed playing in one game.

2. What can you do to avoid being injured when playing sports?
   Sample: I can be sure I put on my safety equipment.

Lab zone: Do the Inquiry Warm-Up *How Do Muscles Work?* Student Lab Manual, p. 81

SUPPORT ALL READERS

Lexile Measure = 870L Lexile Word Count = 808

Prior Exposure to Content: Most students have encountered this topic in earlier grades

Academic Vocabulary: *compare, contrast, infer*

Science Vocabulary: *involuntary muscle, voluntary muscle, smooth muscle, striated muscle*

Concept Level: Generally appropriate for most students in this grade

Preteach With: My Planet Diary "Will's Blog" and Figure 3 activity

Vocabulary
- involuntary muscle
- voluntary muscle
- tendon
- smooth muscle
- cardiac muscle
- striated muscle

Skills
 Reading: Compare and Contrast
Inquiry: Infer

What Muscles Are in Your Body?

TEKS 12B, 12C In this section, you'll examine the different types of muscles that make up the muscular system.

Try to sit without moving any muscles. Can you do it? First, you probably need to breathe, so your chest expands to let air in. Then you swallow. Breathing and swallowing involve muscles, so it is impossible to sit still without any muscle movement.

Involuntary and Voluntary Muscles Some body movements, such as smiling, are easy to control. Other movements, such as breathing, are impossible to control completely. That is because some of your muscles are not under your conscious control. Those muscles are **involuntary muscles.** Involuntary muscles are responsible for other activities such as digesting food. The muscles under your conscious control are **voluntary muscles.** Smiling, writing, and getting out of your seat when the bell rings are all actions controlled by voluntary muscles.

FIGURE 1

Muscle Use

Some muscles are voluntary and others are involuntary.

Relate Text and Visuals Write how the person in each frame is using involuntary and voluntary muscles.

	Frame 1	Frame 2	Frame 3
Involuntary	Sample: digesting food	Sample: heart beating	Sample: breathing
Voluntary	Sample: walking	Sample: holding light	Sample: running

English Language Proficiency Standards

ELPS Listening 2.C.1

Read aloud **Figure 1** to students. Explain that verbs ending in *-ing* also can be used as nouns.

Beginning One by one, name the actions in frames 1–3: *The person eats an apple. Digesting food uses _____ muscles.* Have partners say the word that completes the sentence (*voluntary or involuntary*.)

Intermediate Help students name the actions in each frame. Have partners orally complete the following sentence: _____ *uses (in)voluntary muscles.*

Advanced Give examples of verbs ending in *-ing* that are used as nouns. Have partners use gerunds to complete the activity in **Figure 1.**

Advanced High Have partners use gerunds to complete the activity in **Figure 1.** Then have them list other actions in gerund form, and decide whether each action is controlled by voluntary or involuntary muscles.

Explain

Introduce Vocabulary

To help students understand the terms *voluntary muscle* and *involuntary muscle*, remind students that the English word *voluntary* comes from a Latin word for "free will." Point out that the Latin prefix *in-* means "not."

Lead a Discussion

HOW MUSCLES MOVE Ask each student to hold a science book in one hand. Have students stand and hold the books down at their sides. Ask them to lift the books while feeling their arm muscles with the opposite hand. Ask: **What muscles did you feel contract? What else did you notice?** *(Sample: I felt muscles in the upper arm contracting. I also noticed that fibers stood out through the skin.)* If students mention that they could see cords or fibers that didn't look like muscles, tell them that these are tendons.

21st Century Learning

CRITICAL THINKING Tell students that muscles are found everywhere in the body. Point out that some muscles are voluntary and some are involuntary. Ask: **What is the difference between voluntary and involuntary muscles?** *(Voluntary muscles are under your conscious control, whereas involuntary muscles are not.)* Point out that a person does not have to think all the time about moving voluntary muscles, for example, when walking. However, the person does decide when to start and stop walking. Invite students to imagine a situation in which a flashlight shines into a person's eye. Explain that the light causes the pupil to constrict. Ask: **Do you think this constriction is the result of a voluntary muscle reaction or an involuntary muscle reaction?** *(Involuntary)*

PEARSON Texas.com

Texas Essential Knowledge and Skills

12B Identify the main functions of the systems of the human organism, including the circulatory, respiratory, skeletal, muscular, digestive, excretory, reproductive, integumentary, nervous, and endocrine systems.

12C Recognize levels of organization in plants and animals, including cells, tissues, organs, organ systems, and organisms.

LESSON 6.5

Explain

Teach Key Concepts

Explain to students that the human body contains three types of muscle tissue—skeletal, smooth, and cardiac—some of which are involuntary muscle and some of which are voluntary muscle. Ask: **How are skeletal and cardiac muscles alike?** *(Both have a striated appearance.)* **Where is the smooth muscle located in Figure 2?** *(In the stomach)* **Are skeletal muscles voluntary or involuntary?** *(Voluntary)* **Are smooth muscles voluntary or involuntary?** *(Involuntary)* **Are cardiac muscles voluntary or involuntary?** *(Involuntary)* **How is cardiac muscle different from skeletal muscle?** *(Unlike skeletal muscle, cardiac muscle contracts repeatedly without getting tired.)*

Teach With Visuals

Have students look at **Figure 2.** Ask: **Which type of muscle is located in only one place in the body, and where is this type of muscle located?** *(Cardiac; in the heart)* **Which type of muscle gives you the ability to lift a heavy object with your arms?** *(Skeletal)* **Why is it important that smooth muscle not tire quickly?** *(Sample: Smooth muscles are responsible for helping the body's organs and structures function, so they need to be able to work extensively without tiring.)* Explain to students that striated muscle appears to have stripes when it is seen under a microscope. Ask: **Which two types of muscle appear to have bands or stripes?** *(Skeletal, cardiac)*

Types of Muscle Tissue

Your body has skeletal, smooth, and cardiac muscle tissues. Some of these muscle tissues are involuntary, and some are voluntary.

Skeletal muscles provide the force that moves your bones. A strong connective tissue called a **tendon** attaches the muscle to a bone. Because you have conscious control of skeletal muscles, they are classified as voluntary muscles. In contrast, the inside of many internal body organs, such as the stomach and blood vessels, contain **smooth muscle** tissue. These are involuntary muscles. They work to control certain movements inside your body, such as moving food through your digestive system. The tissue called **cardiac muscle** is found only in your heart. Like smooth muscle, cardiac muscle is involuntary. Look at **Figure 2.**

FIGURE 2

Muscle Tissue

You have three types of muscle tissue: skeletal, smooth, and cardiac.

Classify **In the table, identify the type of muscle tissue in each body structure.**

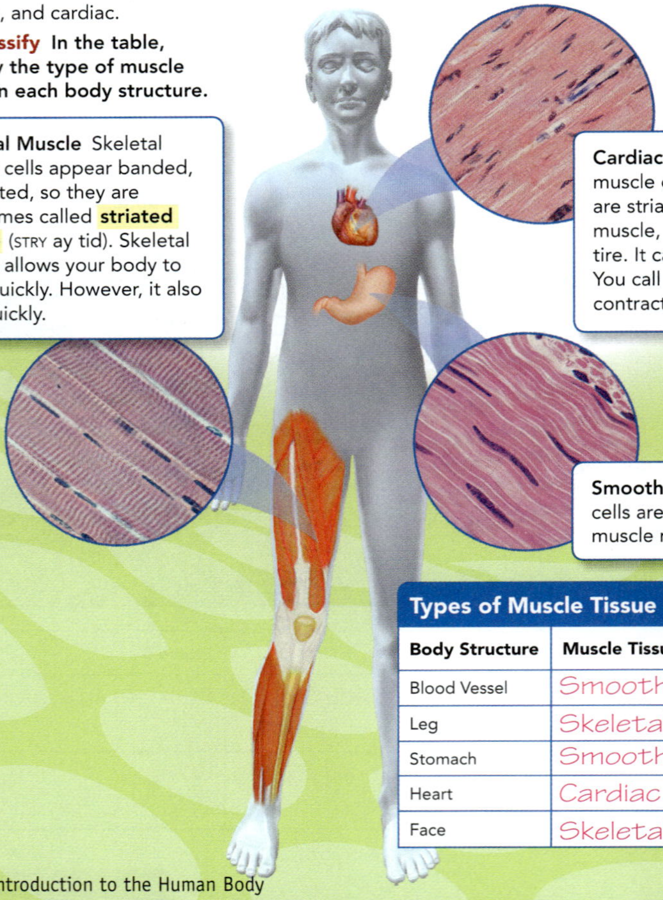

Skeletal Muscle Skeletal muscle cells appear banded, or striated, so they are sometimes called **striated muscle** (STRY ay tid). Skeletal muscle allows your body to react quickly. However, it also tires quickly.

Cardiac Muscle Like skeletal muscle cells, cardiac muscle cells are striated. But unlike skeletal muscle, cardiac muscle does not tire. It can contract repeatedly. You call those repeated contractions heartbeats.

Smooth Muscle Smooth muscle cells are not striated. This type of muscle reacts and tires slowly.

Types of Muscle Tissue

Body Structure	Muscle Tissue
Blood Vessel	Smooth
Leg	Skeletal
Stomach	Smooth
Heart	Cardiac
Face	Skeletal

Skeletal Muscle
- Reacts and tires easily
- Attached to bones
- Voluntary
- Striated

All
- Muscle tissue
- Contract

Smooth Muscle
- Reacts and tires slowly
- Involuntary
- Not striated

Cardiac Muscle
- Does not tire
- Involuntary
- Striated
- Found only in heart

did you know?

Why do you shiver when you get chilled? You shiver when many of your skeletal muscles contract quickly again and again. When your muscles contract, they produce extra heat. So, by shivering, your body produces heat that warms you.

Compare and Contrast In the graphic organizer, write how all three muscle tissues are alike and how each type is different.

Do the Quick Lab *Observing Muscle Tissue.* Student Lab Manual, p. 82

Assess Your Understanding

TEKS 12B, 12C

1a. Identify What is the difference between voluntary and involuntary muscles?
Voluntary muscles are under your control and involuntary muscles are not.

b. Infer Why is it important that cardiac muscle tissue does not tire?
Sample: The heart must beat repeatedly to keep you alive.

got it?

○ I get it! Now I know that the muscles in my body are either involuntary or voluntary, and can be made of skeletal, smooth, or cardiac muscle tissue.

○ I need extra help with See TE note.

247

Support the TEKS

CONNECTIVE TISSUE Explain to students that bones and muscles are not directly connected to one another. Ask: **Which kind of muscle is connected to bones?** *(Skeletal muscle)* **What holds bones and skeletal muscles together?** *(They are attached to each other by strong connective tissue called tendons.)* Tendons allow the muscular and skeletal systems to work together to move the body.

Compare and Contrast Tell students that comparing and contrasting involves examining the similarities and differences between things. When two things are compared, ways in which they are alike are identified. When two things are contrasted, ways in which they are different are identified.

Elaborate

Lab Resource: Quick Lab

L3 **OBSERVING MUSCLE TISSUE** Students will use a microscope to analyze prepared slides of skeletal, smooth, and cardiac muscle. This Quick Lab can be found in the Student Lab Manual, p. 82, and online.

Evaluate

Assess Your Understanding

After students answer the questions, have them evaluate their understanding by completing the appropriate sentence.

RTI Response to Intervention

1a. If students have trouble remembering the definitions of *voluntary muscle* and *involuntary muscle*, **then** have them picture consciously controlling muscles as they raise an arm to *volunteer* an answer.

b. If students have trouble explaining why it is important for cardiac muscle not to tire, **then** remind students that their life depends on their hearts beating continuously.

Differentiated Instruction

L1 **Muscle Movements** Have students wave their hands. Tell them that this movement is voluntary, and point out that they did not wave their hands until their brains sent a message to do so. Then, help students find their pulses. Point to the heart in **Figure 2,** and explain that the pulse is caused by muscles in the heart. Explain that the beating of their hearts is involuntary.

L3 **Tendons** Challenge students to do research to find out about various health issues relating to tendons. Encourage students to make a poster or other large graphic to illustrate the structure and functions of a healthy tendon. The graphic can also contain information about causes and effects of various tendon injuries and conditions.

LESSON 6.5

Explain

Teach Key Concepts

Explain to students that skeletal muscles work in pairs, with one muscle contracting while the other muscle relaxes. Have students refer to **Figure 3** as they reflect on muscles and answer the questions below. Invite students to bend their elbows so that they can feel their muscles contract and relax as shown in the figure. Explain that a muscle can contract, or become shorter. However, it cannot extend, or become longer. After contraction, a muscle relaxes and returns to its original length. Ask: **In what condition is a skeletal muscle at its greatest length?** *(It is relaxed.)* **Why is a muscle's relaxing to its original length not considered lengthening?** *(It can reach its original length when relaxed, but it cannot reach a length greater than this.)* **How does your biceps change as you bend your arm?** *(It gets shorter and thicker and feels firmer.)* **What happens to your arm when your triceps contracts?** *(The arm straightens, and the biceps returns to its original length.)*

Lead a Discussion

DOING WORK Review with students the concept of doing work. Ask: **What is work?** *(Work is force exerted on an object that causes the object to move.)* **How do your muscles do work?** (Muscles exert a force on the bones, which causes them to move.) Challenge students to identify situations in which their muscles do work and situations in which their muscles do not do any work.

Elaborate

Apply It!

L1 Before beginning the activity, review how muscles do work.

Texas Essential Knowledge and Skills

7A Contrast situations where work is done with different amounts of force to situations where no work is done such as moving a box with a ramp and without a ramp, or standing still.

12B Identify the main functions of the systems of the human organism, including the circulatory, respiratory, skeletal, muscular, digestive, excretory, reproductive, integumentary, nervous, and endocrine systems.

12C Recognize levels of organization in plants and animals, including cells, tissues, organs, organ systems, and organisms.

TEKS 7A, 12B, 12C In this section, you'll learn how skeletal muscles work to make the human body move.

How Do Skeletal Muscles Work?

Has anyone ever asked you to "make a muscle"? Like other skeletal muscles, the muscles in your arm do their work by contracting, which means becoming shorter and thicker.

Doing Work The muscles in your body do work by exerting a force on your bones, causing your body to move. For example, when your legs are at rest, your leg muscles are relaxed and no work is done. When a leg muscle contracts, it pulls on your leg bones. Work is done because your leg moves.

Each time you move, more than one muscle is involved. **Skeletal muscles work in pairs. Muscle cells can only contract, not lengthen. While one muscle in a pair contracts, the other muscle in the pair relaxes to its original length.** The biceps and triceps are shown in **Figure 3.**

FIGURE 3

Muscle Pairs

To bend your arm at the elbow, the biceps contracts while the triceps relaxes.

Interpret Diagrams **Tell what happens to each muscle as you straighten your arm.**

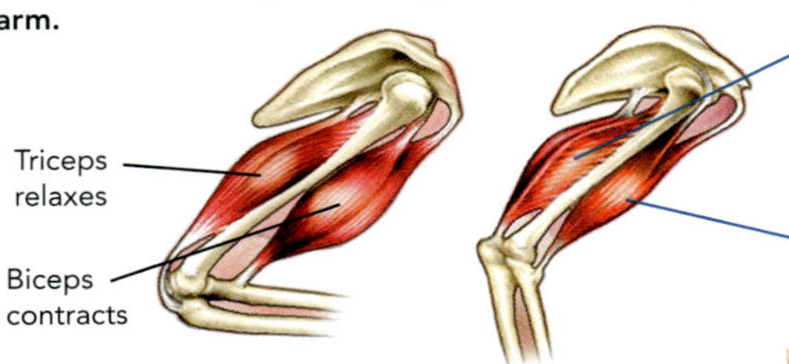

Kerri is sitting still in class when the teacher asks for a volunteer. Kerri raises her hand.

Compare and Contrast Which photo shows a situation in which Kerri's muscles have done work? Explain.

Photo 2. Kerri has raised her arm. Her muscles exerted a force on her bones to get her arm to move.

Keeping Muscles Healthy

Regular exercise is important for maintaining the strength and flexibility of muscles. Exercise makes individual muscle cells grow bigger, so the whole muscle becomes thicker and stronger. Warming up before exercising increases the blood flow to your muscles. Stretching as you warm up helps your muscles become more flexible and prepares them for exercise. Exercise is important even in space, as shown in **Figure 4.**

Sometimes, muscles can become injured. A muscle strain can occur when muscles are overworked or overstretched. After a long period of exercise, a skeletal muscle can cramp, or contract and stay contracted. If you injure a muscle, be sure to follow medical instructions and rest the injured area so it can heal properly.

FIGURE 4

Muscle Loss

Without gravity, astronauts in space can lose muscle mass. Therefore, they need to exercise daily.

CHALLENGE **Explain why a lack of gravity might cause muscles to weaken.**

Sample: You float, so your leg muscles do less work.

Lab zone Do the Quick Lab *Modeling How Skeletal Muscles Work.* Find the lab online.

Assess Your Understanding

TEKS 12B, 12C

2a. Review How do muscles work in pairs?

One muscle contracts while the other muscle relaxes.

b. Make Generalizations Why is it important to exercise both muscles in a pair?

Sample: Both muscles in a pair need to have equal strength.

got it?

O **I get it!** Now I know that skeletal muscles work in pairs.

O I need extra help with See TE note.

Differentiated Instruction

L3 Muscle Pairs Have students observe which muscles contract when they stand or bend one of their knees. Then have them sketch the bones and muscles in the thigh in each position and indicate which muscle is contracted. Display students' sketches in the classroom.

L3 Flexibility Flexibility is the ability to use a muscle throughout its entire range of motion. Flexibility is just as important as having strong muscles. Have students ask a physical education teacher to show them how to correctly perform stretches to improve flexibility. Students can lead the class in performing the stretches.

21st Century Learning

ACCOUNTABILITY To help students review the material in this lesson, have them write ten questions about the muscular system. Suggest that they write a mixture of short answer, fill-in-the-blank, and matching questions. Have partners exchange and answer the questions. Students can work together to reread appropriate passages to find answers.

Lab Resource: Quick Lab 

L2 MODELING HOW SKELETAL MUSCLES WORK Students will create and use models to demonstrate how muscle pairs work together. This Quick Lab can be found online.

Evaluate

Assess Your Understanding

After students answer the questions, have them evaluate their understanding by completing the appropriate sentence.

RTI Response to Intervention

2a. If students have trouble explaining how muscles work, **then** have them review the Key Concept statement.

b. If students have trouble explaining why it is important to exercise both muscles in a pair, **then** have them review **Figure 3** to see opposite muscles in a pair.

Name ______________________ Date ____________ Class ____________

Assess Your Understanding

The Muscular System

What Muscles Are in Your Body?

1a. IDENTIFY What is the difference between voluntary and involuntary muscles?

____

____

b. INFER Why is it important that cardiac muscle tissue does not tire?

____

____

got it?

○ **I get it!** Now I know that the muscles in my body are ______________________

____

____

○ **I need extra help with** ____

____

How Do Skeletal Muscles Work?

2a. REVIEW How do muscles work in pairs?

____

____

b. MAKE GENERALIZATIONS Why is it important to exercise both muscles in a pair?

____

____

got it?

○ **I get it!** Now I know that skeletal muscles work ______________________

____

○ **I need extra help with** ____

____

Place the outside corner, the corner away from the dotted line, in the corner of your copy machine to copy onto letter-size paper.

Name ______________________ Date ____________ Class ____________

Enrich

The Muscular System

Exercise makes individual muscle cells grow wider by stimulating protein synthesis. This causes the muscle to become thicker. Muscles increase in strength as they become thicker. Read the paragraph below and look at the diagram. Then answer the questions that follow in the space provided.

Pumping Iron

Any exercise that makes a muscle try to move an immovable object or to lift a heavy one can work toward increasing that muscle's strength. You can increase the difficulty of the exercise by adding additional weight. The figure below demonstrates a biceps curl, a common strength-training exercise used to build arm muscles. The arrows show the direction of movement, and the circles represent a weight.

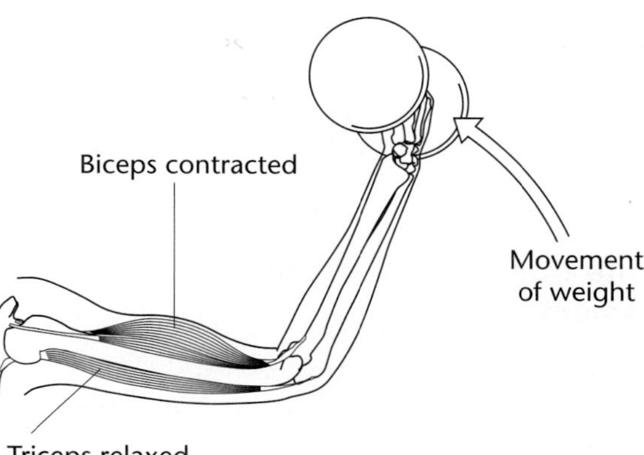

1. How does exercise build muscles?

____

____

2. How can you increase the amount of exercise your arm muscles can do in the exercise pictured above?

____

____

3. How does the exercise shown above demonstrate that skeletal muscles must work in pairs?

____

____

Name ______________________ Date ____________ Class ____________

Lesson Quiz

The Muscular System

Write the letter of the correct answer on the line at the left.

1. ___ What types of muscle tissue does the human body contain?

A skeletal, smooth, and cardiac
B smooth, cardiac, and involuntary
C striated, skeletal, and cardiac
D voluntary, cardiac, and smooth

2. ___ The muscle that is paired with the biceps is called the

A deltoid
B triceps
C pectoralis
D trapezius

3. ___ For which of the following activities are involuntary muscles responsible?

A talking
B smiling
C writing
D breathing

4. ___ Muscle is attached to a bone by a

A bicep
B blood vessel
C tendon
D tricep

If the statement is true, write *true*. If the statement is false, change the underlined word or words to make the statement true.

5. ______________ A muscle strain occurs when muscles are overworked or overstretched.

6. ______________ Voluntary muscles perform essential activities in your body, such as keeping your heart beating and moving food through your digestive system.

7. ______________ Skeletal muscle and cardiac muscle are sometimes referred to as smooth muscle, because of their banded appearance.

8. ______________ Skeletal muscles work in pairs.

9. ______________ The tissue called cardiac muscle is found only in the heart.

10. ______________ Both smooth muscle and cardiac muscle are voluntary.

The Muscular System

Answer Key

Review and Reinforce

Find the worksheet in the Student Workbook.

1. voluntary—skeletal; involuntary—smooth, cardiac
2. react quickly—skeletal ; tire quickly—skeletal
3. While one muscle in a pair contracts, the other muscle in the pair relaxes to its original length.
4. regular exercise, with stretching and warming up beforehand
5. d
6. f
7. c
8. a
9. b
10. e
11. Skeletal muscles are voluntary, striated muscles that provide the force that moves your bones. Skeletal muscles work in pairs. Each time you move, one muscle in a pair contracts while the other muscle in the pair relaxes to its original length.

Enrich

1. Exercise stimulates protein synthesis. This makes the muscle cells grow wider, and causes the muscles to become thicker.
2. Gradually increase the amount of weight being moved.
3. Two muscles are required to perform each movement. As the biceps contract, the triceps relaxes and the arm raises the weight. To move the weight downward, the biceps must relax and triceps contracts. The movement would not be possible without both muscles.

Lesson Quiz

1. A
2. B
3. D
4. C
5. true
6. Involuntary
7. striated
8. true
9. true
10. involuntary

The Integumentary System

 **How does your body work?**

LESSON PACING:
1–2 periods or $\frac{1}{2}$–1 block

Lesson Vocabulary

- epidermis
- melanin
- dermis
- pore
- follicle

Content Refresher

Keratin On the surface of the skin is a layer of dead skin cells made of keratin. The layer forms a barrier between the inside and the outside of the body. It is critical to water and temperature homeostasis and to protection from infectious organisms and toxic substances.

Made by cells in the epidermis, keratin is a sturdy protein that forms sheets of fibers inside the cell. As epidermal cells move upward toward the skin surface, they make increasing amounts of keratin. By the time the cells die, they are essentially flat bags of keratin. Keratin, along with the pigment melanin, protects the skin from the cancer-causing effects of ultraviolet radiation in sunlight.

Keratin is also vital to two specialized skin structures—hair and nails. Cells in each hair bulb produce keratin. The visible hair strand is made up of hard, flat, keratin-rich cells wrapped around a small central space. Nails are made of translucent sheets of keratin produced and released by the cells of the nail bed. As a nail grows, keratin is added at the edge of the nail bed and to the underside of the nail. This process thickens the nail as it lengthens and keeps the nail connected to the living cells of the nail bed.

Lesson Objectives	TEKS	ELPS
Identify the main functions and structures of the integumentary system.	12B, 12C	4.F.1

Texas Essential Knowledge and Skills

12B Identify the main functions of the systems of the human organism, including the circulatory, respiratory, skeletal, muscular, digestive, excretory, reproductive, integumentary, nervous, and endocrine systems.
12C Recognize levels of organization in plants and animals, including cells, tissues, organs, organ systems, and organisms.

English Language Proficiency Standards

ELPS Reading 4.F.1 Use visual and contextual support to read grade-appropriate content area text.

DIFFERENTIATED INSTRUCTION KEY
L1 Struggling Students or Special Needs
L2 On-Level Students L3 Advanced Students

Investigations and Activities

My Planet Diary, **Student Edition,** p. 250

Inquiry: Inquiry Warm-Up, What Can You Observe About Skin?, **PearsonTexas.com**

Introduce Vocabulary, **Teacher's Edition,** p. 251

Teach Key Concepts, **Teacher's Edition,** p. 251

Support the TEKS, Functions of the Skin, **Teacher's Edition,** p. 252

Teach With Visuals, **Teacher's Edition,** p. 252

21st Century Learning, Critical Thinking, **Teacher's Edition,** p. 253

Make Analogies, A Layer of Protection, **Teacher's Edition,** p. 253

21st Century Learning, Communication, **Teacher's Edition,** p. 253

Differentiated Instruction, **Teacher's Edition,** p. 253

Inquiry: Quick Lab, Sweaty Skin, **Lab Manual,** p. 83

TEKS Review

Apply the TEKS, Doing Work, **Student Edition,** p. 254

TEKS Practice, **Student Edition,** p. 256

Texas Test Prep, **Student Edition,** p. 260

Lesson 6.6, **TEKS Preparation and Study Guide Workbook,** p. 64

SHORT ON TIME? To do this lesson in approximately half the time, do the Activate Prior Knowledge activity. A discussion of the Key Concepts will familiarize students with the lesson content. Have students do the Quick Lab. The rest of the lesson can be completed by students independently.

These editable worksheets are available on **PearsonTexas.com.**
Print versions can be found in the **TEKS Preparation and Study Guide Workbook.**

Name ______________ Date ________ Class ________

6.6 The Integumentary System

Key Concept Summary

What Are the Main Functions of the Integumentary System?

The skin is part of the integumentary system, which also includes hair, nails, sweat glands, and oil glands. **The skin has two layers that protect the body. Skin helps regulate body temperature, eliminate wastes, gather information about the environment, and produce vitamin D.** The skin forms a barrier that keeps harmful substances outside the body. Also, the skin keeps important substances such as water and other fluids inside the body. Skin helps the body maintain a steady temperature through perspiration and the enlarging of blood vessels. Perspiration is also responsible for eliminating some waste materials from the body. Nerves in the skin gather information from the environment about pressure, temperature, and pain. Some skin cells produce vitamin D in the presence of sunlight. Together an outer layer and an inner layer perform all the skin's functions. The **epidermis** is the outer layer of the skin, which helps protect your skin. Some cells deep in the epidermis produce **melanin,** a pigment that colors the skin. The **dermis** is the inner layer of the skin, which includes nerves, blood vessels, sweat glands, hairs, and oil glands. **Pores** are openings that allow sweat to reach the surface. Strands of hair grow within the dermis in **follicles.** Oil produced in glands around the follicles keeps the surface of the skin moist and the hairs flexible.

64

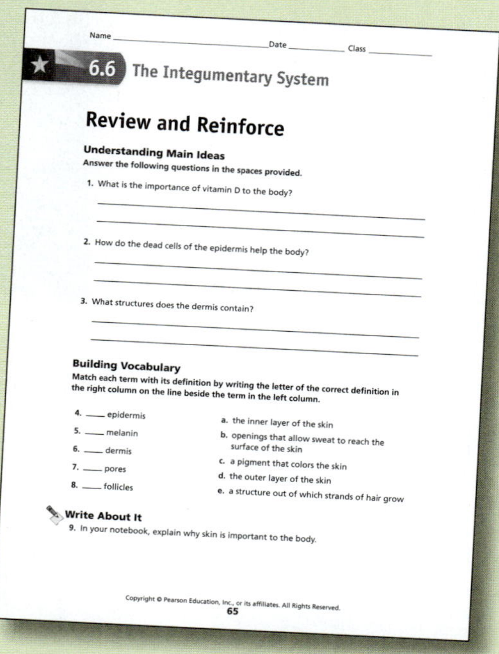

Name ______________ Date ________ Class ________

6.6 The Integumentary System

Review and Reinforce

Understanding Main Ideas
Answer the following questions in the spaces provided.

1. What is the importance of vitamin D to the body?
2. How do the dead cells of the epidermis help the body?
3. What structures does the dermis contain?

Building Vocabulary
Match each term with its definition by writing the letter of the correct definition in the right column on the line beside the term in the left column.

4. ____ epidermis
5. ____ melanin
6. ____ dermis
7. ____ pores
8. ____ follicles

a. the inner layer of the skin
b. openings that allow sweat to reach the surface of the skin
c. a pigment that colors the skin
d. the outer layer of the skin
e. a structure out of which strands of hair grow

Write About It
9. In your notebook, explain why skin is important to the body.

65

Lexile Measure = 900L

LESSON 6.6

The Integumentary System

Establish Learning Objectives

After this lesson, students will be able to:

Identify the main functions and structures of the integumentary system.

Engage

Activate Prior Knowledge

MY PLANET DIARY Read *Would You Like to Be a Skin Doctor?* with the class. Point out that skin cancer is the most common form of cancer in the United States today. Explain to students that one cause of skin cancer is overexposure to the sun's ultraviolet rays. **What do you think a dermatologist might say about spending time outside in the sun?** *(Sample: A dermatologist might advise you to apply sunscreen before spending time in the sun.)*

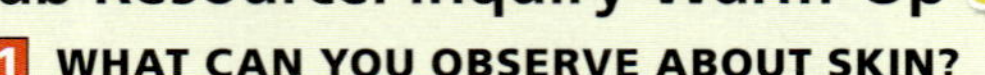

Explore

Lab Resource: Inquiry Warm-Up

L1 WHAT CAN YOU OBSERVE ABOUT SKIN?
Students will observe skin structures to understand perspiration. This Inquiry Warm-Up can be found online.

The Integumentary System

What Are the the Main Functions of the Integumentary System?
TEKS 12B, 12C

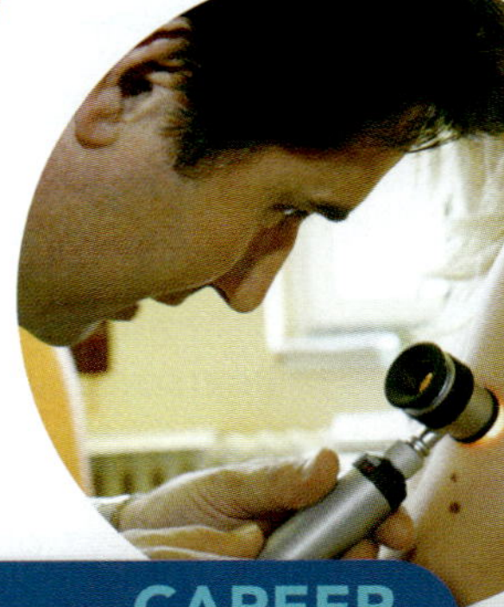

my planet Diary — CAREER

Would You Like to Be a Skin Doctor?

Did you know that there is a special type of doctor who studies and treats skin? This type of doctor, a dermatologist, specializes in caring for skin, hair, and nails. Dermatologists diagnose and treat a variety of skin problems ranging from acne to psoriasis to dangerous cancers.

To become a dermatologist, you need a lot of education. You may spend about ten years in schooling and training after you graduate high school. Then you must pass a certification test. Although becoming a dermatologist is not easy, it can be a rewarding career!

Communicate Discuss the question with a partner. Then write down your answer.

Why do you think dermatologists are important?

Sample: Dermatologists help diagnose and treat skin conditions that may lead to serious illness or even death if left untreated.

Psoriasis

Do the Inquiry Warm-Up *What Can You Observe About Skin?* Find the lab online.

SUPPORT ALL READERS

Lexile Measure = 900L **Lexile Word Count = 882**

Prior Exposure to Content: Most students have encountered this topic in earlier grades

Academic Vocabulary: *cause, effect, observe, relate*

Science Vocabulary: *epidermis, melanin, dermis, pore, follicles*

Concept Level: Generally appropriate for most students in this grade

Preteach With: My Planet Diary "Would You Like to Be a Skin Doctor?" and Figure 2 activity

Vocabulary
- epidermis • melanin • dermis
- pore • follicle

Skills

 Reading: Relate Cause and Effect

Inquiry: Observe

What Are the Main Functions of the Integumentary System?

If an adult's skin were stretched out flat, it would cover an area about the size of a mattress on a twin bed. The skin is part of the integumentary system (in teg yoo MEN tur ee). In addition to the skin, this system includes hair, nails, sweat glands, and oil glands.

TEKS 12B, 12C In this section, you'll identify the roles of the integumentary system in the human body.

ELPS 4.F.1
With a partner, look at the photographs on page 252. Use the photos to help understand the main functions of skin, as you read the text on pages 251–252. Then fill in labels for the photos together.

Functions of the Skin

Your skin helps you in many ways. **The skin has two layers that protect the body. Skin helps regulate body temperature, eliminate wastes, gather information about the environment, and produce vitamin D.**

Protecting the Body The skin forms a barrier that keeps harmful substances outside the body. It also keeps important substances such as water and other fluids inside the body.

Maintaining Temperature The skin helps the body maintain a steady temperature. When you become too warm, like the runner in **Figure 1,** blood vessels in your skin enlarge. This widening of the vessels allows more blood to flow through them and body heat to escape into the environment. In addition, sweat glands produce perspiration in response to excess heat. As perspiration evaporates from your skin, your skin is cooled. When you get cold, blood vessels in your body contract. This reduces blood flow to the skin and helps your body conserve heat.

FIGURE 1
Amazing Skin
As this runner exercises, his skin helps to cool him off.

 **Identify** On the notebook page, write about a time that your skin protected you, maintained your body temperature, or both.

Sample: When I played soccer I perspired, and that helped cool me off. It also helped to keep harmful substances, such as bacteria, out of my body.

Explain

Introduce Vocabulary

To help students understand the terms *dermis* and *epidermis,* point out the identical word part *dermis* as well as the Latin prefix *epi-*. Tell students that this prefix means "outer."

Teach Key Concepts 

Point out to students that the skin's two layers protect the body, and they help the body maintain a steady temperature, get rid of wastes, make vitamin D, and gather information about the environment. Invite students to look at the people around them in the classroom. Ask: **What is the most obvious function of skin?** *(To cover and protect the body)* Remind students that in addition to keeping harmful substances out of the body, skin also keeps water and other fluids in the body. Ask: **How does skin regulate your body's temperature?** *(When you are too warm, blood vessels in the skin get larger and wider and allow heat to move out of your body. Also, you sweat in response to excess heat. Then the perspiration evaporates, cooling the skin.)*

PEARSON Texas.com

LESSON 6.6

English Language Proficiency Standards

ELPS Reading 4.F.1

Have students complete **Figure 2** on page 252.

Beginning Help students name or point to each function of skin. Record answers. Have students point to the picture in **Figure 2** that shows each function.

Intermediate Review and discuss each function in bold type in **Figure 2.** Then have partners match each function to one of the pictures in the figure. Have students switch partners and compare answers.

Advanced Have partners explain each function in bold type in **Figure 2.** Then have them match each function to one of the pictures in the figure.

Advanced High Have partners play a guessing game. One student chooses a picture and gives clues by naming the functions the picture illustrates. The other student guesses the picture.

Texas Essential Knowledge and Skills

12B Identify the main functions of the systems of the human organism, including the circulatory, respiratory, skeletal, muscular, digestive, excretory, reproductive, integumentary, nervous, and endocrine systems.

12C Recognize levels of organization in plants and animals, including cells, tissues, organs, organ systems, and organisms.

LESSON 6.6

Explain

Support the TEKS

FUNCTIONS OF THE SKIN Remind students that skin does more than just protect the body and regulate temperature. Ask: **How is waste eliminated by the skin?** *(Perspiration contains waste materials, and skin eliminates wastes every time a person perspires.)* **What structure within skin allows it to gather information from the environment?** *(Nerves)* **What information can skin gather from the environment?** *(Information about pressure, temperature, and pain)* **How does skin help produce healthy bones?** *(Some skin cells produce vitamin D. Vitamin D is important for healthy bones, because it helps your body absorb calcium from food.)* Emphasize to students the importance of getting enough vitamin D from sun exposure each day. Point out that many adults today are deficient in vitamin D.

Relate Cause and Effect Tell students that a cause makes something happen. An effect is what happens. When students recognize that one event causes another, they are relating cause and effect.

Teach With Visuals

Help students focus their attention on what is happening in each photograph in **Figure 2.** After students have identified the "event" captured in each image, encourage them to think about how the human being's skin relates to this event or set of conditions. Ask: **In the first photograph, what does the person's skin help him or her to feel?** *(Heat)* **In what way is the skin protecting this person's body?** *(The skin is gathering information about the environment and about potential danger.)* **What is happening in the second photograph?** *(The person is perspiring.)* **Which details in the third photograph tell you that the person's skin is exposed to the sun's rays?** *(She is wearing a hat and sunglasses.)*

Teach With Visuals

Ask students to identify the openings in the epidermis *(pores and openings of hair follicles)* in **Figure 3.** Ask: **What are the pores connected to?** *(Sweat glands)* **Where is oil produced?** *(In glands around the hair)* Ask students to note differences between the epidermis and the dermis. *(The dermis is thicker, and it has blood vessels, nerves, hair follicles, sweat glands, and fat.)* Ask: **What are the functions of the dermis?** *(The fat pads protect internal organs and helps keep heat in the body. Sweat glands produce perspiration. Oil helps moisten the skin.)* Ask: **What is the relationship between the hairs and the nerves in the dermis?** *(When something touches or blows against the hairs, the nerves pick up the sensation.)*

Relate Cause and Effect Describe the possible effect of not getting enough sunlight each day.

Sample: The skin would not produce enough vitamin D.

Eliminating Wastes Perspiration contains dissolved waste materials that come from the breakdown of chemical processes. Your skin helps eliminate wastes whenever you perspire.

Gathering Information Nerves in your skin gather information from the environment. They provide information about things such as pressure, temperature, and pain. Pain messages warn you that something in your surroundings can injure you.

Producing Vitamin D Some skin cells produce vitamin D in the presence of sunlight. Vitamin D is important for healthy bones because it helps your body absorb the calcium in your food. Your skin cells need sunlight each day to produce enough vitamin D.

FIGURE 2

Skin at Work

You may not notice that your skin is constantly working.

Complete the activity and answer the questions below.

1. **Interpret Photos** Write below each photo the functions that the skin performs.

Protects the body, gathers information

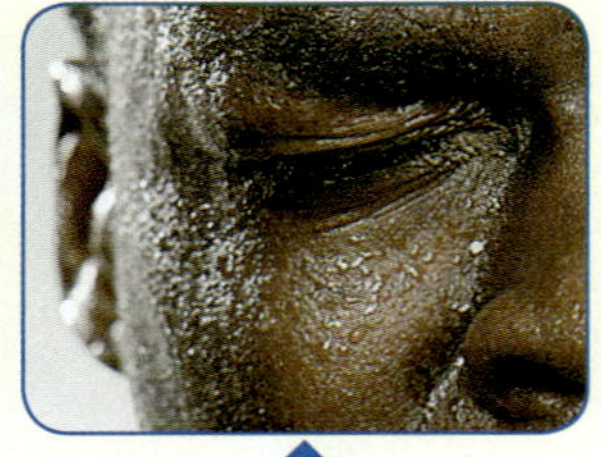

Maintains temperature, eliminates wastes

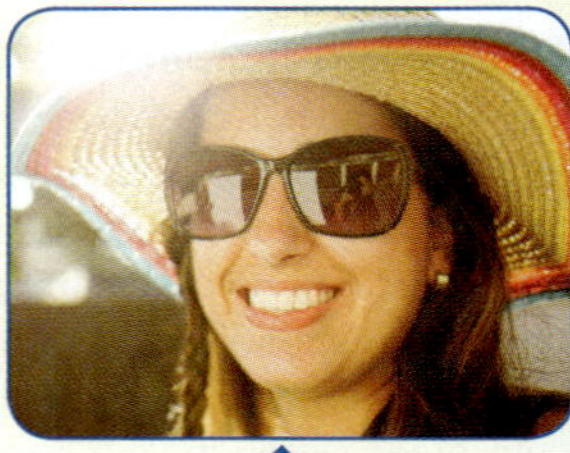

Produces vitamin D, gathers information

2. **Observe** Press down firmly on your arm with your fingertips. Then lightly pinch yourself. What information did you receive?

   I felt pressure when I pressed down and pain when I pinched myself lightly.

3. **CHALLENGE** What might happen if the nerves in your skin did not gather information?

   Sample: I might be hurt badly by something hot before I realized it.

Structures of the Skin

The skin has two main layers, as shown in **Figure 3.** Together, these layers—an outer layer and an inner layer—perform all the skin's functions.

The **epidermis** is the outer layer of the skin. Deep in the epidermis, new cells form. As they mature, they move upward until they die. They then become part of the epidermal surface layer. This surface layer helps protect your skin. Cells stay in this layer for about two to three weeks until they are shed. Some cells deep in the epidermis produce **melanin,** a pigment that colors the skin.

The **dermis** is the inner layer of the skin. It is above a layer of fat. This fat layer pads the internal organs and helps keep heat in the body. The dermis includes nerves, blood vessels, sweat glands, hairs, and oil glands. **Pores** are openings that allow sweat to reach the surface. Strands of hair grow within the dermis in **follicles** (FAHL ih kulz). Oil produced in glands around the follicles keeps the surface of the skin moist and the hairs flexible.

FIGURE 3

Structures of the Skin

Relate Text and Visuals On the lines, write the functions of the epidermis, the nerves, and the sweat gland.

The epidermis protects the body. Nerves gather information. The sweat gland makes sweat that helps maintain temperature and eliminate wastes.

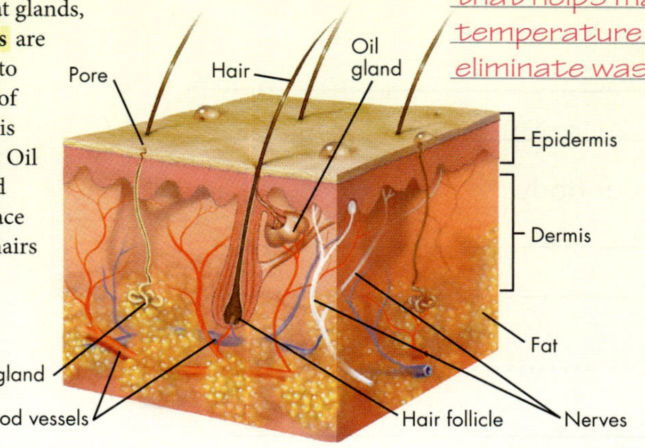

Do the Quick Lab *Sweaty Skin.* Student Lab Manual, p. 83

Assess Your Understanding

TEKS 12B, 12C

1a. Identify How does your skin gather information about the environment?

The nerves in the skin provide information about pressure, temperature, and pain.

b. Recognize Explain how some structures in the skin protect your body.

Sample: A layer of fat keeps heat in the body and pads internal organs.

got it?

○ **I get it!** Now I know that the skin has two layers that protect the body, maintain temperature, eliminate waste, gather information, and make vitamin D.

○ I need extra help with See TE note.

253

Differentiated Instruction

L1 Skin Functions Tell students to gently squeeze one hand with the other. Explain that the nerves in skin help them to feel the pressure. Then have students rub a cool, damp washcloth on their skin to demonstrate that the skin senses temperature. Ask students to describe what sometimes happens to the skin when it feels very cold. *(Sometimes the skin gets "goose bumps.")* Have students dip a finger into a cup of water, and then hold the finger in the air. Ask students to describe how their finger feels. *(Cool)* Explain that this sensation is similar to the feeling people get when they sweat. The sweat, like the water, removes heat.

L1 Science Glossary Pronounce the vocabulary terms *epidermis, dermis, pores,* and *follicles* as you point to them in **Figure 3.** Have students use simple words and phrases to write brief descriptions of these structures.

Elaborate

LESSON 6.6

21st Century Learning

CRITICAL THINKING Remind students that the life cycle of cells in the skin concludes when dead cells become part of the surface layer of the epidermis. Ask: **What is the advantage of having dead cells instead of living tissue make up the outer layer of skin?** *(Living tissue has nerves and blood vessels. You would bleed more easily and feel pain more easily from cuts and pressure if living tissue made up the outer layer of skin.)*

Make Analogies

L1 A LAYER OF PROTECTION To help students understand that the surface layer of skin consists of dead cells, compare this surface layer to a cloth garment such as a shirt. Ask: **What does a shirt do in relation to your body?** *(Sample: It covers the body.)* **In what way does it protect the body?** *(Sample: It keeps the body from various products of the environment, such as the sun's rays, rain, dirt, and cold.)* **How is a shirt like the surface layer of the epidermis?** *(Both are inanimate structures that protect the body.)*

21st Century Learning

COMMUNICATION Have students research the causes of common skin problems and diseases: What kinds of foods and substances help promote healthy skin? What kinds of foods and substances can damage the skin? Students should summarize their findings by writing a script for a public-service announcement about how to maintain healthy skin.

Lab Resource: Quick Lab 

L2 SWEATY SKIN Students will model the cooling effects of the evaporation of sweat. This Quick Lab can be found in the Student Lab Manual, p. 83, and online.

Evaluate

RTI Response to Intervention

1a. If students need help identifying how the skin gathers information about the environment, **then** remind students about the structures that make up the epidermis.

b. If If students have trouble recognizing how some structures in the skin protect the body, **then** encourage students to review the functions of the structures in the dermis from **Figure 3**.

Name ______________________ Date __________ Class __________

Assess Your Understanding

The Integumentary System

What Are the Main Functions of the Integumentary System?

1a. **IDENTIFY** How does your skin gather information about the environment?

____

____

b. **RECOGNIZE** Explain how some structures in the skin protect your body.

____

____

got it?

○ **I get it!** Now I know that the skin has two layers that ______________________

____

____

○ **I need extra help with** ______________________

____

Place the outside corner, the corner away from the dotted line, in the corner of your copy machine to copy onto letter-size paper.

Name ______________________ Date ____________ Class ____________

Enrich

The Integumentary System

Doctors classify burns according to the depth of skin damage. Burns can be classified as first-, second-, and third-degree burns, as shown in the diagram below. Look at the diagram and read the passage. Then answer the questions that follow on a separate sheet of paper.

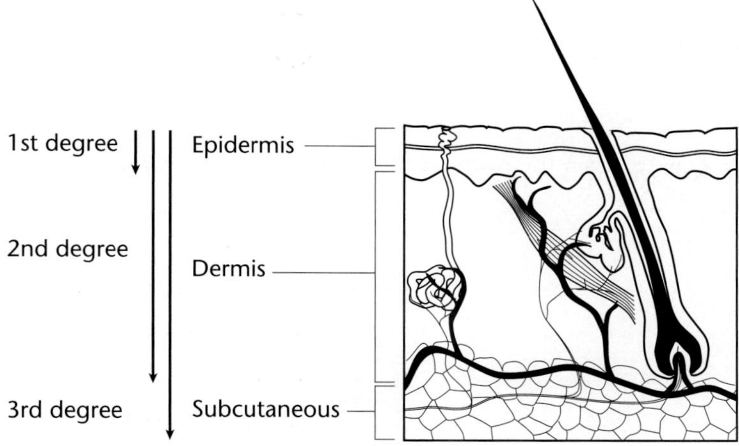

Burns

First-degree burns can be caused by briefly touching a hot object or by coming into contact with hot water or steam. A mild sunburn is also considered a first-degree burn. In a first-degree burn, only the epidermis is damaged. These burns make the skin turn red and swell slightly.

Second-degree burns can be caused by coming into contact with flames, spilling a very hot liquid on yourself, or getting a deep sunburn. In a second-degree burn, both the epidermis and the dermis are damaged. These burns turn the skin bright red and cause blisters. These burns are very painful.

Third-degree burns can also be caused by contact with flames or by spilling hot liquids on yourself. In a third-degree burn, the entire thickness of the skin is damaged, including blood vessels, sweat glands, oil glands, hair follicles, and other skin tissues and structures. These burns are often leathery in appearance and may be red, white, tan, or brown in color. The person feels no pain because even the skin's pain receptors are damaged.

1. Which layer of the skin is affected by a first-degree burn? By a second-degree burn?
2. List at least four parts of the skin that are damaged by a third-degree burn.
3. Because the skin is destroyed in third-degree burns, what would be one possible complication of these burns while the patient is recovering? Explain your answer.
4. Many burns occur in the kitchen. What are two things you can do to prevent burn accidents there?

Name ______________________ Date ____________ Class ____________

Lesson Quiz

The Integumentary System

Write the letter of the correct answer on the line at the left.

1. ___ Which of the following is NOT a function of the integumentary system?

A regulating body temperature

B gathering information

C transporting materials

D eliminating wastes

2. ___ Which of the following is NOT found in the inner layer of skin?

A oil glands

B blood vessels

C pores

D sweat glands

3. ___ Which of the following is a pigment that colors the skin?

A vitamin D

B melanin

C follicle

D pore

4. ___ The surface layer of your skin contains dead skin cells that stay in this layer for about

A one to two hours

B two to three days

C two to three weeks

D one to two months

Fill in the blank to complete each statement.

5. Skin helps the body maintain a steady temp through perspiration and the enlarging of blood vessels.

6. Nerves in the skin gather information from the environment about pressure, temperature, and pain.

7. The dermis is the inner layer of skin.

8. oil produced in glands around the follicles keeps the surface of the skin moist and the hairs flexible.

9. Skin helps eliminate wastes and produce vitamin D.

10. The eprdemis is the outer layer of skin.

The Integumentary System

Answer Key

Review and Reinforce

Find the worksheet in the Student Workbook.

1. Vitamin D is important for healthy bones because it helps the body absorb calcium in food.

2. Dead cells become part of the epidermal surface layer that helps protect your skin.

3. nerves, blood vessels, sweat glands, hairs, oil glands, hair follicles, fat

4. d

5. c

6. a

7. b

8. e

9. The skin's importance involves several functions. Skin protects the body by keeping harmful substances outside the body and keeping important substances inside the body. Skin helps regulate body temperature, eliminate wastes, gather information about the environment, and produce vitamin D.

Enrich

1. the epidermis; the epidermis and dermis

2. Accept all reasonable responses. Students will likely say the epidermis, the dermis, the layer of fat, blood vessels, hair follicles, nerves, sweat glands, oil glands

3. Sample: Infection is one possible complication, because the skin is entirely destroyed, and the skin protects the body from infection.

4. Accept all reasonable responses. Students may say: Always use a pot holder or an oven mitt to remove hot pots from the stove; keep pot handles turned in to prevent pots from being knocked off the stove; keep clothing from coming in contact with the flame or heating element on the stove.

Lesson Quiz

1. C

2. C

3. B

4. C

5. temperature

6. Nerves

7. dermis

8. Oil

9. vitamin D

10. epidermis

Scientific Investigation and Reasoning

This Apply the TEKS feature will enable students to use the scientific investigation and reasoning skills in TEKS 2C, 2D, and 2E and to reinforce the content of TEKS 7A. Students will collect, record, and analyze data in order to contrast situations where work is done with different amounts of force to situations where no work is done.

2C, 2D, 2E, 7A

Doing Work

Have students read about the three situations they are being asked to contrast. Explain that two conditions must exist for work to be done: (1) a force must be applied to an object and (2) the object must move in the same direction in which the force is applied. Ask: **Are you doing work when you lift a stack of books from the floor to a table?** *(Yes)* **Are you doing work if you hold the books in front of you and carry them across the room at a constant velocity? Explain.** *(No; the books are not moving in the direction in which the force is applied.)*

Make sure students understand how the formula for calculating work is derived. Ask: **What two factors determine how much work is done when performing an action?** *(The force applied and the distance the object is moved)* Point out that when force is measured in newtons and distance in meters, the SI unit of work is the newton-meter (N•m). This unit is also referred to as a joule.

Texas Essential Knowledge and Skills

2C Collect and record data using the International System of Units (SI) and qualitative means such as labeled drawings, writing, and graphic organizers.

2D Construct tables and graphs, using repeated trials and means, to organize data and identify patterns.

2E Analyze data to formulate reasonable explanations, communicate valid conclusions supported by the data, and predict trends.

7A Contrast situations where work is done with different amounts of force to situations where no work is done such as moving a box with a ramp and without a ramp, or standing still.

APPLY THE TEKS 2C, 2D, 2E, 7A

DOING WORK

Your class is learning how interactions among your muscular, skeletal, and nervous systems allow you to move your body and other objects. Remember that you do work any time you exert a force on an object that causes the object to move some distance.

Your teacher divides the class into groups and gives each group a 1-kg box. She asks your group to contrast the amount of work done in three different situations: (1) standing still while holding the box, (2) using a force of 10 N to lift the box and put it on a platform that is 2 m high, and (3) using force of 4 N to move the box up a 5-m ramp onto the same platform.

1. **Apply Concepts** The amount of work done on an object can be determined by multiplying the force (in newtons) times the distance. Write the formula that you will use to determine the amount of work done in each situation.
   Work = Force x Distance
2. **Collect and Record Data** Determine the amount of work done in each of the three situations. Record the data in joules (J) in a table like the one shown.

Situation	Work Done (J)
1	*0*
2	*20*
3	*20*

3. **Analyze Data** In which situation was the least amount of work done? Why is this the case?
   Sample: Situation 1 involved no work because the box was not moved at all.

4. **Compare and Contrast** How does the amount of work compare in Situations 2 and 3? How many times more force was required to move the box without the ramp than with it? How much farther was the box moved with the ramp than without it?

The amount of work is the same, 20 N. It required about 2.5 times more force without the ramp. The box was moved about 2.5 times farther with the ramp.

5. **Analyze Data** Remember that a machine makes doing work easier. Explain how the ramp is a machine.

Sample: The same amount of work is done in Situations 2 and 3. The ramp makes it easier to do work because using it requires less force on the box.

Scientific Investigation and Reasoning, Cont.

After students answer Question 4, have them look for patterns in the data. Ask: **What is the relationship between the additional amount of force needed to move the box without the ramp and the additional distance the box was moved with the ramp?** *(They are the same amount.)*

Remind students that a simple machine makes work easier by changing at least one of three factors: the amount of force you exert, the distance over which you exert your force, or the direction in which you exert your force. Ask: **If the ramp reduces the amount of force you have to exert, then what generalization can you make about the distance over which the force is exerted when you use the ramp?** *(The force must be exerted over a greater distance than if no ramp were used.)*

TEKS Practice

Assess Understanding

Have students complete the answers to the TEKS Practice questions. Have a class discussion about what students find confusing. Write Key Concepts on the board to reinforce knowledge.

RTI Response to Intervention

6. If students have trouble comparing and contrasting cells and tissues, **then** have them review **Figure 3** and body organization.

10. If students need help completing the diagram, **then** have them review the terms *stimulus* and *response*.

Alternate Assessment

L1 MAKE A MURAL Have students design and create a six-panel mural that summarizes the chapter content. Each panel should depict the information covered in one lesson. Students should be sure to incorporate all the Key Concepts, vocabulary terms and address the Focus Question. The class can then use the mural to review chapter material.

PEARSON Texas.com

CHAPTER 6

TEKS Practice

TEKS 12B, 12C, 13A, 13B

LESSON 1 Body Organization

1. Bone tissue and fat tissue are examples of
   - a. muscle tissue.
   - b. nervous tissue.
   - c. epithelial tissue.
   - (d.) connective tissue.

2. The cell membrane forms the outside border of a cell.

3. **Identify** Which body system is responsible for transporting oxygen, nutrients, and wastes?
   The circulatory system

4. **Describe** What is an organ system?
   An organ system is a group of organs that work together to carry out certain functions.

5. **Recognize** Why is the job of an organ usually more complex than the job of a tissue?
   Sample: An organ is at a higher level of organization than a tissue.

6. **Compare and Contrast** How is a cell different from a tissue? How are they similar?
   A cell is the basic unit of structure that helps a living thing function. A tissue is a group of similar cells that have the same function. Cells and tissues are both levels of body organization, and both perform specific functions.

LESSON 2 System Interactions

7. Signals from the ________ make skeletal muscles move.
   - (a.) nervous system
   - b. digestive system
   - c. respiratory system
   - d. muscular system

8. Absorption occurs when nutrients move from the digestive system into the bloodstream.

9. **Identify** Which systems' main functions include regulating and controlling body functions in humans?
   The nervous and endocrine systems

10. **Explain** You walk outside from school on a bright, sunny day and immediately shield your eyes with your hand. Use this information to complete the diagram below.

Stimulus
Bright sun

Response
I shield my eyes from the sun

11. **Infer** Your knee is called a hinge joint. It is called this because it bends like the hinge on a door. What are some other examples of joints in your body that work like a hinge?
    Sample: Elbow, finger, or toe are examples of hinge joints.

LESSON 3 Homeostasis

12. Under what circumstances would your endocrine system release adrenaline?
    a. sleep
    b. sudden stress (circled)
    c. absorption
    d. homeostasis

13. Your body systems work together to maintain internal conditions, or homeostasis

14. **Explain** What is the fight-or-flight response? Which body system is responsible for it?
    Sample: It is a response to a possible threat. The endocrine system releases hormones that get your body ready to fight or run from the threat.

15. **Describe** What effect might a problem with your inner ear, such as an infection, have on your body? Explain.
    Sample: I might have trouble keeping balance because structures in the inner ear work with my brain to help me keep my balance.

16. **Apply Concepts** Imagine you are leading a workshop to help students deal with stress. Explain why it is important to reduce stress.
    Sample: Ongoing stress can cause health problems, such as headaches, digestion problems, and heart problems.

LESSON 4 The Skeletal System

17. A soft connective tissue found inside bones is
    a. cytoplasm.
    b. marrow. (circled)
    c. cartilage.
    d. osteoporosis.

18. The vertebrae make up your backbone.

19. **Identify** What are the main functions of the skeletal system?
    Provides shape and support, enables movement, protects organs, produces blood cells, and stores minerals and other materials

20. **Infer** Why do bones need to be both strong and lightweight?
    Sample: Bones need to be able to absorb force without breaking but still be light enough for people to be able to move around easily.

21. **Compare and Contrast** Does an adult human contain more or less cartilage than a human baby? Explain.
    An adult contains less cartilage than a baby because most of it is replaced by bone as we grow.

22. **math!** The smallest bone in the human body is the stapes (or stirrup) in the inner ear, at about 3 mm. The largest bone is the femur, which is about 480 mm in length in an average male. About how many times smaller is the stapes than the femur in the human body?
    About 160 times smaller: $480 \div 3 = 160$

TEKS Practice, Cont.

RTI Response to Intervention

14. If students cannot explain what the fight-or-flight response is, **then** have them describe how they might react if a bear jumped out at them during a hike in the woods.

20. If students need help inferring why bones need to be strong and lightweight, **then** ask them to think about the main functions of the skeletal system.

CHAPTER 6

TEKS Practice, Cont.

RTI Response to Intervention

25. If students have difficulty distinguishing between involuntary and voluntary muscles, **then** ask them to describe what a *volunteer* is.

31. If students cannot explain the effect of a cut, **then** have them look at **Figure 3** to review details of the dermis and epidermis.

L3 WRITING IN SCIENCE Have students write a radio interview with a doctor that explains to listeners how the human body works.

TEKS Practice

LESSON 5 The Muscular System

23. Muscles that help the skeleton move are

a. cardiac muscles.
b. smooth muscles.
(c.) skeletal muscles.
d. involuntary muscles.

24. Skeletal muscles must work in pairs because muscle cells can only contract.

25. Identify Is the act of pouring a glass of water controlled by involuntary muscles or voluntary muscles? Explain.

Sample: Voluntary muscles; the muscles you use are under conscious control.

26. Compare and Contrast Write one similarity and one difference between skeletal muscle and cardiac muscle.

Sample: Both are striated. Skeletal muscle tires quickly, while cardiac muscle does not.

27. Predict What would happen if a tendon in your finger were cut?

Sample: The muscle would not be attached to the bone.

28. Describe When you get a cramp, what is happening to the skeletal muscle?

The muscle is staying contracted, which causes the cramp.

LESSON 6 The Integumentary System

29. A pigment that colors the skin is

a. the dermis.
b. the epidermis.
c. a follicle.
(d.) melanin.

30. Hair grows in follicles.

31. Make Generalizations What skin layers are affected when a cut on your hand bleeds?

The cut affects the epidermis and dermis.

32. Identify What are the main functions of the integumentary system?

It creates a barrier to protect the body, helps maintain temperature, eliminates wastes, gathers information, and produces Vitamin D

33. Interpret Diagrams Label the epidermis and dermis in the diagram below.

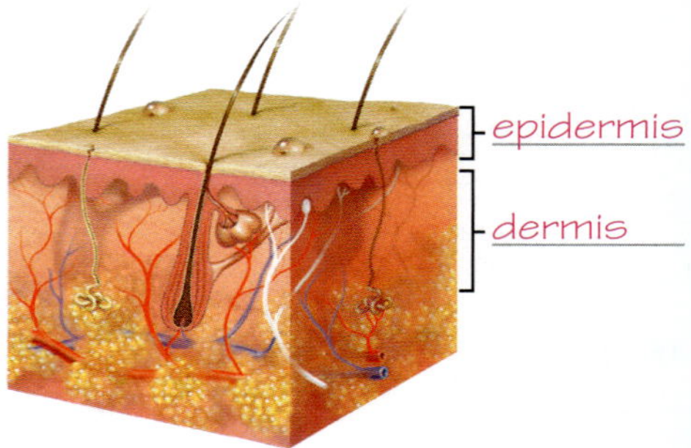

34. Write About It You are out running with your friend on a hot day. Your body begins to warm up. Describe how your skin responds to the excess heat.

See TE rubric.

Write About It **34** Assess student's writing using this rubric.

SCORING RUBRIC	SCORE 4	SCORE 3	SCORE 2	SCORE 1
Describe the skin's response to excess heat	Student describes enlarging of blood vessels in skin to let body heat escape and perspiration evaporating from the skin to cool the body.	Student describes one of the skin's responses to excess heat.	Student describes how the skin responds to excess heat with prompting or assistance.	Student cannot describe how the skin responds to excess heat.

258 Introduction to the Human Body

How does your body work?

35. Describe how the body systems of this boy function during and after the time he eats his lunch. Use at least four different body systems in your answer.

Sample: As the boy eats, his nervous system directs his muscles to move his bones (skeletal system) so he can pick up his food and chew. His digestive system breaks down the food into nutrients. His circulatory system absorbs the nutrients and delivers them to his cells. See TE rubric.

Interactive Science Chapter 6

Lesson 1
In Lesson 6.1 you learned about the levels of organization in the body. These levels consist of cells, tissues, organs, organ systems, and organism.
Supporting TEKS: 12B
TEKS: 12C

Lesson 2
In Lesson 6.2, you learned that muscles, bones, and nerves work together to make your body move. The circulatory, respiratory, digestive, and excretory systems move materials, while the nervous and endocrine systems control body functions.
Supporting TEKS: 12B
TEKS: 12C

Lesson 3
In Lesson 6.3, you learned that all of the human body systems work together to maintain homeostasis and keep the body in balance.
Supporting TEKS: 12B
TEKS: 13A, 13B

Lesson 4
In Lesson 6.4, you learned that your skeleton provides shape and support, enables you to move, protects your organs, produces blood, and stores minerals. Bones are living organs that grow and repair themselves. Joints in the skeletal system allow bones to move in different ways.
Supporting TEKS: 12B
TEKS: 12C

Lesson 5
In Lesson 6.5, you learned about the muscular system. Your body has skeletal, smooth, and cardiac muscle tissues. Some of these tissues are involuntary, and some are voluntary. Skeletal muscles work in pairs because they only contract.
Supporting TEKS: 12B
TEKS: 7A, 12C

Lesson 6
In Lesson 6.6, you learned that the skin has two layers to protect the body. Skin also helps regulate body temperature, eliminate wastes, and produce vitamin D.
Supporting TEKS: 12B
TEKS: 12C

How does your body work?

Assess student's response using this rubric.

SCORING RUBRIC	SCORE 4	SCORE 3	SCORE 2	SCORE 1
Describe how body systems work together to perform a function	Student accurately describes how four systems work together as and after the boy eats.	Student can describe how three systems work together as and after the boy eats.	Student can describe how two systems work together as and after the boy eats.	Student can describe the action of only one system as and after the boy eats.

TEKS Practice, Cont.

How does your body work?

Students should be able to demonstrate understanding of how the human body works. See the scoring rubric below.

Review the TEKS Chapter 6

PARTNER REVIEW Have partners review the definitions of vocabulary terms and quiz each other. Student can read the Key Concept statements and leave out words for their partners to fill in. They also can change a statement so that it is false and then ask their partners to correct it.

CLASS ACTIVITY: CONCEPT MAP Have students work in six groups to develop concept maps to show how the information in this chapter is related. Invite each group to brainstorm in order to identify Key Concepts, vocabulary, definitions, examples, and important details from one lesson in the chapter. Encourage students to write information on sticky notes and to attach them at random on poster board, paper, or the board. Groups can use these notes to develop a concept map that begins at the top with the lesson's Key Concepts. After each group has completed its work, bring the groups together and discuss how the concept maps relate to one another. Ask students to use the following questions to help them organize the information on their sticky notes:

- What are the levels of organization in the human body?
- How do organ systems work both independently and together?
- What are the structures and functions of the different body systems?
- What is homeostasis and how is it maintained?

SAMPLE CONCEPT MAP:

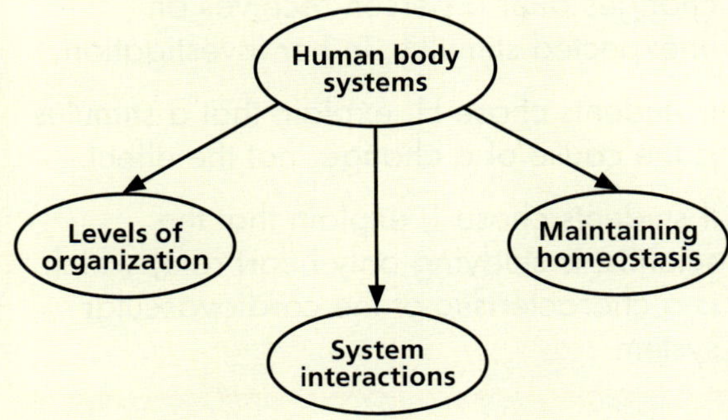

CHAPTER 6

TEKS Practice: Chapter Review

Test-Taking Skills

PAYING ATTENTION TO THE DETAILS Explain to students that two or more answers to a multiple-choice question may be almost identical. If students do not read the answers carefully, then they might select an incorrect answer by mistake. Tell students to read the question and the answer choices carefully before choosing an answer.

Question 1 TEKS 12B

The correct answer is A. When the biceps shortens, it pulls the bones in the upper and lower parts of the arm together.

If students chose B, explain that when muscles contract, they pull on bones.

If students chose C, explain that the muscle in the back of the upper arm—the triceps—straightens the arm when it contracts.

If students chose D, explain that the biceps are responsible for bending the arm, not straightening It. Also, the biceps pulls on bones.

Question 2 TEKS 13A

The correct answer is G. The scientist is comparing heart rates before and after a person receives a stimulus.

If students chose F, explain that the scientist is looking at how heart rate changes after a person receives an unexpected stimulus. in her investigation.

If students chose H, explain that a stimulus is the cause of a change, not the effect.

If students chose J, explain that the scientist is studying only heart rate, which is a characteristic of the cardiovascular system.

★ TEKS Practice: Chapter Review

Read each question and choose the best answer.

1 Your biceps is a muscle in the front part of your upper arm, as shown below. Tendons connect muscles and bones in your arms.

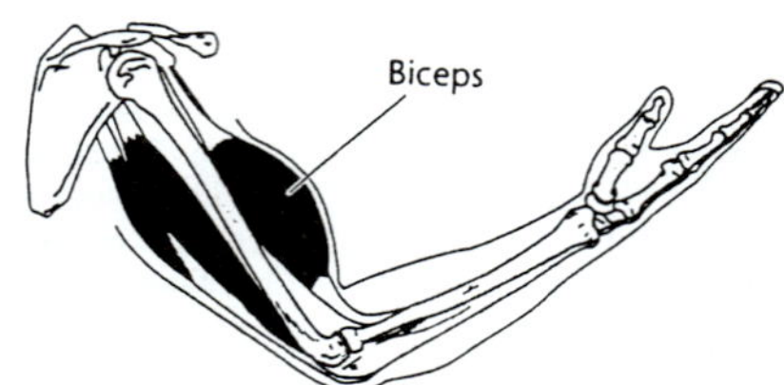

What happens when your biceps contracts, or shortens?

- (A) The biceps pulls on the bones it is connected to and bends your arm.
- B The biceps pushes on the bones it is connected to and bends your arm.
- C The biceps pulls on the bones it is connected to and straightens your arm.
- D The biceps pushes on the bones it is connected to and straightens your arm.

2 A scientist is studying heart rates in people. In her investigation, she measures the heart rate of a person before and after the person receives an unexpected stimulus. Which scientific question could she answer with this investigation?

- F Do people have different heart rates?
- (G) Does heart rate change after a person receives a stimulus?
- H Does a change in heart rate result in a stimulus?
- J Does a stimulus affect body systems other than the cardiovascular system?

★ TEKS Practice: Cumulative Review

3 The Punnett square below summarizes the process by which the gender of a child is determined. The combination XX results in a female child, while XY results in a male child.

	X	Y
X	XX	XY
X	XX	XY

What does this Punnett square indicate about how the gender of a child is determined?

A Sex chromosomes are only passed onto children from the male parent.

B The female sex chromosome is dominant with respect to the male sex chromosome.

C The sex chromosomes passed on by both parents determine the gender of the child.

(D) Only the sex chromosome passed on by the male parent determines the gender of the child.

4 A cell has a low oxygen supply, but has plenty of glucose. Which process can the cell carry out to gather energy?

F Cellular respiration

G DNA replication

(H) Fermentation

J Photosynthesis

If You Have Trouble With . . .				
Question	1	2	3	4
See Lesson	6.5	6.3	4.2	3.2
TEKS	7.12B	7.13A	7.14A	7.7B

TEKS Practice: Cumulative Review

Additional Assessment Resources

Teacher's Edition: Lesson Quizzes, Texas End-of-Year Test Prep A and B
Online assessments

Question 3 TEKS 14A

The correct answer is D. The gender of the child is determined by whether an X or Y chromosome is passed on to the child from the father.

If students chose A, explain that both parents pass on sex chromosomes to offspring.

If students chose B, explain that if this were true, then XY would result in a female child.

If students chose C, explain that the female parent always passes on an X chromosome to offspring.

Question 4 TEKS 7B

The correct answer is H. During fermentation, cells release energy from food molecules such as glucose without using oxygen.

If students chose F, explain that cellular respiration requires oxygen.

If students chose G, explain that DNA replication is needed for cell division. It does not produce energy for cells.

If students chose J, explain that photosynthesis uses light energy to produce glucose and oxygen. It does not release energy for cells to use.

CHAPTER 6

Science Matters

Kids Doing Science

Have students read *Student Athletic Trainers: More Than Counting Exercises.* Point out that because of improved safety rules and protective gear, there is good news in the declining number of sports-related injuries among high school athletes since the 1990s. However, injuries still occur. Athletic trainers deal with many common student sports injuries such as swollen muscles, knee injuries, tendon injuries, sprains, strains, tears, dislocations, pain along the shinbone, and bone breaks. Because student trainers may have to assist in any of these situations, their training includes learning about human body anatomy. They must also learn basic injury evaluations, emergency procedures, taping and treatment of common athletic injuries, injury prevention, and injury rehabilitation. So until students stop playing sports, there will always be a need for qualified athletic trainers. The best preparation is getting experience as a student athletic trainer.

Ask: **Why is knowledge about the skeletal and muscular body systems key for student athletic trainers to understand?** *(Sample: The most frequent student sports-related injuries are to the skeletal and muscular body systems. Examples include swollen muscles, sprains, strains, tears, dislocations, pain along the shinbone, and bone breaks.)* **Why are athletic trainers important to school sports teams?** *(Sample: Because athletic trainers are trained in the anatomy and physiology and know how to diagnose and treat many athletic injuries)*

Texas Essential Knowledge and Skills

12B Identify the main functions of the systems of the human organism, including the circulatory, respiratory, skeletal, muscular, digestive, excretory, reproductive, integumentary, nervous, and endocrine systems.

SCIENCE MATTERS

Kids Doing Science

STUDENT ATHLETIC TRAINERS:

MORE THAN COUNTING EXERCISES

We all know of athletes such as basketball players, figure skaters, or swimmers, who have impressive physical skills. But even the most skilled athletes can use help getting into top form. Athletic trainers are some of the most important people to any sports team or individual athlete. If you are interested in medicine or sports, you don't have to wait until college to learn more. Being a student athletic trainer can give you an inside look into the world of sports medicine.

TEKS 12B

Student trainers gain basic first-aid skills and learn about anatomy and physiology, the study of the body's functions. Student trainers must understand how the skeletal and muscular systems work together. They also learn how to diagnose and treat common athletic injuries. Some student trainers visit college and professional sports training rooms.

Write About It Interview a coach and a trainer at your school. How do they use their knowledge of the muscular and skeletal systems to help athletes train and prevent injuries? How do they help athletes recover from injuries? Write an article for your school newspaper describing the interviews.

262 Introduction to the Human Body

Quick Facts

Concussions that occur during some school sports can put an end to a promising student athlete's career. Zealous athletes often remain silent about any disorientation or nausea they feel during a game because they do not want to be benched. Coaches and medical staff must be keenly aware of any symptoms in their team's players. Studies have shown that concussions are drastically underreported in high school football.

Have students find out more about the dangers of unreported concussions in school sports. Have them write an article for the school newspaper or Web site that alerts students to the dangers of unreported concussions in school sports.

Seeing THE SKELETAL SYSTEM

TEKS 3D

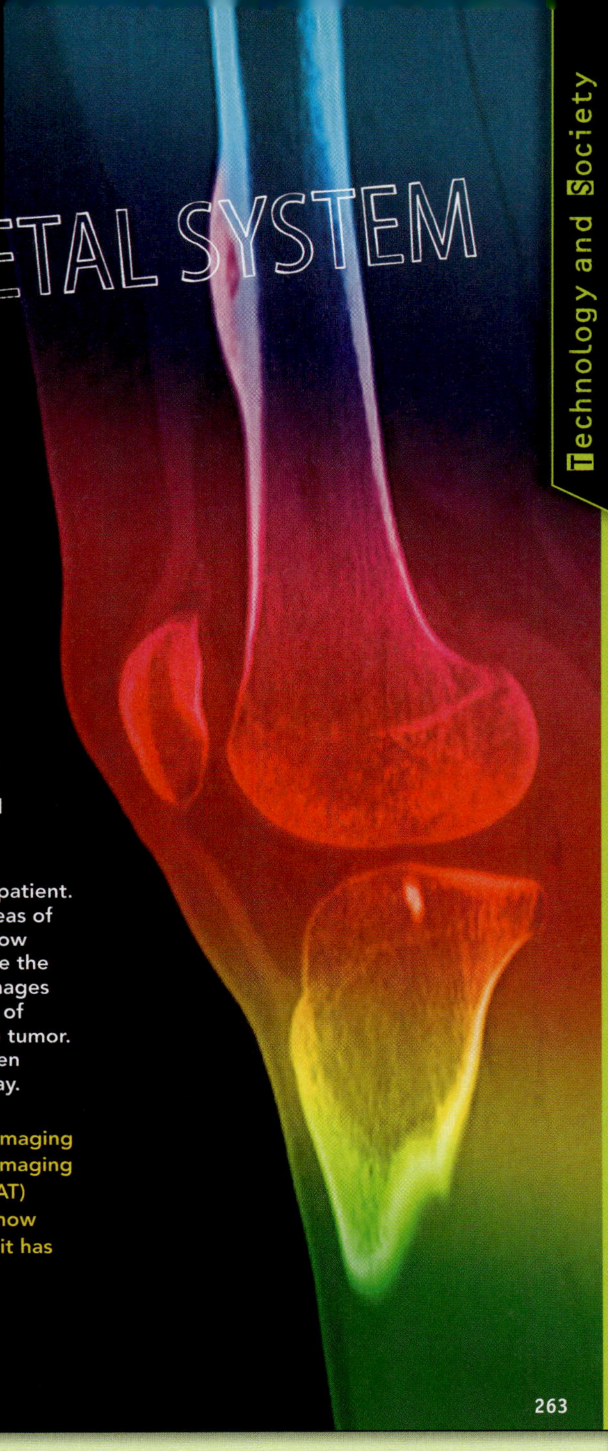

Technology and Society

Have you ever broken a bone? If you have, you may have gotten an X-ray. X-rays are just one example of the powerful tools that doctors use to look inside the human body without surgery.

X-rays can reveal a broken bone, but other tools such as bone scans help doctors determine how a patient's bones are growing. During a bone scan, tiny amounts of radioactive material are injected into the patient's bloodstream. This material contains elements whose molecules have unstable nuclei. As the nuclei of these elements in the radioactive material break down, they release radiation, including gamma rays. The radioactive material accumulates in areas of the bone where growth and cell division are taking place.

After a few hours, a special camera that detects gamma rays is used to scan the patient. The camera shows darker and lighter areas of the patient's bones. The darker areas show where more growth is taking place, while the lighter areas show less growth. These images help doctors figure out whether an area of bone is growing abnormally or if it has a tumor. Bone scans are also used to detect hidden fractures that may not appear on an X-ray.

Research It Research one other medical imaging technique, such as magnetic resonance imaging (MRI) or computed axial tomography (CAT) scans. Write a one-page explanation of how the technology works, and what impact it has had on society in the past 30 years.

CHAPTER 6

Technology and Science

Have students read *Seeing the Skeletal System*. Point out that each type of imaging method reveals different types of bone problems. X-rays show images of the spine and are used to detect spinal fractures. Magnetic resonance imaging (MRI) gives views from many angles and detects back problems such as damage to spinal nerve roots. Computer axial tomography (CAT) is a cross between an X-ray and an MRI. A positron emission tomography (PET) scanner detects radioactive material that has been injected into the patient. Gamma ray detectors then record and produce an image of the area emitting gamma rays.

As students do their research, have them provide examples of medical situations that are best evaluated from the type of medical imaging technique they select.

Ask: **What happens to the radioactive material that is injected into a patient for a bone scan?** *(Sample: Gamma rays accumulate in the areas of the bones where growth and cell division are taking place.)* **What does the camera of a bone scan show?** *(Sample: The camera shows a patient's bones as darker areas where more growth occurred and lighter areas where less growth occurred.)* **What do the bone scan images show?** *(Sample: The images show whether an area of the bone is growing abnormally or has a tumor.)*

English Language Proficiency Standards

ELPS Listening 2.C.1

Read aloud *Seeing the Skeletal System* and point out the words *have you ever.*

Beginning Use *have you ever* to ask questions about tools doctors use: *Have you ever had an X-ray taken? Had a bone scan? Had a blood test? Had an ultrasound?* Discuss students' answers.

Intermediate Model how to use *have you ever* to ask questions about the ways doctors look inside people: *Have you ever had an X-ray?* Have partners ask each other about the tools in the article.

Advanced Display the sentence frame *Have you ever had a ______?* Have partners ask each other about the tools in the article.

Advanced High Have partners use *have you ever* to ask each other about each of the tools in the article.

Texas Essential Knowledge and Skills

3D Relate the impact of research on scientific thought and society, including the history of science and contributions of scientists as related to the content.

CHAPTER 7

Managing Materials in the Body

Chapter TEKS Overview

This chapter focuses on TEKS 12B, 13A, and 13B. Students will explore how materials are managed in the human body by the digestive, circulatory, respiratory, and excretory systems. They also will investigate how these systems help the human body maintain balance in response to external and internal stimuli.

Introduce the TEKS

To pique student interest and introduce TEKS 12B, have students look at the image and read the Focus Question and description. Ask students to explain how blood moves around inside their bodies. Have volunteers share their explanations. Point out that blood is flowing to all parts of our bodies continuously throughout every day and night of our lives. Ask: **What do you think causes blood to flow around your body?** *(Sample: The heart causes blood to flow.)* **Does blood move along certain pathways or does it move randomly?** *(Sample: It moves through veins and arteries.)* **What is blood made of?** *(Accept all responses.)* **What is blood's important function in the body?** *(Accept all responses.)*

Untamed Science Video

BLOOD LINES Before viewing, invite students to discuss what they know about circulation. Then play the video. Lead a class discussion and make a list of questions that this video raises. You may wish to have students view the video again after they have completed the chapter to see if their questions have been answered.

Chapter at a Glance

CHAPTER PACING: 10–14 periods or 5–7 blocks

INTRODUCE THE CHAPTER: Use the Focus Question and the opening image to get students thinking about managing the materials in the body. Activate prior knowledge and preteach vocabulary using the Getting Started pages.

Lesson 1: The Digestive System

Lesson 2: The Circulatory System

Lesson 3: The Respiratory System

Lesson 4: The Excretory System

ASSESSMENT OPTIONS:
Teacher's Edition: Lessons Quizzes, Texas End-of-Year Test Prep A and B
Online assessments

HOW DOES YOUR BLOOD FLOW?

FOCUS ON TEKS 12B

How do systems of the body move and manage materials?

Traffic flows back and forth through a city in all directions, carrying people and goods where they need to go. Like the traffic of a city, your blood flows throughout your body. Some materials "ride" your bloodstream to places in the body where they are used. Other materials get delivered to organs that remove what you don't need. And all the while, your heart keeps things moving.

Infer **What would you like to know about the materials that are moved within your body?**

Sample: I would like to know what kind of materials are used in the body and where. I'd also like to know how wastes are removed.

Watch the **Untamed Science** video to learn more about circulation.

Professional Development Note: From the Author

Most major cities have a public transit system. Metropolitan Washington DC's Metro consists of three major subway lines: The orange line carries passengers traveling northeast and west, the red line serves the north or northwest, and the blue line runs east and south. It's no surprise that the place where these three lines intersect, the Metro Center station, is located in the heart of the capital. It may help some students to think of the cardiovascular system as the body's transit system, transporting materials to the cells and removing waste from the cells in the same way the subway transports people.

Zipporah Miller

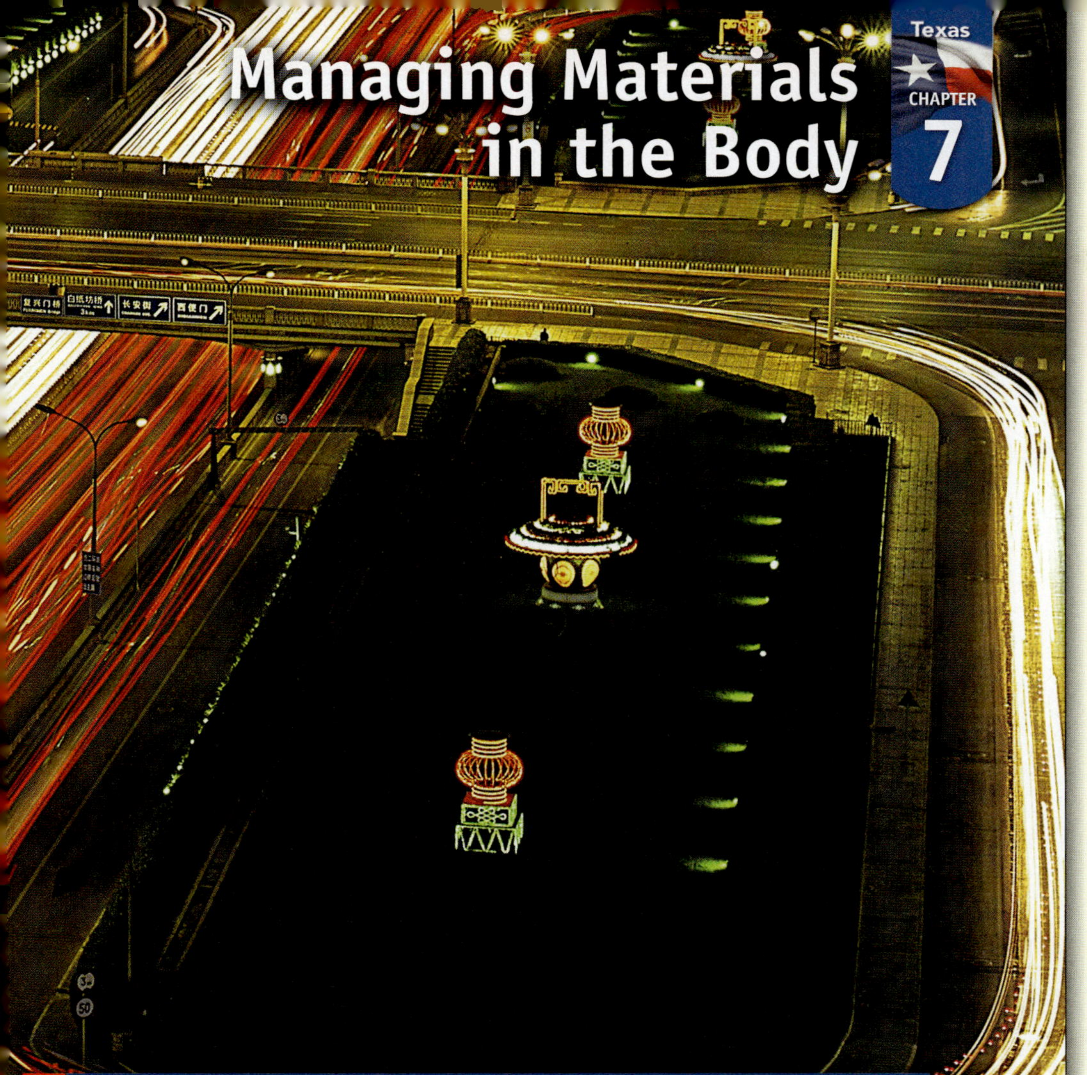

Managing Materials in the Body

Texas CHAPTER 7

Texas Essential Knowledge and Skills

SUPPORTING TEKS: 6A Identify that organic compounds contain carbon and other elements. **6B** Distinguish between physical and chemical changes in matter in the digestive system. **12B** Identify the main functions of the systems of the human organism, including the circulatory, respiratory, digestive, and excretory systems.

TEKS: 2D Construct graphs, using repeated trials and means, to organize data and identify patterns. **2E** Analyze data to formulate reasonable explanations. **3A** In all fields of science, analyze and evaluate scientific explanations by using empirical evidence. **3D** Relate the impact of research on scientific thought and society. **6C** Recognize how large molecules are broken down into smaller molecules such as carbohydrates can be broken down into sugars. **12C** Recognize levels of organization in animals. **12E** Compare the functions of a cell to the functions of organisms such as waste removal. **13A** Investigate how organisms respond to external stimuli in the environment. **13B** Describe and relate responses in organisms that may result from internal stimuli such as vomiting in animals that allow them to maintain balance.

PEARSON Texas.com 265

Spotlight On Technology

Editable Presentation Each lesson includes an editable presentation to help students understand the main functions of the digestive, circulatory, respiratory, and excretory systems.

Flipped Video for Science Show students a Pearson Flipped Video for Science for this chapter so that they can better understand the main functions of the digestive, circulatory, respiratory, and excretory systems.

PEARSON Texas.com

Texas Essential Knowledge and Skills

2D Construct tables and graphs, using repeated trials and means, to organize data and identify patterns.

2E Analyze data to formulate reasonable explanations, communicate valid conclusions supported by the data, and predict trends.

3A In all fields of science, analyze, evaluate, and critique scientific explanations by using empirical evidence, logical reasoning, and experimental and observational testing, including examining all sides of scientific evidence of those scientific explanations, so as to encourage critical thinking by the student.

3D Relate the impact of research on scientific thought and society, including the history of science and contributions of scientists as related to the content.

6A Identify that organic compounds contain carbon and other elements such as hydrogen, oxygen, phosphorus, nitrogen, or sulfur.

6B Distinguish between physical and chemical changes in matter in the digestive system.

6C Recognize how large molecules are broken down into smaller molecules such as carbohydrates can be broken down into sugars.

12B Identify the main functions of the systems of the human organism, including the circulatory, respiratory, skeletal, muscular, digestive, excretory, reproductive, integumentary, nervous, and endocrine systems.

12C Recognize levels of organization in plants and animals, including cells, tissues, organs, organ systems, and organisms.

12E Compare the functions of a cell to the functions of organisms such as waste removal.

13A Investigate how organisms respond to external stimuli found in the environment such as phototropism and fight or flight.

13B Describe and relate responses in organisms that may result from internal stimuli such as wilting in plants and fever or vomiting in animals that allow them to maintain balance.

The following **College and Career Readiness Standards** are covered in this chapter: **I.A.3, I.A.4, I.E.1, III.D.2, VII.A.1, IX.F.1**

CHAPTER 7

Getting Started

Check Your Understanding

This activity assesses students' understanding of how materials circulate and are transported in the body. After students have shared their answers, point out that the materials needed by cells in the body are transported throughout the body by blood in the circulatory system.

Preteach Vocabulary Skills

Invite students to look over the table. Have a volunteer read aloud the heading of the first column, and have two other volunteers read the two verbs and their definitions in that column. Then have different volunteers read aloud the heading and definitions in the second column. Ask students to identify the suffix shared by both terms. Explain that the suffix *-tion* means "the process or act of." Adding this suffix to the root word makes the word into a noun. Have volunteers read the head and the terms in the third column, and ask students to identify the suffix shared by both terms. Explain that the suffix *-atory* means "of, relating to, or connected with the thing specified." Adding this suffix to the root makes the word into an adjective. Clarify the parts of speech with students. Ask: **How do you know the terms in the first column are verbs?** *(Because verbs express actions)* **How do you know the terms in the second column are nouns?** *(Because nouns name abstract ideas)* **How do you know the terms in the third column are adjectives?** *(Because adjectives describe nouns.)*

CHAPTER 7

Getting Started

Check Your Understanding

1. **Background** Read the paragraph below and then answer the question.

Each day, Ken **circulates** from the food pantry to senior centers around the city and then returns to the pantry. He **transports** meals and juice to the seniors and collects their empty bottles. Similarly, your blood circulates materials that the cells in your body need, such as **glucose** and oxygen, and carries away wastes to be excreted from the body.

To **circulate** is to move in a circle and return to the same point.

To **transport** is to carry something from one place to another.

Glucose is a sugar that is the major source of energy for the body's cells.

- What materials does your blood transport to your body cells?

Blood transports oxygen and glucose to body cells.

Vocabulary Skill

Identify Related Word Forms Learn related forms of words to increase your vocabulary. The table below lists forms of words related to key terms.

Verb	Noun	Adjective
respire, *v.* to obtain energy from the breakdown of food molecules	cellular respiration, *n.* the process by which cells obtain energy from the breakdown of food molecules	respiratory, *adj.* concerning respiration
excrete, *v.* to remove or eliminate waste	excretion, *n.* the process by which wastes are removed from the body	excretory, *adj.* concerning excretion

2. **Quick Check** **Fill in the blank with the correct form of *respire*.**
- Obtaining energy from food is a respiratory activity.

English Language Proficiency Standards

ELPS Learning Strategies 1.F

Beginning Use Preview Vocabulary Terms on page 267 to display and read aloud the terms before each lesson. Together scan the lesson looking for clues to meaning in images, labels, and captions. Then give definitions and examples. Have students add the terms to their glossaries along with drawings that illustrate meaning.

Intermediate Before each lesson, display and read aloud the vocabulary terms. Discuss definitions and examples. Have small groups preview the text, look for clues to meaning, and add the terms to their glossaries.

Advanced Provide definitions and examples for each lesson's vocabulary terms. Have partners preview the lesson, locate the terms, and add them to their glossaries.

Advanced High Provide definitions for each lesson's vocabulary terms. Have partners preview the text, locate the terms, and match terms to their definitions.

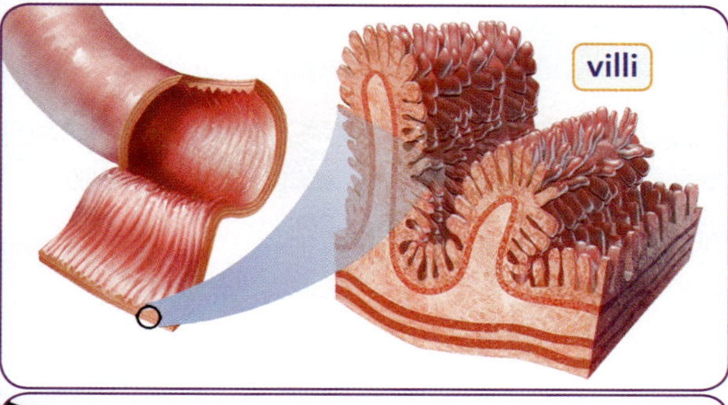

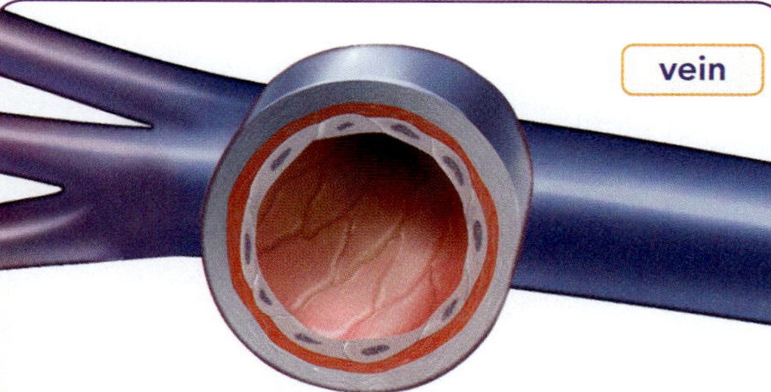

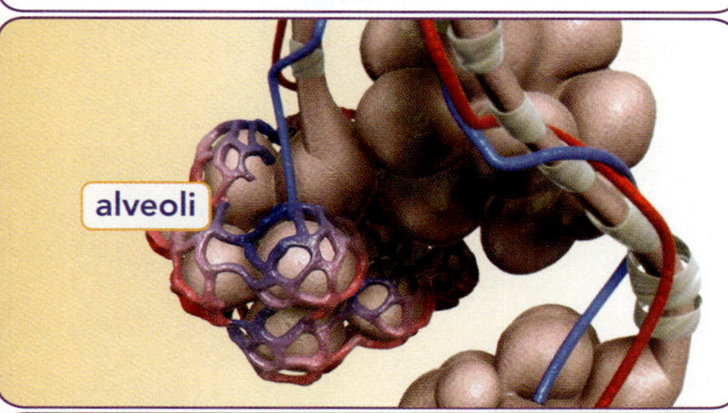

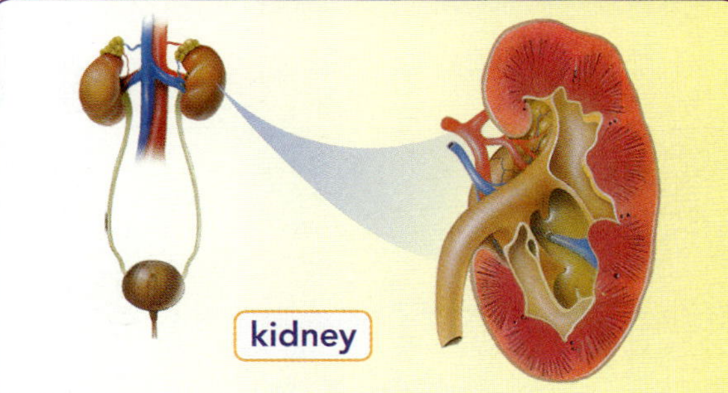

Chapter Preview

LESSON 1

- calorie
- enzyme
- esophagus
- peristalsis
- villi

 Chart
Develop Hypotheses

LESSON 2

- heart
- atrium
- ventricle
- valve
- artery
- aorta
- capillary
- vein
- hemoglobin

Summarize
Observe

LESSON 3

- pharynx
- trachea
- cilia
- bronchi
- lungs
- alveoli
- diaphragm
- larynx
- vocal cords

 Chart
Communicate

LESSON 4

- excretion
- urea
- urine
- kidney
- ureter
- urinary bladder
- urethra
- nephron

Identify the Main Idea
Infer

Preview Vocabulary Terms

Have students work individually to create a personalized glossary to organize the vocabulary terms for this chapter. Be sure to discuss and analyze each term before having students add it to their glossaries. Students may choose to draw an illustration to remind them of the meaning of specific terms. As the class progresses through the chapter, have students return to their glossaries periodically to review vocabulary terms introduced in the earlier lessons.

L1 Have students look at the images on this page as you pronounce the vocabulary word. Have students repeat the word after you. Then read the definition. Use the sample sentence in italics to clarify the meaning of the term.

villi (*VIL eye*) Millions of tiny, finger-shaped structures that line the inner surface of the small intestine. *The villi absorb nutrient molecules from the small intestine and pass them into the blood vessels.*

vein *(VAYN)* The blood vessels that carry blood back to the heart. *A vein has thinner walls than an artery.*

alveoli (*al VEE uh ly*) Compartments in the lungs that allows respiratory gases to exchange with the capillaries. *The thin walls of alveoli allow oxygen to pass into the blood stream.*

kidney (*KID nee*) One of two major organs of the excretory system. *The kidneys' main function is removing urea and other wastes from the blood.*

Academic Vocabulary

Each lesson includes key Academic Vocabulary. See also the Support All Readers box at the start of the lesson.

Lesson 1: chart, develop, hypothesis

Lesson 2: observe, summarize

Lesson 3: chart, communicate

Lesson 4: identify, infer, main idea

The Digestive System

 **How do systems of the body move and manage materials?**

LESSON PACING:
3–4 periods or 1 $\frac{1}{2}$–2 blocks

Lesson Vocabulary

- calorie
- enzyme
- esophagus
- peristalsis
- villi

Content Refresher

Omega-3 Fatty Acids Omega-3 fatty acids are a family of polyunsaturated fats essential to health. One type, alpha-linolenic acid (ALA), is found in flaxseed, walnuts, soybeans, and other plant foods. Two other types, eicosapentaenoic acid (EPA) and docosahexaenoic acid (DHA), are found mainly in cold-water fish such as salmon and halibut. Many scientific studies have shown that some omega-3 fatty acids (specifically, DHA) reduce the risk of coronary heart disease by decreasing the incidence of blood clots and irregular heartbeat, and by lowering blood pressure and triglyceride levels.

Development of some diseases depends on many factors, not just a single nutrient. For example, following a diet low in saturated fat and cholesterol can be one way to reduce the risk of heart disease. Low-fat diets rich in fruits and vegetables (which are low in fat and often contain dietary fiber, vitamin A, or vitamin C) may reduce the risk of some types of cancer.

Lesson Objectives	TEKS	ELPS
Explain why the body needs food and what nutrients it uses.	2D, 6A, 6C	3.G.1
Identify the structures and main functions of the digestive system.	6B, 6C, 12B, 13B	

Texas Essential Knowledge and Skills

2D Construct tables and graphs, using repeated trials and means, to organize data and identify patterns.
6A Identify that organic compounds contain carbon and other elements such as hydrogen, oxygen, phosphorus, nitrogen, or sulfur.
6B Distinguish between physical and chemical changes in matter in the digestive system.
6C Recognize how large molecules are broken down into smaller molecules such as carbohydrates can be broken down into sugars.
12B Identify the main functions of the systems of the human organism, including the circulatory, respiratory, skeletal, muscular, digestive, excretory, reproductive, integumentary, nervous, and endocrine systems.
13B Describe and relate responses in organisms that may result from internal stimuli such as wilting in plants and fever or vomiting in animals that allow them to maintain balance.

English Language Proficiency Standards

ELPS Speaking 3.G.1 Express opinions ranging from communicating single words and short phrases to participating in extended discussions on a variety of social and grade-appropriate academic topics.

DIFFERENTIATED INSTRUCTION KEY
L1 Struggling Students or Special Needs
L2 On-Level Students L3 Advanced Students

LESSON PLANNER 7.1

Investigations and Activities	TEKS Review
My Planet Diary, **Student Edition,** p. 268 Inquiry: Inquiry Warm-Up, Where Does Digestion Start?, **Lab Manual,** p. 84 Introduce Vocabulary, **Teacher's Edition,** p. 269 Teach Key Concepts, **Teacher's Edition,** p. 269 Do the Math!, **Student Edition,** p. 269 Teach With Visuals, **Teacher's Edition,** p. 270 Lead a Discussion, Carbohydrates, Fats, and Proteins, **Teacher's Edition,** p. 270 21st Century Learning, Creativity, **Teacher's Edition,** p. 270 Inquiry: Lab Investigation, Nutrient Identification, **Lab Manual,** p. 85	Apply the TEKS, Observing Enzymes in Saliva, **Student Edition,** p. 304 TEKS Practice, **Student Edition,** p. 305 TEKS Practice: Chapter and Cumulative Review, **Student Edition,** p. 308 Lesson 7.1, **TEKS Preparation and Study Guide Workbook,** p. 68
Teach Key Concepts, **Teacher's Edition,** p. 272 Lead a Discussion, Digestion Begins, **Teacher's Edition,** p. 272 Inquiry: Teacher Demo, Action of Enzymes, **Teacher's Edition,** p. 273 Apply It!, **Student Edition,** p. 273 Support the TEKS, The Stomach, **Teacher's Edition,** p. 274 Differentiated Instruction, **Teacher's Edition,** p. 275 Teach With Visuals, **Teacher's Edition,** pp. 275, 276 Do the Math!, **Student Edition,** p. 276 Lead a Discussion, The Function of the Large Intestine, **Teacher's Edition,** p. 277 Inquiry: Quick Lab, As the Stomach Churns, **PearsonTexas.com**	**SHORT ON TIME?** To do this lesson in approximately half the time, do the Activate Prior Knowledge activity. A discussion of the Key Concepts will familiarize students with the lesson content. Have students do the Quick Lab. The rest of the lesson can be completed by students independently.

These editable worksheets are available on **PearsonTexas.com.**
Print versions can be found in the **TEKS Preparation and Study Guide Workbook.**

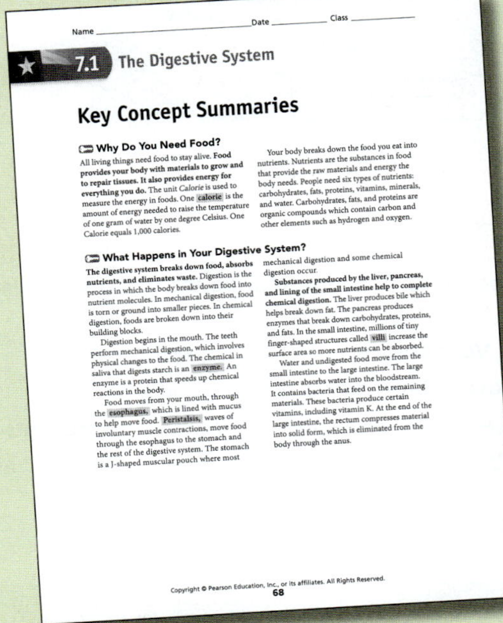

Name ______ Date ______ Class ______

7.1 The Digestive System

Key Concept Summaries

Why Do You Need Food?

All living things need food to stay alive. **Food provides your body with materials to grow and to repair tissues. It also provides energy for everything you do.** The unit *Calorie* is used to measure the energy in foods. One **calorie** is the amount of energy needed to raise the temperature of one gram of water by one degree Celsius. One Calorie equals 1,000 calories.

Your body breaks down the food you eat into nutrients. Nutrients are the substances in food that provide the raw materials and energy the body needs. People need six types of nutrients: carbohydrates, fats, proteins, vitamins, minerals, and water. Carbohydrates, fats, and proteins are organic compounds which contain carbon and other elements such as hydrogen and oxygen.

What Happens in Your Digestive System?

The digestive system breaks down food, absorbs nutrients, and eliminates waste. Digestion is the process in which the body breaks down food into nutrient molecules. In mechanical digestion, food is torn or ground into smaller pieces. In chemical digestion, foods are broken down into their building blocks.

Digestion begins in the mouth. The teeth perform mechanical digestion, which involves physical changes to the food. The chemical in saliva that digests starch is an **enzyme.** An enzyme is a protein that speeds up chemical reactions in the body.

Food moves from your mouth, through the **esophagus,** which is lined with mucus to help move food. **Peristalsis,** waves of involuntary muscle contractions, move food through the esophagus to the stomach and the rest of the digestive system. The stomach is a J-shaped muscular pouch where most mechanical digestion and some chemical digestion occur.

Substances produced by the liver, pancreas, and lining of the small intestine help to complete chemical digestion. The liver produces bile which helps break down fat. The pancreas produces enzymes that break down carbohydrates, proteins, and fats. In the small intestine, millions of tiny finger-shaped structures called **villi** increase the surface area so more nutrients can be absorbed.

Water and undigested food move from the small intestine to the large intestine. The large intestine absorbs water into the bloodstream. It contains bacteria that feed on the remaining materials. These bacteria produce certain vitamins, including vitamin K. At the end of the large intestine, the rectum compresses material into solid form, which is eliminated from the body through the anus.

68

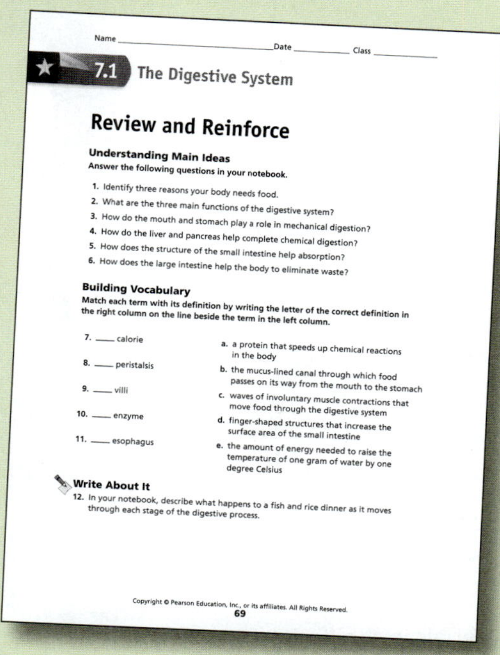

Name ______ Date ______ Class ______

7.1 The Digestive System

Review and Reinforce

Understanding Main Ideas
Answer the following questions in your notebook.

1. Identify three reasons your body needs food.
2. What are the three main functions of the digestive system?
3. How do the mouth and stomach play a role in mechanical digestion?
4. How do the liver and pancreas help complete chemical digestion?
5. How does the structure of the small intestine help absorption?
6. How does the large intestine help the body to eliminate waste?

Building Vocabulary
Match each term with its definition by writing the letter of the correct definition in the right column on the line beside the term in the left column.

7. ____ calorie
8. ____ peristalsis
9. ____ villi
10. ____ enzyme
11. ____ esophagus

a. a protein that speeds up chemical reactions in the body
b. the mucus-lined canal through which food passes on its way from the mouth to the stomach
c. waves of involuntary muscle contractions that move food through the digestive system
d. finger-shaped structures that increase the surface area of the small intestine
e. the amount of energy needed to raise the temperature of one gram of water by one degree Celsius

Write About It

12. In your notebook, describe what happens to a fish and rice dinner as it moves through each stage of the digestive process.

69

Lexile Measure = 900L

LESSON 7.1

The Digestive System

Establish Learning Objectives

After this lesson, students will be able to:

Explain why the body needs food and what nutrients it uses.

Identify the structures and main functions of the digestive system.

Engage

Activate Prior Knowledge

MY PLANET DIARY Read *The Science of Food* with the class. Encourage students to think about the foods they eat every day, as well as their favorite foods. Ask: **What kinds of foods do you eat for a main course in the evenings?** *(Samples: Various meats, fish, pasta, eggs, and salads)* **Which foods do you think of as being "healthy," and which do you think of as being less "healthy"?** *(Accept all responses)*

Explore

Lab Resource: Inquiry Warm-Up

L1 WHERE DOES DIGESTION START? Students will explore where digestion begins in the human body. This Inquiry Warm-Up can be found in the Student Lab Manual, p. 84, and online.

The Digestive System

Why Do You Need Food?
TEKS 2D, 6A, 6C

What Happens in Your Digestive System?
TEKS 6B, 6C, 12B, 13B

MY PLANET DIARY — CAREER

The Science of Food

You know that you need to eat food every day. But did you know that some people study food for a living? Food scientists research and improve the food products you buy at the grocery store. Sometimes, they even think up new foods!

Many food scientists spend a lot of time in a lab. They use what they know about biology and chemistry to test food for nutrition, taste, and shelf life, which is how long the food will last before it spoils. If you like science and food, being a food scientist might be the job for you!

Communicate Discuss these questions with a partner. Then write your answers.

1. Why should food scientists test foods before the foods are sold?
   Sample: Scientists need to find out how long the food will last before it spoils.

2. How might food scientists improve your favorite breakfast food?
   Sample: They might make it contain less sugar so it is better for me; they might make it have a longer shelf life.

Lab zone: Do the Inquiry Warm-Up *Where Does Digestion Start?* Student Lab Manual, p. 84

SUPPORT ALL READERS

Lexile Measure = 900L   Lexile Word Count = 1790

Prior Exposure to Content: May be the first time students have encountered this topic

Academic Vocabulary: *chart, develop, hypotheses*

Science Vocabulary: *calorie, enzyme, esophagus, peristalsis, villi*

Concept Level: Generally appropriate for most students in this grade

Preteach With: My Planet Diary "The Science of Food" and Figure 1 activity

Vocabulary	Skills
• calorie • enzyme • esophagus • peristalsis • villi	Reading: Chart Inquiry: Develop Hypotheses

Why Do You Need Food?

TEKS 2D, 6A, 6C In this section, you'll learn about the nutrients and compounds that the human body obtains from food during the process of digestion.

All living things need food to stay alive. **Food provides your body with materials to grow and to repair tissues. It also provides energy for everything you do.** Exercising, reading, and sleeping require energy. Even maintaining homeostasis takes energy.

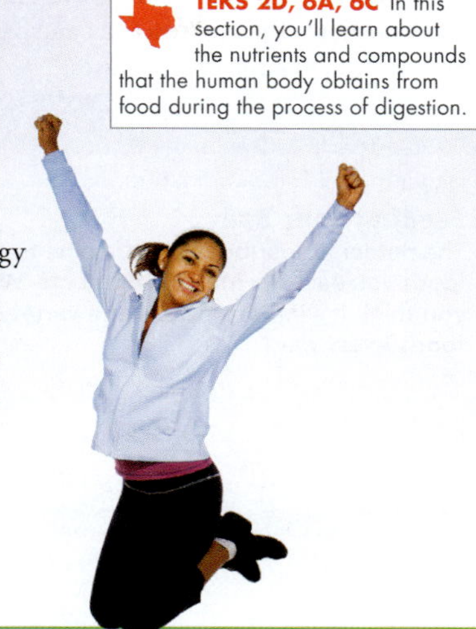

Calories When food is used for energy, the amount of energy released is measured in calories. One **calorie** is the amount of energy needed to raise the temperature of one gram of water by one degree Celsius. The unit *Calorie*, with a capital *C*, is used to measure the energy in foods. One Calorie equals 1,000 calories. Everyone needs a certain number of Calories to meet their daily energy needs. However, the more active you are, the more Calories you need.

do the math!

The U.S. Department of Agriculture recommends that people do about 30 to 60 minutes of physical activity most days. The data table shows the Calories a 13-year-old weighing 45 kilograms burned in 30 minutes of each activity.

Activity	Calories Burned in 30 minutes
Soccer	275
Dancing	115
Walking	75

1. **Graph** Use the data to draw a bar graph.
2. **Name** Write a title for the bar graph.
3. **CHALLENGE** Why does the type of physical activity change a person's dietary needs? Use the graph to explain your answer.

Sample: Some activities, such as soccer, burn more Calories than others, such as walking. The number of Calories a person burns changes how many Calories that person needs to eat to keep a balance in the body.

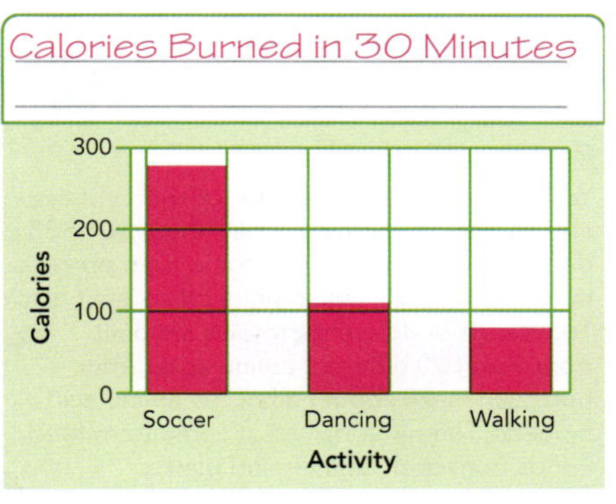

Explain

Introduce Vocabulary

To help students understand that the term *calorie* is a unit of energy, explain that the word derives from the Latin word *calor*, meaning "heat." A calorie is the amount of energy needed to raise the temperature of one gram of water by one degree Celsius.

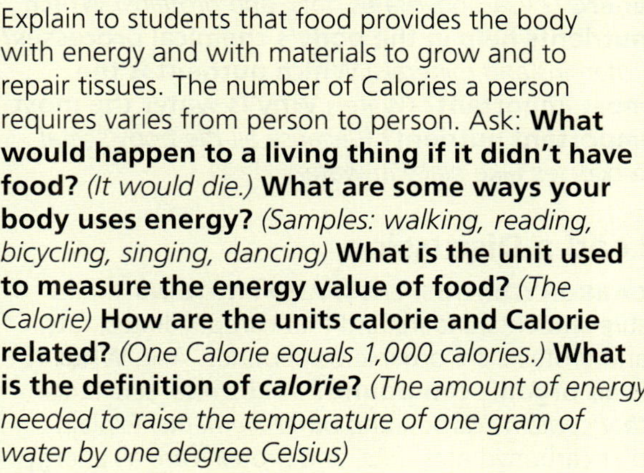

Teach Key Concepts

Explain to students that food provides the body with energy and with materials to grow and to repair tissues. The number of Calories a person requires varies from person to person. Ask: **What would happen to a living thing if it didn't have food?** *(It would die.)* **What are some ways your body uses energy?** *(Samples: walking, reading, bicycling, singing, dancing)* **What is the unit used to measure the energy value of food?** *(The Calorie)* **How are the units calorie and Calorie related?** *(One Calorie equals 1,000 calories.)* **What is the definition of *calorie*?** *(The amount of energy needed to raise the temperature of one gram of water by one degree Celsius)*

Elaborate

Do the Math!

If necessary, help students distinguish the data table from the bar graph. Explain that this data table and this graph express the same information in different ways. Point out that the bar graph allows you to visualize the similarities and differences among the three activities, whereas the data table lets you focus on the precise numbers of Calories burned. Ask: **Where in the data table do the names of activities appear?** (*In the left column*) **Where do these names appear in the bar graph?** (*Along the horizontal axis*) **How many more Calories are burned by 30 minutes of dancing than by 30 minutes of walking?** (*40*) **Which graphic aid did you use to determine the difference?** (*Data table*)

PEARSON Texas.com

LESSON 7.1

English Language Proficiency Standards

ELPS Speaking 3.G.1

Read aloud My Planet Diary on page 268 with students.

Beginning Help students express opinions about food science by scaffolding questions, for example: *Is it important to test foods before they are sold? What could happen to untested food?*

Intermediate Have partners express their opinions by completing the following sentence frames: *Food scientists need to test foods because* _____. *Food scientists could make my* _____ *more* _____ *by* _____.

Advanced Have partners read aloud the text and discuss the questions. Encourage students to elaborate on their ideas.

Advanced High After students discuss the question, have them discuss ways in which food scientists help the public.

Texas Essential Knowledge and Skills

2D Construct tables and graphs, using repeated trials and means, to organize data and identify patterns.

6A Identify that organic compounds contain carbon and other elements such as hydrogen, oxygen, phosphorus, nitrogen, or sulfur.

6C Recognize how large molecules are broken down into smaller molecules such as carbohydrates can be broken down into sugars.

LESSON 7.1

Explain

Teach With Visuals

Direct students' attention to **Figure 1**. Explain that people need six types of nutrients. In order to maintain a balanced state, people must take in a minimum amount of each type of nutrient every day. Ask: **What are the six nutrients people need?** *(Carbohydrates, fats, proteins, vitamins, minerals, and water)* **Which nutrients provide energy?** *(Carbohydrates, fats, and proteins)* **Which nutrients help in the body's chemical processes?** *(Vitamins and minerals)* **Which nutrient is the most important?** *(Water)* **Why is water the most important nutrient?** *(Because all the body's vital processes take place in water)*

Lead a Discussion

CARBOHYDRATES, FATS, AND PROTEINS Make sure students understand that carbohydrates, fats, and proteins are organic compounds. Ask: **What is an organic compound?** *(A chemical substance that contains carbon and other elements)* Explain that carbohydrates, fats, and proteins are all energy sources, but they also play other roles in the body. Ask: **What other role do carbohydrates play in the body?** *(Carbohydrates provide raw material to make cell parts.)* **How do simple carbohydrates and complex carbohydrates differ as energy sources?** *(Simple carbohydrates provide a burst of energy; complex carbohydrates are starches that must first be broken down into sugar molecules. This makes complex carbohydrates a steady long-term source of energy.)* **How do fats and carbohydrates compare as sources of energy?** *(Fats provide more energy per gram than carbohydrates.)* **Besides providing energy, what other roles do fats play in the body?** *(Fats form part of the cell membrane; fatty tissue protects organs and insulates the body.)* **Besides providing energy, what other roles do proteins play in the body?** *(Proteins help the body with growth and tissue repair.)*

21st Century Learning

CREATIVITY Have students research superfoods such as blueberries, spinach, and sweet potatoes. Ask students to create a poster that details the nutrients found in their chosen superfood and how these nutrients contribute to a healthy body. Encourage students to include on their posters several recipes that feature their chosen superfood.

What Nutrients Do You Need? Your body breaks down the food you eat into nutrients. Nutrients are the substances in food that provide the raw materials and energy the body needs. We require six types of nutrients: carbohydrates, fats, proteins, vitamins, minerals, and water.

Nutrients such as carbohydrates, fats, and proteins are organic compounds, or chemical substances that contain carbon and other elements such as hydrogen and oxygen. These compounds are essential to basic life processes.

FIGURE 1

Feeding Your Body

The nutrients your body needs come from the foods you eat. **Make Judgments Why do you think it's important to eat a variety of foods every day?**

Sample: Eating different kinds of food helps you get a balanced diet of nutrients that you need.

Carbohydrates

Carbohydrates (kahr boh HY drayts) are a major source of energy. They also provide raw materials to make cell parts. About 45 to 65 percent of your daily Calories should come from carbohydrates. Simple carbohydrates, called sugars, can give you a quick burst of energy. One sugar, glucose, is the major source of energy for your cells. Complex carbohydrates are made of many linked sugar molecules. Starch is a complex carbohydrate. Potatoes, rice, wheat, and corn contain starches. Your body breaks down starches into sugar molecules.

Fats

Like carbohydrates, fats are energy-containing nutrients. However, 1 gram of fat provides 9 Calories of energy, while 1 gram of carbohydrate provides only 4 Calories. Fats form part of the cell membrane. Fatty tissue also protects your organs and insulates your body. No more than 35 percent of your daily Calories should come from fats.

Proteins

Your body needs proteins for growth and tissue repair. Proteins also can be an energy source. About 10 to 35 percent of your daily Calorie intake should come from proteins. Proteins are made up of small, linked units called amino acids (uh MEE noh). Thousands of different proteins are built from about 20 different amino acids. Your body can make about half of the amino acids it needs. The other half must come from food. Foods from both animals and plants contain protein.

Vitamins and Minerals

Unlike some nutrients, vitamins do not provide the body with raw materials and energy. Instead, vitamins act as helper molecules in your body's chemical reactions. The body can make a few vitamins, such as vitamin D, but foods are the source of most vitamins.

Nutrients that are not made by living things are called minerals. Like vitamins, minerals do not provide your body with raw materials and energy. However, your body still needs small amounts of minerals to carry out chemical processes. For example, you need calcium to build bones and teeth, and you need iron to help red blood cells function. Plant roots absorb minerals from the soil. You obtain minerals by eating plants or animals that have eaten plants.

Water

Water is the most important nutrient because all the body's vital processes take place in water. In addition, water helps regulate body temperature and remove wastes. Water accounts for about 65 percent of the average healthy person's body weight because it makes up most of the body's fluids, including blood. Under normal conditions, you need to take in about 2 liters of water every day to stay healthy.

Do the Lab Investigation *Nutrient Identification.* Student Lab Manual, p. 85

Assess Your Understanding

TEKS 6C

1a. Define What does a Calorie measure?

The amount of energy in foods.

b. Recognize Why are complex carbohydrates like starches considered long-term energy sources?

The human body can break down starches into sugar molecules to provide energy.

c. Apply Concepts What do you think is meant by the phrase "a balanced diet"?

Sample: It means eating a mix of foods every day that gives you the nutrients you need.

got it?

O I get it! Now I know that food provides the body with energy and the materials to grow and repair tissues.

O I need extra help with See TE note.

Differentiated Instruction

L1 Plan a Menu Based on the information in **Figure 1**, have students work in groups to create a day's menu that consists of 45–65% carbohydrates, no more than 35% fat, and 10–35% protein. Encourage students to research which foods fall into these categories. Menus should make allowances for vitamins, minerals, and water.

L3 Simple and Complex Carbohydrates Carbohydrates are made of the elements carbon (C), hydrogen (H), and oxygen (O). Most carbohydrates contain twice as many hydrogen atoms as of carbon and oxygen atoms. Challenge students to research the chemical structure of a simple carbohydrate, such as glucose, and a complex carbohydrate, such as starch. Ask them why complex carbohydrates provide a long-lasting source of energy. *(The sugars in the long-chained complex carbohydrates are absorbed at a slower, steadier rate than the quick burst of energy provided by simple sugars.)*

Lead a Discussion

VITAMINS, MINERALS, AND WATER Explain that vitamins, minerals, and water do not provide energy—only carbohydrates, fats and proteins do. Ask: **Why are vitamins important to the body?** *(They act as helper molecules for the body's chemical reactions.)* **Why are minerals important to the body?** *(They help the body carry out chemical processes.)* **Why is water important to the body?** *(Water helps regulate body temperature and remove waste.)* **How do you get vitamins?** *(The body makes some vitamins, but most come from food.)* **How do you get minerals?** *(You get minerals by eating plants or animals that have eaten plants.)* **What is the main difference between vitamins and minerals?** *(Vitamins are nutrients made by living things; minerals are not made by living things; they come from the soil.)* **How much water should you drink every day?** *(About 2 liters)*

21st Century Learning

INTERPERSONAL SKILLS Tell students that vitamins can be classified into two groups: fat soluble and water soluble. Invite groups of students to research vitamins. Groups should work together to classify the principle vitamins as either fat or water soluble, list foods that contain these vitamins, and explain the different ways the body handles fat and water soluble vitamins. Have groups organize their information on a chart, and display the charts around the room.

Lab Resource: Lab Investigation 

L1 NUTRIENT IDENTIFICATION Students will perform tests to detect starches, sugars, and proteins in foods. This Lab Investigation can be found in the Student Lab Manual, p. 85, and online.

Evaluate

Assess Your Understanding

After students answer the questions, have them evaluate their understanding by completing the appropriate sentence.

RTI Response to Intervention

1a. If students have trouble defining what a Calorie measures, **then** have them review the definition.

b. If students have difficulty recognizing why complex carbohydrates are considered long-term energy sources, **then** have them review Carbohydrates in **Figure 1.**

c. If students struggle to explain "a balanced diet," **then** discuss how to apply the concept of balance to food choices.

LESSON 7.1

Explain

Teach Key Concepts

Explain that the three main functions of organs of the digestive system are digestion, absorption, and elimination. Ask: **What are the organs of the digestive system?** *(Samples: mouth, esophagus, stomach, large intestine, small intestine)* **What is mechanical digestion?** *(Tearing or grinding food into smaller pieces)* **Where does most mechanical digestion take place?** *(In the mouth and stomach)* **What is chemical digestion?** *(The breaking of foods into their building blocks)* If students have studied chemistry, point out that mechanical digestion is a series of physical changes, while chemical digestion is a series of chemical changes in which complex molecules are broken down to make simpler ones. Ask: **What is absorption?** *(The process by which nutrient molecules pass through the wall of the digestive system and move into the blood)* **Where does most absorption take place?** *(In the small intestine)* Tell students that substances produced by the liver, pancreas, and lining of the small intestine help complete chemical digestion.

Lead a Discussion

DIGESTION BEGINS Review with students the distinctions between mechanical and chemical digestion. Have students look over **Figure 2.** Ask: **What is the function of the salivary glands?** *(Producing saliva)* Help students understand that saliva has several important functions. First, saliva adheres chewed particles of food into a ball that is small enough to be swallowed. Saliva also lubricates food, making it easier to swallow. Finally, saliva contains a chemical that can break down starches into sugars. **What are the functions of the tongue?** *(It pushes food toward the teeth, helps mix food with saliva, and assists in swallowing.)* **What is the function of the teeth?** *(Teeth cut, tear, crush, and grind food.)*

Texas Essential Knowledge and Skills

6B Distinguish between physical and chemical changes in matter in the digestive system.

6C Recognize how large molecules are broken down into smaller molecules such as carbohydrates can be broken down into sugars.

12B Identify the main functions of the systems of the human organism, including the circulatory, respiratory, skeletal, muscular, digestive, excretory, reproductive, integumentary, nervous, and endocrine systems.

13B Describe and relate responses in organisms that may result from internal stimuli such as wilting in plants and fever or vomiting in animals that allow them to maintain balance.

TEKS 6B, 6C, 12B, 13B In this section, you'll identify the main functions of the digestive system and what happens during the process of digestion.

What Happens in Your Digestive System?

Your digestive system is about 9 meters long from beginning to end. **Figure 2** shows the structures of the digestive system. **The digestive system breaks down food, absorbs nutrients, and eliminates waste.** These functions occur one after the other in an efficient, continuous process.

Digestion The process by which your body breaks down food into small nutrient molecules is called digestion. In mechanical digestion, bites of food are torn or ground into smaller pieces, which causes physical changes to the food. This kind of digestion happens mostly in the mouth and stomach. In chemical digestion, chemicals break foods into their building blocks, resulting in chemical changes to the food. Chemical digestion takes place in many parts of the digestive system. Substances made in the liver and pancreas help digestion occur.

Absorption and Elimination Absorption occurs after digestion. Absorption is the process by which nutrient molecules pass from your digestive system into your blood. Most absorption occurs in the small intestine. The large intestine eliminates materials that are not absorbed.

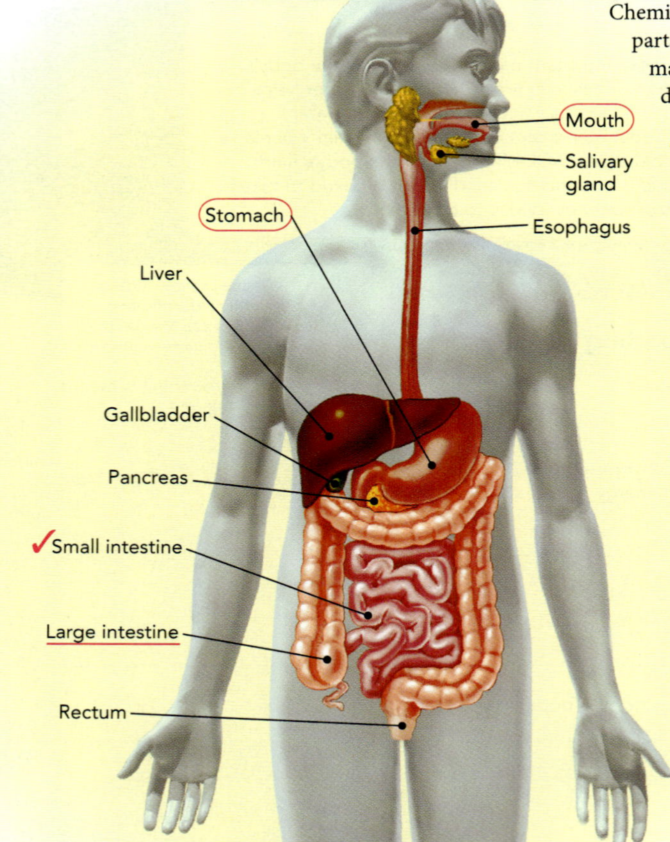

FIGURE 2

The Digestive System

Food passes directly through five organs of your digestive system: the mouth, esophagus, stomach, small intestine, and large intestine.

Identify Circle the name(s) of the organ(s) where mechanical digestion mainly occurs. Check the name(s) of the organ(s) where most absorption occurs. Underline the name(s) of the organ(s) where elimination occurs.

The Mouth

Your mouth is where digestion begins. When you bite off a piece of food, both mechanical and chemical digestion begin inside your mouth. Your teeth and tongue carry out mechanical digestion. Your teeth cut, tear, crush, and grind food into physically smaller pieces. Your tongue pushes food toward your teeth.

As your teeth work, your saliva (suh LY vuh) moistens food into a slippery mass. Saliva is the fluid released by salivary glands when you eat. Saliva contains a chemical that can break down large molecules, such as starches or other carbohydrates, into smaller molecules, such as sugars. This step begins the chemical digestion of your food.

The chemical in saliva that digests starch is an enzyme. An **enzyme** is a protein that speeds up chemical reactions in the body. Your body produces many different enzymes. Each enzyme has a specific chemical shape that enables it to speed up only one kind of reaction. Many different enzymes are needed to complete the process of digestion. **Figure 3** shows how enzymes work.

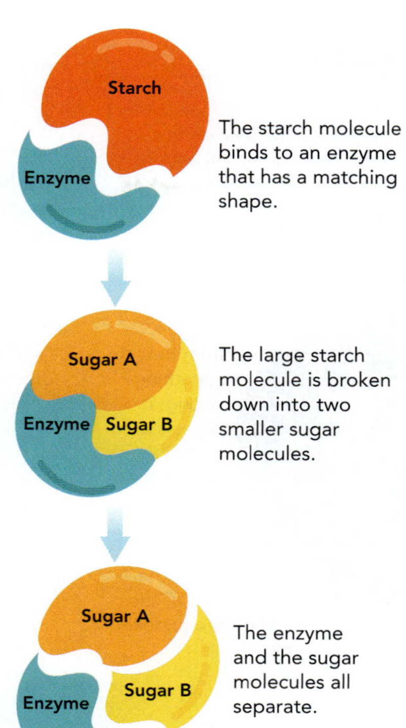

FIGURE 3

How Enzymes Work

Enzymes help break down starches, proteins, and fats.

Recognize Which molecule does not change?

The enzyme does not change.

You have four types of teeth. Each type has a specific function.

1. **Name** Think about eating a carrot. Which type of teeth cuts the carrot into a bite-sized piece? Incisors
2. **Identify** Which teeth at the back of your mouth crush and grind the carrot piece? Premolars and Molars
3. **Interpret Diagrams** When people tear chicken off a bone, they use their pointed teeth called canines.
4. **Summarize** Write about all the teeth people use to eat an apple.

Sample: They cut a bite-sized piece of apple with their incisors. Then they crush and grind the apple with their premolars and molars.

Molars

Premolars

Incisors

Canine

Differentiated Instruction

L1 Word Meanings Help students clarify their understanding of the concept of *mechanical digestion* by focusing their attention on the word *mechanical.* Point out that this word is related to the word *machine.* Explain that a machine is a device with some moving parts that performs work. Then help students recognize that mechanical digestion occurs when parts of the human body such as the tongue and teeth are moved.

L3 Teeth Challenge students to do research to learn more facts and details about the size, shape, number, and functions of the four types of human teeth. Encourage them to include one or more diagrams with detailed labels for all of the upper and lower teeth.

Lead a Discussion

ENZYMES Remind students that, in addition to mechanical digestion, chemical digestion begins in the mouth because saliva contains an enzyme. Ask: **What is the difference between mechanical and chemical digestion?** *(Mechanical digestion involves physical changes in the food. Chemical digestion involves chemical changes in the food.)* **What is an enzyme?** *(A chemical that speeds up chemical reactions)* **What does the enzyme in saliva do?** *(It helps break down large starch molecules into smaller sugar molecules.)*

Elaborate

Teacher Demo 

L3 ACTION OF ENZYMES

Materials meat tenderizer, milk, orange juice, 2 flasks or clear glasses, 2 stirrers

Time 10 minutes

Show students the container of meat tenderizer. If students do not know what it is used for, tell them that it is sprinkled on meats before cooking to make them tender. Explain that meat tenderizer contains papain, an enzyme that breaks down protein. At the beginning of class, place 2 tablespoons of milk into one flask and the same amount of orange juice in another. Add 1 tablespoon of the meat tenderizer to each flask, and stir well with separate stirrers.

Ask: **What will happen to the milk and orange juice?** *(The papain will act on the milk solution, because milk contains protein. The papain will not act on the orange juice, because orange juice does not contain protein.)*

Set aside the flasks until the end of class, and then display both flasks. The milk solution will be thick; the orange juice will be unchanged.

Ask: **What can you say about the action of digestive enzymes based on this demonstration?** *(An enzyme acts on only one type of nutrient.)*

Apply It!

L1 Before beginning the activity, review the photograph and labels carefully. If students have difficulty figuring out what the different kinds of teeth do, suggest that they think about how they bite and chew the food they eat.

LESSON 7.1

Explain

Lead a Discussion

THE FUNCTION OF THE ESOPHAGUS Explain that the mouth leads to the throat, formally called the pharynx. In the pharynx, food and fluid from the mouth join air inhaled through the nose. At the back of the pharynx, the food, fluid, and air are sorted into the esophagus (food and fluid) or the windpipe (air). Explain that the epiglottis is a flap of tissue that seals off the windpipe during swallowing and prevents food from entering the lungs. Ask: **The esophagus is located between which two organs in the digestive system?** *(The mouth and the stomach)* **How does mucus in the esophagus aid digestion?** *(Mucus helps food move easily.)* Large hollow organs of the digestive system contain muscles that help the walls of these organs to move. Ask: **How do the muscles of the esophagus help food to move through the digestive system?** *(Muscles in the esophagus push food toward the stomach.)* **What is this process called?** *(Peristalsis)*

21st Century Learning

CRITICAL THINKING GRAVITY AND DIGESTION Point out that astronauts traveling in space do not feel Earth's gravitational pull on their bodies. Invite students to explain how peristalsis allows astronauts to eat in space. *(Sample: Gravity doesn't affect how food travels from the mouth to the stomach. The muscles in the esophagus contract and push the food down into the stomach.)*

Support the TEKS

THE STOMACH Ask students to think about their own perceptions about the digestive process in their bodies. Ask: **What is one way your stomach lets you know that it is empty?** *(Samples: hunger pangs; stomach "growls" or rumbles.)* Explain that these sounds are caused by peristalsis when the stomach has been empty for some time. Point out that people sometimes experience a sensation of pain, called hunger pangs, when their stomachs have been empty for a long time. Encourage students to look at **Figure 4.** Ask: **How do stomach muscles help the digestive system to perform its function?** *(Layers of muscle contract to churn food and mix it with fluids.)* **What happens to food in the stomach?** *(Food is churned and mixed with fluids, including digestive juices. Proteins in the food are chemically digested. Then the food is released into the next part of the digestive system.)* Have students summarize where mechanical and chemical digestion have taken place to this point. *(Mouth—chemical and mechanical digestion; esophagus—none; stomach—chemical and mechanical digestion)*

did you know?

Our stomachs have an important immune function. Acids in digestive juice can kill harmful microbes and parasites that we ingest with our food. However, some dangerous microbes and toxins, like the norovirus pictured here, are not affected by the acids. When the nervous system detects something wrong, it signals a certain part of the brain. The process of vomiting begins, in which the contents of the stomach are expelled from the body.

To the Stomach Food moves from your mouth through your **esophagus** (ih SAHF uh gus) and then into your stomach. The stomach is a J-shaped muscular pouch where most mechanical digestion and some chemical digestion occur. Mechanical digestion occurs as layers of smooth muscle in the stomach wall contract, producing a churning motion. Chemical digestion occurs as the food mixes with digestive juice. Digestive juice is a fluid produced by cells that line the stomach. It contains the enzyme pepsin that chemically digests proteins into short chains of amino acids.

Food usually stays in your stomach for a few hours until mechanical digestion is complete. Now a thick liquid, the food enters the next part of the digestive system. That is where chemical digestion continues and absorption takes place.

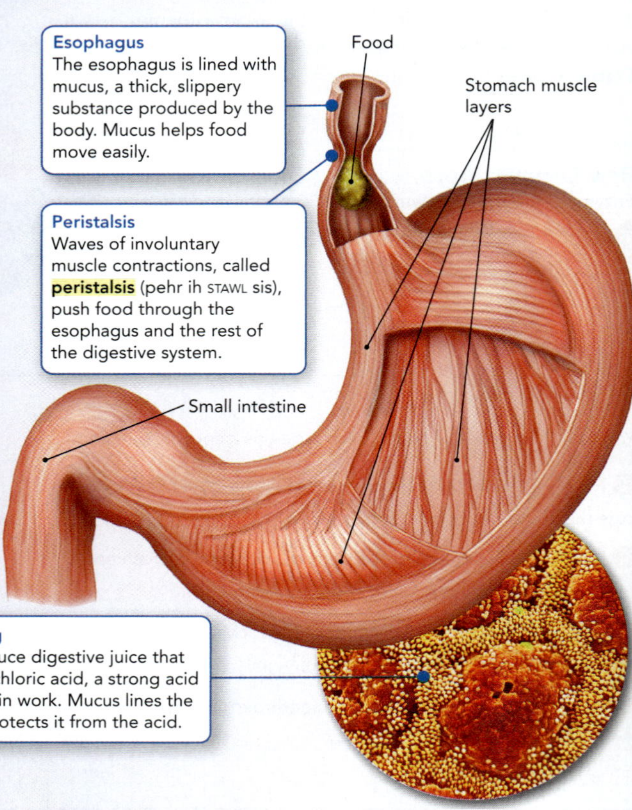

FIGURE 4

The Stomach

The stomach wall has three muscle layers. The microscopic view shows you the cells that line the inside of the stomach.

Classify What type of digestion is aided by the action of stomach muscles?

Sample: Mechanical digestion

The Small Intestine At about 6 meters—longer than some full-sized cars—the small intestine makes up two thirds of the length of the digestive system. The small intestine is the part of the digestive system where most chemical digestion and absorption take place. Its small diameter, from 2 to 3 centimeters wide, gives the small intestine its name.

A great deal happens in the small intestine. When food reaches it, starches and proteins have been partially broken down, but fats have not been digested. **Substances produced by the liver, pancreas, and lining of the small intestine help to complete chemical digestion.** The gallbladder, which stores bile from the liver, and the pancreas send their substances into the small intestine through small tubes.

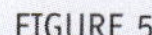

FIGURE 5

Organs of Digestion

The liver, pancreas, and gallbladder aid digestion in the small intestine.

Complete the tasks.

1. **Identify** Fill in the missing labels.
2. **Develop Hypotheses** How may a blockage in the tube between the gallbladder and the small intestine affect digestion?

Sample: Bile may not reach the small intestine, making digestion of fats harder.

Gallbladder
The gallbladder stores bile and releases it into the small intestine.

Liver
The liver has many roles in the body. One job is making bile. Bile physically breaks fats into smaller droplets, so it is not involved in chemical digestion.

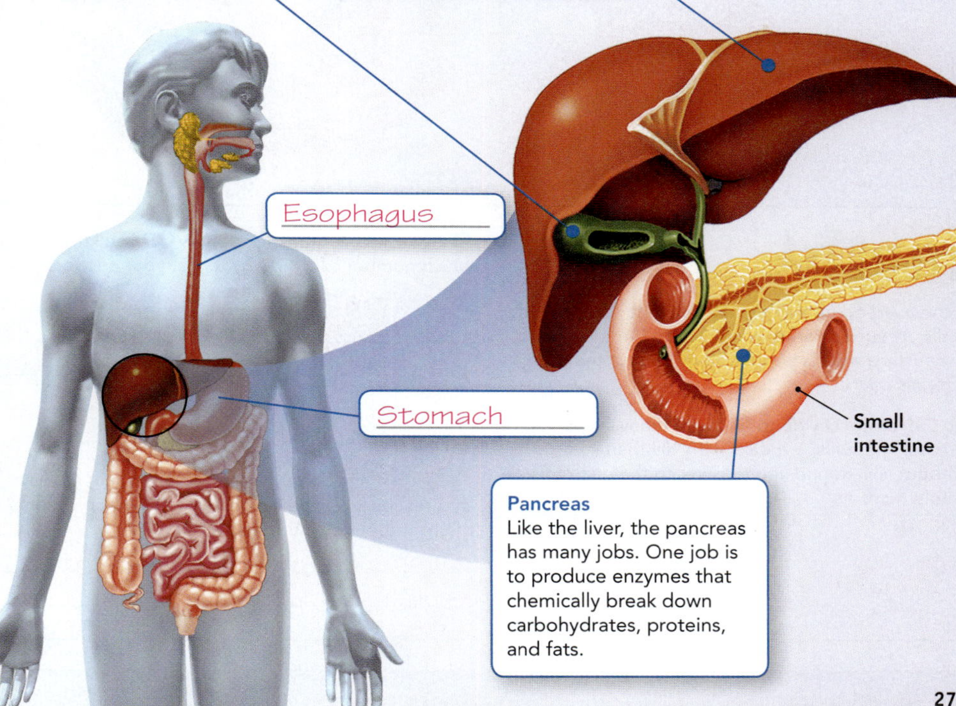

Differentiated Instruction

L1 Table of Digestive Organs Have students draw a table with three columns labeled *Organ, Type of Digestion,* and *What Happens.* As students read this lesson, have them complete the table. Model the first row for them: mouth *(Organ)*; chemical, mechanical *(Type of Digestion)*; teeth cut, tear, crush, and grind food, the tongue mixes food and helps in swallowing, and enzymes in saliva break down starches *(What Happens)*.

L3 Mucus Challenge students to use print and electronic sources to learn more about the composition of mucus, as well as its vital role in the digestive process.

Lead a Discussion

THE SMALL INTESTINE Explain that most chemical digestion and absorption takes place in the small intestine. Ask: **How would you describe the shape of the small intestine?** *(The small intestine is very small and very narrow.)* **How might the shape of the small intestine aid in chemical digestion and absorption?** *(Because it is so long, it has a large surface area, which gives a lot of opportunity for absorption.)* **What is the state of food by the time it arrives in the small intestine?** *(Starches and proteins are partially broken down; fat has not been digested.)* **What does this suggest about how the body digests different types of food?** *(It takes longer for the body to digest fat than to digest starches and proteins.)*

Teach With Visuals

Direct students' attention to **Figure 5.** As you name the organs, including the esophagus and the stomach, have students find them in the illustration. Ask: **What product of the liver aids in digestion?** *(Bile)* **What is the function of bile?** *(It physically breaks down large fat particles into smaller fat particles.)* **What is the function of the gallbladder?** *(It stores bile and releases it into the small intestine.)* **What are the products of the pancreas?** *(Enzymes)* **What do these enzymes break down?** *(Proteins, carbohydrates, and fats)* Remind students that each step in digestion requires a different enzyme. For example, the pancreatic enzyme that digests protein does not affect carbohydrates or fats. If students mention that the pancreas produces insulin, tell them that insulin is not a digestive enzyme and does not enter the small intestine. The pancreas has two separate functions, to produce enzymes for digestion and to produce the hormone insulin.

Develop Hypotheses Help students understand that a hypothesis is one possible explanation or answer to a scientific question. It can be worded as an *If/then* statement.

21st Century Learning

L3 COMMUNICATION Tell students that the presence of insulin causes cells in the liver and other organs to take up glucose from the blood. When insulin is absent, the body uses fat as an energy source. Diabetes is a disease in which the body cannot control the levels of insulin. Assign groups to research either Type 1 or 2 diabetes. Have students prepare and deliver an oral report on the causes, symptoms, and treatment.

LESSON 7.1

Explain

Teach With Visuals

Direct students' attention to **Figure 6**. Tell students the small intestine is involved in mechanical digestion as a result of continued peristalsis, but to a lesser extent than in the mouth and stomach. Remind students most absorption takes place in the small intestine. Ask: **What does the small intestine absorb?** *(Nutrient molecules from digested foods)* **How does the structure of the small intestine aid absorption?** *(The small intestine is folded into millions of villi, making it possible to absorb more nutrients.)* **Where do nutrients go when absorbed by the small intestine?** *(Nutrients pass into the blood vessels, which deliver them to the body's cells.)*

Make Analogies

L1 SURFACE AREA AND ABSORPTION Help students understand that increased surface area helps absorption. Contrast the greater amount of water a terrycloth hand towel can absorb with the lesser amount a smooth dish towel can absorb, even with similar dimensions. Ask: **What is different about the surfaces of these towels?** *(Sample: The hand towel has loops of thread that are raised from the surface. The dish towel is smooth and flat.)* **Which towel do you think contains more fabric?** *(The hand towel)* Point out the loops of thread increase the surface area that can hold water, much as the villi increase the surface area of the small intestine.

Elaborate

Do the Math!

Help students understand that calculating how many times greater the surface area of the small intestine is with villi rather than without it emphasizes one important function of the small intestine in digestion—absorption. Encourage students to round the decimal dimensions of the surface areas before they estimate the difference between the two decimals. Ask: **Which number will be the dividend in your equation?** *($250 \ m^2$)* **Which will be the divisor?** *($0.57 \ m^2$)*

21st Century Learning

INTERPERSONAL SKILLS Have students work in pairs to design a game about the digestive system. The game board should consist of a diagram of the digestive system and questions about its organs that allow players to progress through the body. After developing the game, have pairs trade games and play.

Absorption in the Small Intestine After chemical digestion takes place, the small nutrient molecules are ready for the body to absorb. The structure of the small intestine helps absorption occur. The inner surface of the small intestine is folded into millions of tiny finger-shaped structures called **villi** (VIL eye) (singular *villus*). Villi, shown in **Figure 6,** greatly increase the surface area of the small intestine. More surface area means that more nutrients can be absorbed. Nutrient molecules pass from cells on the surface of a villus into blood vessels and are then delivered to body cells.

FIGURE 6
Villi
Tiny villi line the folds of the small intestine.

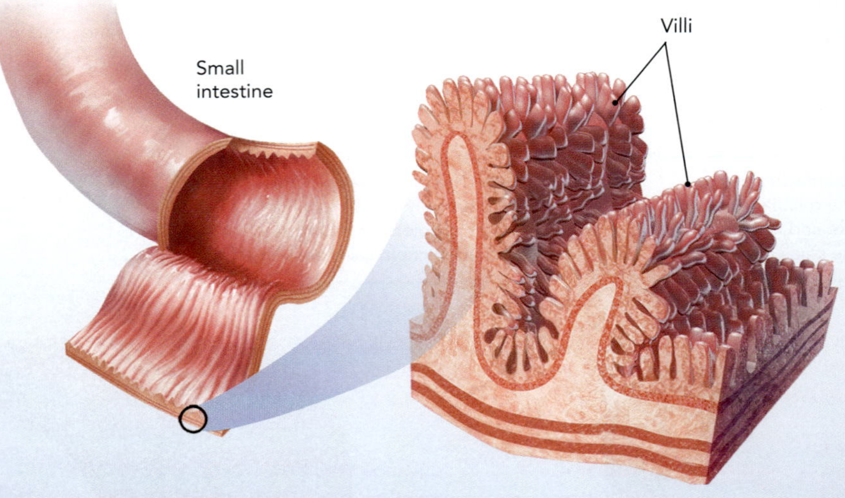

do the math!

If the average person's small intestine had smooth walls, its surface area would be $0.57 \ m^2$. With villi, the surface area is about $250 \ m^2$, about the size of a tennis court.

1 **Calculate** Divide to find how many times greater the surface area is with villi than it is without villi. Round your answer to the nearest whole number.

$$\frac{250 \ m^2}{0.57 \ m^2} = 438.59 = 439$$

2 **Estimate** In Question 1, how did you know which number to divide by to get your answer?

The surface area increases so the answer has to be larger than 250 or 0.57. Dividing by 0.57 gives a larger number.

3 **CHALLENGE** Some people have a wheat allergy that results in villi being destroyed. What problems might these individuals have?

Sample: They might not absorb enough nutrients to stay healthy.

The Large Intestine By the time material reaches the end of the small intestine, most nutrients have been absorbed. The water and undigested food that is left move from the small intestine into the large intestine. The large intestine is the last section of the digestive system. As the material moves through the large intestine, water is absorbed into the bloodstream. The remaining material is readied for elimination from the body.

The large intestine is about 1.5 meters long. It contains bacteria that feed on the material passing through. These bacteria normally do not cause disease. In fact, they are helpful because they make certain vitamins, including vitamin K.

The large intestine ends in a short tube called the rectum. In the rectum waste material is compressed into solid form. This waste material is eliminated from the body through the anus, a muscular opening at the end of the rectum.

Chart Write the sequence of organs that food passes through during digestion.

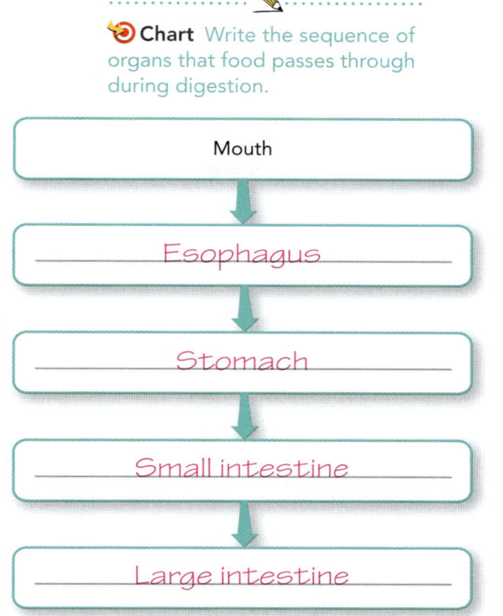

Do the Lab Investigation *As the Stomach Churns.* Find the lab online.

Assess Your Understanding

TEKS 6B, 12B

2a. Define (Chemical/Mechanical) digestion occurs when enzymes break down foods into simpler substances. [Chemical circled]

b. Distinguish Does chewing with your teeth cause physical or chemical changes to the food you eat? Explain.
It causes physical changes because my teeth are only making the pieces of food smaller. It is a change in the appearance of the food, not its chemical makeup.

c. Identify What are the main functions of the digestive system?
The digestive system breaks down food, absorbs nutrients that the body needs, and eliminates wastes from the body.

d. Identify How do villi help the small intestine carry out its function?
Villi increase the surface area for absorption.

got it?

○ **I get it!** Now I know that the digestive system works through the actions of organs that include the mouth, esophagus, stomach, small intestine, pancreas, liver, gallbladder, and large intestine.

○ I need extra help with See TE note.

Differentiated Instruction

L1 Surface Area Give each student two pieces of blank paper of the same size. Ask them to make folds in one of the pieces and place it on top of the flat paper. Students will observe that the two pieces still have the same surface area. Have students describe how the folded paper models the villi in the small intestine. *(The villi allow for greater surface area, yet take up little space in the body.)* If students have difficulty with this concept, ask them to smooth out the folded paper and place it against the flat paper.

L3 Three Sections of the Small Intestine The small intestine can be divided into three subsections: the duodenum, the jejunum, and the ileum. Challenge students to create a diagram of the three sections of the small intestine. Diagrams should describe the function of each section.

Explain

Lead a Discussion

THE FUNCTION OF THE LARGE INTESTINE Explain that the large intestine absorbs water and prepares the remaining material for elimination from the body. Ask: **After nutrients are absorbed, what is left in the food material as it enters the large intestine?** *(Water, bits of undigested food)* **What is left after water is absorbed into the bloodstream?** *(Bits of undigested food)* **What do the bacteria in the large intestine do?** *(Feed on the material and produce vitamins, such as vitamin K.)* **Where does material move after water is absorbed?** (*Into the rectum*) **What happens in the rectum?** *(The waste material is compressed and eliminated through the anus.)*

Chart Tell students that making a chart while reading is a good way to organize information.

Elaborate

21st Century Learning

INFORMATION LITERACY Students may have seen advertisements for yogurt products that claim to aid the digestive system. Challenge students to learn how yogurt is made, what "active cultures" means, and how the bacteria in yogurt aid the digestive system. Encourage students to evaluate the claims.

Lab Resource: Lab Investigation

L1 AS THE STOMACH CHURNS Students will explore how proteins are digested in the stomach. This Lab Investigation can be found online.

Evaluate

Assess Your Understanding

After students answer the questions, have them evaluate their understanding by completing the appropriate sentence.

RTI Response to Intervention

2a. If students have trouble distinguishing between chemical and mechanical digestion, **then** review how *chemical* and *mechanical* apply to digestion.

b. If students have difficulty distinguishing between physical and chemical changes, **then** have them explain what the teeth do to food when it is chewed.

c. If students cannot identify the main functions of the digestive system, **then** have them reread the second Key Concept statement.

d. If students have difficulty describing the role of villi, **then** have them review **Figure 6.**

Name ______________________ Date __________ Class __________

Assess Your Understanding

The Digestive System

Why Do You Need Food?

1a. DEFINE What does a Calorie measure?

____

b. RECOGNIZE Why are complex carbohydrates like starches considered long-term energy sources?

____

c. APPLY CONCEPTS What do you think is meant by the phrase "a balanced diet"?

____

____

What Happens in Your Digestive System?

2a. DEFINE (Chemical/Mechanical) digestion occurs when enzymes break down foods into simpler substances.

b. DISTINGUISH Does chewing with your teeth cause physical or chemical changes to the food you eat? Explain.

____

____

c. IDENTIFY What are the main functions of the digestive system?

____

____

d. IDENTIFY How do villi help the small intestine carry out its function?

____

____

Place the outside corner, the corner away from the dotted line, in the corner of your copy machine to copy onto letter-size paper.

Name ______________________ Date __________ Class __________

Enrich

The Digestive System

Sugar makes food taste sweet, but it also contains a lot of Calories. For this reason, some people use sugar substitutes. Read the passage below. Then use a separate sheet of paper to answer the questions that follow.

Sugar Substitutes

A *sugar substitute* is a chemical that tastes sweet but provides few or no nutrients to the body. Sugar substitutes often taste much sweeter than sugar, and some have no Calories. However, sugar substitutes are not problem-free. Some lose their sweetness over time or at high temperatures; some have an aftertaste; and some have been linked to health problems. Despite these problems, over 100 million people use sugar substitutes in the United States every year.

Saccharin (sa kur uhn), the first sugar substitute, was discovered by a scientist named Constantin Fahlberg in 1879. Saccharin has no Calories and is not digested or absorbed by the body. In 1977, a study showed that rats given saccharin had a greater risk of developing cancer. As a result, all foods containing saccharin carry labels warning that this chemical might be linked to cancer in humans.

Aspartame (AS pahr taym) is another sugar substitute. Unlike saccharin, aspartame contains amino acids, which are digested and absorbed by the body. Although the United States Food and Drug Administration (FDA) has not linked aspartame to any health problems, there is still some debate about its safety.

In 1998, the FDA approved a sugar substitute called sucralose for use in some types of foods. Sucralose is made from sugar, but it has been modified so that the body does not digest or absorb it. One advantage of sucralose is that it does not lose its sweetness at high temperatures, so it can be used in foods that are baked, such as cakes.

1. **The population of the United States was about 270 million in 1998. About what percentage of the population used a sugar substitute?**
2. **Why do you think so many people use sugar substitutes?**
3. **What are some of the advantages and disadvantages of sugar substitutes?**
4. **Which of the three sugar substitutes described above provides the body with nutrients it can use? Explain.**
5. **Look at the ingredients listed on a can of diet soft drink. Does the drink contain a sugar substitute? If so, which one?**

Name ______________________ Date ____________ Class ____________

Lesson Quiz

The Digestive System

Write the letter of the correct answer on the line at the left.

1. ___ The process by which nutrient molecules pass from the digestive system into the blood is called

A absorption

B chemical digestion

C elimination

D mechanical digestion

2. ___ Which mucus-lined organ must food pass through to get from the mouth to the stomach?

A large intestine

B esophagus

C liver

D small intestine

3. ___ Which organ prepares undigested materials for elimination from the body?

A liver

B gallbladder

C pancreas

D large intestine

4. ___ An enzyme speeds up

A absorption

B chemical digestion

C elimination

D mechanical digestion

Fill in the blank to complete each statement.

5. Saliva contains an enzyme that breaks down starches into ____________________.

6. Nutrients such as carbohydrates, fats, and proteins are ____________________ compounds, or chemical substances that contain carbon and other elements such as hydrogen and oxygen.

7. The unit *Calorie* is used to measure the ____________________ in foods.

8. The liver produces ____________________ that helps break down fats.

9. The ____________________ produces enzymes that break down carbohydrates, fats, and proteins.

10. The villi aid the process of absorption by increasing the ____________________ ____________________ of the small intestine.

Place the outside corner, the corner away from the dotted line, in the corner of your copy machine to copy onto letter-size paper.

Place the outside corner, the corner away from the dotted line, in the corner of your copy machine to copy onto letter-size paper.

The Digestive System

Answer Key

Review and Reinforce

Find the worksheet in the Student Workbook.

1. Your body needs food to grow, to repair tissues, and to have energy.
2. The digestive system breaks down food, absorbs nutrients, and eliminates waste.
3. The teeth rip food into smaller bits; contracting stomach muscles create a churning motion to complete mechanical digestion.
4. The liver produces bile to break down fats; the pancreas produces enzymes that break down carbohydrates, fats, and proteins.
5. The small intestine is lined with millions of villi. With increased surface area, the villi can transfer nutrients into the bloodstream.
6. The large intestine absorbs water. Bacteria feed on the remaining materials. The rectum compresses these remaining materials for elimination.

7. e **8.** c **9.** d
10. a **11.** b

12. In the mouth, the teeth mechanically digest fish and rice, while an enzyme in saliva breaks down the starch into sugar molecules. The fish and rice goes into the esophagus where peristalsis moves it into the stomach. The contraction of the stomach muscles continues the mechanical digestion, while the enzyme pepsin breaks down proteins into amino acids. Bile produced by the liver break down fat, while enzymes produced by the pancreas break down carbohydrates, proteins, and fats. In the small intestine, villi transfer nutrient molecules into the bloodstream. The large intestine absorbs water and bacteria feed on undigested materials. The rectum compresses remaining material to be eliminated via the anus.

Enrich

1. about 37% ($\frac{100}{270} \times 100$)
2. Sample: They probably are concerned about the health effects of eating too much sugar.
3. Sample: Sugar substitutes sweeten food without adding Calories. However, they provide few or no nutrients to the body and may have harmful effects on health.
4. Aspartame; it contains amino acids which are digested and absorbed by the body.
5. Accept all reasonable answers. Students will likely say yes and aspartame, as most diet soft drinks contain some form of sugar substitute.

Lesson Quiz

1. A **2.** B
3. D **4.** B
5. sugars **6.** organic
7. energy **8.** bile
9. pancreas **10.** surface area

The Circulatory System

How do systems of the body move and manage materials?

LESSON PACING:
3–4 periods or 1 ½–2 blocks

Lesson Vocabulary

- heart
- atrium
- ventricle
- valve
- artery
- aorta
- capillary
- vein
- hemoglobin

Content Refresher

Transport of Carbon Dioxide Some of the carbon dioxide produced by cells dissolves directly into the bloodstream; some of it binds to hemoglobin. However, most of the carbon dioxide carried by the blood is first converted to carbonic acid by carbonic anhydrase, an enzyme found in red blood cells. The carbonic acid readily dissolves in blood, which contains substances that act as buffers and that prevent the blood from becoming dangerously acidic. In the lungs, the carbonic acid is converted back into carbon dioxide, which diffuses across the capillary walls into air in the lungs.

Lesson Objectives	TEKS	ELPS
Identify the structures and main functions of the circulatory system.	12B, 12C	4.C.4
Describe the characteristics of blood.	3D, 12B, 12C	

Texas Essential Knowledge and Skills

3D Relate the impact of research on scientific thought and society, including the history of science and contributions of scientists as related to the content.
12B Identify the main functions of the systems of the human organism, including the circulatory, respiratory, skeletal, muscular, digestive, excretory, reproductive, integumentary, nervous, and endocrine systems.
12C Recognize levels of organization in plants and animals, including cells, tissues, organs, organ systems, and organisms.

English Language Proficiency Standards

ELPS Reading 4.C.4 Comprehend English language structures used routinely in written classroom materials.

DIFFERENTIATED INSTRUCTION KEY
L1 Struggling Students or Special Needs
L2 On-Level Students L3 Advanced Students

LESSON PLANNER 7.2

Investigations and Activities	TEKS Review
My Planet Diary, **Student Edition,** p. 278 Inquiry: Inquiry Warm-Up, Observing a Heart, **PearsonTexas.com** Teach Key Concepts, **Teacher's Edition,** p. 279 Lead a Discussion, The Heart's Function, **Teacher's Edition,** p. 280 Support the TEKS, The Heart as Transportation Hub, **Teacher's Edition,** p. 280 Differentiated Instruction, **Teacher's Edition,** p. 281 Inquiry: Build Inquiry, Observe Your Heartbeat, **Teacher's Edition,** p. 281 Lead a Discussion, Structures and Functions of Blood Vessels, **Teacher's Edition,** p. 282 21st Century Learning, Interpersonal Skills, **Teacher's Edition,** p. 282 Apply It!, **Student Edition,** p. 283 Inquiry: Quick Lab, Direction of Blood Flow, **Lab Manual,** p. 90	Apply the TEKS, Moving Things Along, **Student Edition,** p. 302 TEKS Practice, **Student Edition,** p. 305 TEKS Practice: Chapter and Cumulative Review, **Student Edition,** p. 308 Lesson 7.2, **TEKS Preparation and Study Guide Workbook,** p. 70
Teach Key Concepts, **Teacher's Edition,** p. 284 Teach With Visuals, **Teacher's Edition,** p. 284 Inquiry: Build Inquiry, Observe Blood Cells, **Teacher's Edition,** p. 284 Make Analogies, Living Components, **Teacher's Edition,** p. 285 Differentiated Instruction, **Teacher's Edition,** p. 285 21st Century Learning, Creativity, **Teacher's Edition,** p. 285 Lead a Discussion, Marker Molecules, **Teacher's Edition,** p. 286 Apply It!, **Student Edition,** p. 287 Inquiry: Quick Lab, Do You Know Your A-B-Os?, **PearsonTexas.com**	**SHORT ON TIME?** To do this lesson in approximately half the time, do the Activate Prior Knowledge activity. A discussion of the Key Concepts will familiarize students with the lesson content. Have students do the Quick Labs. The rest of the lesson can be completed by students independently.

These editable worksheets are available on **PearsonTexas.com.**
Print versions can be found in the **TEKS Preparation and Study Guide Workbook.**

Name ______ Date ______ Class ______

7.2 The Circulatory System

Key Concept Summaries

What Are the Main Functions of the Circulatory System?

The circulatory system, or cardiovascular system, is made up of the heart, blood vessels, and blood. **The circulatory system delivers needed substances to cells, carries wastes away from cells, and helps regulate body temperature. In addition, blood contains cells that fight disease.**

The **heart** is a hollow, muscular organ that pumps blood to the body through the blood vessels. Structurally, the heart has a right side and a left side. Each upper chamber, called an **atrium,** receives blood coming into the heart. Each lower chamber, called a **ventricle,** pumps blood out of the heart. Valves separate the atria from the ventricles. A **valve** is a flap of tissue that prevents blood from flowing backward.

The circulatory system has three kinds of blood vessels. **Arteries** carry blood away from the heart. The **aorta** is the largest artery in the body. From the arteries, blood flows into tiny vessels called **capillaries,** where substances are exchanged between blood and body cells. From capillaries, blood flows into **veins,** which carry blood back to the heart.

What Does Blood Contain?

Blood has four components: plasma, red blood cells, white blood cells, and platelets. Red blood cells are made mostly of **hemoglobin.** Hemoglobin binds with oxygen in the lungs, allowing red blood cells to release oxygen to the body's cells as it travels through the capillaries. Hemoglobin also picks up carbon dioxide in the capillaries and releases it back into the lungs.

White blood cells are the body's disease fighters. Like red blood cells, white blood cells are produced in bone marrow, but white blood cells are larger than red cells and contain nuclei. Platelets are cell fragments that help form blood clots when a blood vessel is cut.

Plasma carries glucose, fats, vitamins, minerals, and chemical messengers that direct body activities. Plasma also carries away carbon dioxide and other waste.

Blood type is determined by marker molecules on red blood cells. There are four major blood types: A, B, AB, and O. In blood transfusions, plasma recognizes blood with "foreign" markers, which can result in dangerous clumping. The Rh factor is a protein in blood. Blood types are either "positive" or "negative" for the Rh factor. In transfusions, it is dangerous to mix Rh positive blood with Rh negative blood.

70

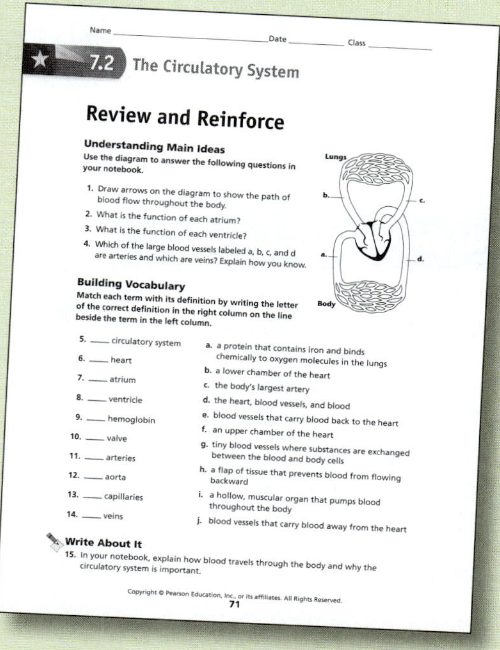

Name ______ Date ______ Class ______

7.2 The Circulatory System

Review and Reinforce

Understanding Main Ideas

Use the diagram to answer the following questions in your notebook.

1. Draw arrows on the diagram to show the path of blood flow throughout the body.
2. What is the function of each atrium?
3. What is the function of each ventricle?
4. Which of the large blood vessels labeled a, b, c, and d are arteries and which are veins? Explain how you know.

Building Vocabulary

Match each term with its definition by writing the letter of the correct definition in the right column on the line beside the term in the left column.

5. ____ circulatory system
6. ____ heart
7. ____ atrium
8. ____ ventricle
9. ____ hemoglobin
10. ____ valve
11. ____ arteries
12. ____ aorta
13. ____ capillaries
14. ____ veins

a. a protein that contains iron and binds chemically to oxygen molecules in the lungs
b. a lower chamber of the heart
c. the body's largest artery
d. the heart, blood vessels, and blood
e. blood vessels that carry blood back to the heart
f. an upper chamber of the heart
g. tiny blood vessels where substances are exchanged between the blood and body cells
h. a flap of tissue that prevents blood from flowing backward
i. a hollow, muscular organ that pumps blood throughout the body
j. blood vessels that carry blood away from the heart

Write About It

15. In your notebook, explain how blood travels through the body and why the circulatory system is important.

71

Lexile Measure = 880L

LESSON 7.2

The Circulatory System

Establish Learning Objectives

After this lesson, students will be able to:

- Identify the structures and main functions of the circulatory system.
- Describe the characteristics of blood.

Engage

Activate Prior Knowledge

MY PLANET DIARY Read *Your Heart, Your Health* with the class. Point out that the heart completes all of these tasks without a person's conscious control, since it contains involuntary muscles. Ask: **Where in the human body is the heart located?** *(In the chest cavity slightly to the left of center)* **How does your heartbeat change when you exercise?** *(It beats faster.)* **When is your heartbeat probably at its slowest rate?** *(During sleep)*

Explore

Lab Resource: Inquiry Warm-Up 

L1 OBSERVING A HEART Students will use a stethoscope to listen to their hearts before and after exercise. This Inquiry Warm-Up can be found online.

The Circulatory System

- What Are the Main Functions of the Circulatory System? TEKS 12B, 12C
- What Does Blood Contain? TEKS 3D, 12B, 12C

MY PLANET DIARY

FUN FACTS

Your Heart, Your Health

Here are some fascinating facts that you may not know about your heart.

- In one year, your heart pumps enough blood to fill more than 30 competition-sized swimming pools!
- A drop of blood makes the entire trip through your body in less than a minute.
- Your heart beats about 100,000 times a day.
- Your heart pushes blood through about 100,000 kilometers of vessels. They would circle Earth more than twice!
- A child's heart is about the size of a fist. An adult's heart is about the size of two fists.

Read the following questions. Write your answers below.

1. Why is it important for a person's heart to be healthy?
   Sample: The heart needs to pump a lot of blood through the body.
2. About how many times does your heart beat in a week? In a year?
   It beats about 700,000 times in a week and 36,500,000 times in a year.

Do the Inquiry Warm-Up *Observing a Heart.* Find the lab online.

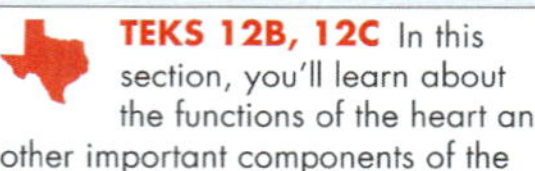
TEKS 12B, 12C In this section, you'll learn about the functions of the heart and other important components of the circulatory system.

What Are the Main Functions of the Circulatory System?

As shown in **Figure 1**, the circulatory system, or cardiovascular system, is made up of the heart, blood vessels, and blood. **The circulatory system delivers needed substances to cells, carries wastes away from cells, and helps regulate body temperature. In addition, blood contains cells that fight disease.**

SUPPORT ALL READERS

Lexile Measure = 880L **Lexile Word Count = 2068**

Prior Exposure to Content: Most students have encountered this topic in earlier grades

Academic Vocabulary: *observe, summarize*

Science Vocabulary: *heart, atrium, ventricle, artery, aorta, capillary, vein*

Concept Level: Generally appropriate for most students in this grade

Preteach With: My Planet Diary "Your Heart, Your Health" and Figure 1 activity

Texas Essential Knowledge and Skills

12B Identify the main functions of the systems of the human organism, including the circulatory, respiratory, skeletal, muscular, digestive, excretory, reproductive, integumentary, nervous, and endocrine systems.

12C Recognize levels of organization in plants and animals, including cells, tissues, organs, organ systems, and organisms.

Vocabulary
- heart
- atrium
- ventricle
- valve
- artery
- aorta
- capillary
- vein
- hemoglobin

Skills
- Reading: Summarize
- Inquiry: Observe

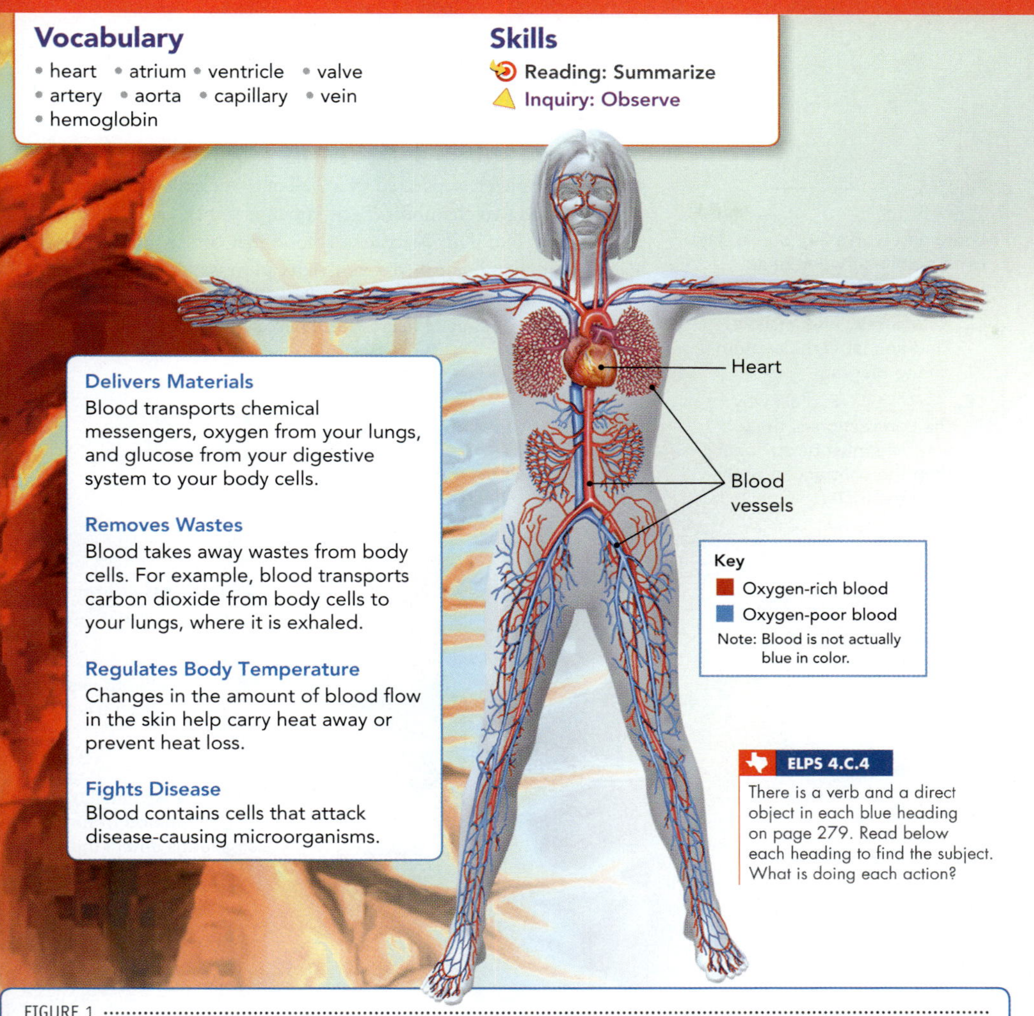

Delivers Materials
Blood transports chemical messengers, oxygen from your lungs, and glucose from your digestive system to your body cells.

Removes Wastes
Blood takes away wastes from body cells. For example, blood transports carbon dioxide from body cells to your lungs, where it is exhaled.

Regulates Body Temperature
Changes in the amount of blood flow in the skin help carry heat away or prevent heat loss.

Fights Disease
Blood contains cells that attack disease-causing microorganisms.

ELPS 4.C.4
There is a verb and a direct object in each blue heading on page 279. Read below each heading to find the subject. What is doing each action?

FIGURE 1

The Circulatory System
Like roads that link all the parts of a town, your circulatory system links all the parts of your body.

 **Use the diagram to answer the questions.**

1. **Infer** What might happen if your circulatory system did not function properly?
   Sample: Cells would not get the materials they need, get rid of wastes, or fight disease.
2. **Pose Questions** After looking at the diagram, write a question that describes one thing you would like to learn about the circulatory system.
   Sample: How does blood get from the oxygen-rich blood vessels to the oxygen-poor blood vessels?

Explain

LESSON 7.2

Introduce Vocabulary
Write the word *artery* on the board. Ask students if they have heard this term used in relation to transportation, which refers to an important route for travel and shipping. Explain that, in terms of the human body, an artery is a blood vessel that carries oxygen-rich blood to all parts of the body.

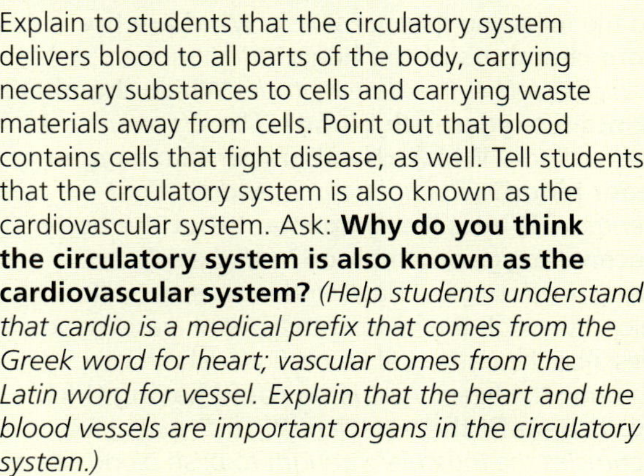

Teach Key Concepts
Explain to students that the circulatory system delivers blood to all parts of the body, carrying necessary substances to cells and carrying waste materials away from cells. Point out that blood contains cells that fight disease, as well. Tell students that the circulatory system is also known as the cardiovascular system. Ask: **Why do you think the circulatory system is also known as the cardiovascular system?** *(Help students understand that cardio is a medical prefix that comes from the Greek word for heart; vascular comes from the Latin word for vessel. Explain that the heart and the blood vessels are important organs in the circulatory system.)*

Teach With Visuals
Directs students' attention to **Figure 1.** Ask: **What are some materials that the circulatory system delivers throughout the body?** *(Chemical messengers, oxygen, glucose)* **How does the circulatory system get rid of carbon dioxide?** *(Carbon dioxide is passed from cells into the blood. Then it is carried to the lungs, where it is exhaled.)* **How does the circulatory system regulate body temperature?** *(By changing the amount of blood flow in the skin, the circulatory system carries heat away or prevents heat loss.)* **How does the circulatory system help fight disease?** *(Blood in the circulatory system carries cells that attack disease-causing microorganisms.)*

PEARSON Texas.com

English Language Proficiency Standards

ELPS Reading 4.C.4

Read aloud the blue headings. Tell students that a direct object is a noun or pronoun that receives the action of a verb.

Beginning Display and read aloud sentence frames based on the headings: _____ *delivers materials,* _____ *removes waste,* and so on. Ask beginners to listen for each missing word as you read aloud page 279.

Intermediate Ask partners to make a three-column chart with the headings *Subject, Verb,* and *Object.* Have partners complete the chart.

Advanced Have partners convert each blue heading to a complete sentence by adding a subject.

Advanced High Have partners write a compound sentence that tells what four things blood does.

LESSON 7.2

Explain

Lead a Discussion

THE HEART'S FUNCTION Explain to students that the heart's function is to push blood through the blood vessels. Point out that the right and the left sides of the heart are the opposite of what students might expect when looking at the page. Explain that the art refers to the right and left sides of the heart, as it would be in a person. Encourage students to refer to **Figure 2** as they consider their answers to the following questions. Ask: **What are the four chambers of the heart?** *(Right atrium, right ventricle, left atrium, left ventricle)* **Which chambers contain oxygen-rich blood?** *(The left atrium and left ventricle)* **Which chambers contain oxygen-poor blood?** *(The right atrium and the right ventricle)* **Why is it so important that the body receive oxygen-rich blood?** *(All the cells in the body need oxygen.)* Have students trace the flow of blood through the heart with a pencil eraser as you describe the pattern of blood flow in both circuits through the heart. Ask: **How could having thick walls benefit the ventricles?** *(Thick walls give ventricles the muscular strength to push blood out of the heart.)*

Support the TEKS

THE HEART AS TRANSPORTATION HUB Point out that the heart is a kind of transportation hub for blood. Ask: **Where does blood come from before it enters the heart's left atrium?** *(The lungs)* **Since the blood came from the lungs, how would you describe this blood?** *(Full of oxygen)* **Where is this blood transported?** *(To the left ventricle and then to all parts of the body)* **Where does blood come from before it enters the heart's right atrium?** *(The body)* **Does this blood have a little or a lot of oxygen?** *(Little oxygen)* **Where is this blood transported?** *(To the right ventricle and then to the lungs)*

21st Century Learning

L3 CRITICAL THINKING Direct students' attention to the ventricles of the heart shown in **Figure 2.** Point out the whitish strands attached to the valves between the atria and the ventricles. Tell students that these strands, called *chordae tendineae,* are tendons that connect the flaps of the valves to the wall of the ventricle. Ask students to suggest the function of these tendons. *(When the ventricle contracts, blood pressure closes the valve. Without the tendons to hold the parts of the valve in place, the valve might be pushed open in the wrong direction, letting blood flow back into the atrium.)*

The Heart Without your heart, your blood would not go anywhere. As **Figure 2** shows, the **heart** is a hollow, muscular organ that pumps blood to the body through blood vessels.

The Heart's Structure The heart has a right side and left side that are completely separated by a wall of tissue called the septum. Each side has two chambers. Each upper chamber, called an **atrium** (AY tree um; plural *atria*), receives blood that comes into the heart. Each lower chamber, called a **ventricle,** pumps blood out of the heart. The pacemaker, a group of cells in the right atrium, sends out signals that regulate heart rate. These signals make the heart muscle contract.

FIGURE 2

The Heart

Your heart works 24 hours a day, resting only between beats.

Complete the activities.

1. **Relate Text and Visuals** Find and label the septum on the diagram.
2. **CHALLENGE** Explain why the contraction of the left ventricle must be stronger than the contraction of the right ventricle.

Sample: The left ventricle must pump blood to the rest of the body.

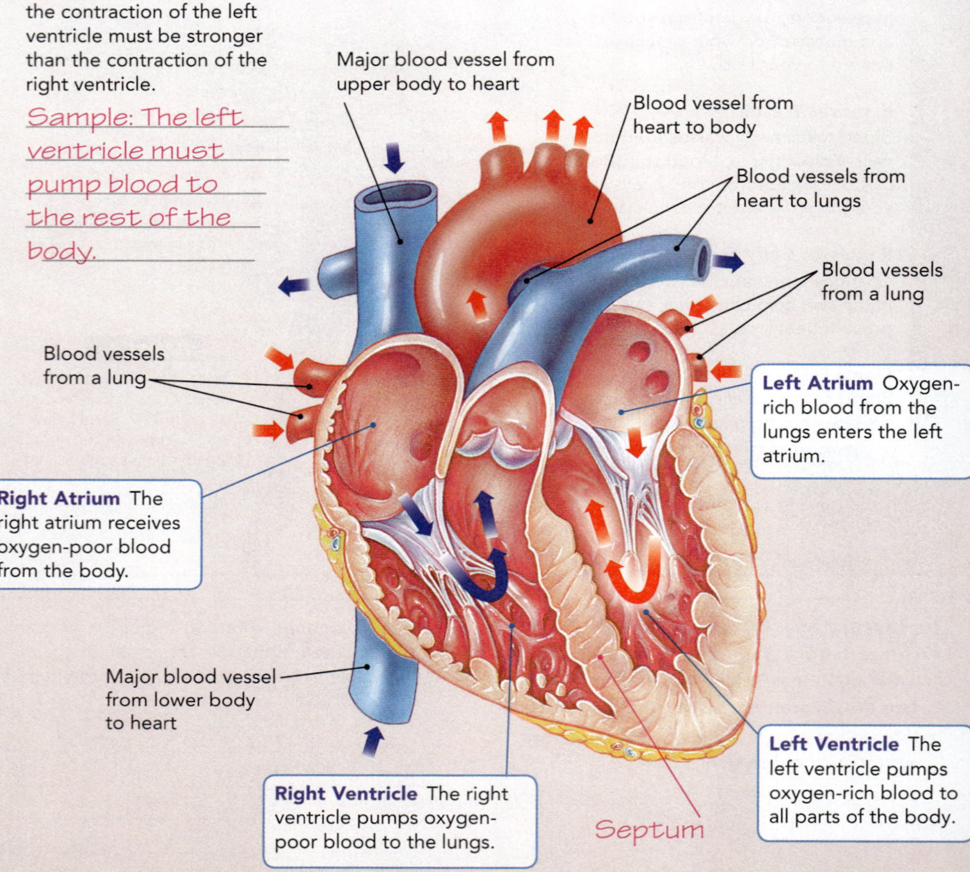

How the Heart Works Valves separate the atria from the ventricles. A **valve** is a flap of tissue that prevents blood from flowing backward. Valves also separate the ventricles and the large blood vessels that carry blood away from the heart.

A heartbeat sounds something like *lub-dup*. First, the heart muscle relaxes and the atria fill with blood. Next, the atria contract, squeezing blood through valves, like those in **Figure 3,** and into the ventricles. Then the ventricles contract. This contraction closes the valves between the atria and ventricles, making the *lub* sound and squeezing blood into large blood vessels. Finally, the valves between the ventricles and blood vessels snap shut, making the *dup* sound. All this happens in less than one second.

FIGURE 3

Heart Valves

Valves control the direction of blood flow through the heart.

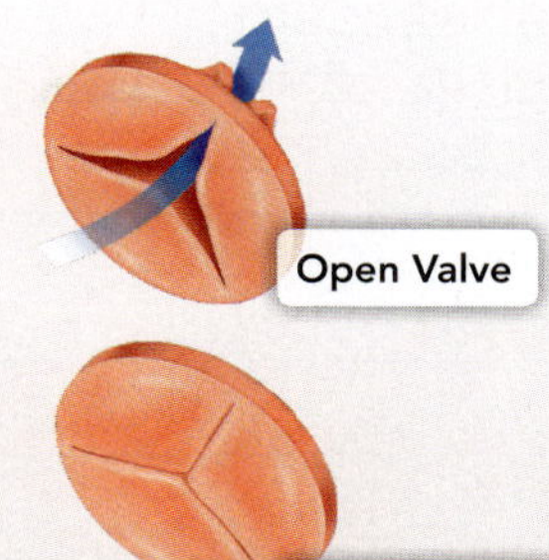

The Path of Blood Flow

As you can see in **Figure 4,** the overall pattern of blood flow through the body is similar to a figure eight. The heart is at the center where the two loops cross. In the first loop, blood travels from the heart to the lungs and then back to the heart. In the second loop, blood travels from the heart throughout the body and then back to the heart.

Your body has three kinds of blood vessels: arteries, capillaries, and veins. **Arteries** carry blood away from the heart. For example, blood in the left ventricle is pumped into the **aorta** (ay AWR tuh), the largest artery in the body. From the arteries, blood flows into tiny vessels called **capillaries.** In the capillaries, substances are exchanged between the blood and body cells. From capillaries, blood flows into **veins,** which carry blood back to the heart.

FIGURE 4

Blood Flow

Your heart can pump five liters of blood through the two loops each minute.

Interpret Diagrams In each box, write where the blood from the heart travels. Then tell where blood travels after it leaves each part listed below.

Right atrium

It enters the right ventricle.

Veins from the body

It enters the right atrium.

Arteries to the lungs

It enters capillaries in the lungs.

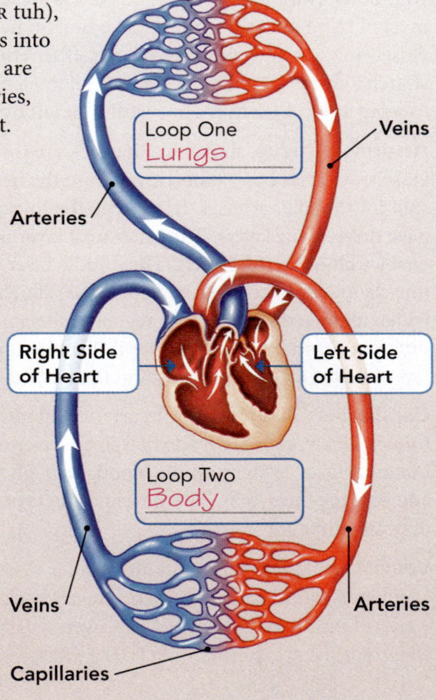

Teach With Visuals

Tell students to look at **Figure 3.** Explain that valves separate chambers from each other and from other parts of the body. Make sure students understand that valves are located between the left atrium and left ventricle, between the right atrium and right ventricle, and between the right ventricle and the artery to the lungs. The valve shown here is between a ventricle and an artery. Point out that valves are like one-way doors in the heart. Ask: **Which way does the blood move through this valve?** *(Up from the left through the open valve)* **What would happen if the blood began to flow backward?** *(The valve would close and prevent backward flow.)*

Lead a Discussion

THREE KINDS OF BLOOD VESSELS Explain that three types of blood vessels are involved in moving blood through the two loops. Ask: **What is the difference between arteries and veins?** *(Arteries carry blood away from the heart; veins carry blood to the heart.)* **What occurs in the capillaries?** *(Substances are exchanged between the blood and body cells.)*

Teach With Visuals

Tell students to look at **Figure 4.** Remind them that oxygen-rich blood is shown in red; oxygen-poor blood is in blue. Explain that while blood passes through the heart twice in completing two loops, the blood from one loop does not mix with the blood from the other loop. In the first loop, blood is moved from the heart to the lungs and back to the heart. In the second loop, blood is moved from the heart throughout the body and back to the heart. Help students locate the flow of blood to the aorta. Have students locate the blood vessels that branch off the aorta. Ask: **Why is the aorta important?** *(It is the largest artery and branches off the aorta carry blood to all parts of the body except the lungs.)* Point out that the large artery shown below the heart is the continuation of the aorta.

Elaborate

Build Inquiry 

L2 OBSERVE YOUR HEARTBEAT

Material stethoscope

Time 15 minutes

Explain that you can use a stethoscope to hear the sound of heart valves closing. Have students use the stethoscope to listen to their own heartbeats and then describe the sound they hear. *(After each student uses the device, use rubbing alcohol to clean the parts of the stethoscope inserted into the ears.)*

Differentiated Instruction

L1 Model a Heartbeat Give each student a beanbag or small, soft ball. Demonstrate how to squeeze the item at a steady rate to approximate 80 beats per minute. Then have students try this for a minute or two. Emphasize that the heart works continuously like this for a person's lifetime.

L3 Pacemakers Explain to students that in the case of disease or accident, an artificial pacemaker can be implanted beneath the skin and connected to the heart by wires. Tiny electrical impulses travel from the battery through the wires and make the heart contract. Have students research artificial pacemakers and present their findings in a poster or visual presentation for the class.

LESSON 7.2

Explain

Lead a Discussion

STRUCTURES AND FUNCTIONS OF BLOOD VESSELS Remind students that the three different kinds of blood vessels have different functions. Their structures help support these functions. Ask: **What type of blood vessel has thick walls?** *(Arteries)* **What is the function of these blood vessels?** *(Carrying blood away from the heart to the body's cells)* **How does the structure of an artery relate to its function?** *(The innermost layer of epithelial tissue helps blood to flow freely. The muscular middle layer of smooth muscle tissue contracts and relaxes, regulating the amount of blood sent to different organs. The outer layer is connective tissue that makes the artery strong yet flexible.)* **What is the smallest type of blood vessel?** *(Capillaries)* **What happens in these thin-walled blood vessels?** *(Materials and wastes are exchanged between the body's cells and blood.)* **How does the structure of the capillaries relate to their function?** *(Capillary walls are thin so that materials can pass easily between them and the body's cells.)* **What type of blood vessel has thinner walls than artery walls?** *(Veins)* **How does the structure of veins relate to their function?** *(The thinner walls allow the veins to carry large amounts of blood from the body's cells back to the heart)*

Teach With Visuals

Tell students to look at **Figure 5.** Ask: **What is the main difference between the structure of arteries and the structures of veins?** *(Arteries are thicker than veins.)* **Why do you think this might be?** *(The extra thickness of arteries is necessary because the force of blood pumped through arteries is stronger than the force of blood flowing through veins.)*

21st Century Learning

INTERPERSONAL SKILLS Have students work in pairs to draw each type of blood vessel. The drawings should include leader lines labeling any parts discussed in the text, as well as captions describing the functions of each type of vessel. Students should read the corresponding text to identify information to include in the drawings. Students can copy and expand upon the illustrations in the text or make original drawings. Encourage students to use different-colored markers or pencils to distinguish the three structures from one another.

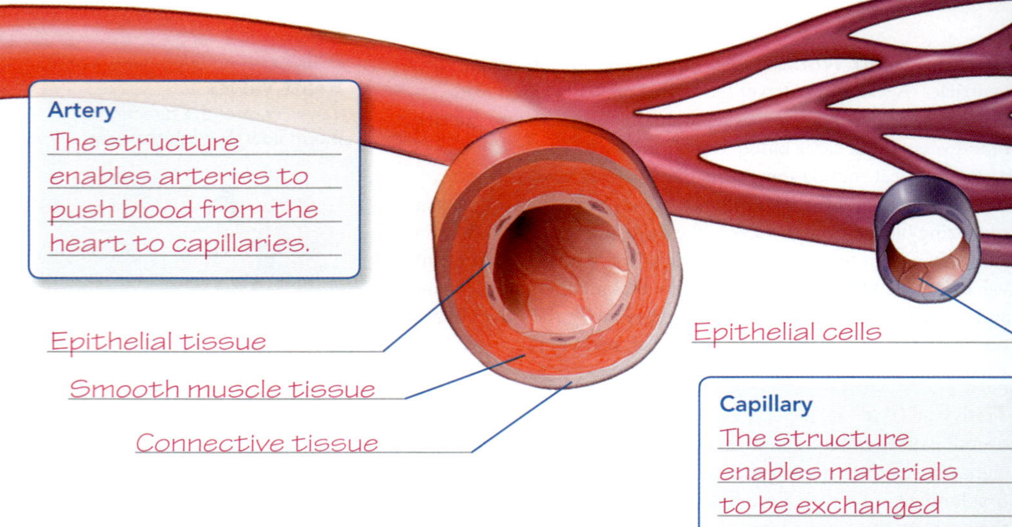

A Closer Look at Blood Vessels

Like hallways in a large building, your blood vessels run through all the tissues of your body. Although some blood vessels are as wide as your thumb, most of them are much finer than a human hair. If all the blood vessels in your body were hooked together end to end, they would stretch a distance of almost 100,000 kilometers. That's long enough to wrap around Earth twice—with a lot left over!

Arteries Arteries, shown in **Figure 5,** are thick-walled, muscular vessels. Arteries carry blood away from the heart. As they do, they split into smaller arteries. In general, the thick, elastic artery walls have three tissue layers. The innermost layer is epithelial tissue that enables blood to flow freely. The middle layer is mostly smooth muscle tissue that relaxes and contracts, allowing the artery to widen and narrow. This layer regulates the amount of blood sent to different organs. The outer layer is flexible connective tissue. These layers enable arteries to withstand the force of pumping blood.

Capillaries Blood flows from arteries into tiny capillaries. The thin capillary walls are made of a single layer of epithelial cells. Materials pass easily from the blood, through the capillary walls, and into the body cells. The waste products of cells pass in the opposite direction.

Veins Capillaries merge and form larger vessels called veins. From capillaries, blood enters veins and travels back to the heart. The walls of veins have the same tissue layers as arteries. However, the walls of veins are thinner than artery walls.

FIGURE 5

Blood Vessels

Read the text before completing the tasks.

1. **Identify** Underline in the text what happens to blood in each kind of vessel.
2. **Interpret Diagrams** In the diagram above, label the parts of each vessel. Then write in each box how the vessel's structure enables it to function.

Professional Development Note — Teacher to Teacher

Demonstration Show students that veins have valves that prevent the backflow of blood. Press 2 fingers over the main vein on the inside of your forearm, about halfway between elbow and wrist, to temporarily stop the blood flowing up it. Massage the blood in the vein upward with your thumb. Take your thumb away, and blood should not leak back down past the valve. Take your fingers away and the vein refills from below as blood once again flows up your arm.

Mrs. Anne Rice
Woodland Middle
SchoolGurnee. Illinois

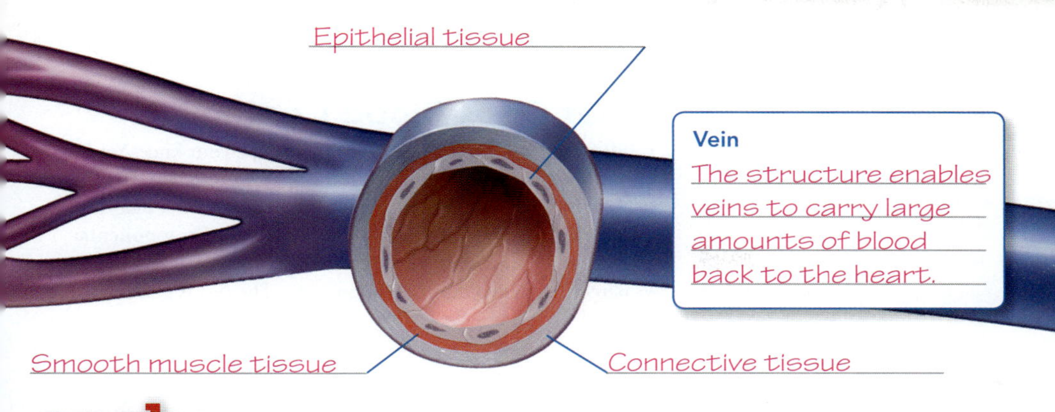

Vein
The structure enables veins to carry large amounts of blood back to the heart.

apply it!

Your pulse results from the alternating relaxation and contraction of arteries as blood is forced through them. Touch the inside of your wrist and find your pulse.

1 Observe How does your pulse feel through your fingertips?

Sample: It feels like a lump is passing through my wrist.

2 Observe How many heartbeats do you count in one minute?

Sample: 65 heartbeats

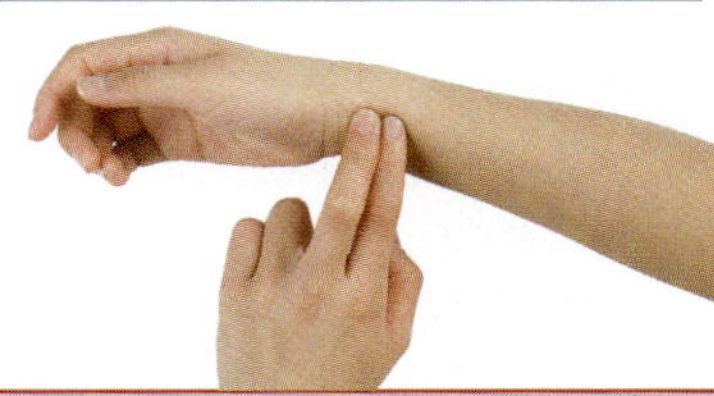

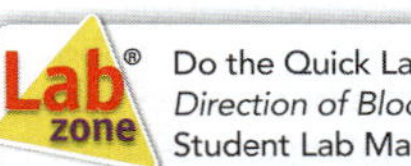
Do the Quick Lab *Direction of Blood Flow.* Student Lab Manual, p. 90

Assess Your Understanding

TEKS 12B, 12C

1a. Identify Blood returning from the lungs enters the heart at the (right atrium/left atrium/right ventricle/left ventricle).

b. Explain In which direction do arteries carry blood?

Away from the heart

c. Draw Conclusions Why is it important for your blood to complete both loops of circulation?

Sample: One loop takes oxygen into my blood and the other loop delivers oxygen to my cells.

got it?

○ **I get it!** Now I know that the circulatory system transports substances to and from cells and regulates body temperature.

○ I need extra help with See TE note.

283

Differentiated Instruction

L3 Coronary Heart Disease Challenge students to do research to learn more about the importance of coronary arteries in the context of coronary heart disease. Students might focus their research on a basic definition of this disease, which is characterized by a blockage or narrowing of the coronary artery by fatty plaques. Such a blockage causes a shortage of oxygen-rich blood to reach the heart muscle. Effects of inadequate levels of oxygen include heart attacks and angina pectoris, both of which are potentially fatal. Encourage students to include information about a variety of ways to prevent or treat coronary heart disease, including diet changes, exercise, not smoking, medications, coronary artery bypass surgery, and balloon angioplasty.

Elaborate

Apply It!

Before they begin the activity, explain to students that heart rate is measured in beats per minute (bpm). The average heart rate for adults at rest is 60 to 100 bpm. In healthy adults, a lower heart rate at rest can indicate a healthier heart. On the other hand, an unusually high or low heart rate may be a symptom of a medical problem. Read the introductory paragraph with students. After they have completed the activity, look at the results as a class. Encourage students to identify the highest and lowest heart rates and calculate the class average. If time permits, have groups graph the class's heart rates.

Observe Explain to students that making observations means using one or more of your senses to gather information.

Lab Resource: Quick Lab 

L2 DIRECTION OF BLOOD FLOW Students will explore how blood can flow against the pull of gravity. This Quick Lab can be found in the Student Lab Manual, p. 90, and online.

Evaluate

Assess Your Understanding

After students have answered the questions, have them evaluate their understanding by completing the appropriate sentence.

RTI Response to Intervention

1a. If students struggle to differentiate between the four sections of the heart, **then** have them review **Figure 2.**

b. If students have trouble identifying the function of the arteries, **then** have them review the function of the three different types of blood vessels.

c. If students have difficulty drawing a conclusion about the two loops of the circulatory system, **then** point out that one of the main functions of the circulatory system is to deliver oxygen-rich blood to the cells.

LESSON 7.2

Explain

Teach Key Concepts

Explain that blood is a complex tissue, consisting of four components: plasma, red blood cells, white blood cells, and platelets. Ask: **What is the function of red blood cells?** *(Delivery of oxygen to body cells)* **How do red blood cells perform this function?** *(The hemoglobin in red blood cells binds to oxygen molecules.)* **What do white blood cells do?** *(Recognize and alert the body to disease-causing organisms, produce chemicals to fight invading organisms, surround and kill the organisms.)*

Teach With Visuals

Tell students to look at **Figure 6.** Ask: **Where is the blood vessel in this figure?** *(the thin, pink structure on the edges.)* **How does the appearance of platelets fit the fact that they are fragments of cells?** *(They are dramatically smaller than white and red blood cells.)*

Build Inquiry 

L2 OBSERVE BLOOD CELLS

Materials prepared slide of human blood (blood smear), microscope

Time 15 minutes

Have students observe human blood under a microscope. Tell them to scan for as many different-looking cells as possible. Point out that the colors they see are due to a stain that is used to make details more visible.

Ask: **Which kind of cell is most abundant?** *(Red blood cells)* **What other kinds of cells can you see?** *(White blood cells)* **How are these cells different?** *(White blood cells are larger and have nuclei. Different kinds of white blood cells have differences in the sizes and shapes of their nuclei.)*

TEKS 3D, 12B, 12C In this section, you'll examine the four components of blood and their functions in the circulatory system.

What Does Blood Contain?

While riding your bike, you fall off and scrape your knee. Your knee stings, and blood oozes from the open wound. You go inside to clean the scrape. As you do, you wonder about what blood is.

Blood is a complex tissue. **Blood has four components: plasma, red blood cells, white blood cells, and platelets.** About 45 percent of the volume of blood is cells. The rest is plasma.

FIGURE 6

Cells in Blood

In addition to red blood cells and white blood cells, blood contains platelets and plasma.

Use the illustration about blood to complete the tasks.

1. **Identify** What is the main function of hemoglobin in the blood?
   Hemoglobin carries oxygen and releases it to the body's cells.
2. **Apply Concepts** What do you think would happen to the number of white blood cells in the body when a person is fighting an infection? Explain your answer.
   The number would increase because many white blood cells would be needed to fight the infection.

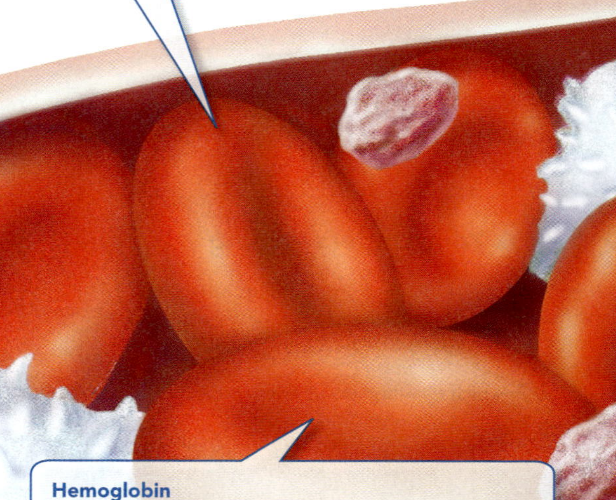

Red Blood Cells

Red blood cells take up oxygen in the lungs and deliver it to cells throughout the body. Red blood cells, like most blood cells, are produced in bone marrow. Mature red blood cells have no nuclei. Without a nucleus, a red blood cell cannot reproduce or repair itself. Mature red blood cells live only about 120 days.

Hemoglobin

A red blood cell is made mostly of **hemoglobin** (HEE muh gloh bin), a protein that contains iron and binds chemically to oxygen molecules in the lungs. Hemoglobin releases oxygen as blood travels through the capillaries. Oxygen makes red blood cells bright red. Without it, the cells are dark red. Hemoglobin also picks up some carbon dioxide produced by cells and releases it into the lungs.

Texas Essential Knowledge and Skills

3D Relate the impact of research on scientific thought and society, including the history of science and contributions of scientists as related to the content.

12B Identify the main functions of the systems of the human organism, including the circulatory, respiratory, skeletal, muscular, digestive, excretory, reproductive, integumentary, nervous, and endocrine systems.

12C Recognize levels of organization in plants and animals, including cells, tissues, organs, organ systems, and organisms.

284 Managing Materials in the Body

Plasma

Most materials transported in blood travel in the plasma. Plasma carries nutrients, such as glucose, fats, vitamins, and minerals. Plasma also carries chemical messengers that direct body activities, such as how your cells use glucose. In addition, plasma carries away most of the carbon dioxide and many other wastes that cell processes produce. Proteins in the plasma make it look pale yellow. Some of these proteins regulate water in the blood. Some help fight disease. Others help to form blood clots.

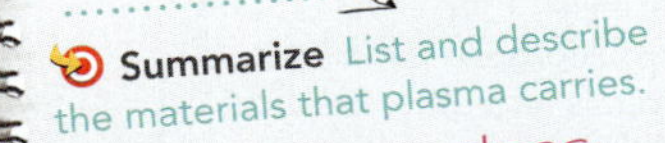

Summarize List and describe the materials that plasma carries.

- Nutrients such as glucose, fats, and vitamins
- Chemical messengers that direct body activities
- Waste products

White Blood Cells

Like red blood cells, white blood cells are produced in bone marrow. White blood cells are the body's disease fighters. Some white blood cells recognize disease-causing organisms, such as bacteria, and alert the body to the invasion. Other white blood cells produce chemicals to fight the invaders. Still others surround and kill the organisms. White blood cells are larger than red blood cells and contain nuclei. They may live for days, months, or even years.

Platelets

Platelets (PLAYT lits) are cell fragments that help form blood clots. When a blood vessel is cut, platelets collect and stick to the vessel at the wound. The platelets release chemicals that produce a protein called fibrin (FY brin). Fibrin weaves a net of tiny fibers across the cut. Platelets and blood cells become trapped in the net, and a blood clot forms.

Make Analogies

L1 LIVING COMPONENTS It may help to personify red blood cells, white blood cells, and platelets. Point out that white blood cells are like a nation's soldiers, that red blood cells are like people transporting food and other valuable cargo all over the country, and that platelets are like emergency medical technicians treating injuries. Ask: **In what way are white blood cells like soldiers?** *(Both fight off threats and invaders.)* **How do red blood cells serve a function that is similar to people transporting valuable cargo?** *(They carry oxygen, a valuable substance, to places all over the body.)* **What do platelets do that resembles the work of emergency medical technicians?** *(When a blood vessel is cut, they attempt to stop the bleeding by forming blood clots.)*

Summarize Tell students that when they summarize, they restate the main idea of a paragraph or passage.

Lead a Discussion

THE ROLE OF PLASMA Point out that about 55 percent of the volume of blood is plasma. Plasma is not a cell. It is a water-like fluid in which red blood cells, white blood cells, and platelets are suspended. Plasma carries many materials that aid the body's cells. Ask: **What are some of the nutrients plasma carries?** *(Glucose, fats, vitamins, and minerals)* **How does plasma help direct the body's activities?** *(By carrying chemical messengers)* **How does plasma help the body eliminate waste?** *(By carrying away most of the carbon dioxide and other wastes produced by cells)* **How do the proteins in plasma help the body function?** *(They regulate water in the blood, help fight disease, and help form blood clots.)*

21st Century Learning

CREATIVITY Invite students to make models of the three cell components of blood—red blood cells, white blood cells, and platelets—using colored modexling clay. Students may refer to **Figure 6** for guidance about representing the appropriate shapes, sizes, and colors of each component. After students have completed their models, have them write an information card for each one that includes the component's name and primary function.

Differentiated Instruction

L1 Red Cells and White Cells To help students distinguish between red and white blood cells, have them create compare/contrast tables or Venn diagrams. They should include words and phrases describing the structure and function of red and white blood cells, as well as the places they are produced in the body.

L3 Research Anemia Tell students that anemia is an illness caused by a shortage of red blood cells or hemoglobin in the body. Invite students to find out more about anemia, especially reasons why some teenagers are prone to anemia *(Samples: They are picky eaters, go on diets to lose weight, or become vegetarians.)* Students might share their research findings in an oral presentation.

LESSON 7.2

Explain

Lead a Discussion

MARKER MOLECULES Explain to students that the presence or absence of marker molecules on the red blood cells determines a person's blood type. Point out that these marker molecules also determine which types of blood the person can safely receive in a transfusion. Encourage students to look carefully at **Figure 7.** Ask: **What type of marker molecule is on type B blood?** *(B marker)* **Why are clumping proteins in type B blood "anti-A"?** *(Clumping proteins recognize markers that are foreign or different. Because type B blood contains B markers, it recognizes A markers as "foreign" and is therefore "anti-A.")* **What would happen if a person with type B blood received type A blood?** *(The type A cells would clump together; the person would get sick and could die.)* **What difference is there between the red blood cells of a person with type A blood and the red blood cells of a person with type AB blood?** *(The red blood cells of a person with type A blood have only one kind of molecule marker—that is, a marker for type A. On the other hand, the red blood cells of a person with type AB blood has molecule markers for type A as well as molecule markers for type B.)* **What difference is there between the red blood cells of a person with type O blood and the red blood cells of people with the other three types of blood?** *(The red blood cells of a person with type O blood have no molecule markers for blood type, whereas the red blood cells of people with A, B, and AB have such molecule markers.)* **Why can a person with type B blood receive type O blood in a transfusion?** *(Type O has no markers for blood type.)*

Transfusions for patients in hospitals and other medical facilities create a constant need for blood. According to the Red Cross, a blood transfusion is needed in the U.S. every two seconds. A typical transfusion of red blood cells is about 3 pints.

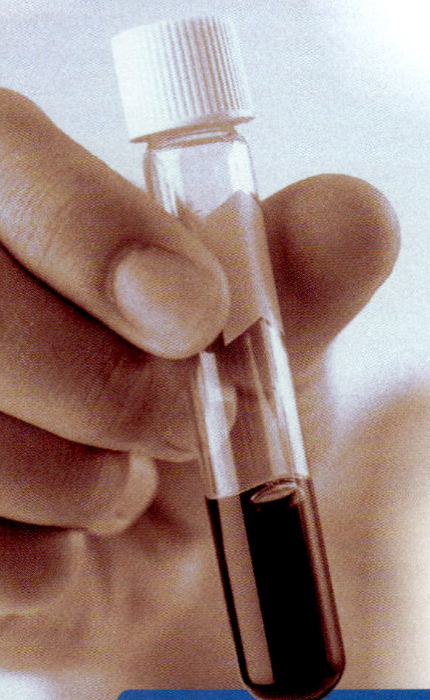

Marker Molecules and Transfusions A blood transfusion is the transfer of blood from one person to another. Most early attempts at blood transfusion failed, but no one knew why. In the early 1900s, a physician named Karl Landsteiner tried mixing blood samples from two people. Sometimes the two blood samples blended smoothly. At other times, the red blood cells clumped together. In a patient, this clumping would clog the capillaries, causing death.

Blood Types Landsteiner identified the four major types of blood: A, B, AB, and O. Blood types are determined by marker molecules on red blood cells. If your blood type is A, you have the A marker. If your blood type is B, you have the B marker. People with type AB blood have both A and B markers. People with type O blood do not have A or B markers.

Clumping proteins in your plasma recognize red blood cells with "foreign" markers that are not your type. The proteins make cells with foreign markers clump together. For example, blood type A contains anti-B clumping proteins that act against cells with B markers. Blood type O has clumping proteins for both A and B markers. In **Figure 7,** you can see all the blood type marker molecules and clumping proteins.

FIGURE 7

Blood Types and Their Markers

Depending on your blood type, you may have certain marker molecules on your red blood cells and certain clumping proteins in your plasma.

Create Data Tables **In the table, label the marker molecules and then identify the clumping proteins.**

Blood Types, Marker Molecules, and Clumping Proteins

Blood Type Characteristic	Blood Type A	Blood Type B	Blood Type AB	Blood Type O
Marker Molecules on Red Blood Cells	A A A A A A	B B B B B B	A B B A A B	
Clumping Proteins	anti-B	anti-A	no clumping proteins	anti-A and anti-B

The marker molecules on your red blood cells determine the type of blood you can safely receive in transfusions. For example, a person with type A blood can receive transfusions of type A or type O blood. But type B blood would cause clumping and would not be safe. Through a process called cross-matching, a patient's blood type is checked so that safe donor types can be determined.

Blood Type	Safe Donor(s)	Unsafe Donor(s)
A	A, O	B, AB
B	B, O	A, AB
AB	A, B, AB, O	None
O	O	A, B, AB

1 **Infer** Use what you know about blood types to complete the table.

2 **Predict** Which blood type may accept safe transfusions from any other blood type? Why?

Type AB because clumping proteins will not recognize A or B markers as foreign

3 CHALLENGE Which blood type is a "universal donor," that is, a blood type that can be used in transfusions to anyone? Explain your reasoning.

Type O can donate to all blood types because it does not have A or B markers.

Rh Factor Landsteiner also discovered a protein on red blood cells that he called *Rh factor.* About 85 percent of the people he tested had this protein. The rest did not. As with blood type, a marker molecule on the red blood cells determines the presence of Rh factor. An Rh-positive blood type has the Rh marker. An Rh-negative blood type does not. Clumping proteins will develop in people with Rh-negative blood if they receive Rh-positive blood. This situation may be potentially dangerous.

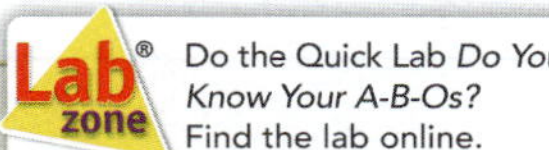

Do the Quick Lab *Do You Know Your A-B-Os?* Find the lab online.

Assess Your Understanding

TEKS 3D, 12B, 12C

2a. **Identify** What is plasma?

It is the liquid part of the blood.

b. **Relate** What did Karl Landsteiner discover and what impact did it have on society?

His discovery of marker molecules on red blood cells made it safer to receive blood.

c. **Relate Cause and Effect** How might a lack of iron in a person's diet affect his or her blood?

Sample: The person might not have enough hemoglobin to make all the red blood cells needed.

got it?

○ I get it! Now I know that blood contains plasma, red blood cells, white blood cells, and platelets.

○ I need extra help with See TE note.

Differentiated Instruction

L1 Karl Landsteiner Tell students that Karl Landsteiner won the Nobel Prize in Medicine in 1930 for his work on the blood groups. Invite students to research Karl Landsteiner and write a brief report on the studies that led to his discovery of the blood groups.

L3 Blood Banks Explain that the American Red Cross and America's Blood Centers control most of the blood banks in the United States. These agencies solicit citizens to donate blood that will be used in emergency transfusions. Invite students to research how blood banks work in the U.S. In their brief reports, students might choose to focus on the price of blood, on the types of procedures that require blood, or on how the need for donated blood often far outweighs the supply.

Lead a Discussion

RH FACTOR Explain that Rh factor is named after the Rhesus macaque, a monkey used by Landsteiner in his experiments with blood. Draw students' attention to a difference as well as the similarities between Rh factor and blood types. Ask: **How are both blood types and Rh factor determined?** *(By the presence or absence of markers on red blood cell molecules)* **Why is it important to match the donor's and recipient's Rh factor during a blood transfusion?** *(If the Rh factor is different, dangerous clumping proteins will develop in the recipient's blood.)* **In what important way are Rh factors and blood types different from one another?** *(Sample: There are four blood types, and two Rh factors.)* If students have previously studied genetics, they may be aware of the pattern of inheritance of A-B-O blood types. Tell them the Rh factor is also inherited, but not controlled by an allele that is part of the A-B-O inheritance pattern.

Elaborate

Apply It!

Before students begin the activity, explain that the *donor* is the blood given. Read the introductory paragraph and review the chart. The Apply It! activity can be made more challenging by adding information about the Rh factor. This exercise may be done as a class discussion, an extension of the initial exercise, or for gifted students to meet the need for differentiated instruction.

Lab Resource: Quick Lab 

L2 DO YOU KNOW YOUR A-B-OS? Students will model interactions between different blood types. This Quick Lab can be found online.

Evaluate

Assess Your Understanding

After students have answered the questions, have them evaluate their understanding by completing the appropriate sentence.

RTI Response to Intervention

2a. If students have difficulty defining plasma, **then** have them review the definition that accompanies **Figure 6.**

b. If students have trouble relating the impact of Landsteiner's discovery, **then** point out he observed that some bloods mixed smoothly; and some clumped.

c. If students struggle to relate iron to red blood cells, **then** have them review **Figure 6.**

Name ______________________ Date __________ Class __________

Assess Your Understanding

The Circulatory System

What Are the Main Functions of the Circulatory System?

1a. IDENTIFY Blood returning from the lungs enters the heart at the (right atrium/left atrium/right ventricle/left ventricle).

b. EXPLAIN In which direction do arteries carry blood?

____

____

c. DRAW CONCLUSIONS Why is it important for your blood to complete both loops of circulation?

____

____

____

got it?

○ **I get it!** Now I know that the circulatory system transports ______________________________

____

○ **I need extra help with** ______________________________

____

What Does Blood Contain?

2a. IDENTIFY What is plasma?

____

b. RELATE What did Karl Landsteiner discover and what impact did it have on society?

____

____

c. RELATE CAUSE AND EFFECT How might a lack of iron in a person's diet affect his or her blood?

____

____

____

Place the outside corner, the corner away from the dotted line, in the corner of your copy machine to copy onto letter-size paper.

Name ______________________ Date __________ Class __________

Enrich

The Circulatory System

Read the passage and study the figure. Then use a separate sheet of paper to answer the questions that follow the diagram.

Heart Murmurs

Sometimes when a doctor listens to a patient's heartbeat, he or she can hear an abnormal flow of blood through the heart. The sound of this abnormal flow is called a *heart murmur.*

Some heart murmurs are caused by blood leaking inside the heart. One type of leak occurs when there is a hole in the wall of tissue that separates the right and left sides of the heart. A hole in this wall causes blood to leak from the left side of the heart into the right side because the left side pumps with more force than the right side. The figure shows a heart with a hole between its ventricles and the direction of blood flow from its left ventricle.

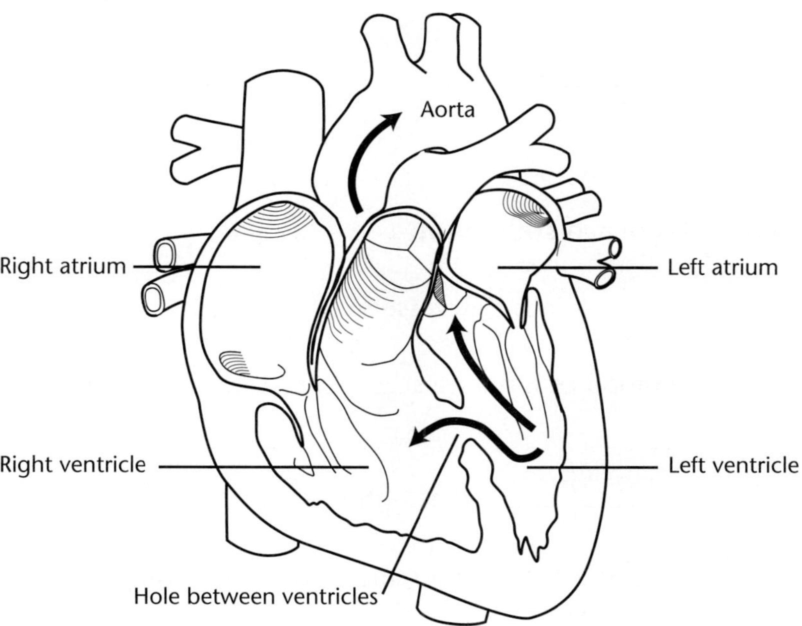

1. In what two directions does blood flow from the left ventricle of the heart in the figure?
2. How does the blood flow in this abnormal heart compare to that in a normal heart?
3. Compare the blood in the right ventricle of the heart in the figure with the blood in the right ventricle of a normal heart. (*Hint:* The wall of tissue normally prevents oxygen-rich blood from mixing with oxygen-poor blood.)
4. How might a hole between the sides of the heart affect the functioning of the circulatory system?

Name ______________________ Date ____________ Class ____________

Lesson Quiz

The Circulatory System

Write the letter of the correct answer on the line at the left.

1. ___ Which of the following helps form blood clots when a blood vessel is cut?

A plasma
B red blood cells
C white blood cells
D platelets

2. ___ A person with type O blood has clumping proteins for

A type A markers
B type B markers
C types A and B markers
D type O markers

3. ___ Oxygen-rich blood from the lungs enters the heart through the

A septum
B right ventricle
C right atrium
D left atrium

4. ___ Which vessel carries blood away from the heart?

A arteries
B capillaries
C veins
D valves

Fill in the blank to complete each statement.

5. From the arteries, blood flows into tiny vessels called ____________________.

6. The hemoglobin in red blood cells releases ____________________ as blood travels through the capillaries.

7. Blood contains ____________________ that fight disease.

8. The ____________________ is a hollow, muscular organ that pumps blood throughout the body.

9. Red and white blood cells are both produced in ____________________ ____________________.

10. The ____________________ is the largest artery in the body.

The Circulatory System

Answer Key

Review and Reinforce

Find the worksheet in the Student Workbook.

1. Students' should have drawn arrows showing blood being pumped out of the right ventricle along *b*, back to the left atrium along *c*, from the left atrium into the left ventricle, out of the left ventricle along *d*, back to the right atrium along *a*, and from the right atrium into the right ventricle.
2. The right atrium receives oxygen-poor blood from the body. The left atrium receives oxygen-rich blood from the lungs.
3. The right ventricle pumps the oxygen-poor blood to the lungs. The left ventricle pumps the oxygen-rich blood to all parts of the body.
4. The blood vessels labeled *b* and *d* are arteries, because they carry blood away from the heart. The blood vessels labeled *a* and *c* are veins, because they carry blood to the heart.
5. d
6. i
7. f
8. b
9. a
10. h
11. j
12. c
13. g
14. e
15. Blood travels through the body in an overall pattern like a figure eight, with the heart at the center where the two loops cross. First, blood travels from the heart to the lungs, where it picks up oxygen, and back. Then blood travels from the heart throughout the body, where it delivers oxygen to the cells, and back to the heart. The circulatory system sends important substances to cells around the body, carries waste products away from cells, and fights disease.

Enrich

1. In the heart in the figure, blood flows from the left ventricle into the aorta and into the right ventricle.
2. In a normal heart, blood only flows from the left ventricle into the aorta.
3. The blood in the right ventricle of a normal heart is oxygen-poor. The blood in the right ventricle of the abnormal heart in the figure contains a mixture of oxygen-poor and oxygen-rich blood.
4. Since oxygen-rich and oxygen-poor blood would mix, the circulatory system might not be able to deliver enough oxygen to body cells.

Lesson Quiz

1. D
2. C
3. D
4. A
5. capillaries
6. oxygen
7. white blood cells
8. heart
9. bone marrow
10. aorta

The Respiratory System

How do systems of the body move and manage materials?

LESSON PACING:
2–3 periods or 1–1 ½ blocks

Lesson Vocabulary

- pharynx • trachea • cilia • bronchi • lungs
- alveoli • diaphragm • larynx • vocal cords

Content Refresher

Control of Breathing Most people are able to control their rates of breathing to some extent. However, if an individual holds his or her breath for a long time, the person would lose consciousness and the brain would take over to resume breathing. Much of the control of breathing is involuntary and is directed by breathing centers in the brain.

It is the concentration of carbon dioxide (CO_2) in the blood, rather than the concentration of oxygen (O_2), that determines the rate of breathing. In general, the higher the blood level of carbon dioxide, the more rapidly a person breathes. For example, the amount of carbon dioxide in the blood increases when a person exercises, which causes the person's breathing rate to increase. In contrast, the amount of carbon dioxide in the blood can fall to abnormal levels when a person hyperventilates, causing the blood acidity to decline. As a result, the brain can misinterpret the blood oxygen level, which can decrease to a dangerous level.

Lesson Objectives	TEKS	ELPS
Identify the main functions and structures of the respiratory system.	12B, 12C, 12E	3.C.4
Explain what happens during breathing and gas exchange.	12B, 12C	

Texas Essential Knowledge and Skills

12B Identify the main functions of the systems of the human organism, including the circulatory, respiratory, skeletal, muscular, digestive, excretory, reproductive, integumentary, nervous, and endocrine systems.
12C Recognize levels of organization in plants and animals, including cells, tissues, organs, organ systems, and organisms.
12E Compare the functions of a cell to the functions of organisms such as waste removal.

English Language Proficiency Standards

ELPS Speaking 3.C.4 Speak using a variety of connecting words with increasing accuracy and ease as more English is acquired.

DIFFERENTIATED INSTRUCTION KEY
L1 Struggling Students or Special Needs
L2 On-Level Students L3 Advanced Students

LESSON PLANNER 7.3

Investigations and Activities	TEKS Review
My Planet Diary, **Student Edition,** p. 288 Inquiry: Inquiry Warm-Up, How Big Can You Blow Up a Balloon?, **PearsonTexas.com** Introduce Vocabulary, **Teacher's Edition,** p. 289 Teach Key Concepts, **Teacher's Edition,** p. 289 Address Misconceptions, Respiration and Breathing, **Teacher's Edition,** p. 289 Lead a Discussion, Breathing Structures, **Teacher's Edition,** p. 290 Teach With Visuals, **Teacher's Edition,** p. 290 21st Century Learning, Critical Thinking, **Teacher's Edition,** p. 290 Differentiated Instruction, **Teacher's Edition,** p. 291 Inquiry: Quick Lab, Modeling Respiration, **Lab Manual, p. 91**	Apply the TEKS, Moving Things Along, **Student Edition,** p. 302 TEKS Practice, **Student Edition,** p. 305 TEKS Practice: Chapter and Cumulative Review, **Student Edition,** p. 308 Lesson 7.3, **TEKS Preparation and Study Guide Workbook,** p. 72
Teach Key Concepts, **Teacher's Edition,** p. 292 Lead a Discussion, The Breathing Process, **Teacher's Edition,** p. 292 Inquiry: Teacher Demo, Air Pressure, **Teacher's Edition, p. 292** Differentiated Instruction, **Teacher's Edition,** p. 293 Apply It!, **Student Edition,** p. 293 Support the TEKS, Gas Exchange in the Alveoli, **Teacher's Edition,** p. 294 Address Misconceptions, Exhaled Gases, **Teacher's Edition,** p. 294 Lead a Discussion, Surface Area, **Teacher's Edition,** p. 295 Inquiry: Lab Investigation, A Breath of Fresh Air, **PearsonTexas.com**	**SHORT ON TIME?** To do this lesson in approximately half the time, do the Activate Prior Knowledge activity. A discussion of the Key Concepts will familiarize students with the lesson content. Have students do the Quick Lab. The rest of the lesson can be completed by students independently.

These editable worksheets are available on **PearsonTexas.com**.
Print versions can be found in the **TEKS Preparation and Study Guide Workbook.**

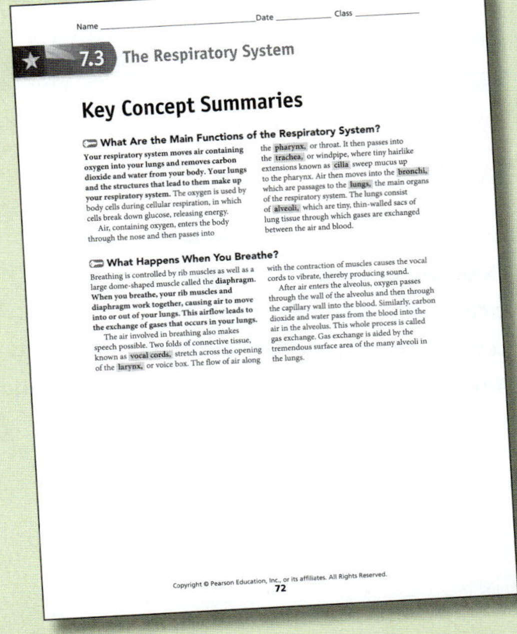

Name ______ Date ______ Class ______

7.3 The Respiratory System

Key Concept Summaries

What Are the Main Functions of the Respiratory System?

Your respiratory system moves air containing oxygen into your lungs and removes carbon dioxide and water from your body. Your lungs and the structures that lead to them make up your respiratory system. The oxygen is used by body cells during cellular respiration, in which cells break down glucose, releasing energy.

Air, containing oxygen, enters the body through the nose and then passes into the **pharynx,** or throat. It then passes into the **trachea,** or windpipe, where tiny hairlike extensions known as **cilia** sweep mucus up to the pharynx. Air then moves into the **bronchi,** which are passages to the **lungs,** the main organs of the respiratory system. The lungs consist of **alveoli,** which are tiny, thin-walled sacs of lung tissue through which gases are exchanged between the air and blood.

What Happens When You Breathe?

Breathing is controlled by rib muscles as well as a large dome-shaped muscle called the **diaphragm. When you breathe, your rib muscles and diaphragm work together, causing air to move into or out of your lungs. This airflow leads to the exchange of gases that occurs in your lungs.**

The air involved in breathing also makes speech possible. Two folds of connective tissue, known as **vocal cords,** stretch across the opening of the **larynx,** or voice box. The flow of air along with the contraction of muscles causes the vocal cords to vibrate, thereby producing sound.

After air enters the alveolus, oxygen passes through the wall of the alveolus and then through the capillary wall into the blood. Similarly, carbon dioxide and water pass from the blood into the air in the alveolus. This whole process is called gas exchange. Gas exchange is aided by the tremendous surface area of the many alveoli in the lungs.

72

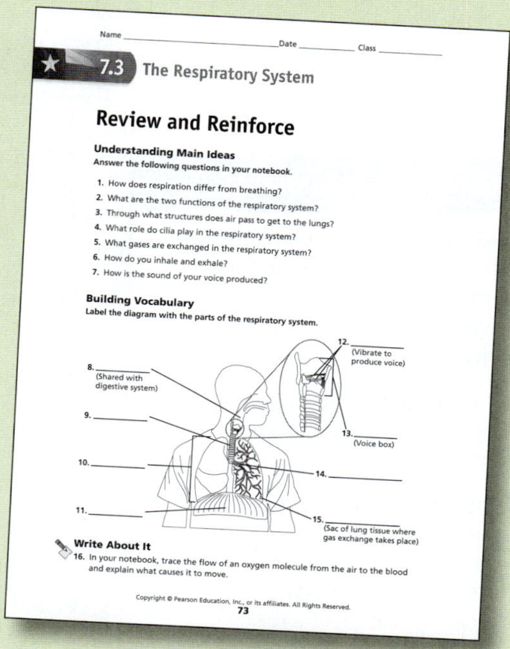

Name ______ Date ______ Class ______

7.3 The Respiratory System

Review and Reinforce

Understanding Main Ideas
Answer the following questions in your notebook.

1. How does respiration differ from breathing?
2. What are the two functions of the respiratory system?
3. Through what structures does air pass to get to the lungs?
4. What role do cilia play in the respiratory system?
5. What gases are exchanged in the respiratory system?
6. How do you inhale and exhale?
7. How is the sound of your voice produced?

Building Vocabulary
Label the diagram with the parts of the respiratory system.

Write About It

16. In your notebook, trace the flow of an oxygen molecule from the air to the blood and explain what causes it to move.

73

Lexile Measure = 910L

LESSON 7.3

The Respiratory System

Establish Learning Objectives

After this lesson, students will be able to:

Identify the main functions and structures of the respiratory system.

Explain what happens during breathing and gas exchange.

Engage

Activate Prior Knowledge

MY PLANET DIARY Read *The Breath of Life* with the class. Remind students that air is a mixture of gases. Invite students to share their ideas about which gases are in air. Ask: **What gases make up a sample of air?** *(Students may name oxygen, carbon dioxide, and nitrogen.)* **Which gas do humans and other animals need to survive?** *(Students may indicate that living things need oxygen to survive.)*

Explore

Lab Resource: Inquiry Warm-Up 

L1 HOW BIG CAN YOU BLOW UP A BALLOON? Students will use a balloon to investigate factors that affect the volume of air a person exhales. This Inquiry Warm-Up can be found online.

The Respiratory System

What Are the Main Functions of the Respiratory System?
TEKS 12B, 12C, 12E

What Happens When You Breathe?
TEKS 12B, 12C

MY PLANET DIARY — MISCONCEPTION

The Breath of Life

Misconception: The only gas you exhale is carbon dioxide.

Actually, about 16 percent of the air you exhale is oxygen. The air you inhale is made up of about 21 percent oxygen. Your body only uses a small portion of the oxygen in each breath, so the unused portion is exhaled.

If a person stops breathing, he or she needs to get more oxygen quickly. As part of CPR (cardiopulmonary resuscitation), a rescuer can breathe into the person's mouth to give unused oxygen to the person. This process is called rescue breathing.

Read the following question. Then write your answer below.

Why would you want to learn to perform rescue breathing?

Sample: I would want to learn it in case I need to save someone's life.

Do the Inquiry Warm-Up *How Big Can You Blow Up a Balloon?* Find the lab online.

TEKS 12B, 12C, 12E In this section, you'll learn about the roles of major organs and structures in the respiratory system.

ELPS 3.C.4 Explain to a partner, what you see in the diagram on page 289. Use connecting words, such as *then, which, that, next,* and *so.*

What Are the Main Functions of the Respiratory System?

In an average day, you may breathe 20,000 times. You breathe all the time because your body cells need oxygen, which comes from the air. **Your respiratory system moves air containing oxygen into your lungs and removes carbon dioxide from your body. Your lungs and the structures that lead to them make up your respiratory system.**

Texas Essential Knowledge and Skills

12B Identify the main functions of the systems of the human organism, including the circulatory, respiratory, skeletal, muscular, digestive, excretory, reproductive, integumentary, nervous, and endocrine systems.

12C Recognize levels of organization in plants and animals, including cells, tissues, organs, organ systems, and organisms.

12E Compare the functions of a cell to the functions of organisms such as waste removal.

SUPPORT ALL READERS

Lexile Measure = 910L Lexile Word Count = 1357

Prior Exposure to Content: Many students may have misconceptions on this topic

Academic Vocabulary: *chart, communicate*

Science Vocabulary: *pharynx, trachea, cilia, bronchi, alveoli, larynx*

Concept Level: Generally appropriate for most students in this grade

Preteach With: My Planet Diary "The Breath of Life" and Figure 3 activity

Vocabulary
- pharynx
- trachea
- cilia
- bronchi
- lungs
- alveoli
- diaphragm
- larynx
- vocal cords

Skills
- Reading: Chart
- Inquiry: Communicate

Breathing and Homeostasis Your body needs oxygen for cellular respiration. During cellular respiration, cells break down glucose, releasing energy. You use energy for activities such as reading this book or playing ball. Your body also uses energy in carrying out processes that maintain homeostasis, such as removing wastes and regulating body temperature.

Breathing gets oxygen into your body. But cellular respiration depends on body systems working together. The digestive system supplies glucose from food. And the circulatory system carries this glucose and the oxygen from the respiratory system to all the cells in the body. The circulatory system also transports a waste product of cellular respiration, carbon dioxide, to the lungs for removal from the body.

FIGURE 1

Systems Working Together
Body systems work together, getting the materials needed for cellular respiration to cells.

Describe System Interactions **Describe how each system provides cells with materials needed for cellular respiration. Then tell how cellular respiration helps the body maintain homeostasis.**

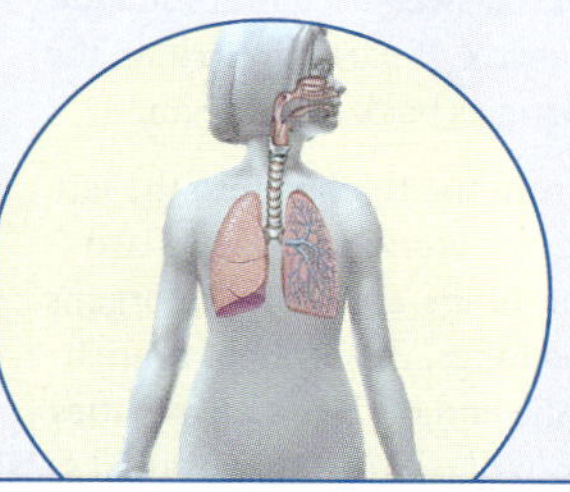

Respiratory System
This system moves air, which contains oxygen, into the body.

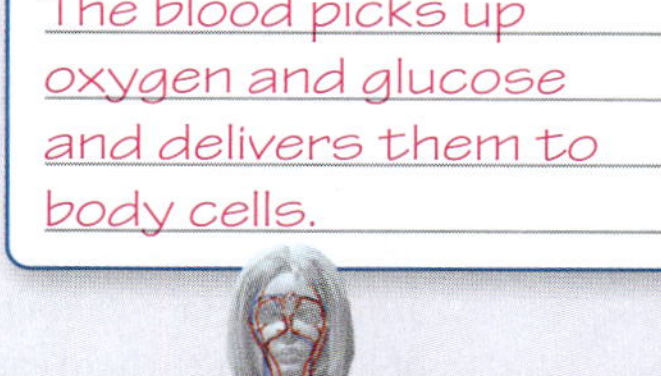

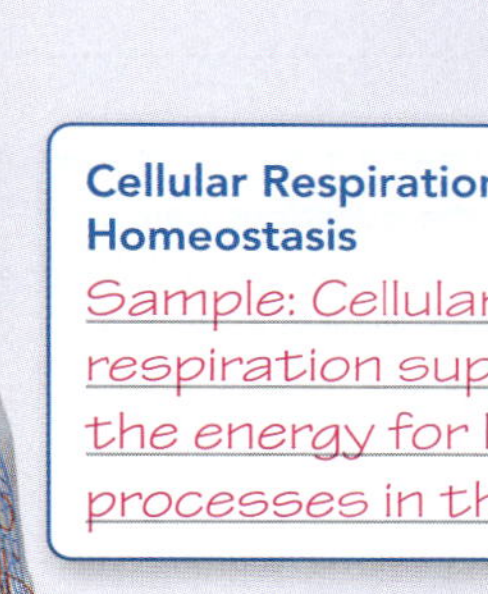

Cellular Respiration and Homeostasis
Sample: Cellular respiration supplies the energy for life processes in the body.

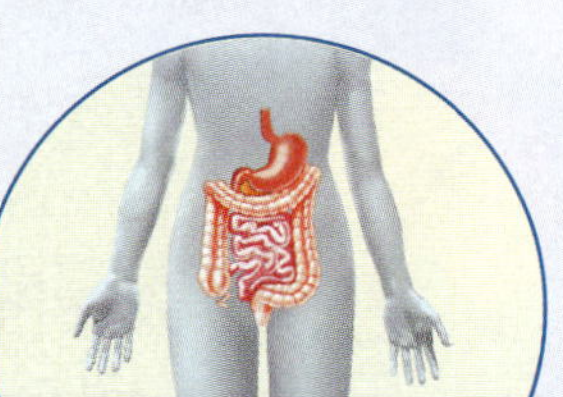

Digestive System
This system provides glucose used in cellular respiration.

English Language Proficiency Standards

ELPS Speaking 3.C.4

Read aloud **Figure 1** with students.

Beginning Read and have students repeat the connecting words from the ELPS activity on page 288. Ask students to listen for connecting words as you complete **Figure 1** orally. Repeat, having students provide the connecting words.

Intermediate Model how to use each connecting word from the ELPS activity on page 288. Then have partners take turns using connecting words as they complete the activity.

Advanced Have partners complete the activity orally before writing their answers.

Advanced High Have partners read the text and complete the activity orally. Encourage them to elaborate with details as they describe how systems work together. Then have students write their answers.

Explain

Introduce Vocabulary

Point out that the plural form of some words ending in *-us* is *-i*. *Bronchi* is plural for *bronchus*. There are two bronchi through which air flows in and out of the lungs. *Alveoli* is plural for *alveolus*. The lungs contain many alveoli, or sacs in which oxygen moves into the blood and carbon dioxide leaves the blood.

Teach Key Concepts

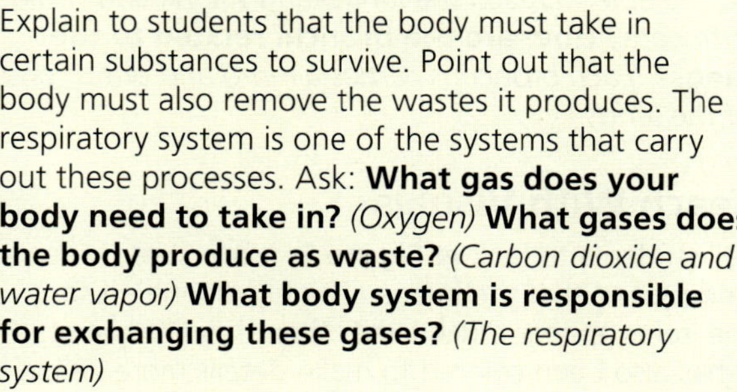

Explain to students that the body must take in certain substances to survive. Point out that the body must also remove the wastes it produces. The respiratory system is one of the systems that carry out these processes. Ask: **What gas does your body need to take in?** *(Oxygen)* **What gases does the body produce as waste?** *(Carbon dioxide and water vapor)* **What body system is responsible for exchanging these gases?** *(The respiratory system)*

Teach With Visuals

Tell students to look at **Figure 1.** Explain that body systems do not work alone. Instead they interact and overlap. Ask: **How are the respiratory and digestive systems similar?** *(They both obtain substances the body needs.)* **How do the respiratory and digestive systems depend on the circulatory system?** *(The circulatory system brings materials from the respiratory and digestive systems to the cells where they are needed.)*

Address Misconceptions

L1 RESPIRATION AND BREATHING Students often confuse the two ways the term *respiration* is used. Students may have heard cellular respiration referred to simply as *respiration*, and they may also have heard of someone who stopped breathing being given *artificial respiration*. Make sure students understand that cellular respiration is a series of chemical reactions that occur within individual cells. Breathing is a physical process that involves moving air into and out of the body. Help them maintain the distinction by always referring to *breathing,* not *respiration,* when discussing the mechanics of air exchange. Ask: **How are breathing and cellular respiration related?** *(Breathing moves air into the lungs, giving the body the oxygen that cells need to carry out cellular respiration.)*

PEARSON Texas.com

LESSON 7.3

Explain

Lead a Discussion

BREATHING STRUCTURES Explain to students that air that is inhaled must travel through several organs on its way to the lungs. Create a flow chart on the board, identifying how each organ processes air on its way to the lungs. Ask: **Into which organ does air pass after it leaves the nose on its way to the lungs?** *(It passes into the pharynx.)* **What might happen if the trachea were blocked?** *(Air would not be able to reach the lungs, and the person could suffocate.)* **How are the bronchi related to the lungs?** *(Each bronchus carries air into and out of one of the lungs.)*

Teach With Visuals

Tell students to look at **Figure 2.** Make sure students understand that the image is highly magnified so the items appear much larger than they really are. It has also been colored to make details more visible. Help students identify each of the items shown. Ask: **What do the cilia resemble?** *(Students might suggest they resemble tiny hairs or even carpet fibers.)* **What do cilia do?** *(Cilia sweep mucus containing particles out of the trachea into the throat where they can be swallowed or coughed out of the body.)*

21st Century Learning

CRITICAL THINKING Direct students' attention to **Figure 2** and ask them to think of other small particles that might get into the respiratory system. Ask: **How do the cilia and mucus help protect the body from diseases?** *(Mucus may trap bacteria or viruses that could harm the body, especially the respiratory system. The cilia move them away from the lungs.)* If students have previously studied the digestive system, ask them to identify the pathway of mucus from the trachea to the stomach. *(Trachea, pharynx, esophagus, stomach)* Point out that saliva and some secretions of the stomach contain substances that act as antibacterial agents.

Substances in the air you breathe can affect your lungs and breathing passages. The Texas Commission on Environmental Quality monitors and reports on air quality for the entire state. The reports include data about particles that can cause allergies and other pollutants that can affect human health.

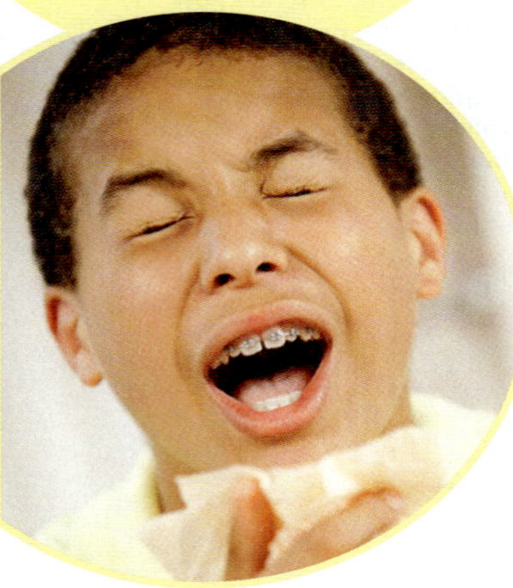

Breathing Structures When you breathe in, air and particles such as pollen and dust move through a series of structures and then into the lungs. You can see these structures—the nose, pharynx, trachea, and bronchi—in **Figure 3**. These structures also warm and moisten the air you breathe.

Nose Air enters the body through the nose or the mouth. Hairs in the nose trap large particles. The air passes into spaces called nasal cavities. Some cells lining the nasal cavities produce mucus, a sticky material that moistens the air and traps more particles.

Pharynx and Trachea From the nose, air enters the **pharynx** (FAR ingks), or throat. Both the nose and the mouth connect to the pharynx. So air and food enter the pharynx. From the pharynx, air moves into the **trachea** (TRAY kee uh), or windpipe. When you swallow, a thin flap of tissue called the epiglottis covers the opening of the trachea to keep food out. Cells that line the trachea have **cilia** (SIL ee uh; singular *cilium*), tiny hairlike extensions that can move together in a sweeping motion. The cilia, like those shown in **Figure 2,** sweep the mucus made by cells in the trachea up to the pharynx. If particles irritate the trachea, you cough, sending the particles back into the air.

Bronchi and Lungs Air moves from the trachea into the left and right **bronchi** (BRONG ky; singular *bronchus*). These two passages take air into the lungs. The **lungs** are the main organs of the respiratory system. Inside the lungs, the bronchi branch into smaller and smaller tubes. At the end of the smallest tubes are **alveoli** (al VEE uh ly; singular *alveolus*), tiny, thin-walled sacs of lung tissue where gases can move between air and blood.

FIGURE 2

Cilia

The photo shows a microscopic view of cilia.

Answer the questions below.

1. **Relate Cause and Effect** How does coughing protect the respiratory system?
   It gets rid of inhaled particles.

2. **CHALLENGE** What might happen if you did not have hairs in your nose and cilia in your trachea?
   Sample: Particles in air could get into my lungs and damage them.

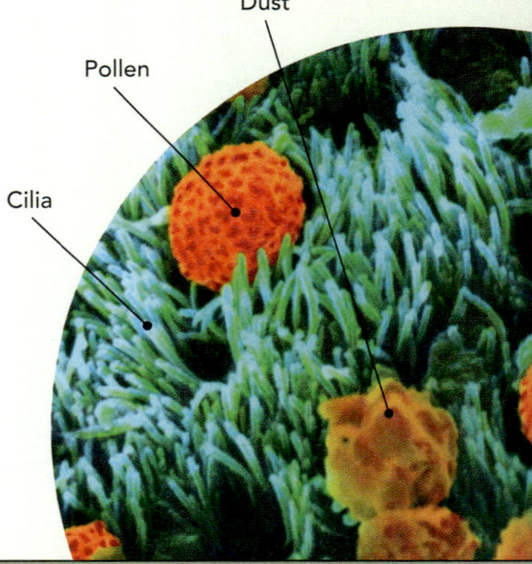

FIGURE 3

Structures of the Respiratory System

Particles in air are filtered out as the air moves through your respiratory system.

Communicate **In your own words, write what each part of the respiratory system does.**

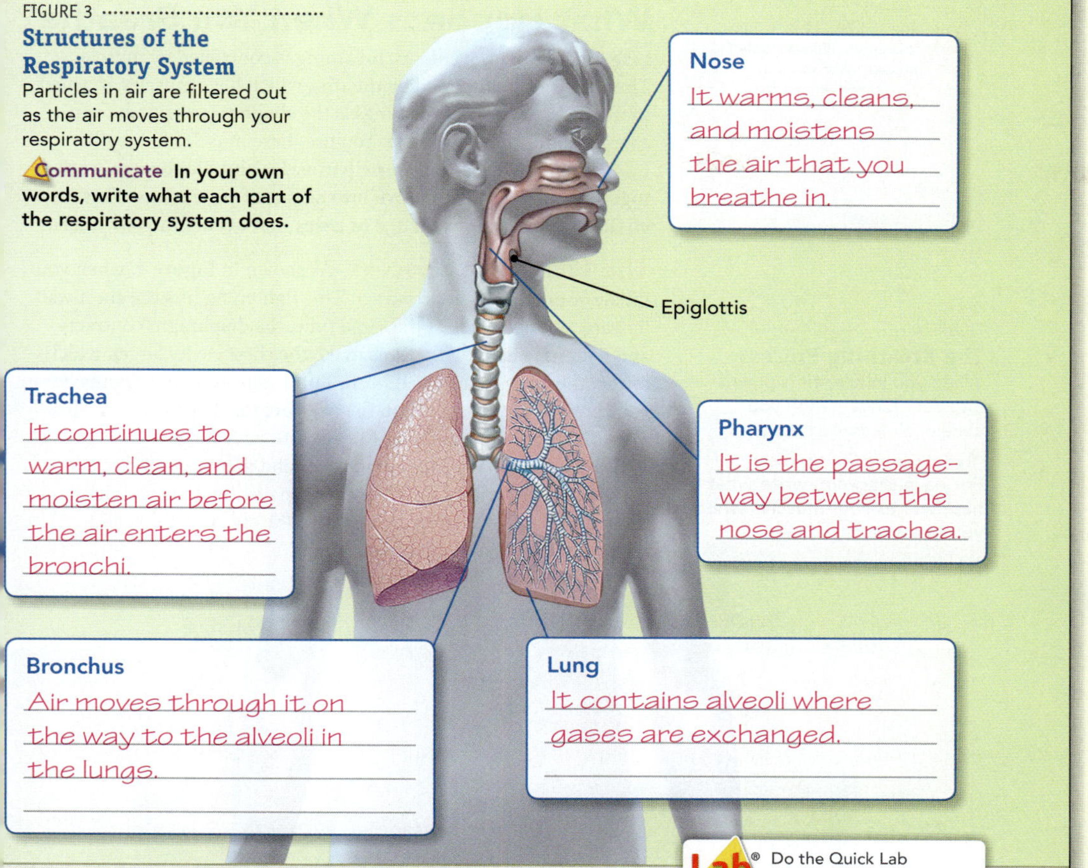

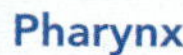

Lab zone Do the Quick Lab *Modeling Respiration.* Student Lab Manual, p. 91

Assess Your Understanding

TEKS 12B, 12E

1a. Identify What happens in the lungs?
In the lungs, there are alveoli, where gases move between the air and the blood.

b. Compare and Contrast How are breathing and cellular respiration different?
Breathing brings oxygen into the body; cellular respiration breaks down glucose in cells.

got it?

○ **I get it!** Now I know that the respiratory system takes in oxygen and removes carbon dioxide from the body.

○ I need extra help with See TE note.

Differentiated Instruction

L1 Pathway for Air Have students write the name of each label in **Figure 3** on an index card. Have them organize the labels by the order in which air passes through the body. Suggest they write a question about each structure on the card. Have them write the answers as they read the section.

L3 Breathing Safely Bring in a gauze mask. Have students work together to explain the purpose of the mask and identify reasons why a person might use it. Ask students to determine why the mask must cover both the nose and the mouth.

Teach With Visuals

Direct students' attention to **Figure 3.** Help them trace the pathways of air from outside the body into the lungs. Ask: **What are the two different ways air can enter the body?** *(Through the nose or through the mouth)* **What happens when you have a cold and your nose is congested?** *(You have to breathe through your mouth.)* **Have you ever laughed or coughed while drinking a beverage? What happens?** *(Most students will have seen or experienced a beverage running out of the nose of someone who has started to laugh in the middle of swallowing. They should identify the pathway as mouth, pharynx, nose.)*

Communicate Tell students that when they communicate, the goal is to share knowledge or information with others so that it can be understood clearly.

21st Century Learning

CRITICAL THINKING Students may have seen a person with a tube in his or her throat. Diseases, such as cancer, or infections, such as croup, may cause damage in the upper airways and block air from getting to the lungs. When an obstruction in the trachea occurs, it can reduce or stop the delivery of oxygen to the lungs. A tracheostomy can be performed to provide an airway and access to remove secretions from the lungs. A tracheostomy is an opening surgically made into the trachea for a breathing tube. Have students find out about patient care after a tracheostomy.

Elaborate

Lab Resource: Quick Lab 

L2 MODELING RESPIRATION Students will identify the structures of the nose and their functions. This Quick Lab can be found in the Student Lab Manual, p. 91, and online.

Evaluate

Assess Your Understanding

After students answer the questions, have them evaluate their understanding by completing the appropriate sentence.

RTI Response to Intervention

1a. If students have difficulty describing what happens in the lungs, **then** have them reread the last paragraph under the head *Breathing Structures.*

b. If students need help contrasting breathing and cellular respiration, **then** help them define each process.

LESSON 7.3

Explain

Teach Key Concepts

Explain that breathing is controlled by muscles. Ask: **What muscles move air in and out of the lungs?** *(The muscles surrounding the ribs and the diaphragm)* **Where are the rib muscles?** *(Ribs surround the lungs and have muscles attached.)* **Where is the diaphragm?** *(The base of the lungs)*

Lead a Discussion

THE BREATHING PROCESS Explain that when you inhale, the rib muscles and diaphragm contract. When you exhale, the rib muscles and diaphragm relax. Ask: **What happens to the chest cavity when the rib muscles and diaphragm contract?** *(It becomes larger.)* **How does this affect air pressure in the lungs?** *(Pressure inside the lungs is less than it is outside the body. Air rushes into the lungs.)* **What happens to the chest cavity when the rib muscles and diaphragm relax?** *(It becomes smaller.)* **How does this affect air pressure in the lungs?** *(Pressure inside the lungs is greater than it is outside the body. Air rushes out of the lungs.)*

Elaborate

Teacher Demo 

AIR PRESSURE

Material balloon **Time** 10 minutes

Blow up a balloon without tying it. Slowly open the inflated balloon and let the air escape.

Ask: **How does the pressure inside the inflated balloon compare to the pressure outside the balloon?** *(The pressure is greater inside the balloon.)* **Why does air rush out of the balloon if you open it?** *(Because the pressure in the balloon is greater than the air pressure outside the balloon)* **How is the balloon like the breathing process?** *(The inflated balloon is like the lungs after inhaling; the higher pressure forces air out of the lungs as it is forced out of the balloon.)*

Texas Essential Knowledge and Skills

12B Identify the main functions of the systems of the human organism, including the circulatory, respiratory, skeletal, muscular, digestive, excretory, reproductive, integumentary, nervous, and endocrine systems.

12C Recognize levels of organization in plants and animals, including cells, tissues, organs, organ systems, and organisms.

TEKS 12B, 12C In this lesson, you'll investigate the process of respiration.

What Happens When You Breathe?

Like other body movements, breathing is controlled by muscles. The lungs are surrounded by the ribs, which have muscles attached to them. At the base of the lungs is the **diaphragm** (DY uh fram), a large, dome-shaped muscle. You use these muscles to breathe. **When you breathe, your rib muscles and diaphragm work together, causing air to move into or out of your lungs. This airflow leads to the exchange of gases that occurs in your lungs.**

The Breathing Process As shown in **Figure 4,** when you inhale your rib muscles contract. This tightening lifts the chest wall upward and outward. At the same time, the diaphragm contracts and flattens. These two actions make the chest cavity larger, which lowers the air pressure inside your lungs. The air pressure outside your body is now higher than the pressure inside your chest. This pressure difference causes air to rush into your lungs.

When you exhale, your rib muscles and diaphragm relax. As they relax, your chest cavity becomes smaller, making the air pressure inside your chest greater than the air pressure outside. As a result, air rushes out of your lungs.

FIGURE 4

The Breathing Process

When you inhale, air is pulled into your lungs. When you exhale, air is forced out.

Interpret Diagrams For each diagram, write what happens to your muscles when you breathe.

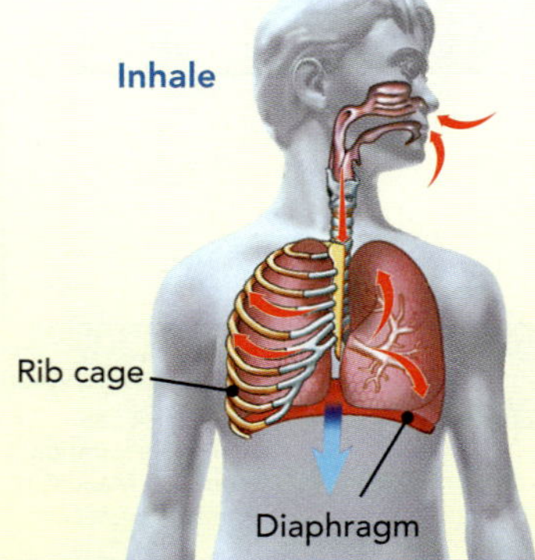

Rib Muscles

The rib muscles contract, expanding the chest cavity.

Diaphragm

The diaphragm contracts and flattens, expanding the chest cavity.

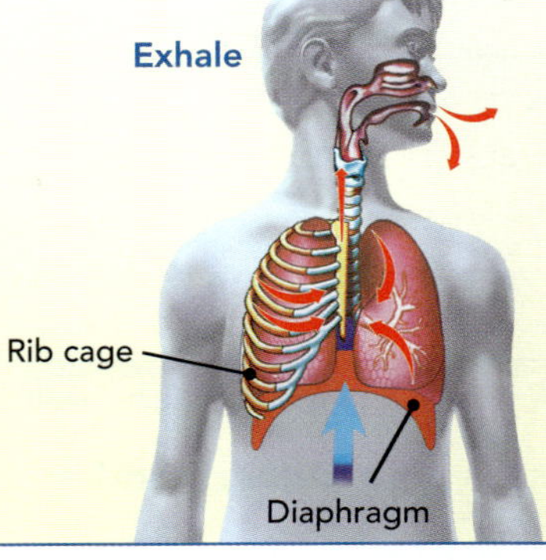

Rib Muscles

The rib muscles relax, reducing the size of the chest cavity.

Diaphragm

The diaphragm relaxes, reducing the size of the chest cavity.

Breathing and Speaking Did you know that the air that moves out of your lungs when you breathe also helps you to speak? Your **larynx** (LAR ingks), or voice box, is located at the top of your trachea. Two **vocal cords,** which are folds of connective tissue, stretch across the opening of the larynx. When you speak, muscles make the vocal cords contract, narrowing the opening 1
as air rushes through. Then the movement of the vocal cords 2
makes air molecules vibrate, or move rapidly back and forth.
This vibration causes a sound—your voice. 3

Chart In the text, underline and number the steps that help you speak. Then write these steps in the graphic organizer.

Sample answers

Step 1

Vocal cords contract, narrowing the opening across the larynx.

Step 2

Air molecules rushing through the narrow opening vibrate.

Step 3

The vibration causes a sound.

apply it!

When you sing, the sounds you make can vary in pitch. That is, you can sing low notes, high notes, and many notes in between. Press your index and middle fingers of one hand gently on the front of your throat.

1 Observe Hum softly, and move your fingers around until you think you have located your voice box. How did humming help you find it?

The humming caused vibrations in my throat that I could feel.

2 Describe With your fingers still in place, sing softly, switching between low notes and high notes. Describe the differences you felt.

I felt stronger vibrations and changes in shape with high notes.

3 Develop Hypotheses How do you think the function of the vocal cords is related to these differences?

Sample: The vocal cords change shape, producing different notes.

Differentiated Instruction

L1 The Breathing Process Ask students to slowly breathe in and out while concentrating on how their diaphragm and rib muscles are contracting and relaxing. Have students look at the structures in **Figure 4** and describe the changes that are occurring.

L3 Deep Breathing Athletic coaches, voice teachers, and yoga practitioners, as well as professionals working in the area of stress management promote the benefits of slow, deep breathing from the abdomen. Have students form small groups to research the various systems of breathing techniques and exercises that are promoted. Then have students present the benefits of deep breathing to the class and demonstrate an exercise or technique that they found most helpful.

Explain

Lead a Discussion

SPEAKING Ask students what they do before they say a long sentence. *(Most students will say they take a deep breath.)* Point out that speaking requires a slow, controlled exhalation. Ask: **What structures are involved in producing sound?** *(Larynx and vocal cords)* **How do they make sounds?** *(Air rushing between the vocal cords makes them vibrate, which causes a sound.)*

Chart Explain to students that a chart helps show how steps in a process fit together. By making a chart of processes, readers can help clarify their understanding.

Elaborate

Apply It!

Before students begin the activity, explain that in music the term *pitch* refers to the way vibrations make a note sound high or low. Review the academic vocabulary. Ask: **What does it mean to observe?** *(It means using one or more of your senses to gather information.)* **What does it mean to describe?** *(It means to use words to give an account of something.)* **What does it mean to develop a hypothesis?** *(It means to come up with a possible scientific explanation for a question.)*

LESSON 7.3

Explain

Support the TEKS

GAS EXCHANGE IN THE ALVEOLI Explain to students that alveoli are like tiny balloons in the lungs. During inhalation, the alveoli fill with oxygen-filled air. That oxygen passes across the walls of the alveoli into the blood in nearby capillaries. At the same time, carbon dioxide and water pass from the blood into the alveoli. Ask: **What happens to the carbon dioxide and water that passes into the alveoli?** *(It is removed from the body during exhalation.)* **What happens to the oxygen passed from the alveoli into the blood?** *(It is carried to cells where it is needed.)* **Which part of blood carries oxygen?** *(Hemoglobin)* **How does the structure of alveoli and capillaries facilitate gas exchange?** *(Both have very thin walls, so that certain materials can pass through them easily.)* **In gas exchange, the respiratory system functions with what other system to manage materials?** *(The circulatory system)*

Address Misconceptions

EXHALED GASES Students often think that they inhale oxygen and exhale carbon dioxide because they know that this is the exchange that takes place in the lungs. Tell students that they inhale air and exhale air, and that air is a mixture of gases. There is no such thing as "a molecule of air." Ask: **What gases make up air?** *(If students have studied the atmosphere, they should recall that air is mostly nitrogen and about 21 percent oxygen, with small amounts of carbon dioxide and other gases.)* **When you inhale, which gases enter your lungs?** *(All the gases in air enter the lungs.)* **Which gas does the body take from the air in your lungs?** *(Oxygen)* **Which gas does the body add to the air in your lungs?** *(Carbon dioxide)* **How is the air you exhale different from the air you inhale?** *(It has less oxygen and more carbon dioxide and water vapor.)* Emphasize that the exhaled air still contains oxygen. Exhaled air is about 16 percent oxygen.

Gas Exchange Air's final stop in its journey through the respiratory system is an alveolus in the lungs. An alveolus has thin walls and is surrounded by many thin-walled capillaries. **Figure 5** shows some alveoli.

Because the alveoli and the capillaries have very thin walls, certain materials can pass through them easily. After air enters an alveolus, oxygen passes through the wall of the alveolus and then through the capillary wall into the blood. Similarly, carbon dioxide and water pass from the blood into the air within the alveolus. This whole process is called gas exchange.

How Gas Exchange Occurs Imagine that you are a drop of blood. You are traveling through a capillary that wraps around an alveolus. You have a lot of carbon dioxide and a little oxygen. As you move through the capillary, oxygen attaches to the hemoglobin in your red blood cells. Carbon dioxide moves into the alveolus. As you move away from the alveolus, you are rich in oxygen and poor in carbon dioxide.

FIGURE 5

Gas Exchange

Gases move across the thin walls of both alveoli and capillaries.

Interpret Diagrams Label each arrow with the gas being exchanged and describe where it is coming from and moving to.

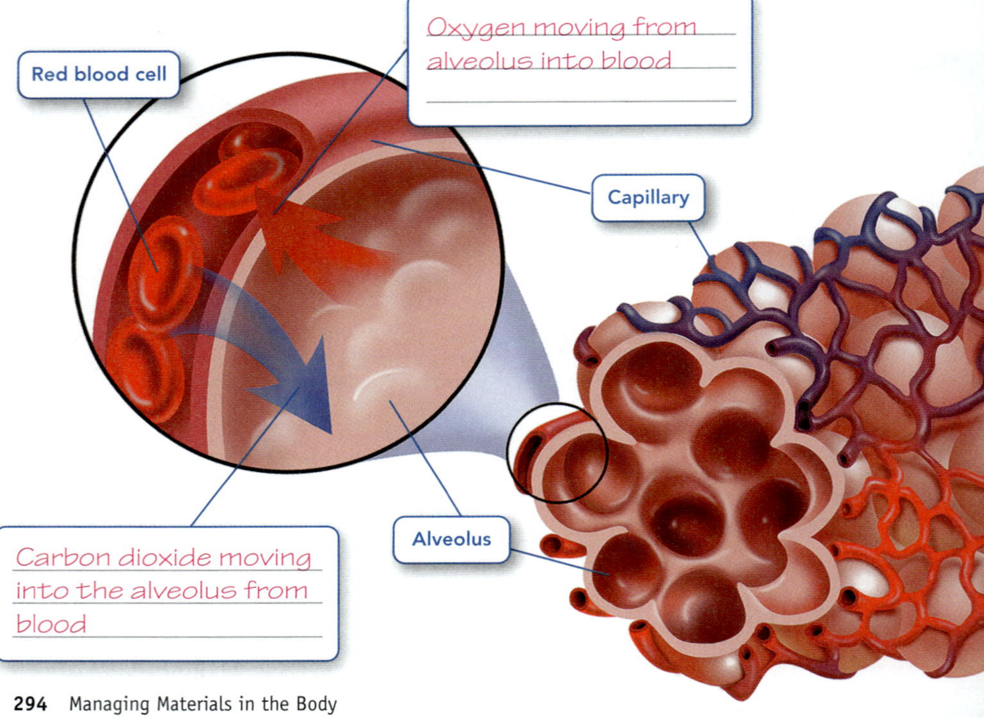

Surface Area for Gas Exchange Your lungs can absorb a large amount of oxygen because of the surface area of the alveoli. An adult's lungs have about 300 million alveoli. Together these alveoli have a surface area of about 100 meters squared (m^2)—the area of the floor in an average classroom! As a result, the alveoli provide a huge amount of surface area for exchanging gases. Therefore, healthy lungs can supply all the oxygen that a person needs—even when the person is very active.

Vocabulary Identify Related Word Forms The verb *absorb* means "to take in." Use this meaning to write a sentence using the noun *absorption*.

Sample: Paper towels take up water through absorption.

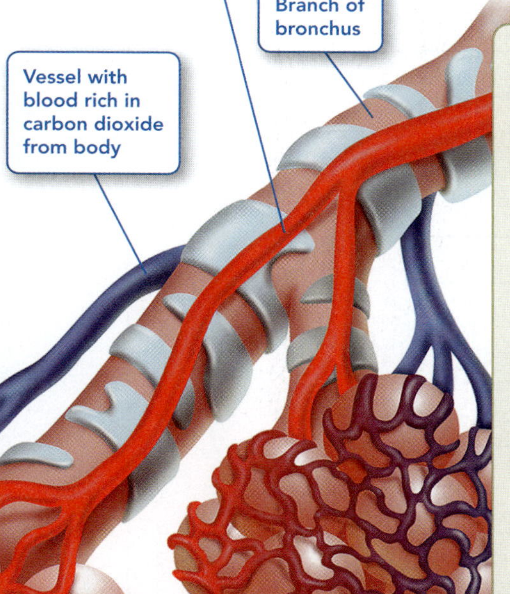

Do the Lab Investigation *A Breath of Fresh Air.* Find the lab online.

Assess Your Understanding

TEKS 12B

2a. **Identify** Where is the larynx located?

It is at the top of the trachea.

b. **Explain** When you inhale, why does air rush into your lungs?

The air pressure inside the chest is lower than the air pressure outside.

c. **Draw Conclusions** How do the alveoli enable people to be very active?

The many alveoli have a large surface area, allowing people to get all the oxygen they need.

got it?

○ **I get it!** Now I know that when I breathe, air moves into and out of my lungs and gas exchange occurs.

○ I need extra help with See TE note.

295

Differentiated Instruction

L1 Cause-and-Effect Chart Encourage students to work in pairs to create a two-column cause-and-effect chart about breathing. Start students off by putting the sentence "The muscles of the rib cage and diaphragm contract" under *Cause*. Students should fill out the rest of the chart ending with the effect, "Carbon dioxide, water vapor, and other gases are exhaled from the lungs."

L3 Investigate Surface Area Ask students to examine the relationship between surface area and the measure of each side of a cube. Have students calculate out how surface area changes each time the length of each side of a cube is doubled. Tell students to present their conclusions using examples.

Lead a Discussion

SURFACE AREA Help students understand that the alveoli are very small. If the lungs were just single large sacs, they would have very little surface area for gas exchange. Ask: **How can the total surface of the alveoli be large if each individual alveolus is so small?** *(There are many tiny alveoli in the lungs.)* Encourage students to make connections between the way surface area works in the respiratory system and the way surface area works in the digestive system. Ask: **By increasing surface area, the alveoli in the lungs are similar to what feature in the small intestine?** *(Villi)*

Make Analogies

L1 THE IMPORTANCE OF SURFACE AREA To understand why it is important for the lungs to have a large surface area, compare the alveoli to the counter at a fast-food restaurant. In the alveoli, oxygen and carbon dioxide are exchanged. At the restaurant, money is exchanged for food. Ask: **What would happen if the restaurant had a small counter with one person taking orders?** *(Samples: Not much food would be served; service would be slow.)* **What would happen if the restaurant had a large counter with many people taking orders?** *(Service would be faster and more food could be served.)*

Elaborate

Lab Resource: Lab Investigation 

L2 A BREATH OF FRESH AIR Students will investigate what causes the body to inhale and exhale air. This Lab Investigation can be found online.

Evaluate

Assess Your Understanding

After students answer the questions, have them evaluate their understanding by completing the appropriate sentence.

RTI Response to Intervention

2a. If students need help locating the larynx, **then** review with them the section titled *Breathing and Speaking.*

b. If students have difficulty relating inhalation to a difference in air pressure, **then** point out that inhaling makes the chest cavity larger, reducing the air pressure inside the body.

c. If students have trouble connecting the alveoli to the amount of accessible oxygen, **then** remind them that the alveoli increase the surface area of the lungs.

Name ______________________ Date __________ Class __________

Assess Your Understanding

The Respiratory System

What Are the Main Functions of the Respiratory System?

1a. IDENTIFY What happens in the lungs?

____

____

b. COMPARE AND CONTRAST How are breathing and cellular respiration different?

____

____

got it?

○ **I get it!** Now I know that the respiratory system ______________________________

____

____

○ **I need extra help with** ______________________________

What Happens When You Breathe?

2a. IDENTIFY Where is the larynx located?

____

b. EXPLAIN When you inhale, why does air rush into your lungs?

____

____

c. DRAW CONCLUSIONS How do the alveoli enable people to be very active?

____

____

got it?

○ **I get it!** Now I know that when I breathe, air ______________________________

____

○ **I need extra help with** ______________________________

Place the outside corner, the corner away from the dotted line, in the corner of your copy machine to copy onto letter-size paper.

Name ______________________ Date __________ Class __________

Enrich

The Respiratory System

Read the passage and study the diagram. Then use a separate sheet of paper to answer the questions that follow.

Analyzing a Spirogram

The volume of air in the lungs increases as a person inhales and decreases as a person exhales. A *spirogram* is a graph of the volume of air in a person's lungs over time. Look at the spirogram shown below. Where the line of the graph moves upward the person is inhaling, and where the line of the graph moves downward the person is exhaling.

The first three breaths on the spirogram show normal breathing. During the fourth breath, however, the person inhaled as much air into his or her lungs as possible. The next two breaths show normal breathing again. Then, during the seventh breath, the person exhaled as much air from his or her lungs as possible.

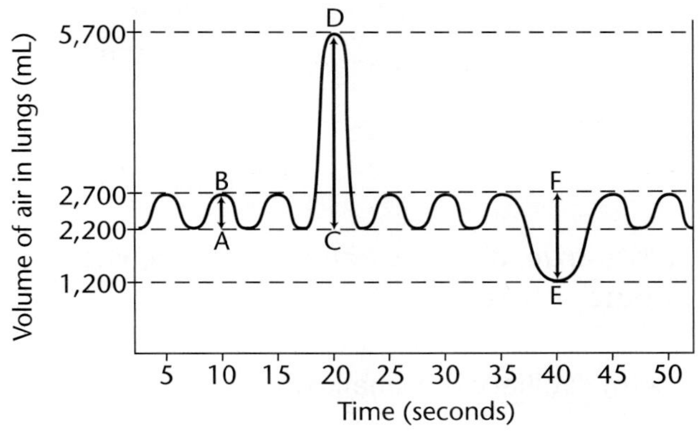

1. During one cycle of breathing, a person begins to inhale, finishes inhaling, then begins to exhale, and finally finishes exhaling. What part of this cycle does a peak on the graph represent? What part of this cycle does a low point on the graph represent?
2. What is the total volume of air in the lungs after the person inhales during a normal breath? What is the volume after the person exhales during a normal breath?
3. What volume of air does the person inhale during a normal breath? This volume is represented by the line AB on the spirogram.
4. What is the maximum volume of air that can be inhaled after a normal exhalation? This volume is represented by the line CD on the spirogram.
5. What is the maximum volume of air that can be exhaled after a normal inhalation? This volume is represented by the line EF on the spirogram.

Name ____________________ Date __________ Class __________

Lesson Quiz

The Respiratory System

Write the letter of the correct answer on the line at the left.

1. ___ Air is exchanged in tiny sacs of lung tissue known as

A trachea

B alveoli

C bronchi

D cilia

2. ___ Fresh air must move through which organ before entering the trachea?

A pharynx

B bronchus

C lung

D alveolus

3. ___ When you breathe, your rib muscles work with the ________ to move air into and out of your lungs.

A larynx

B cilia

C trachea

D diaphragm

4. ___ A person's voice is produced by the vibration of the

A lungs

B pharynx

C diaphragm

D vocal cords

If the statement is true, write *true*. If the statement is false, change the underlined word or words to make the statement true.

5. ____________________ A(n) capillary is a small tube that connects arteries to veins.

6. ____________________Your body uses energy from cellular respiration to carry out processes that maintain homeostasis, such as removing wastes and regulating body temperature.

7. ____________________ Air passes from the nose into the bronchi.

8. ____________________ The trachea is a large, dome-shaped muscle that expands and contracts during breathing.

9. ____________________ The vocal cords stretch across the opening of the larynx.

10. ____________________ The nose produces a sticky material called mucus, which moistens the air and traps particles.

Place the outside corner, the corner away from the dotted line, in the corner of your copy machine to copy onto letter-size paper.

The Respiratory System

Answer Key

Review and Reinforce

Find the worksheet in the Student Workbook.

1. Respiration is the process inside cells through which glucose is broken down using oxygen to release energy. Breathing is the process in which air flows into and out of the lungs.
2. The respiratory system provides oxygen to the body and eliminates carbon dioxide and water from the body.
3. nose, pharynx, trachea, and bronchi
4. Cilia are tiny, hairlike extensions that sweep mucus, along with any particles it might contain, up out of the trachea.
5. Oxygen moves from air to the blood and carbon dioxide moves from the blood into the air.
6. Contraction and relaxation of the diaphragm and rib muscles causes inhalation and exhalation.
7. Sound is produced when air passes over the vocal cords.
8. pharynx
9. trachea
10. lung
11. diaphragm
12. vocal cords
13. larynx
14. bronchus
15. alveolus
16. When the diaphragm and rib muscles contract, the size of the chest cavity increases. This change causes air to flow into the nose. An oxygen molecule in the air will pass through the nose into the pharynx and then into the trachea. Once in the trachea, the oxygen molecule will pass into one of two bronchi, which lead into an alveolus in one of the lungs. The molecule will then pass across the wall of an alveolus and through the wall of a capillary into the blood.

Enrich

1. A peak represents the point where the person has just inhaled and is starting to exhale. A low point represents the point where the person has just exhaled and is starting to inhale.
2. 2,700 mL after inhalation; 2,200 mL after exhalation
3. 500 mL
4. 3,500 mL
5. 1,500 mL

Lesson Quiz

1. B
2. A
3. D
4. D
5. true
6. true
7. pharynx
8. diaphragm
9. true
10. true

The Excretory System

How do systems of the body move and manage materials?

LESSON PACING:
2–3 periods or 1–1 $\frac{1}{2}$ blocks

Lesson Vocabulary

- excretion
- urea
- urine
- kidney
- ureter
- urinary bladder
- urethra
- nephron

Content Refresher

The Formation of Urine The liquid that leaves the blood and enters the capsule in a nephron is called the *filtrate*. At first, the concentration of small molecules (such as glucose) in the filtrate is basically the same as in blood plasma. However, the concentrations of substances in plasma and urine differ greatly. Filtrate is composed of some salts, urea, glucose, amino acids, and water. These differences result from selective reabsorption that occurs as the filtrate travels through the nephron. Amino acids and glucose are moved back into capillaries by active transport. After these substances are removed, water molecules follow them by osmosis. Once the filtrate moves to the collecting tubule, nearly all of it has been reabsorbed. Nitrogenous wastes (which include urea, creatinine, ammonia, and uric acid) are not reabsorbed into the blood.

Lesson Objectives	TEKS	ELPS
Identify the structures and main functions of the excretory system.	2E, 12B, 12C	1.B.1
Explain how excretion contributes to homeostasis.	12B, 13A, 13B	1.B.1

Texas Essential Knowledge and Skills

2E Analyze data to formulate reasonable explanations, communicate valid conclusions supported by the data, and predict trends.
12B Identify the main functions of the systems of the human organism, including the circulatory, respiratory, skeletal, muscular, digestive, excretory, reproductive, integumentary, nervous, and endocrine systems.
12C Recognize levels of organization in plants and animals, including cells, tissues, organs, organ systems, and organisms.
13A Investigate how organisms respond to external stimuli found in the environment such as phototropism and fight or flight.
13B Describe and relate responses in organisms that may result from internal stimuli such as wilting in plants and fever or vomiting in animals that allow them to maintain balance.

English Language Proficiency Standards

ELPS Learning Strategies 1.B.1 Monitor oral language production and employ self-corrective techniques or other resources.

DIFFERENTIATED INSTRUCTION KEY

L1 Struggling Students or Special Needs

L2 On-Level Students L3 Advanced Students

LESSON PLANNER 7.4

Investigations and Activities

My Planet Diary, **Student Edition,** p. 296

Inquiry: Inquiry Warm-Up, How Does Filtering a Liquid Change the Liquid?, **PearsonTexas.com**

Introduce Vocabulary, **Teacher's Edition,** p. 297

Teach Key Concepts, **Teacher's Edition,** p. 297

Do the Math!, **Student Edition,** p. 297

Lead a Discussion, The Stages of Filtration, **Teacher's Edition,** p. 298

21st Century Learning, Critical Thinking, **Teacher's Edition,** p. 298

Inquiry: Build Inquiry, Nephron Filters, **Teacher's Edition,** p. 299

Differentiated Instruction, **Teacher's Edition,** p. 299

Inquiry: Quick Lab, Kidney Function, **PearsonTexas.com**

Teach Key Concepts, **Teacher's Edition,** p. 300

Teach With Visuals, **Teacher's Edition,** p. 300

21st Century Learning, Interpersonal Skills, **Teacher's Edition,** p. 300

Lead a Discussion, Systems Working Together, **Teacher's Edition,** p. 301

21st Century Learning, Accountability, **Teacher's Edition,** p. 301

Differentiated Instruction, **Teacher's Edition,** p. 303

Inquiry: Quick Lab, Perspiration, **Lab Manual,** p. 92

TEKS Review

Apply the TEKS, Moving Things Along, **Student Edition,** p. 302

TEKS Practice, **Student Edition,** p. 305

TEKS Practice: Chapter and Cumulative Review, **Student Edition,** p. 308

Lesson 7.4, **TEKS Preparation and Study Guide Workbook,** p. 74

SHORT ON TIME? To do this lesson in approximately half the time, do the Activate Prior Knowledge activity. A discussion of the Key Concepts will familiarize students with the lesson content. Use the Apply the TEKS activity to help students understand how systems of the body move and manage materials. Have students do the Quick Labs. The rest of the lesson can be completed by students independently.

These editable worksheets are available on **PearsonTexas.com.**
Print versions can be found in the **TEKS Preparation and Study Guide Workbook.**

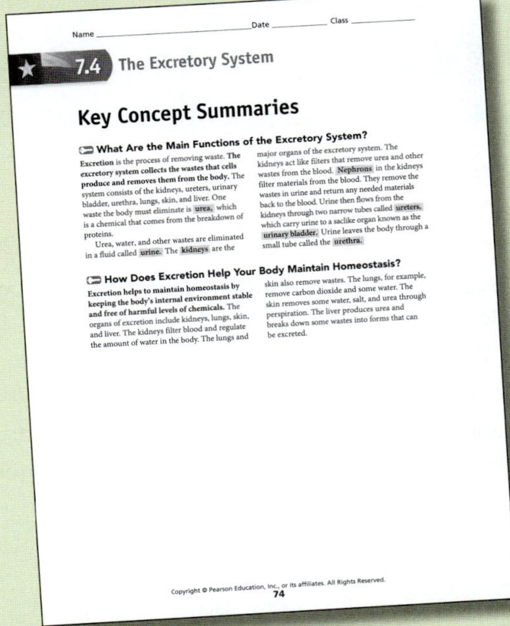

Name ________ Date ______ Class ______

7.4 The Excretory System

Key Concept Summaries

What Are the Main Functions of the Excretory System?

Excretion is the process of removing waste. **The excretory system collects the wastes that cells produce and removes them from the body.** The system consists of the kidneys, ureters, urinary bladder, urethra, lungs, skin, and liver. One waste the body must eliminate is **urea,** which is a chemical that comes from the breakdown of proteins.

Urea, water, and other wastes are eliminated in a fluid called **urine.** The **kidneys** are the major organs of the excretory system. The kidneys act like filters that remove urea and other wastes from the blood. **Nephrons** in the kidneys filter materials from the blood. They remove the wastes in urine and return any needed materials back to the blood. Urine then flows from the kidneys through two narrow tubes called **ureters,** which carry urine to a saclike organ known as the **urinary bladder.** Urine leaves the body through a small tube called the **urethra.**

How Does Excretion Help Your Body Maintain Homeostasis?

Excretion helps to maintain homeostasis by keeping the body's internal environment stable and free of harmful levels of chemicals. The organs of excretion include kidneys, lungs, skin, and liver. The kidneys filter blood and regulate the amount of water in the body. The lungs and skin also remove wastes. The lungs, for example, remove carbon dioxide and some water. The skin removes some water, salt, and urea through perspiration. The liver produces urea and breaks down some wastes into forms that can be excreted.

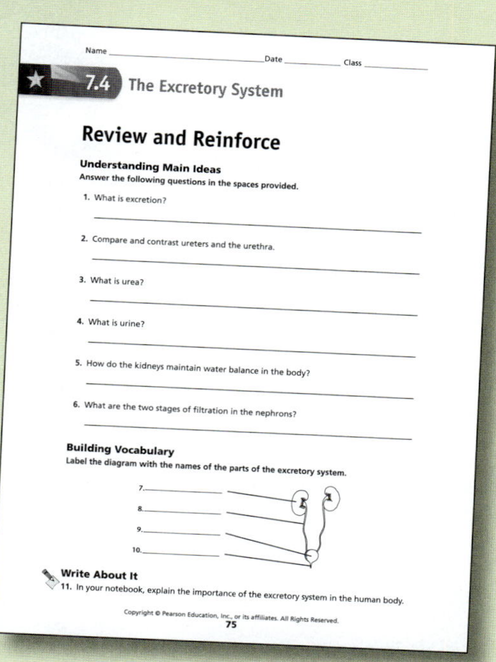

Name ________ Date ______ Class ______

7.4 The Excretory System

Review and Reinforce

Understanding Main Ideas

Answer the following questions in the spaces provided.

1. What is excretion?
2. Compare and contrast ureters and the urethra.
3. What is urea?
4. What is urine?
5. How do the kidneys maintain water balance in the body?
6. What are the two stages of filtration in the nephrons?

Building Vocabulary

Label the diagram with the names of the parts of the excretory system.

7. ______
8. ______
9. ______
10. ______

Write About It

11. In your notebook, explain the importance of the excretory system in the human body.

LESSON 7.4

Lexile Measure = 900L

The Excretory System

Establish Learning Objectives

After this lesson, students will be able to:

- Identify the structures and main functions of the excretory system.
- Explain how excretion contributes to homeostasis.

Engage

Activate Prior Knowledge

MY PLANET DIARY Read *Useful Urine* with the class. Explain that urine is the name for liquid waste that is removed from the body. Ask: **What does it mean to purify a liquid?** *(Students may define this term as cleaning, removing contaminants, or filtering.)* **What is a filter?** *(A device that can take materials out of a substance.)* **Why do astronauts need water?** *(The human body cannot survive for very long without a source of water.)*

Explore

Lab Resource: Inquiry Warm-Up 

L1 HOW DOES FILTERING A LIQUID CHANGE THE LIQUID? Students will filter different materials out of water. This Inquiry Warm-Up can be found online.

The Excretory System

- What Are the Main Functions of the Excretory System?
  TEKS 2E, 12B, 12C
- How Does Excretion Help Your Body Maintain Homeostasis?
  TEKS 12B, 13A, 13B

my planet Diary

FUN FACTS

Useful Urine

You can recycle plastic, glass, and paper. Did you know that urine can be recycled, too? NASA has developed a machine that purifies astronauts' urine. The water that is recovered can be used for drinking, among other things.

Why do astronauts need this kind of machine? Large quantities of water are too heavy to carry into space. So the machine runs urine through a filtering system to remove waste. Then iodine is added to the filtered urine to kill any harmful bacteria. What remains is drinkable water. The astronauts pictured here are celebrating the first batch of recycled water aboard the International Space Station.

Answer the questions below.

1. How else might the astronauts use the filtered urine?
   Sample: They might clean with it.

2. Do you think this system would be useful on Earth? Why or why not?
   Sample: Yes, because the system would help recycle water, which is a limited resource on Earth.

Lab zone: Do the Inquiry Warm-Up *How Does Filtering a Liquid Change the Liquid?* Find the lab online.

SUPPORT ALL READERS

Lexile Measure = 900L **Lexile Word Count = 1069**

Prior Exposure to Content: Most students have encountered this topic in earlier grades

Academic Vocabulary: *identify, infer, main idea*

Science Vocabulary: *excretion, urea, ureter, urethra, nephron*

Concept Level: Generally appropriate for most students in this grade

Preteach With: My Planet Diary "Useful Urine" and Figure 1 activity

Vocabulary
- excretion
- urea
- urine
- kidney
- ureter
- urinary bladder
- urethra
- nephron

Skills
- Reading: Identify the Main Idea
- Inquiry: Infer

What Are the Main Functions of the Excretory System?

TEKS 2E, 12B, 12C In this section, you'll identify the role of organs and structures in the excretory system. You will also analyze data in graphs to formulate reasonable explanations.

The human body faces a challenge similar to keeping your room clean. Just as you must clean up papers that pile up in your room, your body must remove wastes from cellular respiration and other processes. The process of removing waste is called **excretion.**

If wastes were not removed from your body, they would pile up and make you sick. **The excretory system collects the wastes that cells produce and removes them from the body.** The system includes the kidneys, ureters, urinary bladder, urethra, lungs, skin, and liver. Two wastes that your body must eliminate every day are excess water and urea. **Urea** (yoo REE uh) is a chemical that comes from the breakdown of proteins. Excess water, urea, and other wastes are eliminated in a fluid called **urine.**

do the math! Analyzing Data

Urine is made up of water, organic solids, and inorganic solids. The organic solids include urea and acids. The inorganic solids include salts and minerals. The solids are dissolved in the water.

1 **Calculate** Calculate and label on the *Normal Urine Content* graph the percentage of urine that is solids. Calculate and label on the *Solids in Normal Urine* graph the percentage of solids that is urea.

2 **CHALLENGE** What might a sharp decrease in the percentage of water in a person's urine indicate about the health of that person?

Sample: The person might be dehydrated or have other medical problems.

Normal Urine Content

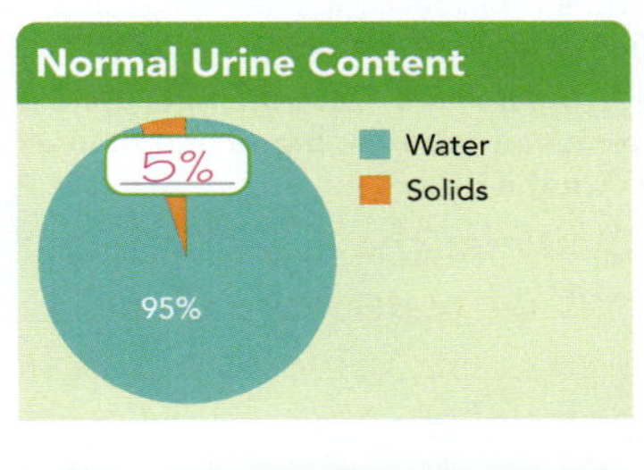

Solids in Normal Urine

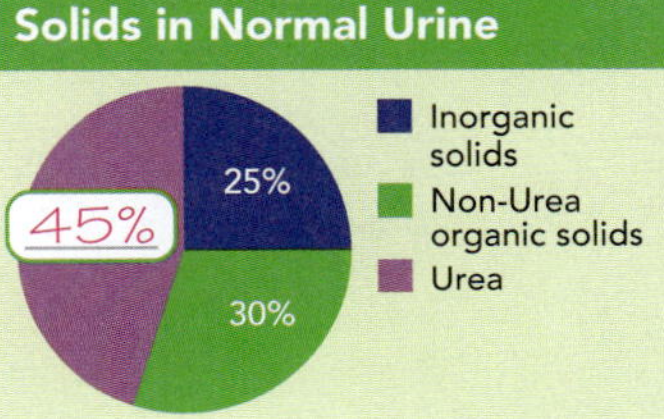

Explain

Introduce Vocabulary

To help students understand the meaning of the term *excretion* in terms of the removal of wastes out of the body, explain that the prefix *ex-* means "out of."

Teach Key Concepts

Explain to students that body cells produce wastes through their normal functioning. The excretory system collects and removes these wastes.
Ask: **What would happen if wastes were not removed from your home?** *(They would pile up and eventually prevent activities in the home from taking place as normal.)* **What would happen if wastes were not removed from the body?** *(They would build up and prevent the cells from functioning properly.)* **Which body structures are involved in excretion?** *(The kidneys, ureters, urinary bladder, urethra, lungs, skin, and liver)* **What is urea?** *(A chemical that comes from the breakdown of proteins)* **How is urea related to urine?** *(Urea is one chemical in urine.)* **What substance makes up most of the urine?** *(Water)*

Elaborate

Do the Math!

L1 Explain that a circle graph is a way to display data that represent parts of a whole. The size of each section of the circle relates to the amount of that substance. Tell students that the top graph compares the water portion of urine to the solid parts. The bottom graph shows the types of solids in the solid part of urine.

PEARSON Texas.com

English Language Proficiency Standards

ELPS Learning Strategies 1.B.1

Read the lesson with students and have them summarize the text in chunks.

Beginning Have students listen as you read aloud a paragraph. Then display simple sentence frames for students to complete, for example: *The excretory system* ______. Provide answer choices as needed. Have several students answer before continuing with the next paragraph.

Intermediate Have partners take turns summarizing each paragraph orally, reminding them to correct themselves and restate if they need to. Have at least one student share a summary before continuing.

Advanced Have partners read the lesson together and summarize each paragraph orally.

Advanced High Have partners read the lesson to each other and summarize each paragraph orally. Ask them to correct their summaries before writing them down.

Texas Essential Knowledge and Skills

2E Analyze data to formulate reasonable explanations, communicate valid conclusions supported by the data, and predict trends.

12B Identify the main functions of the systems of the human organism, including the circulatory, respiratory, skeletal, muscular, digestive, excretory, reproductive, integumentary, nervous, and endocrine systems.

12C Recognize levels of organization in plants and animals, including cells, tissues, organs, organ systems, and organisms.

LESSON 7.4

Explain

Lead a Discussion

THE STAGES OF FILTRATION Remind students that the excretory system removes waste products of cellular respiration and other activities of cells. Ask: **Which organs act like filters?** *(The kidneys)* **How are they like filters?** *(They remove urea and other wastes from the blood but return materials that the body needs to the blood.)* **What is the name for the individual filter units in a kidney?** *(A nephron)* **What happens in the first stage of filtration in a nephron?** *(Both wastes and needed materials are filtered out of the blood.)* **What happens in the second stage?** *(Needed materials are returned to the blood.)* **Why is the second stage important?** *(Sample: It would be wasteful to discard useful materials, such as glucose or water.)*

Teach With Visuals

Tell students to look at **Figure 1.** Students can trace the flow of waste materials from the blood to removal from the body. Ask: **Which organs filter the blood?** *(The kidneys)* **Where do wastes go from the kidneys?** *(They travel through tubes called ureters to the urinary bladder.)* **What is the function of the urinary bladder?** *(It stores urine.)* **Where does urine go from the urinary bladder?** *(It leaves the body through a tube called the urethra.)*

21st Century Learning

CRITICAL THINKING Direct students' attention to **Figure 2** and the nephron capsule. Have students compare the size of the blood vessel leading into the capsule and the size of the blood vessel inside the capsule. Ask: **What happens to the diameter of the blood vessel inside the capsule?** *(It gets smaller.)* **How else does the blood vessel change inside the capsule?** *(It looks like a tangled knot.)* **How do you think these changes affect the blood pressure inside the small vessel?** *(They increase blood pressure.)* **How does this help the nephron function?** *(The increased pressure helps squeeze wastes out of the blood.)*

Structures That Remove Urine **Figure 1** shows the organs that remove urine from the body. Your two kidneys are the major organs of the excretory system. The **kidneys** act like filters. They remove urea and other wastes from the blood but keep materials that the body needs. These wastes are eliminated in the urine. Urine flows from the kidneys through two narrow tubes called **ureters** (yoo REE turz). The ureters carry urine to the **urinary bladder,** a muscular sac that stores urine. Urine leaves the body through a small tube called the **urethra** (yoo REE thruh).

Waste Filtration Each kidney has about one million nephrons. A **nephron** is a tiny filtering factory that removes wastes from blood and produces urine. The nephrons filter wastes in two stages. First, both wastes and needed materials are filtered out of the blood. Next, much of the needed material is returned to the blood, and the wastes are eliminated from the body. Follow this process in **Figure 2.**

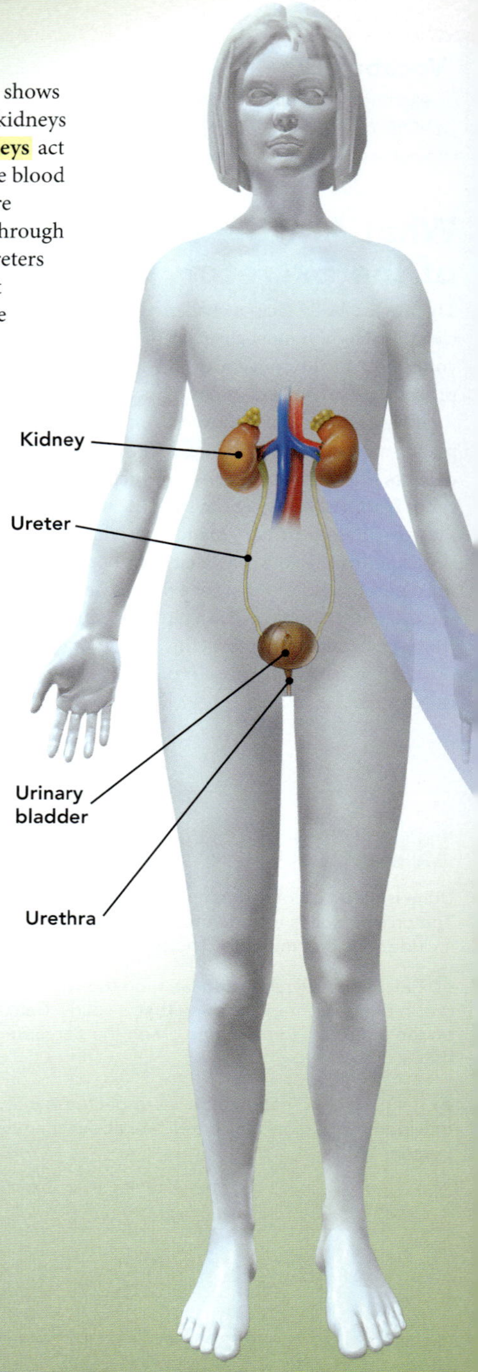

FIGURE 1
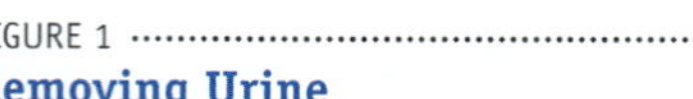

Removing Urine
Urine is produced in the kidneys and then removed from the body.

Summarize Describe how urine is removed from the blood and then eliminated from the body.

Sample: The kidneys filter out wastes from the blood and make urine. It moves into the ureters and collects in the bladder. Urine leaves the bladder and is eliminated from the body through the urethra.

298 Managing Materials in the Body

FIGURE 2

How the Kidneys Work

Most of the work of the kidneys is done in the nephrons.

Infer **In the key, write what each color represents in the diagram. Then explain below why it is important for capillaries to surround the nephron tube.**

Sample: It is important because the nephrons return needed materials to the blood through the capillaries.

Key

- Capillary with oxygenated blood
- Capillary with blood low in oxygen
- Nephron tube

Stage 1

- Blood flows into the cluster of capillaries in the thin-walled, hollow nephron capsule.
- Urea, glucose, and some water are filtered out of the blood and into the capsule.
- These materials then pass into the nephron tube.

Stage 2

- As the material flows through the nephron tube, most of the needed glucose and water move back into the blood through the capillaries.
- Most of the urea and some of the water stay in the nephron tube and become urine.

Nephron capsule

Nephron tube

Do the Quick Lab *Kidney Function.* Find the lab online.

TEKS 12B, 12C

1a. **Identify** The chemical urea comes from the breakdown of proteins.

b. **Draw Conclusions** Why is it important for a kidney to have many nephrons?
Sample: Each nephron is tiny, and all the blood must be filtered.

got it?

○ **I get it!** Now I know that the function of the excretory system is to collect and remove wastes from the body.

○ I need extra help with See TE note.

299

Elaborate

Infer Tell students that when they make an inference, they are using reasoning and prior knowledge to explain or interpret things they observe.

Build Inquiry

L1 **NEPHRON FILTERS**

Materials markers, paper

Time 10 minutes

Remind students that a number of materials enter the kidney, where they are filtered by the nephrons. Have students draw a diagram showing which materials enter the nephron, which materials are returned to blood, and which materials leave the body in urine.

Ask: **Which materials enter the nephron?** *(Urea, water, and glucose)* **Which materials are returned to the blood?** *(Most water and glucose along with a small amount of other materials)* **Which materials leave the body in urine?** *(Urea and other wastes)*

Lab Resource: Quick Lab

L1 **KIDNEY FUNCTION** Students will model the movement of sodium during urine formation. This Quick Lab can be found online.

Evaluate

Assess Your Understanding

After students answer the questions, have them evaluate their understanding by completing the appropriate sentence.

RTI Response to Intervention

1a. If students have trouble naming the chemical, **then** ask students to make a chart showing each of the waste products eliminated from the body.

b. If students need help reaching a conclusion about the number of nephrons, **then** remind them of what they learned about the need for many alveoli in the lungs.

LESSON 7.4

Differentiated Instruction

L1 **Illustrate the System** Give each student several index cards. Have them write a key term on the front of each card, and its definition on the back. Have them refer to the cards as they study **Figure 1** and **Figure 2.**

L3 **Make a Display** Have a group of students work cooperatively to investigate the causes of kidney stones, their treatment, and how the incidence of kidney stones can be reduced. Ask that group members prepare visuals to use as a display to present their findings to the class.

LESSON 7.4

Explain

Teach Key Concepts

The body has many processes that allow its internal environment to remain stable despite changes to the external environment. This condition is called *homeostasis*. For example, the body must maintain certain levels of substances in the blood, such as sugars, salts, and water. The excretory system keeps levels constant by eliminating any extra chemicals, and retaining what the body needs. Ask: **Which organs are involved in eliminating water from the body?** *(Kidneys, lungs, and skin)* **Why is more water reabsorbed into the blood by the kidneys on a hot day?** *(Water is needed to balance the amount of water lost due to perspiration.)*

Teach With Visuals

Tell students to look at **Figure 3.** Explain that the body uses more water during some activities than others. Ask: **Why is it important to drink water before and during physical activities?** *(Students should recognize that not only does the body use water, but it also loses water through sweating.)* **How does the body control its use of water by the amount of urine it produces?** *(The body produces less urine when it needs to retain more water.)*

21st Century Learning

INTERPERSONAL SKILLS Have students dramatize a television interview between a journalist and a doctor about kidney disease and organ donation. The interview should address the effects of malfunctioning kidneys on the human body and what is involved in a kidney transplant. Students can take turns playing the roles of the journalist and the doctor.

Texas Essential Knowledge and Skills

12B Identify the main functions of the systems of the human organism, including the circulatory, respiratory, skeletal, muscular, digestive, excretory, reproductive, integumentary, nervous, and endocrine systems.

13A Investigate how organisms respond to external stimuli found in the environment such as phototropism and fight or flight.

13B Describe and relate responses in organisms that may result from internal stimuli such as wilting in plants and fever or vomiting in animals that allow them to maintain balance.

TEKS 12B, 13A, 13B In this section, you'll explore how the excretory system helps the human body maintain balance and stable internal conditions.

How Does Excretion Help Your Body Maintain Homeostasis?

A buildup of wastes such as urea, excess water, and carbon dioxide can upset your body's balance. **Excretion helps to maintain homeostasis by keeping the body's internal environment stable and free of harmful levels of chemicals.** The organs of excretion include the kidneys, lungs, skin, and liver.

Kidneys As the kidneys filter blood, they regulate the amount of water in your body, helping to maintain homeostasis, or internal stability. Remember that as urine is being formed, needed water passes from the nephron tubes into the blood. The amount of water that returns to the blood depends on conditions both outside and inside the body. For example, on a hot day when you have been sweating a lot and have not had much to drink, almost all the water in the nephron tubes will move back into the blood. You will excrete only a small amount of urine. On a cool day when you have drunk a lot of water, less water will move back into the blood. You will excrete a larger volume of urine. Look at **Figure 3.**

Vocabulary Identify Related Word Forms You know the noun *excretion* means "the process of removing wastes." Use this meaning to choose the correct meaning of the verb *excrete*.

- ○ relating to removing wastes
- ● to remove wastes
- ○ the state of removing wastes

FIGURE 3

Fluid Absorption

These three students have been doing different activities all day.

Relate Text and Visuals Which student will probably produce the least urine? Explain.

Kari is sweating and has not drunk any water, so she will produce the least urine.

Maria has been in classes all morning. She has had nothing to eat or drink.

Kari has been running sprints. She forgot to bring her water bottle.

Mike has been sitting on the bench and drinking water.

Lungs, Skin, and Liver Organs that function as part of other systems in your body also help keep you healthy by excreting wastes. For example, the lungs of the respiratory system remove carbon dioxide and some water when you exhale. The skin, part of your integumentary system, contains sweat glands that produce perspiration. Perspiration consists mostly of water, salt, and a small amount of urea.

The liver, which functions as part of the digestive system, also makes urea from the breakdown of proteins in the body. In addition, the liver breaks down other wastes into forms that can be excreted. For example, the liver breaks down old red blood cells. It even recycles some of their parts. In this way, you can think of the liver as a recycling factory.

Identify the Main Idea In your own words, write about each organ's role in excretion.

Lead a Discussion

SYSTEMS WORKING TOGETHER Explain that organs that function as part of other systems, such as the lungs, skin, and liver, also contribute to excreting wastes. Ask: **What system do the lungs belong to?** *(The respiratory system)* **How do the lungs help the body excrete waste?** *(The lungs remove carbon dioxide and water when you exhale.)* **What system does the skin belong to?** *(The integumentary system)* **How does the skin help the body excrete waste?** *(Sweat glands in the skin produce perspiration, which removes water, salt, and small amounts of urea from the body.)* **What system does the liver belong to?** *(The digestive system)* **How does the liver help the body excrete waste?** *(The liver produces urea from breaking down proteins, breaks down wastes into forms that can be excreted, and recycles materials.)*

Identify the Main Idea Explain to students that the main idea of a paragraph is the topic, central thought, or most important idea in a paragraph. Readers can identify the main idea because most of the sentences in the paragraph support, or give further details about, the main idea.

21st Century Learning

ACCOUNTABILITY Have students write ten true/false statements about the role of the excretory system, the structures of the excretory system, how the kidneys work, and what other organs in the body help the excretory system maintain homeostasis. Then have students exchange papers with a partner. Students should identify true statements and reword false statements to make them true.

Differentiated Instruction

L1 Draw a Diagram Have students prepare a poster showing a simple outline of a human body. Tell students to draw and label the lungs, skin, liver, and kidneys. Ask them to use arrows and labels to show the substances removed by each organ.

L3 Write a Paragraph Have students write a paragraph explaining how wastes are filtered in the kidneys. Ask them to begin by writing two lists—one that includes materials removed from the blood in the kidneys and one that includes materials returned to the blood.

Elaborate

Apply the TEKS

Emphasize that the body's systems do not work independently. Instead, the systems work in harmony with each other to maintain homeostasis, a balanced, relatively stable internal environment. Remind students that the body's cells work best when they have this stable environment, which includes the correct water, glucose, and oxygen levels. Direct students' attention to the illustration of the four body systems. Before students identify the main function of each system, ask: **How does the respiratory system play a role in maintaining homeostasis?** *(The respiratory system conducts gas exchange between the alveoli and the capillaries—transferring oxygen to the blood and picking up carbon dioxide and water to carry out of the body.)* **What is the primary way in which the excretory system helps maintain homeostasis?** *(The excretory system filters urea and other waste from the blood and removes them from the body in the form of urine.)* **How does the circulatory system play a role in maintaining homeostasis?** *(The blood of the circulatory system connects all the other systems. It brings oxygen from the respiratory system and nutrients from the digestive system to the cells; it sends carbon dioxide and water back through the respiratory system, and sends wastes through the digestive and excretory systems.)* **How does the digestive system play a role in maintaining homeostasis?** *(The digestive system breaks down food into nutrients that can be absorbed by the blood and removes digestive waste.)* **Which system is the most important to homeostasis?** *(Accept all well-reasoned answers. Most students should conclude that no system can function properly without the others.)*

Texas Essential Knowledge and Skills

12B Identify the main functions of the systems of the human organism, including the circulatory, respiratory, skeletal, muscular, digestive, excretory, reproductive, integumentary, nervous, and endocrine systems.

Moving Things Along

How do systems of the body move and manage materials?

The systems of the body work together, helping to maintain homeostasis by changing materials and moving them to where they can be used or excreted.

Complete the following tasks.

1. **Identify** For each system, identify its main function and tell what materials are managed or moved.
2. **Describe System Interactions** Which system is the link for moving materials among the other three systems?

The circulatory system

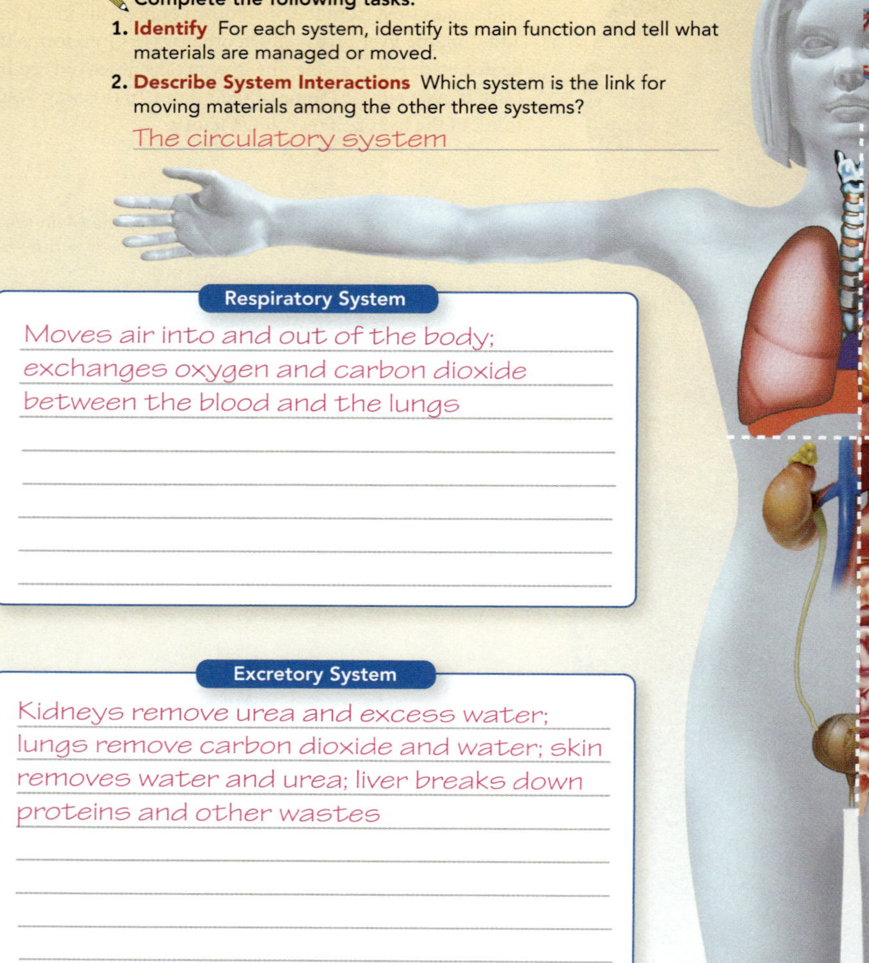

Respiratory System

Moves air into and out of the body; exchanges oxygen and carbon dioxide between the blood and the lungs

Excretory System

Kidneys remove urea and excess water; lungs remove carbon dioxide and water; skin removes water and urea; liver breaks down proteins and other wastes

Circulatory System

Circulates blood, which carries oxygen, nutrients, carbon dioxide, other wastes, and other materials through the body

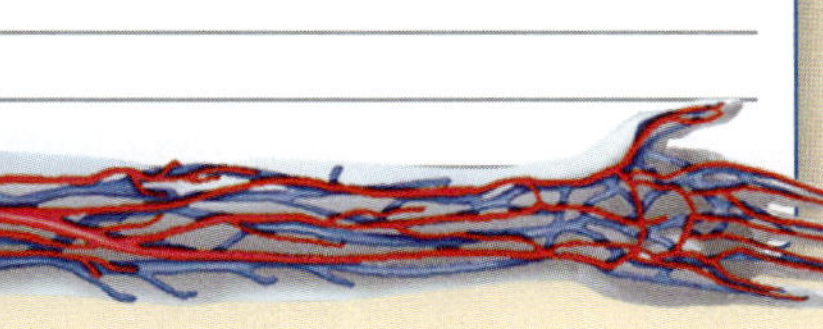

Digestive System

Breaks down food into smaller molecules, which are absorbed into the blood and then used by body cells; also removes digestive wastes

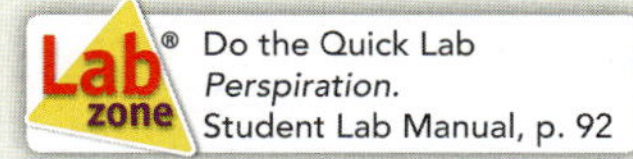

Do the Quick Lab *Perspiration.* Student Lab Manual, p. 92

Assess Your Understanding

TEKS 12B, 12C, 13B

2a. Review How does removing wastes from the body help maintain homeostasis?

The body is protected from a buildup of harmful chemicals.

b. Relate Cause and Effect On a long bus trip, a traveler does not drink water for several hours. How will the volume of urine she produces that day compare to the volume on a day when she drinks several glasses of water? Explain.

She will produce less urine on the bus trip because most of the water in her kidneys will be reabsorbed.

c. Explain How do systems of the body move and manage materials?

Sample: The digestive system moves food and breaks it down into simple substances. The circulatory system moves blood, which carries oxygen, carbon dioxide, and other materials. The respiratory system moves air into and out of the body. The excretory system filters the blood and removes wastes.

got it?

○ **I get it!** Now I know that excretion helps maintain homeostasis in my body by removing wastes and excess water.

○ I need extra help with See TE note.

303

Lab Resource: Quick Lab 

L1 PERSPIRATION Students will investigate how perspiration removes excess water and regulates temperature. This Quick Lab can be found in the Student Lab Manual, p. 92, and online.

Evaluate

Assess Your Understanding

After students answer the questions, have them evaluate their understanding by completing the appropriate sentence.

RTI Response to Intervention

2a. If students have trouble connecting waste removal with maintaining homeostasis, **then** help them recognize that cellular processes naturally produce waste and that too much waste in the body can cause sickness.

b. If students need help relating urine production to drinking water, **then** point out that the kidneys regulate the amount of water in the body and have students review **Figure 3.**

c. If students struggle to describe the way the body's systems move and manage materials, **then** review each of the body's systems independently and help students identify how each system either moves materials, manages materials, or both.

Differentiated Instruction

L1 Chart the Body's Systems Have students create a card for each of the four systems involved in moving and managing materials: the digestive system, the circulatory system, the respiratory system, and the excretory system. On the back of each card, have students identify the ways in which the system moves and manages materials in the body. Have pairs of students quiz each other by reading the backs of the cards and naming the system and/or reading the system and describing the way that system manages and moves materials.

L2 A Narrative of Consumption Point out that the digestive, respiratory, circulatory, and excretory systems all play a role in processing the food we eat. Challenge students to write a paragraph tracking a piece of food through the body and to mention all four systems as they describe the path that item takes from consumption to excretion.

Name ______________________ Date ____________ Class ____________

Assess Your Understanding

The Excretory System

What Are the Main Functions of the Excretory System?

1a. **IDENTIFY** The chemical ____________ comes from the breakdown of proteins.

b. **DRAW CONCLUSIONS** Why is it important for a kidney to have many nephrons?

____

____

got it?

○ **I get it!** Now I know that the function of the excretory system is to ____________

____

○ **I need extra help with** ____

How Does Excretion Help Your Body Maintain Homeostasis?

2a. **REVIEW** How does removing wastes from the body help maintain homeostasis?

____

____

b. **RELATE CAUSE AND EFFECT** On a long bus trip, a traveler does not drink water for several hours. How will the volume of urine she produces that day compare to the volume on a day when she drinks several glasses of water? Explain.

____

c. **EXPLAIN** How do systems of the body move and manage materials?

____

____

____

____

got it?

○ **I get it!** Now I know that excretion helps maintain homeostasis in my body by

____

○ **I need extra help with** ____

Name ______________________ Date ____________ Class ____________

Enrich

The Excretory System

Read the passage and study the table. Then use a separate sheet of paper to answer the questions that follow.

Urinalysis

The table below lists some common urine tests, their possible results, and what these results indicate. Testing for the presence of protein, glucose, ketones, and nitrite involves chemical analysis. These tests are usually performed by dipping into the urine a strip of paper or plastic called a dipstick. There is a specific kind of dipstick for each type of test. If the dipstick changes color, the patient has the condition being tested. Testing for the presence of red and white blood cells is performed by looking at a urine sample with a microscope. Red and white blood cells are visible when magnified if they are present.

Test	Results	Condition Indicated by Test Results
color and texture	pale to dark yellow	normal
	foamy	presence of a protein
	red or red-brown	presence of red blood cells
presence of protein	slight color change in dipstick	kidney disease
	major color change in dipstick	severe kidney disease
presence of glucose	dipstick changes color	diabetes
presence of ketones	dipstick changes color	severe diabetes
presence of nitrite	dipstick changes color	bacteria in urine
presence of red blood cells	cells visible	kidney disease
presence of white blood cells	cells visible	bacteria in urine

1. A patient's urine tests positive for glucose and ketones. What do these results indicate?
2. The dipstick test for protein shows a slight color change. What does this indicate?
3. A patient's urine test shows the presence of nitrite. What does this indicate? What other test could you perform to confirm this result?
4. Dipstick tests for protein, glucose, and nitrite show no color change. What do these results indicate?

Name ______________________ Date ____________ Class ____________

Lesson Quiz

The Excretory System

Fill in the blank to complete each statement.

1. The process of removing waste from the body is called ________________.

2. Water, urea, and other wastes are eliminated in a fluid called ________________.

3. Two ________________ connect the kidneys to the urinary bladder.

4. Nephrons are located within the ________________, which are the two main organs of the excretory system.

5. Urine flows through the ________________ and out of the body.

6. The ________________ makes urea and breaks down old red blood cells.

Write the letter of the correct answer on the line at the left.

7. ___ What is the chemical that comes from the breakdown of proteins and must be eliminated?

A nephron
B salt
C water
D urea

8. ___ What substance is produced by sweat glands in the skin when the body is cooling itself on a hot day?

A carbon dioxide
B perspiration
C urine
D protein

9. ___ In which structure is blood filtered within the kidneys?

A nephron
B ureter
C bladder
D gland

10. ___ What is the process through which the body maintains stable internal conditions?

A gas exchange
B circulation
C homeostasis
D recycling

Place the outside corner, the corner away from the dotted line, in the corner of your copy machine to copy onto letter-size paper.

Place the outside corner, the corner away from the dotted line, in the corner of your copy machine to copy onto letter-size paper.

The Excretory System

Answer Key

Review and Reinforce

Find the worksheet in the Student Workbook.

1. the process of removing waste from the body

2. Each of the two ureters carries urine from a kidney to the bladder. The urethra carries urine outside the body from the bladder.

3. a chemical that is a result of the breakdown of proteins

4. the watery fluid that contains urea and other wastes; produced by the kidneys

5. When the body has more water than needed, the kidneys excrete a large amount of water. When the body needs water, the kidneys produce more concentrated urine so less water is excreted.

6. First the nephrons filter both waste and needed material out of the blood. In the second stage, needed material is returned to the blood and wastes are eliminated.

7. kidney

8. ureter

9. urinary bladder

10. urethra

11. Processes in cells produce wastes. If these wastes build up, the body cannot remain healthy. The excretory system removes wastes from the body. It also maintains homeostasis, which is a relatively stable internal environment.

Enrich

1. The results indicate that the person has a severe condition of diabetes.

2. The results indicate that the person has a kidney disease.

3. Nitrite indicates bacteria in the urine. You could examine the urine under a microscope for the presence of white blood cells. A positive result would confirm the presence of bacteria.

4. The results indicate that the person does not have diabetes, kidney disease, or bacteria in the urine.

Lesson Quiz

1. excretion **2.** urine

3. ureters **4.** kidneys

5. urethra **6.** liver

7. D **8.** B

9. A **10.** C

CHAPTER 7

Scientific Investigation and Reasoning

This Apply the TEKS feature will enable students to use the scientific investigation and reasoning skills in TEKS 2E and 3A to reinforce the content of TEKS 6C. Students will analyze the data from an experiment and evaluate if the data supports a given hypothesis.

Observing Enzymes in Saliva

Have students read about how a group of students collected data to support its hypothesis that enzymes in saliva break down complex carbohydrates like starch into simple sugars. Ask them to identify the dependent and independent variables in the group's experiment. Have them explain why the students ran five trials.

Make sure that students understand the data table before they begin to answer the questions.

Ask: **Why did the students observe the test tubes at 0, 3, and 5 minutes?** *(To see if the iodine changed color, which would indicate that starches are being broken down into sugars)* **Based on the data, what can you conclude happened to the starch in test tube 1?** *(The iodine remained black in the test tube, which indicates that starch is still present. The starch was not broken down into sugars in water.)*

Observing Enzymes in Saliva

A group of students decides to test the hypothesis that enzymes in saliva break down complex carbohydrates like starch into simple sugars. They put a pea-size amount of cracker crumbs into 5 test tubes. Into test tube 1, the students pour 5 mL of water. Next, each of the students added 5 mL of their own saliva to one of the remaining test tubes. Finally, they added a few drops of an iodine solution to each test tube and observed for 5 minutes. Iodine turns from brown to black in the presence of starch.

Test Tube	Color at 0 min.	Color at 3 min.	Color at 5 min.
1	black	black	black
2	black	black	brown
3	black	black	brown-black
4	black	brown-black	brown
5	black	black	brown

Answer the questions below.

1. **Evaluate the Design** Why was the test tube with water included in the experiment?
Sample: Water is the control in the experiment.
2. **Analyze Data** Do the data support the group's hypothesis? Explain your conclusion.
Sample: Yes, the data support the hypothesis. After 5 minutes in saliva, starch in the crackers was broken down into simple sugars.
3. **Draw Conclusions** Based on the results, compare the amount of enzyme in each student's saliva.
Sample: The most enzyme was in Tube 4. Tubes 2 and 5 had average amounts. Tube 3 had the least amount of enzyme.

Texas Essential Knowledge and Skills

2E Analyze data to formulate reasonable explanations, communicate valid conclusions supported by the data, and predict trends.

3A In all fields of science, analyze, evaluate, and critique scientific explanations by using empirical evidence, logical reasoning, and experimental and observational testing, including examining all sides of scientific evidence of those scientific explanations, so as to encourage critical thinking by the student.

6C Recognize how large molecules are broken down into smaller molecules such as carbohydrates can be broken down into sugars.

English Language Proficiency Standards

ELPS Speaking 3.G.2

Read the experiment on this page.

Beginning Clarify meaning as needed. Then help students retell ideas by scaffolding questions 1–3, for example: *Does water have enzymes like saliva? Will water break down starch?*

Intermediate Together, read aloud page 304. Have partners retell ideas by discussing their responses to questions 1–3. Then have students switch partners to share ideas.

Advanced Have partners read aloud the text and discuss the questions. Encourage students to summarize the material.

Advanced High Ask students to summarize the information. Then have them name other substances they could use in the test tubes for this experiment, and tell why they would work.

CHAPTER 7

TEKS Practice

TEKS 6B, 6C, 12B, 12C, 12E, 13B

LESSON 1 The Digestive System

1. Mechanical digestion causes ____________ changes in the food we eat.
   a. nutritional b. energy
   (c.) physical d. chemical

2. Complex carbohydrates can be broken down into sugars.

3. **Interpret Diagrams** How do you think acid reflux, the condition illustrated in the diagram below, affects the esophagus?

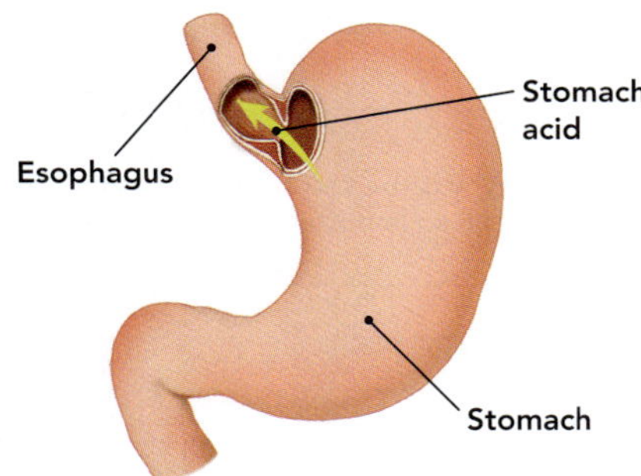

Sample: It causes stomach acid to burn the esophagus.

4. **Recognize** How do enzymes help break down large molecules into smaller molecules?
   Sample: An enzyme helps speed up only one kind of reaction, such as breaking a large starch molecule into smaller sugar molecules.

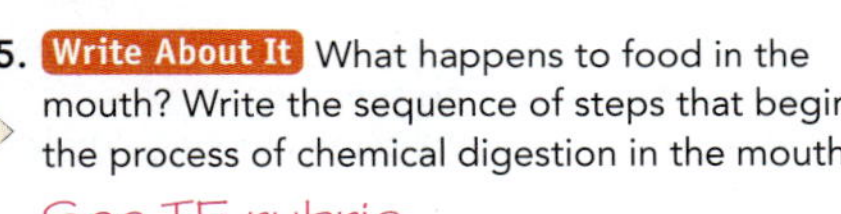

5. **Write About It** What happens to food in the mouth? Write the sequence of steps that begin the process of chemical digestion in the mouth.
   See TE rubric.

LESSON 2 The Circulatory System

6. What structure regulates the direction of blood flow through the heart?
   a. ventricle b. pacemaker
   (c.) valve d. artery

7. The veins in your body carry blood back to your heart.

8. **Name** Which chambers of the heart below are the ventricles? Which chamber receives oxygen-poor blood from the body?

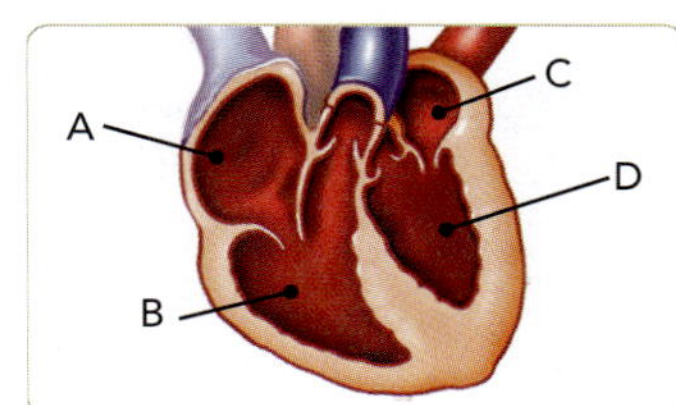

The ventricles are B and D. Oxygen-poor blood enters from the body into A.

9. **Predict** Is it safe for a person with blood type O+ to receive a blood transfusion from a person who has blood type A+? Explain.
   No. A person with type O+ blood will make anti-A clumping proteins, so the match of Rh factor will not matter.

10. **Write About It** People who do not have enough iron in their diets sometimes develop anemia, a condition in which their blood cannot carry a normal amount of oxygen. Write a paragraph to explain why this is so.
    See TE rubric.

PEARSON Texas.com

TEKS Practice

Assess Understanding

Have students complete the answers to the TEKS Practice questions. Have a class discussion about what students find confusing. Write Key Concepts on the board to reinforce knowledge.

RTI Response to Intervention

4. If students cannot recognize how enzymes help break down large molecules, **then** have them review **Figure 3**.

9. If students cannot predict the results of the blood transfusion, **then** have them review the section on Landsteiner's discovery. Remind them that blood is only compatible if it has the same marker molecules and Rh factor.

Alternate Assessment

3D MODEL Challenge student groups or pairs to make a 3D model of the small intestine, the heart, the lungs, or the kidneys. Using a variety of materials such as clay, cardboard, paint, construction paper, tape, glue, and paper maché, students should strive to make their models illustrate how these organs work move and/or manage materials in the body.

PEARSON Texas.com

Write About It 5 Assess student's writing using this rubric.

SCORING RUBRIC	SCORE 4	SCORE 3	SCORE 2	SCORE 1
Explain what happens to food	Explains start of digestion in good detail	Explains with a few details	Partially explains what happens to food in the mouth	Fails to explain what happens to food in the mouth
Sequence steps in chemical digestion	Sequences all the steps of chemical digestion in the mouth	Sequences most of the steps in the process	Sequences some of the steps	Fails to sequence any steps

Write About It 10 Assess student's writing using this rubric.

SCORING RUBRIC	SCORE 4	SCORE 3	SCORE 2	SCORE 1
Explain how poor diet can lead to anemia	Thoroughly explains connection between diet and anemia	Adequately explains connection between diet and anemia	Partly explains connection between diet and anemia	Fails to explain the cause of anemia

CHAPTER 7

TEKS Practice, Cont.

RTI Response to Intervention

14. If students cannot contrast respiration and cellular respiration, **then** review how breathing helps maintain homeostasis.

21. If students cannot recall the role of the kidneys in homeostasis, **then** have them review the kidneys' part in fluid absorption.

22. math!
200 x 356 = 71,200 quarts
2 x 356 = 712 quarts

L3 **WRITING IN SCIENCE** Ask students to write a blog entry that explains to readers how to keep the different systems of the body healthy. In their blogs, students should use what they learned about how the digestive, circulatory, respiratory, and excretory systems manage materials in the body. They should also discuss what nutrients and other materials these systems process.

TEKS Practice

LESSON 3 The Respiratory System

11. Your voice is produced by the

a. pharynx. (**b.**) larynx.
c. trachea. **d.** alveoli.

12. Clusters of air sacs in the lungs are alveoli.

13. Recognize What part of the respiratory system connects the mouth and nose?

It is the pharynx.

14. Compare and Contrast What is the difference between respiration and cellular respiration?

Respiration occurs in the lungs, bringing oxygen into the body and removing carbon dioxide. Cellular respiration occurs inside cells and involves breaking down glucose to release energy.

15. Explain How do mucus and cilia work together to remove dust that enters your nose? How do they differ?

Mucus traps particles in the nose and trachea. Cilia in the trachea sweep the mucus into the pharynx. Mucus traps particles and cilia sweeps them away.

16. Write About It Suppose you are a doctor with patients who are mountain climbers. Write a letter to these patients that explains how gas exchange is affected at the top of a mountain, where air pressure is lower and there is less oxygen than at lower elevations.

See TE rubric.

LESSON 4 The Excretory System

17. Urine leaves the body through the

a. ureters. **b.** nephrons.
c. urinary bladder. (**d.**) urethra.

18. Urine is stored in the urinary bladder.

19. Identify What are the main functions of the excretory system?

The excretory system collects wastes produced by cells in the body and removes these wastes from the body. It helps to maintain homeostasis by keeping the body free of harmful levels of chemicals.

20. Explain Why is the second stage of filtration in a nephron so important?

The second stage allows needed materials like glucose and water to reenter the bloodstream where it can be used by the body to maintain homeostasis.

21. Relate Cause and Effect How do the kidneys help maintain homeostasis?

They adjust the amount of water reabsorbed during excretion to regulate the amount of water in the body.

22. math! Each day, the average person's kidneys filter around 200 quarts of blood. About 2 quarts of urine is produced in the process. How many quarts of blood will the average person's kidneys filter in one year? How many quarts of urine will the average person's kidneys produce in the same amount of time?

See TE note.

Write About It **16** Assess student's writing using this rubric.

SCORING RUBRIC	SCORE 4	SCORE 3	SCORE 2	SCORE 1
Explain effect of low air pressure on breathing	Fully explains why breathing is more difficult	Adequately explains why breathing is more difficult	Partially explains why breathing is more difficult	Fails to explain why breathing is more difficult
Explain effect of low oxygen on breathing	Fully explains why low oxygen causes rapid breathing	Partially explains why low oxygen causes rapid breathing	Describes breathing as more rapid; gives no reason	Incorrectly describes effect of low oxygen on breathing

How do systems of the body move and manage materials?

23. You eat an apple. Describe how the functions of your body systems working together can provide you with energy from the apple.

The digestive system breaks down the apple into sugars and other nutrients. These are absorbed into the bloodstream. Then the circulatory system carries oxygen from my respiratory system and the sugars to my cells. The cells carry out cellular respiration, releasing energy I can use.
See TE rubric.

Interactive Science Chapter 7

Lesson 1
In Lesson 7.1 you learned that the digestive system breaks down food, absorbs nutrients, and eliminates waste. You also learned about mechanical digestion, which causes physical changes in food, and chemical digestion, which results in chemical changes.
Supporting TEKS: 6A, 6B, 12B
TEKS: 6C, 13B

Lesson 2
In Lesson 7.2, you learned that the circulatory system delivers substances to cells, carries wastes away, regulates body temperature, and helps fight illness. You also learned about the four components of blood: plasma, red blood cells, white blood cells, and plasma.
Supporting TEKS: 12B
TEKS: 12C

Lesson 3
In Lesson 7.3, you learned that the respiratory system moves air containing oxygen into your lungs and removes carbon dioxide and water from the body. You also learned how the diaphragm and other muscles work together to cause air to move into and out of the lungs.
Supporting TEKS: 12B
TEKS: 12C, 12E

Lesson 4
In Lesson 7.4, you learned that the excretory system collects the wastes that cells produce and removes them from the body. You also learned how excretion helps maintain homeostasis by keeping the body's internal conditions stable and free of harmful levels of chemicals.
Supporting TEKS: 12B
TEKS: 12C, 13A, 13B

How do systems of the body move and manage materials?

Student answers should describe the roles of the digestive and circulatory systems in providing energy to the body and describe cellular respiration. See the scoring rubric below.

Review the TEKS Chapter 7

PARTNER REVIEW Have partners review the definitions of vocabulary terms and quiz each other. Students can read the Key Concept statements and leave out words for their partners to fill in. They also can change a statement so that it is false and then ask their partners to correct it.

PAIR ACTIVITY: POSTER Have pairs of students create posters to show how the digestive, circulatory, respiratory, and excretory systems all work together to move and manage materials in the body. Provide pairs with poster paper, markers, and other art supplies. Have students identify Key Concepts, key terms, details, and examples, and incorporate this information into their posters. Use the questions below to help students organize the information they should address in their posters.

- Which organs and organs systems bring materials into the body?
- Which organs and organs systems transfer materials across a system?
- Which organs and organs systems carry waste away from cells?
- Which organs and systems break down food into nutrient molecules?
- Which organs and systems process and eliminate waste?

How do systems of the body move and manage materials?

SCORING RUBRIC	SCORE 4	SCORE 3	SCORE 2	SCORE 1
Describe roles of digestive and circulatory systems	Fully describes roles of both systems	Adequately describes roles of both systems	Adequately describes role of one system	Fails to describe role of either system
Describe role of cellular respiration	Fully describes cellular respiration	Adequately describes cellular respiration	Weakly describes cellular respiration	Fails to describe cellular respiration

CHAPTER 7

TEKS Practice: Chapter Review

Test-Taking Skills

ANTICIPATING THE ANSWER Explain to students that they can sometimes figure out an answer to a question before they look at the answer choices. First, they should read the question and answer it in their minds. Then, they should compare their answer with the answer choices and choose the answer that most closely matches their own.

Question 1 TEKS 12B

The correct answer is C. When the diaphragm and rib muscles contract, the chest cavity gets larger and air is drawn into the lungs.

If students chose A, explain that this describes only the respiratory system and not the muscular system.

If students chose B, explain that neck muscles are not involved in bringing air into the body.

If students chose D, explain that the statement describes how the respiratory and circulatory systems work together.

Question 2 TEKS 6B

The correct answer is H. Saliva contains enzymes that begin to break down food molecules through chemical digestion. This model uses water, which does not contain enzymes, so it does not represent the chemical digestion that occurs in the mouth.

If students chose F, explain that absorption takes place mainly in the small intestine, not in the mouth.

If students chose G, explain that this model uses the mashing tool to represent the mechanical digestion performed by teeth in the mouth.

If students chose J, explain that choice G is a correct answer.

★ TEKS Practice: Chapter Review

Read each question and choose the best answer.

1 The diagram shows the respiratory system and part of the muscular system.

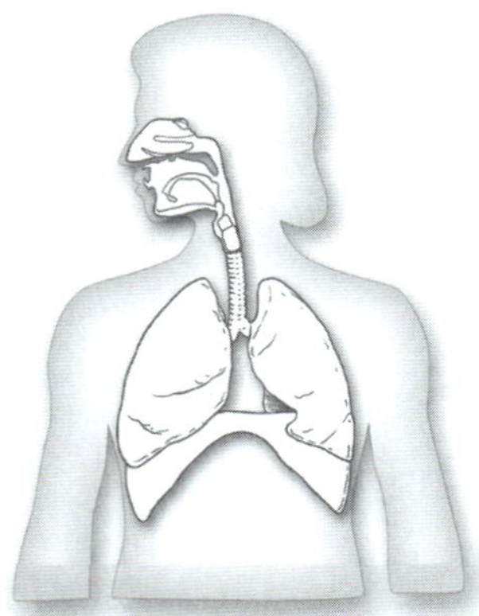

Which of the following is a way that the respiratory and muscular systems work together?

A The bronchi direct air from the trachea into the lungs.

B The neck muscles work to force air through the trachea.

(C) The diaphragm and rib muscles expand the chest cavity, allowing the lungs to take in air.

D Capillaries allow oxygen to pass from the alveoli into the blood.

2 A student wants to model the first stage of digestion, which takes place in the mouth. He places a piece of apple in a bowl. He adds a small amount of water to represent saliva. He crushes the apple and water with a wooden mashing tool to represent how teeth chew food. The resulting mashed apple and water mixture is a model of what food looks like when it is ready to be swallowed. Which of the following is a limitation of this model?

F The model does not represent the absorption that takes place in the mouth.

G The model does not represent the mechanical digestion that takes place in the mouth.

(H) The model does not represent the chemical digestion that takes place in the mouth.

J None of these

★ TEKS Practice: Cumulative Review

3 The drawing is an illustration of an X-ray. X-ray images let doctors see the shapes of a patient's bones without having to open up the patient's body.

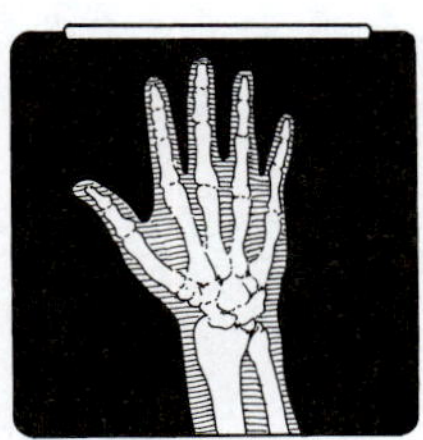

Which of the following is one way that doctors can use an X-ray image?

(A) To examine damage to the skeletal system

B To repair damage to the circulatory system

C To observe how the muscular system moves bones

D To diagnose problems with the integumentary system

4 Dylan drew a pyramid to show levels of organization in an organism as they progress from the simplest at the bottom to the most complex at the top. Dylan cannot remember what level lies between cells and organs.

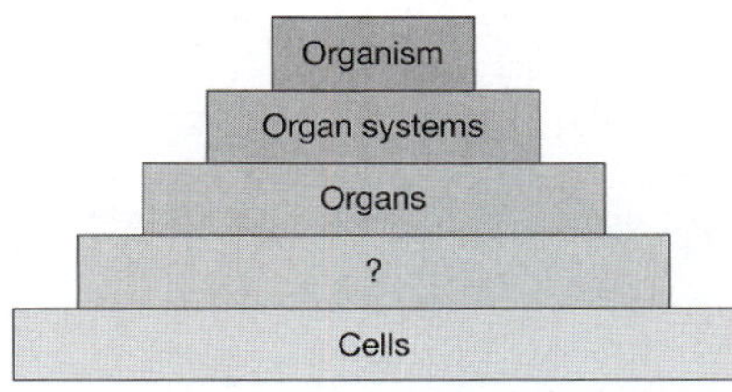

Lauren gives Dylan an example of the correct level of organization in plants to help him remember. Which of the following is the best example of the missing level in the diagram above?

F Flower

G Leaf

H Moss

(J) Phloem

If You Have Trouble With . . .

Question	1	2	3	4
See Lesson	7.3	7.1	6.4	5.1
TEKS	7.12B	7.6B	7.12B	7.12C

TEKS Practice: Cumulative Review

Additional Assessment Resources

Teacher's Edition: Lesson Quizzes, Texas End-of-Year Test Prep A and B
Online assessments

Question 3 TEKS 12B

The correct answer is A. When bones break or crack, an X-ray image can detect damage.

If students chose B, explain that X-ray images show damage; they do not repair damage.

If students chose C, explain that X-ray images are still; they do not show movement.

If students chose D, explain that X-ray images let doctors see through the skin.

Question 4 TEKS 12C

The correct answer is J. Phloem is a type of vascular tissue, and tissue is the missing level of organization in the chart.

If students chose F, explain that flowers may be considered an organ or an organ system.

If students chose G, explain that a leaf is an organ.

If students chose H, explain that moss is a type of organism.

Science Matters STEM

Frontiers of Technology

Have students read *Artificial Blood.* Point out that artificial blood compounds have properties that have several advantages. They are 1) easily available, 2) long-lasting and storable; 3) processed to avoid contamination; and 4) usable for all blood types. However, as the article mentions, they do not perform all the functions that blood does.

Ask: **What kind of job does artificial blood compounds do well?** *(Sample: Artificial blood compounds are effective at carrying oxygen through the blood vessels.)* **What do artificial blood compounds still have to do in order to be a substitute for blood?** *(Sample: In order to do the same job as blood, artificial blood compounds have to carry oxygen, nutrients, hormones, cells that fight disease and infections to all parts of the body, and carry away waste products from the body's cells.)*

SCIENCE MATTERS

Frontiers of Technology

Artificial Blood

TEKS 3D

STEM

Blood is amazing. It carries oxygen, nutrients, and hormones throughout the body, and carries waste products away from cells. If you have a cut, blood cells called *platelets* quickly patch things up. Blood also carries cells that fight diseases and infections.

If patients lose a lot of blood, doctors must rely on donated blood supplies. This donated blood saves a lot of lives, but it is in short supply. So, scientists have been trying to develop compounds that make it possible to use less blood. One such compound is 50 times more effective at carrying oxygen than red blood cells are. This compound also delivers oxygen through blood vessels that are too damaged to allow red blood cells to pass.

Artificial blood compounds have been used to treat a small number of patients, but they can be expensive to make. Also, no compound can perform all the jobs that blood does. However, researchers hope to produce an artificial blood source that is as good as the real thing!

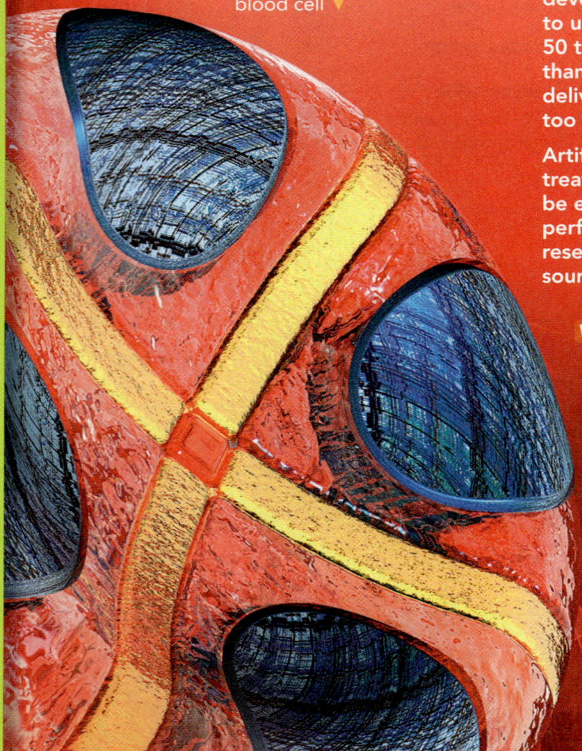

Artist's drawing of an artificial red blood cell ▼

Write About It Scientists have been trying to find many ways to use less donated blood, including artificial blood. Research problems scientists have come across in developing blood replacements, and write an essay describing these problems.

310 Managing Materials in the Body

Quick Facts

The greater the body mass, the more the heart works to get blood with oxygen and nutrients to the body. The Body Mass Index (BMI) is a measurement for assessing body fat in proportion to a person's height and weight. To calculate it, multiply a person's weight in pounds by 703, divide by their height in inches, and divide again by their height in inches—that's the BMI. According to the National Center for Health Statistics: individuals with BMI values under 18.5 are underweight, values from 18.5 to 24.9 are healthy, and values over 25.0 are overweight. The American Heart Association and other health organizations have programs that encourage people to evaluate their health and to follow guidelines to get healthy. Have students form groups and research health organizations in Texas. How do their programs promote health? Have groups report to the class.

Texas Essential Knowledge and Skills

3D Relate the impact of research on scientific thought and society, including the history of science and contributions of scientists as related to the content.

An Olympic Experiment

TEKS 3D

Science and Society

Asthma is a respiratory disease that causes a person's air passages to narrow. People with asthma feel shortness of breath and have trouble getting enough oxygen. For a long time, scientists have known that a link exists between air pollution and asthma. Yet, it is difficult to determine how many cases of asthma are caused by or affected by air pollution. The 1996 Summer Olympics in Atlanta, Georgia, provided a rare chance to explore this question.

During the Olympics, city officials dramatically increased the number of public transportation vehicles in the city. They also limited the number of cars allowed on city roads. These policies reduced the levels of air pollutants that are suspected to trigger asthma attacks.

After the Olympics, researchers analyzed how many children received emergency treatment for asthma and other respiratory problems before, during, and after the Olympics. According to one source, the number of children who received emergency treatment for asthma dropped by 41 percent during the Olympics!

The Atlanta Olympics helped scientists develop a much better understanding of the links between air pollution and asthma. Studies such as these provide strong evidence that reducing air pollution can help us all breathe easier.

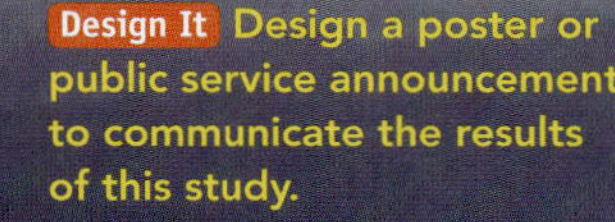

Design It Design a poster or public service announcement to communicate the results of this study.

During the 1996 Summer Olympics, peak ozone levels in Atlanta declined as much as 28 percent. The levels of other pollutants also decreased by between 7 and 19 percent. ▶

311

Science and Society

Have students read *An Olympic Experiment*. Point out that since the Atlanta Olympics report there have been many more scientific studies with findings showing links between pollution and adverse health effects. When the air pollution increases, lung function decreases, directly affecting people with asthma and other types of lung or heart disease. The elderly and children are the most vulnerable to the effects of air pollution. Young lungs in developmental stages and mature lungs with slower healing capacity can become more easily damaged by pollutants and irritant particles. Although overall air quality has improved in the last 20 years, pollution in congested urban areas is still at high levels.

As students prepare their posters or public service announcements, have them include some of the factors that reduced the level of air pollutants such as increased public transportation.

Ask: **After the Olympics in Atlanta, what did the researchers use to analyze the effects of air pollution on asthma occurrences?** *(Sample: The researchers looked at how many children received emergency treatment for asthma before, during, and after the Olympics.)* **What were the researchers findings?** *(Sample: They found a 41 percent drop in asthma treatment during the Olympics.)* **What were some of the factors that contributed to lowering the level of air pollution in Atlanta during the Olympics?** *(Sample: Atlanta city officials increased the number of public transportation vehicles and limited the number of cars in the city greatly reducing the amount of air pollutants.)*

CHAPTER 7

English Language Proficiency Standards

ELPS Listening 2.C.

Help students build vocabulary by using idioms.

Beginning Have beginners listen as you read aloud page 311 and the Design It activity. Discuss two to three idioms about breathing. Have students use an idiom in their posters.

Intermediate Together read page 311 and the Design It activity. Help students find idioms with the words *breathe* and *air*. Then have partners use an idiom in their posters.

Advanced Have partners read page 311 and find idioms with the words *breathe* and *air*. Have them use one idiom in their posters.

Advanced High Have partners read page 311 and find idioms with the words *breathe* and *air*. Ask them to use two or more of the idioms in their posters.

Texas Essential Knowledge and Skills

3D Relate the impact of research on scientific thought and society, including the history of science and contributions of scientists as related to the content.

ACKNOWLEDGMENTS

Staff Credits

The people who make up the ***Interactive Science*** team—representing composition services, core design digital and multimedia production services, digital product development, editorial, editorial services, manufacturing, and production—are listed below:
Jan Van Aarsen, Samah Abadir, Ernie Albanese, Chris Anton, Zareh Artinian, Amy Austin, Bridget Binstock, Suzanne Biron, Niki Birbilis, MJ Black, Nancy Bolsover, Stacy Boyd, Jim Brady, Robin Brandenburg, Katherine Bryant, Michael A. Burstein, Pradeep Byram, Jessica Chase, Jonathan Cheney, Arthur Ciccone, Margaret Clampitt, Brandon Cole, Allison Cook-Bellistri, Rebecca Cottingham, AnnMarie Coyne, Bob Craton, Pat Cully, Chris Deliee, Paul Delsignore, Michael Di Maria, Diane Dougherty, Kristen Ellis, Kelly Engel, Theresa Eugenio, Amanda Ferguson, Jorgensen Fernandez, Kathryn Fobert, Alicia Franke, Louise Gachet, Julia Gecha, Mark Geyer, Meredith Glassman, Steve Gobbell, Paula Gogan-Porter, Jeffrey Gong, Sandra Graff, Robert M. Graham, Adam Groffman, Lynette Haggard, Adam Harter, Christian Henry, Karen Holtzman, Susan Hutchinson, Sharon Inglis, Joan Jacobsen, Marian Jones, Sumy Joy, Sheila Kanitsch, Courtenay Kelley, Chris Kennedy, Toby Klang, Greg Lam, Russ Lappa, Margaret LaRaia, Ben Leveillee, Thea Limpus, Charles Luey, Elsa Machabanski, Dotti Marshall, Kathy Martin, Robyn Matzke, John McClure, Mary Beth McDaniel, Krista McDonald, Tim McDonald, Rich McMahon, Cara McNally, Bernadette McQuilkin, Melinda Medina, Angelina Mendez, Maria Milczarek, Mike Napieralski, Deborah Nicholls, Dave Nichols, Joe O'Day, William Oppenheimer, Jodi O'Rourke, Ameer Padshah, Lorie Park, Celio Pedrosa, Jonathan Penyack, Anita Raducanu, Linda Zust Reddy, Jennifer Reichlin, Stephen Rider, Charlene Rimsa, Walter Rodriguez, Stephanie Rogers, Marcy Rose, Rashid Ross, Anne Rowsey, Hugh Rutledge, Manuel Sánchez, Logan Schmidt, Amanda Seldera, Laurel Smith, Nancy Smith, Ted Smykal, Emily Soltanoff, Cindy Strowman, Dee Sunday, Jennifer Teece, Barry Tomack, Elizabeth Tutsian, Patricia Valencia, Stephanie Van Camp Sarah Vázquez, Stephanie Wallace, Amanda Watters, Molyrhenium Wetterschneider, Christine Whitney, Brad Wiatr, Heidi Wilson, Heather Wright, Tim Yetzina, Rachel Youdelman.

Illustrations

630 Steve McEntee; **630, 631, 632** Stephen Durke.

Photographs

Every effort has been made to secure permission and provide appropriate credit for photographic material. The publisher deeply regrets any omission and pledges to correct errors called to its attention in subsequent editions.

Unless otherwise acknowledged, all photographs are the property of Pearson Education, Inc.

Front Cover Ammit Jack/Shutterstock;
Back Cover DLILLC/Corbis

Title L Creatix/Fotolia; **Title** C jojje11/Fotolia; **Title** Bkgd Larry Landolfi/Science Source/Photo Researchers, Inc.; **Title** Plant Image Studio/Getty Images; **vi–vii** HG Photography/Fotolia; **viii** Banner Brandon Seidel/Fotolia; **ix** NHPA/SuperStock ; **x–xi** JPL/ NASA; **xii** Kim Taylor/Nature Picture Library; **xiii** T Sari ONeal/Shutterstock; **xiii** B Simon Williams/Nature Picture Library; **xiv** Exactostock/SuperStock; **xviii–xix** Andrea Poole/ Fotolia; **xxi** TR Creatix/Fotolia; **xxi** TL jojje11/Fotolia; **xxii** C Creatix/Fotolia; **xxiv** C Image Studio/Getty Images; **1** NASA; **3**T National Geographic Image Collection; **3**TC ©Ken Seet/ Corbis; **4** National Geographic Image Collection; **5** Nature Picture Library; **6**BL ©Manoj Shah/Getty Images; **6**TL Anup Shah/Nature Picture Library; **7** Christoph Becker/Nature Picture Library; **8** Kennan Ward/Corbis; **9** E.D. Torial/Alamy; **10** Sarah Holmstrom/iStockphoto; **11** ©Kurt Lackovic/Alamy; **11**BL Karimala/iStockphoto; **11**CL Stephen Dalton/Photo Researchers, Inc.; **12** Photo Network/Alamy; **13** Gary Woodard/ iStockphoto; **14** ©Ken Seet/Corbis; **15**CR Digital Vision/Getty Images; **15**TR Duncan Walker/iStockphoto; **16–17** Stephen Dorey Creative/Alamy; **17**T Redmond Durrell/Alamy; **18** Sezione di Ziologia "La Specola"; **19** Andy Sands/Nature Picture Library; **21**TR Idamini/Alamy; **22**TL Idamini/Alamy; **23**TR Idamini/Alamy; **25**B Idamini/Alamy; **25**CR US Dept of Energy Genome Programs; **26** D. Hurst/Alamy; **27** Photodisc/ Getty Images; **28** Joe Sohm Visions of America/Newscom; **30**L Ilan Rosen/Alamy; **30**R Bryan Whitney/Photonica/ Getty Images; **31** Kevin Foy/Alamy; **32–33** ©Stephen Frink Collection/Alamy; **34–35** James Balog/Aurora; **36–37**B William Philpott/AFP/Getty Images; **36–37**TR Martin Shields/ Alamy; **41** Lew Robertson/Alamy; **42**L ©Bettmann/Corbis; **42**R Viorika Prikhodko/iStockphoto; **44**Bkgd Ray Roberts/ Alamy; **44**B George Silk/Time & Life Pictures/Getty Images; **45**R Stacy Gold/National Geographic/Getty Images; **46**Bkgd The Macaulay Library at the Cornell Lab of Ornithology; **46–47** Skip Nail/Stockbyte/Getty Images; **46**R John James Audubon/ The Bridgeman Art Library/Getty Image; **48** Stockdale Studios; **49**Bkgd Ekaterina Pokrovskaya/Shutterstock; **49**L Bedo/ Dreamstime LLC; **49**R Pablo631/Dreamstime LLC; **49**C Dell/ Dreamstime LLC; **51** Corbis; **54** ©blickwinkel/Alamy; **54**C Chuck Bargeron **55** Academie des Sciences, Paris/Archives Charmet/ Bridgeman Art Library; **56–57** Nature Picture Library; **59**B iStockphoto; **59**T Nancy Nehring/Getty Images; **60** Michel & Christine Denis-Huot/Science Source/Photo Researchers, Inc.; **61**B Biophoto Associates/Science Source/Photo Researchers Inc; **61**BL Biophoto Associates/Science Source/ Photo Researchers, Inc.; **61**C Nancy Nehring/Stockbyte/Getty Images; **61**CL ©Bon Appetit/Alamy; **61**TR William H. Mullins/ Science Source/Photo Researchers, Inc.; **62–63** Corbis Premium RF/Alamy; **62**BL David Mack/Science Source/Photo Researchers Inc.; **62**TL Michael Abbey/Getty Images; **62**TR Natural Imaging/ Shutterstock; **63**C George Grall/Getty Images; **63**TR Aaron Haupt/Getty Images; **64** NHPA/SuperStock ; **66** Biophoto Associates/Photo Researchers; **67**B ©Nils-Johan Norenlind/AGE Fotostock; **67**TL Steve Gschmeissner/Photo Researchers; **68**C Dr. Jeremy Burgess/Photo Researchers; **68**L Dr. Cecil H. Fox/Photo Researchers; **68**BR Science Museum, London; **68**TR Science and Society Picture Library; **69**B John Walsh/Photo Researchers; **69**CL M. I. Walker/Photo Researchers; **69**CR Perennou Nuridsany/ Photo Researchers; **70**B DK Images; **70**CR Millard H. Sharp/Photo Researchers, Inc.; **70**CL Wes Thompson/Corbis; **70**TL Paul Taylor/ Riser/Getty Images; **70**TR Millard H. Sharp/Photo Researchers; **72–73** ©A. Syred/Photo Researchers; **74** Dr. Torsten Wittmann/ Photo Researchers; **76**TL Alfred Paskieka/Photo Researchers; **76**BR Bill Longcore/Photo Researchers, Inc.; **77**TL Bill Longcore/

ACKNOWLEDGMENTS

Photo Researchers; **78** CNRI/Photo Researchers; **80** Professors Pietro M. Motta & Tomonori Naguro; **81** Biophoto Associates/ Photo Researchers, Inc.; **82**BR SCIENCE SOURCE/Photo Researchers; **82**BL Biophoto Associates/Photo Researchers; **82**TL Profs. P. Motta and S. Correr/Science Photo Library; **82**TR Perennou Nuridsany/Science Source/Photo Researchers Inc.; **84** Tierbild Okapia/Photo Researchers; **85** Tracy Frankel/Getty Images; **86**BR iStockphoto; **86**L David Murray/DK Images; **87** kykydar/Shutterstock; **90** David M. Phillips/Photo Researchers; **91**B Index Stock Imagery/Photolibrary New York/Getty; **91**CR Cultura Limited/Superstock; **92** Perennou Nuridsany/Photo Researchers, Inc.; **93** Steve Gschmeissner/Photo Researchers, Inc.; **94**Bkgd Jim Parkin/Shutterstock; **94**CL Tyler Olson/ Shutterstock; **100**BL David M. Phillips/Photo Researchers ; **100**CL Kim Taylor and Jane Burton/Dorling/DK Images; **101** BARNABAS HONECZY/EPA/Newscom; **102–103** Ocean/ Corbis; **105**B "Kent Wood/Science Source /Photo Researchers Inc; **105**BC Ed Reschke/Peter Arnold/PhotoLibrary Group, Inc./ Getty Images; **105**T Polka Dot Images/Thinkstock; **105**TC Vincenzo Lombardo/Photodisc/Getty Images; **106** David Cook/ blueshiftstudios/Alamy; **107**Bkgd Robbert Koene/Getty Images; **107**BL Adrian Bailey/Aurora Photos; **107**C Polka Dot Images/Thinkstock; **109** Rich Iwasaki/Getty Images; **111** Yuji Sakai/Digital Vision/Getty Images; **112** Pete Saloutos/Flirt/ Corbis; **116** Vincenzo Lombardo/Photodisc/Getty Images; **117** Erik Isakson/Blend Images/Getty Images; **118** George Grall/ National Geographic Image Collection; **119**BL Helmut Gritscher/Peter Arnold/PhotoLibrary, Inc.; **119**BR Eric Bean/The Image Bank/Getty Images; **119**TR ©Heather Angel/Natural Visions/Alamy; **122**B Ed Reschke/Peter Arnold/PhotoLibrary, Inc.; **122**C Ed Reschke/Peter Arnold/PhotoLibrary, Inc.; **122**CR Ed Reschke/Peter Arnold/PhotoLIbrary, Inc.; **123**B Ed Reschke/ Peter Arnold/PhotoLibrary, Inc.; **123**R Ed Reschke/Peter Arnold/PhotoLibrary, Inc.; **123**T Ed Reschke/Peter Arnold/ PhotoLibrary, Inc.; **124** Kent Wood/Photo Researchers, Inc.; **126** Bob Stefko/Botanica/Getty Images; **127**B Olga-i/Shutterstock; **127**T Anup Shah/Getty Images; **132** Steve Byland/Shutterstock; **133** Michelangelo Gratton/Stone/Getty Images; **134** ZSSD/ SuperStock; **137**C Brand X Pictures/Jupiter Images; **137**CL Frank Krahmer/Getty Images; **137**CR Brand X Pictures/Jupiter Images; **137**TCL iStockphoto; **138**B Andrea Jones/Alamy Images; **138**C Bettman/Corbis; **140**R Andrea Jones/Alamy Images; **143**Bkgd Monika Gniot/Shutterstock; **143**TC Herman Eisenbeiss/Photo Researchers, Inc.; **143**TR WildPictures/Alamy; **144** Ron Trumbla/National Weather Service; **145**B Brand X Pictures/Jupiter Images; **146–147** Monika Gniot/Shutterstock; **148**Bkgd Agg/Dreamstime LLC; **149**C iStockphoto; **149**CR Jomann/Dreamstime LLC; **150**C Joel Sartore/National Geographic/Getty Images; **151**BC DK Images; **151**BCL Brand X Pictures/Jupiter Images; **151**BL Frank Krahmer/Getty Images; **151**BR DK Images; **151**C Brand X Pictures/Jupiter Images; **151**CR CreativeAct-Animals Series/Alamy; **152**TC John Daniels/ Ardea; **152**TL Geoff Dann/DK Images; **152**TR Jay Brousseau/ Stone/Getty Images; **153**B Randy Faris/Corbis; **153**BC Michael Melford/National Geographic/Getty Images; **153**BL Luis Carlos Torres/iStockphoto; **153**BR Naile Goelbasi/Taxi/Getty Images; **153**CC Stuart McClymont/Stone/Getty Images; **153**CL Radius Images/Photolibrary, Inc.; **153**CR Blickwinkel/Schmidbauer/ Alamy; **154**B Tomas Bercic/iStockphoto; **154**C Brand X Pictures/Jupiter Images; **154**CL DK Images; **154**FL DK Images; **154**LC Frank Krahmer/Getty Images; **154**R Radius Images/ Photolibrary, Inc.; **154**RC Brand X Pictures/Jupiter Images; **154**TL CreativeAct-Animals Series/Alamy; **155**CL Stuart McClymont/Stone/Getty Images; **155**BL Joel Sartore/National Geographic/Getty Images; **156**BL Science Source/Photo Researcher, Inc. **157**BC Cathlee/iStockphoto; **157**BCL Frank Greenaway/DK Images; **157**BCR Jane Burton/DK Images; **157**BL Eric Isselée/iStockphoto; **157**BR proxyminder/ iStockphoto; **160**CL Andrew Syred/Photo Researchers, Inc.; **164**R Brownstock/Alamy; **164**BL DK Images; **165**CR Colin Milkins/Oxford Scientific (OSF)/PhotoLibrary Group, Inc.; **166**T taelove7/Shutterstock; **166**C CountrySideCollection- Homer Sykes/Alamy; **172**Bkgd AP Images/Chris O'Meara; **172**CL AP Images/E. Skylar Litherland/Sarasota Herald-Tribune; **173**B DK Images; **174**Bkgd Laurent Bouvet/AGE Fotostock; **177**TCR Howard Rice/DK Images; **179**BCR Perennous Nuridsany/Photo Researchers, Inc.; **179**CR Lusoimages/Shutterstock; **182**CL Kjell Sandved/Photo Researchers, Inc.; **184**CR Adrian Davies/Nature Picture Library; **184**BL Philippe Clement/Nature Picture Library; **185**CL BYRON JORJORIAN/SCIENCE PHOTO LIBRARY; **185**R Tom Grundy/Shutterstock; **185**TL Lincoln Rogers/Shutterstock; **186**CL Fletcher & Baylis/Photo Researchers, Inc.; **187**C Lynwood M. Chace/Photo Researchers, Inc.; **187**CR Derek Croucher/Alamy Images; **188–189** Manfred Kage/Peter Arnold/Photo Library, Inc.; **188**TL Peter Hestbaek/Shutterstock; **190**B Pakhnyushcha/ Shutterstock; **194**B Simon Williams/Nature Picture Library; **194**BC Niall Benvie/Nature Picture Library; **194**BCL Kim Taylor/ Nature Picture Library; **194**BCR Nature Picture Library; **194**BL Nature Picture Library; **194**CL Sari ONeal/Shutterstock; **196**B Kim Taylor/Nature Picture Library; **198**CL Maryann Frazier/ Photo Researchers, Inc.; **199**B Mark Turner/Garden Picture Library/Photo Library Group; **200**TL Carole Drake/Garden Picture Library/Photolibrary, Inc.; **201**CL Martin Shields/Alamy; **201**L STILLFX/Shutterstock; **201**R Maryann Frazier/Photo Researchers, Inc.; **202**Bkgd David Noble/Alamy; **202**C Nigel Cattlin/Alamy; **208**Bkgd, CR Rolf Nussbaumer Photography/ Alamy; **209**CR NASA; **213**B Priscilla Gragg/Aurora Open/Glow Images; **213**CB Ed Reschke/PhotoLibrary Group, Inc./Getty Images; **213**T Innerspace Imaging/Photo Researchers; **213**TC Mike Kemp/Rubberball/AGE Fotostock; **214** Mads Abildgaard/ iStockphoto; **215** Dr. Gopal Murti/Science Source; **216**CL Manfred Kage/PhotoLibrary Group, Inc./Getty Images; **216**TL Ed Reschke/Peter Arnold/Getty Images; **216**CR Biophoto Associates /Photo Researchers, Inc.; **220**BL Lebedinski Vladislay/Shutterstock; **221**BL Claro Cortes IV/Corbis; **221**BR Jeff Rotman/Corbis; **222**CL Juice images/PhotoLibrary Group, Inc./Getty Images; **224–225** Photodisc/White/PhotoLibrary Group, Inc./Getty Images; **226** DK Images; **227**CL tedestudio/ iStock; **228**CR O. Burriel/Photo Researchers, Inc.; **229**BR Mike Chew/Corbis; **230–231** Duomo TIPS RF/PhotoLibrary Group, Inc./Getty Images; **232–233** Michael Meisl/Oxford Scientific/ PhotoLibrary Group, Inc./Getty Images; **232**TR Mike Kemp/ Rubberball/AGE Fotostock; **233**TL John Henley/Flirt/Corbis; **233**TR Superstock; **234–235** Michael Wong/Flirt/PhotoLibrary Group, Inc./Getty Images; **236**BL Comstock/Punchstock/Getty Images; **236**BR Nick Calovanis/National Geographic ; **236**TR Bone Clones; **237**BR DK Images; **239**B AGE Fotostock/Corbis; **242**BR moodboard/Corbis; **242**TL JGI/Blend Images/Getty Images; **243**CL Steve Gschmeissner/Photo Researchers, Inc.; **243**CR BSIP/Science Source/Photo Researchers Inc.; **244**B Dan Galic/Alamy; **246**CR Ed Reschke/PhotoLibrary Group, Inc.; **246**CL Eric V. Grave/Photo Researchers, Inc.; **246**CT Science

Photo Library/Photo Researchers; **247** Jason Stitt/Shutterstock ; **249** NASA; **250**B Custom Medical Stock; **250**TL Mauritius/ SuperStock; **251**BR Tom Carter/Alamy; **252**BR Ocean/Corbis; **252**C David Vintiner/Zefa/Corbis; **252**CL Alloy Photography/ Corbis; **252**R Priscilla Gragg/Aurora Open/Glow Images; **254**Bkgd Caspar Benson/fstop/Alamy; **255** Fuse/Getty Images; **259**TR Martin Lee/Mediablitz Images Limited (UK)/Corbis Images; **262**BL Pkruger/iStock International, Inc.; **262**CL Kyle Robertson/Columbus Dispatch/AP Images; **263** PHANIE/Photo Researchers, Inc.; **264** Raga Jose Fuste/AGE Fotostock; **268**B Jonathan Gelber/BlueMoon Stock/AGE Fotostock; **269** Puzant Apkarian/Getty Images; **270**BR Angelo Cavalli/Corbis Images; **271**T Monkey Business Images/Shutterstock; **273**BR Ocean/ Corbis; **274**B Medicimage/Photolibrary; **274**CL Science Photo Library/Alamy; **276** Science Source/Photo Researchers, Inc.; **277** Ron Trumbla/National Weather Service; **278–279** Zephyr/ Photo Researchers, Inc.; **278**C Shelia Terry/Photo Researchers, Inc.; **283**CR Tim Ridley/Dorling Kindersley Limited; **286**CL Photodisc/Alamy; **288**CL Uppercut Images/Masterfile; **290**BR Eddy Gray/Photo Researchers, Inc.; **290**CL BananaStock/ Jupiter Images/Getty Images; **293**CR Anne Ackerman/Digital Vision/Getty Images; **296** AP Photo/NASA; **297** Barry Austin Photography/Riser/Getty Images; **300**BC Steve Skjold/Alamy; **300**BL Seth Joel/Taxi/Getty Images; **300**BR Anderson Ross/ Getty Images; **301**BL Pete Saloutos/Corbis; **304**B jeehyun/ Shutterstock; **304**CR ©1994 Richard Megna/Fundamental Photographs; **304**L Andy Crawford and Tim Ridley/Getty Images; **310** Laguna Designs/Photo Researchers, Inc.; **311**Bkgd Simon Meeds/Alamy; **311**BR Tannen Maury/Alamy; **312–313** AFP/Getty Images; **315**B Dopamine/Photo Researchers; **315**CB Gary Cornhouse/Digital Vision/Alamy; **316** Science Photo Library/Alamy; **317** Larry Dale Gordon/The Image Bank/Getty Images; **318**BR Mile Powell/Allsport Concepts/Getty Images; **318**TL Cre8tive Studios/Alamy; **320**T kolvenbach/Alamy; **322** Gregor Schuster/Ionica/Getty Images; **324**CL ©Image Source/ Corbis; **326**CL Peter Chapwick /Dorling Kindersley, Ltd.; **327**C Michael Nemeth/The Image Bank/Getty Images; **327**CR ©Ocean/Corbis; **328**BL Fuse/Jupiter Images; **328**BR Keith Leighton/Alamy; **329** Sandy Huffaker/Stringer/Getty Images; **330**BR Adrian Brockwell/Alamy; **330**TL Tom Sanders/Aurora/ Getty Images; **334–335** Matt Lange/Southcreek Global/Zuma Press; **335**TC Dream Pictures/Getty Images; **336**C Adrian Arbib/Encyclopedia/Corbis; **336**CL Keystone/Stringer/Hulton Archive/Getty Images; **337**CB David M. Phillips/Photo Researchers; **337**LC Gary Cornhouse/Digital Vision/Alamy; **343** Tig Photo/Alamy; **344** Michelle Del Guercio/The Medical File / Photolibrary, Inc.; **345**L Dopamine/Photo Researchers, Inc.; **345**R Tim Vernon, LTH NHS Trust/Photo Researchers, Inc.; **347** Getty Images; **348** BSIP/Photolibrary, Inc.; **349**L Big Cheese Special/Big Cheese Photo LLC/Alamy; **349**R Tony Freeman/ Photoedit; **350** David Grossman/Alamy; **352**TL Rob Marmion/ Shutterstock; **352–353** Dr. Klaus Boller/Science Source; **352** Custom Medical Stock; **353**CR E. Gueho/Science Photo Library; **353**TR SPL/Photo Researchers; **354** Jose Pedro Fernandes/ Alamy; **355**Bkgd Bruce Heinemann/Stockbyte/Getty Images; **355**C Karen Kasmauski/Corbis; **355**CL Ed Reschke/ PhotoLibrary Group Inc., Inc/Alamy; **356** Science Pictures Ltd./ Photo Researchers; **358** Melianiaka Kanstantsin/Shutterstock; **360** Stem Jems/Photo Researchers; **369** David Joyner/ iStockphoto; **370–371** Billy Currie Photography/Getty Images; **373**CB Pichugin Dmitry/Shutterstock; **373**TL DK Images; **374**B imagebroker/SuperStock; **375**Bkgd Jerome Wexler/Photo Researchers, Inc.; **375**BL Ted Kinsman/Photo Researchers, Inc.; **378**TL, CL, DK Images; **378** BL iStockphoto; **379**T, TL, CC, B DK Images; **379**TCL Judy Ledbetter/iStockphoto; **379**TCR Jerry Young/DK Images; **379**TR Geoff Brightling/DK Images; **379**CL Nicholas Homrich/iStockphoto; **379**CR Frank Greenaway/DK Images; **379**BC iStockphoto; **380**T GlobalP/iStockphoto; **380**TC, BC, CC, BCL, BCR, B DK Images; **382** 2canaries/Alamy; **384** Emma Firth/DK Images; **386** Dr. Paul A. Zahl/Photo Researchers, Inc.; **390** Rolf Nussbaumer/Getty Images; **393**Bkgd Karen Huntt/Photographer's Choice/Getty Images; **393**CB Kenneth M Highfill/Photo Researchers, Inc./Getty Images; **394**B Peter Chadwick/DK Images; **395**C Juergen & Christine Sohns/Getty Images; **396**TC Arco Images GmbH/ Alamy; **396**TL Juan-Carlos Munoz/Peter Arnold/Photolibrary, Inc.; **396**TR Philip Dowell/Dorling Kindersley, Ltd.; **397**B Tim Shepard, Oxford Scientific Films/DK Images; **397**TR mlorenz/ Shutterstock; **398**BL Alan & Sandy Carey/Photo Researchers, Inc.; **398**TL Randy Green/Taxi/Getty Images; **399**B Momatiuk-Eastcott/Corbis; **400** Khoroshunova Olga/Shutterstock; **401**B Hoffmann Photography/AGE Fotostock; **401**CR Sinclair Stammers/Science Source/Photo Researchers, Inc.; **401**TC Danita Delimont/Getty Images; **401**TR Tim Zurowski/All Canada Photos/SuperStock; **402**C Sandy Felsenthal/Corbis; **403** PIER/Stone/Getty Images; **406**BL Jerome Whittingham/ Getty Images; **407**TR, CR arlindo71/iStockphoto; **407**BL Ocean/Corbis; **407**C PhotographerOlympus/iStockphoto; **407** xpixel/Shutterstock; **409** Nature Picture Library; **410** Ivaylo Ivanov/Shutterstock; **411**TR Image Quest 3-D; **411**CL Betsy Gange/Hawaii Dept. of Land and Natural Resources/ Associated Press; **411**CR Rick & Nora Bowers/Alamy; **413**BL Kim Mitchell/Whooping Crane Eastern Partnership; **413**CL Markus Botzek/Corbis Bridge/Alamy; **413**CR Kevin Schafer/ Alamy; **413**CC James Caldwell/Alamy; **413**BC Operation Migration; **413**BR Jason Hahn Photography; **414–415** Wolfgang Amri/Shutterstock; **416** Ilene MacDonald/Alamy; **422–423** Cynthia Farmer/Shutterstock; **423** paulaphoto/ Shutterstock; **430**B Roger del Moral; **432** Mark Conlin/Alamy; **435**BR Stan Osolinski/Corbis Images; **435**CB Cathy Keifer/ Shutterstock; **435**TCR Photoshot/AGE Fotostock; **436**B Andreas Gross/Westend 61/Alamy; **436**BL DK Images; **436**C PoodlesRock/Corbis Images; **437**BC Wardene Weisser/Bruce Coleman Inc./Alamy; **437**BL N. Reed of QED Images/Alamy; **437**CR Enzo & Paolo Ragazzini/Corbis Images; **437**TR Gerard Lacz Images/SuperStock; **438**BR ©Thompson Paul/AGE Fotostock; **438**BL Ryan M. Bolton/Shutterstock; **440**BC Steve Shott/DK Images; **440**BL GK Hart/Vikki Hart/Getty Images; **440**BR Tracy Morgan/DK Images; **441**TC The Art of Animals. co.uk/PetStock Boys/Alamy; **441**TCR Photo-Max/iStockphoto; **441**TL Georgette Douwma/Nature Picture Library; **441**TR Derrell Fowler; **444** Maury Hatfield; **445** David McNew/Getty Images; **446**CL Jim Watt/PhotoLibrary Group, Inc.; **448**BR Anthony Mercieca /Photo Researchers, Inc.; **448**L Photoshot/ AGE Fotostock; **449**CL Steve Gschmeissner/Science Photo Library; **449**TR Philip M. De Renzis/Getty Images; **450**BL Denise Applewhite/Princeton University, Office of Communications; **450**T ©D. & S. Tollerton/AGE Fotostock; **450**TL AGE Fotostock/D. & S. Tollerton; **453**CC Kim Taylor/Nature Picture Library; **453**CL Hemera Technologies/Jupiter Images; **453**CR Joseph Caley/Shutterstock; **453**TC JinYoung Lee/Shutterstock; **453**TL Armando Frazao/iStockphoto; **453**TR Eric Isselee/

Shutterstock; **454**CL Brian Weed/Shutterstock; **454**CR Jesse Kunerth/Shutterstock; **455**CR Joe Mercier/Shutterstock; **455**TR Cathy Keifer/Shutterstock; **456**TR Steve Gschmeissner/Science Photo Library; **457** Mladen Antonov/AFP/Getty Images/ Newscom; **458–459** Nina Leen/Getty Images; **460**BR Tony Wear/Shutterstock; **460**TL John Cancalosi/Peter Arnold/ PhotoLibrary Group, Inc.; **462**BL Ron Cohn, koko.org, the Gorilla Foundation/Newscom; **463** Inski Relanzon/Nature Picture Library; **464**Bkgd Mark Turner/Photo Library, Inc.; **464**BL Linda Freshwaters Arndt/Photo Researchers, Inc.; **465**BR Stan Osolinkski/Corbis Images; **465**TL Robert Muth/ Shutterstock; **465**TR Bob Gurr/AGE Fotostock; **466** AGE Fotostock/SuperStock; **468**BL PRNewsFoto/Perot Museum of Nature and Science/AP Images; **468–469** Mark Carwardine/ Nature Picture Library; **469** Zephyr/Photo Researchers, Inc.; **470**TR Doug Allan/Nature Picture Library; **470**B Kevin Schafer/ age fotostock/Photolibrary, Inc.; **471**TR T. J. Rich/Nature Picture Library; **472**B Giuseppe-R/AGE Fotostock; **474** Nachman et al. 2003 Proc. Nat. Acad. Sci. USA 100: 5268-5273/Smithsonian; **475** Dr. Michael Nachman/Getty Images; **478**BR Boris Djuranovic/iStockphoto; **479**CL Geoff Brightling/DK Images; **480–481** Corbis; **484**B Steve Jacobs/iStockphoto; **484**TR Milos Luzanin/iStockphoto; **485**CR Gerald Hinde/Getty Images; **489**TR Tiburon Studios/iStockphoto; **494**CL Halina/Alamy; **494**CR Philip Kramer/Getty Images; **495**C Jeff Schmaltz/NASA; **495**TL Alison Wright/Getty Images; **495**TR Ralph Lee Hopkins/ Photo Researchers, Inc.; **496**B North Wind Picture Archives/ Alamy; **497**T David Noton Photography/Alamy; **498**C Andrew Ward; **498**CR Ted Spiegel/Corbis; **499**T Plainview/iStockphoto; **501**BC Peter Bowater/Photo Researchers, Inc.; **501**BCL Inga Spence/Photo Researchers, Inc.; **501**BCR Theodore Clutter/ Photo Researchers, Inc.; **501**BL Photoroller/Dreamstime LLC; **502**TL Photodisc/SuperStock; **504**C Mark A. Johnson/Corbis; **504**Bkgd DK Images; **509**Bkgd Greenpeace; **509**CR US Coast Guard Photo /Alamy; **512**BR Rob Marmion/Shutterstock; **512**L Elena Elisseeva /Shutterstock; **518**Bkgd Dave G. Houser/Alamy; **519**BR Courtesy of Melanie Cannon, University of Cincinnati; **519**BR Melanie Cannon, University of Cincinnati; **520–521** Raimund Linke/PhotoLibrary Group, Inc.; **523**B AGE Fotostock; **523**BC Payne Anderson/PhotoLibrary Group, Inc.; **523**T Susan Rayfield/Photo Researchers, Inc; **523**TC Marli Bryant Miller; **524** Rick Bowmer/Associated Press; **527**BL Susan Rayfield/Photo Researchers, Inc; **527**BR Jim Nicholson/Alamy Images; **527**C Deborah Bullen/Travel Ink/Alamy Images; **527**TL Tom Till/Getty Images; **527**TR Corbis; **528**CL Bob Hammerstrom/The Nashua Telegraph/Newscom; **528**TL Associated Press; **529**L Adam Hart-Davis/Photo Researchers, Inc.; **529**T John Elk III/Alamy Images; **532**BR China Daily/Reuters Media; **532**TR CNImaging/ Photoshot/Newscom; **533**TL Chuck Place Photography; **533**R Marli Bryant Miller; **534** Enrique Aguirre/PhotoLibrary Group, Inc.; **536**Bkgd Michael Just/AGE Fotostock; **537**CL Gail Jankus/ Photo Researchers, Inc.; **537**L Gail Jankus/Photo Researchers, Inc.; **537**FL Gail Jankus/Photo Researchers, Inc.; **537**FR Gail Jankus/Photo Researchers, Inc.; **538**BR Payne Anderson/ PhotoLibrary Group, Inc.; **538**L Darwin Wiggett/AGE Fotostock; **539**T Roine Magnusson/PhotoLibrary Group, Inc.; **539**B Aflo/Nature Picture Library; **540**L JPL/NGA/NASA; **540**R Marli Bryant Miller; **542** Chad Ehlers/PhotoLibrary Group, Inc.; **543** Igor Katayev/ITAR-TASS/Landov LLC; **544–545** Radius Images/PhotoLibrary Group, Inc.; **548**Bkgd Landsat Image Mosaic of Antarctica/USGS/NASA; **549** AGE Fotostock; **550** David Nunuk/AGE Fotostock; **552** Photo courtesy Margaret Hiza Redsteer/U.S. Geological Survey; **553** Corbis; **554–555**Bkgd Ermin Gutenberger/iStockphoto; **554**R George Steinmetz/ Corbis; **554**L Nagel Photography/Shutterstock; **555**L Imagebroker/Alamy; **555**R Skip Brown/Getty Images; **556** Q-Images/Alamy; **557** Rusty Dodson/Shutterstock; **558**T Carlo Abbruzzese; **558–559**B Abby Wolters; **558**C Loren M. Smith, Texas Tech University; **559**BL "©Clint Farlinger/Alamy; **559**TR ©Amar and Isabelle Guillen - Guillen Photography/ Alamy; **560** DigitalVues/Alamy; **561** RIRF Stock/Shutterstock; **562** AGE Fotostock America Inc.; **563** Abby Wolters; **565** Associated Press [Wide World Photos]; **566** Corbis; **567**C Str Old/Reuters; **567**R Austin American-Statesman, Ted S. Warren/ Associated Press; **568** Sisters of Charity of the Incarnate Word/ Associated Press; **569** Tony Gutierrez/Associated Press [Wide World Photos] ; **570** USGS; **571** USGS; **578** Newscom; **579** Chris Howes/Wild Places Photography/Alamy; **580–581** NASA; **583**B Madarakis/Shutterstock; **583**CB ©Universal Images Group; **583**T Friedrich Saurer/Alamy; **584** Manfred Kage/ Science Source/Photo Researchers, Inc.; **585** Utah Images / Alamy; **586**B NASA; **586**CL Dennis Hallinan/Venus; **586**CR Loskutnikov/Shutterstock; **586**FL JCElv/Shutterstock; **586**FR Stocktrek Images, Inc./Alamy; **587**CL Dimitar Todorov/Alamy; **587**CR JCElv/Shutterstock; **587**FL Friedrich Saurer/Alamy; **587**FR Friedrich Saurer/Alamy; **588–589** JPL/NASA; **590**BC NASA; **590**CR JPL/NASA; **590**TR NASA; **591**CR Science Source/Photo Researchers, Inc.; **591**TR Science Source/Photo Researchers, Inc.; **592** NASA/Science Photo Library; **592**C NASA/Science Photo Library; **594** NASA /Science Source; **596** Universal Images Group; **597** NASA; **598**BR Photri Inc./ AGE Fotostock; **598**TCR NASA; **599**CL Science Source/ Photo Researchers, Inc.; **599**TR Science Photo Library; **600**B Exactostock/SuperStock; **600**TR NASA; **601**BL JSC/Stanford University/NASA; **601**BR Hank Morgan/Photo Researchers, Inc.; **601**TCR Steven Hobbs/NASA; **602**BC Mark Anderson/ Rubberball/Getty Images; **602**BR Alexander Selektor/ iStockphoto; **602**CL Hugh Threlfall/Alamy; **602**CR Mark Evans/ iStockphoto; **602**TL Madarakis/Shutterstock; **603**BCL NASA Human Spaceflight Collection/NASA; **603**C Mehau Kulyk/ Photo Researchers, Inc.; **603**TR Terry Vine/Getty Images; **603**BL David L. Moore-AK/Alamy; **604**Bkgd NASA/Science Source; **605**Bkgd Tristan3D/Shutterstock; **605**CL Science Source/ Photo Researchers, Inc.; **610**B Orla/Shutterstock; **610**CL Claudio Bertoloni/Shutterstock; **610**CL Shutterstock; **611**B NASA; **611**CR NASA; **616**B Britvich/Dreamstime LLC ; **616**BR Terex/Dreamstime LLC ; **616**T Chiyacat/Dreamstime LLC ; **617** Jupiter/Image Source/Getty Images ; **620**BL Kevin Fleming/ Corbis; **620**TL Brandon Seidel/Fotolia; **622**BL Robert Manella/ Comstock/Corbis; **623** Comstock/Thinkstock; **624**Bkgd H. Lansdown/Alamy; **624**C Frank Greenaway/Dorling Kindersley; **625** Chris Johnson/Alamy; **626** IThinkstock/Getty Images; **626**B Thinkstock/Comstock/Getty Images; **627** Comstock/Thinkstock; **628**B PhotoSky 4t.com/Shutterstock; **628**BL ©Imagebroker/ Alamy; **629**B Ryan M. Bolton/Shutterstock; **629**T Geoff Brightling/DK Images; **630**B Bill Curtsinger/National Geographic Image Collection; **632**BR ©Adrian Sherratt/Alamy; **632**TL ©elfart/Alamy; **633**Bkgd NASA Hubble Space Telescope Collection; **633**TR NASA; **634** Car Culture/Corbis